Studies in Latin American Ethnohistory & Archaeology

Joyce Marcus
General Editor

Volume I	*A Fuego y Sangre: Early Zapotec Imperialism in the Cuicatlán Cañada, Oaxaca,* by Elsa M. Redmond. Memoirs of the Museum of Anthropology, University of Michigan, No. 16 (1983). $15.
Volume II	*Irrigation & the Cuicatec Ecosystem: A Study of Agriculture & Civilization in North Central Oaxaca,* by Joseph W. Hopkins III. Memoirs of the Museum of Anthropology, University of Michigan, No. 17 (1984). $15.
Volume III	*Aztec City-States,* by Mary G. Hodge. Memoirs of the Museum of Anthropology, University of Michigan, No. 18 (1984). $15.
Volume IV	*Conflicts over Coca Fields in XVIth-Century Perú,* by María Rostworowski de Diez Canseco. Memoirs of the Museum of Anthropology, University of Michigan, No. 21 (1988).

MEMOIRS OF THE MUSEUM OF ANTHROPOLOGY
UNIVERSITY OF MICHIGAN
NUMBER 21

Studies in Latin American Ethnohistory & Archaeology

Joyce Marcus, General Editor

Volume IV

Conflicts Over Coca Fields in XVIth-Century Perú

by

María Rostworowski de Diez Canseco

CON LA COLABORACIÓN DEL INSTITUTO DE ESTUDIOS PERUANOS

ANN ARBOR
1988

This series is partially supported by a grant-in-aid No. 4453 from the Wenner-Gren Foundation for Anthropological Research, whose Director of Research, Lita Osmundsen, offered both encouragement and help during the preparation of the grant proposal. Generous funds were also supplied by the Museum of Anthropology, University of Michigan, through the efforts of former Director Richard I. Ford.

Printed in the
United States of America

ISBN 0-915703-13-0

Introduction to Volume IV

by Joyce Marcus

While Mesoamerican ethnohistory and archaeology were treated in the first three volumes in this series, this Memoir constitutes a departure; it kicks off studies in South American ethnohistory and archaeology by examining *Justicia 413*, a lengthy sixteenth-century litigation.

In the past, archaeologists and ethnohistorians have sought to link rich, detailed sixteenth-century documents to important, well-preserved prehispanic remains and long archaeological sequences. Some parts of Latin America are ideal for such a methodology.

As we saw in the last volume in this series (Hodge 1984), such integration of archaeological and ethnohistoric data can provide a good view of prehistoric society. In that case study, the Aztec "empire" (or Triple Alliance headed by three *señoríos)* managed to subjugate 489 towns located in 38 "provinces" containing 15,000,000 people speaking several different languages.

Unlike the Aztec "empire," which was dedicated principally to exacting tribute from its subjects, the expansionist Inca empire concerned itself with tightly controlling its labor force. As Murra (1967, 1980, 1982) has emphasized, *mit'a* (obligatory labor on behalf of the state) was an essential element of Inca state control. The Inca state also engaged people as retainers *(yanacona)*, and resettled people (as *mitmaq* colonists) in other regions for a variety of purposes; the latter group retained sociopolitical rights and ethnic ties to its home base. For example, one of many ethnic groups subject to the Inca were the Canta who had to contribute labor *(mit'a)* to cultivate part of their own lands for the Inca. Further, the Canta lord had to send 50 men to serve in the Inca army, and to supply the Inca with 100 bearers along with 10 men and women to serve as *yanacona*. As we will see below, the Inca also resettled members of the Chaclla ethnic group on *chaupi yunga* (or mid-valley) lands near Quivi, to cultivate coca for the Inca and to serve as messengers and bearers.

The lands of Quivi occur in the Chillón Valley of Perú's central coast, an area with high potential for the integration of archaeology and ethnohistory. Although there are still many significant gaps in our knowledge both of its archaeology and ethnohistory, by integrating current knowledge of the archaeology for the Chillón Valley with relevant ethnohistoric data (especially from the *Justicia 413* document published in this volume, and those listed in Rostworowski 1977:273–74), we hope to show encouraging preliminary results.

In the past few decades Andean ethnohistorians such as John V. Murra, John H. Rowe, Craig Morris, Donald Thompson, and others have tended to concentrate on the *sierra* or highland regions, while María Rostworowski de Diez Canseco and others have focused on the coastal valleys, which include both the coastal *yunga* (ca. 0–600 m) and the mid-valley *chaupi yunga* (ca. 600–1200 m).

Rostworowski has devoted much of her career to developing a *"yunga* ethnohistory." She has done this by discovering, publishing, and analyzing the contents of many documents from Peruvian and Spanish archives, and by subsequent fieldwork in the areas discussed in those documents. The coastal emphasis of Rostworowski's work (e.g. 1961; 1967–1968; 1970; 1972a,b,c; 1973a; 1977; 1978; 1978–1980) and that of others (e.g. Cock 1985; Espinoza Soriano 1963; Hart 1983; Netherly 1977, 1984; Ramírez-Horton 1982) has complemented the highland and Cuzco-centered views.

One of the principal reasons this *yunga* ethnohistory is so important is that we now have data that enable us to understand the nature of diverse coastal polities as they existed before absorption, influence, or subjugation by the expansionist Inca empire based in the highlands. Since the ancient territorial boundaries, political affiliations, and loyalties of numerous *yunga* ethnic groups varied significantly over time, it appears that *sierra* and *yunga* ethnohistories are equally essential for building well-balanced reconstructions of prehispanic Perú.

This volume is divided into three parts. Part I presents a brief synthesis of central coast environmental and archaeological data. These minimal background data are provided so that the reader may view the *Justicia 413* document in a meaningful context; otherwise, it would be more difficult to understand the dynamic interaction and evolution of conflict among the ethnic groups that figure so prominently in the document. Part II provides Rostworowski's very clear exposition of the main themes of the *Justicia 413* document. Part III is the transcription of that document from the Archivo General de Indias de Sevilla, Spain, here published in its entirety for the first time.

In Part I, Jorge Silva and I try to synthesize some of the relevant data that have been amassed for the Chillón Valley by other archaeologists. With caution, the names of ethnic groups and places mentioned in the *Justicia 413* document can sometimes be connected to modern places in the Chillón Valley and the surrounding region. Assessing how far back in time some of the processes mentioned in the document extend (such as inter-ethnic group alliance formation, exchange, and competition) is more problematic. Most scholars only feel comfortable using documents such as the one published here to reconstruct patterns and processes for the later periods (such as A.D. 1000–1500, that is, the Late Intermediate and Late Horizon), while others see no problem projecting them back to even earlier times. After acknowledging that each document and valley presents a different set of circumstances, we cannot offer any hard and fast rules that apply to every situation. Some valleys clearly witnessed dramatic changes during the long span of time from A.D. 1000–1500, while others did not. Additionally, some kinds of data are thought to be more "static" or "dynamic" through time than others; for example, data regarding which products were grown and exchanged between zones are often considered more static, durable, or predictable, than data on political hegemony, which are considered less so. Given the difficulty of projecting ethnohistoric information into the archaeological past, Jorge and I make an effort to synthesize what is known for the Chillón and surrounding valleys so that the document has some archaeological context.

In Part II, María Rostworowski compares the *Justicia 413* document to a gold mine, since ethnohistorians, linguists, archaeologists, and ethnologists can return to it time and again "to mine new data" (e.g. Murra 1972; Rostworowski 1967–1968, 1973b, this volume; Dillehay 1976, 1979). The document provides remarkable information on the long, inter-ethnic group struggle for possession of special lands which, while very limited in extent, were suited ideally for the growing of coca. Two highland ethnic groups (the Chaclla and the Canta) and one lowland group (the Colli) claimed ancestral rights to these lands, which as we have seen were located in the *chaupi yunga* zone of the Chillón Valley near Quivi, known today as Santa Rosa de Quives.

In Perú's central coast—the Chillón, Rímac, and Lurín valleys—it appears that relatively free and peaceful exchanges of products from different zones took place among *curacazgos (señoríos)* and ethnic groups at different times prior to the Inca conquest. However, this was not always the case. The Canta, one of the highland ethnic groups occupying the upper Chillón Valley, wished to expand into the *chaupi yunga* zone, arguing that since their highland water was brought down to the *chaupi yunga* lands for irrigation, they had a right to some lands in that zone. One witness in the *Justicia 413* document recalls a Canta attack on the coastal Colli *curacazgo*, and states that the Canta, upon encountering some resistance, did an about face and retreated to the highlands. Statements regarding expansion into other ecological zones frequently mention highlanders as the intruders and aggressors, and this document is no exception.

With the Inca conquest, the former sociopolitical balance of powers (extant alliances, as well as competitive relationships among *señoríos)* changed dramatically. In addition to the Inca, ethnic groups who appear in the *Justicia 413* document were (1) the Quivi of the *chaupi yunga* (whose coca lands were at the very center of the lawsuit published in Part III); (2) the coastal Colli (whose paramount lord at one time had had control over the Quivi and other *señoríos)*; (3) the highland Chaclla (one of the microethnic groups within the Yauyos stock, who were placed as *mitmaq* colonists on Quivi lands by the Inca state for the purpose of producing coca and fruits); (4) the highland Canta (who received coca from Quivi, Yaso, and Caraguayllo, and whose power increased at the expense of the Atavillos); and (5) the highland Atavillos (whose confederation had included the *curacazgo* of Huamantanga and the Indians of Socos [Xecos], and who occupied the left bank of the Chancay River). The Guancayo, Maca, and Guarauni (Huarabi) were three other Chillón Valley *chaupi yunga* ethnic groups who appear in a 1571 document published by Espinoza (1963).

Thus there were many ethnic groups who had a claim to the lands of Quivi: the Quivi themselves, who lived adjacent to the fields and were the "original occupants"; the Colli, who later dominated the Quivi; the Chaclla, who were placed on the lands after the Inca conquest; and the Canta, who intimidated these Chaclla *mitmaq* and later killed a Chaclla lord and 154 of his men. The Canta, the coastal *Yunga*, and the Quivi *chaupi yunga* occupants all regarded the Chaclla as the intruders, since they were placed on those lands by the Inca. Thus, the *Yunga* testified on behalf of the Canta against the Chaclla.

The arrival of the Spaniards, the defeat of the Inca, and the retreat of the Chaclla *mitmaq* from Quivi lands occasioned an opportunity for the Canta to extend their boundaries and take possession of the lands in Quivi. By A.D. 1549 the Colli were no longer powerful owing to significant population losses, but conflict between the Canta and Chaclla over the coca lands of Quivi continued until the Spaniards intervened. Not satisfied with any of the Spanish solutions, the Chaclla initiated a lawsuit which ran for many years and produced *Justicia 413*.

Access to land where coca (and a range of tropical fruits) could be grown was the goal of the competing Indian groups. However, as we will see, the Spaniards operated throughout the litigation with a variety of Old World concepts such as "land ownership" and "just compensation"—terms which were not only inappropriate to the situation, but also incapable of satisfying the Indian claimants. The concepts of "land ownership" and "just or equal compensation" had no counterpart in Quechua vocabulary or indigenous Indian custom. For example, in A.D. 1549 several Spanish *encomenderos* wanted to put a stop to the inter-ethnic feuding by forcing one ethnic group (the Chaclla) to "sell" its rights to the land to the other party (the Canta) in return for 200 camelids (100 male alpacas, 50 male and 50 female llamas). Since indigenous Andean populations had never "bought" or "sold" land, there was tremendous resistance to this proposed resolution. In the Andean world, neither camelids nor any other items were considered "compensation" for losing one's ancestral lands; thus, the Chaclla continued to claim the fields and access to the coca and fruits that could be grown there.

In Part III we provide a complete transcription of the *Justicia 413* document. This manuscript was first seen in the Archivo General de Indias by Waldemar Espinoza, who drew Murra's attention to it in 1961. In 1967, María Rostworowski received a transcription of the document prepared by Angelina López; later, she lent it to a colleague who never returned it. Knowing that Rostworowski wanted a transcription for further study and eventual publication, Pierre Duviols very kindly sent her a second transcription, prepared by Héléne Moreno, and gave her his generous permission to publish it. Part III consists of the Moreno transcription, which we have copied faithfully; the reader should be advised, however, that we have not had the opportunity to compare the Moreno transcription word for word against the original document in the archives.

In this volume we have tried to meet two goals: (1) to publish this extensive document so that future scholars can use it for their own research purposes; and (2) to analyze the document in preliminary form from two perspectives: a synchronic, ethnohistoric presentation and a diachronic, archaeological perspective.

Acknowledgments

Putting this volume together required the help of many people. Jorge Silva spent hundreds of hours typing the entire text of the *Justicia 413* document onto computer disks and tape, so that the text would not have to be retyped by Ann Arbor typesetters who would have found it a very difficult task. The value of this strategy was that once we had an error-free tape, we knew no additional proofreading would be required, and most importantly, we knew the cost of the book could be kept down so students could afford it. Of tremendous help to us were Charles Hastings and Sally Horvath, particularly during the typing of the text into the computer and the subsequent copying onto disks and tapes.

All the maps and illustrations in this volume were prepared by Kay Clahassey. At every stage of production and preparation, good advice was offered by María Rostworowski and Ramiro Matos M., and we thank them for their help.

Bibliography

Cock, Guillermo Alberto

1985 *From the Powerful to the Powerless: The Jequetepeque Valley Lords in the 16th Century, Peru.* Unpublished Master's thesis. Department of Anthropology, University of California at Los Angeles.

Dillehay, Tom D.

1976 *Competition and Cooperation in a Prehispanic Multi-Ethnic System in the Central Andes.* Unpublished Ph.D. dissertation, Department of Anthropology, University of Texas, Austin.

1979 Pre-Hispanic Resource Sharing in the Central Andes. *Science*, Volume 204, pp. 24–31.

Espinoza Soriano, Waldemar

1963 La Guaranga y la Reducción de Huancayo. Tres Documentos Inéditos de 1571 para la Etnohistoria del Perú. *Revista del Museo Nacional*, Tomo XXXII, pp. 8–80. Lima.

Hart, Elizabeth A.

1983 *Prehispanic Political Organization on the Peruvian North Coast.* Unpublished Ph.D. dissertation, Department of Anthropology, University of Michigan, Ann Arbor.

Hodge, Mary G.

1984 *Aztec City-States.* Studies in Latin American Ethnohistory & Archaeology, Volume III. Memoirs of the Museum of Anthropology University of Michigan, Number 18. Ann Arbor.

Murra, John Victor

1967 La Visita de los Chupachu como Fuente Etnológica. In Visita de la Provincia de León de Huánuco en 1562, por Iñigo Ortíz de Zúñiga, visitador, (John V. Murra, Ed.), Tomo I, pp. 381–406. *Documentos para la Historia y Etnología de Huánuco y la Selva Central*, Tomo I. Universidad Nacional Hermilio Valdizán, Huánuco, Perú.

1972 El 'Control Vertical' de un Máximo de Pisos Ecológicos en la Economía de las Sociedades Andinas. In Visita de la Provincia de León de Huánuco en 1562, por Iñigo Ortíz de Zúñiga, visitador, (John V. Murra, Ed.), Tomo II, pp. 427–76. *Documentos para la Historia y Etnología de Huánuco y la Selva Central*, Tomo II. Universidad Nacional Hermilio Valdizán, Huánuco, Perú.

1980 *The Economic Organization of the Inka State*. JAI Press, Greenwich, Connecticut.

1982 The *Mit'a* Obligations of Ethnic Groups to the Inka State. In *The Inca and Aztec States 1400–1800*, edited by George A. Collier, Renato I. Rosaldo, and John D. Wirth, pp. 237–62. Academic Press, New York.

Netherly, Patricia

1977 *Local Level Lords on the North Coast of Perú*. Ph.D. dissertation, Department of Anthropology, Cornell University. University Microfilms, Ann Arbor.

1984 The Management of Late Andean Irrigation Systems on the North Coast of Perú. *American Antiquity*, Volume 49, Number 2, pp. 227–54.

Ramirez-Horton, Susan

1982 Retainers of the Lords or Merchants: A Case of Mistaken Identity? In El Hombre y su Ambiente en los Andes Centrales, editado por Luis Millones and Hiroyasu Tomoeda, *Senri Ethnological Series*, Number 10, pp. 123–36. National Museum of Ethnology, Osaka, Japan.

Rostworowski de Diez Canseco, María

1961 *Curacas y Sucesiones: Costa Norte*. Librería e Imprenta Minerva, Lima.

1967–1968 Etnohistoria de un Valle Costeño durante el Tahuantinsuyu. *Revista del Museo Nacional*, Tomo XXXV, pp. 7–61. Lima.

1970 Mercaderes del Valle de Chincha en la Epoca Prehispánica: un documento y unos comentarios. *Revista Española de Antropología Americana*, Volumen 5, pp. 135–78. Madrid.

1972a Breve ensayo sobre el señorío de Ychma o Ychima. *Boletín del Seminario de Arqueología*, Número 13. Pontificia Universidad Católica, Lima.

1972b Las etnías del Valle del Chillón. *Revista del Museo Nacional*, Tomo XXXVIII, pp. 250–314. Lima.

1972c El sitio arqueológico de Con Con en el Valle del Chillón: derrotero etnohistórico. *Revista del Museo Nacional*, Tomo XXXVIII, pp. 315–26. Lima.

1973a Urpayhuachac y el "Símbolo del Mar." *Boletín del Seminario de Arqueología*, Número 14. Pontificia Universidad Católica, Lima.

1973b Plantaciones Prehispánicas de coca en la vertiente del Pacífico. *Revista del Museo Nacional*, Tomo XXXIX, pp. 193–224. Lima.

1977 *Etnía y Sociedad: Costa Peruana Prehispánica*. Instituto de Estudios Peruanos, Lima.

1978 *Señoríos Indígenas de Lima y Canta*. Instituto de Estudios Peruanos, Lima.

1978–1980 Guarco y Lunaguaná. Dos Señoríos de la costa sur-central del Perú. *Revista del Museo Nacional*, Tomo XLIV, pp. 153–214. Lima.

Contents

Tables

Figures

PART I
The Chillón Valley "Coca Lands"
Archaeological Background and Ecological Context

Joyce Marcus and Jorge E. Silva

Archaeology and Ethnohistory

Archaeology and ethnohistory are two subfields of anthropological investigation. While they employ different methodologies and frequently reveal non-overlapping data sets, they share a similar goal to reconstruct both the *specific* and *general* processes that characterized the evolution of ancient populations (Marcus 1983:v).

The strength of ethnohistory is that it often permits detailed synchronic reconstructions that involve the interplay of multiple factors, while archaeology's strength has been its diachronic reconstruction of long-term transformations within and among societies. Time depth is provided by archaeology, while ethnohistoric data often supply the motivations of key individuals, and the names, dates, places, and specific events of the past. Ethnohistory provides concrete cases that can sometimes be used to develop testable models about past behavior. How far back those models and patterns can be extended must be determined by archaeologists.

The data gleaned from the archaeological and ethnohistoric records should be considered *complementary*, because rarely are they precisely contemporaneous, nor are they exactly comparable in thematic or geographic coverage. Each set of data should be thought of as a "case study" of a particular situation with its own temporal, geographic, and thematic constraints.

While some archaeologists only turn to ethnohistory as a secondary support—that is, when the ethnohistoric data reinforce their conclusions (and ethnohistorians frequently do the same with archaeology)—it is often far more revealing when archaeological and ethnohistoric data are in conflict. That moment is an invitation to serious investigative work. Research designs should take the areas of conflict as their point of departure, and one goal should be to explain the nature of the conflict; we have in fact been ignoring or glossing over such areas of conflict far too long.

What are the limitations of ethnohistoric data? Unfortunately, such data may have been gathered from as few as one "typical" Indian informant to as many as several highly reliable Indian witnesses. However, all informants have particular biases; inescapably, data are subjective presentations by Indian informants. The Spaniards who recorded the data and took the depositions also had their own set of purposes and biases, some of which resulted from their Western religious education and goals, while others involved a different worldview or a set of concepts and traditions embedded in the sixteenth-century Spanish legal system.

What are the limitations of archaeological data? For non-literate Andean civilizations, the data recovered through excavation cannot supply the names of ethnic groups or ancient villages, towns, and cities, nor the days, months, and exact years when specific events took place. Ethnohistory supplies us with these and other data for the recent past, and thus allows us to flesh out our archaeological reconstructions, particularly for the Late Horizon.

At least one topic—ethnicity—should be mentioned here, because both archaeologists and ethnohistorians have struggled with it, and both provide data on the topic. Using synchronic variation in ceramic and architectural styles, some archaeologists have assumed (1) that such variation was related to the prehistoric establishment and maintenance of ethnic identity, and (2) that such data can be used to map out ethnic groups and their boundaries. Some scholars view such ethnic boundaries as a passive reflection of group differences, while others see stylistic variation as an active system cultivated and maintained by each group. Still other scholars are hesitant to relate ceramic and architectural variability to the maintenance of ethnic differences, and instead prefer to attribute it to environmental, economic, or other "functional" factors.

The names of ethnic groups and the motivations for exchange, competition, and alliance-formation can sometimes be recovered from ethnohistoric data. However, it is still difficult to establish whether entire ethnic groups or segments within these ethnic groups were implementing these strategies. Were segments within these ethnic groups trying to achieve self-sufficiency, or were they quite specialized? At what level of the settlement system was self-sufficiency or specialization operating? For example, a household may be specialized, but the community may be self-sufficient. It is therefore all-important that we isolate the relevant level of society for purposes of analysis and model-building. Between the two extremes or "idealized models" of complete self-sufficiency and full specialization, there is a broad continuum of strategies and adaptations. Such topics are difficult enough if we approach them synchronically with ethnohistoric and ethnographic data, but if we also wish to give the patterns some time depth, and inquire as to their origins, developments, and changes through time, we must work with archaeological data.

Turning now to our particular "case study"—the inter-ethnic group struggle for access to coca lands in the *chaupi yunga* zone, near Quivi, in the Chillón Valley—we must place the protagonists and events mentioned in the document into an historic, ecological, and sociopolitical context. To do that, we must cast our net widely to discuss the regions contiguous to the Chillón Valley (Figure 1); and to understand the present and recent past, we must extend our knowledge into the archaeological record.

Chillón Valley

The Chillón Valley is about four kilometers wide at its mouth, but narrows considerably over the 65-kilometer route inland to the point where it becomes a *quebrada* in the *sierra*. The valley can be divided into several broad zones: (1) the lower valley desert coast or *yunga* (0–600 meters), (2) the middle valley scrub desert or *chaupi yunga* (600–2000 meters), and (3) the upper valley (2000–5000 meters)—the latter including a number of vegetation types from lower sierra thicket, to lower sierra thorny steppe, to mountainous steppe, to humid subalpine plain, to humid mountainous forest, to the rainy alpine tundra (Tosi 1960).

The lower Chillón Valley, which extends from the Pacific Ocean to Huanchipuquio, is contained within the modern Department of Lima (Figure 1). The middle and upper valley zones are part of the modern Province of Canta. The region of the Chaclla ethnic group (people who were part of the macroethnic Yauyos group) falls within the modern Province of Huarochirí (Figure 1).

Let us now look at the lower, middle, and upper valleys.

The Yunga *or Lower Valley Zone*

The *yunga* or "warm land" within the Chillón Valley (Figure 3) includes approximately 10,000 hectares of valley floor land that today, as in the past, is extensively irrigated and farmed, extending from the seashore to Trapiche (elevation of ca. 600 meters). A narrow temperature range (18.6°C on the average) characterizes the coastal desert valley. A low fog (*garúa*) exists on the coast and penetrates as far inland as Trapiche, providing some moisture in the air, but not enough to permit vegetation or agriculture without irrigation (Figure 4).

The Chaupi Yunga *or Middle Valley Zone*

The *chaupi yunga* is the key zone for understanding the *Justicia 413* document, the main focus of this volume. This zone (Figure 2) is geographically, as well as sociopolitically and economically, intermediate between highland and coastal populations.

The Quechua term *chaupi yunga* indicates that the middle valley zone is part of the *yunga* or "warm land," and the word *chaupi*, "middle," indicates that these lands lie between the *sierra* and the coast. The *chaupi yunga* in the Chillón Valley extends from Huarabi to Arahuay on the Arahuay River, and to Yaso on the Chillón River.

Arable land in the *chaupi yunga* is scarce; in contrast to the 10,000 hectares (Figure 5) in the lower valley, the middle valley has only 4000 hectares. As we proceed inland, the valley floor narrows into a canyon, where fertile land is restricted to a strip adjacent to the river. From Macas to Quives, the strip varies in width from .4 to 1 kilometer; between Quives and Pacaybamba, no more than 300 cultivable hectares are available today (Dillehay 1976:52).

Rainfall is ca. 125 to 250 millimeters annually; indigenous vegetation (Tosi 1960:33) includes *Gynerium, Cortaderia, Opuntia, Cereus, Acacia, Schinus*, and *Prosopis*. Temperature averages 17.6°C. Coca and fruits are some of the preferred crops in the *chaupi yunga*.

All settlements along the Chillón and Arahuay valleys are close to the valley-long canal system. At higher elevations each site is connected to one main canal, has feeders from secondary canals, or has access to a small pond, well, or springs.

The Sierra *or Upper Valley Zone*

The highlands or *sierra* begin about 85 kilometers inland, between Yani and Pacaybamba along the Chillón River and near Arahuay on the Arahuay River. This zone begins at an altitude of ca. 2000 meters and reaches 5000 meters (Figure 2).

The lower *sierra* zone (2000–3000 meters) is quite narrow, from 12 to 26 kilometers wide with its eastern limits falling near Huaros, Carhua, and Lachaqui. This rugged zone includes many mountaintop archaeological sites overlooking the valley floor, such as Huamantanga, Huayuacancha, Cantamarca, Purumarca, Kekamarca, and Huancuna. In the next zone upvalley (ca. 3000–3500 meters) potatoes, oca, olluco, and other

Figure 1. Map showing the central coast of Peru.

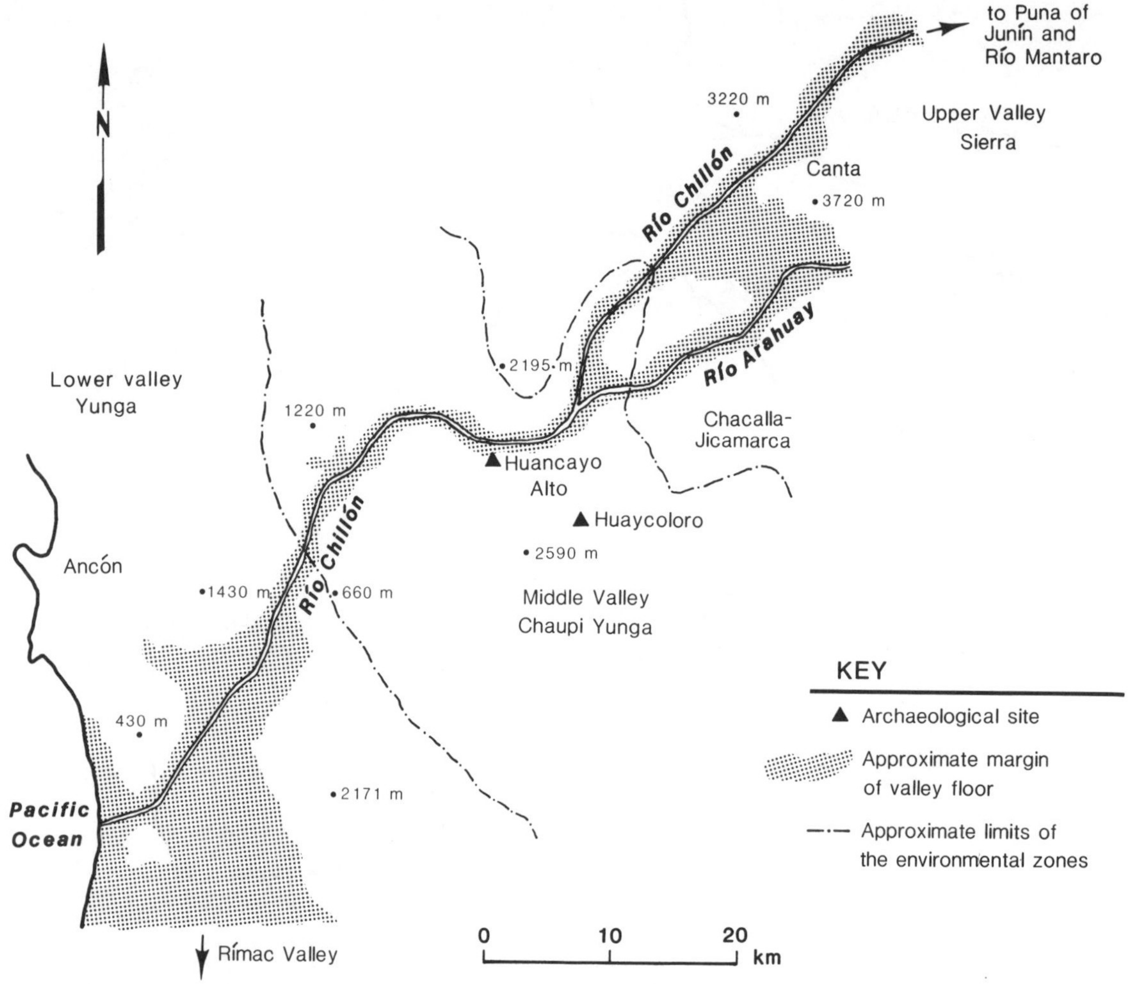

Figure 2. Map showing the three environmental zones within the Chillón River Valley. Note the lower valley (*yunga*), middle valley (*chaupi yunga*), and upper valley (*sierra*) zones as one ascends from sea level in the west to over 3500 meters in the east [redrawn after Dillehay 1977: Fig. 1; 1979: Fig. 1].

crops are grown; camelid meat and wool are other important products.

Inhabitants in the *sierra* were able to acquire desired resources from the *chaupi yunga* and lower valley by means of different strategies: (1) marriage and reciprocal exchange alliances, (2) the establishment of colonies, or (3) conquest (Dillehay 1976, 1979, n.d.).

In the *Justicia 413* document (folio 188v) we learn of a reciprocal exchange alliance from a Guaravi informant who stated that *chaupi yunga* residents had highland Canta as "partners" with whom they exchanged coca for llamas and potatoes. In addition to the Quivi and the Guaravi, for whom we have ethnohistoric data, we might infer that similar reciprocal exchanges linked other *sierra* and *chaupi yunga* groups.

The archaeological data indicate that coastal items also reached the *sierra* (Dillehay 1976). *Spondylus* and other marine shells have been found on the surface of sites in the *chaupi yunga* and in the *sierra*. Evidently the coastal *Yunga* of the Chillón Valley were part of a far-flung trade network linked to the warmer waters of far northern Perú and southern Ecuador.

Other cooperative alliances linked the *sierra* and *chaupi yunga* occupants. For example, the Canta and the Quivi cooperated "to open some *sierra* lakes," that is, to work together for mutual benefit during a drought, a practice mentioned in the *Justicia 413* document (folio 206r). When in a non-cooperative mood, the Canta used the fact that "highland" water was used to irrigate Quivi lands as justification for their raids on lands in the *yunga* zone.

Figure 3. The broad lower sector (*yunga*) of the Chillón Valley.

Figure 4. View of the lower Chillón Valley. In the background the lower valley (*yunga*) begins to narrow as it ascends to the middle valley (*chaupi yunga*) zone.

Figure 5. View of the arable land in the middle Chillón Valley.

The ethnohistoric documents indicate the presence of many ethnic groups in the Chillón Valley and surrounding area—Canta, Chaclla, Colli, Guancayo, Inca, Socos, and Atavillos (Figure 6). At some time most of these either had coca fields in Quivi or were fighting to have access to coca lands.

While there are archaeological and ethnohistoric data that can be integrated to understand the Late Horizon (A.D. 1470–1534), there is much less information (particularly ethnohistoric) available for understanding earlier periods. In fact, there are extremely long periods of time for which we lack both detailed, systematic settlement pattern data and stratigraphic data. With additional excavation at Chillón Valley coastal and highland sites, it should be possible to establish a ceramic sequence to complement that of Huancayo Alto (Dillehay 1979), a site located in the *chaupi yunga* zone. Any evolutionary sequences for the Chillón Valley thus must await additional systematic survey and extensive excavations.

Before looking at the archaeological sequence for the Chillón Valley and the egalitarian and non-egalitarian societies that occupied it, let us review some of the available archaeological evidence for coca use and production.

Coca

The shrub called coca belongs to the genus *Erythroxylon*. Four cultivated varieties containing the alkaline cocaine are known from South América, but they differ significantly in the content of alkaloids (and other chemical elements), and in the anatomy of leaves, bark, fruit, and flowers (Plowman 1984). Each of these varieties was chewed by the indigenous population and designated by a different native term. Coca was used as medicine and stimulation, to allay cold, fatigue, pain, and hunger, as well as for ritual and social purposes.

Various scholars have suggested a brief scenario to explain the origins of coca use and its domestication (Towle 1961; Antonil 1978; Plowman 1979; Bohm, Ganders, and Plowman 1982), in which hunters and gatherers might have first tried the leaves during famines. The beneficial properties of coca were discovered and the leaves were collected. Later developments involved drying the leaves, keeping them in the mouth as a quid, and adding lime to the quid. Entire shrubs might have been transplanted to locations nearer the base camp, perhaps to reduce search area, to increase availability, and to increase shrub density.

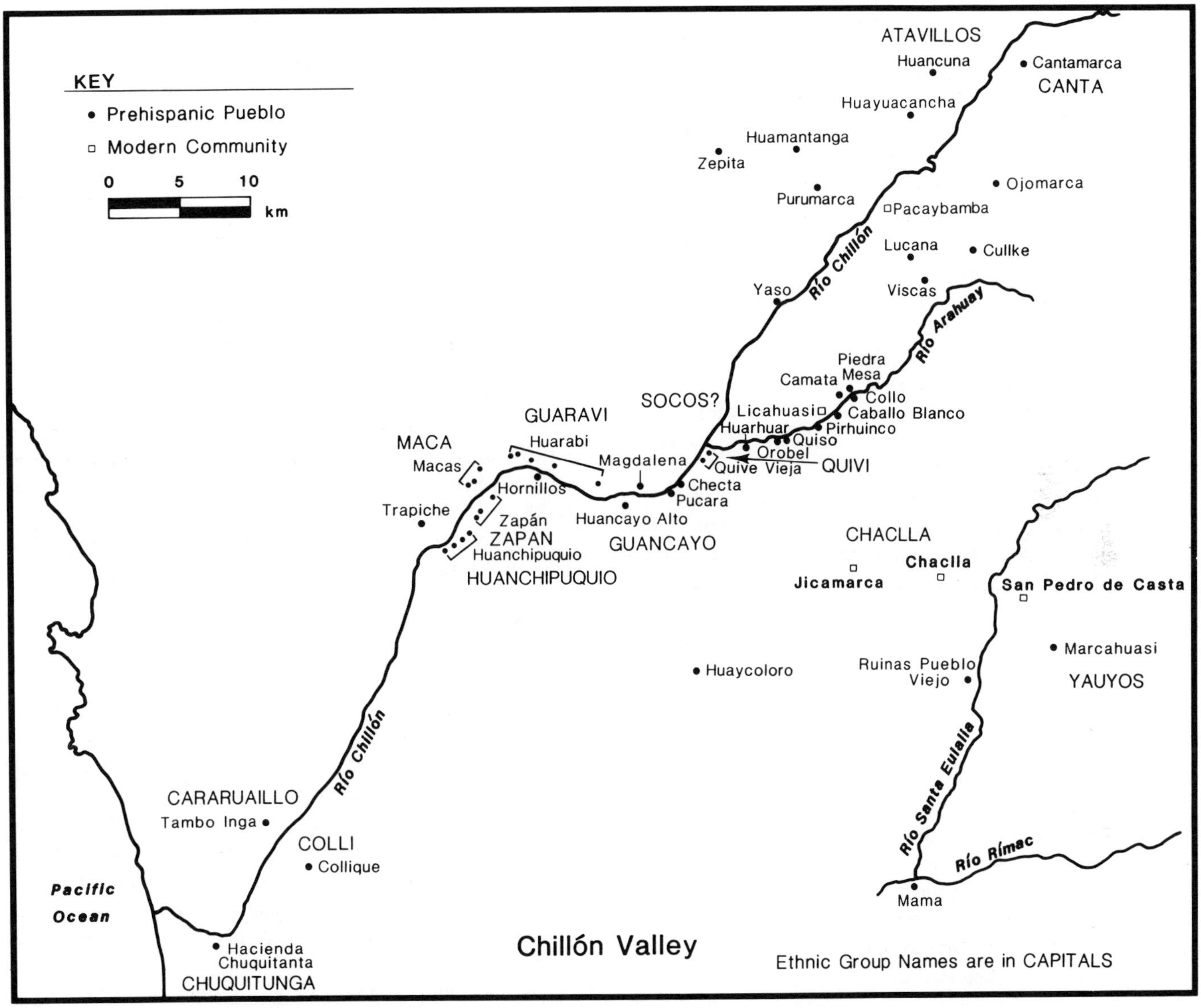

Figure 6. Map showing the approximate locations of many of the sixteenth-century Chillón Valley ethnic groups mentioned in ethnohistoric documents. Additionally, the names of some modern place towns and those of archaeological sites (many of which were occupied in the Late Intermediate and Late Horizon periods) are given. [Redrawn after Dillehay 1976: Fig. 1, 2; Rostworowski 1978: Mapa 1]

Of the four cultivated cocas, only two will concern us here. One is *Erythroxylon coca* var. *coca* [indigenous term, *mamox*], which grows at an elevation from 500 to 1500 meters and is often called *Huánuco* or *montaña coca,* because it still occurs as a wild plant in the eastern Andean montaña (Golte 1968). It may have been grown as early as 7000 B.C., gradually diffusing throughout the eastern Andes from Ecuador to Bolivia. Transplants of this variety to the irrigated Peruvian coast were unsuccessful (Plowman 1984:133).

Another important variety is *Erythroxylon novogranatense* var. *truxillense* [indigenous term, *tupa*]. Significantly, re-examination of prehispanic coca leaves from the coast of Perú has revealed that almost all of them belong to this latter variety. Only one Late Horizon example of montaña coca has been recovered from a coastal context (see Plowman 1984:135, Mortimer 1901). Today Trujillo coca is cultivated on the north coast at elevations from 200 to 1800 meters. It requires some irrigation, although it is a very drought-resistant shrub.

While coca has been reported from Late Preceramic contexts, we still do not know which variety is represented since identifications by botanists were never made. At Culebras, Department of Ancash, during the Late Preceramic 6 period (ca.

2500–1800 B.C.) a gourd vessel and 3 *Mytilus* shell containers filled with lime were reported (Engel 1957:67–68; Bray and Dollery 1983:271). Engel (1963) also reported coca and lime from Asia at 1314 ± 100 B.C. Patterson (1971:195–96) states that coca was cultivated during the Gaviota phase (1900–1750 B.C.) in the floodplains of the Chillón Valley. "Chewed coca quids" from early Initial period levels at the Tank Site at Ancón have also been reported (Cohen 1978:36–37).

Of still greater interest here is the fact that Dillehay (1979) has reported coca from excavations at Huancayo Alto in the *chaupi yunga* zone of the Chillón Valley, not too far from the Quivi lands, which were the focus of the inter-ethnic group struggle in the *Justicia 413* document. Dillehay (personal communication) recovered coca leaves from (1) a storage unit dating to 800 B.C. and (2) Late Intermediate/Late Horizon levels; all leaves were identified by Emma Cerrate and Ramón Ferreyra as Trujillo or coastal coca.

However, from more recent sites and contexts (e.g. Vista Alegre [Rímac Valley], Nazca [Taruga Valley], Monte Grande [Ica Valley]) the archaeological coca leaves have all been identified; all have been assigned to the Trujillo variety.

Plowman (1984:144–45) now believes there is sufficient evidence to suggest that the Trujillo variety was cultivated on the Peruvian coast by 2000 B.C., and he speculates that that variety "could have arisen as early as 4000 B.C." Just as botanists have identified varieties of Peruvian coca, Indians also distinguished between the *tupa* or "small-leaf" coastal Trujillo variety and *mamox*, the "large-leaf" Huánuco or montaña coca (Rostworowski 1973).

From the *Justicia 413* document we know that coca was grown near Quivi, and given present data we might suggest that (1) it was the small-leaf *tupa* or Trujillo variety and (2) that Huancayo Alto had access to this variety.

Now let us turn to the archaeological sequence for the Chillón Valley and surrounding territory.

Preceramic Period (ca. 15,000–1800 B.C.)

Since sites dating to the period from ca. 15,000 to 9000 B.C. are unknown at present for the Chillón Valley, we will turn our attention to sites occupied during the latter part of the Preceramic (Figure 7).

Preceramic occupation seems to be concentrated on the coast, but this apparent pattern may not be a wholly accurate reflection of early settlement. For the most part, intensive survey in zones other than the coast has not been undertaken; however, when limited survey has taken place in such areas it has been difficult to locate early sites. While it is true that it is easier to locate early sites on the coast and that more intensive survey has been conducted there, it may also be the case that occupation there may have been denser.

Preceramic 3: ca. 9000–7000 B.C.

Lanning and Patterson attempted to establish a long Preceramic sequence for the Chillón Valley. Toward that goal, nearly 300 sites were located by them in the area between Ancón and the lower sector of the Chillón River valley (e.g. Lanning 1963, 1965, 1967; Patterson 1966a, 1971; Patterson and Lanning 1964; Patterson and Moseley 1968; see also MacNeish, Patterson, and Browman 1975; Moseley 1975; Bonavia 1979; Quilter 1985).

Lanning located approximately 50 workshops whose stone assemblages he used to define four complexes: Red Zone, Oquendo, Chivateros 1, and Chivateros 2. The first two "industries" were thought to be associated with "undifferentiated gatherers" or "bands." Patterson (1966a) has reported on the limited stratigraphic and radiocarbon evidence he had, arguing that the Oquendo complex occurs between the Red Zone and Chivateros 1, and that the Red Zone dates to ca. 10,000–9000 B.C.

For approximately 20 years the "quarry" or "workshop" sites known as Chivateros have been regarded as the earliest occupations on the central coast (Figure 7). Most attempts to reconstruct the way of life of the people in the Chillón Valley—who produced and used those stone tools—have used the Chivateros data as the basis for reconstructions.

Because these Chivateros sites were located in the *lomas* zone, Lanning emphasized the *lomas*, a hilly area (just inland from the sandy coastal desert) that supports a seasonal vegetation, dependent on continuous exposure to mist and moist air; when the *lomas* are "in bloom" (sometimes for a few months a year, sometimes every few years), they support vegetation that may be suitable occasionally for short-term camelid grazing, but of little use for cultivation or plant collecting for food that is edible by humans. Thus, even though the *lomas* offers an unpredictable resource, it may be useful (1) for helping a herder who is searching for pasturage for his camelids, or (2) for a hunter (who is a member of a band) looking for an area that attracts wild game.

On the basis of his Chivateros data, Lanning concluded that these people were "migratory food gatherers in the lomas areas" (Lanning 1967:56) who concentrated their efforts on the collection and consumption of wild plants, and that "hunting played only a relatively minor role" (Lanning 1967:46). Using the same data, MacNeish, Patterson, and Browman (1975) suggested that hunting was probably a primary activity supplemented by seasonal utilization of the *lomas* (wet season) and the valley floor zones (in the dry season). Still another view held that "highland-based hunters and gatherers apparently migrated down to the lomas seasonally, following game" (Sanders and Marino 1970:34).

Another model was developed by Lynch (1983:110) who

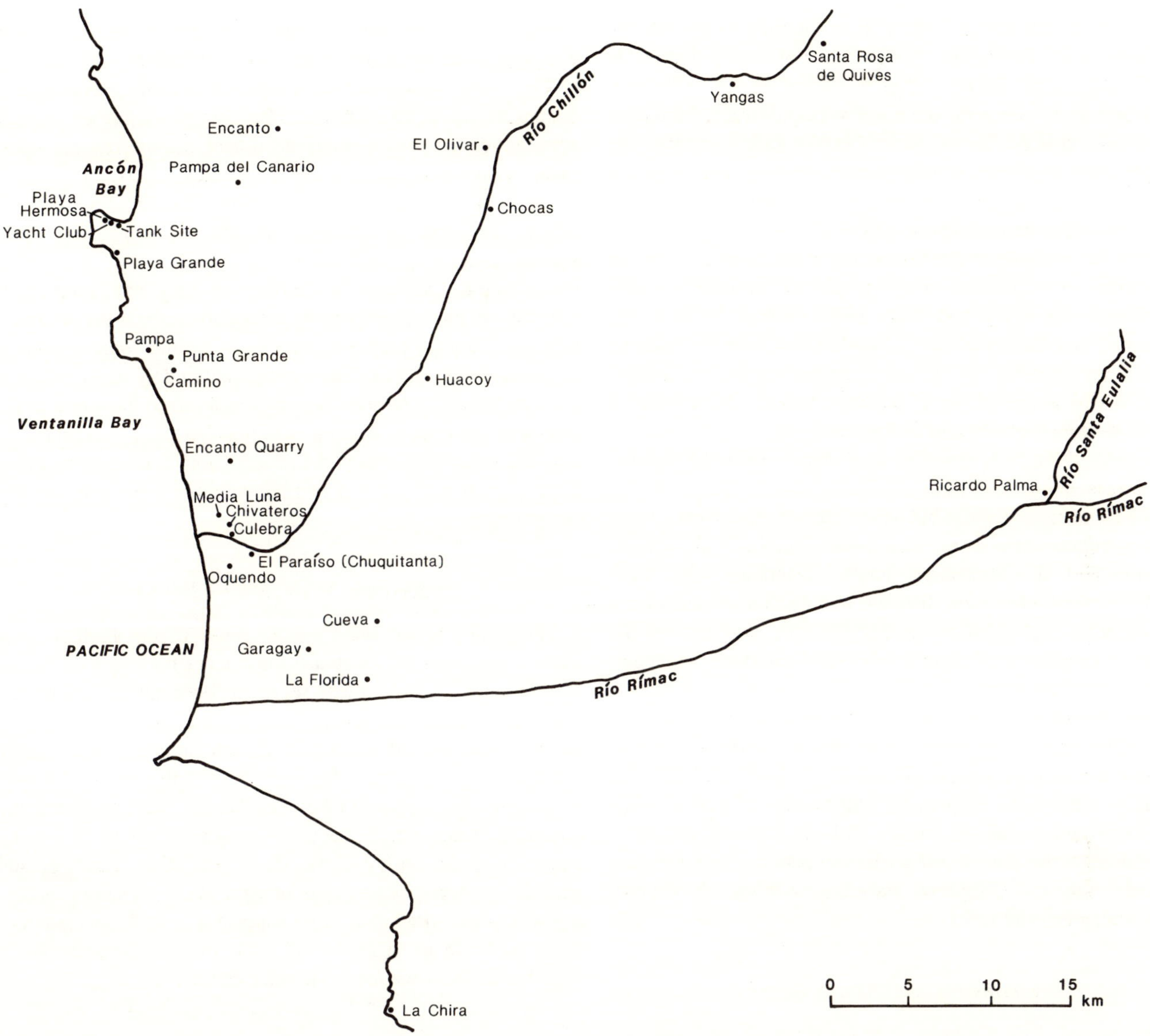

Figure 7. Some of Peru's central coast Preceramic and early pottery-making villages located between Ancón Bay (on the north) and the Rímac River (on the south).

suggested that seasonal transhumance from the highlands to the low altitude zones took place, such that the "coastal occupations represent lateral excursions into habitats that could hardly have provided, even at 9,000 or 10,000 B.C., game and vegetal food resources comparable to those of the montane zones" (Lynch 1983:111). One of the important aspects Lynch emphasizes is that the coastal inhabitants employ different strategies from those at higher elevations. Greater reliance on hunting (meat-eating) characterized the serranos, while gathering (wild plant foods and shellfish) characterized the costeños. In support of this assertion is the fact that the oldest remains from the northern Peruvian coast (the Paiján complex) are not associated with megafauna and that the tooth wear pattern analysis on a young human female (Paiján complex, ca. 8250 ± 180 B.C.) indicates that her diet was "more vegetarian than meaty" (Lynch 1983:115).

The Chivateros complex is considered to be related to that at other coastal sites on the central coast (e.g. Volcán in the Huarmey Valley) and northern coast (e.g. Paiján; cf. Bonavia 1979, 1982). Some scholars have argued it would be more accurate to describe the earliest occupants of the central coast as practicing a subsistence strategy that relied mainly on gathering, fishing, and collecting, rather than hunting. One of the problems, however, is that "the evidence of early gathering in South America is minimal" (Lynch 1983:121).

All attempts to describe the major sociopolitical features and subsistence changes for the lower Chillón Valley societies are fraught with numerous problems related to (1) meager stratigraphic data, (2) absolute chronology, (3) stone tool seriations, and (4) inadequate ecological data.

Since the majority of the Chivateros implements could more accurately be classified as a group of unfinished tools (preforms) (see Craig and Psuty 1968; Parsons 1970; Fung, Cenzano, and Zavaleta 1972; Lynch 1974, 1983; Chauchat 1975; Bonavia 1979, 1982), some scholars question the assignment of Chivateros to the Preceramic period and whether it should be considered a tool industry.

Despite all the problems discussed above, it may be inferred with a reasonable amount of confidence that the band was the unit of social organization, but we still do not know what kind of band societies they were. Are lower Chillón Valley band societies part of a widespread coastal adaptation? Or, is the Chillón coastal adaptation different in important respects from that of other sectors of the coast? We still lack data on the relative importance of plant foods, fish, shellfish, and other foods to the Preceramic diet.

Even though Lanning's work on the Preceramic of the Chillón Valley was a pioneering effort, a better understanding of this era will require a sophisticated research design involving intensive survey and excavation, nutritional analyses of available foods, and ecological studies. This research will enable us to reconstruct the Preceramic settlement patterns, procurement systems, stone tool complexes associated with specific subsistence strategies, and diet.

Preceramic 4: ca. 7000–5500 B.C.

The Arenal, Luz, Canario, Corvina (Pampa), and Encanto phases follow the controversial Chivateros complexes we discussed above. Although the subsistence strategies during Preceramic 4 times are thought to be different or more diversified from those employed in Preceramic 3, the evidence is limited. One of the few things we can say is that the societies continue to be organized as bands.

The Arenal complex (6000–5300 B.C.) includes "pressure-flaked stemmed projectile points...well-made scrapers...chipped stone awls and small pointed tools, and a very few milling stones" (Lanning 1967:48). The type site of El Arenal is located 3 kilometers north of the Chillón Valley; similar material is known from Ancón (Lanning 1963).

A more recent, but related, complex is Luz, which is another complex found on the *lomas*. Lanning reports at least 23 shallow sites that contained gourd rinds (*Lagenaria siceraria*) and a few mortars. The transhumant herders or hunters may have been carrying some water and plant food with them.

The most significant development during Preceramic 4 occurs by ca. 3700-2500 B.C. (Encanto phase) when there is some evidence for an increased use of marine resources and plant cultivation (Lanning 1967; Moseley 1975).

Bands at this time may have employed a diversified strategy, which consisted of utilizing marine resources (e.g. fish, shellfish), hunting, and collecting plants in the *lomas* zone (according to Lanning), or the diversity may be a result of the fact that we have slightly more information about Preceramic 4 than 3. Lanning proposed that now extinct *lomas*—located between Ancón and Chillón—were inhabited by small groups in the wet season. Although Lanning emphasized the *lomas* as the principal place for collecting wild plant foods, he presented no tangible evidence to support this assertion; furthermore, as we have seen, there is next to nothing from this zone that is edible for a human.

Preceramic 5: ca. 5500–4200 B.C.

The type site for the Canario phase (5200–4200 B.C.) is Pampa del Canario, located 6 kilometers northeast of Ancón Bay (Figure 7). Settlement seems to be concentrated on the *lomas* near Ancón. Lanning (1963:49) uses the tools—"milling stones, shaped and unshaped manos, mortars and cobble pestles..."—as evidence for significant subsistence changes. Patterson (1971) argued that diverse habitats were exploited: the *lomas*, the sandy beach, the rocky headland littoral, the coastal desert, and the valley floor. Plant collecting, hunting, and shellfish gathering were apparently the principal strategies during this phase. Patterson also suggested that these resources were used as part of an annual cycle of transhumance which took bands from winter to summer camps.

Preceramic 6: ca. 4200–2500 B.C.

This period is represented by two complexes: the Corvina (4200–3700 B.C.) and the Encanto (3700–2500 B.C.). Only three sites contain the Corvina complex: two in Cerro La Corvina (4 kilometers east of Ancón) and one at Loma Encanto (6 kilometers northeast of Ancón). At Loma Encanto, Lanning (1963:49) found projectile points, and a "fair number of manos and a couple of milling stones, but no mortars or pestles."

The Encanto complex is known from fifteen sites around the Ancón-Chillón area. Five are small workshops while the others are camp sites. Apparently, year-round settlements had been established in this area, and in the Lurín and Chilca valleys (MacNeish, Patterson, and Browman 1975:29). Based on data from the Yacht Club site, a settlement overlooking the Bay of Ancón, and the Chilca site, in the lower Chilca Valley, it was proposed that sea food was the primary subsistence resource, supplemented by squashes (*Cucurbita ficifolia, C. moschata*), and cotton (*Gossypium barbadense*) (Lanning 1967:50). In fact, 23 squash seeds were recovered at the Encanto type site (Moseley 1975). Twined cotton textiles and cotton fish nets were found at the Yacht Club and Chilca sites (Lanning 1967:50).

Obviously, these kinds of remains actually indicate what some of the subsistence strategies were by Encanto times, especially those related to fishing technology as shown by the use of fishing nets (Lanning 1967; Moseley 1975:53).

Preceramic 7: ca. 2500–1800 B.C. *("Cotton Preceramic Stage")*

Qualitative changes apparently did take place during this era, since by the end of this period there are indications that a social hierarchy has emerged. Let us first look at the archaeological evidence before reassessing the models and inferences that have been made by others.

The Pampa phase (2500–2300 B.C.) is named after the type site of Pampa located at the north end of the Ventanilla Bay, on the north margin of the Chillón Valley. Lanning (1967:53) has suggested that the site was a temporary camp, inhabited by people who fished and planted squash who gradually adopted a more marine-oriented subsistence. Although Lanning (1967) and Moseley (1975:116) argue that fishing was the primary subsistence strategy, a series of cultivated plants was found at this site. For example, the oldest wild cotton for the Ancón-Chillón area was discovered at Pampa (Moseley 1975:22). Additionally, such domesticated plants as squash (*Cucurbita ficifolia, C. moschata*) and jack beans (*Canavalia* sp.), as well as wild squash (*C. ecuadorensis*) have been found (Patterson and Moseley 1968; Cutler and Whitaker 1961). Cohen (1977) reported pacae (*Inga feuilli*), achira (*Canna* sp.), guava (*Psidium guajava*), and a legume (*Galactia striala*).

The Playa Hermosa phase (2300–2100 B.C.) is characterized by a continuation of the subsistence strategies of the Pampa phase. However, plant use increased, and new year-round sites are occupied in the Ancón-Chillón area. The Playa Hermosa phase is represented by the Camino and Banco Verde sites (in the Ventanilla area) and the Yacht Club site (in Ancón) (Moseley 1975). Despite the fact that these sites provided remains of a new cultigen (chili pepper, *Capsicum baccatum*) and old cultigens (cotton, gourds, and guava), most archaeologists have concluded that the settlements were marine-oriented.

Conchas phase (2100–1900 B.C.) is known from the Punta Grande site (in Ventanilla) and the Tank Site (in Ancón) (Moseley 1975).

Punta Grande is the larger site, composed of 4 artificial terraces arranged one above another at the base of a hill. The terraces were apparently used as living surfaces. Punta Grande shows heavy reliance on the resources on the sandy littoral zone, even though two new cultigens are present—lima beans (*Phaseolus lunatus*) and lúcuma (*Pouteria lucuma*).

The Gaviota phase (1900–1750 B.C.) may represent an era of increasing sociopolitical complexity as shown by the construction of large civic-ceremonial buildings. The site of El Paraíso (Chuquitanta), located on the south bank of the Chillón River, has provided significant information on the construction techniques of early buildings and on some of the ceremonial activities that took place there (Figure 8). Engel (1966, 1967; see also Moseley 1975) excavated at El Paraíso and suggested that its construction was completed under the direction of some centralized authority.

It has been assumed that El Paraíso's inhabitants engaged in a maritime economy (Moseley 1975). However, a number of cultivated plants were recovered, including cotton, gourds, achira, lima beans, jícama, peanuts (*Arachis hypogaea*), and possibly sweet potatoes (*Ipomoea batatas*).

Although El Paraíso is composed of 9 mounds, they may not all be contemporaneous. Engel excavated Unit I (see Figure 9) which revealed a sequence of 5 construction phases. The most recent construction phase of Unit I yielded a C^{14} date of ca. 2000 B.C. This unit included a special sunken square pit surrounded by 4 circular pits which perhaps served for special civic-ceremonial purposes (see Figure 9).

Lanning (1967) first suggested that El Paraíso was part of the U-shaped building tradition; and more recently, Williams (1971, 1978–1980, 1985) has argued that the site layout of El Paraíso constitutes the oldest evidence for this building type. Doubts about the building shapes and layout of the El Paraíso site have been raised by Scheele (1970) and Silva (1984). More recent work at Units I, II, and IV of El Paraíso by Quilter (1985) suggest that a variety of residential and ceremonial activities took place at this settlement.

El Paraíso has been regarded as a site constructed by a chiefdom (Carneiro 1970; Sanders and Marino 1970). By 2000 B.C. it may be that a series of chiefdoms had emerged in several parts of the central Andes (at Kotosh in the Middle Huallaga River valley, at Pacopampa in the Chotano Basin, and elsewhere). As a focal point within a polity, El Paraíso perhaps integrated settlements within and beyond the lower sector of the Chillón Valley. We still do not know what the internal characteristics of these villages were, the number of non-residents who attended special ceremonies there, the nature of inter-site interaction, nor why such a site as El Paraíso was abandoned ca. 1800 B.C.

Figure 8. View of reconstructed Unit I at El Paraíso (Chuquitanta), a Late Preceramic Period (Late Gaviota phase) site located on the south bank of the lower Chillón River. See also Figure 7. (Photo: J. Silva)

Initial Period (ca. 1750–1000 B.C.)

This period shows significant trends in regard to the economic and sociopolitical development of the Ancón-Chillón area societies. This is a time of intensive cultivation on the floodplains of the lower and middle Chillón River valley. For reasons that need to be explained, fishing apparently was no longer the primary subsistence activity. Instead, a diversified economy in which exchange played an increasingly important role is evident. As part of these trends coca cultivation perhaps became an important item by the beginning of this period. As mentioned above, the earliest evidence for coca comes from Late Preceramic coastal contexts. Early evidence for coca use in the Chillón-Ancón area comes in the form of chewed quids from Initial period levels (ca. 1750 B.C.) at the Tank Site in Ancón (Patterson and Moseley 1968; Cohen 1977:66, 1978:37). Additionally, although coca may have been cultivated at this time in the mid-valley (*chaupi yunga* zone) of the Chillón River valley (MacNeish, Patterson, and Browman 1975:33), sufficient data for that zone are not yet available to say more. Now let us turn to some of the sociopolitical events which took place by this time in the Ancón-Chillón area.

Integrative Role of the La Florida Site

According to Patterson (1971:198), the abandonment of El Paraíso was related to the emergence of La Florida; he suggested that a group of families from El Paraíso apparently joined those who abandoned villages around Ventanilla and together they settled at La Florida (located 1.5 kilometers north of the Rímac River; Figure 7). However, there is no direct archaeological evidence to demonstrate any movement of people from the Chillón to Rímac valleys and this scenario may obscure the processes that brought about the abandonment of El Paraíso and the changes in the Rímac Valley that brought about the emergence of a new political power in that valley.

La Florida and El Paraíso are different in shape, size, and disposition of sacred and non-sacred space. The sacred room of Unit I of El Paraíso is small and designed for the performance of secluded ceremonial activities (Figure 9). The arrangement of the sacred space at La Florida conforms to the more open and public U-shaped layout. In brief, this latter layout exhibits a clear-cut differentiation of secluded and non-secluded areas; the large, open area or plaza is separated from the small, secluded rooms in the west-central, north, and south mounds.

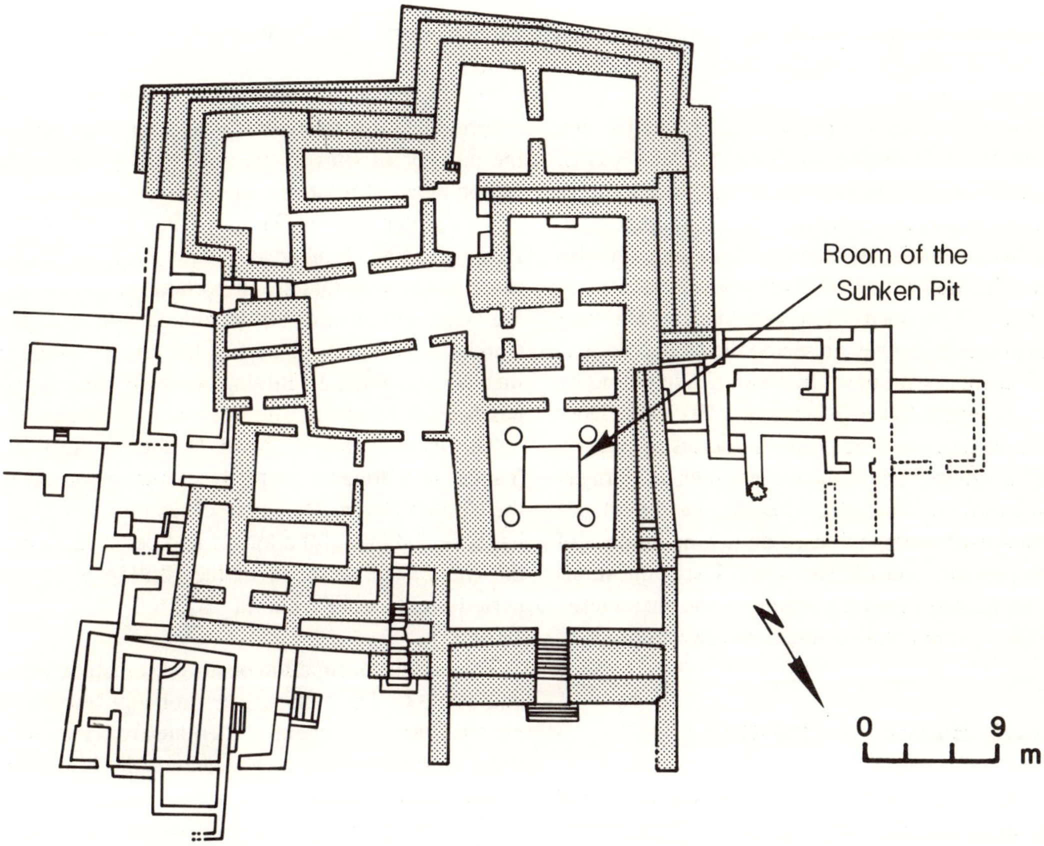

Figure 9. Plan of the reconstructed Unit I at El Paraíso (Chuquitanta). On the north side of the structure a series of steps leads to a square room with a sunken square pit surrounded by 4 circular pits. [Redrawn after Moseley 1975: Fig. 6.5; Quilter 1985: Fig. 1]

Were El Paraíso and La Florida occupied simultaneously (ca. 2000 B.C.), or were they occupied sequentially (as Patterson has suggested)? If the former, was their relationship competitive or did they merely offer rites and services to different populations? These are issues to be addressed by future research.

The construction of the La Florida site apparently co-occurs with other processes: (1) the appearance of pottery for the first time throughout the central coast; (2) the construction of U-shaped public buildings over the central and northern coast of Perú (the one at La Florida is just one example); (3) the apparent emergence of a strong political unit in the Rímac Valley; and (4) the regional importance of La Florida and its apparent control over the Ancón-Rímac area, an area that included several sites (Patterson 1985:65). The fact that similar pottery occurs throughout the Ancón-Rímac area at this time has been used to support the idea of widespread political control.

Although it has been stated that La Florida was abandoned at 1750 B.C., late Early Horizon and early Early Intermediate components were also recovered at the site (Patterson 1985).

Garagay, La Florida, La Salina, Huacoy, Chocas, and Cueva

At more or less the same time the sites of Garagay, La Florida, Huacoy, and Chocas display U-shaped plans. Garagay is located between the Rímac and Chillón valleys (Ravines 1975; Ravines and Isbell 1976); La Florida and La Salina are in the Rímac Valley; Huacoy, Chocas, and Cueva are situated in the Chillón Valley (Figure 7).

At Garagay the Middle Atrium Temple in Mound B was probably the focal point for ritual; it contains polychrome wall friezes that apparently predate Chavín iconography on the central coast (Ravines and Isbell 1976:266). Ravines (1984) has interpreted the principal image of these friezes as a stylized "langosta marina"; Salazar-Burger and Burger (1982:234) interpret it as a spider, which they note was an important theme within the Cupisnique style. Ravines (1984) has identified two clear construction phases in Mound B. At least one of these construction phases in Mound B is perhaps contemporaneous with the La Florida structure.

Garagay and La Florida were probably only two of several

important centers that performed special services and perhaps controlled a number of villages in various parts of the Ancón, Chillón, and Rímac valleys.

The site of Huacoy, located on the south side of the Chillón Valley, also includes a U-shaped structure (Ludeña 1973). Pottery recovered at Huacoy indicates that there is some temporal overlap between Garagay and Huacoy. Other U-shaped structures are known from Chocas and Cueva. These sites are also located on the south side of the Chillón River valley.

Given the number of U-shaped buildings at this time and their distribution, it is possible that we are seeing the archaeological manifestation of "a successful new chiefdom... expanding by accretion as well as internal growth" (Service 1975:79). Does each site with a U-shaped building perform a series of integrative functions for a ramage? Did some centers such as Garagay have more or wider control than others? Do we have evidence for a two- or three-level hierarchy based on (1) the number of public buildings present, and (2) site size? These questions together with other related issues (for example, the characteristics of the residential units) require additional research.

The Early Horizon (ca. 1100–400 B.C.)

Traditionally, this period has been viewed as one of cultural homogeneity throughout central Perú, particularly if it is viewed as the spread of an art style. Additionally, the widespread Chavín style appears as an intrusive style, displacing earlier ones. One of the most frequent explanations for its dispersion is that the style reflects the adoption of a new religion or "cult" which took the feline as its principal image (Lanning 1967).

Chavín-like pottery is known from many coastal sites. Rather than employ the vague term "influence," we might have sufficient evidence to document elite interaction, or chiefly gift-giving. An example of this gift-giving might be the Chavín offerings placed in the Atrium of the Middle Temple at Mound B, Garagay. Garagay may have had enough prestige to have been recognized by the Chavín elite, and the offerings may have been used to establish and reinforce sociopolitical ties.

Some of Garagay's growing prestige may be linked to its remodeling and growth in population during the Early Horizon. A temple was added to the top of the Middle Temple of Mound B. This new addition (called the Late Temple) might have contained Chavín-like depictions, but they are now destroyed. Wing A at Garagay has wall friezes with depictions resembling those from Moxeque, Cerro Blanco, and Caballo Muerto.

Mound A is associated with a sunken circular courtyard. Given a number of similarities between Garagay and other "chiefly" centers elsewhere in Perú, we might suggest that the chiefs of Garagay were in contact with those of other coastal polities. The major concentration of sunken circular courtyards extends from Alto Salaverry (in the north) to Mala (in the south).

For the lower Chillón Valley, Patterson and Lanning (1964) reported that during the Early Horizon there was an increase in the number of sites in the main valley and in the number of hilltop sites, at least one of which is fortified.

While such sites as Garagay (Lanning 1967; Ravines 1984), Huacoy or San Humberto (Ludeña 1970, 1973), and Cueva are three *yunga* sites that are important for coastal residents, there are some settlements further inland that emerge at this time. Chocas is at the lower part of the *chaupi yunga* (at about 500 meters) and some 10 kilometers from Huacoy; since no work has been done at this site not much more can be said.

The *chaupi yunga* site of Huancayo Alto is located near Yangas at Kilometer 56 (at 1050 meters) on the south bank of the Chillón River. Thanks to excavations by Dillehay, the site has revealed one of the longer and better-known sequences for the *chaupi yunga* sector of the Chillón Valley. It includes 7 periods, extending from ca. 800 B.C. to A.D. 1534 (Dillehay 1979:25).

The earliest occupation occurred sometime between 800 B.C. and 500 B.C. The village was probably founded from another *chaupi yunga* settlement, or perhaps from the lower valley. The Early Horizon pottery at Huancayo Alto is similar to that from Garagay and Huacoy.

Different types of construction are known from Huancayo Alto: 3 small *huacas* (artificial mounds); 3 parallel rows of boulders that lead from the largest *huaca* to 2 small mounds 30 meters away; rooms made of river cobbles; and a set of 57 storage rooms laid out in rows on the hillside. The storage complex at Huancayo Alto is a feature not seen elsewhere.

Further upvalley along the Chillón River is the site of Santa Rosa de Quives, which, along with Huancayo Alto, has produced pottery similar to that known from the lower valley. For that reason, it has been argued that some type of exchange characterized *yunga* and *chaupi yunga* relationships (Patterson 1971; MacNeish, Patterson, and Browman 1975; Dillehay 1979).

Since public buildings are concentrated in the *yunga* zone (for example, at Garagay and Huacoy) and in the lower part of the *chaupi yunga* (Chocas), it seems that the *Yunga* ethnic group had control over populations inhabiting the Chillón Valley. As in the previous period the civic-ceremonial centers were the focal points of Early Horizon societies of this valley.

Early Intermediate Period (400 B.C.-A.D. 600)

The Early Intermediate period is often defined as a time of state development throughout the Andes (for example, the Moche state in the north coast, the Lima state in the central coast, etc.) (Lanning 1967; Earle 1972; MacNeish, Patterson, and Browman 1975; Patterson, McCarthy, and Dunn 1982).

An examination of the Early Intermediate period from the perspective of the Chillón Valley archaeological record, indicates that obvious changes occurred during this period in the central coast of Perú. We will present the evidence first; then, we will discuss the socioeconomic characteristics of this period in the Chillón Valley.

The available information indicates that the flat lower sector of the Chillón Valley (Ancón, Ventanilla, Puente Piedra, Carabayllo) was heavily inhabited during this period. From these sites, three ceramic periods have been defined: Ventanilla, Miramar, and Lima (Villar Córdova 1935; Stumer 1953; 1954a,b;1955; Patterson 1966; Patterson and Lanning 1964; Dillehay 1979).

The Ventanilla phase (ca. 250–50 B.C.) is represented by a fishing village located in the lower Chillón Valley which is abandoned in the second century A.D. (MacNeish, Patterson, and Browman 1975:48). Though no Ventanilla-related sites have been discovered in the *chaupi yunga* zone of the Chillón Valley, this phase displays similarities with the Huachipa-Jicamarca C phase of the middle Rímac Valley (Silva, Hirth, García, and Pinilla 1982, 1983).

Following Ventanilla is the Miramar phase defined on the basis of pottery from Miramar in Ancón. Similar pottery was also recovered at Santa Rosa or Playa Grande, and at Huancayo Alto (Stumer 1953; Patterson 1966; Dillehay 1979). This pottery has strong ties with ceramics found in the Chancay, Rímac, Lurín, and Chincha valleys. In fact, it is part of the so-called "White-on-Red Horizon," a pottery style which extended over central Perú at about 100 B.C. (Willey 1948).

The Lima component spans the years from A.D. 200 to 600; it is divided into 9 periods, with the earliest immediately following the Miramar style, and the latest exhibiting attributes that are similar to the Nievería style, usually assigned to Epoch 1 of the Middle Horizon (A.D. 600–650) (Patterson 1966; Menzel 1964). Components of Lima society have been discovered over an area between the Chancay and Lurín valleys, and it is mainly regarded as a coastal polity. In addition to the Maranga mounds in the lower Rímac Valley which are said to represent the type site of this society, at least 16 settlements were reported for the area between the lower Chillón Valley and Ancón (Patterson 1966). The largest settlement is La Uva which includes 13 buildings (Stumer 1954b; Patterson 1966). At present, apart from the two residential sites Playa Grande and Ancón, only two sites with public buildings have been excavated: Media Luna and Culebra (Stumer 1954b,1955; Quilter 1983; Silva, García, Bragayrac, and Morales 1984). Of these, Culebra has provided data on intra-site organization. It is composed of a public building (Figure 10), with at least 3 construction phases, and numerous domestic structures (Figure 11). Excavations conducted in the domestic area indicated that houses were built of cane (Figure 12), stone, or small handmade adobes (Silva, García, Bragayrac, and Morales 1984; Patterson 1966; Stumer 1955). Also, modern structures are built of canes or reed mats (Figures 13–15).

Lima phase remains occur at Huancayo Alto (Dillehay 1979). This site shows an adobe multiroom building constructed in the middle of the *quebrada*, residential stone platforms perhaps used for *kincha* huts, and an oval stone structure which Dillehay (personal communication) says is typical of many highland sites in the Canta area. While Dillehay concludes that Huancayo Alto was inhabited during this time by "a mixture of *chaupi yunga* and highland ethnic groups," it is possible that the same archaeological evidence could be interpreted as reflecting (1) increased exchange between the two areas, or (2) marriages between highlander and *chaupi yunga* residents.

Though information is limited, we can say a few things about the Early Intermediate: (1) sites display public architecture quite different in layout from the U-shaped plan that characterized the previous Initial and Early Horizon periods; (2) population was concentrated in the lower *yunga* sector of the Chillón Valley; (3) the so-called "Cultura Lima" prevailed during a time when the lower valleys were still the centers of power, and when inter-valley contacts were strong; (4) Lima ceramics were less abundant in the *chaupi yunga* and even scarcer at *sierra* sites; and (5) the assignment of "Lima society" to a state level of organization is a controversial topic which needs to be examined using specific types of archaeological evidence which we lack at present.

Middle Horizon (A.D. 600–1000)

Although this is the least-studied period in the Chillón River valley, remains belonging to this period have been recovered in both the *yunga* and *chaupi yunga* zones. Research carried out by Stumer (1954b), Bonavia (1962), and Patterson and Lanning (1964) in the Chillón-Ancón sector indicate that (1) there was a significant occupation of the flat area of Ancón; (2) sites such as Culebra, Media Luna, and Santa Rosa or Playa Grande were virtually abandoned; (3) Early Intermediate period public buildings were used as cemeteries; (4) and the middle and upper valley populations employed a local style which still remains to be defined through stratigraphic excavations.

Information from the *chaupi yunga* zone comes from Huancayo Alto. During this period the *huaca* complex was abandoned and a new stone public building was constructed, its storage complex was enlarged, and camelid bones, wool, and "highland" pottery were more abundant. Based on this evidence, Dillehay concluded that "three ethnic groups" occupied Huancayo Alto and that (1) residents of the multiroom

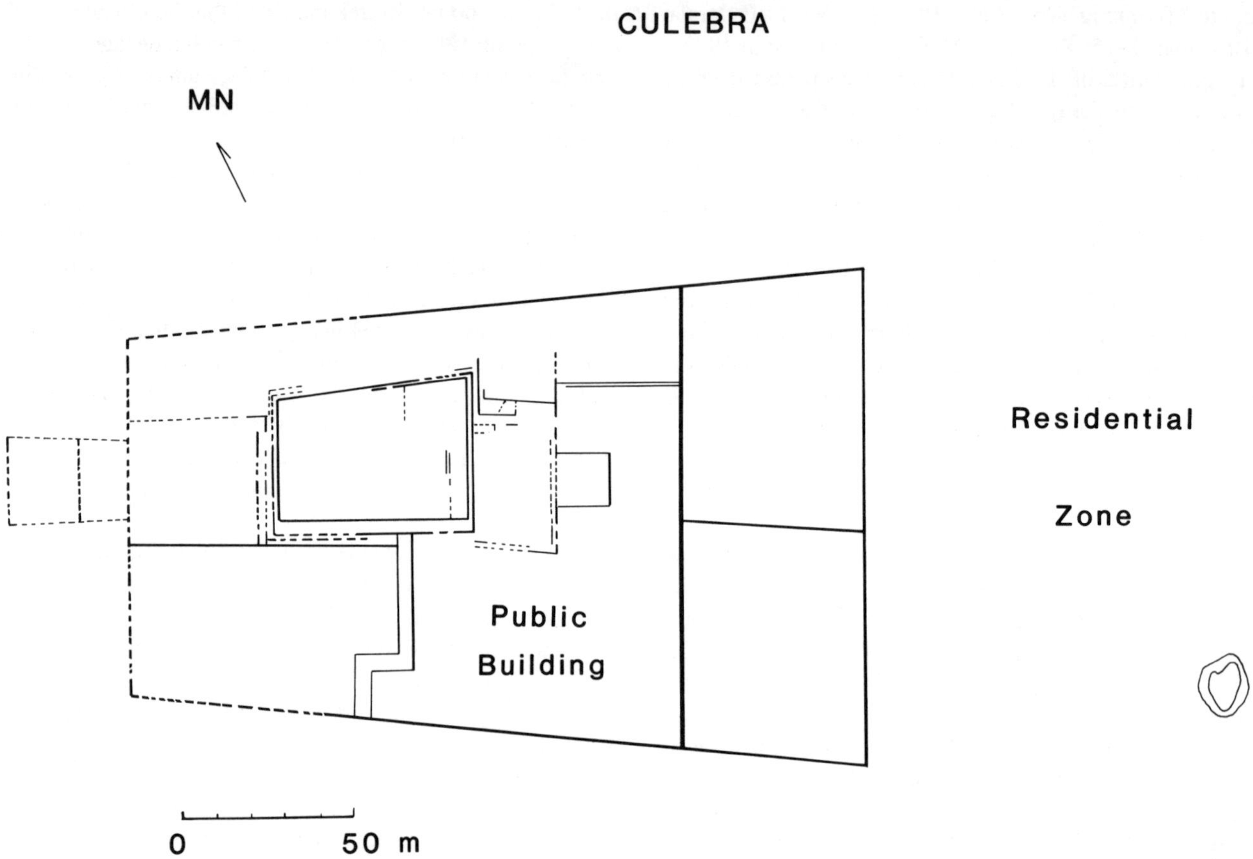

Figure 10. Plan of Culebra's public building and residential zone (drawn from aerial photograph supplied by the Servicio Aerofotográfico Nacional, Las Palmas, Barranco, Lima, Perú). (Plan: J. Silva)

adobe structure were high-status *chaupi yunga* indigenous administrators, (2) residents of the stone rooms were of the lesser ruling elite (of Canta or Junín origins?), and (3) residents of the stone-lined terrace sector were seasonal occupants from the Chacalla-Jicamarca area.

It appears that the *yunga* area of the Chillón Valley was not densely occupied when compared to the Early Intermediate archaeological remains in the *chaupi yunga*. For reasons not well understood, it seems that the lower Chillón Valley was no longer politically important. Instead, the Rímac and Lurín Valleys became powerful, and apparently the population of the central coast was ruled by 2 political centers located in these valleys; of particular significance are Cajamarquilla in the Rímac Valley and Pachacamac in the Lurín Valley.

Late Intermediate Period (A.D. 1000–1476)

Since both the Inca and Spanish policies and institutions can be understood by assessing how each group took advantage of prior intra-group political organization as well as inter-group alliances and enmities, we will try to reconstruct the Late Intermediate situation in some detail. Let us now look at the sociopolitical and economic environment the Inca encountered.

The Chillón lower and middle valleys displayed a concentrated, dense population, while the upper valley had a highly dispersed population organized into loose confederations or a series of *curacazgos* (Dillehay 1976: Chapters 5–6). For the purposes of placing the *Justicia 413* document into perspective, we should enumerate some of the *chaupi yunga* or middle valley characteristics: (1) the greatest concentration of prehispanic sites, (2) the greatest diversity of plants and crops, (3) the capability of producing large coca yields, and (4) a strategic location enabling access to several zones.

In contrast to other well-known valleys, the Chillón Valley at no time saw any single center emerge either to dominate the whole valley or to expand into other valleys. Rather, a mosaic of multiple, semi-autonomous, alternately competitive/cooperative polities characterized by the Late Intermediate and Late Horizon (Dillehay 1976).

The multi-ethnic nature of exchange alliances and the inter-ethnic animosities that existed in the Chillón Valley were

Figure 11. Culebra domestic structure built of cane (*kincha*) dating to the "Lima culture" (ca. A.D. 300). Note the postholes and narrow "channels" that still contain the lower segments of the cane walls in both the foreground and background. For a closeup of the cane wall in the background, see Figure 12. (After Silva, García, Bragayrac, and Morales 1984; photo: J. Silva).

important considerations for the Inca, since the latter needed to secure appropriate allies in the takeover and administration of newly acquired lands.

One pattern that emerges from the archaeological data of the Chillón Valley is that sites on the north bank of the river, especially in the *chaupi yunga*, display greater ceramic affinity with the Chancay Valley to the north (particularly sites near Quebrada Quilca). Although Chancay Black-on-White pottery was recovered from *yunga* and *chaupi yunga* sites, it was most abundant at *chaupi yunga* sites, particularly at Trapiche, Macas, and Huarabi. Many fewer Chancay Black-on-White sherds were found on the southern side of the river, at Collique, Huanchipuquio, Zapán, and Santa Rosa de Quives. Sites on the south side of the valley have ceramics more similar to those of the Rímac Valley sites. Quivi Vieja constitutes the farthest upvalley location to which this ceramic style extended (Dillehay 1976:184).

Earlier we mentioned the topic of ethnicity because archaeologists and ethnohistorians have tried to establish ethnic group boundaries and territories using different sets of data. Archaeologists often use the distribution of synchronic ceramic and architectural styles, while ethnohistorians and ethnographers use names and places mentioned in documents. Both data sets have limitations. Of all the problems that exist, we should mention at least three: (1) ceramic and architectural variation might not be linked to ethnic identity; (2) ethnohistoric data at times may be unreliable, because of the biases of the informants; and (3) ethnohistoric data may correspond to time periods of such short duration that they are difficult or impossible to detect archaeologically.

Notwithstanding, Murra (1972, 1975), Dillehay (1976) and Rostworowski (1972, 1978) have tried to link the names of ethnic groups mentioned in the documents to specific zones and archaeological sites in the Chillón Valley. (See Table 1 for some of these correspondences).

1. Late Intermediate and Late Horizon Yunga *Groups and Sites*

There is evidence for as many as six ethnic groups present in the lower Chillón Valley. These include the Cararuaillo, Colli, Chuquitunga, Sebillay, Sutca, and Inca.

According to documents, the Cararuaillo group occupied part of the *yunga* zone (Figure 6). The name of Carabayllo is still attached to a *yunga* settlement on the north side of the Chillón

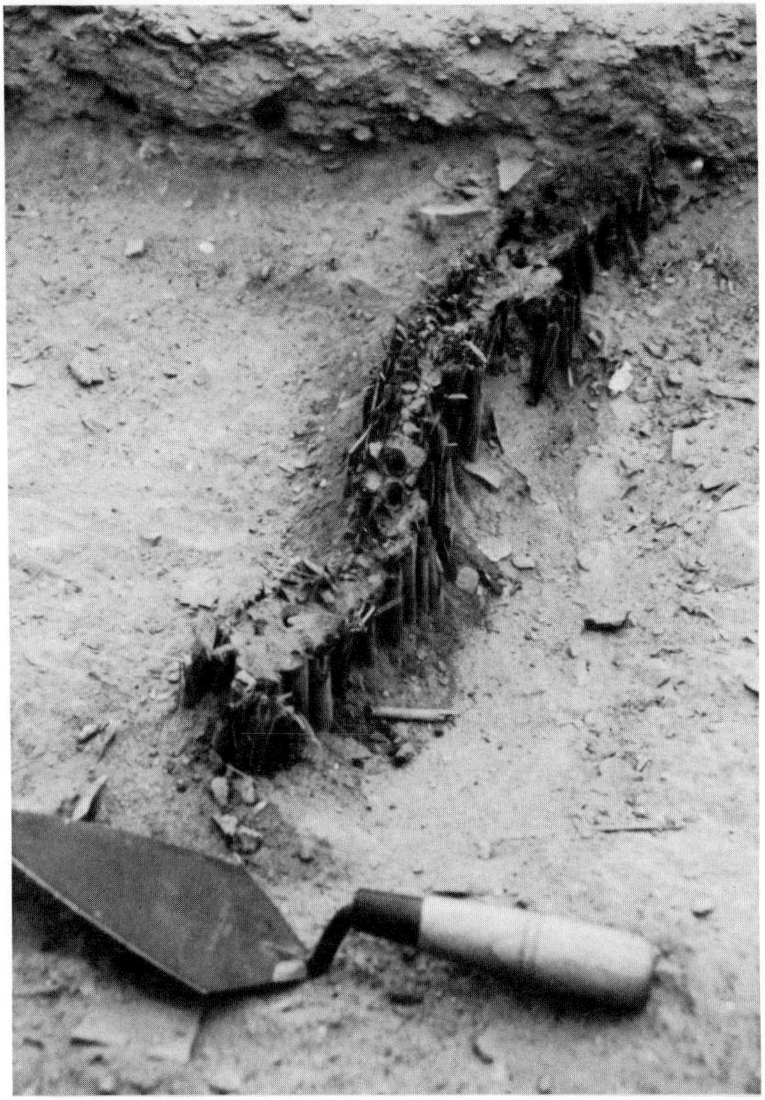

Figure 12. Close-up of a Culebra cane or *kincha*-walled domestic structure dating to the "Lima Culture" (ca. A.D. 300), which shows the lower segments of the canes. (After Silva, García, Bragayrac, and Morales 1984; photo: J. Silva)

River, which covers ca. 1300 hectares at an elevation of 150 meters. In the sixteenth century the *curaca* of Carabayllo was don Juan Quivi; he stated that Carabayllo was part of the lands of don Fernando Nacara, Lord of Collique (Rostworowski 1977:43).

Cobo (1956:301) stated that two "nations" speaking different languages inhabited the village of Carabayllo and the Chillón Valley. One "nation" was Carabayllo, which extended from the town of Carabayllo through the north side of the valley to Chancay; and the second was Pachacamac, which extended from Carabayllo to Pachacamac.

The only large site Dillehay (1976:198) found in his survey of the Carabayllo district was Kara-Hualla, which consisted of rooms of *tapia* and small adobes. This site was called Pueblo

Figure 13. Canes being dried in the middle valley or *chaupi yunga* zone of the Chillón River valley, August, 1986.

Figure 14. Bundles of canes are converted into mats for wall and roof construction. Note house at far right is constructed of these mats. Photo taken in the *chaupi yunga* zone, Chillón River valley, August, 1986.

Figure 15. Another middle valley (*chaupi yunga*) house constructed of mats, Chillón Valley, August, 1986.

Viejo by Villar Córdova (1935:170), who noted the presence of two adobe types: "Maranga" and "Chancay." Accompanying the Maranga and Chancay adobes were Early Intermediate, Late Intermediate, and Late Horizon ceramics (ca. 200 B.C. to A.D. 1550 [Dillehay 1976:202]).

Because the Cararuaillo were identified in ethnohistoric documents as a separate ethnic group in the *yunga* zone of the Chillón Valley, and because the site of Kara-Hualla (Carabayllo) is the largest and most complex set of ruins on the north side of the lower valley, Dillehay (1976:202) identified that site as the local political capital. The smaller, secondary sites yield Early Intermediate, Late Intermediate, and Late Horizon ceramics.

Another *Yunga* group was the Colli. In the *Justicia 413* document several witnesses stated that the Quivi lord was subject to the Colli (or Collique) lord, and that the Collique lord held lands from the ocean to Quivi. Others stated that the Collique lord also had lands in the Rímac Valley.

Dillehay concluded that the Carabayllo were on the north side of the lower valley, while the Colli were located on the south side (Figure 6).

The archaeological site of Collique is located on the south side of the river, between kilometer 16 and 17, and covers 20 to 25 hectares. It is divided into several sectors. A central zone, 1600 × 2000 meters is delimited (1) on its northern side by a wall extending over 1400 meters, which reaches the hillside near the "fort"; (2) on its northwestern side by four large pyramids or *huacas*; (3) on its southwestern side by two hills named Alpacoto and Comicay; and (4) on its eastern side by a hilltop called the "Fortaleza de Collique." Dividing the lower half of this central zone are two subterranean water sources, or *puquios*, and an interconnecting canal system (Rostworowski 1972:271–79; Squier 1877:87–88; Middendorf 1893–1895, II:47; Villar Córdova 1935:171; Dillehay 1976:204–6). Dillehay (1976:212) suggests that a relatively small population lived within this central zone, and that the site functioned as an administrative and religious center for the surrounding villages from at least Late Intermediate to Late Horizon times.

Other coastal groups who occupied the *yunga* zone were the Chuquitanta (were these the Chuquitanga or Chuquitunga?), Sebillay, and Sutca. The first group may have been associated with the cluster of *huacas* not far from the preceramic site of Chuquitanta or El Paraíso. In A.D. 1571, "Chuquitanga" was the name that referred to a *pachaca* (unit of 100 tribute payers) within Guancayo's *repartimiento* (Martínez Rengifo, in Espinoza 1963). In A.D. 1585, Chuquitanta, Sebillay, and Sutca were different ethnic groups in the *reducción* of Carabayllo (Rostworowski 1972:257, 266, 290).

TABLE 1
Possible Correspondences between Ethnic Groups and Archaeological Sites in the Chillón Valley

Ethnic Groups	*Archaeological Sites*
I. *Yunga* or Lower Valley	
Cararuaillo	Carabayllo
Colli	Collique
Chuquitunga	Hacienda Chuquitanta
Sebillay	——
Inca	Tambo Inga
Sutca	——
——	Trapiche
II. *Chaupi Yunga* or Middle Valley	
——	Huanchipuquio
Guancayo	Huancayo Alto
Guaravi	Huarabi Alto
Guaravi	Huarabi Bajo
Maca	Macas
Quivi (Quibi)	Santa Rosa de Quives
Quivi	Quivi Vieja
Sapán (Zapán)	Zapán
——	Checta
Missai (Misayo)	——
——	Yaso
——	Quyso
——	Orobel
——	Magdalena
——	Hornillos
——	Pucara
——	Huarhuar
III. *Sierra* or Upper Valley	
Canta?	Purumarca
——	Huaycoloro
Atabillos (Atavillos)	——
——	Pirhuinco
Canta	Cantamarca
——	Collo
Socos	——
——	Caballo Blanco
Chaclla (Jicamarca)	Jicamarca
Huarochirí	——
Canta?	Piedra Mesa-Camata
Canta?	Lucana
Canta?	Huancuna
Canta?	Huayuacancha
Huamantanga?	Huamantanga
Canta?	Zepitamarca

During the Late Horizon, the Inca occupied Tambo Inga, an archaeological site located 18 kilometers north of Lima on a natural promontory 150 meters west of the Pan American Highway. The small site (ca. 60 × 130 meters) has several high-walled enclosures separated by narrow corridors. Besides being an Inca "waystation," there is ceramic and architectural evidence of pre-Incaic occupation, perhaps extending back to the Middle Horizon.

II. Late Intermediate and Late Horizon Chaupi Yunga *Groups and Sites*

Chaupi yunga ethnic groups included the Huanchipuquio, Maca(s), Guancayo, Zapán, and Guarabi (Guarauni) (Figure 6).

One middle valley site is Trapiche, located on the north side of the Chillón River at about 650 meters (Figure 16); as we ascend into the *chaupi yunga* zone, it is the first large settlement we encounter. Trapiche is located on the east side of the *quebrada* linking the Chillón Valley with the Chancay Valley, and of all the Chillón sites, it reveals more Late Intermediate Chancay Black-on-White pottery than any other. Overlooking Trapiche is a small hilltop "fortress."

Four sites comprise Huanchipuquio, located at about 650 meters elevation on the south side of the valley; each is located on a hillside base or the edge of a *quebrada*. Huanchipuquio is directly across the valley from the site of Trapiche and the south entrance of Quebrada Quilca.

An impressive and highly visible isolated wall is located just east of Buena Vista and about 350 meters to the west of one of the Huanchipuquio sites. While Villar Córdova (1935:188) interpreted the wall as a defensive feature, Dillehay (1976:221) interprets it as a *mojón*, or boundary wall, serving to separate the Huanchipuquio group from the Cararuaillo (Carabayllo) group.

Villar Córdova (1935:173) says that the name Huanchipuquio is derived from *Huancho* (an ethnic group who lived in the Rímac Valley, but extended its territory into Chillón) and *puquio* (water source or spring). It is appropriately named, since there are many *puquios* directly in front of these settlements on the valley bottom land.

Still another town or ethnic group was called Zapán (Rostworowski 1967–1968:36–37; 1972:257, 279–83). One witness in the *Justicia 413* document claims that the Quivi lands adjoined the boundary markers of Zapán, which lay three leagues west of Quivi at an elevation of 750 meters. During the Late Intermediate period, Zapán could have been part of Guancayo whose landholdings now lie between Zapán and Quivi.

Zapán now extends from kilometer 45–48 on the south side of the river, upvalley from the Huanchipuquio sites, but perhaps originally a part of them; in fact, Zapán may once have been the eastern sector of Huanchipuquio. Zapán, like Huanchipuquio, is made up of separate settlements—in this case, three. These settlements display significant variation, which may be partly synchronic and partly diachronic.

Directly across the valley from Guarauni or Huarabi and one kilometer east of one of the Zapán sites is Hornillos, at an elevation of 825 meters. Hornillos has a Late Intermediate occupation and is apparently a secondary center, perhaps the easternmost settlement of Zapán (Figure 6).

According to Martínez Rengifo (in Espinoza 1963:58–69),

Figure 16. Late Intermediate (ca. A.D. 1000-1470) tapia structure at Cerro Temblador, Trapiche, in the *chaupi yunga* or middle valley zone of the Chillón Valley. (Photo: J. Silva)

the Macas *parcialidad* was apparently a dependency of the *cacicazgo* of Guancayo. The ruins called Macas today are on the north side of the valley. Four Macas settlements are at the southern base of Cerro Hormigas along the left margin of the entrance to Quebrada Socos (Trimborn 1970).

The site of Huarabi actually consists of a chain of four settlements on the north side of the valley. Huarabi C is the farthest site upvalley to include buildings of *tapia* construction. Beyond Huarabi C are three stone monoliths measuring 1.8–2.0 meters in height that were apparently set up to establish a boundary line. Local informants thought the boundary was Inca (Late Horizon), but Dillehay (1976:260) recovered Middle Horizon and Late Intermediate pottery in association. Huarabi D, one kilometer upriver, includes no *tapia* walls, although pottery from the Middle Horizon and Late Intermediate periods was found.

During the Late Horizon, *pachacakuna* (each *pachaca* is a unit of 100 tribute payers) of the Maca(s) and Guarauni (Huarabi) groups were political subjects of the lord of Guancayo (Martínez Rengifo 1571). The Guancayo group was a *guaranga*, a unit of 1000 family-heads, who were tribute payers to the Inca. Thus, at least two levels of the tribute-exacting hierarchy existed. All *pachaca* paid the Inca a specified amount in particular items. The *pachacakuna* of the Guancayo and those of their subjects (Maca and Guarauni) were required to give the Inca state various items, including coca in stated units:

> Cinco cestos de coca grandes... Cinco petaquillas de zuara de coca para el ynga.... Y que para el sol le tenían hecha una huertecilla de coca verde la cual cortaban y se la llevaban para ofrecer y quemar al dicho sol. [Martínez Rengifo 1571, in Espinoza 1963:63]

The document states that at Pampavilla and Cancavilla, the Guancayo planted coca on communal fields; the Guarauni maintained a communal field called Corupa for growing coca. For the *parcialidad* of Maca the communal coca field was referred to as "Guaynacapa where they planted coca and picked 50 basketloads of coca in ancient times" (Espinoza 1963:65).

One of the sites apparently occupied by the Guancayo was Huancayo Alto. The main part of the Huancayo Alto site, covering 120 × 500 meters, is located in a *quebrada*. This sector of the site has over 350 stone or *tapia* structures in addition to nearly 100 stone-lined rooms located on the lower slopes of an adjacent hillside. Five specialized "activity zones" for the site center have been identified: (1) administrative units, (2) an elite residential complex, (3) a defensive network, (4) lower class residential zones, and (5) a cemetery (Dillehay

1976). Other activity areas were located on the adjacent hillside and are believed to include an isolated lower class residential zone, some drying terraces, and storage facilities.

Two administrative buildings dominate the site center; they are similar in form, but different in construction. Dillehay (1976:274) suggests that these two administrative structures indicate a dual leadership or some type of moiety system, one being controlled by *chaupi yunga* administrators and the other by highlanders.

After A.D. 900, new architectural construction at Huancayo Alto was limited to a terrace and a few residences within the *quebrada*. During Late Intermediate times, the storage area of the site was used infrequently or abandoned; in support of this conclusion, Dillehay notes the rarity or absence of Late Intermediate pottery in the storage area.

From A.D. 900 until the Inca conquest, Huancayo Alto continued to be an "integrative lattice" where coastal and highland groups overlapped. No urban center or political power emerged there; rather, it seems to have been a center where products were stored, exchanged, and distributed. Dillehay concludes that incoming highlanders willingly yielded political control to the local inhabitants in exchange for access to lower-altitude resources and co-residency at Huancayo Alto (Dillehay 1979:30).

Rather than seeing one ethnic group conquering or dominating another, Dillehay argues that cooperation was the major integrative factor operating in the Chillón Valley. He believes that the increasing importance of cooperation and access to nonlocal resources led to the emergence of Huancayo Alto, a site in the *chaupi yunga* zone, which was apparently occupied by multiple ethnic groups.

We have no evidence that Huancayo Alto paid tribute or sent labor to a centralized state prior to the Inca conquest. However, two *chaupi yunga* ethnic groups (Maca and Guaravi, or Guarauni, and possibly a highland group from the Chacalla region) were at one time "subjects" of the lord of Guancayo, who may have lived at Huancayo Alto. If that were so, Huancayo Alto's extensive storage unit may have involved receipt of goods from "subject" ethnic groups. Since the storage rooms may date as far back as A.D. 200, it also may have been that some of the goods were destined for maintenance of the *huacas*, which were built at an even earlier time (100 B.C.–A.D. 1).

At Huancayo Alto, *tapia* was used for the administrative buildings and elite residences, while stone can be seen in the lower class residences, drying terraces, and storage rooms. Dillehay suggests that the *tapia* reflects the "coastal" tradition and stone the "highland" tradition; he uses differences in construction materials and the presence of apparently nonlocal ceramics and textiles to infer the presence of other ethnic groups.

In front of Huancayo Alto is a bridge at an elevation of ca. 875 meters, and 60 meters southwest of the bridge on the valley floor lies Tambo del Ynga. Residents of the town of Yangas later removed most of the stones of the *tambo* for building materials; the remains now consist of a stone foundation 10 × 10 meters. There Dillehay (1976:338) recovered two Inca star-shaped stone mace heads and several Inca polychrome sherds. Both the bridge and the *tambo* evidently date to the Late Horizon.

The site of Pucara, east of Huancayo Alto and at an elevation of 1000 meters, may be an example of a border settlement separating the Guancayo group from the Quivi. It was occupied from Middle Horizon to Late Intermediate times (A.D. 800–1400). The site layout is similar to upvalley sites like Quyso, Orobel, and Quivi Vieja. The road that connects with Jicamarca enters through the *quebrada* that contains Pucara Alto.

The site of Checta is 1.1 kilometer upvalley from Pucara and 2.3 kilometers downvalley from Quivi Vieja, at an elevation of ca. 1000 meters. During Late Intermediate times, Checta may have been the boundary between the *yunga* and *serranos*. Here the Inca placed members of the Chaclla ethnic group to serve as *chasqui* (messengers, runners) and *mitmaq* (laborers obliged to work for the state). Checta may be close to where the Judcunga boundary was located. The *Justicia 413* document states that the Chaclla *mitmaq* were placed between the boundary marker of Judcunga and a *quebrada*; this *quebrada* was evidently the mouth of the Arahuay Valley. Checta is adjacent to a hill that features a large wall which descends toward the river. Both the hill and wall on the south bank restrict access in this very narrow section of the valley; on the north side, the steep mountains descend near the river. Dillehay states that there is some evidence for dual ethnic groups co-inhabiting the site of Checta, much as he suggested for Huancayo Alto and Pucara Bajo. He further suggests that the large wall at Pucara separated the Quivi from the Guancayo.

The Quivi ethnic group may have occupied Quivi Vieja; among the *yunga* groups, these Quivi figure most prominently in the *Justicia 413* document. Quivi lands are located at the confluence of the Chillón and Arahuay rivers, at about 1150 meters in elevation and 65 kilometers inland. Along the Chillón River the eastern limit of the Quivi domain is near Yaso. Today, Yaso is the boundary between the District of Santa Rosa de Quives and the District of Canta. To the west and on the north side of the river is the site of Magdalena, which Dillehay suggests was another boundary. On the south side, Checta was perhaps the westernmost Quivi settlement.

Quivi probably controlled the lands on the south side of the Chillón and Arahuay rivers. Quivi Vieja is located on the south side of the valley, two kilometers east of Checta and 200 meters south of Hacienda Cabana. It extends along the base of the hills of the modern town of Santa Rosa de Quives. Beneath modern Quivi was a prehispanic settlement, now largely destroyed by earthmoving efforts designed to level the area for construction of a hotel and church.

One of the Huarabi witnesses (*Justicia 413*, folio 123) mentions that boundary markers separated the Huaravi and Quivi lands; those lands at Kilometer 48 (across the valley from Avícola Yangas) may correspond to that boundary line (Dillehay 1976:305). Present-day informants state that Yaso on the Chillón River and Huarhuar on the Arahuay River are the eastern boundaries of the modern district of Santa Rosa de Quives (Dillehay 1976:306).

Before the Inca conquest, the Quivi lord was subject to the Colli Capac who occupied the lower valley. Various witnesses in the document indicated that the Quivi lord "served and paid tribute" to the lord of Collique "before the Ingas came." Murra (1972, 1975) wonders whether the coastal Colli actually placed their own men in Quivi to plant the fields that produced coca, peppers, and fruits, or whether the Colli merely received those products through exchange or tribute.

> In A.D. 1559 the size of the Quivi land was stated to be ...2000 paces long and 300 wide and bordered by a part of the river and a road that comes from Lima to the said town of Canta... [*Justicia 413*, folio 131v]

Dillehay (1976:307) employed a local person to pace out the length and width of the sherd scatter on the flat terrain east of Santa Rosa de Quives, between the river and a hillside road which leads to Canta; interestingly enough, it proved to be 2200 × 340 meters, which corresponds closely to the size given in the *Justicia 413* document.

Based on the ceramics recovered at Quivi, Dillehay (1976:313) believes that diverse highland groups (Chaclla, Canta, Atavillos, Socos, and Inca) and coastal groups (Maranga, Chancay) were either in contact with inhabitants of the Quivi area, or occupied the site from Early Intermediate to Late Horizon times.

Capcha in the *Justicia 413* (folio 44r) stated that the Quybi (Chillón) River formed the boundary between the Atavillos and the Secos. Dillehay, however, suggests that the site of Magdalena, located on the eastern side of a *quebrada* at the base of Cerro Cenicero, may fall within the Socos territory.

The site of Huarhuar (at 1450 meters) is on the south side of the Arahuay River, 3 kilometers east of Quivi Vieja and 2 kilometers west of Orobel and Quyso. Huarhuar may have been another "boundary community," since there is a stone wall which runs perpendicular to the hillside, located about 300 meters upvalley, midway between Huarhuar and Orobel.

Other sites and their possible ethnic affiliation are given by Dillehay; they include Quyso and Orobel located to the east of Quivi at an elevation of 1500 meters. Quyso was a possible "boundary town" in the Canta domain, while Orobel may have been a Chaclla colony.

Modern Orobel is a colony of 30, placed there by Jicamarca to "maintain the boundaries of Huarochirí and Casta." Dillehay's informant (Dillehay 1976:331) stated that he did not have permission to enter and cultivate lands in the Arahuay Valley, because those were the cultivated fields of another group. The Orobel colony cultivated fields between Jicamarca and Collata at a higher elevation. In the sixteenth or seventeenth centuries the communities of Jicama, Culata, Punan, and Chacalla agreed to share the same pasture lands (Rostworowski 1972:299). From April to July the Jicamarca and Collata worked their high altitude lands and grew potatoes; when not working on the high altitude *chacrakuna* they lived in Cullata and Jicamarca.

Yaso, located at kilometer 73 at 1550 meters, was another possible Late Intermediate "border community" located between the Quivi, the Canta, and an Atavillos or Socos "colony."

III. Late Intermediate and Late Horizon Upper Valley Groups and Sites

The Canta, whose homeland is in the headwaters of the Chillón River, occupied parts of the upper Chillón Valley north of the Arahuay River. In the upper valley south of the Arahuay River resided the Chaclla, an ethnic group that was part of the macroethnic Yauyos group, who had their homeland in the headwaters of the Mala and Cañete Rivers. To the north of the Chillón River in the headwaters of the Chancay River lived the Atavillos (see Figure 6).

The Canta region includes the modern settlements of Arahuay, Lachaqui, Carhua, Canta, Pariamarca, Huamantanga, and Huaros. Prehispanic Canta sites include Cantamarca, San Lorenzo (Kekamarca or Lucana), Huancuna, Huamantanga, Zepitamarca, Piedra Mesa, Camata (Collo), Pirhuinco, Cullke, Viscas (Pacramarca), Ojomarca, Huayuacancha, Huantamarca near Carhua, and Purumarca (Figure 6).

Cantamarca is the principal settlement, located on a hilltop overlooking the upper valley (Villar Córdova 1935:298–302; Casana 1976:44–48). One type of architectural construction that characterizes such highland sites is the *kullpi*, a semi-subterranean building in the form of a round tower varying from 4–6 meters in diameter and 1–2.5 meters high, with stylized roof and small rectangular doorway. Entrances are often at ground level.

Cantamarca exhibits significant Late Horizon Inca style architecture, including trapezoidal doors. Other sites with Inca architectural features include Kekamarca, Huancuna, Purumarca, and Huamantanga. The *sierra* sites also display Inca pottery, both plain and polychrome.

At Quyso, Camata, Caballo Blanco, Cantamarca, and Huancuna, Dillehay (1976:357) collected surface remains of marine life including clam (*Mesodesma donacium*), thorny oyster (*Spondylus*), and slipper shell (*Crepidula dilatata*). These remains support documentary evidence which indicates that during peaceful eras, exchange took place between the coastal *Yunga* and the highland Canta. Ceramics of the coastal Nievería style also reached some Canta sites. Such interaction between the highland Canta and coastal sites began at least by Early Intermediate times.

Pirhuinco is a small site at 1750 meters on the south bank of the Arahuay River across from Licahuasi (3.5 kilometers east of Quyso), apparently occupied from Early Intermediate times through the Late Horizon.

Sites in the Arahuay Valley that were controlled by the Canta include Piedra Mesa and Camata (Collo); both are Late Intermediate/Late Horizon sites located on the north side of the river, 4.5 kilometers upvalley from Licahuasi and 2 kilometers upvalley from Caballo Blanco at an elevation of about 2050 meters. Caballo Blanco is located on a spur between Santa Rosa de Quives and Arahuay on the south bank of the Arahuay River. The location is ideal for controlling entry into and exit out of the Arahuay Valley. Caballo Blanco is a single component Inca site.

The Chaclla (a sub-group within the macroethnic Yauyos group) occupied parts of the Provinces of Huarochirí and Yauyos (Rostworowski 1967–1968, 1972). Their settlements included Caballo Blanco, Chuya, Macando, Cullipampa, Huayarcoto, and Kancha-Kancha (Dillehay 1976:366). From Quivi it is two leagues journey to the first Canta population, but it is a two-day trip to the first major Chaclla town. Chaclla *mitmaq* were placed in two areas: near the boundary marker called Judcunga (half a league from Quivi), and within Quivi proper. The Chaclla were an important Yauyos group, but they were subjects of the Huarochirí lord.

Villar Córdova (1935:341–44) discusses the Chaclla-Jicamarca sites of Chuya, Macando, Cullipampa, Huayarcoto, and Kancha-Kancha, sites that exhibit stone architecture and semi-subterranean *kullpi*.

At an elevation of 1200 meters, Huaycoloro is located at the upper end of a *quebrada* between the Chillón and Rímac valleys, in barren desert between rugged mountains some 20 kilometers east of Cajamarquilla. Trails connect Huaycoloro to Huancayo Alto, Pucara, and Quivi in the Chillón Valley, and to Jicamarca and Chacalla in the lower sierra.

Huachoc is located at 1250 meters in the Quebrada Quilca, midway between the Chillón and Chancay valleys. The site consists of 10–15 small, irregularly-shaped rooms with stone walls, and may be of Late Intermediate date. The small group that lived there (ca. 25–50 people) may have been affiliated with a larger group.

The available data allow us to reconstruct the Chillón and adjacent valleys as a mosaic of interacting *curacazgos*. Many of these were relatively small, bounded ethnic polities administered by a *curaca* or "local lord"; these interacting societies are often called *señoríos* or "community kingdoms." A territorially-extensive centralized state did not appear until these *señoríos* were absorbed by the expansionist Inca state which extended into the Chillón Valley during the 1460s.

Before the arrival of the Inca, the highland Chaclla had been in conflict with their neighbors the Canta, and with the coastal Colli, over access to the lands near Quivi (Figures 17–18). These Quivi lands were located in the *chaupi yunga* zone, geographically intermediate between the coastal Colli and the highland groups (Canta and Chaclla) who claimed them. The coastal Colli lord was forced to surrender control of the Quivi lands to the Canta, and to place a boundary marker called Judcunga at Checta to separate highland lands from coastal ones.

When the Inca conquered the valley, they placed the Chaclla on those lands. The competition and struggle among a number of ethnic groups for possession of these Quivi lands foreshadowed the actions and interests of the Inca. All these groups had wanted access to the lands because coca and a range of tropical fruits could be grown on them. By using the highland Chaclla —one of the ethnic groups who had fought for control of the Quivi lands —as *mitmaq*, the Inca were able to achieve the goals of the state and at the same time take advantage of the long-standing desires of the Chaclla.

Inca Impact on the Chillón Valley

Some scholars have argued that following the Inca conquest, the highlands revealed evidence for direct rule by the Inca, while the *yunga* lands were characterized by indirect rule. As evidence for direct rule by the Inca, many writers cite Inca architecture. With regard to the establishment of newly constructed Inca settlements, Morris (1972:401) has noted that state settlements "tend to be particularly large and numerous in areas characterized by certain conditions of ecological and sociopolitical marginality." Morris's comments are particularly pertinent to settlements in the *sierra* and *chaupi yunga* zones. However, rather than revealing many newly-founded sites with Inca architecture, much of the *yunga* region continued to see the construction of buildings in local styles. Further, many of the coastal sites that display the strongest "Inca influence" were those that had been the largest administrative centers prior to Inca conquest (cf. Menzel 1959).

For the late prehispanic periods—especially the Late Horizon (A.D. 1476–1534)—each zone within the Chillón Valley seems to have been affected differently by the Inca expansion (Dillehay 1977). The lower and middle valleys reveal much less overt Inca influence than the upper valley and *sierra* zones. It appears that the Inca were able to exercise indirect control over

Figure 17. Lands near Quivi (Santa Rosa de Quives) on which coca and fruits could be grown.

Figure 18. Another view of land near Quivi (Santa Rosa de Quives) on which coca and fruits could be grown.

the coastal zone from their base of operations in the lower *sierra* by mobilizing their armies there.

To be specific, direct rule characterized the Inca's relationship with the Chaclla and others in the upper valley, while indirect rule (through Collique) was applied to the lower valley. Perhaps a type of indirect rule (administered through Huancayo Alto) also characterized the Inca strategy in governing the middle valley.

The Inca established their base in the upper valley for a variety of reasons. The *Justicia 413* document indicates that the Inca had always feared the coast and were afraid of contracting illnesses and dying there. They also selected the upper valley zone of the Chillón since it was politically marginal, sparsely occupied, and far from the major centers of occupation on the coast and in the middle valley.

Of all the zones, the *yunga* or lower valley reveals the least evidence of Inca occupation. Relying on two lines of evidence—ceramic and architectural remains—Menzel (1959, for the south coast) and Dillehay (1977, for the Chillón Valley) demonstrate that most late coastal sites reveal little or no Inca influence. However, two sites in the Chillón Valley that were occupied before and after the Inca arrival—Tambo Inga and Collique—do reveal some Inca remodeling; the Inca apparently took over these two extant communities and modified them architecturally for their own ends.

Not surprisingly, both plain and painted Inca ceramic wares were most abundant in the lower *sierra* sites in the Chillón Valley, although they were also encountered at Collique and Tambo Inga in the *yunga* zone, and at Huancayo Alto in the *chaupi yunga* zone. The largest number of Inca plain brown wares were found at Caballo Blanco, an Inca site in the Arahuay Valley, located between the *chaupi yunga* and the lower *sierra*.

For much of the sequence, the *chaupi yunga* sites reveal more ceramic similarities with *yunga* sites, although they had significant ceramic affinities with lower *sierra* sites as well (Dillehay 1976:188). Much as the indigenous people classify both the lower and middle valley population and environmental zones as *yunga*, the ceramic evidence also suggests that those two populations were more closely linked than either was to the highlands (if we use ceramic similarity as our sole criterion). This conclusion must be tempered by the fact that *sierra* sites for this valley have been studied much less extensively (cf. Dillehay 1976, n.d.).

One ethnohistoric document that deals with three ethnic groups occupying the *chaupi yunga* zone of the Chillón Valley (the Guancayo, Maca, and Guarauni; see Martínez Rengifo in Espinoza 1963) supplies us with a list of products that these people gave to the Inca each year. In addition to revealing that the Inca benefited economically from control of this mid-valley zone, one of the items specifically mentioned as a tribute item from these groups was coca.

Since coca could be grown in the *chaupi yunga* zone—and only with difficulty elsewhere—we might expect that that zone was a target area for the Inca. In the *chaupi yunga* site of Huancayo Alto, Inca ceramics were recovered in three areas: (1) the administrative sector, (2) the drying terraces, and (3) the storage units nearest the drying terraces (Dillehay 1976:150). The Inca sherds appear to be deliberate imitations of Cuzco types. Additionally, 300 meters to the west of the main part of the site is an isolated building that included only Inca polychrome pottery.

One other *chaupi yunga* site that has produced some Inca ceramics (in addition to local pottery) is Huanchipuquio, which contains a multi-room compound with walls still standing over 2 meters in height. This compound may have been used as a storage facility by the Inca.

When we ascend into the upper valley zone, there is a notable increase both in the number of settlements that were "Inca-influenced" and in the number of newly-built Inca state installations.

Villar Córdova (1935) and Casana (1976) have reported on Purumarca, Cantamarca, Huancuna, Huamantanga, Caballo Blanco, Lucana, and Huayuacancha. Dillehay (1976, 1977) located several large corrals at many of these sites; large llama herds may have been maintained in this zone for state purposes, especially for military operations (Murra 1965) or transport of food.

After having a chance to evaluate extant political conditions, the Inca selected their allies carefully. For example, we have seen that the Inca conquered the macroethnic group, the Yauyos, of which the Chaclla were a part. Although known to be aggressive and warlike, the Yauyos became loyal friends of the Inca. Knowing that for many years the Chaclla, Canta, and Colli had been vying for control of the Quivi coca lands, the Inca stepped in and selected the aggressive Chaclla to serve as their *mitmaq*. The Chaclla and Inca could thereby mutually benefit from the coca lands of Quivi, with the Chaclla receiving additional territory in return for their labor on the Quivi lands.

While we know something about the Inca's relationship with *sierra* groups, we know less about the extent of reorganization that the institutions of coastal polities witnessed when they were incorporated into the Inca empire—especially those coastal polities that were already highly evolved and quite specialized. Additional fieldwork will be needed to address in detail the nature of coastal and *sierra* interrelationships, and each case will have to be examined separately.

Conclusions

A number of important themes have been touched upon in the preceding summary of Chillón Valley archaeology. One of these was community self-sufficiency, thought by some authors to be an Andean ideal. One of the strategies for self-sufficiency that may have been employed in the Chillón Valley, according to Dillehay (1976, 1979), was verticality, whereby a population nucleus sent out a series of "daughter colonies" to exploit a range of ecological zones, each colony still retaining membership, rights, and affiliation with the population nucleus. An alternative strategy, favored by other authors, would be community specialization, accompanied by exchange among communities within a political unit such as a *curacazgo* or within a region.

Which of these two ideals, we may ask, best characterizes the Chillón Valley at different times? or does a mix of strategies somewhere along the continuum between the two result in a better fit with the known evidence? In fact, Dillehay (1976, 1979, n.d.) sees a variety of strategies, of which verticality is but one, and he suggests that the verticality is not the "classic type" analyzed by Murra (1972).

We have seen that prior to the Inca conquest, the Chillón Valley was never unified by a single ruler or ethnic group. In fact, the ethnohistoric record makes it clear that several different ethnic groups occupied areas along the Chillón River (Figure 19) and surrounding terrain. Many of these had access to the coca or coca lands in the *chaupi yunga* zone located near Quivi. We must therefore try to determine if access was (1) direct (for example, by means of verticality) or (2) indirect (by means of alliances among equals, trading partners, community specialization, and/or exchange). We also need to ascertain whether the procuring unit was the household, the community, or the supra-community; whether access was centrally controlled by a *curaca* (local leader) or not; and whether the strategy changed through time. If two contiguous polities are similar in size and political, economic, and military strength, a strategy of interdependency, with some local specialization, might be the appropriate one; however, when two polities of unequal strength and size interact, the stronger, larger polity might achieve self-sufficiency by direct access through the establishment of a system of daughter colonies. In the case of the Chillón Valley, a number of ethnic groups apparently employed a wide variety of cooperative strategies, which probably evolved gradually.

According to Dillehay (1976, 1979), resource-sharing was both facilitated and controlled by multiple ethnic groups co-residing at *chaupi yunga* sites. As an example, he argues that at Huancayo Alto, highlanders lived in peace alongside *chaupi yunga* populations. Dillehay also (1979:30) believes he has sufficient archaeological evidence to demonstrate that a *form* of verticality (but not the "classic type" analyzed by Murra) was initiated there around 800–500 B.C. and carried out by mobile highlanders who descended to the *chaupi yunga* zone to carry out product exchanges under a centralized local authority. By 200 B.C. to A.D. 540, Dillehay (1979:30) says the verticality was in full swing "under the direction of local and highland ethnic leaders housed in administrative buildings," forming some kind of dual administration in Huancayo Alto. One of the key differences between Huancayo Alto's "verticality" and the "classic type" is that Huancayo Alto may have operated in a "pre-state" or "non-state" context in which highlanders were not sent out by a "nucleus," but were resident of their own volition (Dillehay 1976).

Any interpretation of Huancayo Alto as an example of the "classic type" of verticality would be weakened if it could be shown (1) that only *expansionist state-level* political and economic systems implemented verticality; (2) that the highlanders occupying Huancayo Alto did not retain rights or membership in a settlement nucleus in the highlands; (3) that the observed pattern reflects *only* a local *chaupi yunga* population residing there while exchanging goods with highlanders; or alternatively, (4) that it reflects some *chaupi yunga* men taking *sierra* women as wives.

If it is the case that verticality is a strategy that only existed under the political umbrella of a powerful state—and that no such centralized state existed in the Chillón Valley prior to the Inca conquest—then it seems unlikely that Huancayo Alto is an example of the "classic type" of highland-controlled colony, i.e., part of an "archipelago" (Murra 1972). However, if Huancayo Alto was not a highland colony, how might one explain the "highland" ceramics and architecture that Dillehay recovered there?

Dillehay constructs a co-residency model from his Huancayo Alto data, a model in which highlanders co-administered the town and cooperated peacefully with *chaupi yunga* inhabitants. But could not the same set of archaeological data be interpreted in another way—for example, as the result of inter-ethnic group exchange, political or marriage alliances, or contact through pilgrimages? Can we really differentiate inter-ethnic exchange from co-residency without many more archaeological data? Would we not need skeletal evidence to demonstrate two or more biological populations, corresponding to different ethnic groups with different genetic characteristics? The latter would be the kind of data one would like to see to support the argument of co-residency of multiple ethnic groups at *chaupi yunga* sites. In fact, Dillehay (personal communication) has indicated his skeletal data seem to support the presence of a "coastal" and "non-coastal" population.

We should stress, however, that our Late Intermediate and Late Horizon reconstructions of the Chillón Valley—that of a mosaic of small *curacazgos* and multiple ethnic groups—should not be projected blindly into earlier times. Archaeologists must be aware of the temporal limitations inherent in sixteenth-century models, and begin to develop others. A greater diversity of practices, strategies, and institutions almost certainly existed

Figure 19. View of the Chillón River in August, 1986.

prior to the Late Horizon. What this means is that models that might characterize the early periods—but which cannot necessarily be generated from ethnohistoric or Late Intermediate/Late Horizon data—must be reconstructed from the archaeological data of those earlier periods.

Not unexpectedly, we are left with more questions than answers for the periods antecedent to the Late Intermediate and Late Horizon *curacazgos*. Were the multiple ethnic polities of these late periods the result of system breakdown, or of long-term adaptations? How easily and frequently did boundaries change between ethnic groups and polities? How sensitive to ethnicity is stylistic variation in ceramics? Can ethnic groups really be defined on the basis of ceramic complexes? Or is stylistic variation the result of a far more complex mix of variables such as status, class, function, specialization, and manufacturing technique?

We are struck by the fact that there is often more ethnohistoric evidence of highland intrusions onto the coast than of coastal intrusions into the highlands. Is this true only of the later periods, or does it indicate that expansion through conquest and verticality were traditional highland practices, while community specialization, alliance, and regional exchange were more traditional coastal strategies?

Serranos from places such as Canta, Lachaqui, Arahuay, Chaclla, and Jicamarca descended to the *chaupi yunga* to exchange their goods for piedmont and coastal products. It appears that direct contact took place between highland and *chaupi yunga* populations on the one hand, and between *chaupi yunga* populations and *yunga* on the other hand; less direct contact, for the most part, characterizes the interaction between the *sierra* and coastal *yunga* (Dillehay 1976, n.d.).

For the ethnohistoric era, each subregion had at least one principal lord and preeminent ethnic group—the Colli in the *yunga*, the Guancayo in the *chaupi yunga*, and the Canta in the *sierra*. In the *Justicia 413* document (folio 189r), one resident of the *chaupi yunga* described this kind of relationship with the Canta of the *sierra* as follows: "he [*chaupi yunga* resident] is neither a relative, friend, nor enemy of the Indians of Chacalla nor of Canta; but with the Indians of Canta he has a great friendship, and they are partners because he deals with them, selling the coca in exchange for llama and potatoes." Coastal groups such as the Colli presumably controlled marine products, some cotton, and perhaps corn; *chaupi yunga* groups such as the Macas, Quivi, and Guarabi exchanged a range of tree fruits and coca; and *sierra* groups such as the Canta and Chaclla distributed *charqui* and wool, potatoes, olluco, and oca. In most of these transactions, the *chaupi yunga* inhabitants may have been the mediators.

Since the *Justicia 413* document specifically deals with a conflict over lands—and because it is clear that the Spaniards did not understand the indigenous concept of land—we should say a few words about that topic. When the Spaniards ordered that the Quivi coca lands be sold for 200 animals, the Indians claimed there was no precedent for such a sale, since in their experience land had never been sold. The Indian response was puzzling to the Spaniards, who failed to understand that the concept of private ownership of land (which had a very long history in their culture) was virtually absent among the Indians of the Americas. There is perhaps no better illustration of the clash between Old World and New.

For example, among most Mesoamerican Indian groups, such as the Maya, the sale, rent, or transfer of lands also lacked meaning. The Maya did conceive of the *señorío* controlling lands, of lineages possessing rights to land, and of utilized lands being delimited by boundary markers. Utilized land was controlled land; uncultivated land could be utilized by others only if it was uncared for and unprotected—for example, virgin forest never previously cultivated, and without boundary markers or improvements. There were even different levels of rights to land: those of the *señorío*, the town, the *barrio*, the lineage, and the nobility. However, prior to the arrival of the Spaniards, there is no evidence that individual ownership of private property existed, nor evidence that individuals could sell, buy, or transfer land (Landa, see Tozzer 1941; Villa Rojas 1961). A Maya noble could inherit rights to lineage lands, but not private ownership.

The Andean concept of land seems to have been similar to that of the Maya. Cobo (1890–1895: Book 12, Chapter 28) tells us that when the Inca settled or subjugated a town, he set up boundary markers and divided the arable land into three parts: (1) one dedicated to support religion, (2) one to be cultivated to support the ruler, and (3) one reserved for the use of the townspeople. In some provinces, the part that was assigned to religion was the largest, while in others the largest part was assigned to support the Inca ruler. In still others, entire towns and all their territory belonged to the Sun. The lands dedicated to religion might be divided among the Sun, Lightning, or other *huacas*. Townspeople could not "sell" the land assigned to their use.

Some development of private property seems to have taken place during and after the reign of Pachacuti (Rostworowski 1962). However, the growth in private possessions during the reigns of the last Inca rulers leads one to suspect it was a recently-evolving pattern not characteristic of earlier periods. In the case of the Quivi coca lands described in *Justicia 413*, there were fields set aside for the rulers Tupa Yupanqui and Huayna Capac, but also fields for the Sun and the Mamacona.

In summary, we should say that the ethnohistoric data we have utilized in this study provide both advantages and disadvantages. The advantages are that the data inform us of so many details we could not otherwise know—the locations of boundaries in the Chillón Valley, the names of lords, ethnic groups, places, and the dates and motivations for many events. The disadvantages are that we may be tempted not to pursue alternative models suggested by the archaeological data, simply because they have no ethnohistoric corroboration.

Bibliography

Antonil
1978 *Mama Coca*. Hassle Free Press, London.

Bohm, Bruce A., Fred R. Ganders, and Timothy Plowman
1982 Biosystematics and the Evolution of Cultivated Coca *(Erythroxylaceae). Systematic Botany* 7, pp. 121–33.

Bonavia, Duccio
1962 Sobre el Estilo Teatino. *Revista del Museo Nacional*, Tomo XXXI, pp. 43–94. Lima.
1979 Consideraciones sobre el Complejo Chivateros. *Arqueología Peruana* (ed. Ramiro Matos), pp. 65–74. Centro de Proyección Cristiana, Lima.
1982 *Los Gavilanes*. Corporación Financiera de Desarrollo S.A.-Cofide, Editorial Ausonia, Talleres Gráficos S.A., Lima.

Bray, Warwick, and Colin Dollery
1983 Coca Chewing and High-Altitude Stress: A Spurious Correlation. *Current Anthropology*, Volume 24, Number 3, pp. 269–82.

Carneiro, Robert
1970 A Theory of the Origin of the State. *Science*, Volume 169, pp. 733–38.

Casana R., Teodoro
1976 *Restos Arqueológicos de la Provincia de Canta*. Lima.

Chauchat, Claude
1975 The Paiján Complex, Pampa de Cupisnique, Perú. *Ñawpa Pacha* 13, pp. 85–96. Institute of Andean Studies, Berkeley.

Cobo, Bernabé
1890-1895 *Historia del Nuevo Mundo*. Publicada por primera vez por Marcos Jiménez de la Espada, 4 Volúmenes. Sociedad de Bibliófilos Andaluces, Sevilla.
1956 *Historia del Nuevo Mundo*. Biblioteca de Autores Españoles, Volúmenes 91–92. Madrid.

Cohen, Mark N.
1977 *The Food Crisis in Prehistory: Overpopulation and the Origins of Agriculture*. Yale University Press, New Haven.
1978 Archaeological Plant Remains from the Central Coast of Perú. *Ñawpa Pacha* 16, pp. 23–50. Institute of Andean Studies, Berkeley.

Craig, Alan K. and Ncrbert P. Psuty
1968 The Paracas Papers, Studies in Marine Desert Ecology 1, Reconnaissance Report. Occasional Publication Number 1 of the Department of Geography, Florida Atlantic University, Boca Raton.

Cutler, Hugh C., and Thomas W. Whitaker
1961 History and Distribution of the Cultivated *Cucurbits* in the Americas. *American Antiquity*, Volume 26, pp. 469–85.

Dillehay, Tom D.
1976 Competition and Cooperation in a Prehispanic Multi-Ethnic System in the Central Andes. Ph.D. dissertation, Department of Anthropology, University of Texas, Austin.
1977 Tawantinsuyu Integration of the Chillón Valley, Perú: A Case of Inca Geo-Political Mastery. *Journal of Field Archaeology*, Volume 4, pp. 397–405.
1979 Pre-Hispanic Resource Sharing in the Central Andes. *Science*, Volume 204, Number 6, pp. 24–31.
n.d. Estrategias Políticas y Económicas de las Etnías locales del Valle del Chillón durante el Período Prehispánico. *Revista Andina*. Cuzco, Perú.

Earle, Timothy
1972 Lurín Valley, Perú: Early Intermediate Period Settlement Development. *American Antiquity*, Volume 37, Number 4, pp. 467–77.

Engel, Frédéric André
1957 Early Sites on the Peruvian Coast. *Southwestern Journal of Anthropology*, Volume 13, pp. 54–68.
1963 A Preceramic Settlement on the Central Coast of Perú: Asia, Unit I. *Transactions of the American Philosophical Society*, Volume 53, Part 3, Philadelphia.
1966 Le Complexe Précéramique d'El Paraíso (Pérou). *Journal de la Société des Américanistes*, Tome LV-I, pp. 43–95. Paris.
1967 El Complejo El Paraíso en el Valle del Chillón, Habitado Hace Más de 3500 Años; Nuevos Aspectos de la Civilización de los Agricultores del Pallar. *Anales Científicos de la Universidad Agraria*, Volumen V, Número 3–4, Julio-Diciembre, pp. 241–80. La Molina, Lima, Perú.

Espinoza Soriano, Waldemar
1963 La Guaranga y la Reducción de Huancayo. Tres Documentos Inéditos de 1571 para la Etnohistoria del Perú. *Revista del Museo Nacional*, Tomo XXXII, pp. 8–80. Lima.

Fung Pineda, Rosa, Carlos F. Cenzano, and Amaro Zavaleta
1972 El Taller Lítico de Chivateros, Valle del Chillón. *Revista del Museo Nacional*, Tomo XXXVIII, pp. 61–72. Lima.

Golte, Jürgen
1968 Algunas consideraciones acerca de la producción y distribución de la coca en el Estado Incaico. *Actas del Congreso Internacional de Americanistas*, Tomo II, pp. 471–78. Stuttgart.

Lanning, Edward P.
1963 An Early Ceramic Style from Ancón, Central Coast of Perú. *Ñawpa Pacha* 1, pp. 47–59. Institute of Andean Studies, Berkeley.
1965 Early Man in Perú. *Scientific American*, Volume 213, Number 4, pp. 68–76.
1967 *Peru Before the Incas*. Prentice Hall, Inc., Englewood Cliffs, New Jersey.

Ludeña, Hugo
1970 San Humberto: Un Sitio Formativo en el Valle del Chillón (informe preliminar). *Arqueología y Sociedad* 2, pp. 37–51. Universidad Nacional Mayor de San Marcos, Lima.
1973 *Investigaciones arqueológicas en el sitio de Huacoy: Valle del Chillón*. Tesis para optar el grado de Bachiller. Universidad Nacional Mayor de San Marcos, Lima.

Lynch, Thomas F.
1974 The Antiquity of Man in South America. *Quaternary Research*, Volume 4, pp. 356–77.
1983 The Paleo-Indians. *Ancient South Americans*, edited by Jesse D. Jennings, pp. 86–137. W.H. Freeman and Co., San Francisco.

MacNeish, Richard S., Thomas C. Patterson, and David L. Browman
1975 The Central Peruvian Prehistoric Interaction Sphere. *Papers of the Robert S. Peabody Foundation for Archaeology*, Number 7. Phillips Academy, Andover, Massachusetts.

Marcus, Joyce
1983 An Introduction to the Series and to Volume I. In *A Fuego y Sangre: Early Zapotec Imperialism in the Cuicatlán Cañada, Oaxaca*, by Elsa M. Redmond. Studies in Latin American Ethnohistory & Archaeology, Volume I, Memoirs of the Museum of Anthropology University of Michigan Number 16, pp. v-vii. Ann Arbor.

Martínez Rengifo, Juan
1571 La visita de Guancayo, Maca y Guaravni 1571. In La Guaranga y la Reducción de Huancayo, by Waldemar Espinoza Soriano. *Revista del Museo Nacional*, Tomo XXXII, pp. 8–80. Lima.

Menzel, Dorothy
1959 The Inca Occupation of the South Coast of Peru. *Southwestern Journal of Anthropology*, Volume 15, pp. 125–42.
1964 Style and Time in the Middle Horizon. *Ñawpa Pacha* 2, pp. 1–106. Institute of Andean Studies, Berkeley.

Middendorf, Ernst W.
1893-1895 Perú: Observaciones y Estudios del País y Sus Habitantes durante una permanencia de 25 años. Tomo II: La Costa.

Morris, Craig
1972 State Settlements in Tawantinsuyu: A Strategy of Compulsory Urbanism. In *Contemporary Archaeology: A Guide to Theory and Contributions*, ed. Mark P. Leone, pp. 393–401. Southern Illinois Press, Carbondale.

Mortimer, W. G.
1901 *History of Coca*. J. H. Vail, New York.

Moseley, Michael Edward
1975 *The Maritime Foundations of Andean Civilization*. Cummings Publishing Company, Menlo Park, California.

Murra, John V.
1965 Herds and Herders in the Inca State. Man, Culture, and Animals: The Role of Animals in Human Ecological Adjustments. *American Association for the Advancement of Sciences*, Publication Number 78, pp. 184–216.
1972 El 'Control Vertical' de un Máximo de Pisos Ecológicos en la Economía de las Sociedades Andinas. In Visita de la Provincia de León de Huánuco en 1562, por Iñigo Ortíz de Zúñiga, visitador (John V. Murra, ed.), Tomo II, pp. 427–76. *Documentos para la Historia y Etnología de Huánuco y la Selva Central*, Tomo II. Universidad Nacional Hermilio Valdizán, Huánuco, Perú.
1975 *Formaciones Económicas y Políticas del Mundo Andino*. Instituto de Estudios Peruanos, Lima.

Parsons, Mary Hrones
1970 Preceramic Subsistence on the Peruvian Coast. *American Antiquity*, Volume 35, Number 3, pp. 292–304.

Patterson, Thomas C.
1966a Early Cultural Remains on the Central Coast of Perú. *Ñawpa Pacha* 4, pp. 145–53. Institute of Andean Studies, Berkeley.
1966b Pattern and Process in the Early Intermediate Period Pottery of the Central Coast of Peru. University of California Publications in Anthropology, Volume 3, University of California, Berkeley.
1971 The Emergence of Food Production in Central Perú. In *Prehistoric Agriculture*, edited by Stuart Struever, Part III, Chapter 11, pp. 181–207. American Museum Sourcebooks in Anthropology, Natural History Press, Garden City, New York.
1985 The Huaca La Florida, Rimac Valley, Peru. In *Early Ceremonial Architecture in the Andes, a Conference at Dumbarton Oaks, 8–10th October 1982*, edited by Christopher B. Donnan, pp. 59–69. Washington, D.C.

Patterson, Thomas C., and Edward P. Lanning
1964 Changing Settlement Patterns on the Central Peruvian Coast. *Ñawpa Pacha* 2, pp. 113–23. Institute of Andean Studies, Berkeley.

Patterson, Thomas C., and Michael Edward Moseley
1968 Late Preceramic and Early Ceramic Cultures of the Central Coast of Peru. *Ñawpa Pacha* 6, pp. 115–33. Institute of Andean Studies, Berkeley.

Patterson, Thomas C., John P. McCarthy, and Robert A. Dunn
1982 Polities in the Lurín Valley, Perú During the Early Intermediate Period. *Ñawpa Pacha* 20, pp. 61–82. Institute of Andean Studies, Berkeley.

Plowman, Timothy
1979 Botanical Perspectives on Coca. *Journal of Psychedelic Drugs*, Volume 11, pp. 103–17.

1984 The Origin, Evolution, and Diffusion of Coca, *Erythroxylum* spp., in South and Central America. In Pre-Columbian Plant Migration, edited by Doris Stone. Papers of the Peabody Museum of Archaeology and Ethnology, Harvard University, Volume 76, pp. 125–63. Cambridge.

Quilter, Jeffrey
1983 *Cerro de Media Luna: Un Templo del Período Intermedio Temprano en el Valle del Chillón, Perú. Informe Preliminar* presentado al Instituto Nacional de Cultura, Lima.
1985 Architecture and Chronology at El Paraíso, Perú. *Journal of Field Archaeology*, Volume 12, pp. 279–97.

Ravines, Rogger
1975 Garagay: Un Viejo Templo en los Andes. *Textual* 10, pp. 6–12. Revista del Instituto Nacional de Cultura, Lima.
1984 Sobre la Formación de Chavín: Imágenes y Símbolos. *Boletín de Lima*, Número 35, Año 6, Setiembre, pp. 27–45. Editorial Los Pinos, Lima.

Ravines, Rogger and William H. Isbell
1976 Garagay: Sitio Ceremonial Temprano en el Valle de Lima. *Revista del Museo Nacional*, Tomo XLI, pp. 253–75. Lima.

Rostworowski de Diez Canseco, María
1962 Nuevos Datos sobre Tenencia de Tierras Reales en el Incario. *Revista del Museo Nacional*, Tomo XXXI, pp. 130–59. Lima.
1967-1968 Etnohistoria de un valle costeño durante el Tawantinsuyu. *Revista del Museo Nacional*, Tomo XXXV, pp. 7–61. Lima.
1972 Las Etnías del Valle del Chillón. *Revista del Museo Nacional*, Tomo XXXVIII, pp. 250–314. Lima.
1973 Plantaciones prehispánicas de coca en la vertiente del Pacífico. *Revista del Museo Nacional*, Tomo XXXIX, pp. 193–224. Lima.
1977 *Etnía y Sociedad: Costa Peruana Prehispánica*. Instituto de Estudios Peruanos, Lima.
1978 *Señoríos Indígenas de Lima y Canta*. Instituto de Estudios Peruanos, Lima.

Salazar-Burger, Lucy, and Richard L. Burger
1982 La Araña en la Iconografía del Horizonte Temprano en la Costa Norte del Perú. *Beiträge zur Allgemeinen und Vergleichenden Archäologie* 4, pp. 213–53. Bonn, West Germany.

Sanders, William T., and Joseph Marino
1970 *New World Prehistory*. Prentice-Hall, Englewood Cliffs, New Jersey.

Scheele, Harry G.
1970 The Chavín Occupation of the Central Coast of Perú. Unpublished Ph.D. dissertation, Department of Anthropology, Harvard University, Cambridge.

Service, Elman R.
1975 *Origins of the State and Civilization: The Process of Cultural Evolution*. W.W. Norton and Co., Inc., New York.

Silva, Jorge
1984 El Paraíso y los Centros Ceremoniales Tempranos de la Costa Central. In *Arqueología y Sociedad* 10, pp. 1–14. Publicación del Museo de Etnología y Arqueología de la Universidad Nacional Mayor de San Marcos, Casona, Parque Universitario, Lima.

Silva, Jorge, Kenneth G. Hirth, Rubén García, and José Pinilla B.
1982 El Valle del Rímac hace 2,500 Años: Huachipa-Jicamarca. *Boletín de Lima*, Número 21, Año 4, Mayo, pp. 59–68. Editorial Los Pinos, Lima.

Silva, Jorge, Kenneth G. Hirth, Rubén García, and José Pinilla B.
1983 El Formativo en el Valle del Rímac: Huachipa-Jicamarca. *Arqueología y Sociedad* 9, pp. 2–83. Publicación del Museo de Etnología y Arqueología de la Universidad Nacional Mayor de San Marcos, Casona, Parque Universitario, Lima.

Silva, Jorge, Rubén García, Enrique Bragayrac, and Daniel Morales
1984 *Excavaciones Arqueológicas en Culebra, Valle del Chillón (informe de la temporada 1983). Informe Preliminar* presentado al Instituto Nacional de Cultura, Lima.

Squier, Ephraim G.
1877 *Peru: Incidents of Travel and Exploration in the Land of the Incas*. Harper and Brothers, New York.

Stumer, Louis Michael
1953 Playa Grande. *Archaeology*, Volume 6, Number 1, pp. 42–48.
1954a Population Centers of the Rimac Valley of Peru. *American Antiquity*, Volume 20, Number 2, pp. 130–48.
1954b The Chillón Valley of Perú Excavation and Reconnaissance: 1952–1953. *Archaeology*, Volume 7, Number 3, pp. 171–78; Number 4, pp. 220–28.
1955 History of a Dig. *Scientific American*, Volume 192, Number 3, pp. 98–104. New York.

Tosi, Joseph A., Jr.
1960 *Zonas de Vida Natural en el Perú*. Memoria Explicativa sobre el Mapa Ecológico del Perú, Proyecto 39, Boletín Técnico 5. Programa de Cooperación Técnica, Instituto Interamericano de Ciencias Agrícolas de la OEA, Zona Andina, Washington.

Tozzer, Alfred Marston
1941 Landa's Relación de Las Cosas de Yucatán. (A translation edited with notes by Alfred M. Tozzer.) *Papers of the Peabody Museum of Archaeology and Ethnology, Harvard University*, Volume 18. Cambridge.

Towle, Margaret A.
1961 The Ethnobotany of Pre-Columbian Peru. *Viking Fund Publications in Anthropology*, Number 30. Wenner-Gren Foundation for Anthropological Research, New York.

Trimborn, Hermann
1970 Las ruinas de Macas en el Valle del Chillón. *Revista del Museo Nacional*, Tomo XXXVI, pp. 258–66. Lima.

Villa Rojas, Alfonso
1961 Notas sobre la Tenencia de la Tierra entre los Mayas de la Antigüedad. *Estudios de Cultura Maya*, Tomo I, pp. 21–46.

Villar Córdova, Pedro Eduardo
1935 [1982] *Las Culturas Pre-Hispánicas del Departamento de Lima*. [Ediciones Atusparia, Lima].

Willey, Gordon R.
1948 A Functional Analysis of "Horizon Styles" in Peruvian Archaeology. *Society for American Archaeology Memoir*, Volume 13, Number 4, pp. 8–15.

Williams León, Carlos
1971 Centros Ceremoniales Tempranos en el Valle de Chillón, Rímac y Lurín. *Apuntes Arqueológicos* 1, pp. 1–4. Lima.
1978-1980 Complejos de Pirámides con Planta en "U," Patrón Arquitectónico de la Costa Central. *Revista del Museo Nacional*, Tomo XLIV, pp. 95–110. Lima.
1985 A Scheme for the Early Monumental Architecture of the Central Coast of Perú. In *Early Ceremonial Architecture in the Andes, a Conference at Dumbarton Oaks, 8–10th October 1982*, edited by Christopher B. Donnan, pp. 227–40. Washington, D. C.

PARTE I
Los Cocales del Valle del Chillón
Evidencia Arqueológica y Contexto Ecológico

Joyce Marcus y Jorge E. Silva

Arqueología y Etnohistoria

Arqueología y etnohistoria son dos subdisciplinas de la investigación antropológica. Aunque ambas emplean distintas metodologías, proporcionando datos que frecuentemente tratan sobre temas diferentes, ellas comparten metas comunes, es decir reconstruir los procesos *específicos* y *generales* que han caracterizado la evolución de las sociedades del pasado (Marcus 1983:v).

La ventaja de la etnohistoria reside en el hecho que permite a menudo elaborar detalladas reconstrucciones sincrónicas incluyendo la interrelación de múltiples factores; en cambio la capacidad de la arqueología reside en el hecho que provee reconstrucciones y transformaciones de carácter diacrónico en y entre sociedades. Mientras que la arqueología proporciona información temporal, la etnohistoria presenta con frecuencia las motivaciones de individuos específicos, así como los nombres, fechas, lugares y hechos particulares sucedidos en el pasado. De manera pues, que la etnohistoria trata con casos concretos que a veces pueden ser utilizados para proponer y examinar modelos sobre el pasado. Es tarea de los arqueólogos determinar la aplicabilidad de dichos modelos y patrones en la prehistoria.

Los datos recuperados de los registros arqueológicos y etnohistóricos deben ser considerados *complementarios*, dado que rara vez coinciden cronológicamente ni son exactamente comparables tanto en el aspecto geográfico como en el temático. Cada conjunto de datos debe ser pensado como un "caso específico de análisis" vinculado a una situación particular con sus propios componentes temporales, geográficos y temáticos.

Mientras que algunos arqueólogos utilizan la etnohistoria únicamente como fuente secundaria—es decir, cuando el dato etnohistórico confirma sus conclusiones (por supuesto que los etnohistoriadores tratan frecuentemente de la misma manera a la arqueología)—es siempre más revelador cuando los datos arqueológicos y etnohistóricos se contradicen entre sí. Dicho conflicto es una invitación para iniciar serios trabajos de investigación. De modo que los proyectos de investigación deben abordar los temas en conflicto como puntos de partida y uno de los objetivos debe ser explicar la naturaleza del conflicto; de hecho, hace tiempo que hemos venido ignorando o tomando a la ligera dichas áreas de conflicto.

¿Cuáles son las limitaciones de la información etnohistórica? Desafortunadamente, sus datos pueden provenir de un "típico" informante o de numerosos testigos. Sin embargo, todo informante tiene puntos de vista personales; sin duda, la información es una exposición subjetiva de los informantes locales. Los españoles, quienes registraron la información y tomaron las manifestaciones, también tuvieron sus propios intereses y puntos de vista, algunos de los cuales resultaron de sus objetivos y educación religiosa occidental, en tanto que otros tuvieron una perspectiva diferente sobre el mundo, o un conjunto de conceptos y tradiciones enraizados en el sistema legal del siglo dieciséis.

¿Cuáles son las limitaciones del dato arqueológico? Para las civilizaciones sin escritura de los Andes, la información proveniente de las excavaciones no puede proporcionar los nombres de grupos étnicos o ciudades, pueblos o aldeas antiguas, ni los días, meses o los años exactos en que hechos específicos sucedieron. La etnohistoria provee éstos y otros datos sobre el pasado cercano permitiéndonos enriquecer nuestras reconstrucciones arqueológicas, en especial las referidas al Horizonte Tardío.

El tema de la etnicidad debe ser por lo menos mencionado en esta discusión, pues tanto arqueólogos como etnohistoriadores lo han tratado con efervescencia y han proporcionado datos sobre el mismo. A base de la variación sincrónica en la alfarería y los estilos arquitectónicos, algunos arqueólogos han dado por sentado que (1) dicha variación está relacionada al establecimiento y mantenimiento de la identidad étnica en la prehistoria, y (2) que dicha información puede ser empleada para identificar grupos étnicos y establecer sus límites territoriales. Algunos investigadores visualizan dichas fronteras étnicas como parte de una pasiva reflexión de diferencias entre grupos, en tanto que otros ven la variación estilística como un sistema dinámico fomentado y mantenido por cada grupo. Más aún, otros investigadores dudan el asociar la variación alfarera y arquitectónica al mantenimiento de diferencias étnicas, y en lugar de ello prefieren relacionarla a factores ambientales, económicos, u otros de naturaleza "funcional."

Los nombres de grupos étnicos y las motivaciones para el intercambio, competencia y alianzas pueden a veces recuperarse a través del dato etnohistórico. Sin embargo, es aún difícil establecer si es que estas estrategias fueron puestas en práctica por consenso de los miembros de un grupo étnico o solo por segmentos existentes en dichos grupos étnicos. ¿Trataron los segmentos de estos grupos étnicos de alcanzar autosuficiencia? ¿o fueron altamente especializados? ¿A qué nivel del sistema de asentamiento operó la autosuficiencia o la especialización? Por ejemplo, una unidad residencial puede ser especializada pero, sin embargo, la comunidad puede ser autosuficiente. En consecuencia, es del todo importante aislar el nivel relevante de la sociedad con fines analíticos y la construcción de modelos. Entre los dos extremos o "modelos ideales" de autosuficiencia y especialización total existe un amplio continuum de estrategias y adaptaciones. Tales modelos son sumamente difíciles de construir si solo los abordamos sincrónicamente a base de datos etnohistóricos y etnográficos; pero si además deseamos proveer cierta profundidad cronológica a dichos patrones, y descubrir sus orígenes, desarrollo y cambios a través del tiempo, debemos integrarlos con los datos arqueológicos.

Retornando ahora a nuestro "caso de análisis"—la lucha inter-étnica por el acceso a los cocales en la *chaupi yunga*, en Quivi, valle del río Chillón—debemos ubicar a los protagonistas y los sucesos mencionados en el documento en sus respectivos contextos históricos, ecológicos y sociopolíticos. Para ello, es necesario ampliar nuestra red y analizar las regiones contiguas al valle del Chillón. Asímismo, para entender el presente y el pasado inmediato, debemos sumergir nuestro conocimiento en el registro arqueológico.

El Valle del Río Chillón

El valle del río Chillón presenta cerca de cuatro kilómetros de ancho en su desembocadura, estrechándose considerablemente sobre una extensión de 65 kilómetros tierra adentro, punto en donde se convierte en *quebrada*, ya en la sierra. El valle puede dividirse en varias zonas mayores: (1) la costa desértica, valle bajo o *yunga* (0–600 metros), (2) el valle medio, desierto espinoso o *chaupi yunga* (600–2000 metros), y (3) el alto Chillón (2000–5000 metros)—este último incluyendo una serie de tipos vegetacionales tales como Maleza Desértica Subtropical, Estepa Espinosa Montano Bajo, Bosque Húmedo Montano, Páramo muy Húmedo Subalpino y Tundra muy Húmeda Alpino y Formación Nival (Tosi 1960; Mapa Ecológico del Perú, Hoja N° 2 [Sector Centro]).

El bajo Chillón, que se extiende desde el Océano Pacífico hasta Huanchipuquio, se halla en el Departamento de Lima (Figura 1). El medio y alto Chillón son parte de la provincia de Canta. La región del grupo étnico Chaclla (pueblo que fuera parte del grupo macro-étnico Yauyos) se encuentra en la provincia de Huarochirí (Figura 1).

Examinemos seguidamente el bajo, medio y alto Chillón.

La Yunga *o Valle Bajo*

La *Yunga* o "tierra caliente" en el valle del Chillón (Figura 3) incluye aproximadamente unas 10,000 hectáreas de terreno en el piso del valle las que hoy, al igual que en el pasado, se hallan extensamente irrigadas y cultivadas, extendiéndose desde el litoral hasta Trapiche (a unos 600 metros de altura). Esta región presenta temperaturas relativamente uniformes (18.6°C promedio). Existe en esta zona una neblina baja (a veces en forma de *garúa)* que llega hasta Trapiche, produciendo humedad en el aire, aunque no suficiéntemente abundante como para permitir la presencia de flora o agricultura sin riego (Figura 4).

La Chaupi Yunga *o Valle Medio*

La *chaupi yunga* es una zona clave para entender el documento *Justicia 413*, motivo de este volumen. Esta zona es geográfica, sociopolítica y económicamente intermedia entre las poblaciones serranas y costeñas (Figura 2).

El término *chaupi yunga* es una palabra Quechua utilizada para señalar que el valle medio es parte de la *yunga* o "tierra caliente"; a su vez, *chaupi*, "medio," indica que esta zona se encuentra entre la *sierra* y la costa. La *chaupi yunga* en el valle del Chillón se extiende desde Huarabi hasta Arahuay en el río Arahuay, y hasta Yaso en el río Chillón.

Tierras para el cultivo en la *chaupi yunga* son escasas; a diferencia de unas 10,000 hectáreas existentes en el valle bajo, el valle medio tiene sólo unas 4,000 hectáreas. A medida que ingresamos valle adentro, éste se estrecha a modo de un cañón, zona en donde las tierras fértiles se reducen a una franja junto al río. Esta franja varía de .4 a 1 kilómetro de ancho entre Macas y Quives; hoy, los terrenos cultivables entre Quives y Pacaybamba no sobrepasan de 300 hectáreas (Dillehay 1976:52).

Las lluvias anuales oscilan aproximadamente entre 125 y 250 milímetros. La vegetación incluye *Gynerium, Cortaderia, Opuntia, Cereus, Acacia, Schinus* y *Prosopis* (Tosi 1960:33). La temperatura promedio es 17.6°C. Coca y frutales son los cultivos más comunes en la *chaupi yunga*.

Todos los asentamientos humanos a lo largo del Chillón y el Arahuay se hallan cerca del sistema de canales de regadío del valle. En las partes altas cada asentamiento está conectado a un canal principal, el cual recibe agua de otros menores o está asociado a una laguna pequeña, a un pozo, o a un manantial.

La Sierra *o Alto Valle*

El alto valle o la *sierra* empieza a unos 85 kilómetros del litoral, incluyendo el área entre Yani y Pacaybamba en el río Chillón y cerca de Arahuay en el río Arahuay. Esta zona se inicia aproximadamente a 2000 metros de altura ascendiendo hasta los 5000 metros.

La *sierra* baja (2000–3000 metros) es estrecha, de 12 a 26 kilómetros de ancho y limita al este con Huaros, Carhua y Lachaqui. Esta zona escarpada contiene numerosos asentamientos arqueológicos en la cima de los cerros desde donde se domina el piso del valle; ellos son Huamantanga, Huayuacancha, Cantamarca, Purumarca, Kekamarca y Huancuna. En la subsiguiente zona del alto valle (entre unos 3000–5000 metros) se cultivan varios productos figurando papas, oca, olluco y otros; *charqui* y lana son asímismo importantes en esta parte del valle.

Los habitantes de la *sierra* tuvieron acceso a una serie de recursos de la *chaupi yunga* y el valle bajo por medio de diversas estrategias: (1) alianzas de reciprocidad e intercambio y alianzas matrimoniales, (2) el establecimiento de colonias, o (3) conquista (Dillehay 1976, 1979 y n.d.).

El documento *Justicia 413* (folio 188v) ilustra un caso de alianza sobre reciprocidad e intercambio según lo manifestado por un informante de Guaravi quien indicó que los pobladores de la *chaupi yunga* tenían como "socios" a los canteños con quienes intercambiaban coca por llamas y papas. Aparte de los Quivi y los Guaravi, de quienes disponemos datos etnohistóricos, se podría inferir que un patrón de intercambio similar pudo existir entre otros grupos de la *sierra* y la *chaupi yunga*.

Los datos arqueológicos demuestran que productos costeños también fueron llevados a la sierra (Dillehay 1976). Por ejemplo, *Spondylus* y otras especies marinas fueron encontrados en la superficie de sitios de la *chaupi yunga* y la *sierra*. Es obvio que los costeños *Yunga* del valle del Chillón fueron parte de un extenso sistema de intercambio vinculado con las aguas cálidas del lejano norte peruano y el sur ecuatoriano.

Los pobladores de la *sierra* y la *chaupi yunga* establecieron otras formas de alianzas recíprocas. Así, los Canta y los Quivi participaron en la apertura de "...unas lagunas que se fazen alla arriba en la syerra de la nyebe...," es decir, trabajaron mancomunadamente y en mutuo beneficio en tiempos de sequía, hecho que está mencionado en el documento *Justicia 413* (folio 206r). En temporadas de conflicto los Canta argumentaban que el agua de las alturas se usaba para irrigar los cocales de Quivi, hecho que servía para justificar ataques a las tierras *yunga*.

Los documentos etnohistóricos indican la presencia de muchos grupos en el valle del Chillón y en sus alrededores—Canta, Chaclla, Colli, Guancayo, Inca, Socos y Atavillos. Estos grupos tuvieron en cierto momento acceso a los cocales de Quivi o estuvieron disputando el acceso a los mismos.

Mientras que se disponen de datos arqueológicos y etnohistóricos integrables para entender el Horizonte Tardío (1470–1534 d. de C.), existe escasa información (particularmente etnohistórica), para entender los períodos más tempranos. En realidad, existen períodos sumamente extensos sobre los cuales no contamos con información detallada y sistemática sobre los patrones de asentamiento y sobre sus ubicaciones estratigráficas. A base de subsecuentes excavaciones tanto en sitios costeños como serranos del valle del Chillón, será posible establecer una secuencia alfarera para complementar la de Huancayo Alto (Dillehay 1979), un sitio ubicado en la *chaupi yunga*. De modo, pues, que cualquier secuencia evolutiva para el valle del Chillón deberá esperar nuevos reconocimientos sistemáticos y extensas excavaciones.

Antes de examinar la secuencia arqueológica del valle del Chillón así como las sociedades egalitarias y no-egalitarias que lo ocuparon, revisemos brevemente la evidencia arqueológica relacionada con la producción y uso de la coca.

Coca

El arbusto conocido con el nombre de coca pertenece al género *Erythroxylon*. Unas cuatro variedades de coca, conteniendo cocaína alkalina, se cultivan en América del Sur, las cuales presentan diferencias significativas en cuanto se refiere al contenido de alkaloide (y otros elementos químicos), en la configuración de sus hojas, su corteza, sus frutos y sus flores (Plowman 1984). Cada variedad fue *chactada* y tenía su propio nombre. La coca fue utilizada como medicamento y estimulante, para aplacar el frío, la fatiga, el dolor, el hambre, y en eventos sociales y ceremoniales.

Varios investigadores han sugerido que el uso y domesticación de la coca se produjo en un breve escenario (Towle 1961, Antonil 1978, Plowman 1979, Bohm, Ganders, and Plowman 1982), en el que los cazadores y recolectores podrían haber probado las hojas en épocas de hambruna. Así, al descubrirse las propiedades beneficiosas de la coca sus hojas fueron recolectadas. Posteriormente se desarrolló el proceso mismo de secar las hojas y luego *chactarlas* en combinación con cal. Es posible que arbustos completos de coca fueron trasplantados a terrenos ubicados cerca de los campamentos base, quizá como parte de un esfuerzo por reducir el área de búsqueda, aumentar su disponibilidad y su densidad como arbusto.

Para los propósitos de este capítulo nos concentraremos solo en dos de las cuatro variedades cultivadas. Una de ellas es

Erythroxylon coca var. coca [llamada *mamox* en el lenguaje vernacular], la cual crece entre 500 y 1500 metros de altitud conocida comúnmente como Huánuco o coca de la montaña, pues aún existe como planta silvestre en los Andes orientales o montaña (Golte 1968). Esta variedad pudo haberse desarrollado hacia los 7000 años a. de C. difundiéndose gradualmente por los Andes orientales desde Ecuador hasta Bolivia. Los esfuerzos por trasplantar esta variedad a la costa peruana no tuvieron éxito (Plowman 1984:133).

La otra variedad es *Erythroxylon novogranatense* var. *truxillense* [llamada *tupa*]. Significativamente, re-análisis de hojas de coca prehispánicas recuperadas en la costa del Perú señalan que casi todas pertenecen a esta variedad. Unicamente una muestra de coca de la montaña proviene de un contexto costeño del Horizonte Tardío (ver Plowman 1984:135; Mortimer 1901). Actualmente, coca de la variedad Trujillo se cultiva en la costa norte entre 200 y 1800 metros de altitud. Aunque es un arbusto resistente a la sequía su cultivo requiere cierta cantidad de riego.

A pesar que varias muestras de coca han sido encontradas en contextos del Precerámico Tardío, todavía no sabemos qué variedad está representada dado que no fueron identificadas por los botánicos. Una calabaza y 3 moluscos del tipo *Mytilus* conteniendo cal fueron encontrados en el sitio de Culebras, Departamento de Ancash, y asignados al Precerámico Tardío 6 (*ca.* 2500–1800 a. de C.) (Engel 1957:67–68; Bray and Dollery 1983:271). Engel (1963) también da cuenta de muestras de coca y cal provenientes del sitio Asia, cuya antigüedad aproximada es 1314 ± 100 a. de C. A su vez, Patterson (1971:195–96) indica que la coca fue cultivada durante la fase Gaviota (1900–1750 a. de C.) en las terrazas aluviales del río Chillón. Finalmente, Cohen (1978:36–37) da cuenta de restos de "coca *chactada* en forma de bolos" encontrados en los niveles tempranos del período Inicial del sitio el Tanque, Ancón.

De gran importancia es el hecho que Dillehay (1979) recuperó coca durante sus excavaciones en Huancayo Alto, en la *chaupi yunga* del Chillón, no lejos de los terrenos de Quivi, los cuales fueron punto de conflicto inter-étnico según el documento *Justicia 413*. Dillehay (comunicación personal) recuperó hojas de coca en (1) una unidad de almacenamiento fechada hacia los 800 a. de C., y (2) de niveles del Intermedio Tardío y el Horizonte Tardío. Dichas muestras fueron analizadas e identificadas como pertenecientes a la variedad Trujillo o coca costeña por Emma Cerrate y Ramón Ferreyra.

Sin embargo, se han identificado muestra de coca provenientes de sitios y contextos más tardíos (e.g. Vista Alegre [valle del Rímac], Nazca [valle de la Taruga], Monte Grande [valle de Ica]) como pertenecientes a la variedad Trujillo o coca de la costa.

Hoy, Plowman (1984:144–45) considera que existe suficiente evidencia como para sugerir que la variedad Trujillo se cultivó en la costa peruana hacia los 2000 a. de C. indicando también que dicha variedad "pudo crecer hacia 4000 a. de C.". Al igual que las dos variedades de coca peruana identificadas por los botánicos, los indígenas también distinguieron entre la variedad Trujillo *tupa* u "hoja pequeña" y *mamox*, de "hoja grande," variedad Huánuco o coca de la montaña (Rostworowski 1973).

Por el documento *Justicia 413* sabemos que en Quivi se cultivaba coca y considerando la información hoy disponible podríamos señalar que la variedad cultivada en Quivi fue (1) la *tupa* de hoja pequeña o Trujillo, y (2) que Huancayo Alto tuvo acceso a esta variedad.

Retornemos ahora a la secuencia arqueológica del valle del Chillón y el área colindante.

Período Precerámico (ca. 15,000–1800 a. de C.)

Puesto que no se conocen sitios correspondientes al período entre aproximadamente 15,000 y 9,000 a. de C. para el valle del Chillón, nos concentraremos en los asentamientos de la parte tardía del Precerámico (Figura 7).

Al parecer, la ocupación Precerámica se circumscribe a la costa. Sin embargo, este patrón de asentamiento temprano no es necesariamente una reflexión concluyente. Aparte de reconocimientos en la costa no se han hecho estudios en otras zonas del valle del Chillón; de todos modos, cuando éstos se hicieron ha sido difícil localizar sitios tempranos. Mientras es cierto que la mayor parte de la investigación y la ubicación de sitios tempranos en la costa es más fácil, puede ser cierto también que la ocupación humana fue más densa en dicha área.

Precerámico 3: ca. 9000–7000 a. de C.

Edward Lanning y Thomas Patterson intentaron establecer una larga secuencia cronológica para el valle del río Chillón. Como parte de dicho objetivo ellos ubicaron unos 300 sitios en la zona entre Ancón y el bajo Chillón (e.g. Lanning 1963, 1965, 1967; Patterson 1966a, 1971; Patterson and Lanning 1964; Patterson and Moseley 1968; ver también MacNeish, Patterson, and Browman 1975; Moseley 1975; Bonavia 1979; Quilter 1985).

Lanning localizó cerca de 50 talleres a base de los cuales definió cuatro complejos: Zona Roja, Oquendo, Chivateros 1 y Chivateros 2. Las dos primeras "industrias" se pensó estaban asociadas con "recolectores indiferenciados" o "bandas." Patterson (1966a) ha señalado la escasa evidencia estratigráfica y radiocarbónica disponible y arguyó que el complejo Oquendo se ubica entre la Zona Roja y Chivateros 1, agregando que Zona Roja tiene 10,000–9,000 años a. de C.

Por aproximadamente 20 años la "cantera" o "taller" de Chivateros fue considerado como la ocupación más temprana de la costa central (Figura 7). Asímismo, casi todos los intentos por reconstruir el modo de vida de los pobladores del Chillón—quienes fabricaron y utilizaron dichas herramientas líticas—se han basado en los datos de Chivateros.

Lanning consideró que las *lomas* fueron importantes por cuanto los sitios Chivateros se ubican en dicha zona, la cual se compone de suaves colinas, adyacentes al litoral, en donde crece una vegetación estacional a expensas de la garúa y el aire húmedo del ambiente. Cuando las *lomas* se hallan en "florecimiento" (a veces por unos meses o cada varios años) brota una vegetación que es aprovechable ocasionalmente y por corto tiempo para el pastoreo de camélidos. Sin embargo, es de poca utilidad para el cultivo o la recolecta de plantas de valor alimenticio para el ser humano. De modo que a pesar que las *lomas* ofrecen recursos no predecibles pueden ser útiles para (1) ayudar a los pastores quienes buscan pastos para sus rebaños, o (2) para un cazador (como miembro de una banda) quien trata de encontrar una área que atrae a los animales.

Lanning planteó, a base de los datos de Chivateros, que estos antiguos habitantes fueron "recolectores migratorios de las *lomas*" (Lanning 1967:56; traducción nuestra) quienes invirtieron energías en la recolección y consumo de plantas silvestres; en consecuencia, "la caza solamente desempeñó un rol relativamente menor" (Lanning 1967:46; traducción nuestra). Los mismos datos fueron reinterpretados por MacNeish, Patterson, and Browman (1975), postulando que la caza fue probablemente una actividad primordial que se complementó con la utilización estacional de las *lomas* (estación húmeda) y el propio valle (en el verano). Existe además otro planteamiento el cual sostiene que "cazadores y recolectores asentados en la sierra aparentemente descendieron estacionalmente a las *lomas* persiguiendo animales" (Sanders and Marino 1970:34; traducción nuestra).

Lynch (1983:110) propone otro modelo según el cual una transhumancia estacional tuvo lugar entre la sierra y las latitudes bajas de modo que las "ocupaciones de la costa constituyen excursiones laterales a habitats que apenas pudieron proporcionar, aún a 9,000 o 10,000 años a. de C., fauna y flora comparables a la de las montañas" (Lynch 1983:111; traducción nuestra). Uno de los aspectos más importantes que Lynch hace hincapié es el hecho que los habitantes de la costa utilizan otras estrategias, diferentes a las empleadas en la sierra. Los serranos se caracterizaron por depender preponderamente de la caza (consumidores de carne); los costeños en cambio se caracterizaron por ser recolectores (de plantas y de mariscos). Este planteamiento se basa (1) en el hecho que los restos culturales más antiguos de la costa norte del Perú (Complejo Paiján) no están asociados con megafauna; y (2) análisis de los patrones de desgaste de los dientes de una mujer joven (Complejo Paiján, ca. 8250 ± 180 a. de C.) indican que su dieta fue "más vegetariana que carnívora" (Lynch 1983:115).

Se postula que el complejo Chivateros estuvo relacionado a otros sitios costeños de la costa central (e.g. Volcán en el valle de Huarmey) y la costa norte (e.g. Paiján; cf. Bonavia 1979, 1982). Algunos investigadores han sugerido que sería más preciso indicar que los antiguos pobladores de la costa central tuvieron una economía de subsistencia basada principalmente en la recolección y pesca antes que en la caza. Sin embargo, unos de los problemas vinculados a este punto es que "la evidencia sobre la recolección temprana en América del Sur es mínima" (Lynch 1983:121).

Todos los esfuerzos por describir los principales componentes sociopolíticos y los cambios en los patrones de subsistencia sucedidos en el bajo Chillón enfrentan numerosos problemas tales como (1) escasa información estratigráfica ambiguos, (2) cronología absoluta, (3) seriación de las herramientas líticas e, (4) información ecológica insuficiente.

Puesto que la mayoría de herramientas Chivateros son clasificables con toda seguridad en el grupo de artefactos no concluídos (preformas) (ver Craig and Psuty 1968; Parsons 1970; Fung, Cenzano, and Zavaleta 1972; Lynch 1974, 1983; Chauchat 1975; Bonavia 1979, 1982), algunos investigadores cuestionan la ubicación de Chivateros en el período Precerámico así como su propio status como industria lítica.

A pesar de todos los problemas previamente señalados se puede inferir con determinada confianza que la unidad de organización social durante este período fue la banda. Sin embargo, aún no conocemos sus características. ¿Fueron las sociedades de bandas del bajo Chillón parte de un proceso general de adaptación a la costa? o, ¿fue la adaptación al bajo Chillón significativamente diferente a la del resto de la costa peruana? Aparte de estas preguntas no contamos con datos sobre la importancia relativa de plantas comestibles, pescado, mariscos y otros recursos que formaron la base de la dieta precerámica.

Aunque los estudios sobre el Precerámico del valle del Chillón realizados por Lanning constituyen un meritorio y pionero esfuerzo, un claro entendimiento de este período supondrá un diseño sofisticado de investigación el cual deberá considerar intensos reconocimientos y excavaciones, análisis sobre el potencial nutritivo de las plantas consumidas, así como estudios ecológicos que permitan reconstruir los patrones de asentamiento del Precerámico, los sistemas de extracción de recursos, los complejos de herramientas líticas asociadas a específicas estrategias de subsistencia, y estudios sobre dieta.

Precerámico 4: ca. 7000–5500 a. de C.

Las fases Arenal, Luz, Canario, Corvina (Pampa) y Encanto prosiguen al controvertido complejo Chivateros. Si bien las estrategias de subsistencia durante el Precerámico 4 se supone son diferentes o más diversificadas que las del Precerámico 3, la evidencia no es abundante. Todo lo que podemos decir es que la banda sigue siendo la unidad básica de organización social.

El complejo Arenal (6000–5300 a. de C.) incluye "puntas de proyectil pedunculadas trabajadas a presión...raspadores cuidadosamente elaborados... pequeñas herramientas con punta y muy pocas piedras de moler" (Lanning 1967:48). El sitio tipo de esta fase es El Arenal, ubicado 3 kilómetros al norte del río Chillón; material semejante ha sido también encontrado en Ancón (Lanning 1963).

Luz es un complejo relacionado a Arenal pero más reciente y como en el caso anterior también se le encuentra en las *lomas*. Lanning descubrió por lo menos unos 23 sitios con depósitos no profundos los cuales proporcionaron restos de *mate (Lagenaria siceraria)* y unos cuantos morteros. Es probable que los cazadores o "pastores" transhumantes llevaron consigo agua y plantas conjuntamente con el resto del utillaje doméstico.

El logro más significativo durante el Precerámico 4 se produce alrededor de 3600–2500 a. de C. (fase Encanto) y está evidenciado por el uso cada vez más frecuente de los recursos marinos y el cultivo de plantas (Lanning 1967; Moseley 1975).

Las sociedades de bandas de esta época probablemente tuvieron una estrategia de subsistencia diversificada la cual consistió en la utilización de recursos marinos (e.g. pescado, mariscos), caza y recolecta de plantas en las *lomas* (siempre de acuerdo a Lanning), o tal vez esta diversidad se debe al hecho que contamos con una ligera mayor información sobre el Precerámico 4 en comparación al Precerámico 3. Lanning propuso que *lomas* hoy extinctas—entre Ancón y el Chillón—fueron ocupadas en la estación húmeda por pequeños grupos. Aunque Lanning ponderó las *lomas* como el lugar primordial para la recolección de plantas silvestres, no presentó evidencias concretas para apoyar su postulado; además, como ya se ha mencionado, es muy poco lo que se puede aprovechar de esta zona en términos de productos comestibles.

Precerámico 5: ca. 5500–4200 a. de C.

El sitio tipo de la fase Canario (5500–4200 a. de C.) es Pampa del Canario, ubicado a 6 kilómetros y al noreste de la Bahía de Ancón (Figura 7). El asentamiento al parecer se aglutina en las *lomas* aledañas a Ancón. Lanning (1963:49) utiliza las herramientas—"piedras para moler, manos de formas naturales y trabajadas, morteros y machacadores..."—como prueba para indicar que en esta fase se producen cambios significativamente importantes en la subsistencia. Patterson (1971) propuso a su vez que una variedad de habitats fueron utilizados: las *lomas*, la playa, el litoral rocoso, el desierto y el piso del valle. En consecuencia, parecería que las principales estrategias económicas de subsistencia estuvieron representadas por la recolección de plantas, la caza y la recolección de mariscos. Patterson ha sugerido asímismo que estos recursos se utilizaron en el contexto de un ciclo de transhumancia anual por el cual las bandas se desplazaron de los campamentos de invierno a los de verano.

Precerámico 6: ca. 4200–2500 a. de C.

Este período está representado por dos complejos: Corvina (4200–3700 a. de C.) y Encanto (3700–2500 a. de C.). Existen solo 3 sitios de la fase Corvina: 2 en Cerro La Corvina (4 kilómetros al este de Ancón) y 1 en Loma Encanto (6 kilómetros noreste de Ancón). Lanning (1963:49) encontró en Loma Encanto puntas de proyectil y cierta "cantidad de manos y un par de piedras de moler, aunque no hubo morteros o machacadores."

El complejo Encanto fue identificado en 15 sitios situados en la zona de Ancón-Chillón. Cinco de ellos son pequeños talleres mientras que los otros son campamentos. Al parecer, hubo sitios permanentes en esta área así como en los valles de Lurín y Chilca (MacNeish, Patterson, and Browman 1975:29). A base de datos recogidos en los sitios de Yacht Club, situado en la ladera norte del cerro que domina la Bahía de Ancón, y Chilca, en el bajo Chilca, se afirma que los recursos marinos fueron la base de la alimentación, la misma que fue complementada con calabazas (*Cucurbita ficifolia, C. moschata)* y algodón (*Gossypium barbadense)* (Lanning 1967:50). De hecho, 23 semillas de calabazas fueron descubiertas en el sitio tipo de Encanto (Moseley 1975). A su vez, redes para pescar y tejidos de algodón fabricados con la técnica del entrelazado fueron recuperados en los sitios de Yacht Club y Chilca (Lanning 1967:50).

Es obvio que estas evidencias indican cuales fueron realmente algunas de las estrategias de subsistencia durante la época Encanto, sobre todo aquellas vinculadas con la tecnología de la pesca según lo mostrado por el uso de redes de pescar (Lanning 1967; Moseley 1975:53).

Precerámico 7: ca. 2500–1800 a. de C.

Al parecer cambios cualitativos se producen en esta era a juzgar por datos vinculados con la aparición de jerarquías sociales. Antes de presentar los modelos e inferencias propuestos por investigadores que han estudiado esta zona, veamos primero la evidencia arqueológica.

Fase Pampa (2500–2300 a. de C.): Llamada así por haber sido identificada en el sitio del mismo nombre, en el extremo norte de la Bahía de Ventanilla, al norte del río Chillón. Lanning (1967:53) planteó que el sitio fue un campamento temporal habitado por pescadores y cultivadores de calabazas quienes adoptaron paulatinamente una economía eminentemente marina. De todos modos, aunque Lanning (1967) y Moseley (1975:116) señalan que la pesca fue la base económica de subsistencia, el sitio ha proporcionado un conjunto de plantas cultivadas. Así por ejemplo, el sitio La Pampa contiene restos de algodón silvestre más antiguos de la zona Ancón-Chillón (Moseley 1975:22). Además, plantas domesticadas entre las que figuran calabazas (*Cucurbita ficifolia, C. moschata)*, frijoles (*Canavalia* sp.), y calabazas silvestres (*C. ecuadorensis)* han sido encontradas en este sitio (Patterson and Moseley 1968; Cutler and Whitaker 1961). Finalmente, Cohen (1977) identificó en una muestra proveniente de este sitio pacae (*Inga feuilli)*, achira (*Canna* sp.), guava (*Psidium guajava)* y una leguminosa (*Galactia striala)*.

Fase Playa Hermosa (2300–2100 a. de C.): Aunque la base económica de subsistencia es similar a la de la fase anterior, la utilización de plantas se incrementa y aparecen sitios permanentes en el área de Ancón-Chillón. Esta fase está representada por los sitios de Camino y Banco Verde (en Ventanilla) y Yacht Club (en Ancón) (Moseley 1975). A pesar que estos sitios contienen restos de una nueva planta (ají, *Capsicum baccatum)* además de los ya conocidos (algodón, calabazas y guava), muchos arqueólogos señalan que la economía se basó en la utilización de los recursos marinos.

Fase Conchas (2100–1900 a. de C.): Fue definida por materiales encontrados en Punta Grande (Ventanilla) y El Tanque (Ancón) (Moseley 1975). Punta Grande es de mayor tamaño y se compone de 4 terrazas construidas a desnivel y una siguiendo a la otra en la base de una colina. Las terrazas fueron aparentemente utilizadas como superficies de vivienda. Parece que los ocupantes de este asentamiento dependieron de los recursos del litoral aunque dos nuevas plantas se incorporan a la dieta—frijoles lima (*Phaseolus lunatus)* y lúcuma (*Pouteria lucuma)*.

Fase Gaviota (1900–1750 a. de C.): Quizá representa una era de creciente complejidad social a juzgar por la construcción de grandes edificios cívico-ceremoniales. El Paraíso (Chuquitanta), margen sur del bajo Chillón, proporciona información significativa sobre técnicas empleadas en la construcción de los edificios públicos así como sobre las actividades ceremoniales que tuvieron lugar en los mismos (Figura 8). Engel (1966, 1967; ver también Moseley 1975) quien realizó excavaciones en El Paraíso ha señalado que su construcción se produjo bajo la dirección de una autoridad central.

Se ha postulado que la población de El Paraíso se dedicó principalmente a la utilización de los recursos marinos (Moseley 1975). Sin embargo, el sitio ha proporcionado también un conjunto de plantas cultivadas entre las que figuran algodón, calabazas, achira, frijoles lima, jícama, maní (*Arachis hypogaea)* y posiblemente camote (*Ipomoea batatas)*.

Aunque El Paraíso se compone de 9 montículos ellos no necesariamente son contemporáneos. Engel excavó la Unidad I (ver Figura 9) e identificó unas 5 fases de construcción. La más reciente ha sido fechada en ca. de 2000 a de. C. Destaca esta Unidad por presentar un recinto en cuyo piso existe una estructura cuadrangular hundida así como cuatro hoyos situados inmediatamente en el exterior de las esquinas que forman esta estructura. Se atribuye a este recinto funciones cívico-ceremoniales (ver Figura 9).

Lanning (1967) fue quien originalmente sugirió que El Paraíso fue parte de la tradición de edificios con planta en U. Más reciéntemente, Williams (1971, 1978–1980, 1985) indicó que el plano de El Paraíso constituye la evidencia más antigua de los edificios con planta en U. Dudas acerca de la forma de este sitio fueron expresadas por Scheele (1970) y Silva (1984). Por otro lado, excavaciones recientes realizadas por J. Quilter (1985) en las Unidades I, II y IV de El Paraíso sugieren que una variedad de actividades domésticas y ceremoniales tuvieron lugar en este asentamiento.

Se ha sostenido que El Paraíso fue construido por una Jefatura (Carneiro 1970; Sanders and Marino 1970). Es probable que hacia los 2000 a. de C. una serie de jefaturas aparecieron en diversas regiones de los Andes centrales (Kotosh en el valle medio del río Huallaga, Pacopampa en la cuenca del Chotano, etc.). El Paraíso como foco de autoridad probablemente integró un conjunto de poblados sea en el propio Chillón y fuera de él. Aún queda por estudiar las características internas de los asentamientos, la cantidad de no-residentes quienes participaron en ceremonias especiales, y el carácter de las interacciones inter-sitio. Por último, no sabemos porqué El Paraíso fue abandonado ca. de 1800 a. de C.

Período Inicial (ca. 1750–1000 a. de C.)

Este período presenta significativos cambios con respecto al desarrollo económico y sociopolítico de las sociedades que se asentaron en este valle. Esta fue una época de cultivo intensivo en las terrazas aluviales del Chillón medio y superior. Por razones aún no establecidas la pesca dejó de ser la más importante actividad económica de subsistencia. En lugar de ello, se evidencia una economía diversificada en la que el intercambio desempeñó papel preponderante. Es quizá como parte de estos cambios que la coca se convirtió en un importante producto al comienzo de este período. Como se dijera anteriormente, evidencias sobre uso de la coca han sido encontradas en niveles del Precerámico Tardío de la costa. Las evidencias más antiguas sobre uso de la coca en el área Chillón-Ancón se presentan en la forma de *bolos* masticados en niveles del período Inicial (ca. 1750 a. de C.) del Tanque, Ancón (Patterson and Moseley 1968; Cohen 1977:66, 1978:37). Además, si bien la coca pudo haber sido cultivada por este tiempo en el valle medio (*chaupi yunga)* del río Chillón (MacNeish, Patterson, and Browman 1975:33), no podemos extendernos en este punto dado que no existe suficiente información. Retornemos ahora a algunos de los eventos sociopolíticos que tuvieron lugar durante este período en el área Ancón-Chillón.

El Edificio de La Florida y su Rol Integrador

Según Patterson (1971:198), el abandono de El Paraíso se relaciona a la aparición de La Florida; él ha sugerido que un grupo de familias de El Paraíso al parecer se asoció con aquéllas que habían abandonado sus aldeas en las inmediaciones de Ventanilla y todas en conjunto se asentaron en La Florida (a 1.5 kilómetros al norte del río Rímac; Figura 7). Sin embargo, no existe evidencia arqueológica directa para demostrar el desplazamiento de grupos de familias del Chillón al Rímac; este punto de vista podría más bien obscurecer los procesos que dieron lugar al abandono de El Paraíso y los subsecuentes cambios en el Rímac los mismos que permitieron la aparición de un nuevo poder político en este último.

La Florida y El Paraíso son diferentes en forma, tamaño y disposición de los espacios sagrados y seculares. El recinto sagrado de la Unidad I de El Paraíso es pequeño y fue diseñado para realizar actividades ceremoniales de carácter ritual (Figura 9). La organización del espacio ceremonial en La Florida se relaciona en cambio al plano de los edificios en U. En breve, este último presenta una clara separación de los espacios privados y no privados; la gran plaza está separada del pequeño y privado recinto situado en los montículos central-oeste, norte y sur.

¿Fueron El Paraíso y La Florida simultáneamente ocupados (ca. 2000 a. de C.)? o ¿fueron utilizados secuencialmente (tal como ha sido sugerido por Patterson)? Si aceptamos la primera posición, ¿mantuvieron estos sitios una relación competitiva? u ¿ofrecieron realmente ritos y servicios a poblados diferentes? Futuras investigaciones deberán responder estas preguntas.

La construcción de La Florida parece producirse juntamente con otros procesos: (1) la aparición de cerámica en la costa central; (2) la construcción de centros ceremoniales en forma de U en la costa central y norte (el de La Florida es sólo una muestra); (3) el surgimiento de una unidad sociopolítica poderosa en el valle del Rímac; y (4) la importancia regional de La Florida y el probable control que ejerció sobre el área Ancón-Rímac, la misma que incluyó numerosos asentamientos (Patterson 1985:65). El hecho que durante este tiempo existió una alfarería similar a través de esta área sirve para inferir la existencia de un amplio control político.

Si bien se ha indicado que La Florida fue abandonado hacia los 1750 a. de C., este sitio contiene también restos del Horizonte Temprano y el Intermedio Temprano (Patterson 1985).

Garagay, La Florida, La Salina, Huacoy, Chocas y Cueva

Garagay, La Florida, Huacoy y Chocas comparten más o menos al mismo tiempo el plano arquitectónico en U. Garagay se halla entre el Chillón y el Rímac (Ravines 1975; Ravines and Isbell 1976); La Florida y la Salina se encuentran en el Rímac; Huacoy, Chocas y Cueva en el valle del Chillón (Figura 7).

El Atrium del Templo Medio presente en el Montículo B de Garagay parece constituir el recinto ceremonial más importante; sobre las paredes de este recinto existen frisos multicolores que al parecer son anteriores a la iconografía Chavín en la costa central (Ravines and Isbell 1976:266). Ravines (1984) señala que la principal imagen de estos frisos es una langosta marina; por su parte, Salazar-Burger and Burger (1982:234) la identifican con la araña, añadiendo que fue un tema iconográfico importante del estilo Cupisnique. Ravines (1984) descubrió dos fases de construcción en el Montículo B. Por lo menos una de estas fases de construcción es quizá contemporánea con otra similar en el edificio de La Florida.

Es probable que Garagay y La Florida fueron solamente dos de varios edificios públicos importantes dedicados a ofrecer servicios especiales y tal vez controlaron una serie de aldeas presentes en distintos lugares de Ancón, Chillón y el Rímac.

Huacoy, situado en la margen sur del valle del Chillón, también muestra plano en forma de U (Ludeña 1973). La cerámica recogida en este edificio indica que estuvo funcionando al mismo tiempo que Garagay.

Además de los sitios mencionados existen dos edificios más en la margen sur del Chillón: Chocas y Cueva.

Teniendo en cuenta la cantidad y la distribución de los edificios en U durante esta época, es probable que estamos frente a la manifestación arqueológica de "una nueva y exitosa jefatura...que se expande por la incorporación de gente extraña al asentamiento...y crecimiento interno" (Service 1975:79). ¿Estuvo cada sitio en forma de U dedicado a realizar una serie de funciones integradoras de linajes particulares? ¿Tuvieron algunos de estos centros, como Garagay, amplio y mayor control que otros de su época? ¿Existen evidencias sobre dos o tres niveles jerárquicos basados en (1) la cantidad de edificios públicos y, (2) tamaños de los mismos? Obviamente, estas preguntas junto con otros aspectos (por ejemplo, las características de las unidades domésticas) necesitan examinarse.

El Horizonte Temprano (ca. 1000–400 a. de C.)

Tradicionalmente, este período ha sido concebido como uno de homogeneidad cultural en el Perú central, especialmente si es considerado como parte de un estilo artístico en expansión. Además, la difusión del estilo Chavín aparece como un estilo intrusivo que desplaza a otros más antiguos. Una de las explicaciones más frecuentes para la dispersión del estilo Chavín se refleja en la adopción de un nuevo "culto" o religión cuyo tema principal es el felino (Lanning 1967).

En efecto, cerámica similar a Chavín aparece en innumerables sitios de la costa. Sin embargo, creemos que el término "influencia" es ambiguo y es posible que algún día contemos con suficientes datos para demostrar la existencia de interacción entre élites, o intercambios recíprocos de regalos entre los jefes de unidades sociopolíticas diferentes. Un ejemplo que podría ilustrar este planteamiento se apoya en ofrendas Chavín depositadas en el Atrium del Templo Medio del Montículo B de Garagay. Garagay pudo tener tanto prestigio que la élite de Chavín de Huantar no vaciló en reconocerlo y dichas ofrendas pudieron servir para reforzar y establecer interrelaciones sociopolíticas.

El creciente prestigio de Garagay puede reflejarse en el constante remodelamiento de sus recintos asi como al crecimiento demográfico sucedido en el Horizonte Temprano. Así, un nuevo templo se construyó sobre el Templo Medio del Montículo B. Esta estructura (llamada Templo Tardío) pudo haber tenido representaciones similares a Chavín, lamentablemente hoy destruídas. El Montículo A de Garagay también presenta frisos de barro con representaciones que recuerdan a las de Moxeque, Cerro Blanco y Caballo Muerto.

El Montículo A está asociado a un recinto circular semihundido. Al respecto, dado que Garagay exhibe una serie de similitudes con otros centros públicos del Perú, podríamos sugerir que la élite de Garagay estuvo en contacto con otras de la costa. La mayor concentración de recintos circulares semihundidos se extiende desde Alto Salaverry (en el norte) a Mala (en el sur).

Con relación al bajo Chillón, Patterson and Lanning (1964) señalaron que durante el Horizonte Temprano la cantidad de asentamientos aumentó en el valle mismo y en lo alto de los cerros, uno de los cuales pudo ser fortificado.

Mientras que Garagay (Lanning 1967; Ravines 1984), Huacoy o San Humberto (Ludeña 1970, 1973) y Cueva son tres sitios *yunga* importantes para los pobladores de la costa, existen otros asentamientos contemporáneos valle adentro. Chocas, por ejemplo, se halla en la parte baja de la *chaupi yunga* (a unos 500 metros de altura) y a 10 kilómetros de Huacoy; en cualquier caso puesto que estos sitios casi no han sido estudiados es poco lo que podemos decir.

El sitio *chaupi yunga* de Huancayo Alto se encuentra cerca de Yangas, Kilómetro 56 y a 1050 metros de altura, en la margen sur del río Chillón. Gracias a las excavaciones realizadas por Dillehay los datos de este sitio permitieron establecer la secuencia más larga y mejor conocida para la *chaupi yunga* del Chillón. La secuencia en cuestión se compone de 7 períodos cubriendo un lapso que va de ca. 800 a. de C. hasta 1534 d. de C. (Dillehay 1979:25).

La ocupación más antigua se produjo entre 800 y 500 a. de C. y quizá fue fundada por otro asentamiento de la *chaupi yunga* o tal vez por sociedades del bajo Chillón. La cerámica del Horizonte Temprano de Huancayo Alto es semejante a la de Garagay y Huacoy.

Huancayo Alto presenta diversos tipos de arquitectura: 3 *huacas* pequeñas (montículos artificiales); 3 hileras paralelas de piedra que se inician en la *huaca* más grande proyectándose a dos montículos pequeños situados a unos 30 metros de distancia; los cuartos fueron hechos con piedras de río; por último, un conjunto de 57 cuartos para almacenamiento dispuestos uno detrás de otro y situados en la ladera del cerro. El complejo de almacenamiento de este sitio no ha sido aún reconocido en otros sitios.

Ascendiendo el valle del Chillón se encuentra Santa Rosa de Quives, que junto a Huancayo Alto, proporcionó cerámica semejante a la encontrada en el bajo Chillón. Por esta razón, se propuso que una suerte de intercambio caracterizó las relaciones *yunga-chaupi yunga* (Patterson 1971; MacNeish, Patterson, and Browman 1975; Dillehay 1979).

Puesto que los edificios públicos se concentran principalmente en la *yunga* (Garagay y Huacoy por ejemplo) y en la parte baja de la *chaupi yunga* (Chocas), es probable que los *yunga* controlaron las poblaciones del Chillón. Al igual que en el período anterior los edificios públicos fueron el foco principal de las sociedades del Horizonte Temprano asentadas en el Chillón.

Período Intermedio Temprano (400 a. de C.-600 d. de C.)

El período Intermedio Temprano se define frecuentemente como una época de desarrollo estatal en los Andes (por ejemplo, el estado Moche en la costa norte, el estado Lima en la costa central, etc.) (Lanning 1967; Earle 1972; MacNeish, Patterson, and Browman 1975; Patterson, McCarthy, and Dunn 1982).

Una mirada al período Intermedio Temprano desde la perspectiva arqueológica del valle del Chillón revela que durante este período se produjeron cambios socioeconómicos cualitativos en la costa central del Perú. Para ilustrar este punto primero presentaremos la evidencia y luego discutiremos brevemente las características socioeconómicas de esta era en el valle del Chillón.

Según los datos arqueológicos el bajo Chillón (Ancón, Ventanilla, Puente Piedra y Carabayllo) estuvo densamente poblado, habiéndose definido tres fases alfareras: Ventanilla, Miramar y Lima (Villar Córdova 1935; Stumer 1953, 1954a,b; 1955; Patterson 1966; Patterson and Lanning 1964; Dillehay 1979).

La fase Ventanilla (ca. 250–50 a. de C.) está representada por un asentamiento de pescadores situado en el bajo Chillón, el cual fue abandonado en la segunda centuria de nuestra era (MacNeish, Patterson, and Browman 1975:48). Aunque no se han descubierto sitios relacionados a Ventanilla en la *chaupi yunga* del Chillón, esta fase exhibe semejanzas con la fase

Huachipa-Jicamarca C del Rímac Medio (Silva, Hirth, García and Pinilla 1982, 1983).

Miramar se superpone a Ventanilla y fue definida a base de cerámica encontrada en la localidad de Miramar, Ancón. Alfarería similar fue encontrada igualmente en Santa Rosa o Playa Grande y en Huancayo Alto (Stumer 1953, Patterson 1966, Dillehay 1979). Es evidente que la cerámica de Miramar se vincula con la encontrada en los valles de Chancay, Rímac, Lurín y Chincha. De hecho, la fase Miramar es parte del denominado "Blanco sobre Rojo," un estilo de cerámica que se extendió a través del Perú central hacia los 100 a. de C. (Willey 1948).

La modalidad Lima se ubica más o menos entre 200 y 600 de nuestra era y fue dividida en 9 fases: la primera inmediatamente después de Miramar y la última con atributos similares a Nievería, un estilo fundamentalmente asignado a la Epoca 1 del Horizonte Medio (600–650 d. de C.) (Menzel 1964; Patterson 1966). Componentes de la sociedad Lima fueron descubiertos sobre una área que se extiende entre Chancay y Lurín y es principalmente conocida como una expresión cultural de la costa. Aparte de los montículos Maranga en el bajo Rímac, sitio que se dice constituye el asentamiento tipo de esta sociedad, por lo menos 16 asentamientos han sido descubiertos en el área de Ancón y el bajo Chillón (Patterson 1966). La Uva es el asentamiento más grande e incluye 13 edificios (Stumer 1954b; Patterson 1966). Actualmente, además de los sitios residenciales de Playa Grande y Ancón, solo dos edificios públicos fueron estudiados: Media Luna y Culebra, ambos en el bajo Chillón (Quilter 1983; Stumer 1954b, 1955; Silva, García, Bragayrac, and Morales 1984). De éstos, Culebra proporcionó datos sobre su organización interna compuesta por un edificio público (Figura 10), que por lo menos tiene 3 fases de construcción, y numerosas estructuras domésticas (Figura 11). Las excavaciones realizadas en el área residencial pusieron al descubierto casas construidas con cañas (Figura 12), casas construidas con piedras y adobes modelados a mano (Stumer 1955; Patterson 1966; Silva, García, Bragayrac, and Morales 1984). Hoy, la antigua área residencial de este sitio muestra otra vez estructuras construidas reciéntemente con cañas y esteras (Figuras 13–15).

Restos de cerámica Lima en la *chaupi yunga* del Chillón están representados por los encontrados en Huancayo Alto (Dillehay 1979). Este sitio fue construido en la parte media de la *quebrada* y se compone de un edificio de adobes con numerosos cuartos, plataformas de piedra—que quizá sirvieron para construir estructuras de *kincha* para uso residencial—así como una estructura de piedra de forma oval que según Dillehay (comunicación personal) tipifica numerosos asentamientos de las serranías de Canta. Mientras que Dillehay concluye señalando que Huancayo Alto fue habitado en este tiempo por "una combinación de grupos étnicos de la *chaupi yunga* y de la sierra," es posible que la indicada evidencia arqueológica descrita líneas arriba podría interpretarse como el reflejo de (1) un creciente intercambio entre dichas zonas, o (2) matrimonios entre pobladores de la *chaupi yunga* y la sierra.

Aunque la información es escasa podemos presentar algunas ideas sobre este período: (1) los sitios exhiben arquitectura pública cuyo plano es distinto al modelo en U de los períodos Inicial y Horizonte Temprano; (2) la población estuvo concentrada en la parte baja de la *yunga* del Chillón; (3) la denominada "Cultura Lima" se desarrolló durante una época en que la parte baja de los valles fueron aún centros de poder y cuando los contactos intervalle fueron preponderantes; (4) la cerámica Lima fue menos abundante en la *chaupi yunga* y aún más escasa en sitios de la *sierra*; y (5) asignar la "sociedad Lima" a la categoría de organización estatal es debatible y tema de controversia que requiere examinarse a la luz de evidencias arqueológicas específicas que por ahora no las tenemos.

Horizonte Medio (600–1000 d. de C.)

Aunque este período es el menos estudiado en el valle del Chillón, algunos restos se han encontrado en la *yunga* y la *chaupi yunga* de este valle. Los estudios realizados por Stumer (1954b), Bonavia (1962), Patterson y Lanning (1964) revelan que (1) hubo una significativa ocupación en Ancón; (2) sitios tales como Culebra, Media Luna y Santa Rosa (o Playa Grande) fueron prácticamente abandonados; (3) los edificios públicos del período anterior fueron convertidos en cementerios y; (4) es posible que las poblaciones del medio y alto Chillón utilizaron una alfarería que aún debe definirse mediante excavaciones estratigráficas.

Huancayo Alto proporciona datos sobre este período para la *chaupi yunga*. Es durante esta época que el complejo de la *huaca* fue abandonado y se construyó un nuevo edificio público de piedras, el complejo de almacenes fue ampliado y se incrementó la presencia de camélidos, lana y cerámica *serrana*. Tomando como base estas evidencias Dillehay propuso que Huancayo Alto fue ocupado por "tres grupos étnicos," agregando que (1) el edificio de adobes (compuesto por innumerables recintos) fue ocupado por la élite administrativa local de la *chaupi yunga*, (2) que las estructuras de piedra fueron ocupadas por personas de menor status (quizá provenientes de Canta o Junín?), y (3) los habitantes de la terraza de piedra fueron ocupantes temporales procedentes del área Chacalla-Jicamarca.

Al parecer la *yunga* del Chillón no fue densamente ocupada si se la compara con los restos del Intermedio Temprano presentes en la *chaupi yunga*. Por razones aún no explicadas el bajo Chillón perdió importancia política durante el Horizonte Medio. En lugar de ello, los valles del Rímac y Lurín se convirtieron en centros de poder y tal vez la población de la costa central fue gobernada por centros políticos localizados en estos valles. Es en este contexto que tiene particular importancia el asentamiento de Cajamarquilla en el valle del Rímac y Pachacamac en el valle de Lurín.

Período Intermedio Tardío (1000–1476 d. de C.)

Dado que los procedimientos e instituciones Inca e Hispánica son entendibles evaluándolos a través de la forma cómo aprovecharon en beneficio propio la organización política de los grupos étnicos así como las alianzas y enemistades interétnicas, trataremos de reconstruir con cierto detalle el caso del Intermedio Tardío. Consecuentemente, veamos el ambiente sociopolítico y económico que fue encontrado por el Inca.

Mientras que el poblamiento del bajo y medio Chillón es denso y aglutinado, el poblamiento del alto Chillón es disperso y organizado en confederaciones o *curacazgos* (Dillehay 1976:Capítulos 5–6). A fín de ubicar en perspectiva al documento *Justicia 413* debemos enumerar algunas de las características de la *chaupi yunga* o valle medio; ellas son: (1) la gran concentración de sitios prehispánicos, (2) la gran diversidad de plantas y cultivos, (3) la capacidad productiva de grandes cantidades de coca, y (4) su ubicación estratégica que le permite el acceso a varias zonas.

A diferencia de otros valles, el Chillón no tuvo en ningún momento un centro de poder como para dominar el valle o extender su dominio a otras regiones. En lugar de ello, el Horizonte Tardío y el Intermedio Tardío se caracterizan por la presencia de un múltiple mosaico de unidades sociopolíticas semi-autónomas que alternadamente competían o se prestaban apoyo cooperativo (Dillehay 1976).

La naturaleza multi-étnica de alianzas e intercambio así como los conflictos inter-étnicos sucedidos en el Chillón fueron aspectos importantes para la administración Inca por cuanto era necesario contar con aliados apropiados en el dominio y administración de las nuevas tierras incorporadas al incanato.

El patrón que surge del dato arqueológico recuperado en este valle es que los sitios ubicados en su margen norte, principalmente en la *chaupi yunga*, presentan mayor afinidad con el valle de Chancay, en el norte (particularmente con sitios cerca de Quebrada Quilca). Aunque cerámica Chancay Negro sobre Blanco ha sido encontrada en sitios *yunga* y *chaupi yunga*, ésta fue más abundante en esta última zona, particularmente en Trapiche, Macas y Huarabi. Unicamente escasos fragmentos de cerámica Chancay Negro sobre Blanco fueron encontrados en la margen sur del Chillón, especialmente en Collique, Huanchipuquio, Zapán y Santa Rosa de Quives. Los sitios situados en el lado sur del Chillón contienen alfarería similar a los sitios del Rímac. Quivi Vieja, en el alto Chillón, es el sitio más alejado hasta donde se extendió este estilo (Dillehay 1976:184).

Anteriormente habíamos mencionado el tema de la etnicidad por cuanto arqueólogos y etnohistoriadores han tratado de establecer el territorio y los límites de los grupos étnicos utilizando varias clases de datos. Los arqueólogos emplean con frecuencia la distribución sincrónica de estilos alfareros y arquitectónicos, en tanto que los etnohistoriadores y etnógrafos utilizan nombres y lugares mencionados en los documentos. Ambas aproximaciones tienen sus limitaciones. Mencionemos por lo menos tres de los innumerables problemas sobre este punto: (1) la variación alfarera y arquitectónica no necesariamente se vincula con identidad étnica alguna; (2) la información etnohistórica puede en ciertos momentos no ser útil debido a los puntos de vista personales de los informantes; y (3) el dato etnohistórico puede corresponder a épocas de tan corta duración que es difícil o imposible detectarlo arqueológicamente.

Sin embargo, Murra (1972, 1975), Dillehay (1976) y Rostworowski (1972, 1978) han tratado de relacionar nombres de grupos étnicos mencionados en los documentos a zonas y sitios arqueológicos específicos en el valle del Chillón (el Cuadro 1 ilustra dichas correspondencias).

1. Grupos y Sitios Yunga de los períodos Intermedio Tardío y el Horizonte Tardío

Por lo menos existe evidencia sobre unos seis grupos étnicos del bajo Chillón. Estos son Cararuaillo, Colli, Chuquitunga, Sebillay, Sutca e Inca.

Según los documentos el grupo Cararuaillo se asentó en la *yunga* (Figura 6). El nombre de Carabayllo aún designa un asentamiento *yunga* en el lado norte del Chillón, el mismo que cubre alrededor de 1300 hectáreas y se halla a una altitud de 150 metros. El *curaca* de Carabayllo en el siglo XVI fue don Juan Quivi quien indicó que Carabayllo fue parte de las tierras de don Fernando Nacara, señor de Collique (Rostworowski 1977:43).

Cobo (1956:301) indicó que Carabayllo y el Chillón fueron habitados por dos "naciones" que hablaban lenguas diferentes. Una de ellas fue Carabayllo la cual se extendió desde Carabayllo hasta el lado norte del valle de Chancay. La segunda fue Pachacamac la cual se extendió desde Carabayllo hasta Pachacamac.

Kara-Hualla es el único sitio grande que Dillehay (1976:198) descubrió en su reconocimiento del distrito de Carabayllo, componiéndose de cuartos hechos con *tapial* y adobes pequeños. Villar Córdova (1935:170) denominó este sitio con el nombre de Pueblo Viejo, notando la presencia de dos tipos de adobe: "Maranga" y "Chancay." Cerámica del Intermedio Temprano, Intermedio Tardío y Horizonte Tardío (ca. 200 a. de C. a 1550 d. de C. [Dillehay 1976:202]) fue encontrada junto a los adobes tipo Maranga y Chancay.

Puesto que, conforme a los documentos etnohistóricos, los Cararuaillo fueron identificados como un grupo étnico aparte de la *chaupi yunga* del Chillón, y considerando que el sitio de

CUADRO 1

Grupos Etnicos	*Sitios Arqueológicos*
I. La *Yunga* o Valle Bajo	
Cararuaillo	Carabayllo
Colli	Collique
Chuquitunga	Hacienda Chuquitanta
Sebillay	——
Inca	Tambo Inga
Sutca	——
——	Trapiche
II. La *Chaupi Yunga* o Valle Medio	
——	Huanchipuquio
Guancayo	Huancayo Alto
Guaravi	Huarabi Alto
Guaravi	Huarabi Bajo
Maca	Macas
Quivi (Quibi)	Santa Rosa de Quives
Quivi	Quivi Vieja
Sapán (Zapán)	Zapán
——	Checta
Missai (Misayo)	——
——	Yaso
——	Quyso
——	Orobel
——	Magdalena
——	Hornillos
——	Pucara
——	Huarhuar
III. La *Sierra* o Alto Valle	
Canta?	Purumarca
——	Huaycoloro
Atabillos (Atavillos)	——
——	Pirhuinco
Canta	Cantamarca
——	Collo
Socos	——
——	Caballo Blanco
Chaclla (Jicamarca)	Jicamarca
Huarochirí	——
Canta?	Piedra Mesa-Camata
Canta?	Lucana
Canta?	Huancuna
Canta?	Huayuacancha
Huamantanga?	Huamantanga
Canta?	Zepitamarca

Kara-Hualla (Carabayllo) es el sitio más grande y más complejo de la margen norte del Chillón, Dillehay (1976:202) afirma que este sitio fue la capital política local. Los sitios secundarios o más pequeños han proporcionado cerámica de los períodos Intermedio Temprano, Intermedio Tardío y Horizonte Tardío.

Los Colli fueron otro grupo *Yunga*. Varios testigos mencionados en el documento *Justicia 413* manifestaron que el señor de Quivi estuvo bajo el dominio del señor de Colli (o Collique) y que este último tenía tierras que se extendían desde el litoral hasta Quivi. Otros testigos dijeron además que el señor de Collique también tuvo tierras en el valle del Rímac.

Dillehay concluye señalando que los Carabayllo se asentaron en la margen norte en tanto que los Colli en la margen sur del valle del Chillón (Figura 6).

Collique se ubica entre los Kilómetros 16 y 17 de la Avenida Tupac Amaru y cubre unas 20–25 hectáreas. Se divide en varios sectores entre los que destaca una zona central de 1600 x 2000 metros de lado la misma que está delimitada por (1) por una pared en su flanco norte extendiéndose unos 1400 metros de longitud prolongándose hasta cerca de la fortaleza, (2) por 4 pirámides grandes o *huacas* en el noroeste, (3) por dos cerros denominados Alpacoto y Comicay en el lado suroeste, y (4) por la propia "fortaleza de Collique" en el este. La sección baja de la zona central se halla dividida por dos fuentes de aguas subterráneas o *puquios* y un sistema de canales (Squier 1877:87–88; Middendorf 1893–1895, II:47; Villar Córdova 1935:171; Rostworowski 1972:271–79; Dillehay 1976:204–6). Dillehay (1976:212) sugiere que la zona central fue ocupada por una población relativamente pequeña y que el sitio en su conjunto funcionó como un centro administrativo y religioso que controló una serie de asentamientos durante el Intermedio Tardío y el Horizonte Tardío.

Otros grupos costeños que ocuparon la *yunga* fueron los Chuquitanta (¿fueron éstos los Chuquitanga o Chuquitunga?), Sebillay y Sutca. El primero pudo estar asociado al grupo de *huacas* que se hallan no lejos del sitio precerámico de Chuquitanta o El Paraíso. Por el año 1571 de nuestra era el nombre de "Chuquitanga" se refería a una *pachaca* (unidad de 100 tributarios) en los confines del *repartimiento* de Guancayo (Martínez Rengifo, en Espinoza 1963). Por el año de 1585 Chuquitanta, Sebillay y Sutca fueron grupos étnicos diferentes en la *reducción* de Carabayllo (Rostworowski 1972:257, 266, 290).

Tambo Inga, localizado a unos 18 kilómetros al norte de Lima y a unos 150 metros al oeste de la carretera Panamericana, fue construído sobre un promontorio rocoso de baja altura. Este sitio, que fuera ocupado por los Incas tiene unos 60 x 130 metros de tamaño y se halla encerrado por gruesos paredones separados por estrechos corredores. Además de ser un tambo Inca, existen evidencias de ocupación pre-Inca que quizás retroceden hasta el Horizonte Medio.

II. Grupos étnicos y Asentamientos en la Chaupi Yunga *durante el Intermedio Tardío y el Horizonte Tardío*

Entre los grupos étnicos de la *chaupi yunga* se incluye Huanchipuquio, Maca(s), Guancayo, Zapán y Guarabi (Guarauni) (Figura 6).

Trapiche es un sitio del valle medio y se halla en el lado norte del río Chillón a unos 650 metros de altura (Figura 16); este es el asentamiento más grande que se puede encontrar a medida que ascendemos en dirección a la *chaupi yunga*. Trapiche se halla justo en el lado este de la *quebrada* que une el Chillón con el valle de Chancay y de todos los sitios del Chillón es el que presenta mayor cantidad de cerámica Chancay Negro sobre Blanco. Existe una "fortificación" en el cerro adyacente que domina el asentamiento de Trapiche.

Huanchipuquio comprende cuatro sitios y se halla a unos 650 metros de altura en la margen sur del valle. Los sitios se localizan en la falda de los cerros o al borde de una *quebrada*. Huanchipuquio está justo al frente de Trapiche y la entrada sur de la Quebrada Quilca.

Al este de Buena Vista y ca. de 350 metros al oeste de uno de los sitios de Huanchipuquio existe una pared sumamente impresionante y visible. Mientras que Villar Córdova (1935:188) atribuyó funciones defensivas a esta pared, Dillehay (1976:221) ha señalado que se trata de un *mojón*, o lindero, que sirvió para separar a los Huanchipuquio de los Cararuaillo (Carabayllo).

Villar Córdova (1935:173) indica que Huanchipuquio deriva de los vocablos *Huancho* (nombre de un grupo étnico que se asentó en el valle del Rímac pero cuyo territorio se extendió hasta el Chillón) y *puquio* (fuente de agua o manantial). Al respecto, el nombre de *puquio* es apropiado dado que existen numerosos manantiales en las inmediaciones de los sitios presentes en el piso del valle.

Otro poblado o grupo étnico es Zapán (Rostworowski 1967–1968:36–37; 1972:257, 279–83). Un testigo, quien está mencionado en el documento *Justicia 413*, declaró que las tierras de Quivi delimitaban con los linderos de Zapán, los cuales se hallan a unas 3 leguas al oeste de Quivi y a unos 750 metros de altura. Es posible que Zapán fue parte de Guancayo durante el Intermedio Tardío; hoy las tierras de Huancayo se encuentran entre Zapán y Quivi.

En la actualidad Zapán se extiende desde el Kilómetro 45 hasta el 48 (carretera a Canta), en el lado sur del valle, continuando río arriba hasta alcanzar los sitios Huanchipuquio, aunque tal vez fue originalmente parte de ellos. De hecho, Zapán pudo ser alguna vez el sector oriental de Huanchipuquio. Zapán, al igual que Huanchipuquio, se compone de tres asentamientos separados los cuales exhiben una significativa variación debido en parte a causas sincrónicas y en parte diacrónicas.

Directamente al frente de Guarauni o Huarabi, a 1 kilómetro de uno de los sitios Zapán y a unos 825 metros de elevación se halla Hornillos. Este sitio presenta una ocupación del Intermedio Tardío y aparentemente es un sitio de segunda categoría, quizá el asentamiento más oriental de Zapán (Figura 6).

Según Martínez Rengifo (en Espinoza 1963:58–69), la *parcialidad* de Macas al parecer dependía del *cacicazgo* de Guancayo. Las ruinas que hoy conocemos con el nombre de Macas se encuentran en la margen norte del Chillón. Asímismo, existen cuatro asentamientos Macas en la base sur del Cerro Hormigas y al borde izquierdo del paso a la Quebrada Socos (Trimborn 1970).

El sitio de Huarabi es realmente un grupo de cuatro asentamientos situados en el lado norte del valle. Huarabi C se halla valle arriba y tiene edificios construidos con *tapiales*. Más allá de este sitio existen tres monolitos de 1.8–2.0 metros de alto los cuales aparentemente fueron colocados para marcar límites territoriales. Al respecto, informantes locales opinaron que estos hitos territoriales fueron puestos por los Incas (Horizonte Tardío); sin embargo, Dillehay (1976:260) encontró en el lugar cerámica del Horizonte Medio y el Intermedio Tardío. Huarabi D, a 1 kilómetro río arriba, no presenta *tapiales* en su construcción pero sí cerámica del Horizonte Medio y el Intermedio Tardío.

En el Horizonte Tardío las *pachacakuna* (cada *pachaca* corresponde a una unidad de 100 tributarios) de Maca(s) y Guarauni (Huarabi) estuvieron políticamente controladas por el señor de Guancayo (Martínez Rengifo 1571). El grupo Guancayo fue una *guaranga*, o unidad de 1000 cabezas de familia, quienes fueron tributarios del Inca. Consecuentemente, hubo por lo menos una jerarquía de tributarios repartida en dos niveles. Todas las *pachacakuna* tributaban al Inca una cantidad específica de bienes. En cambio las *pachacakuna* de Guancayo y aquéllas que estaban bajo el control de este grupo (Maca y Guarauni) debían tributar al Inca varios bienes, incluyendo coca en cantidades específicas:

> Cinco cestos de coca grandes....Cinco petaquillas de zuara de coca para el ynga.... y que para el sol le tenían hecha una huertecilla de coca verde la cual cortaban y se la llevaban para ofrecer y quemar al dicho sol. [Martínez Rengifo 1571, en Espinoza 1963:63]

El documento indica que los Guancayo sembraron coca en terrenos comunales de Pampavilla y Cancavilla; los Guarauni mantenían un terreno comunal llamado Corupa para sembrar coca. El terreno comunal para la siembra de coca en la *parcialidad* de Maca fue relacionado a "Guaynacapa en donde sembraron coca y recogieron 50 cestos de coca en tiempos pasados" (Espinoza 1963:65).

Huancayo Alto fue aparentemente uno de los sitios ocupados por los Guancayo. La parte principal de este asentamiento, que cubre 120 x 500 metros, se encuentra en la *quebrada*. Esta zona del sitio tiene más de 350 estructuras de piedra o *tapial* además de ca. de 100 cuartos hechos de piedra localizados en la ladera baja de un cerro adyacente. Por lo menos cinco "zonas de actividad" especializadas se han identificado en este asentamiento: (1) unidades administrativas, (2) un complejo residencial ocupado por la élite, (3) un sistema de defensa, (4) zona de unidades domésticas ocupadas por gente de menor status y, (5) un cementerio (Dillehay 1976). Además, otras áreas de actividad fueron identificadas en la ladera de un cerro adyacente y se cree que contiene una zona doméstica aislada la cual fue ocupada por familias que no pertenecían a la élite, algunas terrazas para secar productos, y depósitos o almacenes.

El centro del asentamiento se compone de dos edificios administrativos; aunque exhiben similar forma las técnicas de construcción utilizadas son distintas. Dillehay (1976:274) sugiere que estas dos estructuras administrativas reflejan una administración bipartita o una suerte de *moiety*, uno de ellos controlado por administradores *chaupi yunga* y el otro por administradores *serranos*.

Después de los años 900 de nuestra era, la construcción de otros edificios en Huancayo Alto se redujo a una terraza y unas cuantas unidades domésticas en la *quebrada*. Asímismo, los depósitos fueron abandonados o esporádicamente utilizados durante el Intermedio Tardío. Este enunciado se apoya, según Dillehay, en la escasa o nula presencia de cerámica del Intermedio Tardío en la zona de los almacenes.

A partir de los años 900 de nuestra era hasta la conquista de la zona por los orejones cusqueños, Huancayo Alto continuó desempeñando el rol de "foco integrador" en donde convergieron grupos costeños y serranos. Sin embargo, no surgió centro urbano o poder político alguno en este lugar; más bien, parece que fue un centro en donde los productos fueron almacenados, intercambiados y distribuidos. Dillehay concluye señalando que los *serranos* permitieron que los habitantes locales tuvieran el control político a cambio del acceso a recursos situados valle abajo y residir en Huancayo Alto (Dillehay 1979:30).

Dillehay argumenta que la cooperación, en vez de la conquista o el dominio de un grupo sobre otro, fue el principal factor de integración en el valle del Chillón. Agrega que la importancia cada vez más creciente de la cooperación y el acceso a recursos no-locales permitieron el surgimiento de Huancayo Alto, un sitio en la *chaupi yunga*, que fue aparentemente ocupado por diversos grupos étnicos.

No contamos con evidencias que prueben que Huancayo Alto tributó o suministró fuerza de trabajo *mitmaq* a una organización estatal centralizada antes de la conquista Inca. Sin embargo, el señor de Guancayo tuvo bajo su dominio a dos grupos *chaupi yunga* (Maca y Guaravi [Guarauni] y posiblemente un grupo serrano de la región de Chacalla) quienes probablemente residieron en Huancayo Alto. Si éste fue el caso, el extenso complejo de almacenes de Huancayo Alto pudo incluir la recepción de bienes proporcionados por los grupos que se hallaban bajo el control del señor de Huancayo Alto. Puesto que la fecha de los almacenes retrocede a los 200 años de nuestra era, cabe la posibilidad que algunos de los bienes fueron destinados para mantener las *huacas*, las mismas que fueron construidas en épocas tempranas (100 a. de C.-1 d. de C.).

Mientras que los edificios administrativos y las residencias de la élite de Huancayo Alto se hicieron con *tapiales*, las unidades domésticas de la población, las terrazas para secar productos y los almacenes fueron construidos con piedras. Para Dillehay el *tapial* simboliza la tradición "costeña" y las piedras la "serrana." Para inferir la presencia de otros grupos étnicos en Huancayo Alto Dillehay utiliza las diferencias en los materiales de construcción y la existencia de tejidos y cerámica supuestamente extrañas al sitio.

Al frente de Huancayo Alto existe un puente a ca. de 875 metros de altura y en cuya cercanía (60 metros al suroeste) se encuentra Tambo del Ynga. La mayor parte de las piedras de este *tambo* fueron retiradas y utilizadas como materiales de construcción por los pobladores de Yangas. Hoy solo queda un cimiento de 10 x 10 metros. Dillehay (1976:338) encontró en este sitio dos *porras* en forma de estrella y varios tiestos polícromos Inca. Tanto el puente como el *tambo* son evidentemente de la época Inca.

Pucara, un sitio al este de Huancayo Alto y a unos 1000 metros de altura, puede tratarse de un asentamiento fronterizo que separó a los Guancayo de los Quivi. Fue ocupado desde el Horizonte Medio hasta el Intermedio Tardío (800–1400 d. de C.) y su plano es semejante a sitios del alto valle tales como Quyso, Orobel y Quivi Vieja. Pucara Alto está en la *quebrada* y a la vera del camino que conecta la zona de Jicamarca.

Checta se halla río arriba y a 1.1 kilómetro de distancia con respecto a Pucara y a 2.3 kilómetros de Quivi Vieja, a unos 1000 metros de elevación. Checta pudo ser durante el Intermedio Tardío el asentamiento límite entre *yungas* y *serranos*. El Inca puso en este sitio gente Chaclla para servir como *chasquis* (mensajeros) y *mitmaq* (individuos obligados a trabajar para el estado). Checta se encuentra cerca del límite conocido con el nombre de *Judcunga*. Según el documento *Justicia 413* los *mitmaq* Chaclla fueron colocados entre el hito de *Judcunga* y una *quebrada*; ésta última fue al parecer la desembocadura de la *quebrada* (o río) de Arahuay. Checta está junto a un gran cerro que contiene una gran pared que desciende hasta el río. Tanto el cerro como la pared, en la margen sur, restringen el acceso a esta parte sumamente estrecha del valle; a su vez, en el lado norte, las montañas con declive pronunciado descienden hasta muy cerca del cauce del río. Dillehay indica que existe cierta evidencia sobre dualismo y co-residencia de varios grupos étnicos en Checta, en la forma como ha sido sugerido para Huancayo Alto y Pucara Bajo. El sugiere además que el gran muro de Pucara separó a los Quivi de los Guancayo.

El grupo étnico Quivi pudo haber ocupado Quivi Vieja; éste es el grupo que mayormente destaca entre todos los grupos *yunga* en el documento *Justicia 413*. Las tierras de Quivi se hallan en la confluencia de los ríos Chillón y Arahuay, a unos 1150 metros de altura y a 65 kilómetros del Océano Pacífico. El límite oriental de los dominios de Quivi se encuentra cerca de Yaso. Hoy, Yaso es el límite que se separa los distritos de Santa Rosa de Quives y Canta. Magdalena, ubicado al oeste y en la margen norte del valle, constituye según Dillehay otro límite. Checta, en el lado sur, fue quizá el asentamiento más occidental de Quivi.

Es probable que Quivi controló tierras en la margen sur de los ríos Chillón y Arahuay. Quivi Vieja se encuentra en el lado sur del valle, a dos kilómetros al este de Checta y a 200 metros al sur de la ex-Hacienda Cabana. Se extiende siguiendo la base de los cerros que rodean la actual población de Santa Rosa de Quives. Santa Rosa de Quives ha sido construido sobre un poblado prehispánico que hoy se halla ampliamente destruído debido a los trabajos de nivelación del terreno para la construcción de un hotel y la iglesia.

Uno de los testigos Huarabi (ver folio 123 de *Justicia 413)* menciona que las tierras de Huaravi y Quivi estaban separadas por hitos o marcadores territoriales. Las tierras que se hallan en el Kilómetro 48 (al frente de Avícola Yangas) pueden corresponder a dicha demarcación (Dillehay 1976:305). Informantes de la localidad indican que Yaso en el Chillón y Huarhuar en el río Arahuay son los límites orientales del actual distrito de Santa Rosa de Quives (Dillehay 1976:306).

Antes de la conquista Inca, el señor de Quivi estuvo bajo el dominio del Colli Capac, asentado en el bajo Chillón. Varios testigos mencionados en el documento indicaron que el señor de Quivi "servía y tributava" al señor de Collique "antes de la llegada de los Ingas." Murra (1972, 1975) se pregunta si los Colli costeños realmente pusieron su propia gente en Quivi para cultivar coca, ají y frutales, o si es que los Colli recibieron dichos productos mediante intercambio o tributo.

En 1559 d. de C. el tamaño de la tierra de Quivi se estableció que era:

> ...dos myll pasos de largo y tresçientos de ancho e tienen por linderos//por una parte el rrio y por la otra camyno rreal que viene de Lima al dho pueblo de Canta... [*Justicia 413*, folios 131r-131v]

Dillehay (1976:307) empleó una persona de Quivi para establecer en pasos el largo y el ancho del área (que contenía tiestos en superficie), al este de Santa Rosa de Quives, y entre el río y la ruta que conduce por la base del cerro a Canta; fue sumamente interesante comprobar que las dimensiones del área medida proporcionaron 2200 x 340 metros, la cual se aproxima muy cercanamente a las dimensiones expresadas en el documento *Justicia 413*.

Dillehay (1976:313), tomando como base la cerámica recogida en Quivi, sugiere que numerosos grupos serranos (Chaclla, Canta, Atavillos, Socos e Inca) y costeños (Maranga, Chancay) estuvieron sea en contacto con los habitantes de Quivi, u ocuparon el sitio desde el Intermedio Temprano hasta el Horizonte Tardío.

Capcha expresó, según *Justicia 413* (folio 44r), que el río Quybi (o Chillón) fue el límite entre Atavillos y Secos. Dillehay sugiere, sin embargo, que el sitio de Magdalena, situado en el lado oriental de una *quebrada* al pié del Cerro Cenicero, puede hallarse en el territorio de los Secos.

El sitio de Huarhuar (a unos 1450 metros de altura) está en el lado sur del río Arahuay, 3 kilómetros al este de Quivi Vieja y 2 kilómetros al oeste de Orobel y Quyso. Existe en el lado norte de este río un gran complejo de terrazas en cuya superficie se encontraron tiestos del Intermedio Tardío e Inca. Huarhuar pudo haber sido otra "comunidad fronteriza," debido a una pared de piedras que se desplaza perpendicularmente a la ladera del cerro y localizada ca. de 300 metros río arriba y a mitad del camino entre Huarhuar y Orobel.

Dillehay provee los nombres de otros sitios y sus probables filiaciones étnicas. Ellos son Quyso y Orobel situados al este de Quivi y a unos 1500 metros de altura. Quyso fue probablemente un "pueblo fronterizo" en el dominio de Canta en tanto que Orobel pudo ser una colonia Chaclla.

Orobel moderna es una colonia de 30 individuos puestos por los Jicamarca a fín de "mantener los límites de Huarochirí y Casta." El informante de Dillehay (1976:331) indicó que no tuvo permiso para entrar y cultivar tierras en el valle de Arahuay puesto que dichos terrenos eran cultivados por otro grupo étnico. La colonia Orobel tuvo campos de cultivo a mayor altitud entre Jicamarca y Collata. Las comunidades de Jicama, Culata, Punan y Chacalla acordaron, en los siglos XVI o XVII, compartir los pastos de la región (Rostworowski 1972:299). Así, entre abril y julio los Jicamarca y Collata trabajaron y cultivaron papas en terrenos de las alturas; cuando no trabajaban las *chacrakuna* en las alturas ellos vivían en Cullata y Jicamarca.

Yaso, situado en el Kilómetro 73 y a 1550 metros de altura, fue otra probable "comunidad de frontera" situada entre los Quivi, los Canta y una "colonia" Atavillos o Socos durante el Intermedio Tardío.

III. Grupos y Sitios del Alto Valle durante el Intermedio Tardío y el Horizonte Tardío

Los Canta, cuya tierra natal se encuentra en las cabeceras del río Chillón, ocuparon partes del alto Chillón, al norte del río Arahuay. Los Chaclla vivían también en el alto valle al sur del río Arahuay. Este grupo fue parte del grupo macro-étnico Yauyos cuyo territorio incluyó las cabeceras de los ríos Mala y Cañete. Los Atavillos ocuparon a su vez las cabeceras del río Chancay, al norte del río Chillón (Figura 6).

La región Canta incluye los asentamientos modernos de Arahuay, Lachaqui, Carhua, Canta, Pariamarca, Huamantanga y Huaros. Entre los sitios prehispánicos Canta figuran Cantamarca, San Lorenzo (Kekamarca o Lucana), Huancuna, Huamantanga, Zepitamarca, Piedra Mesa, Camata (Collo), Pirhuinco, Cullke, Viscas (Pacramarca), Ojomarca, Huayuacancha, Huantamarca cerca de Carhua y Purumarca (Figura 6).

El principal asentamiento es Cantamarca y se ubica sobre una colina desde donde se domina el alto valle (Villar Córdova

1935:298–302; Casana 1976:44–48). El tipo arquitectónico que predomina en estos sitios es el *kullpi*, una estructura semi-subterránea en forma de una torre redondeada que varía de 4 a 6 metros de diámetro y 1–2.5 metros de alto, mostrando además techo estilizado y pequeños accesos rectangulares situados generalmente a ras del suelo.

Cantamarca contiene una significativa arquitectura de estilo Inca, incluyendo accesos de forma trapezoidal. Otros sitios que presentan elementos arquitectónicos Inca son Kekamarca, Huancuna, Purumarca y Huamantanga. Los sitios de la *sierra* también exhiben cerámica llana y polícroma de estilo Inca.

En la superficie de Quyso, Camata, Caballo Blanco, Cantamarca y Huancuna, Dillehay (1976:357) encontró restos marinos entre los que se cuentan almejas (*Mesodesma donacium)*, ostras (*Spondylus)* y conchas (*Crepidula dilatata)*. Estos restos apoyan la evidencia documental según la cual los *yungas* y los Canta establecieron relaciones de intercambio en tiempo de paz. Cerámica costeña del estilo Nievería fue también encontrada en sitios Canta. Consecuentemente, dicha interacción costeña-serrana pudo empezar durante el Intermedio Temprano, o quizá antes.

Pirhuinco es un sitio pequeño situado en la margen sur del río Arahuay y a 1750 metros de altura, al frente de Licahuasi (8 kilómetros al sureste de Quivi y 3.5 kilómetros al este de Quyso). Al parecer fue ocupado desde el Intermedio Temprano hasta el Horizonte Tardío.

Los sitios del valle de Arahuay controlados por los Canta son Piedra Mesa y Camata (Collo); ambos contienen restos del Intermedio Tardío y el Horizonte Tardío y se hallan en el lado norte del río, a 4.5 kilómetros río arriba de Licahuasi y 2 kilómetros río arriba de Caballo Blanco a una elevación de ca. 2050 metros. Caballo Blanco, un sitio Inca, se localiza sobre un espolón entre Santa Rosa de Quives y Arahuay en la margen sur del río Arahuay. Su ubicación es ideal para controlar la entrada y la salida del valle de Arahuay.

Los Chaclla (un sub-grupo del grupo macro-étnico Yauyos) ocuparon partes de las provincias de Huarochirí y Yauyos (Rostworowski 1967–1968, 1972). Entre los asentamientos Chaclla figuran Caballo Blanco, Chuya, Macando, Cullipampa, Huayarcoto y Kancha-Kancha (Dillehay 1976:366). Desde Quivi toma un viaje de dos leguas para llegar al primer asentamiento Canta; en cambio, se necesita viajar dos días para llegar al primer poblado Chaclla. Los *mitmaq* Chaclla fueron ubicados en dos áreas: cerca del hito o marca llamado Judcunga (a 1/2 legua de Quivi) y, dentro del propio Quivi. Los Chaclla fueron un grupo Yauyos muy importante pero estuvieron bajo el dominio del señor de Huarochirí.

Villar Córdova (1935:341–44) describe los sitios Chaclla-Jicamarca de Chuya, Macando, Cullipampa, Huayarcoto y Kancha-Kancha, los cuales tienen arquitectura de piedra y *kullpi* semi-subterráneos.

Huaycoloro se halla a una elevación de 1200 metros en el extremo superior de la *quebrada* entre los ríos Chillón y Rímac y sobre terreno árido y escarpado, a unos 20 kilómetros al este de Cajamarquilla. Huaycoloro está conectado mediante una serie de caminos con Huancayo Alto, Pucara y Quivi en el valle del Chillón y con Jicamarca y Chacalla en la sierra baja. Huaycoloro estuvo también relacionado a la región Chaclla-Jicamarca y a las aisladas unidades residenciales sobre la falda del cerro en Huancayo Alto.

Huachoc se halla en la Quebrada Quilca y a 1250 metros de altura, a mitad del camino entre los valles del Chillón y Chancay. El sitio que puede pertenecer al Intermedio Tardío consiste de 10–15 recintos pequeños y de forma irregular construidos con piedras. El pequeño grupo que ocupó este lugar (ca. 25–30 personas) pudo estar afiliado a uno mayor.

Los datos disponibles nos permiten reconstruir el Chillón y los valles adyacentes como un mosaico de *curacazgos* en interacción. Muchos de éstos fueron relativamente pequeños, otros fueron unidades étnicas territoriales administradas por un *curaca* o "señor local"; estas unidades políticas de interacción se conocen frecuentemente con el nombre de *señoríos* o "reinos." No hubo un estado centralizado y territorialmente extenso sino hasta que estos *señoríos* fueron absorbidos por el expansionista estado Inca el cual penetró al valle del Chillón en la década de 1460.

Antes de la llegada de los Incas, existía entre los Chaclla, sus vecinos de Canta y los Colli del bajo Chillón un permanente conflicto por el control de tierras cerca de Quivi (Figuras 17–18). Estas tierras se localizaban en la *chaupi yunga*, una zona geográficamente intermedia entre los Colli de la costa y los grupos serranos (Canta y Chaclla) quienes alegaban legítimo derecho sobre las tierras de Quivi. El señor costeño de Quivi fue forzado a ceder el control de las tierras de Quivi a los de Canta y a colocar un hito territorial en Checta llamado Judcunga con el propósito de separar tierras serranas y costeñas.

Cuando el Inca conquistó el valle, ellos pusieron a los Chaclla en las tierras de Quivi. La competencia y la lucha entre los varios grupos étnicos por la posesión de las tierras de Quivi precipitaron las acciones e intereses de los Incas. Estos grupos pretendían acceso a estas tierras por cuanto reunían inmejorables condiciones para el cultivo de la coca y frutas tropicales. Mediante el uso de los Chaclla—uno de los grupos étnicos que luchaba por controlar las tierras de Quivi—como *mitmaq*, los Incas resolvieron los objetivos del sistema estatal y al mismo tiempo tomaron ventaja del tan esperado deseo de los Chaclla.

El Impacto Inca en el Valle del Chillón

Algunos investigadores han sostenido que subsecuentemente a la conquista Inca, la sierra se caracterizó por presentar evidencias sobre gobierno Inca directo, en tanto que el gobierno de las tierras *yunga* fue indirecto. Muchos investigadores usan la arquitectura Inca como prueba de gobierno directo. Con relación a la fundación y construcción de nuevos asentamientos Inca, Morris (1972:401) notó que los asentamientos estatales "tienden a ser particularmente grandes y numerosos en áreas caracterizadas por ciertas condiciones de marginalidad ecológica y sociopolítica." Los comentarios de Morris son particularmente pertinentes para los asentamientos de la *sierra* y la *chaupi yunga*. Sin embargo, en lugar de mostrar innumerables sitios expresamente fundados y construidos en neto estilo Inca, la mayor parte de la región *yunga* continuó levantando construcciones en puro estilo local. Además, muchos sitios costeños que exhiben la más nítida "influencia Inca" fueron centros administrativos importantes antes de ser conquistados por los cusqueños (cf. Menzel 1959).

Para los períodos prehispánicos tardíos—especialmente el Horizonte Tardío (1476–1534 d. de C.)—cada zona al interior del Chillón parece que fue afectada de modo diferente por la expansión Inca (Dillehay 1977). El bajo y medio Chillón muestran influencia Inca menos visible que el alto valle y la *sierra*. Al parecer el Inca tuvo la habilidad de ejercer control indirecto de la costa desde su base de operaciones en la baja *sierra* mediante el desplazamiento de su ejército.

Para ser específicos, el gobierno directo caracterizó la relación del Inca con los Chaclla y otros grupos del alto valle, en tanto que en el bajo valle se aplicó el gobierno indirecto (a través de Collique). Es posible que un tipo de gobierno indirecto (administrado desde Huancayo Alto) caracterizó también la estrategia Inca en el gobierno del valle medio.

El Inca estableció su base en el alto valle por diversas razones. El documento *Justicia 413* indica que el Inca tuvo siempre temor de la costa y tenía miedo de contraer enfermedades y morir en dicha región. También seleccionó el alto Chillón por cuanto era políticamente marginal, ocupado de modo disperso, y lejos de los grandes centros poblados de la costa y el valle medio.

De todas las zonas, la *yunga* o el bajo Chillón contiene menos evidencia de ocupación Inca. Menzel (1959, para la costa sur) y Dillehay (1977, para el Chillón) usan dos tipos de evidencia—cerámica y arquitectura—para demostrar que la mayoría de los sitios tardíos de la costa presentan escasa o ninguna evidencia de influencia Inca. Sin embargo, dos sitios en el valle del Chillón—Tambo Inga y Collique—que fueron ocupados antes y después de la llegada de los Incas muestran remodelamiento Inca. Al ejercer control sobre estos sitios el Inca los remodeló conforme a sus propios intereses.

No es de extrañar entonces que tanto cerámica llana como polícroma de estilo Inca son más frecuentes en sitios de la *sierra* baja del Chillón, aunque también tiestos semejantes fueron encontrados en Collique y en Tambo Inga, en la zona *yunga*, y en Huancayo Alto, en la *chaupi yunga*. La muestra más grande de cerámica Inca marrón llana fue descubierta en Caballo Blanco, un sitio Inca en el valle de Arahuay, situado entre la *chaupi yunga* y la *sierra* baja.

Los sitios *chaupi yunga* exhiben, a través de casi toda la secuencia cronológica, más afinidades alfareras con los sitios de la *yunga*, aún cuando también se observan fuertes similitudes con sitios de la *sierra* baja (Dillehay 1976:188). Del mismo modo como los pobladores de la localidad clasifican como *yunga* a la población y al ambiente del valle bajo y medio, la evidencia alfarera sugiere también que estas dos poblaciones estuvieron mucho más relacionadas en comparación a la sierra (si usamos la cerámica como único indicador). Esta conclusión debe tomarse con cautela pues los sitios de la *sierra* (para el valle del Chillón) fueron escasamente estudiados (cf. Dillehay 1976, n.d.).

Existe un documento etnohistórico que trata sobre tres grupos étnicos que ocuparon la *chaupi yunga* del Chillón (los Guancayo, los Maca y los Guarauni; ver Martínez Rengifo en Espinoza 1963) y provee una lista de productos que estos grupos tributaban anualmente al Inca. Asímismo, además de revelar que el Inca se benefició económicamente al controlar el valle medio, el documento indica específicamente que la coca fue uno de los productos que estos grupos tributaban al Inca.

Puesto que la coca sólo se puede cultivar en la *chaupi yunga*—y con suma dificultad en otras zonas—es de suponer que dicha zona fue de especial interés para el Inca. Cerámica Inca fue encontrada en tres áreas de Huancayo Alto: (1) en el sector administrativo, (2) en las terrazas para secar productos y (3) en las unidades de almacenamiento cerca de las citadas terrazas (Dillehay 1976:150). Los tiestos Inca parecen ser imitaciones deliberadas de los tipos cusqueños. Además, a 300 metros al oeste de la parte principal del sitio existe un edificio aislado del resto del conjunto en donde únicamente cerámica polícroma Inca fue encontrada.

Huanchipuquio es otro sitio *chaupi yunga* que contiene cerámica Inca (además de la local) la cual fue encontrada en un complejo de cuartos con paredes de hasta 2 metros de altura. Este complejo pudo haber sido empleado como depósito por el Inca.

A medida que ascendemos al valle se observa un notable aumento en la cantidad de asentamientos con "influencia Inca" y en el número de edificios construídos por los Incas.

Villar Córdova (1935) y Casana (1976) dan cuenta de sitios Inca en Purumarca, Colli, Cantamarca, Huancuna, Huamantanga, Caballo Blanco, Lucana y Huayuacancha. Dillehay (1976, 1977) ubicó varios corrales grandes en muchos de estos sitios; es posible que grandes rebaños de llamas fueron mantenidos en esta zona para propósitos estatales, especialmente para operativos militares o transporte de alimentos (Murra 1965).

Los Incas seleccionaron sus aliados con sumo cuidado luego de examinar las condiciones políticas existentes. Así por ejemplo, hemos visto que el Inca conquistó al grupo macro-étnico Yauyos siendo los Chaclla parte de dicho grupo. Aunque los Yauyos eran conocidos como agresivos y guerreros ellos fueron amigos leales del Inca. El Inca al enterarse de la vieja disputa por el control de las tierras de Quivi entre los Chaclla, Canta y Colli, intervino y seleccionó a los agresivos Chaclla para servir como *mitmaq*. Los Chaclla y orejones pudieron en consecuencia beneficiarse mutuamente con los cocales de Quivi, los primeros recibiendo otros territorios como retribución a su trabajo en las tierras de Quivi.

Mientras que conocemos algo sobre las relaciones del Inca con los grupos de la *sierra*, sabemos menos sobre la reorganización impuesta a las instituciones locales de la costa al ser incorporadas al imperio Inca—sobre todo en cuanto se refiere a las unidades sociopolíticas costeñas que ya habían logrado alto grado de desarrollo y especialización. Es necesario, pues, estudios adicionales a fín de establecer en detalle el carácter de las relaciones costeñas y *serranas*; el análisis deberá hacerse por separado en cada caso.

Conclusiones

El presente resumen sobre la arqueología y la etnohistoria del valle del Chillón ha puesto de relieve un conjunto de temas importantes. Uno de ellos ha sido el tema de la comunidad autosuficiente, considerada por algunos autores como el ideal andino. Según Dillehay (1976, 1979) la verticalidad fue una de las estrategias de dicho ideal en el valle del Chillón. Para ser breves, verticalidad se materializa cuando un núcleo poblacional dispone el envío de "colonias hijas" a fín de explotar una serie de pisos ecológicos; cada colonia retiene su filiación cultural y sus respectivos derechos en sus lugares de origen. Otros autores han descubierto, sin embargo, datos que ilustran una estrategia alternativa distinta a la descrita, la cual tiene que ver con la existencia de comunidades especializadas, acompañadas de relaciones recíprocas de intercambio entre comunidades pertenecientes a una unidad sociopolítica (llámese *curacazgo)* o a través de una región.

Nos preguntamos, ¿cuál de estos dos ideales caracterizan mejor el valle del Chillón a través del tiempo? ó ¿quizá la combinación de ambas estrategias en algún punto de la secuencia se ajusta mejor a la evidencia? De hecho, Dillehay (1976, 1979, n.d.) ve una variedad de estrategias, en que la verticalidad es sólo una de ellas, y propone que la verticalidad en el Chillón no corresponde al "tipo clásico" descrito por Murra (1972).

También hemos visto que antes de la conquista Inca el valle del Chillón nunca estuvo integrado mediante un sólo grupo étnico o señor local. En realidad, los registros etnohistóricos revelan claramente que distintos grupos étnicos se asentaron a lo largo del Chillón y en sus alrededores. Muchos de estos grupos tuvieron acceso a la coca o los cocales de la *chaupi yunga*, cerca de Quivi. En consecuencia, debemos tratar de precisar si el acceso fue (1) directo (por ejemplo, mediante verticalidad) o (2) indirecto (por medio de alianzas entre grupos de individuos, socios, comunidades especializadas y/o intercambio). También necesitamos establecer si es que la unidad de trabajo fue la familia nuclear, la comunidad o la supra-comunidad; si es que el acceso estuvo centralmente controlado por un *curaca* (jefe local) o no; y si es que la estrategia cambió a través del tiempo. Asímismo, si dos unidades sociopolíticas vecinas son similares en tamaño y en poder político, económico y militar, una estrategia de interdependencia, con cierta especialización local, podría ser la más apropiada; sin embargo, cuando dos unidades políticas de poder y tamaño diferentes establecen relaciones, el grupo más fuerte se halla en condiciones de lograr autosuficiencia mediante el acceso directo y a través del establecimiento de un sistema de colonias. En cuanto al Chillón se refiere, una serie de grupos étnicos aparentemente hizo uso de una amplia variedad de estrategias cooperativas las mismas que probablemente se desarrollaron gradualmente.

Según Dillehay (1976, 1979) el compartir los recursos fue facilitado y controlado por numerosos grupos étnicos que co-habitaban en sitios de la *chaupi yunga*. Dillehay pone como ejemplo a Huancayo Alto e indica que este sitio fue ocupado por gente de la sierra y la *chaupi yunga*. El mencionado autor (1979:30) también señala que cuenta con suficiente evidencia arqueológica para demostrar que una *forma* de verticalidad (no el "tipo clásico" de Murra) se inició en dicho asentamiento hacia 800–500 a. de C. y consistió en el desplazamiento de la sierra a la *chaupi yunga* para intercambiar productos mediante el control de una autoridad local centralizada. Dillehay agrega (1979:30) que la verticalidad estuvo en pleno apogeo "bajo la dirección de líderes locales de la sierra quienes ocupaban edificios administrativos" formando cierta clase de administración dual en Huancayo Alto. Una de las diferencias fundamentales entre la verticalidad de Huancayo Alto y el "tipo clásico" es que Huancayo Alto pudo haber operado en un contexto "pre-estatal" o "no-estatal" en el cual los *serranos* no fueron enviados como colonos por un núcleo central sino más bien ellos residieron voluntariamente en dicho lugar (Dillehay 1976).

Cualquier interpretación de Huancayo Alto como ejemplo del "tipo clásico" de verticalidad carecerá de firmeza si se demuestra que (1) solamente sistemas políticos y económicos *expansionistas de nivel estatal* pusieron en práctica la verticalidad; (2) que los serranos que ocuparon Huancayo Alto no mantuvieron sus derechos y su filiación a un asentamiento central situado en la sierra, (3) que el patrón examinado refleja *solamente* una población local *chaupi yunga* que vivía en dicho lugar mientras intercambiaba productos con los de la sierra; o alternativamente, (4) que varones de la *chaupi yunga* tomaron mujeres de la *sierra* como esposas.

Si se da el caso en el que la verticalidad es una estrategia que solo existió bajo la sombrilla política de un estado poderoso—y que no hubo estado centralizado en el valle del río Chillón antes de la conquista Inca— entonces parece improbable que Huancayo Alto es un ejemplo del "tipo clásico" de colonia controlada por la sierra, i.e., parte de un "archipiélago" (Murra 1972). Sin embargo, si Huancayo Alto no fue una colonia serrana ¿Cómo se podría explicar la presencia de cerámica y arquitectura *serrana* recuperada por Dillehay en dicho lugar?

Dillehay presenta un modelo de co-residencia a base de los datos encontrados en Huancayo Alto, un modelo en el que los serranos co-administraron el poblado y cooperaron pacíficamente con los pobladores de la *chaupi yunga*. Pero acaso no puede interpretarse la misma información de otra manera—por ejemplo, ¿como resultado de intercambio inter-étnico, alianzas políticas o matrimonios o contacto mediante peregrinaciones? ¿Podemos realmente distinguir intercambio inter-étnico de co-residencia sin otros datos arqueológicos? ¿No sería necesario contar con evidencias de entierros y esqueletos para demostrar la presencia de dos o más grupos biológicamente distintos y correspondientes a diferentes grupos étnicos con características genéticas diferentes? Esta última sería el tipo de datos que quisiéramos observar para apoyar el argumento de co-residencia por diversos grupos étnicos en sitios de la *chaupi yunga*. Al respecto, Dillehay (comunicación personal) indicó que los datos de entierros parecen apoyar la presencia de pobladores "costeños" y "no-costeños."

Debe enfatizarse, sin embargo, que nuestras reconstrucciones sobre el Intermedio Tardío y el Horizonte Tardío del valle del Chillón—aquellas relacionadas con el mosaico de pequeños *curacazgos* y múltiples grupos étnicos —no deben proyectarse ciegamente hacia el pasado. Los arqueólogos deben ser conscientes de las limitaciones temporales asociadas a los modelos del siglo XVI y más bien lo que se debe hacer es desarrollar otros. Realmente, una gran diversidad de prácticas, estrategias e instituciones existió antes del Horizonte Tardío. Esto significa que los modelos que podrían caracterizar a los períodos tempranos—pero que no necesariamente pueden desarrollarse mediante la información etnohistórica o los datos del Intermedio Tardío/ Horizonte Tardío—deben formularse a partir de los datos arqueológicos correspondientes a dichos períodos tempranos.

No sin sorpresa, nos encontramos ante más preguntas que respuestas para los períodos que anteceden a los *curacazgos* del Intermedio Tardío y el Horizonte Tardío. ¿Fueron las múltiples etnías de estos períodos tardíos resultado de un sistema de división en grupos pequeños o un largo proceso de adaptación? ¿Cuán fácil y con qué frecuencia cambiaron los límites entre grupos étnicos y unidades políticas? ¿Hasta qué grado es la variación estilística alfarera sensible a la etnicidad? ¿Es realmente posible definir grupos étnicos a partir de los complejos alfareros? ó, ¿es la variación estilística resultado de una combinación mucho más compleja de variables tales como status, clase, función, especialización y técnicas de manufactura?

Es sorprendente el hecho que a menudo existe más evidencia etnohistórica de intrusiones serranas a la costa que a la inversa. ¿Es éste aspecto solo cierto para los períodos tardíos? ó ¿indica que la expansión mediante conquista y verticalidad fueron prácticas serranas tradicionales, mientras que comunidades especializadas, alianzas e intercambios regionales fueron estrategias tradicionalmente costeñas?

Los *serranos* de Canta, Lachaqui, Arahuay, Chaclla y Jicamarca descendieron a la *chaupi yunga* para intercambiar sus productos por otros de la costa y de piedemonte. Parece que por un lado hubo contacto directo entre pobladores de la sierra y la *chaupi yunga*, y por otro lado, contactos entre las poblaciones *yunga* y *chaupi yunga*; relaciones menos directas caracterizaron en la mayoría de casos la interacción cultural entre la *sierra* y la *yunga* (Dillehay 1976, n.d.).

En cuanto a la era etnohistórica, cada región tuvo al menos un señor principal y una etnía preeminente—los Colli en la *yunga*, los Guancayo en la *chaupi yunga* y los Canta en la *sierra*. Un poblador de la *chaupi yunga* describe este tipo de interrelación en el documento *Justicia 413* (foja 189r) del siguiente modo:

> ...e [residente *chaupi yunga*] que no es pariente amigo ni henemygo yntimo de los yndios de Chacalla ny de Canta mas de que con los yndios de Canta tyene gran amystad e son companeros porque trata con ello este t° en bendelles coca y los de Canta obejas y papas...

Los grupos costeños tales como los Colli probablemente controlaron productos del mar, algodón y quizá maíz; los grupos *chaupi yunga* tales como Macas, Quivi y Guarabi intercambiaron una serie de frutas y coca; los grupos de la *sierra* tales como Canta y Chaclla distribuyeron *charqui* y lana, papas, olluco y oca. Los *chaupi yunga* pudieron ser mediadores en la mayoría de estas transacciones.

Puesto que el documento *Justicia 413* trata conflictos sobre tierras—y porque es claro que los españoles no entendieron el concepto indígena de tierra—debemos ofrecer unos breves comentarios sobre dicho tópico. Cuando los españoles dispusieron que los cocales de Quivi se vendan por 200 animales, los indígenas afirmaron que no había precedente alguno sobre ventas por cuanto de acuerdo a sus experiencias las tierras nunca se vendieron. La respuesta indígena confundió a los españoles quienes no lograron entender que el concepto de propiedad privada de la tierra (concepto que tenía una larga historia en el viejo mundo) fue virtualmente desconocida entre los indígenas de América. Tal vez no existe mejor ejemplo que éste para ilustrar el choque cultural entre el viejo y el nuevo mundo.

Por ejemplo, la venta, arrendamiento o transferencia de tierras no tuvo significado alguno entre la mayoría de sociedades Mesoamericanas, tales como los Maya. En la sociedad Maya los *señoríos* fueron quienes controlaron las tierras, los derechos sobre las mismas eran ejercidos por los linajes; además, las tierras utilizadas fueron marcadas con hitos. Tierra utilizada fue tierra controlada; la tierra no cultivada podía ser empleada por otros únicamente si se hallaba abandonada y sin protección—por ejemplo, el bosque vírgen nunca antes cultivado y sin hitos de marcación o modificados. Aún más, los derechos sobre la tierra tenían diferentes niveles: aquéllos del *señorío*, del pueblo, del *barrio*, del linaje y de la nobleza. Sin embargo, antes de la llegada de los peninsulares, no se encuentran evidencias de propiedad privada individual, ni evidencias que los individuos podían vender, comprar o transferir tierras (Landa, ver Tozzer 1941; Villa Rojas 1961). Un noble Maya podía heredar derechos sobre las tierras de los linajes pero no tenía derecho a poseerlas como bien individual.

El concepto Andino de tierra parece haber sido similar al de los Maya. Cobo (1890–1895:Libro 12, Capítulo 28) nos dice que cuando el Inca construyó o incorporó un asentamiento a su dominio, las tierras cultivables fueron marcadas con hitos y divididas en tres partes: (1) una dedicada a la religión, (2) otra para sostener al gobernante, y (3) una reservada para el uso de la gente común. En algunas provincias la parte asignada a la religión fue la más grande, mientras que en otras la parte más grande fue destinada a sostener al Inca. En otras partes poblados enteros incluyendo su territorio pertenecieron al Sol. Las tierras dedicadas a la religión podían dividirse entre el Sol, el Relámpago u otras *huacas*. Los pobladores no podían "vender" la tierra destinada para su uso.

Parece que cierto desarrollo sobre propiedad privada tuvo lugar durante y después del gobierno de Pachacuti (Rostworowski 1962). Sin embargo, el crecimiento de la posesión privada durante el reinado de los últimos Incas nos hace sospechar que fue un patrón que apareció tardíamente y no tiene parangón en períodos anteriores. En el caso de los cocales de Quivi descritos en *Justicia 413*, hubo terrenos separados para Tupa Yupanqui y Huayna Capac pero también terrenos para el Sol y la Mamacona.

En resumen, debemos decir que la información etnohistórica que hemos utilizado en este estudio tiene ventajas y desventajas. Las ventajas se reflejan en el hecho que los datos nos informan sobre innumerables detalles que de otro modo no podríamos conocer—la ubicación de hitos territoriales en el Chillón, los nombres de los *curacas*, nombres de las etnías, nombres de lugares y las fechas y motivaciones sobre muchos eventos. Las desventajas consisten en que podemos ser tentados a no plantear modelos alternativos sugeridos por los datos arqueológicos, simplemente porque no son corroborables etnohistóricamente.

[*Véase p. 30 para la bibliografía de este capítulo*]

PART II
Prologue

María Rostworowski de Diez Canseco

Instituto de Estudios Peruanos

> *"the aforementioned lands of Quibi were so good and so much coca was picked in them..."*
>
> [statement of Alonso Paucar, subject of don Xpoual, *cacique principal* of Chacalla (folio 145v)]

Justicia 413 is an extensive document in the Archivo General de Indias in Seville, Spain, published here for the first time by the University of Michigan Museum of Anthropology thanks to the support of Dr. Joyce Marcus. This document is valuable for the study of prehispanic Perú for many reasons.

First, its early date (A.D. 1558–1570) means that it provides us with information from a period before the reforms imposed by Viceroy Toledo, a time when indigenous populations still remained in their original locations and before the *Ordenanzas* which irrevocably altered the Indian lifeway.

Second, this lawsuit—which pits one rival ethnic lord against another—was the result of enmities and other motives (see below) that were very different in their purposes from those of the Spaniards. The Spaniards simply did not understand the reason for the tenacious fight and subsequent lawsuit among Indian lords for some relatively poor, rocky, and not very extensive lands. The stubborn fight for possession of that land led the Indians to take their appeal to the Consejo de Indias without their even considering the exorbitant cost it would exact in terms of the number of years the litigation lasted—litigation that would end by leaving both parties poor and ruined. Perhaps because the Indians were persuaded by their Spanish representatives that they might be successful in their lawsuit, the Indians continued, feeling they would ultimately be favored if the litigation were prolonged, first in the Real Audiencia of Los Reyes, and later on in Spain.

The information one may glean from litigation such as the one we are discussing here differs from and complements the information contained in the *crónicas* and *relaciones*. The primary motivation for writing the *crónicas* was the necessity of establishing the rights of the Spanish Crown in American lands, and, at the same time, revealing in them the usurpation and consequent lack of rights of the Inca in the conquest of their states. On the other hand, the information one obtains from documents such as *Justicia 413* deals with local problems, restricted to a single zone. In these testimonies, we gain information that permits not only the reconstruction of events at a specific place, but also the possibility of a different perspective on the past, through the information provided on the *"provincias"* that formed the Tahuantinsuyu. With this type of data one can gain a local or regional perspective, as opposed to that of the state as a whole.

The first Peruvianist who took the initiative of studying such documents and carrying out related fieldwork was John Murra with his studies on the Chupaychu (Chupaychos) of Huánuco, using the 1562 *Visita* of Iñigo Ortíz de Zúñiga. His contributions (1964, 1967, 1975) were important and form the basis for future work.

For our part, we have dedicated ourselves for many years to the investigation of prehispanic coastal ethnohistory by studying the documents in Peruvian and Spanish archives. Later on we compared the data obtained from the documents with data gathered from fieldwork in such a way as to understand better the *expedientes* and to evaluate their content. It is extremely worthwhile to go over the region of interest carefully with documents and maps in hand, then speak with the people who live in the area, who often remember the ancient place names that figure in the *expedientes* and can clarify more than one problem of interest.

One of the *expedientes* that served as the beginning of our investigations was *Justicia 413*, the document published here. When a document is rich in content, it becomes a gold mine to which scholars with different concerns can return again and again to exploit. The same thing occurs with the *crónicas*, to which one can return frequently and in which one always finds new data.

The researcher trying to use this document possibly will find the sixteenth-century terms difficult and cumbersome, and the text repetitive. Nevertheless, Joyce Marcus and I decided that it was preferable to maintain the integrity of the original manuscript rather than introduce changes in spelling or punctuation, and run the risk of altering the meaning. Additionally, linguists interested in sixteenth-century Spanish will find this document a source for their studies.

Location of Quivi

The original Quivi was located upriver in the Chillón Valley near the present town of Santa Rosa de Quives, at kilometer 64 on the highway that links Lima with Canta. According to local information, the remains of the ancient settlement were destroyed when land was leveled to create a flat area near the modern church, as well as in the construction of a hotel. The name of "Santa Rosa" was given to the place in honor of the patron saint of Lima, but formerly in the era of the Toledo *reducciones* the village received the name of San Juan de Quives.

The valley of the Chillón River, together with those of Rímac (Lima) and Lurín, form what is considered the central coast.

The History of the Valley

In order to understand why this long lawsuit exists, and to appreciate the degree of rancor and accumulated hatred and enmity that are evident among the groups in litigation, we need to reconstruct the sociopolitical environment of the central coast valleys in prehispanic times. In attempting to understand the past, we employ a great number of manuscripts from Peruvian and Spanish archives; combining all the data is somewhat like working out a puzzle that can provide us with an idea of what occurred in the fifteenth century when the Inca conquered the region, and later when the Spaniards arrived (Rostworowski 1972, 1977).

During the Late Intermediate period (which is to say, before the formation of the Inca state), the Andean territories—and with this term we are referring to both the coast and the highlands—were divided into macroethnic groups that varied in power and frequently were in conflict among themselves. The central coast included two macroethnic groups. The first was situated in the valley of the Chillón River and formed a *señorío* to which we will return later. The second, called Ychma or Ychmay, extended in both of the lower valleys of Rímac and Lurín, forming a *señorío* later called "Pachacamac" by the Inca. Numerous small *curacazgos* made up these valleys, but they remained under the hegemony of the religious center of Ychma or Ychmay (see Rostworowski 1981–1982:63, *Probanza* de don Gonzalo, *curaca* of Lima, witness of Hernando Llaxaguailas, *curaca* of Pachacamac).

In the highlands of both valleys lived the Yauyos, warlike and aggressive people who nevertheless achieved a friendship with the Inca, a situation that favored their acquisition of large pieces of land later on; for example, the Yauyos helped subdue and wipe out the rebels who were involved in some local uprisings against the Cusqueños (see Avila 1966).

The Yauyos

The Yauyos, who consisted of a series of ethnic groups or "tribes" that were related to each other through kinship, expanded across the highlands of the Cañete and Mala rivers until they adjoined the Canta on the north (Rostworowski 1978–1980). In 1586 Dávila Briceño, *corregidor* of the region, took on the job of reducing the 200-plus indigenous villages into 39 Spanish towns, changing the ancient political boundaries and creating new territorial demarcations. According to this new realignment, which was part of a land grant administered at the time by the Spanish Crown, the "province" of Yauyos took its name from one of the ethnic groups located to the south. In accord with the traditional Andean system the zone was divided into Anan Yauyos formed by the Manco and Larao groups (inhabitants of the highlands of the Cañete River), and the Lurín Yauyos, whose territories extended along the western slopes of the highlands of the Mala, Lurín, and Rímac rivers (Dávila Briceño 1881, Relaciones Geográficas de Indias, Volume 1; *Justicia 413*, folio 262, mentions Anan Yauyos and Hurin Yauyos).

Nevertheless, in the 1586 map that accompanies the 1881 edition of the Relaciones Geográficas de Indias, it seems that the Anan and Lurín Yauyos together constituted a macroethnic group. In the 1586 map (shown in Figure 21 at end of book) the *"reducción"* of "Santa María de Jesús de Guadochirí" is shown surrounded by a circle that indicates it was the major town and residence of the local lord. Additionally, according to the document, the "province" of the Yauyos apparently was divided into four parts—two of Anan, two of Lurín—a territorial division that reproduces the quadripartition that was the basis of the Andean organizational model of space (Rostworowski, in press).

The oral tradition of one of the Yauyos groups, possibly that of the Checa, was compiled by the eradicator of idolatries Francisco de Avila, the priest in the village of San Damián at the end of the sixteenth century. This collection of stories and myths, written in Quechua, is one of the most important documents on ancient Perú. The epic tone of the narrative is in the tradition of a people that conquered others in a series of waves thereby occupying new lands and throwing the original inhabitants out of them.

It is possible that the expansion of the Yauyos was initiated immediately following the end of Huari hegemony at the close of the Middle Horizon. Everything seems to indicate that this epoch was characterized by migratory movements of people who were free from the restrictions formerly imposed by the Huari. It was a time in which no macroethnic group imposed its will on the inhabitants of the highlands; an example of this aforementioned migration is the transhumance of the brothers Ayar mentioned in the Inca myth, which deals with the movements of agricultural peoples desirous of obtaining good lands for cultivation.

The later Inca policy of administration, however, prohibited its subjects from wandering from place to place; the state assigned them to a location so that the administrative structure would not be destroyed. Only the state had the right to implement mass movements of people, who were known as *mitmaq*, and they were placed in localities that served the interests of the Cusco government.

In earlier articles the narrative of the informants of Avila has been analyzed in greater detail, and we refer the reader to those for additional information. Here we will only review the events in the myths that reveal the methods employed by the Yauyos to gain possession of new lands and settlements (Rostworowski 1977, 1978). The motivation for the expansion of the Yauyos groups may have been an increase in their population which forced them to look for new lands for cultivation. Their expansion and advances were stopped by other inhabitants belonging to the Atavillos macroethnic group, who resided in the highlands of the present-day Department of Lima. The Atavillos included various subgroups, among them the Canta, who also figure in the document upon which we are commenting.

In Avila there is mention of an earlier epoch when different coastal groups occupied the *sierra* regions of their respective valleys, a time that preceded the great expansion of the Yauyos. This situation presumably occurred during the hegemony of Huari, and witnessed the apogee of the coastal sanctuary of Ychma or Ychmay (Pachacamac). The lands that were painstakingly cultivated by the coastal people (for example, there is mention that even the steep lands were cultivated by means of terraces and canals) were the motivation for the greed of the Yauyos, a warlike and combative people, who together with their god Pariacaca conquered the *yunga* lands.

Avila mentions first the battle of Pariacaca against Guallallo, and later on his fight with the female *huaca* of Manañamca; after her flight in the direction of the sea, Manañamca was replaced by a highland deity who was Pariacaca's sister, named Chaupiñamca.

The informants of Avila (Chapter 25) told him that the coastal people called Colli were thrown out by the wind and most of them died; only a few of the Colli succeeded in establishing themselves in the Chillón Valley. Their major towns in the highlands were Yarutini, Huayquijusa, and Colli, all conquered by the Checa. This information allows us to link up the "mythical" Colli to the later inhabitants of Collec in the Chillón Valley; the defeat of the Colli meant the abandonment of their highland possessions and a retreat to their coastal habitat. The text of Avila insists that "the Yunga all lived the same way," i.e., "shared a similar lifestyle" (Chapter 9). The *Yunga* ethnic group of Lima also had their towns in the *sierra*, and Dávila Briceño mentions one of them, named Limac.

The overlap of the names of the *ayllus* and places mentioned by Avila with those in the documents from the archives is impressive. These names can still be verified through fieldwork in that area. The persistence of the oral tradition and the survival of ancient names in that zone are incredible.

The "sons" of Pariacaca who commanded and conquered the lands of the *Yunga* were Chucpaico, Chancharuna, Huariruna, Utcochuco, Tutayquiri, Huarquinri, Aren, and Male. Later the last "son," named Pachacayro, appeared (Avila 1966:63).

The hero that dominated the Rímac Valley (Valley of Lima) was Tutayquiri, and the informants of Avila described him as a very powerful "lord" who vanquished all of his opponents and advanced through the *quebradas* of the valley until he reached a black hill in Pariachi; just as Manco Capac did, he carried with him a staff of gold that he nailed into the mountain as a sign of his taking possession of the place. Tutayquiri and his army occupied the left side of the valley, while the right side was conquered by another branch of the Yauyos composed of the Chaclla and the Carampoma—perhaps led by other mythical "sons" of Pariacaca. These Chaclla established themselves in the mountainous regions and contiguous parts of the highlands located between the Rímac and Chillón valleys. The *quebrada* of Jicamarca was the natural route to the Rímac River.

The Yauyos, once having conquered the land, began to allot or distribute among themselves the fields (or *chacras)* and the houses of the defeated *Yunga*, occupying from then on the property and possessions of the coastal people (Avila 1966:141).

We do not know how much time passed between (1) the establishment of the "tribes" of the Yauyos all along the length of the western side of the mountainous region of the central coast, and (2) the arrival of the Inca armies. Apparently the people of Huarochirí did not occupy the lowlands of the valleys; nevertheless, it is possible that this disinterest was only apparent. In order to conquer the coast, it was not necessary to have an army present; it was sufficient to control the upstream intakes of the irrigation canals. During the Late Intermediate period neither the Rímac nor Lurín valley reveals any evidence of fortifications or defenses, such as one finds in other coastal valleys, which would have protected them against possible attacks coming from the mountains (Rostworowski 1985).

The *Yunga* were easy prey if they were cut off from their supply of water from the rivers and aqueducts. In another article

(Rostworowski 1978–1980) we indicated how the inhabitants of the *señorío* of Guarco (Cañete) possessed a whole system of defenses to protect their lands by means of fortifications, including an enclosing wall that assured the protection of their fields. The same can be confirmed for the *curacazgo* of Collec in the Chillón Valley.

Additionally, in the documents it is frequently stated that highlanders were afraid of installing themselves on the coast because of their fear of getting sick and dying. Due to this they preferred to remain in the zone called the *chaupi yunga*, "costa media," where they enjoyed a dry, sunny climate. In manuscripts mentioning the *chaupi yunga* region of the Rímac Valley one finds numerous references to highlanders coinhabiting the zone with the *Yunga*. (It is a separate topic of research from the one we are discussing here to assess the degree of influence exercised by the highlanders on the central coast during the period designated as Late Intermediate; that research lies in the archaeological domain.)

The document of *Justicia 413* names Tupac Yupanqui as the first Inca ruler to dominate the region. Later on we will examine the relevant data referring to the Inca time period (Late Horizon), but first let us indicate the different groups of Yauyos who occupied the zone, and point out its political divisions. In order to do that, we must consult the Spanish administrative documents.

In Colonial times the *curaca principal* of Huarochirí in the Lurín Yauyos governed a number of *guaranga* (a unit approximately equivalent to 1000 households). Dávila Briceño, in charge of reducing the indigenous population into villages created by the Spanish administration, gave the following 5 *guaranga* to the lord of Huarochirí: Quinti, Langasica, Chaucarima, Colcaruna, and Checa. Nevertheless, these 5 *guaranga* correspond to the organizational patterns imposed by the Spaniards and *do not* represent the former indigenous jurisdictions. In a manuscript on the *Retasa* for Huarochirí dated to 1577 (archive of the Museo Nacional de Historia, not catalogued) 6 *guaranga* are mentioned, instead of 5: Quinte, Colcaruna, Xica, Checa, Chaucarima, and Sisicaya. In this list the *guaranga* of Langasica is omitted, and in its place, two others are named, those of Xica and Sisicaya. In the same document it is said that the last *guaranga* (Sisicaya) was composed of *Yunga* from the middle of the Lurín Valley, a group that was detached from the *señorío* of Pachacamac in order that it be annexed to the Yauyos *guaranga* of Chaucarima (see Figure 20).

In the document of the Museo Nacional de Historia, the Xica were included within the Huarochirí macroethnic group, and other information in *Justicia 413* support that affiliation. With the European conquest and the creation of the system of *encomiendas*, the division of the macroethnic groups into various *repartimientos* frequently proceeded in compliance with the requirements of granting a large number of *encomiendas*; for that reason, at times it becomes very difficult to reconstruct the prehispanic political divisions.

The boundaries of the Xica or Jicamarca included, on the south, the right bank of the Santa Eulalia or Cayao River as far as Mama, and continued along the right bank until Huampaní. In the valley of the Chillón River, the same group's boundary was the *quebrada* of Arahuay on the left bank of the river. The Inca ruler Tupac Yupanqui assigned the Yauyos (in the role of *mitmaq*) to new lands, *yunga* lands in the Chillón Valley, whose location we will see later on.

Further, the *guaranga* of Xica or Jicamarca included (besides the Xica), the Chacalla or Chaclla, the Cullata or Collata, and the Punan. All of them constitute what in various documents are referred to at times as the "Chaclla" or "Jicamarca" (Table 2).

Two different *guaranga* of these same Yauyos were those of Carampoma and of Casta; their lands extended along the upper Chillón Valley on the left bank of the river. Later, due to the very low population in this area, these *guaranga* were restructured. Finally, another group of the Yauyos, composed of the *guaranga* of Picoy and Yaucha, established themselves in Matucana and in San Mateo de Huanchor (Espinoza 1983–1984).

There is substantial documentation concerning the *guaranga* of Jicamarca in the Archivo General de la Nación. At the present time, the *guaranga* of Jicamarca includes a Comunidad Campesina recognized in 1929 (Expediente N°. 97–70, Ministerio de Trabajo; for the boundaries see Archivo General de la Nación, Escritura N°. 1175, year 1883; I had the opportunity to consult these documents when they were sent to the archives for transcription as part of a lawsuit that the *comunidad* brought against the government). The community of Chaclla was recognized separately from that of Jicamarca in 1933 (Expediente N°. 1550, Ministerio de Trabajo); the whole Archivo de Comunidades has since disappeared, taken away by SINAMOS, and today its whereabouts are unknown.

In a lawsuit begun in 1973, the members of the *comunidad* of Jicamarca claim in their own name all the ancient domain of Jicamarca and that of the Chaclla on the basis of "títulos supletorios." The extent of the claimed lands is approximately 100,000 hectares (many uncultivated) and includes the present districts of Ate, Carabayllo, Chosica, the San Cristóbal mountain, the zone of the shooting range and barracks of the army, Huarhuar, and the *quebrada* of Río Seco. All of these lands are located in the valleys of Rímac and Chillón.

The Canta

The data that we have on the Canta in *Justicia 413* are supplemented by the "Cuaderno" of the *visita* to Canta which

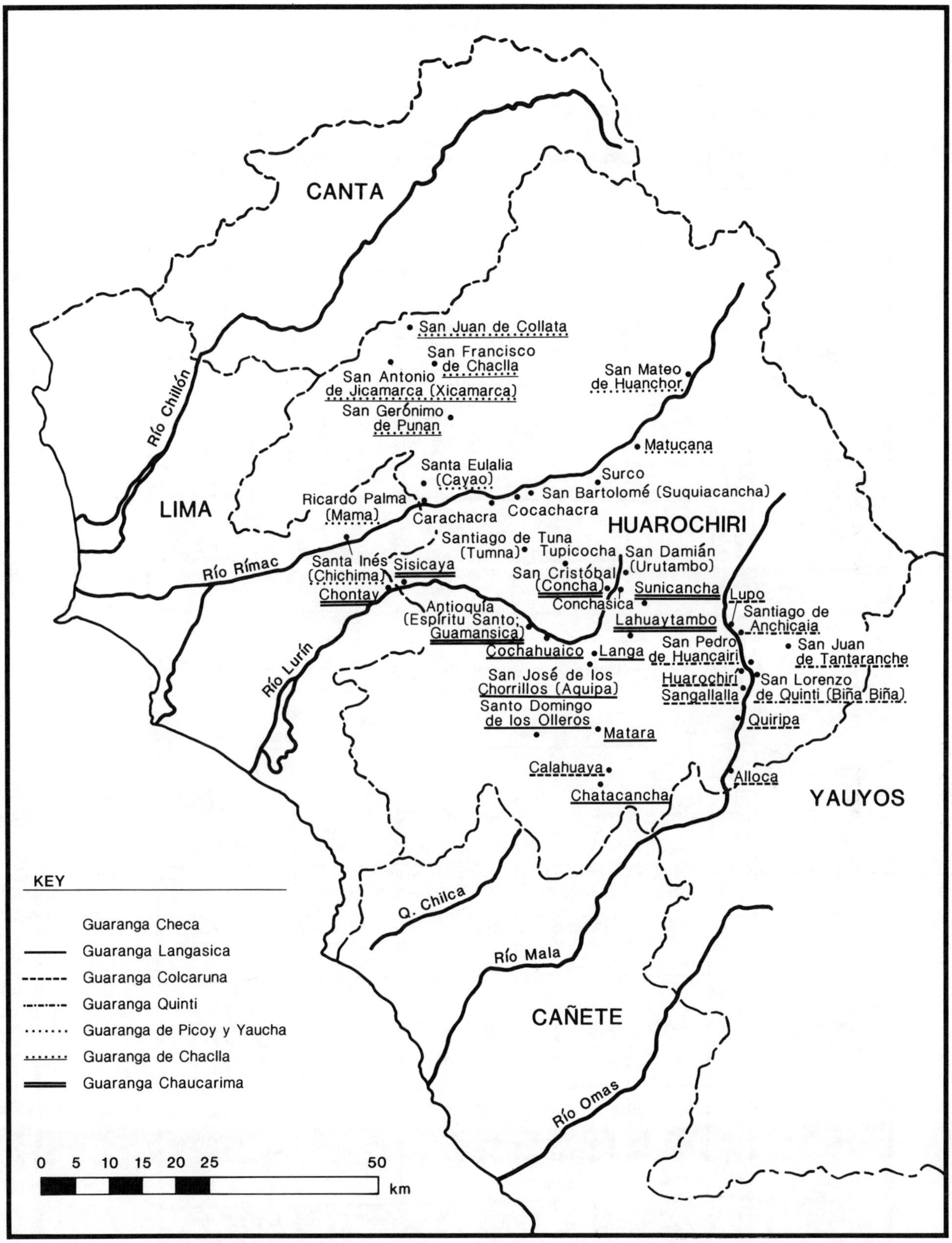

Figure 20. Colonial communities with their indigenous names that formed part of the Huarochirí macroethnic group. Through analysis of the documents, it has also been possible to supply some of the names of the *ayllus* that formed the various *guaranga* (see Rostworowski 1978: 109–122).

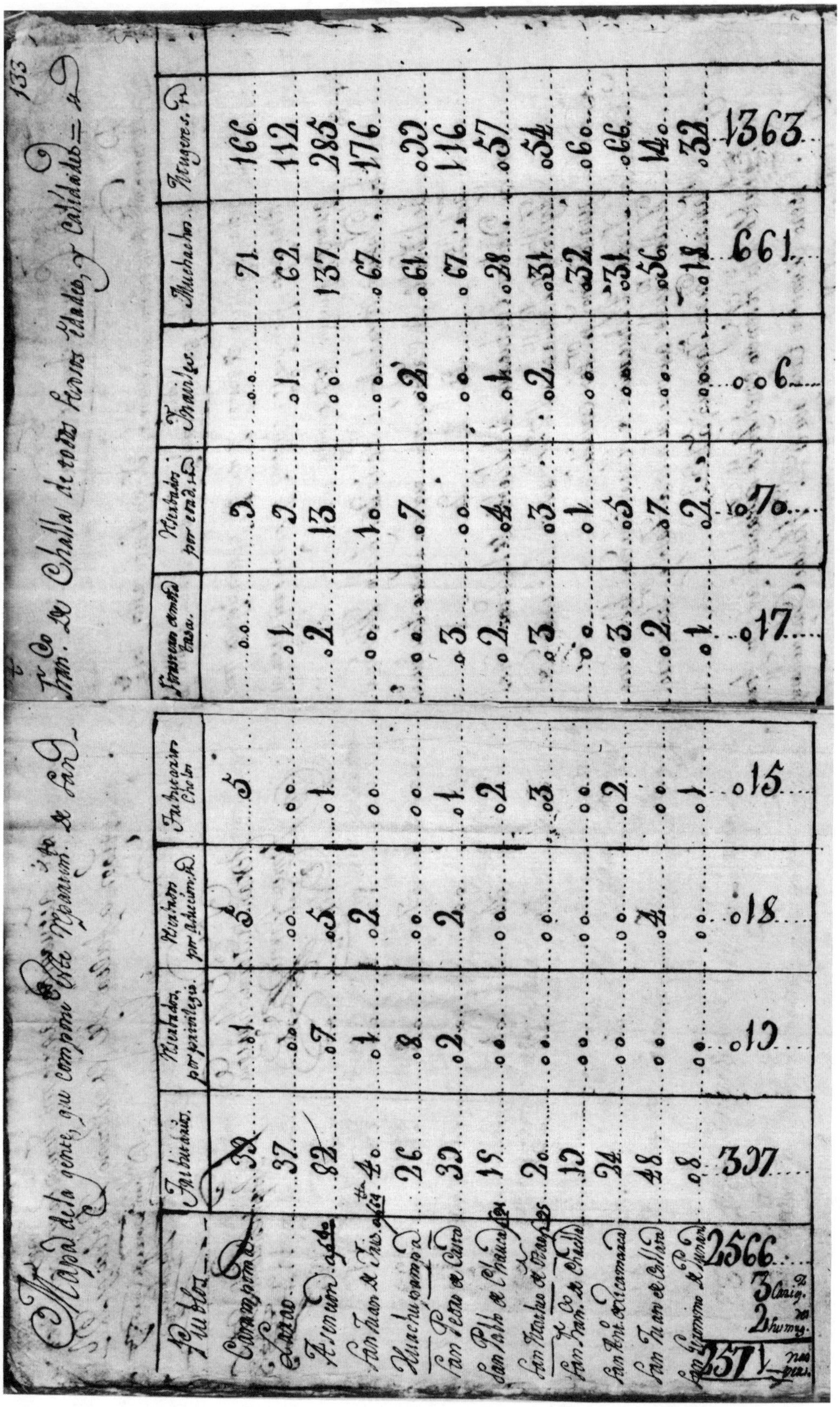

Mapa de la gente que compone este Repartimiento de San Francisco de Challa de todos sexos, estados y calidades

Pueblos	Tributarios	Reservados por privilegio	Reservados por ocupación	Tributarios cholos	Forasteros sin tierra ni casa	Reservados por edad	Frailes	Muchachos	Mugeres
Chaupimarca	33	1	5	5	0	3	0	71	166
Lacsao	37	0	0	0	1	3	1	62	112
Aromuera	82	7	5	1	2	13	0	137	285
San Juan de Chisnayo	40	1	2	0	0	1	0	67	176
Huachuyacampa	26	8	0	0	0	7	2	61	33
San Pedro de Casta	33	2	2	1	3	0	0	67	116
San Pablo de Chaucayán	15	0	0	2	2	4	1	28	57
San Francisco de Llanchas	20	0	0	3	3	3	2	31	54
San Francisco de Challa	13	0	0	0	0	1	0	32	60
San Antonio de Lleamanca	24	0	0	2	3	5	0	31	66
San Juan de Collana	48	0	4	0	2	7	0	56	14
San Jerónimo de Punán	8	0	0	1	1	2	0	18	32
2566	337	10	18	15	17	70	6	661	1363

3 Caciques. 2 Alcaldes. 2571 personas.

Table 2. Indians comprising the *repartimiento* of San Francisco de Challa (Chaclla), 1752. (Rostworowski 1972:300; photo by A. Guillén).

was ordered by *licenciado* Pedro de La Gasca in 1549, followed by a revisit in 1553 (see Rostworowski 1978, Appendix II). Through these testimonies we know that the *curacazgo* of the Atavillos was located on the left bank of the Chancay River as far as the right bank of the Chillón River, and that it was divided into an upper and lower area. In prehispanic times this area was occupied by a macroethnic group known as the Atavillos, which included within its *señorío* other groups called the Canta, Piscas, Huamantanga, and Secos or Xecos. Given its wealth, this *señorío* awakened the interest of Francisco Pizarro, who wanted to take possession of it for himself although he already had other *encomiendas* in Yucay, Cusco, Huaylas, and Chimu Conchuco, as well as the two small possessions of Lima and Chuquitanta. Upon the death of Pizarro, the Atavillos were divided into several *encomiendas*. The supporters of Almagro delivered Canta to a Saavedra who benefited from it for only a few months, and later it passed on to the son of the Marqués, don Gonzalo. Then, beginning in 1549, Canta was an *encomienda* that was given to Nicolás de Ribera, el mozo.

With the arrival of the Colonial period, the Atavillos lost their ancient importance, surrendering it to the Canta. Two of the reasons for the change were Canta's proximity to the city of Los Reyes, and its location right on the route linking Lima with Cerro de Pasco and Huánuco. The two routes of penetration from the central coast into the interior of the country were those of Canta on the one hand, and on the other hand the road from Lima through Sisicaya, Huarochirí, and Jauja which led to Cusco. (The central route through the valley of the Rímac River is a modern creation.)

The first *visita* to Canta was carried out in the village of Quivi on July 29, 1549, in a meeting with the *curacas*. The inspectors did not take the time to travel throughout the *señorío*, preferring instead to stay only in Quivi. Diego Flores, representative of Nicolás de Ribera, was in charge of the meeting. In this first *visita*, 8 Canta *ayllus* were counted, a number that conforms to the Andean pattern. The second *visita* was begun in the town of Mallo or Mayo on June 18, 1553, and it lasted for 20 days. The men encharged with carrying out this *visita* were friar Antonio de Figueroa of the Dominican order, and clergyman Gómez Hernández. This second *visita* produced many more detailed data than those of the first, since the clergymen traveled to all the villages in the *señorío*. Thus it is valuable and interesting to compare both *visitas*. Both give the names of the *curacas* and the villages they governed. The *ayllus* that constituted the *señorío* of Canta were Canta, Racas, Rocha, Lachaqui, Copa, Esquibamba, Carua, and Visca.

The prehispanic boundaries of the *curacazgo* of Canta do not coincide with the boundaries of the modern Canta province. Its lands were located in a linear pattern along the left bank of the Chillón River, embracing several ecological zones from the warm *chaupi yunga* to the cold *puna*. In the high altitude zones its territories extended from the Cordillera of La Viuda up to Caruacayán, through the valley where the Chillón River originates and where it adjoins the Colonial hacienda of Corpacancha, located in the province of Lima (Archivo General de la Nación, Property Titles, File 500, year 1774).

During the Inca period the Canta were neighbors of the small *Yunga curacazgo* of Quivi, and it is also during this period that the Canta received an allotment of lands following the defeat suffered by the *Colli Capac* of Collec, to whom the Quivi were subject. The Arahuay River constituted the southern boundary of the Canta lands, separating them from those of the Chaclla.

The outstanding feature of the Canta settlement pattern lies in the existence of 16 hamlets that were inhabited for short periods of time, hamlets in which certain specific, communal tasks were carried out. When the work was completed, the hamlets remained deserted; jobs were rotated according to a plan developed beforehand. Both *visitas* mention this method as rotation strategy; the earlier *visita* is the one that provides more data on this topic. Later in the Colonial period, various impositions brought about the deterioration of the system. The *visita* of 1549 named 16 hamlets in addition to the villages declared for each *ayllu* that belonged to their respective *parcialidades*. These hamlets, and their tasks, where known, are as follows:

Paron Marca Cambis (weavers of fine cloth, or *cumbis)*
Quiby chacara
Caruacaya (shepherds)
Chacalla ("bird hunters")
Cancha Cancha (sandal makers)
Choccha Calla (uninhabited)
Apio chacara
Aragaco (sandal makers)
Guaro chacara
Yaca chacara
Achim (potters)
Cato (unoccupied by *mitimaes*)
Chochacara
Copi chacara
Mallo chacara
Cazmo chacara

One can see in this list a functional division whereby different activities were carried out in temporally occupied hamlets. If we note the presence of the word *chacara* (as in Chochacara, the name of a field which apparently grew a particular cultigen or crop, possibly for a religious end), eight of the names refer to cultivated fields. All we know about these fields are that they were located in different environmental zones, and that they were worked communally by all the *ayllus*. What we do not know is whether the crops belonged to the *parcialidades* or if they only benefited the *curacas* or the *huacas*. In the documents on the eradication of idolatry in Cajatambo in the seventeenth century it is often stated that each *huaca*, according to its importance, had the benefit of lands that were set aside for it and were cultivated by the common people. Its crops were destined for the celebration of its fiesta, sacrifices to the *huaca*, and preparation of beverages for the faithful.

The other hamlets served in craft production, which demonstrates that specialized groups of artisans or craftsmen did not exist. According to information in the manuscript, the first hamlet, called Parón Marca, was for the manufacture of textiles, a place where fine clothing *(cumbis)* was prepared. The visitors stated that in addition to weaving clothing in Parón Marca (Pariamarca), each *ayllu* wove clothing to satisfy its own personal needs.

By carrying out fieldwork in the region we were able to locate some of the hamlets mentioned in the documents; others remain to be investigated, because we did not ascend to the *puna*. For additional data we refer the reader to an earlier publication (Rostworowski 1978); here we will only discuss the places mentioned in the *Justicia 413* document.

A hamlet whose "specialty" seems very surprising is Chacalla, which was composed of men who were "paxareros." It is possible that these Chacalla were *mitmaq* whose job was to prepare *charqui*, or dehydrated bird meat, as a product of the hunt. In the *visita* carried out in the *curacazgo* of Guancayo, a place located downriver in the Chillón Valley and adjoining that of Quivi, it states that in the time of the Inca the inhabitants delivered as tribute some 10 *"petaquillas"* (small baskets) of dried birds made up as "small pieces of *charqui*" (Espinoza 1963; Rostworowski 1972a). The hunting of fowl, doves, and "perdices" for the preparation of dried meat is mentioned by Guamán Poma (1936: folios 205, 207). "Perdices" were abundant in the Canta region, and, in the 1549 *visita*, the Indians said that they delivered to Nicolás de Ribera 400 "perdices" in tribute.[1]

As for the place designated as Chacalla, an informant in Arahuay indicated that it was located in a *quebrada* "to the right of the town of Licahuasi."

Among the eight special *chacras* is that of Quiby *chacara*, named by the visitors as a town of *Yunga*, in which all the *parcialidades* of Canta were represented. This is a typical case of a highland enclave or "archipelago" established in the *yunga* zone following defeat in war. The peculiarity of the Quivi zone was, as we will see later and in much more detail, that of being a special region for the cultivation of coca and ají *(Capsicum)*, i.e., crops that are either impossible or difficult to grow in high altitude zones. Nevertheless, the Canta had other lands suitable for coca that possibly were theirs before the Inca conquest; they were in the town of Cararuayllo, located in the *parcialidad* of Esquibamba in the low *quebrada* of Arahuay, also occupied by *Yunga* people who cultivated coca (Rostworowski 1978:255).

[1] In Perú today the term "perdiz" usually refers to one of several native tinamous (*Nothoprocta* spp.), since the partridge is an Old World genus. [Editor's note]

Such was the situation of the Canta in 1549, before the lawsuit involving the Chaclla. In order to understand better the circumstances surrounding the litigation, it is necessary to indicate what happened to the coastal peoples in the same valley.

The *Yunga* of Collec

The existence of the ancient *Yunga señorío* of Collec or Collic (called Collique by the Spaniards) was unsuspected until our investigations, because the *crónicas* gave no indication of its existence. However, a series of manuscripts eventually brought to light the sociopolitical situation of the Chillón River valley immediately before the Inca conquest of the region (Rostworowski 1972, 1977). We worked hard to reconstruct the history of Collec, first using the documents in the archives and then, in strenuous fieldwork, searching for the towns, canal intakes, springs, aqueducts, fortifications, boundaries, and landmarks named in the many different documents. In my outings to the valley I could always rely on the enthusiastic collaboration of the late Josefina Ramos de Cox, director of the Seminario de Arqueología at the Catholic University in Lima, who died before her time.

Before the Inca conquest of the coast, the *curacazgo* of Collec extended from the ocean to the *chaupi yunga* zone and included the *curacazgo* of Quivi; numerous small *señoríos* were included inside its boundaries, and the Collec "chief" or principal lord was called the *Colli Capac*. The lower valley or *Yunga* macroethnic group included the following *señoríos*: Collec, Chuquitanta, Seuillay, Comas, Sutca, and Carabayllo. In the upriver region, on the left bank, it included Sapan, Guancayo, and Quivi; and on the right bank were the Macas, Guarauni, and Missai (Rostworowski 1977; Trimborn 1970). The final phase of the Late Intermediate period seems to have been the boom time of the *señorío* of Collec, and its holdings then also included a part of the Rímac Valley. In fact, a witness of the *probanza* of don Gonzalo, *curaca* of Lima in A.D. 1555, declared that an Indian *auca* (equivalent to a soldier or enemy) called Colli forcibly took possession of a certain part of the lands of the Rímac Valley. He said these events took place when the father of the old Taulichusco was the *curaca* of the *señorío*, when the city of Los Reyes was founded, i.e., in A.D. 1535. The witness added that "there is no memory of the Indians of Colli" (Rostworowski 1981–1982, Apéndice:120).

For some time previously the Collec also had had encounters with the Canta, who coveted their coca lands. A witness, reporting on one of these attacks, described a time when the well-armed Canta descended by the river bed, passed the settlement of Quivi without harming its *curaca* Chaumecaxa, and advanced four leagues in the direction of Collec. The witness said that on that occasion the Canta obtained through annexation the village of Quivi, which was "courteously" yielded to them by the *Colli Capac* (folio 211r); this seems to be improbable, since the coastal "chief" felt himself sufficiently strong to oppose the Inca.

Nevertheless, in spite of the enmity between coastal and highland peoples, in times of drought both groups forgot their grudges and hatreds and together addressed themselves to the "sierra de la nieve," work jointly carried out that brought water to the Quivi River (folio 206r).

We know little about the macroethnic groups in Tahuantinsuyu and the role that they played in the state, although a fundamental fact is the presence of *curacas* at lower levels of the hierarchy who were dominated by more powerful lords. We still lack major studies regarding the links and obligations that tied all these lesser lords to their superiors. The *Justicia 413* document provides some information on that topic, which we will analyze when we deal with the data the *expediente* offers us.

We have briefly outlined the situation on the central coast up to the arrival there of the armies of Tupac Yupanqui. The Inca offered the *Colli Capac* the establishment of a reciprocal relationship that would tie the lords of the macroethnic groups to the *Sapa Inca* of Cusco, but the Collec lord—accustomed to resisting the attacks of his neighbors—refused all deals, defended as he was in his fort-palace which was surrounded by a high wall. Inside the enclosing walls he had extensive lands and two sources of water that allowed him to cultivate his fields. Because a long siege did not worry him, he decided to oppose the new conquerors.

We do not know the details of the battle, his defeat, nor his later death. In his place, the Inca ruler named a *curaca* from the social category *yanacon yanayacu* (Rostworowski 1977). The "chiefs" of this status were not part of the system of reciprocity, and thus it was one way for the sovereign to avoid the continuous "requests" and "gifts" that the relationship normally required, and which the Inca was permanently obliged to renew. An additional piece of support is given by Cabello de Valboa (1952:371) who states that Huayna Capac gave preference to the *yanayacu curacas*, especially after the Cusco nobles let him fall from his litter during the war against the Carangues in the north.

The defeat of the Collec was the prelude to their annihilation and the end of their *señorío*. First, the punishment of the Inca decimated the population; later, the already low population could not rebound from the Spanish Conquest. In Colonial times, there were many causes for the decline of population on the coast and among these one could list epidemics, civil wars among Spaniards, and the construction of the viceregal towns. As a result of all these, the inhabitants of Collec disappeared; the many censuses and *visitas* reveal a very small population.

On May 13, 1571, Juan Martínez Rengifo carried out a *visita* to the Indians of Collec which was ordered by Viceroy Toledo; on that occasion he was informed of the former existence of 10 *pachaca*, of which only 15 tribute-payers in total remained. On the eighteenth day of the same month, Martínez Rengifo continued his inspection upriver, visiting the small *señoríos* of Guancayo, Macas, and Guarauni (Espinoza 1963). In 1586, 15 years later, in an investigation carried out by order of the viceroy Fernando de Torres Portugal, count of Villar, it was discovered that of all the ancient inhabitants of Collec only the *curaca*, don Hernando Nacar, and two young men survived. Given the circumstances, it is understandable that the Spanish chroniclers did not find out about the existence of the *señorío* of Collec.

The *Yunga* of Quivi

The data that we possess on the Quivi are provided by the *Justicia 413* document. According to what is stated by all the parties in the lawsuit, the Quivi were *Yunga* people, and before the Inca Conquest they were part of the *señorío* or *curacazgo* of Collec, whose "lord" carried the title of *Colli Capac*. His important status was confirmed by witnesses. As we indicated above, the Collec macroethnic group included many small *curacazgos*, and according to the statements of witnesses, all the inhabitants from Quivi to the ocean were considered *Yunga* (Rostworowski 1972, 1977).

The *curaca* of Quivi paid tribute in "coca, corn, ají, and other things" to the *Colli Capac*. We do not know if the delivery of those products was the result of harvests on lands possessed by the *Colli Capac* in Quivi, or if it was a gesture of subordination on the part of an inferior to a superior.

When the Inca armies appeared, the *Colli Capac* decided to defend himself; at the same time, the *curaca* of Quivi went to the aid of his lord with armed men (folio 198v). These data allow us to assume that both situations represent the terms of subordination between lords at different levels of the hierarchy, and they throw an interesting light on the ties of ongoing reciprocity in the Andean world. The relationship between the *curaca* of Collec and that of Quivi can be dimly seen through the sporadic incursions of the highlanders into the lands of the *Yunga*. The document upon which we are commenting provides data on some of these attacks, and we will see that the palace of the *Colli Capac* was fortified for that reason.

As one of the consequences of the victory of Tupac Yupanqui, 100 Chaclla *mitmaq* were installed in the lands of Quivi, dedicated to cultivating coca for the Inca ruler in addition to serving as *chasqui* or messengers (folio 284r). That number was increased several times both by the same Tupac, and by Huayna Capac and Huascar, who took for themselves a large expanse of lands and assigned new *mitmaq* to work them.

As a result of the new boundaries a noble or "orejón" named Apar Yupanqui arrived from Cusco (folio 149r), and the former boundary markers were removed and transferred from Chuquicoto to Judcunga, a distance of half a league from Quivi (folio 211r). It is possible that the lands between the former boundary and the new one were granted to the Canta, bringing them closer to Quivi (Rostworowski 1978).

The motivation for the introduction of the Chaclla into the Chillón Valley by the Inca was the fact that all the Yauyos were friends, servants, and personal confidants of the Inca, information that is confirmed by the testimony of Avila's informants (1966, 1980). These friendly relationships between the Inca and Chaclla motivated the cessation of the traditional hostilities between the Quivi and the Canta, so that the latter pair could present a unified front against the intruders.

In the document under study here, references are found indicating that the Chaclla belonged to the region of Mama (see Figures 6–7) and were forbidden to intervene in the valleys that did not belong to them. (Mama was located in what is now known as Ricardo Palma near Chosica, folio 189v; see Figure 6.)

The four *ayllus* that formed the *curacazgo* of the Chaclla were represented by *mitmaq* in Quivi. These were enclaves placed by the Inca in accord with the highland system of archipelagos (folio 121v).

After the *Yunga* defeat at the hands of the Inca, the *curaca* of Quivi (named Chumbiquivi) returned to his lands, and a little later he was accused of conspiring against the health of Tupac Yupanqui through use of sorcery of a *huaca* (folio 198r, 198v). The Inca ruler became aware of the machinations of the *curaca*, sent for him, and along with his men he was led to Cusco where they were put to death. Only a few women and children were spared, which explains why in 1557 the Quivi population was so low (see the statement of the *curaca* Felipe Taulichumbi of Quivi, folio 207v). Later on the Canta attributed these acts to the Chaclla, an unlikely situation since they were friends and servants of the Inca ruler and had been favored by him in receiving new lands (folio 219v). Perhaps the Canta just wanted to put the Chaclla in a bad light.

Such was the situation reigning in the upper valley of the Chillón River. And here we will make one clarification: a river in prehispanic times did not carry just one name from its source to its mouth, but changed names according to the lands that it passed through. For that reason, the terms "Quivi River" or "Canta River" are used here in accord with the places through which the river meandered; "Chillón" is a modern designation.

Events Leading Up to the Filing of the Lawsuit

Upon the disappearance of the Cusco government after the Spanish Conquest, during the civil wars among the Spaniards and up to the establishment of the viceregal order, the Canta and Chaclla continued killing each other, motivated by intense passions in which each argued its right to the aforementioned coca fields.

While the two were airing their differences and complaints the rebellion of Manco II broke out, and the Inca lay siege to Los Reyes. The Canta took the side of the Cusqueños or Inca, but the Chaclla refused to aid the Inca because of their ongoing enmity with their Canta rivals. When the generals Ylla Topa and Quiso Yupanqui appeared, the Canta accused the Chaclla of having "two hearts" or dual loyalties, one that sided with Manco II and the other that served the Spaniards.

In reprisal (and not wanting to unite the Chaclla with the Canta against the Spaniards), Guamán Yauri had Vilca Poma, *curaca* of the Chaclla, brought to him through deceit and trickery, and he killed him with blows on his back and shoulders delivered with a club (folios 25v, 25r) (although other witnesses state that he was only wounded). Next the *curaca* ordered the unwary Chaclla to bring provisions for the Inca; but when they arrived at Canta they were put in an enclosure and 150 bearers of supplies were killed. Later the Canta apologized to Ylla Topa for the massacre, offering the explanation that the Chaclla had sent messengers to the Spaniards to warn them that the Inca "chiefs" were present (folios 150r, 150v). A witness named Juan Merlo Chauca, Indian of Huarochirí, confirmed that the Spaniards killed Quiso Yupanqui during the siege of the city.

In 1549, during the stay of the *licenciado* Pedro de La Gasca in Perú, the *encomenderos* Nicolás de Ribera el mozo (of the Canta) and Francisco de Ampuero (of the Chaclla), wanted to impose peace between the *curacas*. They decided that the Chaclla should sell their lands to the Canta for the price of 200 "sheep of this land" (an expression used to indicate the New World domestic camelids, e.g. llamas and alpacas). The Chaclla resisted the deal, since they did not understand the concept of "selling land," and they did not want to receive the animals that Ampuero proposed. In spite of the opposition of the *curaca*, Francisco Vilca Poma of the Chaclla, the agreement was carried out. In view of the cacique's refusal to take the camelids, Ampuero ordered that the animals should be taken to his house in Los Reyes, where he sold some and sent others to the slaughterhouse. Later, pressured to confess his actions, he admitted to having taken them, together with Father Godoy.

On finding out about these events, a *curaca* named Ninavilca of the macroethnic group of Huarochirí sent for Vilca Poma, in order to have him explain the situation and to reprimand him for the actions, which had left Ninavilca and his people without lands. Additionally, the *curaca* of Huarochirí, who had his own lands in Quivi, threatened that he would go up against the Canta with an army composed of all the Yauyos (folio 16v). But times had changed, and the ethnic lords could not mete out justice by their own means, a situation that made the Chaclla desperate. In looking for a way to abolish the agreement, they initiated the lawsuit.

The Trial

The Chaclla initiated the litigation in the Real Audiencia of Los Reyes in 1558; the trial lasted until 1567, with the verdict eventually going in favor of the Chaclla. The judges based their verdict on an important consideration for the Spanish Crown: a policy that the *mitmaq* should stay on the lands to which they had been assigned before the arrival of the Spaniards. Had the Spanish Crown established the precedent of permitting the *mitmaq* to move from their assigned lands, an unending number of trials and lawsuits would have resulted which would have put the entire country in chaos.

The Canta, advised by their Spanish consultants of the content of the viceregal *Ordenanzas*, tried to deny the fact that the Chaclla were *mitmaq* placed by the Inca in Quivi; instead, the Canta insisted that the Chaclla intervention in the valley was just a consequence of the *capacocha* ceremony (see below). As a result, the document provides invaluable data on these rites, which we will analyze later on in this chapter.

Later, the Quivi or the *Yunga* intervened, alleging that the lands were originally theirs and that they had been stripped of them by the Chaclla. Until that point, the Quivi had remained witnesses on behalf of the Canta, without claiming their ancient pre-Inca possessions. This may also have been a stratagem advised by the Spanish attorney, in order to show that the lands had formerly belonged to the Quivi.

The testimonies of the *curacas* and noble *Yungas* are interesting, such as those provided by (1) don Felipe Taulichumbi of Quivi (folio 207v); (2) the *curaca* don Antonio Chumbiquivi of Sapan, a small *curacazgo* adjoining that of Guancayo (folio 54r); (3) don Pedro, *curaca* of Guancayo (folio 183r); and (4) don Francisco Yauri, *curaca* of Collique (folio 54r).

In spite of the long *probanzas* and their stubbornness, the Canta were unable to prove their rights to the lands under litigation (folio 383r). The president and attorneys of the Real Audiencia pronounced their sentence in front of the attorneys for the Indians: it involved the Canta returning the lands to the Chaclla, and the Chaclla returning the 200 camelids to the Canta. Not content with the verdict, the Canta continued to present written testimony and asked for more *probanzas*. Under these circumstances, Fray Gaspar de Carbajal advised the Real Audiencia to impose permanent silence on both parties; even so, the Indians did not pay attention to the advice, deciding instead to go on with the lawsuit (folio 415v). Eventually, the Viceroy Toledo ordered the end of the lawsuits and cases among the Indians in the Real Audiencia, because after all accounts had been made, the Indians remained very poor, the trials cost them much more than they gained, and the Spaniards considered them a nuisance (Rostworowski, in press).

In July, 1567, in the public court of justice Juan de Arrendolaza and Diego de Ocampo were notified that both parties should exhibit clay "models" of their lands (folio 365v). Don Felipe Taulichumbi, *curaca* of Quivi, said he had used such a model on a previous occasion to explain in a general way the disposition of the lands that had been stolen (folios 365r, 367r). For their part, the Chaclla showed "paintings," but they also were asked to show a clay model of the valley. They pleaded that they could not do so, because although they had formerly brought to the Audiencia such a model prepared according to descriptions provided by their elders, it had since fallen to pieces (folios 365r, 365v).

This information on the indigenous custom of making clay models is extremely interesting, because it confirms data provided by Sarmiento de Gamboa (1943, Chapter 39) on the custom of preparing models of a conquered valley in order that the Inca ruler be apprised of the situation and make the changes that seemed necessary. In reference to the "paintings" of the Chaclla, Betanzos (1968: Chapters 10, 16) relates an incident in which the Cusco nobles appeared before Yupanqui in order to offer him the *mascapaicha* (insignia of rulership), only to find him busily painting the changes he wanted to introduce to Cusco.

In accord with the royal decrees issued by the Real Audiencia, a delivery of 200 camelids was made in the town of Quivi on June 28, 1567, by the Chaclla *curaca*, don Juan Bautista Caxa Guaraca, who was hoping to recover his lands. In order to foil the proceedings, the Canta *curaca* did not appear, but many of his noblemen were present at the event, and they said he was to be found in Los Reyes.

While the local Spanish *alguacil* or "constable" wanted to make the delivery, the Canta refused to cooperate, insisting that their *curaca* was absent and hence unable to participate in the transfer. In view of this problem, the *alguacil* wanted to establish proof of the Canta's refusal, so he ordered that the camelids be left in Quivi. Then the *curaca* of the Chaclla proceeded to take possession of the lands; following the Spanish custom for demonstrating proof of ownership, don Juan Caxa Guaraca walked around the fields, cut limbs from the trees, threw stones, and removed clods of dirt. The *alguacil* that was present then ordered that many stone and clay markers be set up in order to mark the boundaries. Don Juan Caxa Guaraca declared to the scribe that he was also taking possession of the waters that irrigated the Quivi lands, in order that he not have trouble later in asserting this right (folio 337v).

After barely an hour of the ceremony had transpired, the Canta suddenly attacked, entering the scene screaming and uttering threats, and in quick succession they destroyed the boundary markers. Upon hearing the uproar, the *alguacil*, scribe, and interpreter came out of the *tambo* (from *tampu*, waystation) and were able to confirm that Pedro Guari, Cristóbal Caruavilca, and Gerónimo Caruavilca (followed by the other Indians) were responsible for the attack. As a result of this lack of respect for authority, the nobles were taken to the prison of the Real Audiencia at Los Reyes (folio 338v). In the Audiencia there followed new depositions and confessions by the Canta prisoners, and from these declarations we learn not only that they had knocked down the boundary markers, but also that they had damaged the irrigation canals, which might cause future floods for the inhabitants of Quivi (folio 348r).

On August 20, 1567, the scribe returned to Quivi accompanied by Friar Gaspar de Carbajal in order to complete the delivery of the camelids to the Canta (folio 386v). When the litigants were called together on the following day, the Canta declared that they had never received the animals, nor did they know of their whereabouts; and they added that the herd did not consist of llamas, but rather of alpacas, and that the latter were sick with *caracha* (a mange caused by the mites *Sarcoptes scabiei* or *Psorotes comunis)* .[1] Additionally, they demanded that the Chaclla go in search of the animals and return with them (folio 387r). Long discussions followed and claims were made by both parties; on August 27, the litigants concluded by bringing together a herd of camelids that were carefully examined by the scribe in the presence of friar Gaspar de Carbajal. Having counted the herd, he confirmed that there were only 113 animals, including both females and males (folio 388v), of which 18 had *caracha*; hence they could not be accepted. The Chaclla offered to replace them, but the Canta continued to demand a delivery of 200 healthy animals, just as they had been sent to Ampuero in 1549.

In several testimonies, the Indians asked the *preceptor* to recommend in the proceedings that the irrigation canals of Quivi be protected against damage in the future, because in the heat of the discussions the aqueducts had been broken, and the course of the water had been altered (folio 419r). Both parties had mutually harassed each other by cutting off the water, by pulling out the coca, and by destroying the harvests.

On the thirtieth of August, Friar Gaspar de Carbajal, appearing before the scribe, ordered the replacement of the 50 markers that bordered the lands won by the Chaclla (folio 396v). The authorities proceeded to arrange a new delivery of the lands, but this time it was contested by don Felipe Taulichumbi and Alonso Caguar, *Yunga* noblemen of Quivi.

This late intervention of the *Yunga* was but a new twist to prolong the lawsuit and obstruct the granting of the lands to the Chaclla. The coastal people accused Friar Gaspar of being partial because of the anger he felt as a result of an incident with a Canta nobleman (folio 412r). They repeated on this occasion the destruction brought about by the Canta, breaking the canals and knocking over boundary markers. As had occurred before, they provided testimony and evidence as to the extent of the damages incurred (folio 429v). On the twenty-first of October, the scribe Pedro de Herrazti was commissioned to verify the extent of the damages that had occurred in Quivi. The Chaclla complained that the coca shrubs had been yanked out and many pacay trees were uprooted as well (folio 434r).

In spite of all the claims by the Canta, the Real Audiencia ratified its verdict, awarding the Quivi lands to the Chaclla. Understandably displeased with this verdict, the Canta—inspired by ancient and ancestral passions—decided to appeal to the Consejo de Indias in Spain. This probably was the result of poor advice from the Spanish representatives, who made no effort to help the Canta understand the meaning of the verdict. Thus on December 19, 1569, the lawsuit was presented in Madrid and on March 6, 1570, Gaspar de Zárate delivered a petition as a "second appeal" of the *sentencia de revisita* to the Real Consejo (folio 7r).

The Coca Fields of Quivi

It is necessary to explain why the lands of Quivi, rocky and not very extensive, awakened among the indigenous population such impassioned desires of possession, expressed not without considerable persistence and hatred. Friar Gaspar de Carbajal, together with Friar Pedraza, in February of 1559 carried out a "data collection trip" in the name of the king and by decree of the Real Audiencia. They measured the land in litigation which amounted to an area 2000 paces in length by 300 paces in width. Its boundaries were, at its width, the river and the royal road of Lima; on the west, the land adjoined the *curacazgo* of Guancayo, and on the east it adjoined a *quebrada* that descended diagonally in the land of the Secos on the opposite side of the river from Quivi. In monetary terms, Carbajal estimated the value of the land at 600 to 700 pesos, but he explained that the indigenous population assessed its value in a different way, because the lands were ideal for the cultivation of coca (folio 131r).

Coca was a basic element in Andean culture, not only for chewing, but also for offerings, ceremonies to the *huacas*, curing rituals, and divination. If coca leaves were not available in the proper zone of the *chaupi yunga*, they had to be obtained from the distant and fairly inaccessible *montaña* region. Thus, that was the reason such great value was placed on the *chaupi yunga* or costa media, a fringe of land in the coastal valleys that has an elevation between 300 and 1200 meters above sea level, and extends from Trujillo in the modern Department of La Libertad as far south as Arica.

The importance of having access to lands ideal for the cultivation of coca came up in the testimony of Juan Merlo Chauca, Indian of Huarochirí, who declared (folio 30r) that he was sent

[1]The Spaniards were evidently unaware that alpacas are notoriously difficult to raise at low elevations. [Editor's note]

by his *curaca* Ninavilca to visit the *chacras* of coca that the aforementioned cacique possessed in Quivi. This information not only indicates the value of such coca fields, but also that the *hatun curaca* of a macroethnic group had available to him lands in the *señoríos* that were subordinate to him.

In our work we found a series of documents referring to the existence, in prehistoric times and in the beginning of the viceroyalty, of coca fields in the costa media of the *yunga* valleys (Rostworowski 1973, 1977; for additional detailed information we refer the reader to these works). Nevertheless, we were lacking botanical evidence for the type of coca cultivated in that zone in prehispanic times.

As a result of our research, we found that the word "coca" comes from the Aymara language *(Aru* or *Jaqí)*. In Quechua, two terms existed for the plant—*tupa* for the coastal variety which has smaller leaves, is aromatic, and has a lower alkaloid content, and *mamox* for the montaña variety.

During our visit to the Regional Museum of Ica, señor Alejandro Pezzia kindly allowed us to open a certain number of *chuspa* (small bags for coca) found in archaeological excavations in Ica; some pertained to the Middle Horizon and others to the Inca (Late Horizon) period. Additionally, we were able to take a sample for analysis. Some time later I met Timothy Plowman, then a botanist at Harvard University dedicated to the study of the varieties of coca. In his research Plowman (1979a, 1979b) recognized the coastal coca of Perú as belonging to the variety *Erythroxylon novogranatense* var. *truxillense*, which is different from the montaña variety *Erythroxylon coca* var. *coca*.

We can now add an interesting piece of data that was revealed in a recent letter from Tom Dillehay. He found coca leaves dated by C^{14} to the Chavín period in Guancayo Alto, a site located in the mid-valley zone of the Chillón Valley. Also, in Quivi Viejo he recovered remains of coca leaves and lime dated to 750 B.C. These data indicate a very old tradition of coca use on the coast.

One final piece of information is needed to explain why the indigenous population, in their fit of hatred, pulled up pacay and guava trees. From the thesis of Anibal Orihuela Noli (1953) on the modern cultivation of coca in the Department of La Libertad, Trujillo, we find out that the coca plantings grow under the protective shade of the pacay *(Inga feuilli)*.

Finally, the disappearance of the major part of the coca fields on the coast was due to the evangelical zeal of the Spanish religious zealots who saw in coca a "diabolical cult" (Arriaga 1968).

The Capacocha

The document of *Justicia 413* is also a source of regional information on the *capacocha*, a ceremony that is far from completely understood and requires much more research.

According to Albornoz (1967:37), the Indians not only called the rites and festivals by that name, but also used the term to refer to the clothing worn by the "idols" or *huacas*. Cieza de León (1943: Chapter 29) dedicates a chapter to the *capacocha* and refers to it as a group of ceremonies whose purpose was to reunite all the important *huacas* of Tahuantinsuyu once a year in the capital, accompanied by numerous followers. In the great plaza of Aucaypata in Cusco the idols were brought together, and in solemm, special rites each *huaca* issued its prophecy in turn. The purpose of these prophecies was to predict the results of wars undertaken by the Inca, his health, the fertility of the earth, the future harvests, or whether the year would be abundant or full of misery.

On such occasions the Indians performed large sacrifices at which they immolated thousands of domestic camelids. After the festivities were over, the *huacas* were returned to their towns bearing offerings that were called *capacocha*. The *huacas* that had hit upon the correct predictions were later rewarded and held in high regard, but those that had been wrong were not so honored.

Cieza de León, possibly even referring to the very document upon which we are commenting, says that soon after the establishment of the Audiencia in Los Reyes, and after La Gasca had returned to Spain in 1549, some lawsuits were initiated in Lima which mentioned the rites of the *capacocha*. Given the designated dates, there is no doubt that these lawsuits fall within the time period of the litigation between the Canta and Chaclla, and surely this litigation had repercussions in Lima. The passion that both parties showed in the dispute for some poor lands, held in low esteem by the Spaniards, must have stunned Cieza de León.

According to other writers, the *capacocha* included sacrifices not only of llamas, but also of humans. Sarmiento de Gamboa (1943: Chapter 13) relates that Manco Capac instituted such sacrifices in order to honor the *huaca* of Huanacauri after the birth of Sinchi Roca. The offering consisted of two children, one male and one female. The inclusion of human sacrifice in the *capacocha* is also supported by statements in Guamán Poma (folios 228, 247, 251, and 259), Cabello de Valboa (1952), Betanzos (1968: Chapters 11, 17), and Estete and Acosta (1942: Book 5, Chapters 4, 7). In the document upon which we are commenting, however, only the blood of camelids mixed with ground-up or crushed shell *(chaquira)* is mentioned.

Duviols (1976) has thoroughly treated the theme of the *capacocha*, and to him we defer for additional information. Duviols is right on the mark when he says that the *capacocha* included the most solemn ceremonies in the life of the Inca realm, with a great number of individuals from the entire "empire" participating. Duviols supposes that there existed at least two types of *capacocha*: some cyclical, and others *ad hoc*. According to the document published here, the last two *capacocha* that took place in Tahuantinsuyu and passed through

Quivi were carried out for Huascar and his wife, the *Coya* (folios 195r, 230v). This reference shows the importance that a female sovereign had in sociopolitical matters, perhaps as a representative of the female gender.

Duviols states that there were different kinds of journeys associated with the *capacocha*—some that were local and other that were long-distance, some that involved traveling in a circle (concentric), and others that went out like the spokes of a wheel (radial). The departure from Cusco was made, according to Duviols, by the four *suyu* or spatial divisions of Cusco in a ceremony similar to that of the *Citua* (Molina 1943), whose purpose was purification through casting evil and illnesses out of the capital.

The data on the *capacocha* in *Justicia 413* are purely local, and have the advantage of providing information at the level of the *provincia*. The Canta were the ones who told all the stories about the episodes that took place during the *capacocha*, with the purpose of explaining how the Chaclla had introduced themselves into Quivi. Finding out through their attorneys that the Crown would not allow *mitmaq* population movements (and assuming that the Spaniards would also honor the rights of those who occupied lands at the time of the Spanish Conquest), the Canta devised another story to explain the presence of Chaclla in the land under litigation (folio 63r).

In the document, various *capacocha* celebrated during the reign of Huayna Capac are mentioned, besides the two carried out in the time of Huascar. The rite consisted of offering the blood of camelids, and such animal sacrifices must have been held with some frequency, since the document emphasizes that the offering had to remain liquid, and that there existed the constant danger that the contents of the vessels would spill. The blood was carried in small gourds (*mates)* or clay pots; when the bearers arrived in high places near a *huaca*, they put the gourd or vessel in a sling, and as if they were throwing a stone, they hurled it on the adoratory. If the bearer did not hit the mark, he was killed on the spot. The same fate was meted out to the messenger if a part of the offering was lost or spilled on the soil, even if it be but a drop.

The Canta maintained that on different occasions the Chaclla took advantage of the fact that they were transporting the offerings in order to trespass beyond boundary markers, placing themselves inside the lands of the Canta in Quivi (folio 245v). While they were carrying the gourds of blood they went along saying "get out of the way, get out of the way, *capacocha, capacocha*...up to here has arrived my *capacocha*, up to here are my lands..." (folio 250r). The Canta became very enraged at the Chaclla's impudence, and on one occasion a Canta named Coqui Guaranga hurled himself on the Chaclla gourd carrier in order to impede his way; in this sudden attack, the sacrificial blood was spilled. Punishment was quick in coming, and the guilty one who spilled the blood was hanged and buried on the spot where it happened.

Through these data we see the existence of different types of *capacocha*, covering different routes and perhaps meeting different objectives. A herder of the Inca once stated that he witnessed a *capacocha* taking place in the *puna* of Pumpu (or Bombón near Junín), and said that the blood was being carried from Cusco to Quito; it was taken from boundary to boundary by different ethnic groups located along the route. In the Pumpu case, the Guanca (Huanca) people were the ones carrying it when seen by the herder. These data confirm the fact that the inhabitants of a region carried the offering within that region, and when they arrived at their territorial limits, the sacrifice was delivered to their neighbors. On another occasion, the Yauyos of Chaclla received an offering from the Jaujas and carried it to their boundary with the Canta.

It is difficult to understand why a *capacocha* was passing through Quivi territory, so far off the main route to Quito, Ecuador, unless we are dealing with different *capacocha*. For example, there is mention of one of these offerings being brought to the *curacazgo* of Guancayo, whose inhabitants in turn carried it to the sea. Another piece of information indicates that it was brought to a river in order to wash the blood. Undoubtedly, the documents refer to different *capacocha*, with varied purposes. It is possible that some *capacocha* were not destined for the north but directed to the sea as we have seen, following the route from Cusco to Jauja, and from there by way of Huarochirí to Quivi and Collec and on to the ocean. Perhaps the different routes corresponded to different *huacas* that had to be honored. Since *Mamacocha*, the sea, was an important deity in the Andean realm, it is not surprising that offerings were sent to her from Cusco.

To sum up, in these lines of introduction we have indicated some of the information contained in the document published here. We have not exhausted the topics here, but we hope that by providing access to such a valuable resource we will contribute to future studies.

Chronological Sequence of the Events in *Justicia 413*

The long lawsuit includes numerous *probanzas* and *averiguaciones* of the witnesses presented by both parties. The contradictions, the different versions presented, and all the repetition make the document difficult to read. In order to facilitate the reading of the document, we offer a chronological sequence of the essential events.

1) In the first epoch, before the conquest of the Inca, the small *curacazgo* of Quivi formed part of the *Yunga* macroethnic group of Collec.
2) The indigenous population of Canta coveted the Quivi lands, brought together an army, passed through Quivi, and advanced four leagues downvalley in the direction

of the town of Collec (folio 211r). The Canta asserted that on this occasion, the Collec lord ceded to them the Quivi *curacazgo*.

3) In view of the threat of the Inca arrival on the central coast, the Quivi lord met with his "war people" and came to the aid of the *Colli Capac*. The coastal people were conquered by the Cusqueños (this assertion conflicts with, and denies, the Canta version of an earlier submission of the Quivi to the Canta).
4) With the arrival of the Inca, the situation in the valley changed; a group of Chaclla, of the macroethnic Yauyos group, was installed in Quivi in the capacity of *mitmaq* in order to cultivate the coca on lands designated by the Inca.
5) Right after the Spanish Conquest, the valley experienced a second transformation. Afraid of and perhaps threatened by the Canta, the Chaclla retreated to the mountains, abandoning their lands in Quivi.
6) Manco II rebelled, the Canta decided to aid the Inca, and the Cusco leaders—Illa Topa and Quisu Yupanqui—passed through Canta. The Chaclla did not want to collaborate with them. The Canta took advantage of this fact to set up an ambush for the *curaca* Vilca Poma, who was assassinated along with his 150 Chaclla burden-carriers.
7) In the time of *licenciado* Pedro de La Gasca, in A.D. 1549, the *encomenderos* Francisco de Ampuero (of the Chaclla) and Nicolás de Ribera el mozo (of the Canta) hit upon a solution to end the conflict. They attempted to resolve the situation by ordering the Chaclla to sell their lands to the Canta for the price of 200 domestic camelids.
8) The Chaclla resisted, having no concept for the "sale" of land; they refused to receive the animals that Ampuero was forcing on them. Finally, the *encomendero* took the animals for himself.
9) The *curaca* of the macroethnic group of Huarochirí met with the Chaclla *curaca* in order to explain what happened. The Chaclla became desperate.
10) In 1558, the Chaclla initiated the lawsuit; the *probanzas* took place; but the testimonies were filled with contradictions. The Real Audiencia decided in 1567 in favor of the Chaclla, based on the principle that the *mitmaq* who occupied lands at the time of the Spanish Conquest should continue to occupy those lands and should not be moved. (This verdict is understandable because if the Spaniards had decided otherwise, it would have given rise to a never-ending stream of lawsuits among the Indians that the Crown could not control.)
11) Finally, in December, 1569 the lawsuit was filed in Madrid, to be presented to the Real Consejo by Gaspar de Zárate, as a request in a second appeal of the *sentencia de revisita*.

[*Bibliography for this section begins on page 81*]

PARTE II
Prólogo

María Rostworowski de Diez Canseco

Instituto de Estudios Peruanos

"...las dichas tierras de Quibi como heran tan buenas y se coxía en ellas mucha coca..."

[declaración de Alonso Paucar, sujeto a don Xpoual, cacique principal de Chacalla (foja 145v)]

El voluminoso documento del Archivo General de Indias de Sevilla—que en esta ocasión publica la Universidad de Michigan gracias al apoyo de la Dra. Joyce Marcus—es valioso para el estudio del Perú pre-hispánico por muchas razones.

Primeramente, su temprana fecha (1558–1570) nos permite tener acceso a noticias anteriores a las reformas impuestas por el virrey Toledo, cuando los poblados indígenas aún permanecían en sus lugares de origen y no se habían dictado las "Ordenanzas" que impulsarían la desestructuración andina.

En segundo lugar, el juicio se efectuaba entre jefes étnicos siguiendo los odios y motivos indígenas, muy distintos en sus propósitos a los intereses de los grupos españoles. Los hispanos no comprendían la razón de la tenaz lucha entablada entre los naturales por unas tierras relativamente pobres, pedregosas y de reducida extensión. El empecinamiento por su posesión llevó a los indígenas a apelar al Consejo de Indias, sin mirar en el gasto exorbitante que significaba para ellos mantener durante tantos años un litigio que terminaría por dejar a ambas partes pobres y arruinadas. Aunque, es posible también que fueran persuadidos por sus representantes españoles a continuar el juicio, pues ellos se verían favorecidos con prolongar la situación de beligerancia, primero en la Real Audiencia de Los Reyes y luego en España.

La información que se puede lograr de testimonios como el que comentamos aquí, difiere y enriquece las noticias contenidas en las crónicas y relaciones. El motivo principal para que se escribieran estas crónicas fue la necesidad de probar los derechos de la Corona española en tierras Americanas y, a la vez, señalar en ellas la usurpación y la carencia de derecho de los Incas en la conquista de sus estados.

En cambio las noticias que se pueden obtener de los documentos son datos sobre problemas locales, restringidos a una zona. En los testimonios aparece información que permite no solamente reconstruir los sucesos del lugar, sino que posibilita un enfoque distinto del pasado, a través de las noticias de las "provincias" que formaban el Tahuantinsuyu. Con este tipo de datos se percibe lo regional bajo un punto de vista opuesto al del estado.

El primer peruanista que tuvo la iniciativa de estudiar documentos y llevar al campo la investigación fue Murra, con sus trabajos en Huánuco sobre la visita de Iñigo Ortiz de Zúñiga a los Chupaychos de 1562. Sus aportes fueron importantes y sentaron las bases para futuras investigaciones (1964, 1967, 1975).

Por nuestra parte, nos dedicamos por largos años a investigar la etnohistoria de la costa pre-hispánica a través de la documentación de archivos peruanos y españoles; más tarde confrontamos las noticias obtenidas en los testimonios en un posterior trabajo de campo, como manera de comprender mejor los expedientes y valorar su contenido. Es muy enriquecedor recorrer la región investigada con los documentos y mapas a la mano, hablar con la gente del lugar, que con frecuencia recuerda los nombres antiguos que figuran en los expedientes y pueden aclarar más de un problema.

Uno de los expedientes que sirvió para el inicio de nuestras investigaciones fue el de Justicia 413 que se publica ahora. Cuando un testimonio es rico en su contenido, se convierte en una cantera a la cual pueden acudir estudiosos con distintas preocupaciones. Lo mismo sucede con las crónicas a las que se vuelve frecuentemente y en las que siempre se hallan datos nuevos.

El investigador encontrará posiblemente difíciles y engorrosos los términos usados en el siglo XVI, unidos a la tendencia repetitiva del texto. Sin embargo juzgamos con Joyce Marcus que, era preferible mantener el manuscrito sin introducir cambios ortográficos, ni puntuación pues se corría el peligro de deformar el sentido. Por otra parte los linguístas interesados en el idioma castellano del XVI, hallarán en este documento una fuente para sus estudios.

Ubicación de Quivi

El lugar indígena de Quivi se hallaba en el valle del Chillón, río arriba, cerca del actual pueblo de Santa Rosa de Quives en el km. 64 de la carretera que une Lima con Canta.

Según información local, los restos del antiguo asentamiento fueron destruídos al nivelarse el terreno para la explanada, cercana a la moderna iglesia y en la edificación del hotel. El nombre de Santa Rosa le fue dado en honor a la santa limeña, pero anteriormente, época de las reducciones toledanas, la aldea recibió el nombre de San Juan de Quives.

El valle del río Chillón junto con los de Lima y el de Lurín forman lo que se considera como la costa central.

La historia del valle

Para comprender el por qué del largo juicio y el grado de rencores y odios acumulados entre los grupos en litigio, necesitamos reconstruir la situación sociopolítica de los valles de la costa central en tiempos prehispánicos. En el esfuerzo por conocer el pasado, empleamos un gran número de manuscritos de archivos peruanos y españoles y, juntando las noticias tal como un rompe-cabeza, pudimos hacernos una idea de lo que sucedió en el siglo XV, cuando la conquista Inca a la región y más adelante con el arribo de los españoles (Rostworowski 1972, 1977).

Durante el Intermedio Tardío, es decir antes de la formación del Estado Inca, los territorios andinos—y con este término nos referimos tanto a la costa como a la sierra—se dividían en macro-etnías más o menos poderosas y con frecuencia en conflicto entre ellas.

La costa central comprendía dos macro-etnías. La primera se situaba en el valle del río Chillón y formaba un señorío sobre el que volveremos más adelante, la segunda, llamada Ychma o Ychmay se extendía en los dos valles bajos de Lima y de Lurín formando un señorío llamado posteriormente por los Incas como Pachacamac. Numerosos pequeños curacazgos se dividían ambos valles, pero permanecían supeditados a la hegemonía del centro religioso de Ychma o Ychmay (ver Rostworowski 1981–82:63, *Probanza* de Don Gonzalo, *curaca* de Lima, Testigo de Hernando Llaxaguailas, *curaca* de Pachacamac).

En la sierra de ambos valles vivía el grupo de los Yauyos, gente agresiva y guerrera que, sin embargo, logró la amistad de los Incas, hecho que favoreció su adquisición de mayores tierras, pues en algunas sublevaciones locales contra los cusqueños ayudaron a la sumisión y al aniquilamiento de los rebeldes (ver Avila 1966).

Los Yauyos

Los llamados Yauyos comprendían una serie de bandas emparentadas entre sí y se expandieron desde las cabeceras serranas de los ríos de Cañete y Mala hasta colindar con los Canta al norte (Rostworowski 1978–80).

En 1586, Dávila Briceño, corregidor de la región, asumió la tarea de reducir las doscientas y tantas aldeas indígenas a treintainueve pueblos españoles, dividiendo los antiguos límites políticos y creando nuevas demarcaciones territoriales. Según éste, la "provincia" de los Yauyos (ver Figura 20 al final del libro) tomó el nombre genérico de un grupo sureño de ellos mismos, encomendados en aquel entonces a la Corona. De acuerdo con el sistema andino la zona se dividía en Anan Yauyos formados por los grupos de Manco y Larao, habitantes de las serranías del río Cañete y los Lurín Yauyos, cuyos territorios se extendían por la sierra marítima de los ríos de Mala, Lurín y Rímac (Dávila Briceño 1881, *Relaciones Geográficas de Indias* Tomo 1; *Justicia 413* foja 262 menciona a los Anan Yauyo y Hurin Yauyos).

Sin embargo en el mapa de 1586 que acompaña la edición de 1881 de las Relaciones, parece que los Anan y los Lurín Yauyos conformaban una macro-etnía. En el mencionado mapa la reducción de "Santa María de Jesús de Guadochiri" está rodeada de un círculo que indicaba el pueblo principal y la residencia del poder local (ver Figura 21 al final del libro). Además la "provincia" de los Yauyos figura dividida en cuatro partes, dos de Anan y dos de Lurin, demarcación que reproduce la cuatripartición del espacio y que fue la base del modelo organizativo andino (ver Rostworowski, en prensa).

La tradición oral de uno de los grupos Yauyos, posiblemente de los Checa, fue recopilada por el extirpador de idolatrías que fue Francisco de Avila, doctrinero del pueblo de San Damián, a finales del siglo XVI. Este conjunto de relatos y de mitos escritos en quechua es uno de los documentos más importantes sobre el antiguo Perú. El tono épico del relato es la tradición de un pueblo que por oleadas conquistó y ocupó nuevas tierras, echando de ellas a sus primeros habitantes.

Es posible que la expansión de los Yauyos se iniciara a raíz del decaimiento de la hegemonía Huari al final del Horizonte Medio. Todo parece indicar que esa época se caracterizó por movimientos migratorios de pueblos que se vieron libres de las restricciones impuestas anteriormente por los Huari. Un tiempo que ninguna macro-etnía imponía su voluntad entre los habitantes de las serranías; un ejemplo de dicho vagar está mencionado en la transhumancia de los hermanos Ayar del mito Inca, migraciones de pueblos agrícolas deseosos de conseguir buenas tierras de cultivos.

La búsqueda de nuevos territorios fue posteriormente prohibida por los Incas, que no podían permitir se les estropease sus cuadros administrativos. Sólo el Estado se atribuyó el derecho del traslado masivo de los pueblos bajo el nombre de *mitmaq*, para cumplir los intereses del gobierno cusqueño.

Ya en otros trabajos hemos analizado la narrativa de los informantes de Avila y a ellos nos remitimos para mayores detalles; aquí sólo repasaremos—a grandes rasgos—los acontecimientos, narrados en forma de mitos, que muestran los métodos empleados por los Yauyos para posesionarse de nuevos asentamientos (Rostworowski 1977, 1978).

El motivo de la expansión de los grupos Yauyos fue posiblemente un aumento demográfico que los obligó a buscar nuevas tierras de cultivo. Sus avances fueron detenidos por otros habitantes, pertenecientes a la macro-etnía de los Atavillos, de las serranías del actual departamento de Lima. Ellos comprendían varios grupos, entre los que se hallaban los Canta, que intervienen en el documento que comentamos.

En la información de Avila se menciona una época anterior a la gran expansión de los Yauyos, durante la cual los diferentes grupos costeños ocupaban la sierra de sus respectivos valles. Situación que se mantuvo durante la hegemonía Huari y marcó la época del mayor apogeo del santuario costeño de Ychma o Ychmay (Pachacamac). Las tierras esmeradamente cultivadas por los costeños (hay mención que hasta los precipicios eran cultivados) por medio de andenerías y de canales hidráulicos fueron motivo de la codicia de los Yauyos, pueblo aguerrido y combativo que, junto con su dios Pariacaca emprendió la conquista de las haciendas yungas.

Avila nombra primero la lucha de Pariacaca contra Guallallo, y luego con la huaca femenina de Manañamca reemplazada, después de su huída hacia el mar, por la divinidad serrana de Chaupiñamca, hermana de Pariacaca.

Los informantes de Avila (cap. 25) contaban que los costeños llamados Colli huyeron arrasados por el viento y murieron, y sólo algunos lograron establecerse en el valle del Chillón. Sus principales pueblos en las serranías fueron Yarutini, Huayquijusa y Colli, conquistados por los Checa. Esta noticia permite relacionar a los míticos Colli con los posteriores habitantes de Collec en el Chillón: la derrota significó para los Colli el abandono de sus posesiones serranas y el repliego a su habitat costeño. El texto de Avila insiste que "la vida de todos los yungas era una sola" (cap. 9); también los *yunga* de Lima tuvieron sus pueblos en la sierra, y Dávila Briceño nombra a uno de ellos llamado Limac.

La comprobación y el cotejo de los nombres de los ayllus, y lugares mencionados por Avila, con los documentos de archivos es impresionante; datos que se verifican más aún en el trabajo de campo. La persistencia de la tradición oral en la zona es increíble.

Los "hijos" de Pariacaca que capitanearon y conquistaron las tierras de los *yunga* fueron Chucpaico, Chancharuna, Huariruna, Utcochuco, Tutayquiri, Huarquinri, Aren y Male; después de un tiempo salió un último "hijo" llamado Pachacayro (Avila 1966:63).

El héroe que dominó el valle de Lima fue Tutayquiri y, los informantes contaron que se trató de un jefe poderoso que venció a todos sus oponentes y avanzó por las quebradas de Lima hasta alcanzar un cerro negro en Pariachi y, tal como lo hiciera Manco Capac, llevaba una vara de oro que clavó en la montaña como señal de su toma de posesión.

Tutayquiri y su ejército ocuparon la margen izquierda del valle, mientras la banda derecha fue conquistada por otra rama de los Yauyos compuesta por los Chaclla y los Carampoma, capitaneados quizá por algún otro de los míticos "hijos" de Pariacaca.

Estos Chaclla se establecieron en las serranías y las partes contiguas a la sierra, entre los valles de Lima y el del Chillón, siendo la quebrada de Jicamarca la ruta natural hacia el río Rímac.

Los Yauyos una vez conquistada la tierra empezaron a repartirse los campos, las chacras y las casas que fueron de los vencidos yungas, ocupando de ahí en adelante las haciendas y posesiones de los costeños (Avila 1966: 141).

No sabemos cuánto tiempo pasó entre el establecimiento de las bandas de los Yauyos a lo largo de las serranías marítimas de la costa central y el arribo de los ejércitos inca. Al parecer la gente de Huarochirí no ocupó las tierras bajas de los valles; sin embargo, es posible que el desinterés fuese tan sólo aparente: para conquistar la costa no se necesitaba de la presencia de un ejército, bastaba con tener acceso al control de las bocatomas y de los canales de riego. Tanto el valle de Lima como el de Lurín no muestran haber dispuesto durante el Intermedio Tardío, de fortificaciones y defensas contra posibles ataques provenientes de las serranías, como sí se pueden encontrar en otros valles costeños (Rostworowski 1985).

Los yungas eran presa fácil si se les cortaba el suministro del agua de sus ríos y acequias. En otro trabajo señalamos cómo los habitantes de Guarco (Cañete) poseían todo un sistema de protección de sus tierras por medio de fortificaciones y de una muralla envolvente que aseguraba la defensa de sus campos. Lo mismo podemos afirmar para el curacazgo de Collec, en el valle del Chillón.

Por otro lado en los documentos se menciona con frecuencia el temor que sentían los serranos de instalarse a vivir en la costa por miedo a enfermar y morir. Debido a ello preferían quedarse en lo que llamaban el *chaupi yunga* o costa media, donde disfrutaban de un clima seco y soleado. En los manuscritos sobre dicha región del valle de Lima se hallan numerosas referencias a serranos co-habitando la zona junto con los yungas. Es tema aparte de investigación el grado de influencia ejercida por los serranos sobre la costa central durante el período Intermedio Tardío; y que corresponde al terreno de la arqueología.

El documento de Justicia 413 nombra a Túpac Yupanqui como el primer Inca que dominó la región. (Más adelante retornaremos a examinar las noticias referentes a la época Inca, para primero señalar a los diferentes grupos de Yauyos que ocuparon la zona y sus divisiones políticas. Para el caso tenemos que valernos de los documentos administrativos españoles).

En tiempos coloniales, el curaca principal de Huarochirí de los Lurín Yauyos gobernaba un número de *guaranga* (cifra hipotética de mil hogares). Dávila Briceño, encargado de reducir a los naturales a aldeas creadas por la administración española, señaló a este señor cinco *guaranga* o sea la de Quinti, Langasica, Chaucarima, Colcaruna, y Checa. Sin embargo estas cinco *guaranga* respondían a los patrones organizativos impuestos por los hispanos y *no representaban* la auténtica jurisdicción anterior.

En un manuscrito sobre la *Retasa* para el mismo lugar fechada de 1577 (Archivo Museo Nacional de Historia s/n) se mencionan seis *guaranga* en lugar de cinco: Quinte, Colcaruna, Xica, Checa, Chaucarima y Sisicaya. En esta lista quedó omitida la *guaranga* de Langasica y, en su lugar, se nombró a dos, la Xica y la de Sisicaya. En el mismo documento se decía que la última *guaranga* estaba compuesta por habitantes *yunga* del valle medio de Lurín, grupo desprendido seguramente del Señorío de Pachacamac, para ser anexado a la guaranga Yauyos de Chaucarima.

En el documento del Museo Nacional de Historia, los Xica estaban comprendidos dentro de la macro-etnía de Huarochirí y diversas noticias del documento de Justicia 413 apoyan esta versión. Con la conquista europea y la creación del sistema de encomiendas, se procedió frecuentemente a la división de las macro-etnías en varios repartimientos para complacer las exigencias de otorgar un mayor número de encomiendas; por eso es que se hace a veces difícil reconstruir las divisiones políticas pre-hispánicas. Los linderos de los Xica o Jicamarca comprendían por el sur la margen derecha del río de Santa Eulalia o Cayao hasta Mama y continuaba siempre por la banda derecha hasta Huampaní. Por el valle del río Chillón, sus fronteras eran la quebrada de Arahuay en la banda izquierda del río. El Inca Túpac Yupanqui les donó nuevas tierras en calidad de *mitmaq*, cuya ubicación veremos más adelante.

Ahora bien, la *guaranga* de Xica no era otra que la de Jicamarca que comprendía además de los Xica a los Chacalla o Chaclla, a los Cullata o Collata y a los Punan. Todos ellos conformaban lo que en diversos documentos nombraban a veces como Chaclla o Jicamarca (Cuadro 2).

Dos *guaranga* distintas de los mismos Yauyos fueron las de Carampoma y de Casta y sus terrenos se extendían por el valle alto del Chillón en la margen izquierda del río. Debido a la baja demográfica posterior, estas *guaranga* fueron reestructuradas. Por último un grupo más de los Yauyos compuesto por las *guaranga* de Picoy y de Yaucha se establecieron en Matucana y en San Mateo de Huanchor (Espinoza 1983–1984).

Sobre la *guaranga* de Jicamarca hay numerosa documentación en el Archivo General de la Nación, y en tiempos actuales ellos integran una Comunidad Campesina reconocida en 1929 (Expediente Nro. 97–70, Ministerio de Trabajo; para sus linderos ver AGN Escritura nro. 1175, año 1883; Tuve oportunidad de consultar algunos documentos cuando fueron remitidos al archivo para su transcripción para el juicio que la comunidad sostenía entonces con el gobierno). La Comunidad de Chaclla fue reconocida separadamente de la de Jicamarca en 1933 (Expediente Nro. 1550 Ministerio de Trabajo Archivo de Comunidades que desapareció, llevado por SINAMOS y hasta hoy se desconoce su paradero).

En un juicio entablado en 1973 por los miembros de la comunidad de Jicamarca, reclamaban en nombre propio todos sus antiguos dominios, y de los Chaclla, a base de títulos supletorios. La extensión de tierras pretendidas abarcaba aproximadamente cien mil hectáreas (muchas eriazas) e incluían los distritos actuales de Ate, Carabayllo, Chosica, el cerro San Cristóbal, la zona del polígono de Tiro y el cuartel del ejército, Huarhuar, la quebrada del Río Seco; todas estas tierras situadas en los valles de Lima y el Chillón.

Los Canta

Las noticias que tenemos sobre los Canta en Justicia 413 se enriquecen con el "Cuaderno" de la visita a dicho lugar, ordenada por el Licenciado Pedro de La Gasca en 1549, seguida de una revisita en 1553 (ver Rostworowski 1978, Apéndice II).

A través de esos testimonios sabemos que el curacazgo de los Atavillos se situaba en la banda izquierda del río Chancay hasta la margen derecha del río Chillón y se dividía en alto y bajo. En tiempos prehispánicos se trataba de una macro-etnía compuesta por el señorío de los Atavillos propiamente dicho, además de los Canta, Piscas, Huamantanga y Secos o Xecos. Por sus riquezas este Señorío despertó la ambición de Francisco Pizarro que quería tomarlo para sí, además de las encomiendas que ya poseía en Yucay, en el Cusco, Huaylas, Chimú Conchuco y de las dos pequeñas posesiones de Lima y Chuquitanta.

Al fallecer Pizarro, los Atavillos fueron divididos en varias encomiendas. Los partidarios de Almagro entregaron Canta a un tal Saavedra quien la gozó sólo unos meses, y más adelante pasó al hijo del Marqués, don Gonzalo.

A partir de 1549, Canta fue encomienda de Nicolás de Ribera, el mozo.

Con el advenimiento de la Colonia, los Atavillos perdieron su antigua importancia cediéndola a Canta. Una de las razones del cambio se debió a su proximidad con la ciudad de Los Reyes y el estar situada en la ruta que unía Lima con Cerro de Pasco y Huánuco. Las dos rutas de penetración de la costa central con el interior del país fueron la de Canta por un lado, y el camino de Lima por Sisicaya, Huarochirí y Jauja que se dirigía al Cusco. La ruta central por la cuenca del río Rímac es un camino moderno.

La primera visita a Canta se realizó en el asiento de Quivi, el 29 de julio de 1549, con reunión de los curacas, pero los visitadores no recorrieron el señorío pues prefirieron permanecer sólo en Quivi; el acto estuvo a cargo de Diego Flores, representante de Nicolás de Ribera. En esta primera visita, se contaron ocho ayllus canteños, número que sigue el patrón y modelo andino. La segunda visita se inició en el pueblo de Mallo o Mayo, el 18 de junio de 1553 y duró veinte días. Los encargados de llevarla a cabo fueron Fray Antonio de Figueroa, de la orden de los dominicos y el clérigo Gómez Hernández. Más prolijos que los primeros, los religiosos recorrieron todos los pueblos; y es interesante cotejar ambos documentos.

Ambas visitas incluyen los nombres de los curacas y de los pueblos que gobernaban. Los ayllus o parcialidades que comprendían el Señorío de Canta fueron: Canta, Racas, Rocha, Lachaqui, Copa, Esquibamba, Carua y Visca.

Los linderos prehispánicos del curacazgo de Canta no coinciden con los límites de la actual provincia. Sus tierras se situaban en la margen izquierda del río Chillón en forma alargada, abarcando varios pisos ecológicos, desde el cálido *chaupi yunga* a la frígida *puna*. En la gran altura se ampliaban sus territorios y se extendían tras la cordillera de La Viuda hasta Caruacayan, por la cuenca del nacimiento del río Chillón y colindaba con la hacienda colonial de Corpacancha, situada en la provincia de Lima (AGN Títulos de Propiedad, cuaderno 500, año 1774).

En época inca los Canta eran vecinos del pequeño curacazgo yunga de Quivi, y es entonces cuando recibieron una porción de tierras, tras la derrota sufrida por el Colli Capac de Collec, al cual estaban sometidos los Quivi. Por el sur colindaban con el río Arahuay que los dividía de los Chaclla.

La nota llamativa de la organización de los Canta radica en la existencia de dieciséis pueblos habitados sólo temporalmente para realizar en ellos ciertas faenas específicas y comunales. Cuando terminaban el trabajo, las aldeas quedaban desiertas, rotando sus ocupaciones según un plan previsto de antemano.

Ambas visitas mencionan este procedimiento, y la más temprana es la que proporciona mayores noticias: posteriormente las imposiciones coloniales deterioraron el sistema. El visitador de 1549, nombra los 16 villorios, además de los pueblos declarados para cada ayllu y que pertenecían a sus respectivas parcialidades. Ellos eran:

Paron Marca-Cambis
Quiby chacara
Caruacaya-obejeros
Chacalla-paxareros
Cancha Cancha-alpargateros
Choccha Calla-despoblado
Apio chacara
Aragaco-alpargateros
Guaro chacara
Yaca chacara
Achim-olleros
Cato-despoblado de mitimaes
Chochacara
Copi chacara
Mallo chacara
Cazmo chacara

De la lista mencionada se observa una división en las funciones de las aldeas temporales; ocho son nombradas como chacras contando con Chochacara, lo que indica que en ellas se ocupaban de un determinado cultivo, posiblemente con un sentido religioso. Lo único que sabemos sobre esos campos es que se hallaban en distintos medio-ambientes y que eran trabajados en común por todos los ayllus, sin precisar si los productos pertenecían a las parcialidades o si sólo se beneficiaban de ellas los curacas o las huacas.

En los documentos sobre la extirpación de la idolatría en Cajatambo del siglo XVII, es frecuente encontrar la noticia de que cada *huaca*, según su importancia disfrutaba de tierras propias, cultivadas por el común de la gente de la población. Sus frutos se destinaban para la celebración de su fiesta, sacrificios a la *huaca* y preparación de bebidas para los fieles.

Los otros villorios servían para ejecutar trabajos artesanales, lo que demuestra que no existían grupos de artífices especializados. La primera aldea llamada Paron Marca según noticias contenidas en el manuscrito, servía para la manufactura de textiles y se reunían para preparar ropa fina de *cumbis*. Los visitadores informaron que, además de tejer ropa en Pariamarca, también hacían lo propio en cada *ayllu* para sus necesidades personales.

Con el trabajo de campo llegamos a ubicar algunos de los villorios indicados en los documentos; y otros quedaron por investigar, pues no subimos a la *puna* y, nos remitimos a nuestra publicación (1978). Sólo mencionaremos aquí los lugares indicados en el testimonio de *Justicia 413*.

Una aldea especializada cuya particularidad es bastante sorprendente es la de Chacalla, dedicada a los "paxareros." Es posible que estos Chacallas fueron *mitmaq*, cuya ocupación era preparar *charqui*, carne deshidratada, como producto de la caza. En la *visita* realizada al *curacazgo* de Guancayo, lugar situado río abajo en el valle del Chillón y colindante con el de Quivi ellos manifestaron que en tiempo del Inca sus habitantes entregaban como tributo unas diez *"petaquillas"* de pájaros secos hechos *charqui* pequeño (Espinoza 1963; Rostworowski 1972a).

También Guamán Poma (1936 foja 205, 207) en la sexta y

séptima *visitas* de los varones menciona la cacería de aves, palomas y perdices para la preparación de carnes deshidratadas. Las perdices abundaban en la región de Canta; y en la *Visita* de 1549, los naturales dijeron que a Nicolás de Ribera le entregaban 400 perdices de tributo.[1]

En cuanto al lugar denominado como Chacalla, un informante en Arahuay indicó que se encontraba en una *quebrada* a la derecha del pueblo de Licahuasi.

Entre las ocho chacras especiales, se hallaba la de *Quiby chacara*, nombrada por los visitadores como un pueblo de *yungas*, en el cual todas las *parcialidades* de Canta estaban representadas. Se trata de un típico caso de enclave o de "archipiélago" serrano establecido en tierras costeñas, después de una derrota bélica. La particularidad de la zona de Quivi era, como veremos más adelante, y con más detalle, la de ser una región especial para el cultivo de plantaciones de cocales, además de producir ají (*Capsicum*) cultivos imposibles o difíciles en las tierras altas.

Sin embargo, los Canta disfrutaban de otras tierras adecuadas para la coca que posiblemente eran suyas antes de la conquista inca; se trataba del pueblo de Cararuayllo situado en la parcialidad de Esquibamba, en la *quebrada* baja de Arahuay, habitada también por gente *yunga* que les beneficiaba la coca (Rostworowski 1978: 255).

Tal era la situación de los Canta en 1549, antes del inicio del juicio con los Chaclla. Para comprender mejor las circunstancias en torno al litigio, es forzoso señalar lo que sucedía con los pueblos costeños en el mismo valle.

Los *Yungas* de Collec

La existencia del antiguo *señorío yunga* de Collec o Collic llamado Collique por los españoles había permanecido ignorada hasta nuestras investigaciones, pues las *crónicas* no dan cuenta de su presencia.

Una serie de manuscritos dieron cuenta de la situación sociopolítica del valle del río Chillón en tiempos inmediatos a la conquista Inca a la región (Rostworowski 1972, 1977). La historia de Collec la fuimos descubriendo afanosamente, primero, en los documentos de archivos y, luego, en árduo trabajo de campo, buscando los pueblos, bocatomas, fuentes, acequias, fortificaciones, linderos e hitos nombrados en los muy diversos testimonios. En las salidas al valle contamos con la entusiasta colaboración de Josefina Ramos de Cox, Directora del Seminario de Arqueología de la Universidad Católica, tempranamente fallecida.

Antes de la conquista inca a la costa, el *curacazgo* de Collec se extendía desde el mar a lo largo del valle en dirección a la sierra, incluyendo el *curacazgo* de Quivi. Numerosos pequeños *señoríos* estaban comprendidos dentro de sus fronteras, cuyo jefe era el *Colli Capac*. La macro-etnía *yunga* comprendía en el valle bajo, a Collec, Chuquitanta, Seuillay, Comas, Sutca y Carabayllo, mientras río arriba se situaban en la margen izquierda Sapan, Guancayo y Quivi y, en la banda derecha se contaban Macas, Guarauni y Missai (Rostworowski 1977; Trimborn 1970). La última fase del Intermedio Tardío fue el tiempo de auge del Señorío de Collec y, sus posesiones comprendían también una parte del valle de Lima. En efecto, un testigo de la probanza de don Gonzalo, curaca de Lima en 1555, manifestó que un indio *Auca* (vale decir soldado o enemigo) llamado Colli se adueñó por la fuerza de cierta parte de las tierras de los Lima. Dichos sucesos tuvieron lugar en tiempo del padre del viejo Taulichusco, curaca del *señorío* cuando la fundación de la ciudad de Los Reyes en 1535. El testigo añadió que: "de los yndios de Colli ya no hay memoria" (Rostworowski 1981–82, Apéndice: 120).

De tiempo atrás los Collec sostenían también encuentros con los Canta que ambicionaban sus tierras de coca. Un testigo narró uno de aquellos ataques, cuando los Canta fuertemente armados bajaron por el cauce del río, pasaron el asiento de Quivi sin hacer daño a su curaca Chaumecaxa y avanzaron cuatro leguas en dirección de Collec. Manifestó el testigo que en dicha oportunidad consiguieron anexarse el pueblo de Quivi, cedido gentilmente por el *Colli Capac* (foja 211r); hecho poco probable puesto que el jefe costeño se sentía lo suficientemente fuerte como para oponerse a los Incas.

Sin embargo y a pesar de la enemistad entre costeños y serranos, en tiempos de sequía olvidaban rencores y odios y todos juntos se dirigían a la "sierra de la nieve" para ejecutar trabajos que permitían conducir el agua hasta el río Quivi (foja 206r, 286v).

Poco es lo que conocemos sobre las macro-etnías en el Tahuantinsuyu y el rol que ejercían en el Estado, aunque un hecho fundamental es la presencia de curacas de menor jerarquía dominados por señores más poderosos. Faltan aún mayores investigaciones sobre los lazos y obligaciones que unían entre sí a todos los jefes menores con sus superiores. El documento de *Justicia 413* suministra ciertas noticias que analizaremos al tratar los datos que nos ofrece el expediente.

En pocas palabras hemos mencionado la situación de la costa central cuando se hicieron presentes los ejércitos de Túpac Yupanqui. El Inca ofreció al *Colli Capac* el establecimiento de la reciprocidad que ataba a los señores de las macro-etnías con el *Sapa Inca* del Cusco, pero el señor de Collec, habituado a resistir los ataques de sus vecinos, resguardado como lo estaba en su fortaleza-palacio, circundado por una alta muralla, se negó a todo trato. Dentro del periplo de sus defensas disponía de

[1]En el Perú de hoy el término "perdiz" se refiere a una de las numerosas especies indígenas de *tinamous* (*Nothoprocta* spp.), pues la verdadera perdiz proviene del Viejo Mundo. [Nota de la Editora]

amplias tierras y de dos fuentes de agua que le permitían cultivar sus campos. Un largo asedio no le inquietaba, así que decidió oponerse a los nuevos conquistadores.

No conocemos los detalles de la lucha, de su derrota, ni de su posterior muerte. En su lugar el Inca nombró a un curaca de la categoría social *yanacon yanayacu* (Rostworowski 1977). Los jefes de dicho *status* no formaban parte del sistema de la reciprocidad y era una manera, para el soberano, de librarse de los contínuos "ruegos" y "dádivas" que el modelo llevaba consigo; y obligaba permanentemente al Inca a renovar estos pedidos. Más aún, según Cabello de Valboa (1952:371) Huayna Capac dió preferencia a los curacas *yanayacu* después que los nobles cuzqueños lo dejaron caer de sus andas, durante la guerra contra los Carangues en el norte.

La derrota de los Collec fue el preludio de su aniquilamiento y del fin de su *señorío*; el castigo del Inca debió diezmar su población, que no pudo sobreponerse posteriormente a la baja demográfica que siguió a la conquista española. Numerosas fueron las causas, en tiempos coloniales, de la disminución poblacional en la costa, y entre éstas se puede nombrar las epidemias, las guerras civiles entre españoles y la edificación de las ciudades virreinales. A consecuencia de todo aquello, los habitantes de Collec desaparecieron, y los diversos censos y *visitas* muestran a una muy disminuida población.

El 13 de mayo de 1571, Juan Martínez Rengifo realizó una visita a los naturales de Collec por orden del virrey Toledo; en dicha ocasión el visitador fue informado de la antigua existencia de diez pachacas, de las cuales sólo quedaban en total quince tributarios. El 18 del mismo mes, Martínez Rengifo continuaba río arriba su visita a los pequeños *señoríos* de Guancayo, Macas y Guarauni (Espinoza 1963). En 1586, quince años después, en una averiguación emprendida por orden del virrey Fernando de Torres Portugal, conde del Villar, de los antiguos habitantes de Collec no quedaba más que el curaca don Hernando Nacar y dos mozos. Dada las circunstancias, es comprensible que no se supiera nada de la antigua existencia de este *señorío*.

Los *Yungas* de Quivi

Las noticias que poseemos sobre Quivi provienen del documento de *Justicia 413*. Según el decir de todos los involucrados en el juicio, los Quivi era gente *yunga* y, anteriormente a la conquista Inca, formaban parte del *curacazgo* de Collec, cuyo jefe llevaba el nombre genérico de *Colli Capac* y tenía un *status* importante; los testigos afirmaron que lucía "las orejas largas." Como lo señalamos más arriba, la macro-etnía de Collec comprendía a varios pequeños *curacazgos* y, según el decir de los testigos, todos sus habitantes desde Quivi hasta el mar eran gente *yunga* (Rostworowski 1972, 1977).

El curaca de Quivi tributaba "coca, maíz, ají y otras cosas" al *Colli Capac* y no sabemos si esos productos eran la entrega del fruto de las cosechas de las tierras poseídas por *Colli Capac* en Quivi o un gesto de subordinación y de reconocimiento de un inferior hacia su superior.

Cuando aparecieron los ejércitos incas, el *Colli Capac* decidió defenderse; y en esa oportunidad el curaca de Quivi acudió con gente armada al socorro de su señor (foja 198v). Estas noticias permiten suponer que ambas situaciones representaban los términos de subordinación entre jefes de distintas jerarquías y confieren una interesante luz sobre los lazos de reciprocidad habituales en el ámbito andino. Sin embargo la relación entre el curaca de Collec y el de Quivi se veía empañada por las esporádicas incursiones de los serranos a tierras *yungas*. El documento que comentamos trae noticias de varios de estos ataques y, por supuesto no en vano la sede del *Colli Capac* estaba fortificada.

La victoria de Túpac Yupanqui trajo como consecuencia para Quivi, la instalación en sus tierras de cien *mitmaq* Chaclla dedicados a cultivar coca para el soberano, además de servir de *chasque* o mensajeros (foja 284r). Su número se aumentará varias veces, tanto por el mismo Túpac como Huayna Capac y Huascar que tomaron para sí una mayor extensión de tierras y designaron a nuevos *mitmaq*.

Para los efectos de los nuevos linderos, llegó del Cusco un orejón llamado Apar Yupanqui (foja 149r) y las señales anteriores fueron removidas de su lugar y trasladadas de Chuquicoto a Judcunga, distante media legua de Quivi (foja 211r). Es posible que las tierras entre el hito anterior y el nuevo fueran otorgadas a los Canta que se aproximaron así al lugar de Quivi (Rostworowski 1978).

El motivo de la introducción de los Chaclla al valle se debió a que todos los Yauyos eran amigos, criados y privados del Inca; noticia confirmada por los testimonios de los informantes de Avila (1966, 1980). Esta medida motivó el cese de las tradicionales hostilidades entre los Quivi y los Canta y, entre ambos, hicieron frente común contra los intrusos.

En el documento se hallan referencias de que los Chaclla pertenecían a la región de Mama y no debían intervenir en el valle que no les correspondía (Mama se situaba en lo que actualmente se conoce como Ricardo Palma, cerca de Chosica, foja 189v).

Los cuatro *ayllus* que formaban el *curacazgo* de los Chaclla estaban representados en Quivi en calidad de *mitmaq*. Se trataba de enclaves puestos por el Inca de acuerdo con la organización serrana de archipiélagos (foja 121v).

Después de la derrota *yunga* ante el Inca, el curaca de Quivi llamado Chumbiquivi retornó a sus tierras y poco después fue acusado de conspirar a través de los hechizos de una *huaca* contra la salud de Túpac Yupanqui (fojas 198r, 198v). Enterado el soberano de las maquinaciones del curaca, envió por él y fue conducido junto con sus hombres al Cusco, donde les dieron muerte. Sólo unas pocas mujeres y niños se salvaron, motivo por el cual era escasa la población *yunga* de Quivi (ver la declaración del curaca de Quivi, Felipe Taulichumbi, foja 207v).

Más adelante, durante el juicio, los Canta achacaron estos episodios a los Chaclla, hecho poco probable por ser ellos amigos y criados del soberano y favorecidos con las nuevas tierras (foja 219v). Quizá sea que los Canta quisieron dar una imagen adversa de los Chaclla.

Tal era la situación reinante en el valle alto del río Chillón—y aquí haremos la aclaración de que un río en tiempos prehispánicos no llevaba un sólo nombre desde su nacimiento hasta su desembocadura, sino que cambiaba de apelativo, según las tierras que atravesaba. Por eso se decía río de Quivi o río de Canta, de acuerdo con los lugares por donde discurría; Chillón es una designación moderna.

Los episodios anteriores al inicio del juicio

Al desaparecer el gobierno cusqueño después de la conquista hispana, mientras se sucedían las guerras civiles entre los españoles y hasta que se asentó el orden virreinal, los Canta y los Chaclla se mataban entre ellos, movidos por intensas pasiones en las que cada cual alegaba sus derechos a las plantaciones de coca.

En esas diferencias y querellas estaban cuando estalló la rebelión de Manco II y el Inca ordenó poner cerco a Los Reyes. Los canteños se plegaron a los cusqueños no así los Chaclla por oposición y enemistad con sus rivales. Cuando aparecieron los generales Ylla Topa y Quiso Yupanqui, los Canta acusaron a los Chaclla de tener "dos corazones," uno a favor de Manco II y otro al servicio de los hispanos.

En represalias, por no querer los Chaclla unirse a los Canta contra los españoles, Guaman Yauri mandó traer con engaños a Vilca Poma, curaca de Chaclla, y lo mató a golpes de porra en la espalda (fojas 24v y 25r); aunque otros testigos afirmaron que sólo fue herido. Luego el curaca de Canta ordenó a los incautos Chaclla traer bastimentos para el Inca y, cuando llegaron a Canta los metieron en un corral y mataron a los 150 cargadores. Más adelante los canteños se excusaron ante Ylla Topa por la masacre so pretexto que los Chaclla habían enviado mensajeros a los españoles para dar aviso de la presencia de los jefes inca (fojas 150r, 150v). Un testigo llamado Juan Merlo Chauca natural de Huarochirí, afirmó que los españoles mataron a Quiso Yupanqui durante el asedio de la ciudad.

En 1549, cuando la estadía en el Perú del *licenciado* Pedro de La Gasca, los *encomenderos* Nicolás de Ribera el mozo—de los Canta—y Francisco de Ampuero—de los Chaclla—quisieron imponer la paz entre los curacas y decidieron que los Chaclla vendieran sus tierras a los Canta por el precio de doscientas "ovejas de la tierra," vale decir llamas.

Los Chaclla se resistieron al trato pues no entendían lo que significaba *vender* y no quisieron recibir las "ovejas" que les proponía Ampuero. A pesar de la oposición del curaca Francisco Vilca Poma de Chaclla, el acuerdo se cumplió. Ante la negativa del cacique de hacerse de las llamas, Ampuero ordenó se las llevasen a su casa en Los Reyes, donde vendió unas y envió otras a la carnicería. Más tarde, presionado por confesar su acción, admitió habérselas llevados junto con el padre Godoy.

Enterado de los sucesos, el curaca de la macro-etnía de Huarochirí, llamado Ninavilca envió por Vilca Poma para que explicara la situación y reprenderle por acciones que le dejaban a él y a su gente sin sus tierras. Además el curaca de Huarochirí tenía chacras propias en Quivi y amenazó que iría con un ejército compuesto por todos los Yauyos contra los Canta (foja 16v). Pero los tiempos habían cambiado y los jefes étnicos no podían hacer cumplir la justicia por sus propios medios, es así que los desesperados Chaclla buscaron la manera de abolir el convenio e iniciaron el juicio.

El juicio

Los Chaclla iniciaron el litigio ante la Real Audiencia de Los Reyes en 1558, juicio que se prolongó hasta 1567 con el fallo a favor de los Chaclla. Los oidores basaban su veredicto sobre un hecho que debía ser contemplado por la Corona, en el sentido que las tierras en posesión de *mitmaq*, al momento de la conquista española, permanecían inamovibles. No se podía sentar el precedente de mudanzas en el derecho colonial, de lo contrario hubiera dado pie a un sin número de juicios y litigios que llevarían al país al caos.

Los Canta informados a través de sus consultores españoles de las Ordenanzas virreinales, trataron de negar el hecho de los Chaclla eran *mitmaq* puestos por los Incas en Quivi, e insistieron en su intromisión en el valle a consecuencia de las ceremonias de la *capacocha*. El documento trae en efecto invalorables noticias sobre estos ritos, que analizaremos por separado.

Tardíamente intervinieron los Quivi, o sea los *yungas*, alegando que las tierras eran suyas y que fueron despojados de ellas por los Chaclla. Hasta entonces habían permanecido como testigos de los Canta, sin reclamar sus antiguas posesiones preincaicas. Es posible también que fuese una estratagema aconsejada por el procurador español, para indicar que anteriormente las tierras fueron primero de los Quivi.

Son interesantes los testimonios de los curacas y principales *yunga*, como don Felipe Taulichumbi de Quivi (foja 207v), el curaca don Antonio Chumbiquivi de Sapan—pequeño *curacazgo* colindante con el de Guancayo (foja 54r), don Pedro curaca de Guancayo (foja 183r) o don Francisco Yauri, curaca de Collique (foja 54r).

A pesar de las largas *probanzas* y del empecinamiento de los Canta, ellos no llegaron a probar sus derechos a las tierras en litigio (foja 383r). El presidente y oidores de la Real Audiencia pronunciaron su sentencia ante los procuradores de los naturales; y consistió en que los Canta debían devolver las tierras a los Chaclla y estos últimos retornar las doscientas "ovejas." No contentos con el veredicto, los Canta continuaron presentando escritos y pidieron más *probanzas*. En tales circunstancias, fray Gaspar de Carbajal aconsejó a la Real Audiencia imponer el perpetuo silencio a las partes, pues los indígenas, con sus pasiones, hacían caso omiso de las provisiones reales (foja 415v).

Más adelante el virrey Toledo ordenó terminasen los pleitos y causas entre indígenas en la Audiencia, por que al fin de cuentas, quedaban los naturales muy pobres y gastados en los juicios (ver Rostworowski, en prensa, Tasa de Capachica de 1575).

En el mes de julio de 1567, estando en audiencia pública, se notificó a Juan de Arrendolaza y a Diego de Ocampo a que ambas partes exhibieran moldes, vale decir maquetas de sus tierras (foja 365v).

Don Felipe Taulichumbi, curaca de Quivi dijo haber mostrado en otra oportunidad una maqueta "en forma gruesa y demostrativa de sus tierras de las que fueron despojados" (fojas 365r, 367r).

Por su parte los Chaclla enseñaron pinturas; pero se les pidió un molde del valle, a lo que alegaron haber traído anteriormente a la Audiencia un modelo confeccionado de acuerdo con el decir de los ancianos, pero que ya se había hecho pedazos (fojas 365r, 365v).

Estas informaciones sobre la costumbre indígena de confeccionar moldes de barro es sumamente interesante porque confirma las noticias dadas por Sarmiento de Gamboa (1943, cap. 39) sobre la costumbre de preparar maquetas de un valle conquistado para que el soberano se enterara de la situación e hiciese en él los cambios que le parecían necesarios. En referencia a las pinturas, Betanzos cuenta un episodio en que los nobles cusqueños se presentaron ante Yupanqui para ofrecerle la *mascapaicha*, y le encontraron ocupado pintando las alteraciones que deseaba introducir en el Cusco (1968, cap. 10 y 16).

De acuerdo con la ejecutoria real emitida por la Real Audiencia, se procedió en el pueblo de Quivi—28 de junio de 1567—a la entrega por el curaca de Chaclla, don Juan Bautista Caxa Guaraca, de las "ovejas" para poder recuperar sus tierras. Para frustrar el procedimiento, el curaca de Canta no apareció y sus numerosos principales presentes al acto dijeron encontrarse en Los Reyes.

Al querer el *alguacil* efectuar la entrega, se negaron los canteños, alegando la ausencia del curaca y no poder intervenir en la transmisión. Ante el obstáculo surgido, el *alguacil* hizo constar la negativa de los principales y ordenó dejar las llamas en Quivi. Luego se procedió a la toma de posesión de las tierras por el curaca de Chaclla, y según la costumbre española y en demostración de propiedad, don Juan Caxa Guaraca se paseó por los campos, cortó ramas de los árboles, lanzó piedras, arrancó terrones. El *alguacil* presente, ordenó se levantaran diversos mojones de piedra y barro para marcar los hitos. El curaca don Juan Caxa Guaraca declaró ante el escribano que tomaba también posesión de las aguas que regaban las tierras de Quivi, para no ser inquietado más adelante en su derecho (foja 337v).

Apenas había transcurrido una hora de la ceremonia, cuando los canteños irrumpieron en la escena dando de gritos y profiriendo amenazas y, en rápida acción, desbarataron los mojones. Al oír la algarrabía, salieron del tambo donde se encontraban, el *alguacil*, el escribano y el intérprete y constataron la actitud de Pedro Guari, Cristóbal Caruavilca y Gerónimo Caruavilca seguidos de otros indígenas. Por desacato a la autoridad, los principales fueron apresados y conducidos a Los Reyes, a la cárcel de la Real Corte (foja 338v).

En la Audiencia se sucedieron nuevos escritos y confesiones de los canteños presos en Lima; y de las declaraciones se entiende que no solamente habían derribado los hitos; además habían quebrado los canales hidráulicos, con la consecuencia de aniegos y perjuicios para los pobladores de Quivi (foja 348r).

El 20 de agosto de 1567, retornó el escribano a Quivi acompañado por Fray Gaspar de Carbajal, para que se cumpliese la entrega de las "ovejas" a los Canta (foja 386v). Al día siguiente, reunidos los litigantes, los Canta declararon que no habían recibido las "ovejas" ni sabían de su paradero; y añadieron que el rebaño no se componía de llamas sino de pacos y que estaban enfermos con *caracha* o sarna (*Sarcoptes scabiei, Psorotes comunis*).[1] Además exigieron a los Chaclla que fuesen en busca de los animales y los trajesen de vuelta (foja 387r). Siguieron largas discusiones y pretensiones de ambas partes y, el 27 de agosto, los litigantes terminaron por reunir un hato de "ovejas" y pacos que fueron atentamente examinados por el escribano en presencia de Fray Gaspar de Carbajal. Contado el rebaño constataron la existencia de 113 "ovejas" hembras y machos (foja 388v) de las cuales 18 tenían *caracha* y no se podían admitir. Los Chaclla ofrecieron reemplazarlas, pero los Canta discutían exigiendo recibir las doscientas llamas juntas, tal como las habían remitido a Ampuero en 1549.

En varios escritos los indígenas pedían al preceptor que se recomendase en los autos que los canales de riego de Quivi no fuesen dañados porque en el acaloramiento de las discusiones quebraban las acequias o desviaban el curso de las aguas (foja 419r). Ambos bandos se hostigaban mutuamente cortando los cursos del agua, arrancando los cocales y destruyendo las cosechas.

[1] Es obvio que los españoles no sabían que es sumamente difícil criar alpacas en las tierras bajas. [Nota de la Editora]

El 30 de agosto, el padre Gaspar de Carbajal, ante la presencia del escribano, ordenó se repusiesen los cincuenta mojones que deslindaban las tierras ganadas por los Chaclla (foja 396v) y se procedió a una nueva entrega de las tierras en litigio, siendo esta vez contradicho por don Felipe Taulichumbi y por Alonso Caguar, principales *yunga* de Quivi.

La tardía intervención de los *yunga* era una nueva treta para prolongar el juicio y estorbar el otorgamiento de las tierras a los Chaclla. Los costeños acusaban a fray Gaspar de parcialidad por el enojo que tuvo en una oportunidad con un principal de Canta (foja 412r).

Se repitieron en esa oportunidad los destrozos ocasionados por los Canta, quienes rompieron los canales y derrumbaron los hitos. Como sucediera anteriormente se hicieron otras informaciones sobre los daños causados (foja 429v).

El 21 de octubre, el escribano Pedro de Herrazti fue comisionado para constatar los daños ocurridos en Quivi, y los Chaclla se quejaron que los cocales fueron arrancados junto con numerosos árboles de pacay (foja 434r).

A pesar de todos los reclamos de los Canta, la Real Audiencia ratificó su veredicto de entregar las tierras de Quivi a los Chaclla. No contentos los Canta con el fallo, y movidos por sus ancestrales pasiones, decidieron apelar al Consejo de Indias en España. Esto, probablemente por consejo de sus representantes españoles, que asusaban a los Canta; en lugar de hacerles comprender el sentido del fallo. Es así que el 19 de diciembre de 1569, el juicio se presentó en Madrid y el 6 de marzo de 1570, Gaspar de Zárate entregó una petición en segunda suplicación de la *sentencia de revisita*, al Real Consejo (foja 7r).

Las plantaciones de coca de Quivi

Es preciso explicar el porqué las tierras de Quivi, pedregosas y poco extendidas despertaban, entre los naturales, tan grandes y apasionados deseos de posesión, expresados no sin odio y persistencia. El padre Gaspar de Carbajal, junto con fray Pedraza, realizaron en febrero de 1559 una información en nombre del rey y por mandato de la Real Audiencia. Ellos midieron las chacras en litigio: 2,000 pasos de largo por 300 pasos de ancho; y sus linderos eran, a lo ancho del río y el camino real de Lima, mientras por el oeste con una *quebrada* que bajaba al sesgo de la banda opuesta a Quivi en tierra de los Secos.

En dinero, Carbajal estimó su valor en 600 a 700 pesos, pero aclaró que los naturales las estimaban de otra manera por tratarse de tierras especiales para el cultivo de cocales (foja 131r).

La coca es un elemento básico en la cultura andina no sólo como masticatorio sino para las ofrendas rituales, las ceremonias a las *huacas*, en las curaciones y como adivinatorio. De no disponer de sus hojas en la zona costeña apropiada, había que conseguirla de las regiones selváticas, lejanas y de difícil acceso. De ahí la gran estima de la región del *chaupi yunga* o costa media, franja de tierras de los valles costeños de una elevación que iba de los 300 metros sobre el nivel del mar a unos 1,200 m., que se extendía desde Trujillo en el actual departamento de la Libertad, hasta Arica.

La importancia de tener acceso a tierras aptas para el cultivo de cocales se traduce en el testimonio de Juan Merlo Chauca, natural de Huarochirí quien declaró (foja 30r) haber sido enviado por su curaca Ninavilca a *visitar* las chacras de coca que dicho cacique poseía en Quivi. Esta noticia no sólo indica el valor de tales plantaciones, sino que el *Hatun Curaca* de una macro-etnía disponía de tierras en los *señoríos* subordinados a él.

En nuestra investigación hallamos una serie de documentos referentes a la existencia, en tiempos prehispánicos y el inicio del virreinato, de plantaciones de cocales en la costa media de los valles *yunga* (Rostworowski 1973 y 1977, para mayor información nos remitimos a esos trabajos). Sin embargo nos faltaba la confirmación botánica del tipo de coca que se cultivaba en tiempos prehispánicos en dicha zona.

En la investigación encontramos que la voz *coca* proviene del idioma aymara (*aru* o *jaqí)*. En quechua existían dos términos para la planta: *tupa* para la variedad costeña de hojas menudas, aromáticas y de menor contenido de alcaloides, y *mamox* para la coca cultivada en la región selvática.

En una *visita* al Museo Regional de Ica, el señor Alejandro Pezzia tuvo la amabilidad de permitirme abrir un cierto número de *chuspa* o bolsitas para coca halladas en excavaciones arqueológicas en Ica, que pertenecían unas al Horizonte Medio y otras al período Inca. Incluso, nos llegó a entregar parte de su contenido.

Tiempo después conocí a Timothy Plowman, botánico de la Universidad de Harvard, dedicado entonces al estudio de las variedades de coca. En sus investigaciones, Plowman (1979a, 1979b) reconoció la coca de la costa peruana como perteneciente a la variedad de *Erythroxylon novogranatense var. truxillense* a diferencia de la especie selvática *Erythroxylon coca var. coca*.

Como noticia interesante añadiremos que en una comunicación recibida (carta), Tom Dillehay me manifestó haber hallado en sus excavaciones en el valle medio, en Guancayo Alto, hojas de coca cuyos datos de Carbono-14 indicaron pertenecer al período Chavín. También en Quivi Viejo, excavó restos de cal y de hojas de coca fechados de 750 a. C. Estas noticias indican una muy antigua tradición y presencia de esta variedad de coca en la costa.

Una última explicación se refiere al porqué los indígenas, en sus arranques de odio quemaban o cortaban los árboles de pacay y de guavas. En la tesis de Anibal Orihuela Noli (1953) sobre el cultivo moderno de la coca en la zona del departamento de La

Libertad, Trujillo, nos enteramos que las plantaciones de coca crecen bajo la sombra protectora de los pacay (*Inca feuilli*).

La desaparición de la mayor parte de las plantaciones de cocales en la costa se debió al celo evangelizador de los eclesiásticos que veían en la coca un culto diabólico (Arriaga 1968).

La Capacocha

El documento de *Justicia 413*, es una fuente de información regional sobre lo que fue la *capacocha*, ceremonia que está lejos de ser esclarecida y que necesita mayor investigación.

Según Albornoz (1967:37) llamaban *capacocha* no sólo a los ritos y fiestas de ese nombre, sino a los vestidos de las *Huacas*. Cieza de León (Señorío 1943, cap. XXIX) le dedicó un capítulo y se refiere a un conjunto de ceremonias cuyo fin era el de reunir una vez al año todas las *Huacas* importantes del Tahuantinsuyu en la capital, acompañadas de un numeroso séquito y de servidores.

En la gran plaza de Aucaypata en el Cusco, se congregaban los ídolos y, en solemnes y especiales ritos, las *Huacas* emitían, cada una por su turno, sus oráculos. Profetizaban en torno a los resultados de las empresas guerreras del Inca, sobre su salud, la fertilidad de la tierra, el futuro de las cosechas, o si el año sería de abundancia o miseria.

Para la ocasión hacían grandes sacrificios y solemnes ceremonias, en las que inmolaban miles de "ovejas de la tierra." Pasado el tiempo de las festividades, las *Huacas* emprendían el regreso a sus pueblos, cargadas con las ofrendas que llamaban *capacocha*. Las *Huacas* que acertaban en sus augurios eran posteriormente premiadas y tenidas en mayor estima, no así las que se habían equivocado.

Cieza de León añade un dato curioso, es posible que referente al documento que comentamos, pues decía que poco después de establecida la Audiencia en Los Reyes y de haber retornado La Gasca a España, se iniciaron en Lima unos juicios que mencionaban los ritos de la *capacocha*. No cabe duda, por las fechas indicadas, que se trató del litigio entre los Canta y los Chaclla y seguramente tuvo resonancia en Lima. La pasión que en la disputa mostraron ambas partes por unas tierras poco estimadas por los españoles debió asombrar a este cronista.

Según otros, la *capacocha* comprendía sacrificios no sólo de llamas, sino humanos. Sarmiento de Gamboa (1943, cap. 13) cuenta que Manco Capac instituyó esos sacrificios para honrar la *huaca* de Huanacauri, después del nacimiento de Sinchi Roca. La ofrenda consistía en dos niños, macho y hembra. El carácter de sacrificio humano de la *capacocha* es sostenido también por Guaman Poma (fojas 228, 247, 251 y 259); Cabello de Valboa (1952), Betanzos (1968, cap. 11 y 17); Estete y Acosta (1942, lib. 5, cap. 4 y 7). En el documento que comentamos se menciona sólo sangre de camélidos mezclada con conchas molidas *chaquira*.

Duviols (1976) ha tratado bastante exhaustivamente el tema de la *capacocha* y a él nos remitimos para mayor información. Acertadamente dice que fueron unas de las ceremonias más solemnes de la vida del incario, ya que en ellas intervenía un gran número de individuos de todo el "imperio." Supone que existieron por lo menos dos tipos de *Capacocha*: unas *cíclicas* y otras *excepcionales*. Según el testimonio que publicamos, las dos últimas que se dieron en el Tahuantinsuyu y que pasaron por Quivi fueron la una para Huascar y la otra para su mujer la *Coya* (fojas 195r, 230v). Esta referencia muestra la importancia que tenía una soberana, en el aspecto socio-religioso, quizá como representante del género femenino.

Plantea Duviols diversos mecanismos de proyección radial o concéntrico, urbanos o territoriales de las *capacocha*. La partida del Cusco se hacía, según él, por los cuatro *suyu* (división espacial del Cusco), en una ceremonia parecida a la de la *Citua* (Molina 1943) cuyo fin era una purificación y echar los males y enfermedades fuera de la capital.

Las noticias referentes a la *capacocha* de *Justicia 413*, son puramente locales y tienen la ventaja de dar una información a nivel de la *provincia*. Fueron los canteños quienes narraron los episodios surgidos durante las *capacocha*, con el fin de explicar la introducción de los Chaclla en Quivi. Enterados por sus procuradores que la Corona no admitía cambios poblacionales entre los *mitmaq* y asumían el respeto a las posesiones de tierras tal como se hallaban al tiempo de la conquista, idearon dar otra explicación a la presencia de los Chaclla en las tierras en juicio (foja 63r).

En el documento se mencionan varias *capacocha* celebradas durante el gobierno de Huayna Capac, además de las dos efectuadas en tiempo de Huascar. El rito consistía en sangre de camélidos, y debieron ser frecuentes los sacrificios de los animales para que la ofrenda permaneciese líquida, ya que existía el peligro de que su contenido se derramara. Se le transportaba en pequeños mates u ollas y, cuando llegaban a lugares agrestes cercanos a una *huaca*, ponían el mate o la vasija en una honda y como quien lanza una piedra, la arrojaban al adoratorio. En caso que el encargado no acertara en su puntería, lo mataban en el mismo lugar. Igual suerte corría el mensajero si una parte de la ofrenda se volcaba al suelo, aunque fuese una gota.

Los canteños relataron cómo en distintos episodios, los Chaclla aprovecharon el hecho de portar los sacrificios, y pasaron los linderos señalados, adentrándose en tierras de los Canta en Quivi (foja 245v). Mientras llevaban los mates iban diciendo: "aparte, aparte, capacocha capacocha...hasta aquí llega mi capacocha, hasta aquí es mi tierra" (foja 250r).

Los Canta se enfurecían por tamaña desfachatez y, en una oportunidad, uno de ellos llamado Choqui Guaranga se abalanzó sobre el portador de los mates para impedirle el paso, y, en la embestida, se derramó la sangre. El castigo no se hizo esperar y el culpable fue ahorcado y enterrado en el lugar de los hechos (fojas 252r, 263r y otras).

Por esas noticias se sabe de la existencia de diversos tipos de *capacocha*, cubriendo distintas rutas y, quizá, objetivos. Un "ovejero" del Inca contó haber presenciado una *capacocha* estando en la *puna* de Pumpu y que era conducida desde el Cusco a Quito; la llevaban de término a término los diferentes grupos étnicos situados en la ruta. En el caso en cuestión, la traía gente *guanca*.

Los datos confirman el transporte de la ofrenda por los habitantes de una región y que llegados a sus límites territoriales, el sacrificio era entregado a los vecinos con quienes colindaban. En una ocasión los Yauyos de Chaclla la recibieron de los Jauja y llevaron la ofrenda hasta el mojón límite con los Canta.

Es difícil comprender por qué una *capacocha* pasaba por Quivi, lugar fuera del camino a Quito; a menos que se tratara de *capacocha* diferentes. Existe mención a una de esas ofrendas conducida hasta el *curacazgo* de Guancayo, cuyos habitantes la llevaron al mar. Otra noticia indica que era conducida a un río para lavar la sangre. Indudablemente se trataba de *capacocha* distintas unas de otras, con fines varios.

Es posible que algunas *capacocha* no eran destinadas al norte sino que se dirigían al mar, siguiendo la ruta del Cusco a Jauja y de ahí por Huarochirí a Quivi y Collec hasta el océano. Quizá los distintos caminos correspondían a diferentes *Huacas* que había que honrar; en ese caso Mamacocha (el mar) era una importante deidad en el ámbito andino y no sorprende que se le enviaran ofrendas desde el Cusco.

En fin, en estas líneas hemos señalado algunas noticias contenidas en el documento publicado. No hemos agotado los temas contenidos en el testimonio y es de esperar que el acceso a tan valiosa fuente sea un aporte para futuros estudios.

Secuencia cronológica de los sucesos en *Justicia 413*

El largo juicio comprende numerosas *probanzas* y averiguaciones a los testigos presentados por ambas partes. Las contradicciones y las distintas versiones exhibidas, así como las repeticiones, hacen difícil la lectura del documento. Para facilitarla, ofrecemos una secuencia cronológica de los sucesos esenciales.

1) En una primera época, anterior a la conquista de los Incas, el pequeño *curacazgo* de Quivi, formaba parte de la macro-etnía *yunga* de Collec.
2) Los naturales de Canta ambicionaban la posesión de las tierras de Quivi y, reunieron un ejército, pasaron por Quivi y avanzaron cuatro leguas valle abajo en dirección del pueblo de Collec (foja 211r). Los Canta afirmaron que en dicha ocasión el señor de Collec les cedió el *curacazgo* de Quivi.
3) Ante la amenaza del arribo de los Incas a la costa central, el señor de Quivi concurre con su gente de guerra al socorro del *Colli Capac*. Los costeños son vencidos por los Cusqueños; esta afirmación anula la versión canteña de una sumisión previa de los Quivi a Canta.
4) Con la presencia Inca, la situación del valle cambia; un grupo de Chaclla, de la macro-etnía de los Yauyos es instalada en Quivi en calidad de *mitmaq* para cultivar las plantaciones de cocales en tierras designadas para el Inca.
5) A raíz de la conquista española, la situación del valle sufre una segunda transformación. Atemorizados los Chaclla, quizá amenazados por los Canta, se retiran a sus serranías y abandonan sus tierras de Quivi.
6) Manco II se rebela, los Canta toman partido por el Inca, los capitanes cusqueños Illa Topa y Quisu Yupanqui pasan por Canta. Los Chaclla no quieren colaborar con ellos. Los Canta aprovechan para tender una emboscada al curaca Vilca Poma quien es asesinado junto con ciento cincuenta cargadores Chaclla.
7) En tiempos del Licenciado Pedro de La Gasca en 1549, los *encomenderos* Francisco de Ampuero, de los Chaclla, y Nicolás de Ribera-el mozo, de los Canta, encuentran una solución para terminar el conflicto. Intervienen y creen apaciguar la situación ordenando a los Chaclla la venta de sus tierras a los Canta por el precio de doscientas "ovejas de la tierra," vale decir llamas.
8) Los Chaclla se resisten y no entienden lo que significa "vender," no quieren recibir las "ovejas" que les impone Ampuero. El encomendero se beneficia con ellas y se lleva las llamas para sí.
9) El curaca de la macro-etnía de Huarochirí, convoca al curaca de Chaclla para que explique lo sucedido. Los indígenas se desesperan.
10) Los Chaclla inician el juicio en 1558, las *probanzas* se suceden, los testimonios se llenan de contradicciones. La Real Audiencia se pronuncia en 1567 a favor de los Chaclla bajo el concepto basado en que la posesión de la tierra de los *mitmaq* se mantenía tal como estaba al momento de la conquista española. Este fallo es explicable; ya que de haber sido de otro modo, daría pié a un sin fin de juicios entre indígenas, que la Corona no podría controlar.
11) En diciembre de 1569 el juicio se halla en Madrid, para ser presentado al Real Consejo por Gaspar de Zárate en calidad de petición en segunda suplicación de la sentencia de revisita.

Bibliografía

Acosta, Fray José de
1940 [1590] *Historia Natural y Moral de las Indias*. Edición preparada por Edmundo O'Gorman. Fondo de Cultura Económica. México.

Albornoz, Cristóbal de
1967 [1580] La Instrucción para descubrir todas las Guacas del Pirú y sus Camayos y Haziendas. *Journal de la Société des Américanistes*, Tome LVI-1, pp. 7–39. Paris. [ver Duviols]

Arriaga, Fray Pablo José de
1968 [1621] *Extirpación de la Idolatría del Perú*. Biblioteca de Autores Españoles, Vol. 209. Ediciones Altas, Madrid.

Avila, Francisco de
1966 [1598?] *Dioses y Hombres de Huarochirí*. Traducción del quechua por José María Arguedas, Instituto de Estudios Peruanos. Lima.
1980 ver Gerald Taylor

Betanzos, Juan de
1968 [1551] *Suma y Narración de los Incas*. Biblioteca de Autores Españoles, Vol. 209. Ediciones Atlas, Madrid.

Cabello de Valboa, Miguel
1951 [1586] *Miscelánea Antártica*. Instituto de Etnología, Universidad Nacional Mayor de San Marcos, Lima.

Cieza de León, Pedro de
1943 [1550] *Del Señorío de los Incas*. Ediciones Argentinas "Solar", Buenos Aires.

Dávila Briceño, Diego
1881 Descripción y Relación de la provincia de los Yauyos toda Anan Yauyos y Lorin...En *Relaciones Geográficas de Indias*, ed. M. Jiménez de la Espada, Tomo I, pp. 61–78. Madrid.

Dillehay, Tom D.
1976 *Competition and Cooperation in a Prehispanic Multi- Ethnic System in the Central Andes*. Unpublished Ph.D. dissertation, University of Texas, Austin.
1979 Prehispanic Resource Sharing in the Central Andes. *Science*, Vol. 204, pp. 24–31.

Duviols, Pierre
1967 "Un inédit de Cristóbal de Albornoz: La instrucción para descubrir todas las Guacas del Pirú y sus camayos y haziendas". *Journal de la Société des Américanistes*, Tome LVI-1, París.
1976 La Capacocha. *Allpanchis Phuturinqa*, Volumen IX, pp. 11–57, Cusco, Perú.

Espinoza Soriano, Waldemar
1963 La Guaranga y la reducción de Huancayo. *Revista del Museo Nacional*, Tomo XXXII, pp. 8–80, Lima.
1983-1984 Los Señoríos de Yaucha y Picoy en el abra del medio y alto Rímac (Siglos XV y XVI). *Revista Histórica*, Tomo XXXIV. Lima.

Estete, Miguel de
(ver Fernández de Oviedo, Gonzalo)

Fernández de Oviedo y Valdés, Gonzalo
1945 [1549] *Historia General y Natural de las Indias*. Editorial Guaranía, Asunción, Paraguay.

Guaman Poma de Ayala, Felipe
1936 [1615] *Nueva Corónica y Buen Gobierno*. Edición Facsimilar, París.

Molina, Cristóbal de
1943 [1575] *Fábulas y ritos de los Incas*. Los Pequeños Grandes Libros de Historia Americana, Serie 1, Tomo IV, pp. 7–92. Lima.

Murra, John V.
1964 Una apreciación etnológica de la Visita. *Visita Hecha a la Provincia de Chucuito Por Garci Diez de San Miguel en el Año 1567*, pp. 421–2. Casa de la Cultura, Lima.
1967 La Visita de los Chupachu como fuente etnológica. *Visita de la Provincia de León de Huánuco en 1562, Iñigo Ortiz de Zúñiga, visitador*, (John V. Murra, ed.) pp. 381–406. Huánuco, Perú.
1975 *Formaciones Económicas y Políticas del Mundo Andino*. Instituto de Estudios Peruanos, Lima.

Orihuela, Aníbal
1953 *Tesis sobre la coca*. Biblioteca de la Universidad Nacional Agraria, "La Molina", Lima.

Ortiz de Zúñiga, Iñigo
1967, Visita de la Provincia de León de Huánuco en 1562, 2 Volúmenes.
1972 [1562] *Documentos para la Historia y Etnología de Huánuco y la Selva Central*, ed. John V. Murra. Universidad Nacional Hermilio Valdizán, Huánuco, Perú.

Plowman, Timothy
1979a Botanical Perspectives on Coca. *Journal of Psychedelic Drugs*, Volume 11, pp. 103–117.
1979b The identity of Amazonian and Trujillo Coca. *Botanical Museum Leaflets*, Vol. 27, pp. 45–68.

Relaciones Geográficas de Indias
1881-1897 Editadas por Marcos Jiménez de la Espada. (Publicaciones del Ministerio de Fomento, 4 tomos, Madrid).

Rostworowski de Diez Canseco, María
1972a Las Etnías del valle del Chillón. *Revista del Museo Nacional*, Tomo XXXVIII, pp. 250–314. Lima.
1972b Breve ensayo sobre el Señorío de Ychma o Ychima. *Boletín del Seminario de Arqueología*, Número 13, Pontificia Universidad Católica, Lima.
1973 Plantaciones prehispánicas de coca en la vertiente del Pacífico. *Revista del Museo Nacional*, Tomo XXXIX, pp. 193–224. Lima.
1977 *Etnía y Sociedad. Costa Peruana Prehispánica*. Instituto de Estudios Peruanos, Lima.
1978 *Señoríos indígenas de Lima y Canta*. Instituto de Estudios Peruanos, Lima.
1978-1980 Guarco y Lunaguaná. Dos señoríos prehispánicos de la Costa sur central del Perú. *Revista del Museo Nacional*, Tomo XLIV, pp. 153–214. Lima.
1981-1982 Dos probanzas de don Gonzalo, Curaca de Lima (1555–1559). *Revista Histórica*, Tomo XXXIII. Organo de la Academia Nacional de Historia. Lima.
(en prensa) La Tasa de Capachica de 1575. *Revista Histórica*, Tomo XXXV, Organo de la Academia Nacional de Historia. Lima.
1985 Paper given at Dumbarton Oaks Symposium entitled "The Northern Dynasties: Kingship and Statecraft in Chimor", co-chaired by Michael E. Moseley and María Rostworowski.

Sarmiento de Gamboa, Pedro
1943 [1572] *Historia General llamada Indica*. Editores Emecé. Buenos Aires.

Spalding, Karen
1984 *Huarochiri. An Andean Society under Inca and Spanish Rule*. Stanford University Press, Stanford, California.

Taylor, Gerald
1980 *Rites et Traditions de Huarochirí*. L'Harmattan, Paris.

Trimborn, Hermann
1970 Las ruinas de Macas en el valle del Chillón, Lima. *Revista del Museo Nacional*, Tomo XXXVI, pp. 258–266. Lima.

Documentos Citados

Archivo del Museo Nacional de Historia s/n: Retasa de Huarochirí
Archivo del Ministerio de Trabajo: Expediente N°. 97–70
Archivo General de la Nación (sigla usada AGN)
Sección: Títulos de Propiedad, cuaderno 500, año 1774
Documentos sobre Jicamarca Escritura N°. 1175
Escritura N°. 1550 Comunidad Campesina de Chaclla.

PARTE III

Justicia 413, Archivo General de Indias, Sevilla
Año de 1567, Lima

Principió en 5 de octubre de 1558 y se concluió en 24 de Abril de 1570.

El pr[al], caziques e indios del rrepartimi[o] de Chacalla, encomendado en Fran[co] de Ampuero con:

El prinçipal, caziques e indios del rrepartimiento de Canta, encomendado en Nicolas de Rribera, el mozo a que salió despues el pr[al] y los ynd[o]s yungas de Quivi; intentando terceria de dominio sobre las tierras,

sobre

que se les restituyeran las tierras y chacaras de coca del valle de Quivi, junto al rio de este nombre el qual alegaban partia los terminos de ambos repartimientos que decian poseian de imemorial en tiempo de los Yngas, y antes de la entrada de los españoles, y que despues les usurparon los de Canta, matando en tiempo de Fran[co] Pizarro a su cazique pr[al] Vilcacapcha con mas de 150 indios que tenian en las dhas tierras, porque eran fieles al servicio del rrey y los españoles y no quisieron rebelarse con el Inga, que se alzó en aquel rreino: y que despues de esta violencia las han retenido a pretexto de un concierto, que decian haver hecho con el cazique; el qual no lo hizo sino violentado por su encomen-//dero Fran[co] de Ampuero que fue quien firmo y recibió 200 ovejas en que se transigieron: lo resistio hasta mas no poder, protesto a su modo; pero aun quando lo huviera consentido no pudiera perjudicar a sus indios ni sucesores, haviendo sido engañados enormisimamen[te].

Los de Canta alegaban que ellos eran los antiguos poseedores de las tierras, que caian en sus limites desde los tiempos mas remotos y rrepartim[o] de los Ingas; y que se las usurparon los de Chacalla en tiempo de Guascar Inga, sucesor de Guainacapac, que aquel mandó hacer por si y por su muger, en los que era rito salir unos ind[o]s con ciertos mates, u ollas pequeñas llenas de sangre de las reses que havian sacrificado, dando voces, y haciendo termino suio hasta el parage donde llegaban; y si algun indio hacia derramar aq.[a] sangre o parte de ella, o la derramaban los que la conducian, eran muertos de garrote, o ahogados y enterrados en el mismio sitio; y con esta malicia se aposesionaron de aquella tierra, intimidando con la pena a los de Canta, pues mataron a 5 de ellos (al f. 84 estan los nonbres) que intentaron resistirlo y derramaron la sangre: que despues de muertos ag[o]s yngas, se entraron otra vez en sus tierras, hasta los tpos del presid[te]. Gasca en que puesto pleito por los de Chacalla, mando dho presidente se concertasen, y con efecto a presencia de las partes, los encomenderos y muchos ind[o]s viejos hicieron un papel de // transacion que presentaron al fol. 6 en que por evitar litigios, aunque era suio el derecho, y no valian tanto las tierras, dieron 200 ovejas, con que quedaron en pacifica posesion, y otras cosas en su favor.

Despues el cazique o pr[al] inmediato de la poblacion de Quivi, salió intentando terceria de dominio fol. 313 y alegando ser suias en propiedad aquellas tierras y no poderle perjudicar el convenio que huviesen hecho los de Canta, en cuia jurisdicion se decia estaban (y no se averiguó porque en las visitas y demarcaciones ya los agregaban a Canta ya a Chacalla, y ya a unos y otros) ni tenian otras tierras ni habitaciones ni mas sementeras con que mantenerse.

De este pleito como negocio peculiar de indios y privilegiado, conoció la Rreal Aud.[a] de Los Rreyes; y alegado y probado lo expuesto por dhos ind[o]s de Chacalla y Canta (f. 14 y 32) con instancia de estos (f. 49) sobre mas prueba por no poder bajar a deponer los ind[o]s serranos a causa de la calor del verano, y quera de haver ido los de Chacalla a coger de mano armada sus sementeras (f. 5l); se dio sentencia en 10 de enero de 1559 (f. 60) absolviendo al cazique e indios de Canta de la demanda puesta por los de Chacalla.

Suplicaron estos (f. 6l) y alegado de nuevo con mas extension lo que queda dho y otras cosas por ambas partes; y pidiendo vistas de ojos, planes y //demarcaciones, se mandó por auto de 23 de diciembre del dho año (f. 64 v[to]) que cada parte nombrase por la suia persona que fuese a la vista de ojos, y a tomar los dhos de los indios antiguos comarcanos segun mas les conviniese; y fuese en su menor daño y perjuicio.

Seguido articulo sobre probar hechos nuevos y no sobre lo ya articulado (f. 67): incidencias sobre daños en las sementeras; recusaciones y oposiciones sobre es[o]s (f. 69 /88/y 280/desde el f. 67) y comisionados nombramy[o]s de lenguas o interpretes; y señalado para la averiguacion y vista de ojos al provincial de S[to] Domingo Fr Gaspar de Carbajal, se hizo esta ynform[on](f. III) y probaron de nuevo las partes hasta el folio 278 y se dio 2[a] sentencia en II de julio de dho a[o] de 1559 (f. 301), revocando la I[a] y condenando a los indi[o]s de Canta a que volviesen las tierras del pleito a los de Chacalla, restituyendo estos las 200 ovejas del concierto; y se mando dar egecutoria a los de Chacalla, con fianzas en 14 del mismo julio (f. 302).

Suplicaron los de Canta para ante la Rr[l] Persona y se concedió la suplicacion en el dho dia 14, dentro del término y con arreglo a la ordenanza (f. 303 y su v[ta]).

Suplican los de Chacalla de que se someta este recurso a los de Canta (f. 304), porque debia oirse sumariam^{te} como pleito de indios, y las tierras no valian la cantidad que era necesa-//ria para la 2ª suplicas^{n} y que quando esta se otorgase, debia ser sin fianzas, porque juraban que no las tenian, y las tierras eran bastante hipoteca; y se mandó fuese en prov^{n} a la Persona R^{l} p^{a} si huviese o no lugar de concederla (dho f^{o} v^{to}).

Articulo sobre volver las ovejas o su precio (desde el fol. 305), segun valian quando se entregaron, pretendiendo los de Canta se les volviesen con los partos y pospartos, y las mejoras de las tierras, y alla mandose a compensar con los frutos que han cogido, (se mandó pagar en ovejas por auto f. 306 v^{to}) y otras incidencias de reconocim^{o}s y quexas de entrarse unos en tierras de otros, en que se dieron varias providencias medias, y mando despachar la egecutoria a los de Chacalla (fol. 311 v^{to}), sin fianzas 1° jun^{o} 1567.

En 19 de junio de 1567 pusieron la demanda de terceria de dominio los principales e ind^{o}s yungas de Quivi (f. 313), y mas adelante suplicaron en caso necesario de las sentencias dadas entre terceros y que se hiciese averiguas.^{es} e informas.^{es} (f. 314 y 315).

Siguiose la instancia con las incidencias y quexas de las entregas de las 200 ovejas, a que fueron comisionados (desde el f. 382) y volvio el provinz^{l} de S^{to} Domingo: de los daños que se hacian en las sementeras (sobre lo que y el atentado de arrancar los mojones puestos por los comisionados de la Aud.^{a} e impedir el uso de las aguas huvo dos criminalidades) (f. 316 y 411 y 422) y de despachar la egecutoria a los de Chacalla, de que al fin fue dada sent.^{a} en 19 de junio de 1567 y dada ynterim providencia en 4 de sett^{re} de dho 1567 (f. 318 / f. 409 v^{to}) para que cogiesen (los de Canta y Quivi) la coca y frutos de los arboles que havian plantado, pero que no los destruyesen ni arrancasen: ultimam^{te} por sentencia de 18 de nov^{re} de dho año (f. 455) dada con todos los litigantes, se mandaron guardar las de vista y revista y lo hecho y proveido en su egecusion por el p^{e} Gaspar de Carbajal y por el rreceptor Pedro de Herrasti: y que dando 25 ovejas mas cumplian los indios de Chacalla; y dieron por libres a las partes de lo demas demandado, pedido y acusado por unas contra otras: y se mandó guardar lo proveido en 22 del mismo nov^{re} sin embargo de la suplicaz^{n} interpuesta por los de Chacalla.

Ramo separado.

En prosecucion de la 2ª suplicacion intentada por los ind^{o}s de Canta y que se les concedió en 14 de julio de 1559 fol. 303 y su v^{ta} citad.^{s} se presentaron en el Consejo R^{1} de las Indias en Madrid a 4 de marzo de 1570 y ante la R^{1} Persona en Cordova a 13 del mismo marzo; y despachada comision a dho Consejo en 29 y aceptada; se declaró por sentencia de 21 de abril de dho 1570; no haver grado de 2ª suplicacion.

Los y^{o}s de Chacalla con los de Canta sobre las tierras de Quibi / en que ay sentençia de vista e rrevista y dada ex.^{a} y asimesmo esta aqui acomulado el pleito que trataron los dhos y^{o}s sobre el prençipal e y^{o}s de Quibi.

Ay autos de vista e rrevista.

secretario Avendaño

Los yndios de Chacalla con los
yndios de Canta sobre las tierras de
Quibi. De lo qual ay sentençia de vista
y rrevista y dada ex.^{a} la qual esta aqui.

- Lo que aora se trata sobre dezir a quienes an de ser sujetos los dichos yndios de Quibi a Canta o a Chacalla que es con don Felipe Taulichunbi y sus yndios . Pareçe que el dicho don Felipe y los y^{o}s a el sujetos estan visitados por de Canta y lo mismo por de Chacalla; como por ellas pareçe vease en este proçeso las petiçiones que el don Felipe tiene dadas que son las dobladas.
- Pareçe estar visitados estos yndios de Quibi por sujetos a Chacalla que es el principal don Felipe.
- Pareçe asimesmo que Alonso Manuel de Anaya en la visita que hizo del rrepartim^{o} de Canta por sujetos a Canta al pueblo de Quibi.

Don Philipe Taurichunbi prinçipal del pueblo de Quivi dize que perteneçe a Chacalla y a el ser subjeto su pueblo y por executoria y estar judiciada y dada posesion de sus tierras y los de Canta piden lo contrario y que son a ellos sujetos y que estan visitados por de Canta por Al^{o} Manuel de Anaya aunque P^{o} Ruiz los visito por de Chacalla.

7 de octubre de 1558. Avendaño

/1r/ En Madrid a nuebe de diz^{e} de myll y quy^{o}s y sesenta y nueve a^{o}s.Se resçivio este p^{o}ceso juntam^{te} con otros.

Muy p^{o}s ss^{r}

Don Xristoval Bilcapoma caçique prinçipal de Chacalla / y don Felipe prinçipal del dho rrepartim^{o} / y Frs^{co} Yndio. Por nosotros y en n^{o} de los demas y^{o}s y prinçipales del dho rrepartim^{o} / demandamos al caçique e y^{o}s de Canta / y dezimos que tenyendo y poseyendo nosotros y nros antepasados unas tierras y chacaras de coca que se dizen de Quibi / desde el tpo de los yngas pasados de tpo ynmemorial a esta pte / y las teniamos al tpo que los españoles entraron en esta trra los dhos caçiques e yndios en tpo del marques don Fran^{co} Piçarro por engaño prendieron a un caçique que entonçes era nro que se dezia Vilcacapcha / y lo mataron y juntamen^{te} con el çiento y çinq^{ta} otros y^{o}s que tenian las dhas tierras / y muerto que los ovyeron les tomaron las dhas tierras / lo qual hizieron porque el dho nro caçique e yndios servian a los españoles y estavan bien con ellos / y no se alçaban con el inga quando se alço en este rreyno / y aunque les huviesen pedido muchas vezes las dhas tierras no nos las han querido bolver / antes dizen tener un conçierto y desistim^{to} dellas que dizen abelles hecho un caçique que dizen haver sido nro / el qual nunca tal hizo a lo menos por su boluntad / y aunque lo hiziera en el fuymos todos los caçiques y^{o}s y prinçipales del dho rrepartim^{o} engañados enormisimam^{te} / y nos deve ser deshecho el dho engaño de manera que nosotros ayamos las dhas tierras y quedemos con ellas como cosa que es nra.

/1v/ A V. Alt^{a} pedimos y suplicamos que declarando las dhas tierras por nras nos las mande bolber y entregar librem^{te} como // cosa que es nra y de nros antepasados con los frutos y rreņtas que han llevado y podido llevar hasta la rreal rrestituçion / y si conçierto alguno paresçiere que nos pueda perjudicar declarando ser en el engañados enormisiman^{te} lo rresçinda y anule conforme a dr^{o} sobre que pedimos justiçia y costas y para ello etc.

El licen^{do} M^{o} Lusio

En los Reyes çinco dias del mes de ottubre de myll e quy^{o}s e çinquenta e ocho años ante los señores oydores en audi^{a} se rrepresento esta demanda e los dhos señores mandaron dar dello traslado a la otra parte para que rrespondan a ello lo qual paso. Presente Françisco de la Torre pr. de la otra pte e al qual se notifyco.

Diego Muñoz

/2r/ Muy p^{o}s ss^{r}

Fran^{co} de la Torre en nonbre del caçique e yndios del rrepartimi^{o} de Canta / encomendado en Niculas de Ribera en el pleyto que contra ellos trata los caçiques e yndios del rrepartim^{o} de Mama encomendado en Fran^{co} de Anpuero /digo que en la dha cabsa / tengo negada la demanda puesta a mys ptes / e alegado otros e pr^{do} otros rrecabdos de los quales agora de nuevo hago pr^{on} contra la demanda que agora se pone a los dhos mys ptes.

A B. Alt. pido y supl^{co} los aya por pr^{dos} en la dha cabsa sobre que pido justiçia.

Fr^{co} de la Torre

En los Reyes a siete dias del mes de otubre de myll e quy^{o}s e çinquenta y ocho años ante los señores presidente e oydores en Audiençia Rreal / la presento el cont^{do} con la dha demanda e los dhos señores la mandaron poner en el proçeso y rreçebieron la causa a prueva con tr^{no} de seis dias.

Fran^{co} de.....

Notifique este dho dia haga prueva a los dhos Fran^{co} de la Torre e Alonso Moreno prs de las dhas ptes.

Diego Muñoz

/3r/ Muy pod^{e}s señores

Fr. de Di^{o} Dabila por los yndios de Canta.
Mill y quatroçientos [*sic*] y sesenta y quatro años.

Fran^{co} de la Torre en nonbre de don Her^{do} caçique prinçipal del rrepartim^{o} de Canta y de sus yndios y prinçipales rrespondiendo a una demanda que fue puesta a mys partes por parte de don Xristoval Vilcapoma caçique prinçipal del rrepartim^{o} de Chacalla y de sus y^{o}s en que en efeto pide a mys partes las trras de la chacara e balle de Quibe que esta junto al camino por donde suben a Canta diziendo pertenesçerle y que çierto conçierto de que hazen rrelaçion que se hizo sobre las dhas trras no debe baler por ser hecho por su encomendero y que por aquello le deben ser rrestituydas las dhas tierras segun que esto y otras cosas mas largamente en la dha demanda se contiene un tenor e avido aqui por rresumydo. La niego en todo y por todo segun y como se pone e digo que mys partes deben ser dadas por libres e quitas de la dha demanda por lo siguiente:

- Lo primero porque la dha demanda es puesta por no parte y contraparte no obligada por todo lo demas gr^{l} que es aqui por espreso y por no ser puesta en tpo ni en forma.
- /3v/ Lo otro porque las dhas trras son de mys partes que las tubieron e poseyeron desde antes de Topa Yupanqui y la color que las dhas partes contrarias tienen para pedir fue porque en aquella parte y lugar Guaynacaba puso çiertos mytimaes de los yndios de los pueblos de la parte contraria por çiertas causas que le mobieron y en la dha trra le señalo lo que abian de labrar para el dho Guaynacaba y despues que los españoles entraron // en este rreyno y çeso la sujeçion del ynga y no avia para que estar alli los dhos mitimaes se bolvieron a sus tierras y ansi mys partes lo tornaron a labrar como cosa suya e propia que era y es y desde entonçes aca siempre lo an tenido y poseydo y teniendolo y poseyendolo las partes contrarias pidieron las dhas tierras a vro pr^{te} el lic^{do} Gasca y pusieron a pleito a mys partes sobre las dhas trras y andando en el dho pleito el dho bro presidente mando a los encomenderos de mys partes y de la parte contraria que fuesen a ber las dhas trras y procurasen de los conçertar y ansi las partes contrarias y mi parte fueron a ver las dhas tierras y estando en la dha diferençia y debate se binieron a conçertar y hazer transaçion en el dho pleito / y no debiendo mys partes cosa alguna por quytarse del dho debate dieron a las partes contrarias todo lo cont^{do} en esta transaçion y conçierto de que hago presentaçion por que no los mobiesen mas pleitos sobre las dhas trras por lo qual aunque las dhas trras fueran de las dhas partes contrarias y tubieran dr^{o} alguno a ellas aquel avian perdido e no podian pedir mas las dhas trras ni poner la demanda que ponen quanto mas siendo como son de mys partes las dhas trras y no teniendo como no tienen las dhas partes contrarias dr^{o} alguno a ellas. Por las quales rrazones y cada bna dellas pido y sup^{co} a V. Alt^{a} mande dar por libre e quyto a las dhas mys partes. Sobre que pido justiçia y costas y el of^{o} de V. Alt^{a} ynploro.

El licen^{do} de Leon Fr^{co} de la Torre

/4r/ En la çibdad de los Reyes en siete dias del mes de abril de mill e qui^{o}s e çinquenta e ocho años ante los señores oydores en Abdiençia Real. La presento el dho Fran^{co} de la Torre en el dho nonbre e los dhos señores mandaron dar tr^{do} a la otra pte e que rresponda a la prim^{a} abdiençia.

Fran^{co} de...

/5r/ Don Her^{do} caçique de Canta

Muy pod. ss^{or}

Presenta el conçierto

Fran^{co} de la Torre / en nonbre de don Hr^{do} e de los demas prinçipales del pueblo de Canta / en el pleyto con don Xristobal Bilcapoma / sobre las tierras / hago pr^{on} ante Bra Alt. deste conçierto que se hizo con los dhos mys partes / por las ptes contr^{as}.

A Bra. Alt. pido e supl^{co} lo aya por pr^{do} e lo m^{de} poner en el proçeso de la cabsa sobre que pido justicia.

Fr^{co} de la Torre

En los Reyes en treze de mayo de myll e quy^{o}s e çinquenta y ocho a^{o}s ante los señores presidente e oydores est^{do} en Audiençia Real. La presento el p^{e} y los dhos señores lo obieron por presentado e que se ponga en el proçeso e mandaron dar traslado a la otra parte y que rresponda a la primera aud^{a}.

Diego Muñoz

/6r/ Conçierto sobre las trras de Quibi entre don Fran^{co} de Chacalla y Guaman Yaure de Canta.

Digo por don Fran^{co} señor del pueblo de Mama y chacla que sirvo don Francisco de Anpuero y lo tengo por señor y amo y le sirvo como su caçique que es verdad que por cuanto entre bos Guaman Yari señor del pueblo de Canta y pueblos y prinçipales a el sujetos y teneys por amo y le serbis a Niculas de Ribera y por quanto entre bos el dicho Guaman Yabre y mi don Fran^{co} teniamos diferençia sobre çiertas tierras mucho tpo a que se dizen Quibi que confinan con los yndios de Guancayo en el mismo rrio y por arriba azia la sierra con yndios de Canta / y los dhos nros amos Fran^{co} de Anpuero y Niculas de Ribera por nos hazer bien y quitarnos de diferençia y pasion binieron ambos a dos juntamente a las trras del dicho Quibi y todos juntos ellos y nosotros nos conçertamos en esta manera que me diesedes bos el dicho Guaman Yabre dozientas obejas y que yo el dicho don Fran^{co} os doy las dichas trras de Quibi para bos y para bros deçendientes y desenbaraçadas y digo que agora ni en ningun tpo por mi ni yndio mio ni mi amo os sera puesto entrebalo alguno y porque es berdad que rrezebi las dichas duçientas obejas de que me doy por contento y pagado a toda mi boluntad os doy esta carta firmada de mi nonbre y de mi amo Fran^{co} de Anpuero y porque yo no se firmar rroge al padre Garçia de Godoy firmase por mi. Testigos que fueren presentes a todo lo sobredicho / nros amos Fran^{co} de Anpuero y Niculas de Ribera y el padre Lorenço de Godoy y Diego Dabila y Pedro Ruiz de Barahona y Diego de Flores y en lo que toca a las trras fuera del rrio azia la parte de Chacalla se partieron desta manera que la trra que cae de los yndios de Canta a do esta vna + para habaxo azia Guancayo son del dicho Guaman Yabre y de la + para arriba del dicho don Fran^{co} y porque es berdad lo firmaron de sus nonbres. Fecho sabado a veynte y seys dias del mes de henero de myll e qui^{o}s y quarenta y nuebe años.

Nicolas de Ribera

Francisco de Anpuero

Por Fr^{o} Diego de Abila

/7r/ Don Xristobal Vilcapoma

Muy p^{o} s^{r}

20 dias de tr^{no} mas comun.

A^{o} Moreno en nonbre de don Xristobal Bilcapoma caçique de Chacalla e de don Felipe e Fran^{co} e los demas y^{o}s de su rrepartimiento en el pleyto con el caçique e y^{o}s del rrepartimiento de Canta sobre las tierras e chacaras de Quybi e con Fran^{co} de la Torre pr. en su nonbre digo que en el termino de los seys dias conque B. Alt. rr^{o}pr. esta cavsa es brebe y en ellos mys ptes no pueden hazer su probança.

Pido y suplico a B. Alt^{a} me conçeda otros veynte dias de mas termino atento que son autores e para ello etc.

Otro sy pido e sup^{co} a B. Alt^{a} mande a Fran^{co} de la Torre pr. de los dhos caçiques e y^{o}s de Canta que ponga en esta cabsa el poder que tiene de los dhos y^{o}s e para ello etc.

Al^{o} Moreno

En los Reyes onze de otubre de myll e quy^{o}s e çinquenta e ocho a^{o}s en audiençia pu^{ca} ante los señores presydente e oydores. La presento el p^{e} e los dhos señores le conçedieron el dho termyno que pide comun a las ptes e que oy en todo el dia el dho Fran^{co} de la Torre ponga el poder que tiene de los yndios de Canta lo qual notifico que estuvo presente.

Diego Muñoz

/8r/ En 6 de otubre de 58

Receptoria a Al^{o} Moreno
de los indios de Chacalla.

En la çibdad de los Reyes juebes seis dias del mes de otubre de myll e qui^{o}s e çinquenta e ocho a^{o}s ante el muy mag. señor Fran^{co} caçique prinçipal e de horden en esta dha çibdad por su mag^{t} y en presençia de mi Juan Garçia de Nogal escrivano pu^{co} e del num^{o} desta dha çibdad e de los t^{o}s de yuso scryptos. Paresçio presente don Xristoval Bilcapoma caçique prençipal del rrepartim^{o} de Chacalla encomendado en Fran^{co} de Anpuero vz^{o} y rregidor desta çibdad e don Felipe prençipal del dho rrepartim^{o} e don Pedro e don Diego y Anton Paria Guaman y Albaro Paico y Niculas Macata y Juan Lobres Capcha y M^{yn} Pillo yndios e naturales del dho rrepartim^{o} por si y en boz y en nb. de los demas yndios de el dho rrepartim^{o} del dho caçique e dieron por lengua de Juan Osorio lengua e yntepetre del dho don Xristobal caçique que el tiene neszecidad de ser probeido de un curador ad liten para un pleito e demanda que quieren poner al caçique prençipal de Canta y a los yndios del dho rrepartim^{o} y a su encomendero por tanto que pide a su md le probea e mande que lo sea Al^{o} Moreno pr^{or} que esta presente e pide costas.

/8v/ El dho señor alc.^{de} pregunto al dho Alonso Moreno si lo quiere ser y estar el qual dixo que si y el dho señor alc.^{de} tomo rr^{o} del juramento por dios nro señor e por las palabras de los santos evangelios e por una señal de cruz tal como esta + en que puso su mano dr^{a} en forma de dr^{o} que husara bien e fielmente del off^{o} de curador ad liten de el dho caçique e yndios e que seguira sus pleitos e cabsas e los feneszera e acabara e no los dejara yndefensos e donde su consexo no bastare / lo tomara de letrado e personas savias que se lo sepan dar e donde biere su daño se lo quitara e rretraera e donde biere su bien se lo allegara y adquirira y en todo hara aquello que bueno y deligente curador de ley es obligado a fazer e a la fuerza e conclusion del dho juramento dixo si juro e amen e si algun daño por duda o menoscavo a dho caçique e yndios le viniese por su culpa e negligençia que en tal casso el lo pagara por su persona e los que para ello dixo que obligava e obligo e dio juntamente consygo para su fiador a Pero Lopez de Mediano que pres^{te} estava el qual dixo que salia e sale por tal fiador de el dho Al^{o} Moreno.

/9r/ E ambos a dos de mancomun la boz de bno e cada vno dellos por si e por el todo rreu^{do} las leyes...y el autentico pres^{te} de...de... y el beneff^{o} de la division y el rremedio de la yncursion y las demas leyes que hablan en rrazon de la mancomunidad como en ellas se contiene e debajo de la dha mancomunidad prometieron e se obligaron por sus personas e bienes que el dho Alonso Moreno cumplira todo lo que tiene prometido sin faltar cosa alg^{a} so pena que lo pagaran por sus personas e bienes e para que lo cumplan a la paga e cumplim^{o} de lo susodho dieron poder cumplido a qualesquier juezes e just. de Su Magestad de qualquier fuero e jur^{on} que sean para que les conpelan a ello como por sent^{a} dif^{a} contra ellos dada e pasada en cosa juzgada y sobre ello rrenunçiaron todas qualesquier leyes fueros y dr^{o}s que en su fabor e contra lo que dho es //

/9v/ sean que ellos non valgan en esta rrazon en jui^{o} ny fuera del y espeçialmente rrenunçiaron la ley e rregla del dr^{o} en que dizque general rrenunçiaçion de ley es fha non vala en termyno de lo qual otorgue la presente carta en el dia e mes e año arriba dho syendo t^{o}s presentes: G^{o} de Luque e Ju^{o} osorio y Rr^{do} de La Fuente y los otorgantes lo firmaron de sus nonbres ante mi Joan Garcia de Nogal escrivano pu^{co}.

E luego el dho senor alc.^{de} dixo que disçernia e disçernyo el dho cargo de curador ad liten de el dho caçique e yndios en el dho Al^{o} Moreno que estava presente e le dio poder cumplido qual de dr^{o} en tal caso se rrequyere generalmente para que en nonbre de el dho caçique e yndios pueda hazer e poner todas las demandas pedimy^{o}s rrequerimyentos execuçiones prisiones ventas de bienes e rremates dellos querellar e protestar conbenir rreconbenir y rresponder negar e conoszer y presentar //

/10r/ testigos probanças escripturas e testimy^{o}s y hazer qualesquier jur^{o}s en su anyma diziendo verdad e los deferir en las partes contrarias e tachar e contradeçir los t^{o}s que las otras partes presentaren ansi en dhos como en personas e hazer todos los demas autos e deligençias ansi judiçiales como estrajudiçiales e concluir e çerrar rrazones e pedir e oir sent^{a} o sent^{as} ansi ynterlocutorias como difin^{as} e las que por ellas se dieren consentir e de las en contr^{o} apelar e suplicar e seguir el apelaçion e sup^{on} alli e do con dr^{o} se deba seguir o dar quien las siga e fazer todos los otros e demas autos e deligençias que convengan porque quan cunplido poder es neszesario para lo que dho es otro tal y ese mysmo y tan cunplido el dho senor alc.^{de} dixo que le dava e dio al dho Al^{o} Moreno con todas sus ynçidençias e dependençias e con libre e general admynistraçion e si es neszesario rrelebaçion //

/10v/ lo rrelebaba en forma de dr^{o} e le dio el dho poder para poder sostituir en un procurador e dos o mas y aquellos rrebocar e de nuevo poner con la clausula de rrelebaçion como dicho es;en todo lo qual dixo que ynterponia e ynterpuso su abtoridad e decreto judiçial tanto quanto puede y con dr^{o} debe y lo firmo de su nonbre siendo t^{o}s los dhos G^{o} de Luque e Ju^{o} Osorio e Hr^{do} de La Fuente / Fran^{co} Talabera ante my Joan Garcia de Nogal escrivano pu^{co}.

Yo Joan Garçia de Nogal escrivano pu^{co} susodho presente fui a lo que dho es y lo fize escrivir e fize aqui mi signo en testimy^{o} de verdad.

Joan Garcia de Nogal
escrivano pu^{co}

El Licen.
M^{o} Luzio

Muy p^{o}s s^{or}

/11r/ Alonso Moreno en nonbre de don Xpoval Bilcapoma caçique prençipal del rrepartimi^{o} de Chacalla e don Felipe por sy e por los demas sus y^{o}s en el pleyto con el caçique de Canta sobre las tierras e chacarras e con Fran^{co} de la Torre pr. en su nonbre digo que en esta cavsa tienen mys ptes de presentar por testigos a don Pedro caçique de Guancayo / y otro yndio biejo que se dize Pariasca su sujeto / y a don Luys caçique de Guarabiri y a Caxa Llabre su sujeto del rrepartimi^{o} de la del contador Caçeres [*sic*] / y a don Fran^{co} Yaybe / caçique prençipal de Collique de B. Alt. los quales saben muy bien como las dhas tierras son de los dhos mys ptes e dizen que porque los y^{o}s cantas los an amenazado que no lo digan no lo an de dezir y ansy no quyere benyr dezir sus dhos. Pido e sup^{co} a B. Alt. mande dar su mandami^{o} para que los dhos caçiques e y^{o}s de los dhos rrepartimyentos vengan luego a dezir sus dhos el qual dho mandami^{o} V. Alt. mande que hable con qualquier español que estubiere en los dhos rrepartim^{o}s para que se lo notifique mandandoles so pena que benga luego sobre que pido just. e para ello etc.

Al^{o} Moreno

/12r/ Don Xristobal Bilcapoma

Muy p^{o}ss s^{o}s
Los t^{o}s se examynen aunque sea p^{do} el tr^{no}.

Al^{o} Moreno en n^{e} de don X^{o}val Bilcapoma y sus y^{o}s del rrepartimi^{o} de Chacalla en el pleito con don Hr^{do} caçique de Canta y sus y^{o}s. Digo que el tr^{no} de la prueva es pasado y no esta fecha pu^{on} en la causa ni pedida; y agora an venido dos testigos a mi pte que no an podido antes ser abidos.

A V. Al. pido y supp^{co} los mande examinar no enbargante que este pasado el ter^{no} de la prueva lo qual pido se haga del off^{o} y para saber la verdad e como mejor de dr^{o} oviere lugar y pido justicia e para ello etc.

El Licenc^{do} Al^{o} Moreno
M^{o} Luzio

En Los Rreyes tres de noviembre de myll e quy^{o}s e çinq^{ta} e ocho años ante my la presento Alonso Moreno en n^{e} de su pte con protestaçion de la rretificar para la primera avdi^{a}.

Diego Muñoz

/12v/ En Los Rreyes quatro de otubre de myll e quy^{o}s e çinquenta y ocho a^{o}s ante los señores presydente e oidores desta Rreal Aud^{a} estando haziendo aud^{a} de petiçiones la presento el dho pr. y se rretifico en la dha present^{on} que esta fha e los dhos señores probeyeron en esta pet^{on} lo que esta probeydo en la petiçion de adelante.

Diego Muñoz

/13r/ Don Hr^{do} de Canta

Muy poders. s^{or}
Pu^{an} a que se tomen dos t^{o}s.

Fran^{co} de La Torre en nonbre de don Hr^{do} s^{or} de Canta en el pleyto con don Xpoval Bilcapoma / e con A^{o} Moreno en su nonbre sobre las trras digo que el tr^{no} p^{o}vat^{o} es pasado.
A Bra Al. pido e suplico mande hazer publicaçion de t^{o}s en la dha cabsa sobre que pido just. etc.

Fran. de La Torre

En Los Rreyes quatro de novienbre de myll e quy^{o}s e çinquenta e ocho a^{o}s ante los señores presydente e oydores la presento el dho pr. en nonbre de su pte e los dhos señores mandaron que juren e se tomen e examynen los dos t^{o}s que el dho Alonso Moreno dize aunque sea pasado el termyno probatorio e jurados e tomados sus dhos se haga la dha publicaçion que el dho Ffran^{co} de La Torre en nonbre de su pte pide.

Diego Muñoz

Muy p^{o} S^{or}

/1r/ Gaspar de Çarate en n^{e} de don Hernando Libia Guaranga y don Al^{o} Macuri e don Pedro Yauri y don Fran^{co} Guaraca y don Fran^{co} Yanacuri y don Diego Tunguisa e don Pedro Poma yndios caçiques prinçipales del pueblo de Canta de las provinçias del peru. Digo que mys partes an tratado pleito en la rreal avdiençia de la çiudad de los Rreyes con los yndios de Chacalla sobre çiertas tierras que llaman de Quibi el qual se a traido rremytido por la dha rreal audien^{a} ante V. Al^{a} / porende yo en nombre de los dhos mys partes hago presentaçion deste poder que dellos tengo para me mostrar parte en este pleito. Y pido y supp^{co} a V. Al^{a} mande que se me de el proçesso para hazer las diligençias que combengan y alegar de la just^{a} de mys ptes e para ello etc.

Gaspar de Çarate

/2r/ En Madrid a tres de març̧o de myll y quy^{o}s y setenta años lo presento Gaspar de Çarate esc. de su pte

Sepan quantos esta carta de poder e sostituçion vieren como yo Diego de Valençia hgo v^{o} de la villa de Valejo en nonbre de don Her^{do} Libia Guaranga y don Al^{o} Macuri y don Pedro Yauri y don Fran^{co} Guaraca y don Fran^{co} Yanacuri y don Diego Tunguisa y don Pedro Poma caçiques prinçipales de Canta rrepartimy^{o} encomendado en Nycolas de Rribera vezino de la çibdad de los Rreyes y por vyrtud del poder por los susodhos por sy y en nonbre de los demas prinçipales e yndios del dho rrepartimy^{o} a ellos suxetos a my otorgado en la dha çibdad de los Rreyes a veynte e syete dias del mes de otubre del año de myll e qui^{o}s e sesenta e ocho a^{o}s el tenor del qual es el syguiente:

Sepan quantos esta carta vieren como nos don Her^{do} Libia Guaranga y don Al^{o} Macuri y don Pedro Yauri y don Fran^{co} Guaraca y don Fran^{co} Yanacuri y don Diego Tunguisa e don Pedro Poma / caçiques prençipales de Canta rrepartimy^{to} encomendado en Nycolas de Rribera bez^{o} desta çibdad por nos y en nonbre de los demas prinçipales e yndios del dho rrepartimy^{o} a nos suxetos por quien prestamos voz y cauçion de rrato que estaran y pasaran por lo que en virtud desde poder hiçiere so espresa obligaçion que haçemos de nras personas e vs e muebles e rraizes abidos y por aber otorga^{o}s y conozemos que damos e otorgamos todo nro poder conplido segun que lo nos abemos y tenemos e segun //

/2v/ que mejor y mas cunplidamente lo podemos y debemos dar e otorgar e de dr^{o} mas puede y deve baler a vos el padre Diego de Valençia que soys presente y a Di^{o} Dabila vez^{o} desta çibdad que soys ausente e anvos a dos juntamente e a cada uno de por sy ynsolidim generalmente para en todos mys pleitos e causas çebiles e crimynales que nos abe^{o}s e tenemos y esperamos aver y tener y mober con qualesquyer persona o personas de qualquier estado e condiçion que sean y las tales personas contra nos asy en demandando como en defendiendo y espeçial para que podays seguir un pleito que tratamos con los yndios de Chacalla encomendados en Fran^{co} de Anpuero v^{o} desta çibdad sobre çiertas tierras que ba rremetido a su mag^{t} y a su rreal consejo de yndias y seguillo y feneçillo por todas ynstançias hasta conclusion e pedir tasaçion de costas e syendo nesçesario en la dha rrazon podays pedir e demandar rresponder negar y conoçer defender pedir y rrequerir y querellar y protestar testimy^{o} e testimy^{o}s e pedir e tomar e para

dar e presentar t^{o}s e provanças escriptos y escrituras e tachar lo en contr^{o} present^{o} y façer juramentos ansy de calunia como de sesorio e para rrecusar e poner sospecha en qualesquyer juezes y test^{o}s y otras personas e concluyr e çerrar rrazones e pedir e oyr sentençia e sentençias ansy ynterlocutorias como difinytibas y de las en su favor dadas consentir y de las en contr^{o} apelar y suplicar para //
/3r/ alli o do con dr^{o} debieredes y seguir la tal apelaçion e suplicaçion hasta la final conclusion e pedir tasaçion de costas sy las hubiere y para que en nros nonbres podays pedir qualesquier mrdes que a nro dr^{o} conbengan y sacar las .... y probisyones que sobre ello nos dieren duplicadas y enbiarnoslas a nra costa e rriesgo e para pedir benefiçio de rrestituçion yn yntegrin e para que podais haçer e hagais todos los demas autos e diligençias judiçiales y estrajudiçiales que conbengan de se haçer e que nos hariamos e haçer podriamos presentes syendo aunque sean tales y de tal calidad que segun derecho rrequerian nra presençia personal e para que podays haçer y sostituyr un pr^{or} dos o mas y los rrebocar quando quisyere quedando ansy este poder prençipal e quan conplido e bastante poder e es nesçesario para lo susodho otro tal y eso mysmo lo otorgamos todos los susodhos con sus ynçidençias anexidades y conexidades y con libre e general admynystr^{on} y bos rrelebamos en forma de dr^{o} para lo aver por firme lo que en nro nonbre hiçieredes obliga^{o}s nras personas e vienes muebles e rraizes avidos e por aver el qual dho poder otorgaron los dhos caçiques prinçipales susodhos por lengua de Sebastian de Leon yndio ladino sobrino del dho caçique don Her^{do} natural que es de Canta el qual les declaro el dho poder y lo que en el se contenia y dixeron por la dha lengua //
/3v/ que lo otorgavan e otorgaron como en el se convi^{e} en testimy^{o} de lo qual otorgaron la presente carta ante my el presente escry^{o} pu^{co} e t^{o}s de yuso scriptos en el rreg^{o} de la qual lo firmo de su nonbre el dho don Pedro Poma e por que los demas no supieron tomar a su rriesgo lo firmaron e^{o} desta carta a los quales dhos otorgantes yo el presente escry^{o} doy fee que conozco que es fho en los Rreyes a veynte e syete dias del mes de otubre de myll e qui^{o}s e sesenta e ocho años t^{o}s que fueron presentes Br^{no} Martinez e Diego Goije e Juan Ochoa estantes en esta çibdad don Pedro Poma por t^{o} e a rruego de los otorgantes Jhoan Ochoa / Al^{o} de Balençia escry^{o} pu^{co} e yo Alonso de Balençia escry^{o} de su mag^{t} e pu^{co} del mand. desta çibdad de los Rreyes presente fuy en uno con los dhos testigos al otorgamy^{o} desta carta e la fiz escribir e fize aqui este myo sygno tal en testimonyo de verdad Alonso de Valençia escry^{o} pu^{co}.

Los escry^{o}s de su mag^{t} pu^{cos} del mand. desta çibdad de los Rreyes del peru aqui contenidos damos fee que Al^{o} de Balençia de quien ba sygnado e firmado este poder es tal escry^{o} de su mag^{t} pu^{co} del man^{o} desta dha çibdad como por el se nonbra a las scripturas de autos que ante el an pasado y pasan se a dado y da entera fee y carta en juiçio y fuera del / fho en los Rreyes a dos de dizienbre de myll e qui^{o}s e sesenta e ocho a^{o}s Fran^{co} de la Bega escry^{o} pu^{co}

Esteban Peres escry^{o} pu^{co}
Al^{o} Herre escrivano pu^{co}

Por vyrtud del qual dho poder que de suso ba yncorporado y del usando yo el dho Diego de Valençia en el dho nonbre digo que sostituyo doy e otorgo todo my poder como yo le tengo para todo lo conv^{te} en el dho poder y para cada una cosa y pte della a Gaspar de Çarate pr. en la corte y Consejo Rreal que esta ausente y le doy el mysmo poder quan cumplido yo le tengo de los susodhos syn rreserbaçion alguna con todas sus ynçidençias y dependençias anexidades y conexidades y para que sera çierto e firme este poder y sostituçion y lo que en virtud del fuere ffecho y no sera rrebocado ny contradho en manera alguna obligo las personas e vienes a my obligados y le rreliebo en forma de dr^{o} segun que yo lo soy en testimy^{o} de lo qual lo otorgue en la dha villa de Valejo ante Pedro Lucas escrivano de su mag^{t} e^{s}. della a veynte e seys dias del mes de henero de myll e qui^{o}s y setenta años syendo presentes por t^{o}s e vieron firmar su nonbre en el rreg^{o} al dho Diego de Balençia al qual yo el dho escriv^{o} doy fee que conozco Bernardino de Gordejuela y Ant^{o} de Estrada estudiantes y Vernabe Ruiz estantes en esta dha villa Diego de Balençia paso ante my Pedro Lucas

ba entre renglones y dependençias .vala / va testado / consejo. no vala / e yo el dho Pedro Lucas escrivano pu^{co} susodho presente fuy a lo que dho es con los dhos t^{o}s e fize aqui my signo en testimonyo de verdad.

C. R. M.

/6r/ Los yndios del rrepartimiento de Canta que es en las provinçias del peru se presentan ante V. Md en grado de segunda suplicaçion de çierta sentençia de rrevista dada y pronunçiada contra ellos y en favor de los indios del rrepartimiento de Chacalla por el presidente y oydores de la real audiençia y chançilleria de la çiudad de los Reyes en el pleito que ante ellos entre las dhas partes se a tractado en la dha Real Audiençia sobre las tierras del valle de Quibi por la qual en efecto condenaron a los dhos yndios de Canta a rrestituçion de las dhas tierras en çierta firma segun en la dicha sentençia se contiene de la qual suplicaron segunda vez para ante vra persona real y les fue otorgada la dha segunda suplicaçion y para prosecuçion della se a traido el proçesso del dho pleito y esta presentado en vro Real Consejo de Indias como pareçe por este testimonio que presentan signado en publica forma / Supp^{an} a V. Md los mande rreçebir en el dho grado de suplicaçion y dar su comission y provision rreal para que los del dho Consejo de las Indias vean el dicho proçesso y le sentençien y determinen y hagan justiçia en el dho grado y para ello etc.

yo la presente Gaspar de Çarate

/6v/ En la çibdad de Cordoba a treze dias del mes de março de mill y quinientos y setenta años yo Xristoval de Leon escriv^{o} de su mag^{t} ley la petiçion desta otra parte contenida a su mag^{t} del rrey don Felipe nro s^{or} y abiendola entendido dixo y rrespondio . Yo lo oy a lo qual fueron testigos don Fadrique Enrrico de Rribera presidente del Consejo de las Hordenes de su mag^{t} y don Pedro Manuel de la camara de su mag^{t} y Sebastian de Santoyo y Domingo de Zuazo ayudas del y yo el dho Xristobal de Leon escriv^{o} susodho fuy presente a lo que dho es y fize aqui my signo en testimy^{o} de verdad

Xristoval de Leon

/7r/ En la villa de Madrid a seis dias del mes de março de myll e qui^{o}s y setenta años en el Consejo Rreal de las Yndias de su mag^{t} presento una petiçion Gaspar de Çarate en nombre de los caçiques prinçipales y yndios del rrepartimyento de Canta su tenor de la qual es como se sigue:

Muy Poderoso Señor. Gaspar de Çarate en nombre de los caçiques prinçipales e yndios del rrepartimyento de Canta en el pleito que an tratado en la vra rreal avdiençia de los Rreyes contra los yndios del rrepartimyento de Chacalla sobre las tierras del balle de Quibi me presento ante V. Al^{a} en grado de segunda suplicaçion de la sent^{a}. de rrevista en el dho pleito dada y pronunçiada por el presidente e oydores de la dha rreal avdiençia por la qual en efecto rrevocaron la sent^{a}. de visita en este dho pleito dada e condenaron a los dhos mys partes en rrestituçion de las dhas tierras sobre que es pleito segun que mas largamente en la dha sent^{a}. se contiene a que me rrefiero y afirmadome en la suplicaçion que por mys partes fue ynterpuesta en la dha rreal audiençia / Pido y supp^{co} a V. Al^{a} que de los mysmos autos la mande rrevocar e confirmar la dha sentençia de vista en favor de mys partes y para ello mande cometer este pleito y causa al vro Rreal Consejo de las Yndias adonde se a traido y esta el proçesso oreginal de la causa del qual hago rrepresentaçion para el dho efecto e para ello etc.

- Otro si en caso que sea neçesario hazerse la dha psent^{on} ante vra rreal persona por hallarse //
/7v/ al presente fuera desta corte donde vro Rreal Consejo rreside y en la çiudad de Cordova / Pido y suplico a V. Al^{a} mande que al pie de esta petiçion se me de fee y testimonyo de la presentaçion della e de como el dho proçesso esta traido y presentado en vro Rreal Consejo de las Yndias para lo presentar ante vra rreal persona y suplicar que se me de la comysion en tal caso neçesaria en forma y para ello etc. Gaspar de Çarate

La qual vista por los señores del dho Consejo mandaron que se le dieze a la parte de los dhos yndios la dha fee para el efecto que la piden y en cumplimyento de lo qual yo Diego de Enzinas escrivano de su mag^{t} a cuyo cargo esta el exerçiçio de just^{a} del dho Consejo hize sacar y saque del proçesso del pleito que esta en el dho Consejo y se trato en la Avdiençia Rreal de los Rreyes entre los dhos yndios de Canta y los de Chacalla las sentençias de vista y rrevista que en el se dieron y pronunçiaron entre las dhas partes su tenor de las quales con las pronunçiaçiones dellas es como se sigue:

En el pleito que es entre partes de la una los caçiques e yndios del rrepartimy^{o} de Chaclla y Alonso Moreno su procurador y de la otra el caçique e yndios de Canta y Francisco de la Torre su procurador sobre las tierras de Quibi.

Fallamos atento los avtos y meritos de este proçesso //
/8r/ que devemos absolber y absolvemos al caçique e yndios de Canta de la demanda contra ellos puesta por parte de los dhos caçiques e yndios del Chaclla y por esta nra sent^{a}. definitiva juzgando asi lo pronunçiamos y mandamos syn costas.

El marques. El doctor Brauo de Saravia. El liçen^{do} Mercado de Peñalosa. El doctor de Cuenca.

Pronunçiose esta dha sent^{a}. difinitiva por los dhos ss presidente y oidores que en ella firmaron sus nonbres est^{do} aziendo audi^{a} rreal pu^{ca} en la dha çiudad de los Rreyes en diez dias del mes de hen^{o} de myll e qui^{o}s e çinq^{ta} e nueve a^{o}s psentes los dhos Fran^{co} de la Torre y Al^{o} Moreno procuradores de las ptes a quien se notifico.

Diego Muñoz

En el pleito entre ptes de la una los yndios de Chaclla y de la otra los yndios de Canta y Al^{o} Moreno y Fran^{co} de la Torre sus prs fallamos atento lo nuevam^{te} alegado e p^{o}vado que devemos rrevocar y rrevocamos la sent^{a}. en esta cavsa dada por nos el presidente e oidores de esta Rreal Audi^{a} de que por parte de los dhos yndios de Chaclla fue suplicado e haziendo eñ el caso lo que de just^{a} deve ser hecho condenamos a los yndios de Canta a que buelban y rrestituyan las dhas tierras de Quibi sobre que es este pleito a los dhos yndios de Chaclla con que prim^{a}mente los dhos yndios de Chaclla buelban y rrestituyan a los dhos yndios de Canta las dozientas ovejas que por el conçierto en esta causa presentado pareçe que rreçivieron dellos y por esta nra sent^{a}. juzgando ansi lo pronunçiamos e mandamos en grado de rrevista

El marques. El doctor Brauo de Saravia. El doctor Mercado de Peñalosa. El doctor de Cuenca.

Dada e pronunçiada fue esta sent^{a} por los dhos señores psidente e oydores de esta Rreal Audiençia haziendo audi^{a} pu^{ca} en los Rreyes a honze dias del mes de julio de myll e qui^{o}s e çinquenta e nueve años estando presentes los procuradores de las dhas ptes a los quales lo notifique.

Diego Muñoz

/8v/ Fecho y sacado corregido y conçertado fue este dho traslado con las dhas sentençias oreginales que estan en el dho proçesso en la villa de Madrid a seis dias del dho mes de março del dho año syendo psentes por t^{o}s Sebastian de Santander procurador del num^{o} de esta corte y Martin de Aroztegui y Pedro de Geruria estantes en esta corte Yo el dicho Diego de Enzinas escrivano sobre el auto presente fuy en uno con los dhos t^{o}s al corregir deste dho t^{do} con las dichas sent^{as}. originales el qual va çierto y verdadero y en fee dello y de que por parte de los dhos yndios de Canta esta suplicado segunda vez para ante la persona real de su mag^{t} lo firme de my nonbre e signe con my signo que es tal en testimy^{o} de verdad.

Di^{o} de Enzinas

/9r/ Don Phelippe por la graçia de dios Rey de Castilla de Leon de Aragon de las dos Seçilias de Jhlm de Navarra de Granada de Toledo de Valençia de Galizia de Mallorcas de Sevilla de Çerdeña de Cordova de Corçega de Murçia de Jaen de los Algarbes de Algezira de Gibraltar de las yllas de Canaria de las Yndias yslas y tierras firme del mar oçeano conde de Flandes y de Tyrol est^{do} ansi los del nro Consejo Rreal de las Indias saved que pleito se a tratado ante el presidente y oydores de la nra Audiençia Real que rreside en la çiudad de los Reyes de las provinçias del piru entre los caçiques e indios del repartimiento de Chacla de la una parte y el caçique e indios de Canta que son en las provinçias del piru de la otra sobre çiertas tierras y sobre las otras causas y razones en el proçesso del dicho pleito contenidas en el qual por los dichos nro presidente y oidores fue dada sentençia de revista por la qual rebocaron la por ellos dada en vista y mandaron bolver y restituir las tierras sobre que hera el dicho pleito a los dhos caçiques e indios del repartimiento de Chacla con que los dichos indios volviesen a los de Canta dozientas obejas segun en la dicha sentençia de revista mas largo se contiene y agora la parte de los dichos indios de Canta nos a hecho relaçion que por su parte fue suplicado segunda vez de la dicha sentençia de revista para ante nra real persona conforme a las nuebas leyes por nos hechas para la buena gobernaçion de las Indias en que se da la orden que se a de tener en interponer las segundas suplicaçiones de las causas que se trataren en las nras audiençias de las dichas nras Indias el qual se presento ante nra real persona en el dicho grado de segunda suplicaçion y nos suplico mandasemos nombrar juezes que viesen el dicho pleito en el y lo determinasen conforme a las dichas leyes o como la nra merçed fuesse y nos confiando de vosotros acordamos de os cometer y por la presente os cometemos la dicha causa y os mandamos que veais el proçeso del dicho pleito y atento el tenor y forma de las dichas leyes le libreis y determineis por vra sentençia como allaredes por justiçia para lo qual vos damos poder cumplido con todas sus inçidençias y dependençias anexidades y conexidades dada en Cordova a syete dias del mes de março de mill y qui^{o}s y setenta años

yo el Rey

Yo Fran^{co} de .... secr^{o} de su mag^{t} Real. La fize escrivyr por su mandado

/9v/ El lic^{do} don Gomez Çapata Gonçalez

Reg^{dor} Ochoa de Aguirre M^{yn} de Ramoyne

En la villa de Madrid a çinco dias del mes de abril de mill e qui^{o}s y setenta años yo Fran^{co} de Valmaseda escrivano de su mag^{t} que sirvo el ofiçio de la secretaria del Consejo de las Yndias de pedimy^{o} de Alonso de Herrera en nonbre y como procurador de los yndios de Canta ley y notifique la comision rreal desta otra parte escripta a los señores del dho Consejo Rreal de las Yndias estando en consejo los quales la obedeçieron con el acatamy^{o} devido y rrespondieron que estaban prestos de hazerlo por ella se les manda y en çertinidad dello lo firme de my nonbre.

Fran^{co} de Valmaseda

/10r/ En la bylla de Madrid a veynte y vn dias del mes de abril de myll e quy^{o}s y setenta años los señores del Consejo Real de las Yndias avyendo visto el pleito de los yndios de Canta con los yndios de Chacalla en las p^{o}vinçias del peru que al dho Consejo vyno en grado de segunda suplicaçion dixeron que declaravan e declararon en el dho pleito no aver el dho grado de la segunda suplicaçion ynterpuesta por los dhos yndios de Canta y ansi le p^{o}nunçiaron y mandaron

ante mi Fran^{co} de Sopando

En Madrid a veynte y uno de abril de dho año lo notifique a Al^{o} de Herrera y Gaspar de Çarate en n^{e} de sus ptes y en sus personas los quales dixeron que lo oyen.

/11r/ Sepan quantos esta carta de poder bieren como yo don Joan Baptista Caxa Guaraca caçique prinçipal del rrepartimyento de Chacalla encomendado en Françisco de Ampuero vezino desta dha çibdad e yo Joan Llacxa Vilca e Joan Tanpta xueca e Martin Yavri prinçipales del dho rrepartimyento por nosotros y en nonbre de los demas prinçipales e yndios del dho rrepartimyento por quyen prestamos boz e cavçion de rrato para que estaran y pasaran por lo que por birtud deste poder se hiziere e actuare otorgamos e conoçemos por llengua [*sic*] de Joan Osorio que esta presente que damos nuestro poder cunplido libre e llenero bastante qual de derecho en tal caso se rrequyere con libre e general admynistraçion para lo que en el contenydo a bos el padre Hernando de Ribera clerigo rresidente en esta dha çibdad que estays avsente como si fuesedes presente generalmente para en todos mys pleytos cavsas e negoçios movidos e por mover que nosotros e los dhos yndios abemos y tenemos y esperamos aber y tener y mover con qualesquyer yndios e caçiques e otras //

/11v/ personas que sean en qualquyer manera como si en demandando como en defendiendo e otro si bos damos el dho poder para que en nuestro nonbre e de los dhos prinçipales e yndios podays pedir a su mag. y senores de su Real Consejo de las Yndias que sean servidos de nos hazer las mercedes que en nuestro nonbre les pidieredes e suplicardes conforme a la ynstrucçion y memoria que para ello llevays e como mejor bierdes e os paresçiere que conviene para nuestro aumento e conservaçion e hazer librar y despachar las provisiones çedulas e sobre cartas que sobre ello sean neçesarias en publica forma de poder de qualesquyer secretarios e otras personas ante quyen se probeyere e nos las enbiar duplicadas para que se cumpla e aya efecto lo en ellas contenydo e otros para que podays rrecusar qualesquyer secretarios e otras personas ante quyen se proveyere lo de contrario e para que podays rrecusar e poner sospecha en qualesquyer letrados juezes y escribanos e les probar //

/12r/ las tales rrecusaçiones e las jurar y para contradezir todo aquello que contra nosotros se pidiere ynpetrare [*sic*] y ganare e pedir costas e las ber tasar e jurar e para que siendo neçesario en rrazon de lo que dho es podays pareçer en juizio ante su mag^{t} e otras qualesquyer susjustias e pedir e demandar rresponder defender negar e conoçer e hazer en nuestra anyma qualesquyer juramentos ansi de calunya como de çisorio e otro qualquyer que conbenga e presentar qualesquyer testigos e provanças escritos y escrituras e otro qualquier genero de prueba e abonar la tal provança tachar e contradezir lo que contra nos se presentare e con cluyr e çerrar rrazones pedir e ber dar sentençias ansi ynterlocutorias como difinytibas e las consentir e apelar para alli e donde con derecho devays e para que podays hazer qualesquyer provanças ad perpetuam rey memoriam e de otra manera que mejor a nuestro derecho e al pro e utilidad de los dhos yndios e naturales del dho pueblo e //

/12v/ rrepartimyento de Chacalla conbengan de se hazer e sean neçesarias lo qual rremytimos e dexamos a vro parecer e boluntad por el buen zelo e yntençion que hemos conoçido de vos en el deseo de obrar que aveys tenydo en hazer bien a los naturales deste rreyno e fechar las dhas provanças las podays sacar de poder de qualesquyer escribanos e otras personas en cuyo poder esten e pedir todo aquello que mas nos conbenga ynformando a su mag^{t} e a los dhos senores de su Real Consejo de las Yndias del estado y ser de nuestras tierras y naturales para que con mas voluntad sea servido de nos otorgar las merçedes que por nosotros y en nonbre de los demas yndios e prinçipales del dho rrepartimyento les pidierdes e demandardes e hazer todos los demas abtos e diligençias judiçiales y extrajudiçiales que en rrazon de lo que dho os convengan de se hazer e que nosotros por lo que nos toca y en el dho nonbre hariamos e hazer podriamos si presentes fuesemos porque quan cumplido poder como abemos y tenemos y en el dho n^{e} le podemos dar e otorgar //

/13r/ otro tal y ese mysmo lo damos e otorgamos a bos el dho Hernando de Rribera segun dho es con todas sus ynçidençias e dependençias anexidades e conexidades e para que podays sostituyr un procurador dos o mas a los quales damos el mysmo poder que a bos e a vos e a ellos rrelevamos segun derecho e para lo aber por firme obligamos nuestras personas e bienes abidos e por aber en testimonyo de lo qual otorgamos esta carta ante el escribano e testigos de yuso escritos que ffecho en la dha çibdad de los Reyes estando en ella el Abdiençia Real de su mag^{t} a beynte e ocho dias del mes de março año del señor de myll e quynientos e sesenta e nueve anos e porque los dhos otorgantes a los quales yo el dho escribano doy fee que conozco. Dixeron que no sabian escrebir por ellos lo firmo a su rruego un testigo desta carta. Testigos que fueron presentes //

/13v/ a lo que dho es Xristoval de Hojeda e Fran^{co} Hernandez e Alonso Myguel e Joan Osorio este es en esta çibdad por testigo. Françisco Hernandez va t^{do} del no vala

E yo Ambrosio de Moscoso escry^{o} de su mag^{t} e secret^{o} pu^{co} que fuy presente a lo que dho es con los dhos otorgantes e ts lo escrevi segun que ante my paso e porende fize aqui este mio signo tal en test^{nio} de verdad.

Ambr^{o} de Moscoso

dr^{o}s un peso

Nos los escriv^{o}s pu^{co}s e de su mag^{t} que aqui firmamos nuestros nonbres damos fee a los que la presente vieron como Ambrosio de Moscoso escriv^{o} de de [*sic*] cuya mano va signada e firmada esta escritura de poder esta el escriv^{o} de su mag^{t} como en ella se nonbra e a sus escrituras e abtos que ante el pasan e se otorgan se les da entera fee y credito en juyzio e fuera del como a escrituras signadas e firmadas de mano de tal escriv^{o} de lo qual damos la presente a fecha en los Reyes a veynte e nueve de março de myll e qui^{o}s e sesenta e nueve a^{o}s.

Pedro de Valverde
escriv^{o} pu^{co}

Joan Garçia de Nogal
escr^{o} pu^{co}

Estevan Perez
escr^{o} pu^{co}

/14r/ Sost^{on}

En la çibdad de los Reyes quatro dias del mes de junyo de myll e quy^{o}s e sesenta e nuebe a^{o}s ante my el dho Ambrosio de Moscoso escriv^{o} pareçio presente el padre Hernando de Ribera clerigo e dixo que en su lugar y en n^{e} de don Juan Baptista Caxaguaranca caçique prinçipal del rrepartimy^{o} de Chacalla e de los demas sus partes contenydos el poder desta otra pte le sostituye e sostituye en Al^{o} de Herrera pr de corte en don Pedro Martynez clerigo thesorero de la yglesia desa çibdad de la Plata y en cada uno dellos ynsolidim para en todas las cosas y casos en el contenydos sin eçeptar ny rreservar en si cosa alguna e para ello les çedio el dho poder segun lo tiene de los dhos sus ptes con todas sus ynçidençias e dependençias anexidades e conexidades e los rreleve segun es rrelevado e obligo los bienes a el por el dho poder obligados e firmela de su n^{e}.

Sebastian de Leon
en fee de lo qual fize aqui este mio signo

Niculas Garçia Hrrez
tal en test^{nio} de verdad

Hr^{do} de Rivera

Ambr^{o} de Moscoso
escriv^{o} de su mag^{t}

don X^{o}val Vilcapoma

dio don Xristoval Vilcapoma
seys p^{o}s de su pte para en pte de
pago de derechos.

Por las preguntas syguientes sean preguntado los t^{o}s que son e fueren presentados por pte de don Xristoval Vilcapoma caçique e de don Felipe e de Fran^{co} Yndio e los demas y^{o}s del rrepartimi^{o} de Chacalla en el pleito con el caçique e y^{o}s del rrepartimiento de Canta sobre las tierras de Quybi.

1 - Primeramente sean preguntado sy conosçen a las ptes e sy tienen notiçia de las tierras e chacaras de Quybi sobre que es este pleito.

2 - Yten sy saben etc. que los caçiques y^{o}s e prençipales del rrepartimi^{to} de Chacalla an tenido e poseydo por suyas e como suyas las dhas tierras e chacaras de Quybi desde el tpo de los yngas pasados e por suyas an sydo tenidas de tanto tpo que memoria de onbres no es en contr^{o}. Digan lo que saben.

3 - Yten sy saben etc. que al tpo que los españoles entraron en este rreyno los dhos y^{o}s de Chacalla tenian e poseyan las dhas tierras e chacaras de Quybi e las tubieron e poseyeron como señores dellas syn contradiçion de persona alguna mucho tpo despues.

4 - Yten sy saben etc. que los dhos caçiques e y^{o}s de Canta en el tpo del marques don Fran^{co} Piç^{o} enbiaron a llamar a vn caçique que entonçes hera de Chacalla que se llamava Vilcacapcha y porque servia a los españoles e no se alçaba como ellos a trayçion y

estando el seguro lo mataron e juntamente con el çiento e çinquenta otros y^{o}s que abitavan en las dhas tierras e chacaras de Quybi.

5 - Yten si saben etc. que los dhos y^{o}s de Canta despues de aver muerto al dho caçique de Chacalla les toma-//
/14v/ ron por fuerça las dhas tierras e chacaras de Quybi y se las an tenido usurpadas y las tienen agora syn otro titulo ny cabsa. Digan lo que saben.

6 - Yten si saben etc. que por henero del año pasado de qui^{o}s e quarenta e nuebe que fue quando el presydente Gasca estava en este rreyno valia cada obeja de la tierra en esta çibdad a quatro pesos e no mas.

Digan los t^{o}s el valor de las dhas obejas.

7 - Yten sy saben etc. que las dhas tierras e chacaras de Quybi valian por el dho mes de henero del año pasado de qui^{o}s e quarenta e nuebe y al presente valen mas de ocho myll pesos de buen oro porque son muchas tierras y muy buenas y en ellas muchas caxas [*sic*] de coca y de mayz e aji e de otros ap^{o}bechamientos e matenymy^{tos}.

Digan los t^{o}s lo que valen e pueden valer las dhas tierras o chacaras de Quybi.

8 - Yten sy saben etc. que todo lo susodho es pu^{ca} boz e fama. Ponense por pusyçiones al caçique e prençipales del rrepartimyento de Canta.

El Licenc^{do} M^{o} Luzio Al^{o} Moreno

En los Reyes onze de otubre de myll e quy^{o}s e çinquenta e ocho a^{o}s en audiençia pu^{ca} ante los señores presydente e oydores. La presento el dho pr. en n^{e} de su pte e los dhos señores la ovieron por presentada quanto avia lugar de dr^{o} e mandaron que por el digan e declaren los t^{o}s que presentare e que las ptes absuelvan las dhas preguntas conforme a la ley e so la pena della.

Diego Muñoz

/15r/ En los Reyes doze dias del mes de otubre de myll e quy^{o}s e çinquenta e ocho años ante my Diego Muñoz escr^{o} de camara desta Real Audi^{a} Alonso Moreno en n^{e} de don Xristoval Bilcapoma caçique de Chaclla e los demas caçiques e prinçipales del dho rreptimy^{o} presento por t^{o}s para en esta causa a vn yndio que dixo llamarse Xristoval Malcachagua e a Diego Chunbicaxa naturales del pueblo de Guarochiri subjetos a don Sebastian caçique e asy mysmo presento por testigo a otro yndio que dixo llamarse H^{yn} Tocari natural de Atunyabyo los quales e cada uno dellos por lengua de Joan de Alvarado ynterprete desta Real Audi^{a} juraron por dyos nro s^{or} e sobre la señal de la cruz en forma de derecho e dixeron sy juramos e amen e prometieron dezir verdad.

Diego Muñoz

En los Reyes treze dias del dho mes de ot^{e} del dho año el dho Al^{o} Moreno en los dhos nonbres presento por t^{o}s a un yndio que dixo llamarse don P^{o} Xuyo e a otro que dixo llamarse don P^{o} Guacho natural de Ayabiri los quales juraron como xristianos que dixeron ser por lengua del dho Joan de Alvarado e prometieron dezir verdad.

Diego Muñoz

En los Reyes quatorze dias del mes de otubre del dho año el dho Alonso Moreno en los dhos nonbres presento por testigo para en esta causa a un negro que dixo llamarse R^{o} de Ampuero esclavo que fue de Fran^{co} de Ampuero e al presente es lybre el qual juro sobre la señal de la cruz en forma de dr^{o} dixo si juro e amen e prometio dezir verdad.

Diego Muñoz

En los Reyes veynte e dos dias del dho mes de ot^{e} del dho año el dho Al^{o} Moreno en los dhos nonbres presento por t^{o}s en esta causa a Alonso y GuamanMango yndio natural que dixo ser del pueblo de Mango en los Yanayabyos el qual juro sobre la señal de la cruz en forma de dr^{o} por lengua del dho Joan de Alvarado e prometio dezir verdad.

Diego Muñoz

En los Reyes veynte e quatro dias del dho mes de ot^{e} del dho año el dho A^{o} Moreno en los dhos nonbres presento por t^{o}s en esta causa a un yndio que dixo llamarse Juan Carvaguilca natural del pueblo de Yabcha e a otro yndio que dixo llamarse P^{o} Calco natural del pu^{o} de Pucurucha los quales juraron por la dha lengua segun de suso.

Diego Muñoz

/15v/ En los Reyes quatro dias del mes de noviembre de myll e quy^{o}s e çinquenta e ocho años el dho Alonso Moreno en los dhos nonbres presento por t^{o}s en esta causa a Xristoval de Leon el qual juro sobre la señal de la cruz en forma de derecho e dixo si juro e amen e prometio dezir verdad.

Diego Muñoz

En los Reyes quatro dias del mes de novi^{e} del dho año el dho Alonso Moreno presento por t^{o}s en esta causa en nonbre de sus ptes a un yndio que dixo llamarse Joan Merlo Chauca natural del pueblo de Guarochiri subjeto al caçique don Sebastian el qual dho yndio testigo juro por lengua del dho Joan de Alvarado ynterprete sobre la señal de la cruz en forma de dr^{o} e dixo sy juro o amen e prometio dezir verdad.

Diego Muñoz

En que los dhos t^{o}s de suso jurados dixeron en esta causa por las preguntas porque fueron presentados por t^{o}s / por si cada uno apartadamente en le sig.^{e}

/16r/

Probança de don X°val Vilcapoma
en el pleyto con el caçique e yndios
de Canta sobre las tyerras.

juro en 12 de ot^e
Christobal Malcachagua indio
nral de Guadacheri de 35 / no le toca.

T° el dho Xristoval Malcachagua yndio natural del pueblo de Guarocheri subjeto de don Sebastian caçique del dho pueblo aviendo jurado en forma de dr° e seyendo preg^do por las preguntas del ynterrogatorio por lengua de Juan de Alvarado ynterprete desta Real Audiençia dixo e depuso lo sig^e:

1 - A la primera pregunta dixo que conosçe al dho don Xristoval Vilcapoma caçique de Chacalla e a don Felipe Vilca Yanat e al dho Fran^co Yndio con todos en la pregunta e conosçe a algunos yndios e prinçipales del dho pueblo de Chacalla / e conosçe a don Hernando caçique de Canta e a algunos yndios e prinçipales del dho rrepartimy° e tiene notiçia de las tyerras e chacaras de Quibi por que las a visto y est^do en ellas.

Preg^do por las preguntas generales dixo que no se acuerda de la hedad que tiene mas de que por su aspeto paresçe ser de hedad de treynta e çinco años poco mas o m°s. E que no es pariente ny enemygo de nynguna de las ptes ny que a sido atemorizado ny sobornado ny le an dado cosa alguna para que diga la mentira y el contr° de la verdad ny le tocan nynguna de las preguntas generales de la ley que le fueron dhas.

2 - A la segunda pregunta dixo que lo que sabe desta pregunta es que en tpo de Guascar Ynga hera caçique del dho pueblo de //
/16v/ Chacalla Vilcapoma que hera antes que los xristianos vinyesen a esta tierra / el qual yva algunas vezes a Guarocheri de donde este t° es natural a visitar al caçique de dho pueblo de Guarocheri que se llamava Nynaguilca padre de don Sebastian caçique que es del dho pueblo al presente. El qual llebava maçorcas de mayz e coca e agi e otras cosas de comyda para el dho caçique Nynaguilca / e dezia: esto traygo de my chacara Quybi e que en tpo del Ynga tenya el dho caçique de Guarocheri chacaras en el dho Quibi y siendo este t° muchacho de doze o treze años por mandado del dho caçique Nynavilca fue donde el dho pueblo de Guarocheri a Quybi a ver e visitar las chacaras que alli tenya el dho caçique / e andandola visitando e myrando vioeste t° que las dhas chacaras que alli avia llamado Quibi heran de los dhos yndios de Chacalla las quales son yendo el rrio abajo a mano yzquierda e de la otra pte del rrio heran de los yndios de Canta e desde entonçes a esta pte los unos yndios e los otros an tenido e poseydo las dhas trras e chacaras e que este testigo oyo dezir a çiertos yndios de Chaclla vinyendose al dho Nynabilca como los yndios de Canta tanbien le querian tomar sus trras al tpo del alçamyento de los naturales e que el dho Nynabilca dixo que vendiesen sus trras e chacaras a los yndios de Canta e si mas molestias e daños les hizieren yran todos los yndios yabyos contra los yndios de Canta e asy despues aca se quedaron los dhos yndios de Canta //
/17r/ con las trras de la vna pte del rrio e con las de la otra e las an tenydo de los andenes para abaxo en el llano que se llama Quybi e que es todo lo que sabe desta pregunta.

3 - A la terçera pregunta dixo que este t° que antes que los españoles entrasen en esta tra los dhos yndios de Chaclla tenyan e poseyan las dhas trras de Quibi que es el dho llano despues de los andenes e las tenyan e poseyan como señores dellas sin contradiçion de los dhos yndios de Canta.

4 - A la quarta pregunta dixo que oyo dezir lo en ella contenydo estando en Guarochiri que muchos yndios lo vinyeron a dezir.

5 - A la quinta pregunta dixo que dize lo que dho tiene en la segunda pregunta e asy mysmo oyo dezir que despues de la muerte del dho caçique e yndios que los yndios de Canta an tenydo e poseydo las dhas trras.

6 - A la sesta pregunta dixo que no la sabe porque no se acuerda dello.

7 - A la septima pregunta dixo que este testigo a visto que las dhas trras de Quybi son buenas e le paresçe que en el tpo que la pregunta dize valyan tres myll e quy°s o quatro myll p°s porque a visto que en ellas ay muchos arboles de frutos e son buenas chacaras e trras las de Quybi hazia la pte de abaxo e la de arriba de andenes e pedregales e que esto es lo que sabe desta pregunta.

8 - A la octava pregunta dixo que lo que dho e declarado tiene //
/17v/ de suso es pu^co e notorio a este t° e es la verdad segun que lo tiene declarado para el juramento que hizo e no firmo este t° ny el dho ynterprete porque dixeron que no sabian.

Diego Muñoz

Di° Chunbicaxa nral de Guarocheri
de 34 / no le toca.

T^{o} el dho Diego Chunbicaxa natural del pueblo de Guarochiri subjeto que dixo ser a don Sebastian caçique prinçipal del dho pueblo aviendo jurado segun forma de derecho e siendo preguntado por lengua del dho Joan de Albarado ynterprete por las preguntas del dho ynterrogatorio dixo e depuso lo siguyente:

1 - A la primera pregunta dixo que conosçe al dho don Xristoval Bilcapoma caçique prinçipal de Chaclla e no conosçe a los demas de la pregunta. Preguntado por las preguntas g^{es} dixo que no sabe la hedad que tiene e paresçio por su aspecto ser de hedad de treynta e quatro años poco mas o menos e que no es pariente ny enemygo de nynguna de las ptes ny le va ynteres en esta causa ny le an dado ny prometydo que le daran cosa alguna para dezir este su dho ny le tocan las demas generales e que vença la pte que tuviere just^{a}.

2 - A la segunda pregunta dixo que este testigo oyo dezir a Bilca Poma caçique de Chaclla padre del dho don Xristoval e a otros yndios del dho Chaclla como las dhas trras de Quyvi heran suyas //
/18r/ e de sus yndios e syendo vn dia este t^{o} muchos dias a al tpo del alçamyento desta trra fue por mensagero a Chaclla e bio que las dhas trras de Quyvi las sembravan e benefyçiavan los yndios de Chaclla como suyas e que esto es lo que sabe desta pregunta.

3 - A la terçera pregunta dixo que este testigo oyo dezir a Nynabilca caçique de Guarochiri e a otros yndios viejos como las dhas trras heran de los dhos yndios de Chaclla antes que los españoles entrasen en esta trra e al tpo que entraron en ella las tenyan e que las tenyan e senbravan como suyas e dize lo que dho tiene en las preguntas antes desta.

4 - A la quarta pregunta dixo que lo en ella contenydo lo oyo dezir al dho Nynabilca caçique de Guarochiri e a otros muchos yndios viejos que dello dezian tener notiçia e por pu^{co} e notorio lo trattavan como cosa que çierto avia pasado.

5 - A la quinta pregunta dixo que lo en ella contenydo lo a oydo dezir por muy pu^{co} e notorio a los yndyos de Chaclla o a otros yndyos yabyos porque este t^{o} no le a visto.

6 - A la sesta pregunta dixo que al tpo que la pregunta dize valia cada vna oveja grande desta trra a quatro e a çinco p^{o}s e corderos e pacos cada cabeça a tres e a quatro p^{o}s o este t^{o} lo vendyo asy en aquel tpo e lo vio vender asy a otros yndios e esto es lo que sabe desta pregunta.

/18v/ 7 - A la septima pregunta dixo que le paresçe a este t^{o} que las dhas trras al tpo que la pregunta dize e al presente podian e podran valer tres myll e dozientos o tres myll e trezientos p^{o}s porque la mytad de las dhas trras por hazia abaxo es llano e donde ay arboles de fruttas e coca e agi e trra llana e del medyo arriba es andenes e pedregales e por esto e por lo ay visto le paresçe que pueden las dhas trras valer el presçio que dho tiene.

8 - A la octava pregunta dixo que lo que dho e declarado tiene de suso es verdad segun que lo a declarado e no firme ny el dho Joan de Alvarado ynterprete porque dixeron que no sabian e el dho Joan de Alvarado dixo aver hecho la dha ynterpretaçion bien e fyelmente sin añadir ny quytar della cosa alguna.

Diego Muñoz

Mrtin Tocaxi nral del pueblo
de Atunyavyo en cabeça de su mag^{t}
de 36 / no le toca.

T^{o} el dho M^{yn} Tocari natural del pueblo de Atunyauyo que en cabeça de su mag^{t} esta aviendo jurado segun forma de derecho e siendo preguntado por las preguntas del dho ynterrogatorio dixo e depuso lo syguyente:

1 - A la primera pregunta dixo que conosçe a los dhos don Xristoval Bilcapoma caçique de Chaclla e al dho don Felype prinçipal e conosçe a casi todos los yndios de Chaclla de los aver visto e no conosçe a las otras ptes e a visto las dhas trras de Quyvi porque las a visto una vez.
/19r/ Preguntado por las preguntas g^{es} dixo que no sabe la hedad que tiene paresçio por su aspeto ser de hedad de treynta e seys años poco mas o menos e que no es pariente ny enemygo de nynguna de las dhas ptes ny le va ynteres en esta causa ny le tocan las demas generales que a los demas t^{o}s desta causa se an preguntado e que vença la pte que toviere just^{a}.

2 - A la segunda pregunta dixo que este t^{o} a oydo dezir a muchos caçiques e yndios de Ananyavyo e Urinyavyo que dezian que las dhas trras e chacaras de Quyvi heran antiguamente en el tpo de Ynga e al tpo que los españoles entraron en esta trra heran propias de los yndios de Chaclla e las senbravan e benefyçiavan e como tales suyas las avian senbrado e benefyçiado e que en tpo del marques este t^{o} fue a las dhas chacaras de Quybi e vio que las senbravan e benefyçiavan los yndios de Chaclla como cosa suya sin que nyngunos yndyos se lo contradixesen.

3 - A las tres preguntas dixo que dize lo que dho tiene en la pregunta antes desta porque asy lo vio.

4 - A la quarta pregunta dixo que lo en ella contenydo lo oyo por muy pu[co] e notorio que asy avia pasado e lo dezian los caçiques e yndyos de Vrinyavyo e Ananyavyo.

5 - A la quynta pregunta dixo que asy mysmo lo en ella contenydo le a oydo dezir por muy pu[co] e notorio como dho tiene en la pregunta antes desta.

/19v/ 6 - A la sesta pregunta dixo que este t[o] vio que en el tpo que la pregunta dise valia vn carnero grande e una oveja grande quatro p[o]s cada vno y los corderos y pacos valia cada uno a dos e a tres pesos. Sabelo este t[o] porque a los dhos preçios vendio en aquel tpo este t[o] del dho ganado de la trra en esta çivdad.

7 - A la setima preg[ta] dixo que no la sabe mas de que una vez que las vio lo pareçe que pueden valer las dhas tierras tres myll e ochoçientos e noveçientos pesos que esto es lo que sabe e le pareçe.

8 - A la octava preg[ta] dixo que lo que dho tiene es berdad e lo que sabe para el juram[o] que hizo e no firmo porque no sabia ny el dho ynterpetre el qual dixo aver hecho la dha ynterpretaçion fielm[te] syn añadir ny quytar en ella cosa alg[a].

Diego Muñoz

Don P[o] Juyo curaca de los Ananyauyos
en cabeça de su mag[t] / de 40 / no le toca
juro en 13 de ot[e].

T[o] el dho don Pedro Xuyo yndio curaca que dixo ser de çiento yndios que dixo ser del pueblo de Tamara en los Ananyavyos que esta en cabeça de su mag[t] aviendo jurado segun forma de derecho e siendo preguntado por las preguntas del dho ynterrogatorio dixo e depuso lo siguy[e]:

1 - A la primera pregunta dixo que conosçe al dho don Xristoval Bilcapoma caçique de Chaclla e a otros yndios del dho rreptm[o] de que no tiene particular memoria e no conosçe a los caçiques de Canta e tiene notiçia de las trras de Quybi por las aver visto al tpo del alçamyento de los naturales desta trra.

/20r/ Preguntado por las preguntas g[es] dixo que no sabe la hedad que tiene e psçio por su aspecto de hedad de quarenta años poco mas o menos e que no es pariente ny enemygo de ninguna de las dhas ptes ny le va ynteres en esta causa ny le an dado ny prometido que le daran cosa alguna e que no le tocan las demas generales e que vença la pte que toviere just[a].

2 - A la segunda pregunta dixo que este t[o] a oydo dezir a Nynabilca caçique de Guarochiri e a otro que se llamava Xacxalaqui e a otros muchos caçiques e yndios de los de Guarochiri e Avchas e Chaclla por muy pu[co] e notorio como las dhas trras de Quyvi heran de los dhos yndios de Quyvi e lo avian sydo de mucho tpo a esta pte e que las tenyan e poseyan por suyas e como suyas.

3 - A la terçera pregunta dixo que al tpo que ovo el alçamyento de los naturales este t[o] vino a poner çerco sobre esta çiudad con vn yndyo ynga que se llama Quyçu Yupanguy al qual mataron en esta çiudad los españoles e despues deste este t[o] fue e yendo a su trra paso por las dhas trras e chacaras de Quyvi donde estevo vn dia e vio que las dhas trras son como corre el rrio a la mano yzquyerda e de la otra pte son las trras de Canta e que el dho su caçique deste t[o] que se llamava Xacxa Llaqui estando en las dhas trras de Quyvi e señalo e dixo: estas son las trras de los yndyos de Chaclla e las avian tenydo de mucho tpo antes e que entonçes las tenyan sin contradiçion alguna e asy lo vio este t[o] estonçes e despues lo oyo asi dezir por muy pu[co] e notorio.

/20v/ 4 - A la quarta pregunta dixo que lo en ella contenydo lo oyo dezir estando en esta çiudad e que se venyan a quexar los yndios de Chaclla al dho Nynabilca caçique prinçipal de Guarochiri e al caçique Racra Llaquy en que vio este t[o] que los yndios que se quexavan dezian que los yndios de Canta les avian muerto a los de Chaclla que heran los que se quexavan çiento e çinq[ta] otros yndios de los que estavan en Chaclla e con ellos al dho Vilcacapcha que hera primo hermano del dho Bilcapoma e que estonçes les tomaron las dhas trras por fuerça contra la voluntad de los yndyos de Chaclla / e que los dhos caçiques rrespondieron a los yndyos que se quexaron que estonçes no heŗa tpo de entender en ello e que despues que estoviese acabado e paçifico el rreyno entenderian en ello e que esto es lo que sabe desta preg[ta].

5 - A la quynta pregunta dixo que dize lo que dho tiene en la pregunta antes desta e que es verdad e es muy pu[co] e notorio que los yndyos de Canta despues aca an tenydo las dhas trras de Quyvi sin tener otro titulo ny causa ny rrazon para ello ny aviendo sido suyas ny de sus antepasados.

6 - A la sesta pregunta dixo que lo que della sabe es que al tpo que la pregunta dize valya en esta çiudad vna oveja o un carnero grande a quatro p[o]s e un cordero e un paco lo vno e lo otro valya a tres p[o]s e asy los vendyo este t[o] e vio que los yndios lo vendieron en el dho tpo e conque no valya mas prezçio el dho ganado porque estonçes avia mucho ganado de la trra.

/21r/ 7 - A la septima pregunta dixo que este testigo a visto como dho tiene las dhas trras e chacaras de Quyvi en que ay chacaras de coca e many e aji e de mayz e arboles de ffruta e son muy buenas hasia la pte de abaxo que es llano e lo otro es andenes e pedregales e que los yngas e caçiques antiguamente no acostumbravan vender en aquel tpo las trras e que aora al paresçer deste t[o] podran valer las dhas trras tres myll e nueveçientos p[o]s e que no sabe çierto si valen mas o menos.

8 - A la ottava pregunta dixo que lo que dho e declarado tiene de suso es pu[co] e notorio a este t[o] e es la verdad segun que lo tiene declarado para el juramento que hizo e no firmo porque dixo que no sabia ny el dho ynterprete el qual dixo aver hecho la dha ynterpretaçion bien e fyelmente sin añadir ny quytar en ella cosa alguna.

Diego Muñoz

Pedro Guacho nral de Ayabiri
de los Ananyauyos en cabeça de su mag[t].

T[o] el dho Pedro Guacho yndio natural que dixo ser del pueblo de Ayaviri de los Ananyavyos que esta en cabeça de su mag[t] subjeto a don Diego Curupaullo caçique prinçipal de los dhos Ananyavyos aviendo jurado segun forma de derecho e siendo preguntado por el dho ynterrogatorio dixo e depuso lo siguy[e]:

1 - A la primera pregunta dixo que conosçe al dho don Xristoval Bilcapoma caçique prinçipal de Chaclla e a otros yndyos del dho rreptmy[o] de cuyos nombres no tiene memoria e conosçe asy mysmo a don Hr[do] Gua-//
/21v/ man Yavre caçique prinçipal de Canta e no conosçe a otros yndios mas de al dho caçique e tiene notiçia de las dhas trras e chacaras de Quybi porque las a visto e estado en ellas.

Preguntado por las preguntas g[es] dixo que no sabe la hedad que tiene e paresçio por su aspetto ser de hedad de quarenta e tres años poco mas o menos e que no es pariente ny enemygo de ninguna de las dhas ptes ny lo an dado ny prometido que le daran por dezir este dho cosa alguna ny le va ynteres en esta causa e que en ella no le va ynteres [*sic*] e que vença la pte que toviere just[a].

2 - A la segunda pregunta dixo que antes que en esta trra entrasen los españoles oyo este t[o] dezir a Nynabilca caçique de Guarochiri e Racra caçique de Ananyauyo e a otros muchos caçiques e yndios de aquella provinçia como dezian que las dhas trras de Quyvi heran de los yndios de Chaclla e como suyas las tenyan e poseyan e sembravan e gozaban della.

3 - A la terçera pregunta dixo que lo que della sabe es que este t[o] vio que despues que los españoles entraron en esta trra los caçiques e yndios de Chaclla tenyan e poseyan las dhas trras de Quyvi como suyas e yendo en ellas una vez donde estovo en las dhas trras de Quyvi tres dias se las vio senbrar e gozar dellas e estovo [*sic*] fue çierto tpo antes que los yndyos se alçasen.

/22r/ 4 - A la quarta pregunta dixo que lo contenydo en ella lo oyo dezir a los dhos caçiques Nynabilca e a otros muchos yndios de Ananyauyo e Urinyauyo que asy avia pasado como la pregunta lo dize lo qual oyo estando en esta çiudad despues de muerto don Fran[co] Piçarro de sy a vn mes o dos.

5 - A la quinta pregunta dixo que lo en ella contenydo lo a oydo dezir a los dhos caçiques e yndyos que dho tiene en la pregunta antes desta lo que hera muy notorio entrellos.

6 - A la sesta pregunta que al tpo que la pregunta dize [*sic*] valya en esta çiudad vn carnero o vna oveja desta trra grande quatro p[o]s e los corderos e pacos a dos e a tres p[o]s / e esto lo sabe asy porque estonçes avia mucho ganado desta trra e este t[o] lo trajo a vender a esta çiudad e lo vio vender como dho tiene.

7 - A la septima pregunta dixo que en tpo de los yngas no se acostumbrava en esta trra a vender las chacaras e que despues que los españoles entraron en ella que se venden le paresçe a este t[o] que valdran las dhas chacaras e trras de Quyvi quatro myll p[o]s por ser trras donde avia coca e muchos arboles de frutta e agi e otras comydas e que la mayor pte dellas es llano e buena trra e del medyo para arriba es andenes e pedregales.

/22v/ 8 - A la ottava pregunta dixo que lo que dho e declarado tiene es verdad segun como lo sabe e a visto e oydo e entendydo para el juramento que hizo e no fyrmo porque dixo que no sabia e el dho Joan de Alvarado ynterprete dixo aver hecho la dha ynterpretaçion bien e fyelmente sin añadir ny quytar della cosa alguna e no firmo porque dixo que no sabia.

Diego Muñoz

R[o] de Ampuero Moreno fue esclavo de
Fran[co] de Ampuero encomendero de los yndios
de 43 / no le toca.
juro en 14 de ot[e] ante my.

T^{o} el dho Rodrigo de Ampuero Moreno esclavo que fue de Fran^{co} de Ampuero e al presente es lybre testigo presentado por pte del dho don Xristoval Bilcapoma aviendo jurado segun forma de derecho e siendo preguntado por las preguntas del dho ynterrogatorio dixo e depuso lo siguy^{e}:

1 - A la primera pregunta dixo que conosçe a las dhas ptes e yndios de Chaclla e Canta e a los mas dellos desde que mataron al marques don Fran^{co} Piçarro e tiene notiçia de las trras de Quyvi.

Preguntado por las preguntas g^{es} dixo que es de hedad de quarenta e tres años poco mas o menos e que no le va ynteres en esta causa ny le tocan las demas generales e que vença la pte que toviere just^{a}.

2 - A la primera [*sic*] pregunta dixo que lo que della sabe es que al tpo que el presydente Gasca yba al Cuzco estando en Guamanga sytio al Cuzco contra Gonçalo Piçarro los caçiques de Chaclla su tio del dho don Xristoval pidie-//
/23r/ ron al dho presydente que les hiçiese bolber sus trras que se las avia quytado el dho Gonçalo Piçarro e dadolas a los yndios de Canta e las avian tenydo e heran suyas del tpo de los yngas e que se las hiziese bolber e el dho presydente se las mando bolber e que las dexasen los yndios de Canta e que asy se las bolbieron las dhas trras e chacaras de Quyvi e este testigo estuvo en ellas que le pusieron ally don Fran^{co} su tio del dho don Xristoval e estando en ellas e senbrandolaslas [*sic*] e queriendolas los dhos yndios de Chaclla benefiçiar vinyeron Guaman Yabre e otros muchos yndios de Canta e hecharon fuera dellas a este t^{o} e a los yndios de Chaclla e descalabraron a muchos dellos e a este t^{o} le amarraron e se quedaron con las dhas trras de Quyvi e esto por asy de ay dos o tres meses poco mas o menos despues que se las avian de entregar a los dhos yndios de Chaclla los quales en aquella sazon avian començado ya a coger la coca e benefyçiar las dhas trras e que al tpo que ovo la dha diferençia e les quytaron las dhas trras vinyeron ally los dhos Guaman Yavre e Marcapoma caçique de Chaclla e entrellos trattaron por donde yvan los mojones e termynos de las dhas trras e se conçertaron e quedaron ally e este t^{o} se fue arriba e no vio mas.

3 - A la terçera pregunta dixo que este t^{o} yendo a las dhas trras pocos dias despues que mataron al marques don Fran^{co} Piçarro fue muchas vezes a las dhas trras de //
/23v/ Quyvi e vio en ellas yndios mytimaes de Chaclla e mytimaes de Canta que estavan rrebueltos los unos con los otros e en aquel tpo las tenyan por suyas los dhos yndios de Chaclla e siendo este t^{o} esclavo del dho Fran^{co} de Ampuero yba muchas vezes a las dhas trras e lo vio asy como dho tiene.

4 - A la quarta pregunta dixo que lo en ella contenydo lo oyo este t^{o} dezir a Marcapoma tio del dho don Xristoval Bilcapoma estando en las dhas trras e le mostro la pte donde avia muerto a los dhos yndios.

5 - A la quinta pregunta dixo que dize lo que dho tiene en la segunda pregunta deste ynterrogatorio.

6 - A la sesta pregunta dixo que en el tpo que la pregunta dize vio este t^{o} vender en est çiudad cada cabeça de oveja o carnero grande a quatro p^{o}s e lo que hera pequeño valya a menos presçio e tanbien los a conprado este t^{o} en el campo en el dho tpo a menos presçio de los dhos quatro p^{o}s.

7 - A la septima pregunta dixo que sabe que las dhas trras de Quybi son muy buenas donde ay coca e arboles de todas fruttas desta trra e otras sementeras e que valen mucho para los yndios por ser muy buen valle porque este t^{o} no sabe lo que valen.

8 - A la octava pregunta dixo que lo que dho e declarado tiene de suso es pu^{co} e notorio a este t^{o} e es la verdad segun que lo tiene declarado para el juramento que hizo e siendole leydo este susodho en el se rretifico e afirmo y no firmo porque dixo que no sabia.

Diego Muñoz

/24r/ Al^{o} Mrin indio de los Anayauyos
encomen^{o} en doña Mª Martel
de 40 / no le toca.
juro en 22 de otu^{e}.

T^{o} el dho yndio que dixo llamarse Alonso Martin y antes que fuese xristiano Guaman Mango natural que dixo ser del pueblo de Mango en los Yananyabyos encomendado en doña Maria Martel sujeto al caçique don Fran^{co} Guaynacama t^{o} presentado por el dho Al^{o} Moreno en n^{e} de los caçiques e y^{o}s de Chacalla sus ptes aviendo jurado segun forma de dr^{o} como xristiano que dixo ser y siendo preg^{do} por las preguntas del dho ynterrgat^{o} por lengua del dho Ju^{o} de Alvarado ynterprete dixo e declaro lo siguiente:

1 - A la primera preg^{ta} dixo que conosçe a los caçiques prençipales de Chacalla y Canta y a don Felipe y a Fran^{co} Yndio y a otros y^{o}s que son de ambas partes e que tiene notiçia de la dhas trras de Quybi sobre que es este pleyto.

Preguntado por las preg^{tas} generales dixo no saber su hedad paresçio por su aspeto ser de hedad de quarenta años poco mas o menos y que no es pariente ny enemygo de nynguna de las ptes ny le an dado ny prometido que le daran cosa alguna para que en este caso diga el contrario de la verdad ny le an amenazado ny atemorizado para ello e que vença la pte que tuviere just^{a}.

2 - A la segunda preg^{ta} dixo que lo que della sabe es que syendo este t^{o} muchacho antes que los españoles entrasen en esta trra oyo este t^{o} dezir a Nynabilca caçique de Guadacheri y a Vilcapoma caçique de Chacalla y a otros muchos caçiques e //
/24v/ prinçipales de aquella comarca en como las dhas tierras de Quyvi eran de los dhos y^{o}s de Chacalla y les tenyan e poseyan como suyas y asy vio este t^{o} que el dho Vilcapoma caçique de Chacalla yendo a ver y visytar al dho Nynavilca caçique de Guaracheri llevava coca y aji y otras frutas que le dava de presente y quando se las dava le oya dezir este t^{o} que aquello era de las tierras e chacaras de Quivi y que despues que los españoles entraron en esta trra vio este t^{o} al tpo que se alçaron los naturales della como los dhos y^{o}s de Chacalla tenyan e poseyan las dhas tierras de Quyvi y este t^{o} lo bio asy vinyendo con los yngas de tpo que vinyeron de guerra para echar los españoles desta çivdad e tierra e que esto es lo que sabe.

3 - A la terçera preg^{ta} dixo que dize lo que dho tiene en la pregunta antes desta.

4 - A la quarta pregunta dixo que lo que della sabe es que al tpo que dho tiene que vino con los dhos yngas al dho alçamy^{o} este t^{o} bino con un caçique de Canta que se llamava Bilca Maqui y estando en las dhas tierras de quivi bio como el dho caçique dixo muy enojado como por que los caçiques e y^{o}s de Chacalla no querian venir con ellos contra los españoles syno serbillos. El dho caçique les hizo arrancar los cocales e otros arboles y comidas que alli avia diziendo: comed vosotros esto pues estos y^{o}s no quyeren venir a serbir al ynga e que estando estavan alli algunos y^{o}s de Chacalla y que despues desto al tpo que los yngas bolvieron //
/25r/ desta çibdad despues del dho alçamy^{o} estando en un pueblo que se dize Carvacaya del rrepartimy^{o} de Canta de alli los yngas y el caçique de Canta que se llamava Guaman Yaure enbiaron a llamar a Vilcapoma caçique de Chacalla para que alli vinyese y el dho caçique Vilcapoma no quiso benir y enbio a otro caçique que se llama Vilcacapcha el qual como vino alli le mataron los y^{o}s y este t^{o} lo oyo asy dezir el mysmo dia que aconteçio porque yba delante con otros y^{o}s que yban camynando y que despues desto este t^{o} vio que Guaman Yavre caçique de Canta mato a Vilcapoma caçique de Chacalla y que despues de avello muerto le vio este t^{o} mandar que fuesen çiertos y^{o}s a matar los y^{o}s que estavan en las chacaras de Quyvi y alli mataron al dho bilcacapcha y a mas de çiento y^{o}s con el y de alli adelante se quedaron con las dhas tierras de Quibi e que esto es lo que sabe desta preg^{ta}.

5 - A la quinta preg^{ta} dixo que dize lo que dho tiene en la preg^{ta} antes desta y que despues aca que los dhos y^{o}s de Canta mataron al caçique e y^{o}s de Quivi se an quedado con las dhas trras no syendo suyas ny de sus antepasados e que esto es lo que sabe desta pregunta.

6 - A la sesta preg^{ta} dixo que al tpo que la preg^{ta} dize vio este t^{o} que en esta çiudad se vendia un carnero e una oveja grande cada bno en quatro p^{o}s y los medianos a tres y los pequeños a dos y este t^{o} lo vio vender asy y lo vendio.

/25v/ 7 - A la septima preg^{ta} dixo que las dhas trras de Quibi al tiempo que este t^{o} las vio / vio que eran muy grandes chacaras e mucha tierra y en ellas muchas comydas y arboles de fruta e que no sabe lo que vale e que esto es lo que sabe desta preg^{ta}.

8 - A la otava preg^{ta} dixo que lo que dho e declarado tiene es verdad e lo que sabe deste caso para el juramento que hizo y el dho Juan de Alvarado ynterprete dixo aver hecho la ynterpretaçion del dho testigo bien y fyelmente sin añadir ny quitar dello cosa alguna e no lo firmo porque no sabia ny el dho ynterprete.

Diego Muñoz

Ju^{o} indio del pueblo Yaucha
encom^{o} en Jhrmo de Silba /
de 40 / no le toca.
juro en 24 de otubre.

T^{o} el dho yndio que dixo llamarse Juan e antes que fuese xristiano Carvaguilca natural que dixo ser del pueblo Yaucha encomendado en Yeronymo de Silva subjeto al caçique don Geronymo Pomachagua testygo presentado por el dho Alonso Moreno en n^{e} de los yndios de Chacalla sus ptes aviendo jurado por lengua de Juan de Alvarado en forma de drcho como xristiano que dixo ser e syendo preguntado por las preguntas del dho ynterrogatorio por la dha lengua dixo e depuso lo sig^{e}:

1 - A la primera pregunta dixo que conosçe a los caçiques prinçipales de Chacalla y al dho don Felipe e Fran^{co} Yndio e a otros yndios del dho pueblo e asi mysmo conosçe al caçique e prinçipales de Canta e a algunos yndios del dho rrepartimy^{o} e tiene notiçia //
/26r/ de las dhas tyerras e chacaras de Quybi porque a estado en ellas / e sobre que es este pleyto.

Preguntado por las preguntas generales de la ley dixo no saber su hedad. Paresçio por su aspeto ser de hedad de quarenta años poco mas o menos e que no es pariente ny enemygo de ninguna de las partes ny le an dado ny prometido que le daran cosa alguna porque en esta causa diga el contrario de la verdad ny le an amenazado ny atemorizado para ello o ffue aperçibido diga la verdad syno que sera castigado sy dize la mentira.

2 - A la segunda pregunta dixo que lo que sabe es que antes que los xristianos entrasen en esta trra este t^{o} oyo dezir a Pomachagua caçique del pueblo de Yaucha de donde es natural e al caçique de Guadacheri que se llamava Nynaguilca e a otros caçiques e prinçipales de los pueblos de Ananyauyo e Hurinyauyo que las dhas trras e chacaras de Quybi heran del caçique e yndios de Chacalla e de sus antepasados e que este t^{o} en tpo de Guascar Ynga por mandado de un capitan suyo que se dezia Andaguarca governador del balle de Xauxa por su mandado vino al dho valle e chacara de Quybi por coca e agi para el dho ynga y estuvo en ella dos dias e vide que los dhos yndios de Chacalla estavan en la dha chacara senbrando e benefiçiando la coca que en ella avia / e que despues de venydos los xristianos al tpo del alçamy^{o} de los naturales este t^{o} venya con el ynga llamado Quyçu Yupangui y otro Yanqui Yupangui contra los xristianos llegaron al valle de Lati e de alli vinyeron a esta çibdad y estando aquy contra los dhos españoles murio el uno de los yngas e por su muerte se rretiraron e fueron huyendo todos los yndios e llegaron el dho ynga y este t^{o} //
/26v/ e muchos yndios que yban con el al dho valle e chacaras de Quybi e vide que estavan alli los yndios de Chacalla como de antes este t^{o} los avia visto la primera vez que alli estuvo e que tenyan la dha chacara e trras por suyas e como suyas e que este es lo que sabe desta pregunta.

3 - A la terçera pregunta dixo que dize lo que dho tiene en la pregunta antes desta a que se rrefyere.

4 - A la quarta pregunta dixo que lo que sabe es que en tpo que entraron los españoles en esta tierra estando en el dho su pueblo llego vn yndio de Chacalla a dar aviso a su caçique Pomachagua como los yndios de Canta e caçiques del avian muerto ciento e cinquenta e tres yndios de Chacalla que estavan en la chacara e tierras de Quybi porque servian a los españoles e no se querian alçar como ellos se avian alçado con el ynga contra los dhos españoles e que con los dhos yndios mataron al dho caçique llamado Vilcacapcha con su muger e hijos e asy fue pu^{co} e not^{o} e que este es lo que sabe.

5 - A la quynta pregunta dixo que es pu^{co} e notr^{o} que los dhos yndios de Canta despues que mataron a los dhos yndios de Chacalla que estavan en las dhas tyerras e chacaras de Quybi las tomaron por fuerça e contra su voluntad e las an tenydo e poseydo por fuerça e que hasta agora las tienen e poseen e que esto es lo que sabe.

6 - A la sesta pregunta dixo que este t^{o} sabe que quando el presydente Gasca estuvo en esta çibdad valia una oveja grande e un carnero quatro p^{o}s / e un cordero / o paco tres p^{o}s e dos p^{o}s y este t^{o} lo bio asi vender a muchos yndios e por esto lo sabe porque lo vio.

/27r/ 7 - A la setima pregunta dixo que este t^{o} dize lo que dho tiene en la segunda pregunta e que no sabe lo que las dhas tierras en aquel tpo valian ny lo que al presente valen porque este t^{o} no sabe vender ny conprar chacaras ny esta acostumbrado a ello / mas de que como dho tiene en el dho tpo que estuvo en las dhas chacaras de Quybi vido que avia muy buenas trras e chacaras de coca e agi e arboles frutales e otras comydas que en ellas se davan e que esto es lo que sabe desta preg^{ta}.

8 - A la otava pregunta dixo que lo que dho tiene es la verdad e lo que deste caso sabe para el juramento que hizo y el dho Juan de Alvarado ynterpetre dixo aver declarado e ynterpetado bien e fyelmente lo que el dho yndio a rrespondido e dho a lo que le a sido pregunt^{do} syn añadir ny quytar dello cosa alguna e es la verdad para el juramento que tiene hecho e no firmo ny el dho testygo porque no supieron.

Diego Muñoz

P^{o} Calco yndio natural de Pucurucha
encom^{o} en Hr^{do} R^{o} Nyno
paresçio de 60 años / no le toca.

T^{o} el dho Pedro Calco yndio natural del pueblo de Pucuracha subjeto a don Hr^{do} caçique del dho pueblo encomendado en el Hr^{do} R^{o} Nyno testigo presentado por pte de los dhos yndios de Chaclla aviendo jurado segun forma de derecho por lengua de Joan de Alvarado ynterprete desta Real Audiençia por las preguntas del dho ynterrogatorio dixo e depuso lo siguyente:

1 - A la primera pregunta dixo que conosçe a los caçiques de Canta e a los dhos caçiques de Chaclla don Xristoval Bilcapoma e don Felipe //
/27v/ e Fran^{co} e otros yndios del dho rreptmy^{o} e que tiene notiçia de las dhas tierras de Quyvi por aver estado en ellas çinco vezes.

Preguntado por las preguntas g^{es} dixo que no se acuerda la hedad que tiene. Pareçio ser de sesenta años poco mas o menos e que no le an dado ny prometido que le daran nynguna de las ptes por que diga el contrario de la verdad ny le an atemorizado e que no tiene pte en las dhas trras mas de que dira la verdad como lo supiere e no le tocan las demas g^{es} e que vença la pte que toviere just^{a}.

2 - A la segunda pregunta dixo que lo que della sabe es que en tpo de Guaynacapa este t^{o} fue tres o quatro vezes a las dhas trras de Quyvi a rrescatar ovejas con los yndios serranos e vio que los yndios de Chaclla tenyan e poseyan las dhas trras e las senbravan e cogian el frutto dellas como cosa suya e estando en las dhas trras oyo dezir a los yndios contiguos e muy viejos como tenyan aquellas trras desde muy antiguo e las avian tenydo asy sus antepasados e que esto es lo que sabe desta pregunta.

3 - A la tercera pregunta dixo que dize lo que dho tiene en las preguntas antes desta e que oyo dezir a los yndios de Chaclla vinyendo el pueblo deste t^{o} de cuyos nonbres no se acuerda como al tpo que la pregunta dize tenyan e poseyan las dhas trras de Quyvi e que las tenyan e poseyan como suyas e no vio ny supo ny entendio que se las contradixesen otros yndios ny persona alguna e que esto es lo que sabe desta pregunta.

/28r/ 4 - A la quarta pregunta dixo que lo en ella contenydo le oyo dezir a los caçiques de Chaclla llamados Bilcapoma e de otros nonbres e a los caçiques de algunos yndios de Canta diziendo los de Canta: nosotros avemos muerto çiento e tantos yndios de los que estavan en Quyvi e que este oyo dezir asy quando volbio el ynga con los caçiques de Chaclla deste valle a la syerra despues del alçamyento e que los avian muerto porque los dhos yndios de Chaclla e Quyvi servian a los españoles e no se avian alçado como ellos y no avian hecho lo que el ynga les mandava e que esto es lo que sabe desta pregunta.

5 - A la quinta pregunta dixo que despues que los dhos yndios de Canta tomaron las dhas trras a los de Chaclla este t^{o} no a ydo a las ver ny a estado en ellas pero a oydo dezir que las an tenydo hasta aora a muchos yndyos de cuyos nonbres no tiene memoria al presente.

6 - A la sesta pregunta dixo dixo [*sic*] que al tpo que la pregunta dize este t^{o} vio que valya en esta çiudad una cabeça de oveja grande vn peso e medyo e otras que valyan menos e este t^{o} las conpro para comer en Quyvi al dho presçio con los yndios serranos e que esto es lo que sabe desta pregunta.

7 - A la septima pregunta dixo que las dhas trras e chacaras de Quyvi son buenas e muchas trras donde ay e se pueden senbrar
muchas comydas e tienen arboles de frutta e que no sa-//
/28v/ ve lo que pueden valer porque nunca a visto vender trras.

8 - A la ottava pregunta dixo que lo que dho e declarado tiene de suso es verdad para el juram^{o} que hizo e no firmo porque dixo que no sabia e en ello se rretifyco e el dho Joan de Alvarado ynterprete dixo que a hecho la dha ynterpretaçion bien e fyelmente sin añadir ny quytar dello cosa alguna e no lo firmo porque dixo que no sabia.

Diego Muñoz

Juro en quatro de novi^{e} ante my
Xristoval de Leon de 40 a^{o}s / no le toca.

T^{o} el dho Xristoval de Leon testigo presentado por el dho Alonso Moreno en n^{e} de los dhos sus ptes aviendo jurado segun forma de derecho e siendo preguntado por las preguntas del dho ynterrogatorio dixo e depuso lo siguy^{e}:

1 - A la primera pregunta dixo que conosçe a los caçiques prinçipales de ambos rreptmy^{o}s e a algunos prinçipales e yndios dellos porque a estado en los dhos rreptmyentos muchas vezes e tiene notiçia de las trras e chacaras e trras [*sic*] de Quyvi porque las a visto e estado en ellas.

Preguntado por las preguntas generales dixo que es de hedad de quarenta años poco mas o menos e que no le tocan las generales ny alguna dellas e que vença la pte que toviere just^{a}.

2 - A la segunda pregunta dixo que lo que della sabe es que este t^{o} puede aver diez años poco mas o menos que este t^{o} fue a las
dhas trras de Quyvi con Anpuero e Nyculas de Ribera e Alonso Davila a meter en paz e conçertar //
/29r/ los caçiques de Chaclla con los de Canta sobre la diferençia que tenyan en rrazon de las dhas trras de Quyvi e se avian descalabrado e este t^{o} sito a vno dellos e estando en el mysmo asyento de Quyvi uvo conçierto entre ellos en que los de Canta diesen a los de Anpuero que son los de Chaclla dozientas cabeças de ganado desta trra e que con ellas dexasen las dhas trras e se quytasen de rruido e estonçes este testigo entendyo que los dhos yndios de Chaclla pretendian que heran suyas las dhas trras de Quyvi e por tales las defendian e esto es lo que sabe desta pregunta.

3 - A la terçera pregunta dixo que dize lo que dho tiene en la pregunta antes desta [*sic*] e no sabe otra cosa.

4 - A la quarta pregunta dixo que no la sabe.

5 - A la quinta pregunta dixo que dize lo que dho tiene en las preguntas antes desta e que cree que si los dhos yndios de Canta poseen las dhas trras de Quyvi al presente es por aver dado a los de Chaclla las dhas ovejas e no sabe otra cosa.

6 - A la sesta pregunta dixo que este t° oyo dezir en aquella sazon a vn mayor^mo de Fran^co de Anpuero que no se acuerda de su n^e e a otro su criado que se llamava Molyna que se vendian estonçes las cabeças de ganado desta trra a quatro p°s e medyo e a quatrto e a menos e otras avia a mas presçio e que esto sabe desta pregunta.

7 - A la septima pregunta dixo que no la sabe porque no myro en las dhas trras quando las vio ny tener espiriençia dello ny saber la quantidad de trras que son.

8 - A la ottava pregunta dixo que lo que dho tiene de suso es la verdad segun que lo a declarado para el juram° que hizo e en ello se rretifyco e firmolo de su n^e.

Xristoval de Leon Diego Muñoz

/29v/ Juro en quatro de novi^e
Joan Merlo Chauca yndio del pu° de Guarochiri subjeto
a don Sebastian caçique / paresçio de 60 años / no le toca.

T° el dho yndio que dixo llamarse Joan Merlo Chauca natural del pueblo de Guarochiri subjeto a don Sebastian caçique prinçipal del testigo presentado por pte de los dhos caçiques e yndios de Chaclla aviendo jurado segun forma de derecho e siendo preguntado por las preguntas del dho ynterrogatorio dixo e depuso lo siguy^e:

1 - A la primera pregunta dixo que conosçe al dho don Xristoval Bilcapoma caçique prinçipal del Chaclla e a don Felipe e a otros prinçipales e yndyos del dho rreptmy° e conosçio a Guaman Yaure caçique prinçipal de Canta e al que al presente lo es no lo conosçe e conosçe a algunos yndios de Canta de vista e tiene notiçia de las dhas trras e chacaras de Quybi porque a estado en ellas e las a visto e estado en ellas tres vezes que fue a ellas por mandado de Nynabilca caçique prinçipal de Guarochiri.

Preguntado por las preguntas generales dixo que no sabe la hedad que tiene. Paresçio ser de hedad de mas de sesenta años e que no le an dado ny prometido que le daran cosa alguna ny atemorizado ny le va ynteres en esta causa ny es pariente ny enemygo de nynguna de //
/30r/ las partes ny le tocan las demas generales e que vença la pte que toviere just^a.

2 - A la segunda pregunta dixo que lo que della sabe es que siendo este t° moço antes que entrasen los españoles en esta trra este t° que tres vezes por mandado de Nynabilca caçique de Guarochiri a visitar e myrar las chacaras que en el valle de Quyvi tenya el dho Nynavilca para coca e vio estonçes que donde se juntan los dos rrios a la mano yzquyerda es de los yndios de Canta e a la mano derecha son las chacaras de Quyvi de los yndios de Chaclla e todo aquello se llama Quyvi e que estavan en las dhas trras de Quyvi de la pte de Chaclla dos prinçipales con çiento yndio que el uno dellos se llama Vilcacapcha e otro Carvacapcha los quales con los dhos çiento yndios benefiçiavan las dhas trras e las senbravan e cogian el fructo dellas de coca e mayz e agi e otras comydas e asy lo vio este t° tres años en cada uno dellos una vez que fue a las dhas trras e que luego entraron los españoles en esta trra e no fue mas a las dhas trras e esto fue en tpo de Guaynacapa ynga por cuyo mandado este t° vino desde Xauxa a ellas por coca e agi e yuca para el dho ynga e oyo dezir a los yndios antiguos mas viejos o mas ancianos que desde una sierra que se llama Tayaca Xaly hasta Quyvi llegavan los mojones o termynos de Chaclla e que las dhas trras de Quyvi de la pte de la mano derecha era de los dhos yndios de Chaclla e nunca oyo dezir ny entendyo otra cosa en contrario hasta a poco tpo a esta pte que oyo dezir que los yn-//
/30v/ dios de Canta pretendian las dhas trras de Quyvi e que esto es lo que sabe desta pregunta.

3 - A la terçera pregunta dixo que dize lo que dho tiene en la pregunta antes desta e que es asy verdad que quando los españoles entraron en esta trra los dhos yndios de Chaclla tenyan e poseyan las dhas trras de Quyvi.

4 - A la quarta pregunta dixo que lo que della sabe es que al tpo que los naturales deste rreyno se alçaron contra los españoles que en el avia este t° vino con Quyçu Yupanguy e Ylla Topa e otros muchos yngas para esta çiudad e que pasando por el valle de Quyvi vio este t° como por mandado del ynga Quyçu Yupanguy un yndyo dio al dho Vilca Capcha muchos palos con bna porra e quando para morir e hecho esto se vinyeron a esta çiudad para el dho alçamy° e despues que bolbieron los dhos yngas pasando por las dhas trras de Quyvi hallo ally al dho Vilcacapcha e lo llevo consigo un quarto de legua e en vn arroyo que se llama Malluc le cortaron la cabeça los yndios de Canta e este t° por ser su conosçido e que hera su caçique le enterro en vn agujero e que despues de hecho esto vio como los yndios de Canta començaron a matar a los yndios de Chaclla que estavan en las dhas trras de Quyvi e este t° de myedo que lo matasen con ellos se huyo hazia su trra con el dho Bilcapoma e Hurcopoma e otros yndios a que llamose e otro dia por la mañana vino ally un mensajero del ynga a dezir que fuesen muchos yndios a servir el ynga e asy fueron çiento e çinquenta yndios e con ellos los dhos Vilcapoma //
/31r/ e Urcopoma e oyo dezir despues otro dia adelante como los yndios de Canta avian muerto los dhos çiento e çinq^ta yndios que fueron a Quyvi por mandado del ynga e le oyo dezir a muchos yndyos de los Yauyos e de Canta los quales dhos çiento e çinq^ta yndios heran de los que estavan en las dhas trras de Quyvi por mytimaes e que desde entonçes se quedaron los dhos yndios de Canta con las dhas trras de Quyvi e las an tenydo hasta aora e que esto lo a asy oydo dezir a los yndios de Chaclla e que esto es lo que sabe desta pregunta.

5 - A la quinta pregunta dixo que dize lo que dho tiene en las preguntas antes desta e que nunca las dhas trras de Quyvi fueron de los yndios de Canta sino las que ellos se tenyan como lo tiene declarado en la segunda preg^{ta}.

6 - A la sesta pregunta dixo que lo que della sabe es que este t^{o} por el tpo que la pregunta dize vio este t^{o} [*sic*] que en esta çiudad valia vna oveja grande quatro p^{o}s e un paco dos p^{o}s e dos p^{o}s e medyo e este t^{o} lo vendyo a este presçio en esta çiudad e lo vio vender asy a otros yndyos.

7 - A la septima pregunta dixo que en tpo de los yndios pasados antes que los españoles entrasen en esta trra no se vendian las chacaras e por esto no sabe lo que valen las dhas trras de Quyvi e que sabe que son buenas trras e muchas para coca e agi e mayz e otras comidas e tienen muchos arboles de frutta e otros aprovechamyentos e que esto es lo que sabe desta pregunta.

/31v/ 8 - A la ottava pregunta dixo que lo que dho e declarado tiene de suso e es la verdad segun que lo tiene declarado para el juramento que hizo e no firmo porque dixo que no sabia / la qual dha declaraçion hizo este t^{o} por lengua de Joan de Alvarado ynterprete desta Real Avdi^{a} que dixe averla hecho bien y fyelmente sin añadir ny quytar dello cosa alguna e no firmo porque dixo que no sabia.

Diego Muñoz

/32r/ El caçique e y^{o}s de Canta.

Fran^{co} de la Torre en n^{e} de los yndios de Canta con los de Chaclla. pagaron los yndios seys p^{o}s para çinqu^{ta} deste pleyto.

Por las preguntas siguyentes sean preguntado los t^{o}s que son e fueron presentados por pte del caçique e yndios de Canta encomendados en Nycolas de Ribera en el pleyto que trata con los yndios de Chaclla sobre las tierras del balle de Quybi que estan juntas a las tierras e pueblo de los y^{o}s del dho valle de Canta.

1 - Primeramente si conosçen a las ptes y a cada vna dellas y tienen notiçia del valle e tierras que dizen de Quybi que estan en los tr^{o}s de los dhos pueblos de Canta y de quanto a esta pte a; y si conosçieron al presydente Gasca e conosçieron a los yngas pasados Topa Ynga Yupangui y Guaynacapa / e los demas señores naturales que an sydo deste rreyno; y si conosçen a Fran^{co} de Anpuero que tiene en encomyenda los y^{o}s de Chaclla y conosçen a nycolas de Rribera que tiene en encomyenda a los dhos yndios de Canta; y si tienen notiçia de vn rrio que se llama el rrio de Quybi que parte los termynos del dho rrepartym^{o} de Canta y tienen carta de vna escritura de conçierto que esta firmada del dho Rribera y del dho Ampuero y este ba en esta cabsa que pido se muestre a los t^{o}s.

2 - Yten si saben etc. que las dhas tierras que dizen de Quybi estan en todos los termynos e pueblos del dho rrepartymy^{o} de Canta y sus ayllos e parçialidades porque el dho rrio que dizen de Quybi parte los dhos tr^{o}s de Canta con los y^{o}s que estan de la otra pte del rrio de Quybi que son Atavillos e otras parçialidades de y^{o}s los quales dhos tr^{o}s tienen los dhos y^{o}s de Canta desde que poblaron donde agora estan poblados e de tpo ynmemorial a esta pte sin aver avido en ello contradiçion alguna sino que son termynos conosçidos suyos. Digan lo que saben.

3 - Yten si saben etc. que las dhas tierras e chacaras de Quybi desde vno y dos y diez e veynte e treynta e çinq^{ta} y mas años a esta pte que memoria de hombres no es en contrario siempre fueron tierras e chacaras de los dhos yndios de Canta y su caçique los quales siempre las an tenydo y poseydo por suyas y como suyas labrandolas y senbrandolas y haz^{do} dellas como cosa suya propia / hasta que pasando por las dhas tierras Guaynacapa / y paresçiendole fertiles las poblo de mytimaes y asy al tpo que el dho Guaynacapa yba a las p^{o}binçias de Quyto lo qual puede aver treynta y çinco o treynta y seys a^{o}s.
Digan lo que saben.

4 - Yten si saben etc. que al dho tpo que el dho Guaynacapa paso por las dhas trras e pueblos de Canta cuyas heran/el dho Guaynacapa / por paresçelle fertiles mando poner e puso çiertos yndios mytimaes que las labrasen para el y asy fueron ally por mytimaes çiertos y^{o}s yabyos de los de Chaclla e los quales m^{do} que las labrasen y senbrasen para el dho ynga y asy lo hizieron quatro o çinco o seys a^{o}s fasta que los españoles entraron en estos rreynos. Digan lo que saben.

5 - Yten si saben etc. que despues que entraron los españoles en este rreyno que fue poco tpo despues que el dho ynga puso los dhos mytimaes en las dhas trras de Quybi los dhos y^{o}s yabyos que asy puso por mytimaes dexaron despobladas las dhas tierras de Quybi e se fueron a sus pueblos y el caçique e y^{o}s de Canta tornaron a tomar e poseer las dhas trras de Quybi y asy las an tenydo y poseydo desde entonçes hasta agora sembrandolas y labrandolas y haz^{do} dellas como de cosa suya propia y como lo heran de antes que pusiese los dhos mytimaes el dho Guaynacapa segun lo que saben.

/32v/ 6 - Yten si saben etc. que tenyendo e poseyendo los dhos y^{o}s de Canta las dhas tierras de Quybi como cosa suya puede aver nueve años poco mas o menos que moviendo pleyto los dhos y^{o}s de Chaclla a los dhos y^{o}s de Canta sobre las dhas trras diz^{do} que les pertenesçian por rrazon de aver sydo çiertos y^{o}s suyos mytimaes en ellas y defendiendose el dho caçique e yndios de Canta ante el presydente Gasca y mandando el dho presydente Gasca que fuesen a ver las dhas tierras e p^{o}curasen de conçertar a las dhas partes los dhos y^{o}s de Canta se conçertaron con los dhos y^{o}s de Chaclla por rredimyr su vejaçion en que les davan como les dieron dozientas ovejas por que se dexasen del dho pleyto segun consta por el conçierto que tengo presentado en esta cabsa que pues sea mostrado a los t^{o}s. Digan lo que saben.

7 - Yten si saben etc. que los dhos yndios de Canta y su caçique dieron a los dhos y^{o}s de Chaclla y a su caçique en cumplim^{o} del dho conçierto las dhas dozientas ovejas y se las dieron y entregaron a ellos mesmos y a el dho su caçique e desde entonçes sea los dhos yndios de Canta an tenydo e poseydo las dhas tierras sin contradiçion ny pleyto alguno. Digan lo que saben.

8 - Yten si saben etc. que todo lo susodho es pu^{co} e notorio e pu^{ca} boz e fama y ponense por pusiçiones a los dhos yndios de Chaclla para que las juren e declaren conforme a la ley e so la pena della.

El licen^{do} de Leon　　　　Fr^{o} de la Torre

Añadida

Yten si saben etc. que los dhos yndios de Chaclla tienen sus tierras e tr^{o}s muy apartadas de las dhas tierras de Quyvi e de la otra pte del dho rrio de Quyvi pasados vnos çerros que estan entre el dho rrio de Quyvi e pueblos de Chaclla como es notorio por manera que nynguna continuydad ny nynguna apariençia de ser suyas ny tener derecho a las dhas tierras de Quybi tienen los dhos y^{o}s de Chaclla por ser como es cosa tan distinta y apartada de las dhas tierras de Quybi los dhos pueblos de Chaclla. Digan lo que saben.

Diego Muñoz

En los Reyes catorze de otubre de myll e quy^{o}s e çinquenta e ocho a^{o}s en audi^{a} pu^{ca} ante los señores presydente e oydores / presento el dho pr. este ynterrogat^{o} de preguntas e los dhos señores le ovieron por presentado quanto avia lugar de drcho y hera ynpertynente e que los t^{o}s que presentare digan lo que saben.

Diego Muñoz

/33r/ En la çibdad de los Reyes en dyez y siete dias del mes de otubre de myll e quy^{o}s e çinq^{ta} e ocho años ante los señores oydores en Abdiençia Real Fran^{co} de la Torre en nonbre del caçique de Canta en el pleyto con los yndios de Anpuero sobre las trras / presento por testigo a Di^{o} Dabila del qual se tomo e rresçibio juram^{o} en forma de dr^{o} so cargo del qual prometio de dezir verdad.

Fran^{co} de ....

En la çiudad de los Reyes beynte e dos dias del mes otubre [*sic*] de myll e quy^{o}s e çinq^{ta} e ocho años Fran^{co} de la Torre en n^{e} del caçique e y^{o}s del rrepartimy^{o} de Canta presento por t^{o}s a Pedro Ambate / e a Domyngo Pariguana / e a don Xristoval ayur / e a don Juan Cabcha Yauri / e a don Diego Chunbi Quivi / e a don Felipe Taurichunbi / e a don Fran^{co} caçique de Culli / los quales por lengua de Ju^{o} de Alvarado juraron por dios nro señor sobre la señal de la cruz en forma de dr^{o} diziendo: sy juramos e amen e prometieron de dezir verdad.

Diego Muñoz

/33v/ En los Reyes diez e siete dias del mes de otubre de myll e quy^{o}s e çinq^{ta} e ocho años ante los señores oydores en audiençia de rr^{al} el dho Fran^{co} de la Torre en nonbre de los dhos yndios sus ptes presento por t^{o}s en esta causa a vn honbre que dixe llamarse Diego Florez el qual juro por dyos nro señor e sobre la señal de la cruz en forma de derecho e dixo: sy juro e amen e prometio dezir verdad.

Diego Muñoz

En los Reyes este dho dia el dho Fran^{co} de la Torre en el dho nonbre presento por t^{o}s en esta cavsa a vn yndyo que dixo llamarse Yayvi caçique de Collyque el qual juro por lengua de Joan de Alvarado e prometio dezir verdad en forma de derecho.

Diego Muñoz

En los Reyes veynte e seys dias del dho mes de ot^{e} del dho año el dho Fran^{co} de la Torre presento por testigo en esta causa a Nyculas de Ribera el moço vs^{o} desta çiudad el qual juro sobre la señal de la cruz en forma de derecho e dixo: si juro e amen e prometio dezir verdad.

Diego Muñoz

/34r/ En los Reyes treynta e un dias del dho mes de otubre de myll e quy^{o}s e çinq^{ta} e ocho años el dho Fran^{co} de la Torre presento por t^{o}s en esta causa a vn yndio que dixo llamarse Capcha natural del pueblo de Atapillo el qual por lengua del dho Joan de Alvarado ynterprete juro sobre la señal de la cruz en forma de derecho e dixo: si juro e amen e prometio dezir verdad.

Diego Muñoz

/35r/ Probança del caçique e yndios de Canta
en el pleyto con los yndios de Chacalla.

Juro en 17 de otu^{e} ante el s^{er} dotor Cuenca
Diego Flores de 40 a^{o}s / no le toca.

T^{o} el dho Diego Flores estante en esta çiudad testigo presentado por pte de los yndios de Canta aviendo jurado en forma de derecho e seyendo preguntado por las preguntas del ynterrogatorio dixo lo sig^{e}:

1 - A la primera pregunta dixo que conoçe al caçique de Canta que al presente es e a algunos prinçipales e yndios del que estan encomendados en el dho Nyculas de Rribera e conosçe al caçique e a algunos yndios e prinçipales de Chacalla / e tiene notiçia del valle e tierras de Quyvi sobre que es este pleyto las quales estan en termynos del dho pueblo de Canta de ocho años a esta parte poco mas o menos e conosçe al lic^{do} Gasca presydente que fue destos rreynos / e conosçe al dho Ffran^{co} de Anpuero encomendero de los dhos yndios de Chacalla / e a Nyculas de Rribera / e tiene notiçia del dho rrio de Quyvi del conçierto que la pregunta dize.

Preguntado por las preguntas generales dixo que es de hedad de quarenta años poco mas o menos e que no es pariente ny enemygo de nynguna de las ptes ny le ba ynteres en esta causa ny le tocan nynguna de las preg^{tas} gen^{ales} de la ley e que vença quyen tuviere just^{a}.

2 - A la segunda pregunta dixo que lo que sabe de la pregunta es que de ocho a^{o}s a esta pte poco mas o menos que aqueste t^{o} sabe e tiene notiçia del dho rrepartimy^{o} de Canta a visto rresidiendo en el dho rrepartimy^{o} que los caçiques e yndios //
/35v/ de Canta an tenydo e poseydo las dhas tierras de Quibi e gozado dellas como suyas e que por esto sabe que estan en los termynos de Canta e porque todo aquello por alli a ellas comarcano es de los dhos yndios de Canta sin aver otras de otra pte entre ellos e que esto es lo que sabe desta pregunta / e que asi mysmo sabe que de la otra parte del rrio son yndios atavillos de la encomyenda de M^{yn} Piçarro que confynan con los yndios de Canta.

3 - A la terçera pregunta dixo que dize lo que dho tiene en la pregunta antes desta e no sabe otra cosa.

4 - A la quarta pregunta dixo que dize lo que dho ti^{e} e no sabe otra cosa.

5 - A la quynta pregunta dixo que no la sabe.

6 - A la sesta pregunta dixo que lo que della sabe es que puede aver ocho a^{o}s poco mas o menos que estando este t^{o} en el dho rrepartimy^{o} de Canta en el dho valle de Quyvi fueron alli el dho Fran^{co} de Anpuero e Nyculas de Rybera e Diego de Avila e otros españoles que no tiene memoria de sus nonbres al presente / y estando en las mysmas tierras de Quybi conçertaron a los caçiques de Canta con los de Chacalla que estavan alli presente / en que / por que los de Chacalla se apartasen del drcho que pretendian a las dhas tierras de Quyvi e de la diferençia e pleyto que sobre ellas tenyan les dyesen los caçiques e yndios de Canta / dozientas ovejas desta tierra / y se las dieron / y las rresçibieron y este t^{o} se las entrego por los dhos yndios de Canta / e con ellas quedaron contentos los vnos y los otros / e siendole mostrado el dho conçierto de que en esta pregunta se haze mynçion //
/36r/ dixo que no se acuerda averlo visto escrivir ny mas de lo que dho tiene e que esto es lo que sabe de la pregunta.

7 - A la setima pregunta dixo que dixe lo que dho tiene en la pregunta antes desta e que despues aca de aver rr^{do} los dhos caçiques e yndios de Chacalla las dhas ovejas por rrazon del dho conçierto a visto que los dhos yndios de Canta an tenydo e poseydo las dhas tierras de Quyvi senbrandolas e cogendolas los frutos dellas syn contradiçion de los dhos yndios de Chacalla hasta agora e que esto es lo que sabe desta pregunta.

Preg^{do} por la preg^{ta} añadida dixo que a bisto este t^{o} que Chacalla donde los yndios tienen su asiento e pueblos es apartado de Quybi çinco leguas poco mas o menos y en medio esta una sierra e que en el tpo que dho tiene que a que esta en el dho rrepartimyento de Canta no vio ny a visto en las dhas tierras de Quybi ny en todo aquel valle yndios de Chacalla / por lo qual e por estar las dhas tierras de Quybi en las tierras e valle de los dhos yndios de Canta le paresçe a este t^{o} que las dhas tierras de Quybi seran de los dhos yndios de Canta e que esto es la verdad e lo que deste caso sabe e siendole leydo su dho se rretifico en el para el juramento que hizo e firmolo de su n^{e}.

Diego Muñoz Di^{o} Flores

/36v/ Don Fran^{co} Yayvi caçique de Cullique
encom^{o} en Muñoz de Avila.
Paresçio de 40 a^{o}s / no le toca.

T^{o} el dho don Fran^{co} Yayvi caçique del rrepartimy^{o} de y^{o}s de Cullique encomendado en Muñoz de Avila t^{o} presentado por pte del caçique e y^{o}s del rreptimy^{o} de Canta aviendo jurado e siendo preguntado por el tenor del dho ynterrogt^{o} dixo lo sygui^{e}:

1 - A la primera preg^{ta} dixo que conoçe a las ptes e ti^{e} notiçia de las tierras en ella contenydas que son las de Quibi que estan junto a Canta de mucho tpo a esta pte e conosçio al presydente de la Gasca e oyo nombrar a Guaynacapa y a Topa Ynga Yupangue e a otros yngas señores naturales destos rreynos porque fueron el y sus antepasados yanaconas yanayacos de los dhos yngas e conoçe a los dhos Nyculas de Ribera e Fran^{co} de Ampuero e tiene notiçia del dho rrio de Quyvi que es de Canta.

Preguntado por las generales dixo que no sabe la hedad que tiene / pareçio por su aspeto ser de quarenta años poco mas o menos e que no le tocan las generales que a los demas se an preg^{do} e que vença quyen tuviere just^{a}.

2 - A la segunda pregunta dixo que lo que della sabe es que siendo este t^{o} mochacho en tpo de los yngas antes que los españoles entrasen en esta trra vio como los y^{o}s de Canta tenyan çierta pte de las tierras de Quivi y tenyan echado sus mojones y çerca de lo que era suyo y asy mysmo los y^{o}s de Chacalla y los de Guancayo y los de Culli que eran quatro parçialidades y cada bna dellas tenia su pedaço señalado y amojonado por los yngas las quales se rriegan parte dellas por açequias e otras por xagueyes y asy se las vio tener e poseer este t^{o} en dos vezes que fue a ellas yendo a llebar la .... de su caçique y asy mysmo se las vio tener e poseer despues que entro el marques don Fran^{co} Piçarro en esta trra e que el dho rrio de Quibi esta en las tierras de Canta e que los Atavillos estan muy apartados e que esto que dho ti^{e} lo a visto asy despues que se acuerda e que si este t^{o} estuviese en las dhas tierras //
/37r/ el amostraria los termynos e mojones de cada vna de las dhas parçialidades asy como se las vio tener e poseer e que en tiempo del marques tuvieron çierta diferençia sobre las dhas tierras e que es todo lo que sabe desta pregunta.

3 - A la terçera preg^{ta} dixo que dize lo que dho ti^{e} en la preg^{ta} antes desta y que este t^{o} oyo dezir que el dho Guaynacapa tomo çierta pte de las dhas tierras de Quivi que era de los yndios de Canta y de los naturales del dho Quyvi y la dio a los dhos y^{o}s de Chacalla que los puso alli por mytimaes y para ello enbio un prinçipal suyo que se las quytase y las diese a los dhos y^{o}s de Chacalla y que quando se las quytavan lloravan los dhos y^{o}s de Canta y este t^{o} lo oyo asy dezir a sus padres y mas ançianos y era muy pu^{co} y notorio entrellos y que despues que el marques entro en esta tierra Guaman Yabre caçique prençipal de Canta dixo a los dhos y^{o}s de Chacalla que tenyan las dhas tierras de Quyvi que se fuesen dellas a sus trras y las dexasen aquellas que heran suyas y de sus pasados e que esto es lo que sabe desta preg^{ta}.

4 - A la quarta pregunta dixo que dize lo que dho tiene en la preg^{ta} antes desta.

5 - A la quinta preg^{ta} dixo que dize lo que dho tiene en las preg^{tas} antes desta y que al tiempo que el dho Guaman Yabre echo de las dhas tierras a los dhos y^{o}s de Chacalla se quedo con ellas y se las an tenydo despues aca el y sus y^{o}s y por ello dieron el dho Guaman Yabre y sus y^{o}s a los dhos y^{o}s de Chacalla dozientas cabeças de ganado por conçierto que entrellos huvo.

6 - A la sesta pregunta dixo que dize lo que dize lo que [*sic*] dho tiene en la preg^{ta} antes desta y que el dho Guaman Yabre caçique prençipal de Canta dixo a este t^{o} //
/37v/ como los dhos y^{o}s de Chacalla avian movido pleyto a el y a los dhos sus y^{o}s sobre las dhas tierras de Quivi y que por que se apartasen del pleyto avian dado a los dhos y^{o}s de Chacalla dozientas ovejas e que esto sabe desta preg^{ta}.

7 - A la septima pregunta dixo que dize lo que dho tiene en la preg^{ta} antes desta e que no sabe otra cosa.

8 - A las ocho preg^{tas} dixo que dize lo que dho ti^{e} en las preguntas antes desta lo qual es verdad para el juram^{o} que hizo y asy lo declaro el dho don M^{yn} ynterprete el qual dixo aver hecho e declarado la dha ynterpretaçion sin aver añadido ny quit^{do} en ello cosa alguna e no firmo el dho t^{o} ny el dho ynterprete porque no sabian.

Diego Muñoz

Di^{o} de Avila de 40a^{o}s
no le toca.

T^{o} el dho Diego Davila testigo presentado por pte de los dhos yndyos de Canta aviendo jurado segun forma de drcho e seyendo preguntado por las preguntas del dho ynterrogatorio dixo lo siguiente:

1 - A la primera pregunta dixo que conosçe al caçique e yndios de Canta encomendados en Fran^{co} de Anpuero e que tiene notiçia de las tierras de Quybi de diez años a esta pte e que conosçio al presydente Gasca e que no conosçio a los dhos yngas e asy mysmo conosçe al dho Ffran^{co} de Anpuero e al dho Nyculas de Ribera encomenderos de los dhos yndios e que tiene notiçia del dho rrio de Quybi e del dho conçierto present^{do} en esta causa que le fue mostrado.
/38r/ Preguntado por las preguntas generales de la ley dixo que es de hedad de quarenta años poco mas o menos e que no es pariente ny enemygo de nynguna de las ptes ny le va ynteres en la causa ny le toca nynguna de las preguntas generales de la ley e que vença el pleyto quyen tuviere just^{a}.

2 - A la segunda pregunta dixo que lo que sabe es que este t^{o} a estado en el dho pueblo de Quybi e Canta e que por la pte de hazia Chacalla a visto que no ay syno piedras syn planta ny se puede plantar por ser toda pedregosa / e por la otra pte del rrio donde estan los yndios de Canta esta todo plantado de charas [*sic*] de coca e arboleda asi frutales como de otros arboles donde tienen sus chacaras los yndios de Quybi e Canta e otros yndios subjetos a Canta / e que estan mas çerca las dhas tierras de Quybi a Canta que no a Chacalla y en el mysmo rrio e valle de Canta e que por esto le paresçe a este t^{o} que las dhas tierras de Quybi estan en termy^{o}s del dho rrepartimy^{o} de Canta e por ser tan pocas que todas ellas las que estan en el mysmo pueblo de Quybi y las chacaras de coca no harian veynte fanegas de mayz en senbradura sy se senbrasen / e porque lo a oydo dezir a los dhos yndios de Canta e asi se da a entender viendolo por vista de ojos e que esto es lo que sabe e le paresçe desta pregunta.

3 - A la terçera pregunta dixo que dize lo que dho tiene en la pregunta antes desta e lo demas no sabe.

4 - A la quarta pregunta dixo que dize lo que dho ti^{e} a que se rrefyere.

5 - A la quynta preg^{ta} dixo que se rrefyere a lo que dho tiene.

6 - A la sesta pregunta dixo que lo que sabe es que puede aver el tpo cont^{do} en las preg^{tas} poco mas o menos este t^{o} salio desta çibdad con Nyculas de Ribera el moço e otras personas e fueron al dho pueblo de Quybi diziendo el dho Ribera que yva a conçertar a su caçique //
/38v/ con el de Chacalla sobre çiertas diferençias que tenyan sobre unas tierras de coca por m^{do} del presydente Gasca / e que a lo mysmo yva Anpuero amo de los yndios de Chacalla / e llegados al dho pueblo de Quybi llegaron alli el caçique de Canta que se llamava Guaman Yaure / y el de Chacalla que se dezia Marcapoma e sus prinçipales e yndios antiguos y estando en el dho pueblo quatro e çinco dias vio que los dhos yndios se juntavan para tratar de las dhas tierras e vio que los yndios de Chacalla dezian que vn pedaço de tierra que ellos señalavan heran suyas que se las avia dado el ynga y los de Canta dezian que heran suyas e no de los de Chacalla que si ellos alli labraron alguna cosa hera por mytimaes de Guaynacapa que los avia alli puesto para que le senbrasen coca e otras cosas pero que las tierras no heran suyas e sobre esto estuvieron los dhos dias que no se conçertavan porque el caçique de Canta no queria nyngun conçierto e ynportunaçion del dho Nyculas de Ribera e de otros españoles e deste t^{o} que alli estavan. Vino a conçierto con el caçique de Chacalla en que le dio dozientas ovejas por el dho pedaço de tierra e vnas casas las quales tenyan los dhos yndios de Chacalla hazia el dho pueblo de Chacalla / e dello se hizo çierta escritura de conçierto entre los dhos caçiques la qual este t^{o} escrivio e firmo su nonbre de su letra el qual le fue mostrado a la qual se rrefyere e que esto es lo que sabe.

7 - A la setima pregunta dixo que lo que sabe es que despues de fho el dho conçierto cont^{do} en la pregunta antes desta este t^{o} vio que el dho caçique de Canta ynbio çiertos yndios prinçipales e a un español que se dize Diego Fflores a un poblezuelo que seria dos o tres leguas de alli a entregar las dhas dozientas ovejas al dho caçique de Chacalla / el qual dho caçique de Chacalla ynbio a vn prinçipal con çiertos yndios para que las rresçibiesen e donde a dos dias llegaron yndios de Canta e de Chacalla adonde este t^{o} estava con carta del dho Diego Flores de como avia entregado las dhas dozientas ovejas a los yndios de Chacalla / e los yndios que en aquella sazon estavan en Quybi dexaron las dhas tierras e se fueron dellas e el dho caçique e yndios //
/39r/ de Canta se metieron en ellas e las an tenydo e poseydo hasta agora / e que despues de lo susodho este t^{o} quexarse al dho caçique de Canta Guaman Yaure por aver dado las dhas ovejas / porque dezia que las tyerras heran suyas e que contra rrazon las avia dado; e que esto es lo que sabe desta pregunta.

8 - A la otava pregunta dixo que dize lo que dho tiene en las preguntas antes desta.

Preguntado por la pregunta añadida dixo que dize lo que dho tiene en la segunda preg^{ta} e que esta es la verdad e lo que deste caso sabe para el juramento que hizo; e se rretifyco en lo demas que dho tiene e firmolo de su n^{e}.

Diego Dabila Diego Muñoz

Diego yndio de 60 a^{o}s / no le toca.
juro en 25 de ot^{e}.

T^{o} el dho Diego yndio que antes que fuese xristiano se llamava Chunbiquyvi natural que dixo ser de Quyvi subjeto al caçique de Canta que se llama Guaman Yabre encomendado en el dho Nyculas de Ribera testigo presentado por pte de los dhos caçiques e yndios de Canta aviendo jurado segun forma de derecho e siendo preguntado por las preguntas del dho ynterrogatorio dixo e depuso lo syguy^{e}:

1 - A la primera pregunta dixo que conosçe a los caçiques e prinçipales de ambos rreptmy^{o}s e a muchos yndios dellos e que tiene notiçia de las dhas trras de Quyvi por ser como es natural dellas que estan en ter^{no} de Canta desde que se acuerda e conosçio al dho presydente Gasca e conosçio a Guayna-//
/39v/ capa e a su hijo Guascar Ynga señores naturales que fueron desta trra e conosçe a los dhos Fran^{co} de Anpuero e Nyculas de Ribera el moço e tiene notiçia del dho rrio de Quyvi que parte los termynos de Canta con Guancayo e que a oydo dezir del dho conçierto.

Preguntado por las preguntas g^{es} dixo que es de hedad de sesenta años poco mas o menos e siendole hechas las demas preguntas g^{es} que a los demas t^{o}s yndios dixo que no le an dado ny prometido que le daran cosa alguna ny para ello le an puesto temores algunos e que tiene sus casas e trras dentro de las dhas trras de Quyvi e le va ynteres en esta causa pero que aunque le va ynteres en ella dira la verdad.

2 - A la segunda pregunta dixo que las dhas trras e chacaras de Quyvi caen e son en los termynos de Canta lo qual parte los termynos con los yndios de Collyque que son de la otra pte de Quyvi los quales dhos ter^{nos} an tenydo e tienen los dhos yndios de Canta desde tpo ynmemorial a esta pte sin aver tenydo en ello contradiçion alguna en todo el tpo que este t^{o} se los a visto tener e poseer ny a oydo dezir que la ayan tenydo.

3 - A la terçera pregunta dixo que dize lo que dho ti^{e} e ay en la pregunta antes desta y que a visto de muy antiguo que los dhos yndios de Canta an senbrado las dhas trras y gozadolas hasta que Guainacava yendo a Quito y pasando por las dhas trras puso en ellas çiertos mytimaes de los yndios yaiyus de anbas parçialidades y alli estuvieron hasta que entraron los españoles en esta trra lo qual puede aver el tpo que la preg^{ta} dize poco mas o m^{o}s.

/40r/ 4 - A la quarta pregunta dixo que dize lo que dho ti^{e} en la pregunta antes desta y que los yndios mitimaes que ansi el dho Guainacapa dexo en las dhas trras y mando que en ellas estuviesen algunos de ellos fueron de los de Chacalla / y que lo que labravan y senbravan hera para el ynga lo qual hizieron çierto tpo hasta que el dho Guainacapa murio y que es todo lo que sabe desta preg^{ta}.

5 - A la quynta pregunta dixo que lo que della sabe es que este t^{o} vio que despues que el dho Guainacapa murio luego los dhos mitimaes yauyos de los de Chaclla dexaron las dhas trras y se tornaron a sus pueblos y los de Canta se bolbieron a las dhas trras de Quivi y las an tenido y gozado dellas como lo solian hazer antes que el dho Guainapa [*sic*] pusiese en ellas los dhos mitimaes.

6 - A la sesta preg^{ta} dixo que puede aber el tpo que la preg^{ta} dize poco mas o menos que este t^{o} vio que Niculas de Ribera el moço y Anpuero fueron a las dhas trras de Quivi y con ellos los caçiques y prinçipales y otros yndios de Chacalla y de Canta y estando en las dhas trras de Quivi se conçertaron en que los dhos caçiques de Canta dieron a los de Chacalla dozientas obejas por que se apartasen de lo que abian pedido sobre las dhas trras de Quivi diz^{do} ser suyas y este t^{o} vio como rreçibieron las dhas obejas y a todo ello se hallo pres^{e}.

7 - A la setima preg^{ta} dixo que dize lo que dho ti^{e} en las preg^{tas} antes desta y que es verdad que este t^{o} vio como los caçiques y yndios de canta dieron y entregaron a los mysmos caçiques de Chacalla las dhas doz^{as} obejas y las rresçibieron y despues aca las an tenydo y poseydo syn contradiçion alg^{a} ni fuile mover pleyto sobre ello y que esto es lo que sabe desta pregunta.
/40v/ 8 - Syendo preg^{do} por la preg^{ta} añadida dixo que la trra y terminos de Chacalla es muy apartado de los Canta porque tienen en tr^{o}s vna sierra de manera que no estan juntos ni apartados mucho lo bno de lo otro y ansi lo a visto este t^{o} por abello andado y porque por esto lo sabe y que esto es ansi verdad y lo que deste caso sabe para el juram^{o} que hizo y en ello se rretifyco y afirmo y el dho Ju^{o} de Albarado ynterpetre dixo aver hecho la dha declaraçion y ynterpretaçion fyelm^{te} syn aver quitado ni añadido en ella cosa alguna so cargo del juram^{o} que ti^{e} fho y no firmo ni el dho t^{o} porque dixo que no sabian firmar.

Diego Muñoz

Juro en 25 de otu^{e} Joan Capucha Yaule
del pueblo de Urco del rrepartrmi^{o} de Canta
encomendado en Nyculas de Ribera el moço
paresçio de 55 a^{o}s / no le toca.

T^{o} el dho Juan Cabpucha Yaule natural del pueblo que se llama Urco del rrepartimi^{o} de Canta encomendado en Niculas de Ribera el moço sujeto al caçique de Canta t^{o} presentado por el dho Fran^{co} de la Torre en n^{e} de los dhos caçiques y yndios del dho rrepartimi^{o} de Canta aviendo jurado segun forma de dr^{o} syendo preg^{do} por las preg^{tas} del dho ynterrogatorio por lengua de Ju^{o} de Alvarado ynterprete dixo e depuso lo sigui^{e}:

1 - A la primera preg^{ta} dixo que conoçe a los caçiques y prinçipales de anbos rrepartimy^{o}s y algunos yndios del rrepartimy^{o} de Chacalla y a los yndios del de Canta por ser este t^{o} nral del y tiene not^{a} de las dhas trras y chacaras de Quyvi porque a estado en ellas e//
/41r/ que esta en los termy^{o}s de Canta y conosçio a Guainacapa señor natural desta trra y a Guascaringa su hijo y conosçe a los dhos Fran^{co} de Anpuero y Niculas de Ribera y que el dho rrio de Quivi es adentro de los termy^{o}s de Canta y que se acuerda de la escritura de conçierto que la preg^{ta} dize porque se hallo pres^{e} quando se hizo.

Preg^{do} por las preg^{tas} gen^{les} dixo que no sabe la hedad que tiene y paresçio por su aspeto ser de çinq^{ta} y çinco años poco mas o m^{o}s y que es natural del dho rrepartimi^{o} de Canta y que querria que vençiese su caçique pero que no enbargante esto dira la verdad y que para dezir este dho no a sido atemorizado ny enduzido por el dho su caçique ni otra persona.

2 - A la segunda preg^{a} dixo que dize lo que dho ti^{e} en la prim^{a} preg^{ta} y que es verdad que las dhas trras y chacaras estan dentro de los termy^{o}s del dho rrepartimi^{o} de Canta como la preg^{ta} dize y que el dho rrio de Quivi pasa por dentro de los termy^{o}s de Canta y que pasado los dhos termy^{o}s de Canta entran los atabillos que son yndios de la encomyenda de M^{yn} Piçarro y que los dhos caçiques y yndios de Canta y sus pasados an estado poblados donde agora estan de tpo ynmemorial a esta pte y nunca este t^{o} a visto lo en contr^{o}.

3 - A la terçera preg^{ta} dixo que despues que este t^{o} se acuerda lo a bisto ser ansi como la preg^{ta} lo dize en el qual tpo los caçique e yndios de Canta an tenido y poseydo las dhas trras senbradolas y gozando dellas y que este t^{o} vio que pasando el dho Guainacapa por las dhas trras //
/41v/ de Quibi quando yba a Quito paresçiendole que eran buenas las quito a los dhos yndios de Canta y puso en ellas mitimaes de los yndios ananyayus / y urinyayus y entre ellos de los dhos yndios de Chacalla y desde entonçes las tuvieron hasta que los hecharon de ally.

4 - A la quarta pregunta dixo que dize lo que dho tiene en la preg^{ta} antes desta y que lo que los dhos mitimaes senbravan no sabe para quyen hera los quales estarian en las dhas trras haziendo esto el tpo que la preg^{ta} dize poco mas o m^{o}s hasta que los españoles entraron en esta trra.

5 - A la quynta preg^{ta} dixo que este t^{o} vio que despues que los españoles entraron en este rreino los dhos mitimaes y Chacalla y yauyus dexaron las dhas trras dexaron las dhas trras [*sic*] y las desmanpararon [*sic*] y se fueron a sus trras y ansi se entraron en ellas y las tornaron a tomar los caçiques y yndios de Canta como cosa suya y las an tenido y poseydo como cosa suya.

6 - A la sesta preg^{ta} dixo que lo que della sabe es que en tpo del presidente Gasca vio este t^{o} que fueron a las dhas trras de Quivi los dhos Niculas de Ribera y Fran^{co} de Anpuero y el p^{e} Diego de Avila y los caçiques de Chacalla y Canta y otros yndios y estando en las dhas trras las personas susodhas conçertaron a los dhos caçiques en la diferençia que tenian sobre ellas en que los de Canta dieron a los de Chacalla por que se apartasen dello y //
/42r/ de no pedir las dhas trras y que las dexasen dozientas cabeças de ganado desta trra çien cabeças de carneros y çien cabeças de obejas y se las vio entregar y bio como las rresçibieron y hizieron vn conçierto sobre ello y este t^{o} lo vio escrevir que fue como lo ti^{e} dho en esta preg^{ta}.

7 - A la setima preg^{ta} dixo que dize lo que dho ti^{e} en la preg^{ta} antes desta y a visto que despues de hecho el dho conçierto a esta pte an tenido los dhos yndios de Canta las dhas trras de Quibi sin contradiçion alguna.

8 - Syendo preguntado por la preg^{ta} añadida dixo que los dhos yndios chacalla tienen su asyento y trras y pueblos muy apartados de las chacaras de Quivi que esta en m^{o} vn gran çerro y de la otra pte estan las dhas trras de Canta y que dize lo que dho ti^{e} y es la verdad para el juram^{o} que ti^{e} fho fuele leydo su dho / rretificose en el y no firmo porque dixo que no sabia y el dho Ju^{o} de Alvarado ynterpetre dixo aber fho la dha ynterpetraçion fyelm^{te} sin añadir ni quitar cosa alguna y no firmo porque no sabia escrevir.

Diego Muñoz

Nyculas de Ribera vs^{o} /de 50 a^{o}s
los y^{o}s de Canta son deste t^{o} / dira verdad.

T^{o} el dho Nyculas de Ribera vz^{o} desta çibdad testigo presentado por pte de los ca-//
/42v/ çique e yndios de Canta aviendo jurado segun forma de derecho e siendo preguntado por las preguntas del dho ynterrogatorio dixo e depuso lo sig^{e}:

1 - A la primera pregunta dixo que conosçe a los caçiques e prinçipales de los dhos dos rreptmy^{o}s e conosçio a los padres de los dhos caçiques e conosçe a muchos de los yndios dellos e tiene notiçia del valle e trras de Quyvi que estan dentro de los termynos de los yndios de Canta de lo qual tiene notiçia de onze años a esta pte poco mas o menos e conosçio al dho presydente Gasca e a los yngas señores desta trra que a avido de veynte e siete años a esta pte que fue desde Atabalypa aca e conosçe al dho Fran^{co} de Anpuero este t^{o} es el dho Nyculas de Ribera en quyen estan encomendados los dhos yndios de Canta.

Preguntado por las preguntas g^{es} dixo que es de hedad de çinq^{ta} años poco mas o menos e que los dhos yndios de Canta tiene este t^{o} encomendados porque por esto no dira el contrario de la verdad e que no le tocan las demas g^{es} e que vença la pte que toviere just^{a}.

2 - A la segunda pregunta dixo que el dho rrio de Quyvi adonde estan las dhas trras parte lymytes con los yndios de M^{yn} Piçarro e yndyos de Canta e que las dhas trras de Quyvi sobre que es este pleyto estan en el dho valle cabe el dho rrio metidas en medyo de las trras de los yndios de Canta porende paresçe ser suyas de los dhos yndyos//
/43r/ de Canta e que este t^{o} los a visto tener e poseer las dhas trras a los dhos yndios de Canta del dho tpo de los dhos onze años a esta pte que a que tiene notiçia de las dhas trras paçificamente sin contradiçion alguna.

3 - A la terçera pregunta dixo que dize lo que dho tiene en la pregunta antes desta e que della no sabe otra cosa.

4 - A la quarta pregunta dixo que no la sabe.

5 - A la quinta pregunta dixo que dize lo que dho tiene en las preguntas antes desta e della no sabe otra cosa.

6 - A la sesta pregunta dixo que lo que della sabe es que el dho presydente Gasca mando a este t^{o} que fuese a las dhas trras de Quyvi e procurase de conçertar a los yndios de Canta con los yndios de Chaclla e los conçertaron a los vnos con los otros sobre la diferençia que entre ellos avia de las dhas trras e las vieron e pasearon e fue el conçierto que los caçiques e yndios de Canta diesen a los caçiques e yndios de Chaclla dozientas ovejas e por ellas se dexasen de pedir lo que pretendian tener de derecho a las dhas trras e sobre ello hizieron el conçierto que en esta causa esta presentado que es el que por my el que le fue mostrado que esta firmado de su nonbre e por tal su firma rreconosçio la qual dize en el dho conçierto Nyculas de Ribera e hecho el dho conçierto todos con-//
/43v/ formes se escrivye e firmo de las firmas que en el estan las quales este t^{o} vio hazer a los que las firmaron y hecho esto enbio este t^{o} vn moço suyo para que entregasen las dhas dozientas ovejas los caçiques e yndios de Canta a los caçiques e yndios de Chaclla los quales fueron luego e se las entregaron e el dho caçique de Canta dixo estonçes que para que el avia de conprar las trras que heran suyas e por que quedasen paçificos e de ally adelante no tuviesen mas diferençia por hazer lo que el dho presydente le encargo sobre ello les rrogo a los dhos yndios de Canta que diesen a los de Chaclla las dhas dozientas ovejas e se las dieron e entregaron como dho es e que asy a visto que despues aca del dho conçierto los dhos yndios de Canta an tenydo e poseydo las dhas trras senbrandolas e cogiendo los fruttos dellas syn contradiçion alguna e que esto es lo que sabe desta preg^{ta}.

8 - [*sic*] A la ottava pregunta dixo que lo que dho e declarado tiene de suso es verdad e lo que deste caso sabe para el juramento que hizo e syendole leydo este su dho en el se rretifyco e firmolo de su n^{e}.

Nicolas de Ribera — Diego Muñoz

Juro en 31 de ot^{e}
Capacha natural del pu^{o} de Atabillo
encom^{o} en Her^{do} de Montenegro vs^{o}
de mas de 60 a^{o}s / no le toca.

T^{o} el dho yndio testigo que dixo llamarse Capcha natural del pueblo de Atabillo //
/44r/ encomendado en Hr^{do} de Montenegro vs^{o} desta çiudad / aviendo jurado en su ley por lengua de don M^{yn} Yndio ynterprete desta R^{l} Abdiençia e siendo preguntado por el dho ynterprete por las preguntas del dho ynterrogatorio dixo e depuso lo sig^{e}:

1 - A la primera pregunta dixo que conosçe a los caçiques e prinçipales de Canta e a algunos yndios e a los de Chaclla no los conosçe / e tiene notiçia de las trras e chacaras de Quyvi que estan en los termynos de Canta despues que se acuerda e asy mysmo se acuerda de Guaynacapa e de otros yngas señores que fueron desta trra e conosçe al dho Nyculas de Ribera el moço e no al dho Fran^{co} de Anpuero.

Preguntado por las preguntas gen^{ales} dixo que no sabe la hedad que tiene e paresçio por su aspeto ser de hedad de mas de sesenta años e que no le tocan las gen^{ales} que a los demas t^{o}s yndios se an preguntado e que vença la pte que toviere just^{a}.

2 - A la segunda pregunta dixo que las dhas trras e chacaras de Quyvi estan en los termynos de Canta e de sus ayllos e parçialydades e que el rrio que se llama de Quybi pte los tr^{nos} con los Atavillos e con los Secos los quales dhos termynos an tenydo de tpo ynmemorial a esta pte e despues que este t^{o} se acuerda sin aver avido otra cosa en contrario e que esto es lo que a visto e sabe desta pregunta.

3 - A la terçera pregunta dixo que este t^{o} siendo muchacho vio que en tpo de Guaynacapa los yndios de Canta tenyan e poseyan //
/44v/ las dhas trras por suyas e como suyas senbradolas e gozandolas e que despues que Guaynacapa paso por las dhas trras que no se acuerda quanto tpo puede aver yendo este t^{o} a las dhas chacaras por coca vio en ellas puestos mor mitimaes los yndios yauyos de Chaclla que tenyan un pedaço de las dhas trras de Quyvi e estonçes oyo dezir como el dho Guaynacapa los puso ally por mytimaes por coca camayos e los vio en las dhas trras en aquella sazon e otras vezes de ally adelante.

4 - A la quarta pregunta dixo que lo en ella contenydo le oyo este t^{o} dezir a los yndios yauyos mytimaes que estavan en las dhas trras yendo a ellas a rrescatar coca que asy lo avia mandado el dho Guaynacapa como la pregunta lo dize e que estuvieron los dhos mytimaes en las dhas trras çiertos años no sabe quantos.

5 - A las çinco preguntas dixo que lo que dellas sabe es que los dhos mytimaes yavyos de Chaclla dexaron las dhas trras despues que los españoles entraron en esta tierra de a ya çierto tpo que no se acuerda quanto por que los dhos yndios de Canta les dieron dozientas ovejas desta trra e que si no se las dieran no se las dexaran las dhas trras e que esto le dixeron los yndios de Canta e que si no les dieran las dhas dozientas ovejas pelearan e rrenyeran sobre ello los yndios de ambos rreptmy^{o}s e asy desta manera despues aca las an tenydo e poseydo e tie-//
/45r/ nen e poseen las dhas trras por suyas e como suyas.

6 - A la sesta pregunta dixo que dize lo que dho tiene en la pregunta antes desta e que della no sabe otra cosa.

7 - A la septima pregunta dixo que dize lo que dho tiene en la pregunta antes desta e que despues aca de hecho el dho conçierto a visto este t^{o} tener e poseer a los dhos caçiques e yndios de Canta las dhas trras de Quyvi e que esto es lo que sabe desta pregunta.

8 - A la ottava pregunta dixo que lo que dho tiene de suso es pu^{co} e notorio a este t^{o} e es la verdad e lo que deste caso sabe para el juram^{o} que hizo e no fyrmo porque dixo que no sabia e el dho don M^{yn} ynterprete dixo aver hecho la dha ynterpretaçion fielmente sin añadir ny quytar dello cosa alguna e no firmo porque dixo que no sabia.

Diego Muñoz

/46r/ Poder de los de Canta.

En los Reyes a primero dia del mes de otubre de myll y quiny^{o}s y çinq^{ta} y ocho años ante el muy mg^{o} señor Sebastian Cherino de Loaisa corregidor en esta çibdad por su mag^{t} por ante my Ju^{o} de Padilla es^{o} pu^{co} desta çibdad paresçio presente don Hr^{do} Guaman Yaure caçique prinçipal del rrepartimy^{o} de Canta por si y en nonbre de los yndios y prinçipales a el subjetos; por lengua de Juan de Albarado lengua de la Real Avdiençia e dixo que por q^{to} el trata y entiende tratar pleitos ante la justiçia de se mag^{t} en esta çibdad y ante la Real Audençia [*sic*] y por ser como es menor de veinte y çinco años e no puede paresçer en juiçio que pedia e pidio a su md le de liçençia e facultad para que pueda dar y otorgar su poder e poderes a la persona e personas que le paresçiere que haran sus neg^{o}s e pleitos; y el dho señor corregidor bisto su pedimy^{o} e como por su aspeto el dho don Fr^{do} paresçe menor de veinte y çinco a^{o}s dixo que como a dr^{o} a lugar daba e dio liçençia e facultad al dho don Fernando Guaman Yaure caçique para que por ante qualquier / o qualesquier es^{o} e es^{o}s pueda haçer e otorgar qualquier / o qualesquier escripturas de poder e los dar a qualesquier personas para sus pleitos e causas segun e como pide en forma e para que valgan e hagan fee en juyçio y fuera del dixo que ynterponya e ynterpuso su autoridad e decreto judiçial tanto quanto puede y con dr^{o} deve y lo firmo de su nonbre siendo t^{o}s el secr^{o} Sebastian Sanches de Merlo e Gr^{mo} de Silba //
/46v/ y el padre Gr^{mo} M^{yn} clerigo / Sebastian Cherino de Loaisa. Paso ante my Juan de Padilla es^{o} pu^{co}. Luego el dho don Fr^{do} Guaman Yaure por lengua del dho Ju^{o} de Albarado dixo que daba y dio / otorgaba y otorgo todo su poder cunplido libre llenero bastante segun que mejor y mas cunplidam^{te} lo puede y deve dar y otorgar y de dr^{o} mas puede y deve valer a Fran^{co} de la Torre pr. de causas en la Real Audençia [*sic*] desta çibdad generalmente para todos los pleitos e causas çeviles e crimynales mobidos y por mober que tiene e tubiere contra todos e qualesquier persona / o personas de qualquier estado e condiçion que sean y que las tales persona y personas las tienen e tubieren contra el en qualquier manera / en los quales dhos pleitos y cada vno dellos demandando e defendiendo pueda paresçer e paresca ante su mag^{t} e ante los señores presidentes e oydores de la dha Real Audiençia e ante todas e qualesquier justiçias e juezes que de los dhos pleitos puedan e deban conosçer e pedir e demandar querellar e rresponder defender negar e conosçer pleito e pleitos contestar el çesiones e definisiones poner alegar rrequerir enplaçar protestar conbenir rreconbenir testimy^{o} o testimy^{o}s pedir e sacar e jurar qualesquier juram^{o}s de calunya e de çesorio / berdad diziendo / e pedir sean fechos e poner articulos e pusiçiones y los confesar y negar los de en contrario e rrecusar y poner sospecha en qualesquier juezes y es^{o}s y la jurar con debida solenidad e pedir e demandar benef^{o} de rrestituçion yn yntegrim y las cosas y casos que //
/47r/ le convengan e fazer probanças e presentar t^{o}s y escripturas y otro qualquier genero de prueba e pedir publicaçion de t^{o}s y abonar sus t^{o}s eprobanças e tachar e contradezir las en contr^{o} en sus dhos y personas y ganar qualesquier mds y probisiones que le conbengan de poder de qualesquier es^{o}s e personas que las tengan y concluir y çerrar rrazones e pedir sent^{ia}/ e sentençias ynterlocutorias y difinitibas y las dadas en su favor consentir; e las en contrario apelar e suplicar e las seguir e acabar por todas ynstançias y dar quien las siga e acabe y en juiçio y fuera del pueda hazer e haga todas las otras cosas / tantos judiçiales y estrajudiçiales que conbengan y sean menester de se fazer e que haria siendo de hedad cunplida presente seyendo aunque aqui no se declaren ny espaçefiquen [*sic*] e rrequieran su presençia personal; que tan cunplido y bastante poder como el dho don Fr^{do} abia y tenya para lo que dho es daba e dio al dho Fran^{co} de la Torre con sus ynçidençias e dependençias anexidades e conexidades y con aprobaçion y rretificaçion de todo lo que por el y en su nonbre obiere fecho e pedido ansi en el pleito contra don Diego Astamiche y consortes suyos como en otros qualesquier y lo susodho fizo y otorgo con obligaçion de su persona y bienes de aver por firme lo que en su nonbre fuere fecho por el dho Fran^{co} de la Torre. Fecho en el dho dia mes y año susodho / siendo presentes por testigos: Gr^{mo} de Silba. Y el padre Gr^{mo} Martin clerigo. Y Sebastian Sanches de Merlo y porque el dicho don Hr^{do} dixo que no sabia escrivir por el a su rruego lo //
/47v/ lo [*sic*] firmo el dicho padre Gr^{mo} M^{yn} en el rregistro desta carta al qual dicho otorgante yo el dicho es^{o} doy fe que conozco Gr^{mo} M^{yn}. Ba testado; e porende aquy myo signo en testimy^{o} de verdad.

Joan de Padilla

/48r/ Los y^{o}s de Chacalla

Muy pr^{o}s ss^{r}

Al^{o} Moreno en nonbre del caçique e y^{o}s de Chacalla en el pleito con los y^{o}s de Canta sobre las tierras y chacaras de Quibe. Digo que visto por v. al^{a} el proçeso de la causa fallara por mi parte probado todo aquello que a su dr^{o} conbenya y con bastante numero de t^{o}s antiguos y de mucha hedad y que depusieron con todas las calidades y rrequisitos del dr^{o} y la pte contr^{a} nyng^{a} cosa provo que escluyo la yntençion y provanças de mi parte y sus t^{o}s son ylegitimos y partes formales en la causa y tienen otros defetos y tachas que notrm^{te} rresaltan de sus dhos y dipusiçiones.

Por que pido y sup^{co} a b. al^{a} que pronunçiando la yntençion de mys ptes por paguada faga en esta causa segun tengo pedido y just^{a} la qual pido y costas y para ello etc.

El licenc^{do} sr Luzio — Al^{o} Moreno

En los Reyes doze de novienbre de myll e quy^{o}s e çinquenta e ocho a^{o}s en audi^{a} pu^{ca} de petiçiones la presento el p^{e} e los dhos ss^{r} mandaron dar tr^{do} a la otra pte e que rresponda a la primera audiençia lo qual paso. Presente Ffran^{co} de la Torre pr. contrario a quien se notyfico; el qual dixo que afirmandose en lo que dho y alegado tiene concluya e concluye y el dho Al^{o} Moreno asimesmo concluyo e por los dhos s. visto ovieron la dha causa por conclusa difynytivam^{te}.

Diego Muñoz

/48v/ Tasaçion de costas

- Al secretario de la pte de los yndios de Chaclla onze p^{o}s e dos tomines. — 11 p^{o}s 2 ts
  a pagado seys p^{o}s; deve la rresta.
- A la pte de los yndios de Canta
  diez e seys p^{o}s — 16 p^{o}s.
  a pagado seys; deve diez.
- Al rrelator de cada pte sacando
  rr^{on} treynta e seys ts. — 4 p^{o}s 4 ts

Diego Muñoz

Pago don Hernando el caçique de Canta de su parte quatro pesos y medio en quinze de nobienbre de 1558 a^{o}s. Pago tanbien el otro caçique don Xristoval Vilcapoma otros quatro p^{o}s e medio en diez e seys de nobienbre de 1558 a^{o}s.

En los Reyes beynte e ocho dias del mes de novienbre de myll e quy^{o}s e çinq^{ta} e ocho años los señores presydente e oydores en acuerdo de just^{a} nonbraron por terçero para la determynaçion desta causa al lic^{do} Hidalgo al qual mandaron haga para ello la solenydad del juram^{o} que en tal caso se rrequyere.

Diego Muñoz

En los Reyes veynte e nueve dias del mes de novi^{e} del dho año en audi^{a} ante los señores psydente e oydores el dho lic^{do} Hidalgo juro como terçero en esta causa haziendo la señal de la cruz en forma de derecho e prometio de la servir con diligençia e dar boto e paresçer en ella con formula de derecho e guardar secreto.

Diego Muñoz

/49r/ Los y^{o}s de Canta

Muy p^{o}s ss

Fran^{co} de la Torre en nonbre de don Fran^{co} e de los demas señores e prençipales yndios de Canta en el pleyto con los yndios de Chacalla sobre las trras de Quybi / digo que por Vra Alteza mandados ber y esamynar los t^{o}s en la dha cabsa presentados hallara que la parte contraria no provo cosa alguna que le aproveche e los t^{o}s que presento son singulares e no contestas definendo oydas e creençias banas no dan rrazon sufiçiente de sus dhos e depusiçiones demas de lo qual son todos yndios yabyos que pretenden ynteres en esta cabsa e my pte provo todo lo que provar debia y le conbenya para aver vitoria y si este pleyto no fuera hecho en el verano donde corre tanto rriesgo en esta çibdad los yndios serranos provara con ynfinyto numero de test^{o}s ser cosa notoria todo lo cont^{do} en las preg^{tas} de my ynterrogatorio conque se quytara de dubda este dho pleyto y asi pido e ssup^{co} a Bra Alteza que la dha provança de suspenda o se difiera hasta tanto que cese el verano / haga tpo frio en esta çiudad para que se sepa la verdad e Bra Alteza mande enbiar un jues rreceptor para que haga la ynformaçion en las propias trras de Quybi sobre que pido just^{a} y el of^{o} de Bra Alteza ymploro para lo qual si es nesçesario pido rrestituçion yn yntegrim y juro en forma que no la pido de maliçia y pido just^{a}.

El licen^{do} de Leon — Fr^{co} de la Torre

Otro si por quanto aliende de lo que dho tengo demas de que es neçesario para hazer la dha provança conviene que una persona sin sospecha vaya a ver las dhas trras en que parte estan y asimesmo para que hagan la dha provança e conste de la verdad a Vra Alteza//
/49v/ pido e suplico mande que vaya vna persona a costa de la parte cayda ante quyen se haga la dha ynformaçion por los mesmos ynterrogatorios de las partes y asi mesmo ponga la descripçion de la trra para que se determyne la cabsa e se haga just^{a} la qual pido.

El licen^{do} de Leon — Fr^{co} de la Torre

En los Reyes quynze de novienbre de myll e quy^{o}s e çinquenta e ocho a^{o}s en audi^{a} de petiçiones ante los señores presydente e oydores la presento el p^{e} e los dhos señores ovieron la dha cabsa por conclusa e mandaron que se trayga para la determynar.

Diego Muñoz

/50r/ Relaçion de la provança de t^{o}s presentados por pte de don Xristoval Bilcapoma caçique de don Felipe e de Fran^{co} Yndio e los demas yndios del rrepartimy^{o} de Chacalla en el pleito con el caçique e yndios del rrepartimy^{o} de Canta sobre las tierras de Quivbi [*sic*].

1 - Primeramente sean preguntados si conosçen a las ptes e si tienen notiçia de las tierras e chacarras de Quibi sobre que es este pleito.

2 - Yten si saben que los caçiques e yndios e prinçipales del rrepartimy^{o} de Chacalla an tenydo e poseydo por suyas e como suyas las dhas tierras e chacarras de Quibi / desde el tpo de los yngas pasados e por suyas an sido tenidas a tanto tpo que memoria de honbres no es en contrario. Digan lo que saben.

T^{o} Xristoval de Leon de cuarenta años / no le toca / dixo que lo que della sabe es que este t^{o} puede aver diez años poco mas o menos que este t^{o} fue a las dhas tierras de Quibi con Ampuero e Niculas de Ribera e Alonso de Avila a meter en paz e conçertar los caçiques de Chaclla con los de Canta sobre la diferençia que tenian en rrazon de las dhas tierras de Quibi e se avian descalabrado e este t^{o} curo a vno dellos e est^{do} en el mesmo asiento de Quibi / obo conçierto entrellos en que los de Canta diesen a los de Ampuero que son los de Chaclla doçientas cabeças de ganado desta tierra e que con ellas dexasen las dhas tierras e se quitasen de rruido e entonçes este t^{o} entendio que los dhos y^{o}s de Chaclla pretendian que heran suyas las dhas tierras de Quibi e por tales las defendian e que esto es lo que sabe desta preg^{a}.

T^{o} Xristoval Vilcachagua yndio natural de Guadacheri de treynta e çinco a^{o}s / no le toca / dixo que lo que sabe desta pregunta es que en tpo de Guascar Ynga hera caçique del dho pueblo de Chacalla Vilcapoma que hera antes que los españoles vinyesen a esta tierra el qual yba algunas vezes a Guarocheri de donde este t^{o} es natural a visitar al caçique del dho pueblo de Guarocheri que se llamava Nyna Guylca padre de don Sebastian caçique que es del dho pueblo al presente al qual llevava maçorcas de mayz e coca e agi e otras cosas de comyda para el dho caçique Nynaguilca e deziale: te traygo de mi chacara Quybi e que en tpo del dho ynga tenia el dho caçique de Guarocheri chacaras en el dho Quibi / y siendo este t^{o} muchacho de doze o treze años por mandado del dho caçique Nynabilca fue desde el dho pueblo de Guarocheri a Quibi a ver e visitar las chacaras que alli tenia el dho caçique / e andandola visitando e myrando bio este t^{o} que las dhas chacaras que alli avia llamado Quibi / heran de los dhos yndios de Chacalla las quales son yendo el rrio abaxo a mano yzquierda e de la otra pte del rrio heran de los yndios de Canta e desde entonçes a esta pte los vnos yndios y los otros an tenido e poseydo las dhas tierras e chacaras e que este t^{o} oyo dezir a çiertos yndios de Chacalla viniendose al dho Nynavilca comos los yndios de Canta tanbien le querian tomar sus tierras al tpo del alçamy^{o} de los naturales e que el dho Ninabilca dixo que vendiezen sus trras e chacaras a los yndios de Canta e si mas molestias e daños la hizieren yran todos los yndios yauyos contra los yndios de Canta (roto) ... pues aca se quedaron los dhos yndios de Canta con (roto)...de la vna pte del rrio e despues de la otra e las (roto)...los andenes para abaxo en el llano que se llama (roto)...lo que sabe desta preg^{ta}.

/50v/ T^{o} Rodrigo de Anpuero Moreno fue esclavo de Fran^{co} de Anpuero encomendero de los yndios de quarenta e tres años / no le toca dixe que lo que della sabe es que al tpo que el presydente Gasca yba al Cuzco est^{do} en Guamanga que yba al Cuzco contra Gonçalo Piçarro los caçiques de Chacalla su tio del dho don Xristoval pydieron al dho presydente que les hisiese bolver sus tierras que se las avia quitado el dho G^{o} Piçarro / e dadoles a los yndios de Canta / e las abian tenido e heran suyas del tpo de los yngas; e que se las hiziese bolver / e el dho presydente se las mando bolver / e que las dexasen los yndios de Canta e que asi se las bolvieron / las dhas tierras e chacaras de Quibi / e este t^{o} estovo en ellas que le pusieron alli con don Fran^{co} su tio del dho don Xristoval estando en ellas e senbrandolas e queriendolas los dhos yndios de Chacalla benefiçiar binieron Guaman Yabre e otros muchos yndios de Canta e echaron fuera dellas a este t^{o} e a los yndios de Chacalla e descalabraron a muchos dellos e a este t^{o} le amarraron e se quedaron en las dhas tierras de Quibi / e esto paso asi de ay a dos o tres meses poco mas o menos despues que se las avian fho entregar a los dhos yndios de Chacalla los quales en aquella sazon avian començado ya a coger la coca e benefiçiar las dhas trras vinieron alli los dhos Guaman Yaure e Marcapoma caçique de Chacalla e entrellos trataron por donde yban los mojones e tr^{es} de las dhas tierras e se conçertaron e quedaron alli e este t^{o} se fue arriba e no vio mas.

T^{o} Alonso M^{yn} yndio de los Ananyabyos encomendado en doña Maria Martel de quarenta años / no le toca / dixo que lo que della sabe es que siendo este t^{o} mochacho antes que los españoles entrasen en esta trra oyo este t^{o} dezir a Nynabilca caçique de Guadacheri / y a Bilcapoma caçique de Chacalla y a otros muchos caçiques prinçipales de aquella comarca que como las dhas tierras de Quibi heran de los dhos yndios de Chacalla e las tenian e poseyan como suyas y asi vio este t^{o} que el dho Bilcapoma caçique de Chacalla yendo a ver e bisitar al dho Ninavilca caçique de Guadacheri llevava coca y agi y otras cosas y frutas que le davan de presente y quando se las dava le oya dezir este t^{o} / que aquello hera de las tierras e chacaras de Quibi / y que despues que los españoles entraron en esta trra vio este t^{o} al tpo que se alçaron los naturales della como los dhos yndios de Chacalla tenian e poseyan las dhas trras de Quibi y este t^{o} vido ansi viniendo con los yngas al tpo que binyeron de guerra para echar los españoles desta çibdad e tierra e que esto es lo que sabe.

T^{o} Joan yndio del pueblo Yaucha encomendado en Gr^{mo} de Silva de quarenta años / no le toca / dixo que lo que sabe es que antes que los xristianos entrasen en esta trra este t^{o} oyo dezir a Pomachagua caçique del pu^{o} de Yavcha de donde es natural e al caçique de Guadacheri que se llamava Nynaguilca e a otros caçiques e prinçipales de los pueblos de Ananyauyo o Hurinyauyo [*sic*] que las dhas tierras e chacara de Quibi heran del caçique e yndios de Chacalla e de sus antepasados e que este t^{o} en tpo de Guascar Ynga por mandado de vn capitan suyo que se dezia Andaguarca governador del valle de Xauxa por su mandado vino al dho valle e chacara de Quibi por coca e agi para el ynga y estuvo en ella dos dias e vido que los dhos yndios de (roto)...estavan en la dha chacara senbrando e benefiçiandola (roto)...e que despues de benidos los xristianos al (roto) naturales; este ts^{o} benia con el ynga llamado (roto)...otro Yanqui Yupangui contra los xristianos//

/51r/ llegaron al valle de Lati e de alli vinieron a esta çiudad y estando aqui contra los dhos españoles murio el uno de los yngas e por su muerte se rretiraron e fueron huyendo todos los yndios e llegaron el dho ynga y este t^{o} / e muchos yndios que yban con el al dho valle e chacaras de Quibi / e vido que estaban alli los yndios de Chacalla como de antes este tº los avia visto la primera bez que alli estovo e que tenian la dha chacara e trras por suya e como suyas; e que esto es lo que sabe desta preg^{ta}.

Tº Pº Calco yndio natural de Pucurucha encomendado en el liçençiado Rodrigo Nyno. Paresçio de sesenta aºs / no le toca / Dixo que lo que desta sabe es que en tpo de Guaynapa [*sic*] este tº fue tres o quatro vezes a las dhas tierras de Quibi a rrescatar obejas con los yndios serranos e vio que los yndios de Chacalla tenian e poseyan las dhas tierras e las senbraban e cogian el fruto dellas / como cosa suya e estando en las dhas tierras / oyo dezir a los yndios antiguos e muy biejos como tenian aquellas tierras desde muy antiguo e las avian tenido ansi sus antepasados e que esto sabe desta pregunta.

Tº Joan Merlo Chauca / yndio del pueblo de Guadacheri subjeto a don Sebastian caçique / paresçio de sesenta aºs / no le toca / vease por el oreginal a fojas treynta ay dos tºs mas de oyda.

3 - Yten si saben que al tpo que los españoles entraron en este rreyno los dhos yndios de Chacalla tenian e poseyan las dhas tierras e chacaras de Quybi e las tubieron e poseyeron como señores dellas sin contradiçion de persona alguna mucho tpo despues.

Tº Xristoval Malcachagua. Dixo este tº que antes que los españoles entrasen en esta trra los dhos yndios de Chacalla tenian e poseyan las dhas tierras de Quybi que es el dho llano despues de los andenes e las tenian e poseyan como señores dellas sin contradiçion de los dhos yndios de Canta.

Tº don Pº Xuyo curaca de los Ananyauyos en cabeça de su mag^{t} dixo que al tpo que ovo el alçamyº de los naturales este tº vino a poner çerco sobre çiudad con vn yndio ynga que se llama Quiçu Yupangui al qual mataron en esta çiudad los españoles e despues desto este tº fue huyendo a su tierra e paso por las dhas tierras e chacaras de Quibi donde estovo vn dia e vio que las dhas tierras son como corre el rrio a la mano yzquierda e de la otra pte son las tierras de Canta e que el dho su caçique deste tsº que se llamava Xacxallaqui estando en las dhas tierras de Quibi e señalo e dixo: estas son las tierras de los yndios de Chacalla e las avian tenido de mucho tpo antes e que entonçes las tenian sin contradiçion alguna e ansi lo vio este tº entonçes e despues le oyo ansi dezir por muy pu^{co} e notorio.

Tº Pº Guacho natural de Ayabiri de los Ananyauyos en cabeça de su mag^{t} dixe que lo que della sabe es que este tº vio que despues que los españoles entraron en esta tierra los caçiques e yndios de Chacalla tenian e poseyan las dhas tierras de Quibi como suyas; e yendo a ellas vna bez donde estovo en las dhas tierras de Quibi tres dias se las bio senbrar e gozar dellas e esto fue çierto tpo antes que los yndios se alçasen.

Tº Rodrigo de Anpuero Moreno dixo que este tº yendo a las dhas tierras pocos dias despues que mataron al marques don Fran^{co} Piçarro fue muchas vezes a las dhas tierras de Quibi e vio en ellas yndios mytimaes de Chacalla e mytimaes de Canta que estavan rrebueltos los unos con los otros e en aquel tpo las tenian por suyas los dhos yndios de Chacalla e siendo este tº esclavo del dho Fran^{co} de Anpuero yba muchas vezes a las dhas tierras e lo vio ansi como dho tiene.

Tº Joan Merlo yndio dixo que dize lo que dho tiene en las preguntas antes desta e que es ansi mesmo que quando los españoles entraron en esta tierra los dhos yºs de Chacalla tenian e poseyan las dhas tierras de Quibi.

/51v/ Tº Pº Calco yndio natural de Pucurucha encomendado en el lic^{do} Rodrigo Nyno dixo que dize lo que dho tiene en las preguntas antes desta e que oyo dezir a los yndios de Chacalla biniendo al pueblo deste tº de cuyos nonbres no se acuerda como al tpo que la pregunta dize tenian e poseyan las dhas tierras de Quibi e que las tenyan e poseyan como suyas e no vio ni supo ny entendio que se las contradixesen otros yndios ny persona alguna e que esto es lo que sabe desta pregª.

Tº Dº Chumbicaxa natural de Guadacheri dixo que este tº oyo dezir a Nynabilca caçique de Guadacheri e a otros yndios biejos como las dhas tierras heran de los dhos yndios de Chacalla antes que los españoles entrasen en esta tierra e al tpo que entraron en ella las tenian e que las tenian e senbravan como suyas e dizo lo que dho tiene en las preguntas antes desta. No ay mas.

4 - Yten si saben que los dhos caçiques e yndios de Canta en el tpo del marques don Fran^{co} Piçarro enbiaron a llamar a un caçique que entonçes hera de Chacalla que se llamaba Bilcacapcha y por que servia a los españoles e no se alçava como ellos a trayçion y estando el seguro lo mataron e juntamente con el çiento e çinquenta e tres yndios avitaban en las dhas tierras e chacaras de Quibi.

Tº Alº Martin yndio subjeto a doña Maria Martel dixo que lo que della sabe es que al tpo que dho tiene que vino con los dhos yngas al dho alçamyº este tº vino con un caçique de Canta que se llamava Bilcamaqui / y estando en las dhas tierras de Quibi vio como el dho caçique dixo muy enojado como porque los caçiques e yndios de Chacalla no querian venyr con ellos contra los españoles syno serbillos / el dho caçique les hizo arrancar los cocales e otros arboles y comydas que alli avia diziendo: "come [*sic*] vosotros esto pues estos yndios no quieren venyr a serbir al ynga" e que entonçes estavan alli algunos yndios de Chacalla e que despues desto al tpo que los yngas bolvieron desta çiudad despues del dho alçamyº est^{do} en un pueblo que se dize Carvocaya del rrepartimiº de Canta de alli los yngas y el caçique de Canta que se llamaba Guaman Yabre enbiaxon a llamar a Bilcapoma caçique de Chacalla para que alli binyese y el dho Bilca Capcha al qual como vino alli lo mataron los yºs y este tsº lo oyo dezir asi mysmo lo que acontesçio porque yba delante con otros yndios que yban camynando y que despues desto este t^{o} vio que Guaman Yaure caçique de Canta mato a Vilcapoma caçique de Chacalla / y que despues de avello muerto / le vio este t^{o} le vio [*sic*] muerto y mandar que fuesen çiertos y^{o}s a matar los y^{o}s que estaban en las chacaras de Quibi y alli mataron al dho Vilca Capcha y a mas de çien y^{o}s con el y de alli adelante no quedaron con las dhas tierras de Quibi e que esto es lo que sabe desta preg^{a}.

T^{o} Joan yndio encomendado en G^{mo} de Silva dixo que lo que sabe es que en tpo que entraron los españoles en esta trra estando en el dho su pueblo llego un yndio de Chacalla a dar aviso a su caçique Pomachagua como los yndios de Canta e caçique del avian muerto çiento e çinquenta e tres yndios de Chacalla que estavan en la chacara e tierras de Quibi porque serbian a los españoles e no se querian alçar como ellos se avian alçado con el ynga contra los dhos españoles e que con los dhos yndios mataron al dho caçique llamado Bilcacapcha / con su muger e hijos e ansi fue pu^{co} e notorio e que esto es lo que sabe.

T^{o} P^{o} Calco yndio natural de Pucurucha encomendado en el liçençiado Rodrigo Niño dixo que lo en ella contenido lo oyo dezir a los caçiques de Chacalla llamados Bilcapoma e de otros nonbres e a los caçiques e a los //
/52r/ yndios de Canta disiendo los de Canta: "nosotros avemos muerto çiento e tantos yndios de los que estavan en Quibi" e que esto oyo dezir asy quando bolvio el ynga con los caçiques de Chacalla deste valle a la sierra despues del alçamy^{o} e que los avian muerto porque los dhos yndios de Chacalla e Quibi sirbian a los españoles e no se avian alçado como ellos y no avian hecho lo que el ynga les mandava. E que esto es lo que sabe desta preg^{a}.

T^{o} Juan Merlo yndio dixo que lo que della sabe es que al tpo que los naturales deste rreyno se alçaron contra los españoles que en el avia este t^{o} vino con Quiçu Yupangui e Ylla Topa e otros muchos yngas para esta çiudad e que pasando por el valle de Quibi vio este t^{o} como por mandado del ynga Quiçu Yupangui un yndio dio al dho Bilcacapcha muchos palos con una porra e quedo para morir / e hecho este se binyeron a esta çibdad para el alçamy^{o} e despues que bolvieron los dhos yngas pasando por las dhas tierras de Quibi hallo alli al dho Bilcacapcha / e lo llevo consigo un quarto de legua e en un arroyo que se llama Malluc le cortaron la cabeça los yndios de Canta e este t^{o} por ser su conosçido le enterro en un agujero e que despues de hecho esto vio como los yndios de Canta començaron a matar a los yndios de Chacalla que estavan en las dhas tierras de Quibi e este t^{o} de myedo que no lo matasen con ellos se huyo hazia su tierra con el dho Bilcapoma e Hurcopoma e otros yndios aquella noche e otro dia por la mañana vino alli un mensajero del ynga a dezir que fuesen muchos yndios a serbir al ynga e ansi fueron çiento e çinquenta yndios e con ellos los dhos Vilcapoma e Urcopoma e oyo dezir despues otro dia adelante como los yndios de Canta avian muerto los dhos çiento e çinquenta yndios que fueron a Quibi por mandado del ynga e lo oyo dezir a muchos yndios de los yauyos e de Canta los quales dhos çiento e çinquenta yndios heran de los que estavan en las dhas tierras de Quibi por mitimaes e que desde entonçes se quedaron los dhos yndios de Canta con las dhas tierras de Quibi e las an tenido hasta agora. E que esto lo a asi oydo dezir a los yndios de Chacalla e que este es lo que sabe desta preg^{a}.

T^{o} Rodrigo de Anpuero negro dixo que lo en ella contenido lo oyo este ts^{o} dezir a Marcapoma tio del dho don Xristoval Bilcapoma / estando en las dhas tieras [*sic*] o le mostro la pte donde avia muerto a los dhos yndios.

T^{o} P^{o} Guacho natural de Ayabiri / dixo que lo contenido en ella lo oyo dezir a los dhos caçiques Nynabilca e a otros muchos yndios de Ananyavyo e Urinyavyo que asi avia pasado como la pregunta lo dize lo qual estando en esta çibdad despues de muerto el marques don Fran^{co} Piçarro de ay a un mes o dos.

T^{o} don p^{o} Yuyo en cabeça de su mag^{t} dixo que lo en ella contenido lo oyo dezir estando en esta çiudad e que se benian a quexar los yndios de Chacalla al dho Nynabilca caçique prinçipal de Guadocheri / e al caçique que Xacxallaqui [*sic*] / en que vio este t^{o} que los yndios que se quexaban / dezian que los yndios de Canta les avian muerto a los de Chacalla / que heran los que se quexavan / çiento e çinquenta e tres yndios de los que estavan en Chacalla e con ellos al dho Vilcacapcha que hera primo her^{no} del dho Vilcapoma e que entonçes las tomaron las dhas tierras por fuerça contra la voluntad de los yndios de Chacalla e que los dhos caçiques rrespondieron a los yndios que se quexaron / que entonçes no hera tpo de entender en ello e que despues que estuviese acabado e paçifico el rreyno entenderian en ello e que esto es lo que sabe desta preg^{a}. Ay mas t^{o}s de avello oydo dezir por muy pu^{co} e not^{o} y a y^{o}s caçiques.

/52v/ 5 - Yten si saben que los dhos yndios de Canta despues de aver muerto al dho caçique e yndios de Chacalla les tomaron por fuerça las dhas tierras e chacarras de Quibi y se las an tenido usurpadas y las tienen agora sin otro titulo ni cabsa. Digan lo que saben.

T^{o} don Pedro Yuyo dixo que dize lo que dho tiene en la pregunta antes de esta e que es verdad e es muy pu^{co} e not^{o} que los yndios de Canta despues aca an tenido las dhas tierras de Quibi sin tener otro titulo ni causa ni rrazon para ello ni aviendo sido suyas ny de sus antepasados.

T^{o} Alonso M^{yn} yndio dixo que dize lo que dho tiene en la pregunta antes desta e que despues aca que los dhos yndios de Canta mataron al caçique e yndios de Quibi se an quedado con las dhas tierras no siendo suyas ni de sus antepasados e que esto es lo que sabe desta preg^{a}.

Tº Joan yndio encomendado en G^{mo} de Silva dixo que es pu^{co} e notº que los dhos yndios de Canta despues que mataron a los dhos yndios de Chacalla que estavan en las dhas tierras e chacaras de Quybi las tomaron por fuerça e contra su voluntad / e las an tenido e poseydo por fuerça e que hasta agora las tienen / e poseen / e que esto es lo que sabe.

Tº Pº Calco yndio natural de Pucurucha encomendado en el lic^{do} Rodrigo Nyño dixo que despues que los dhos yndios de Canta tomaron las dhas tierras a los de Chacalla / este tº no a ydo a las ver ni a estado en ellas pero a oydo dezir que las a [*sic*] tenido hasta agora a muchos yndios de cuyos nonbres no tiene memoria al presente.

Tº Joan Merlo yndio dixo que dize lo que dho tiene en las preguntas antes desta e que nunca las dhas tierras de Quibi fueron de los yºs de Canta sino las que ellos se tenian como lo tienen [*sic*] declarado en la segª pregª. Ay mas tºs de oydas.

6 - Yten si saben que por el henero del año pasado de mil e quarenta e nueve que fue quando el presydente Gasca estava en este rreyno valia cada obeja de la tierra en esta çiudad a quatro pºs e no mas. Digan los tºs el valor de las dhas ovejas.

Tº Dº Chunbicaxa dixo que al tpo que la pregunta dize valia cada una oveja grande desta tierra a quatro e çinco pºs / e corderos e pacos cada cabeça a tres e a quatro pºs e este tº lo vendio asi en aquel tpo e lo vio vender asi a otros yndios e que esto es lo que sabe desta pregª.

Tº Martin Caxi [*sic*] en cabeça de su mag^{t}d dixo que este tº vio que en el tpo que la pregunta dize valia un carnero grande e una oveja grande quatro pesos cada uno / y los corderos y pacos valia cada uno a dos e a tres pesos. Sabelo este tº porque a los dhos presçios vendio en aquel tpo este tº del dho ganado de la tierra en esta çiudad.

Tº don Pº Yuyo dixo que lo que della sabe es que al tpo que la pregª valia en esta çiudad una obeja e un carnero grande quatro pesos e un cordero e un paco lo uno o lo otro valia a tres pºs e asi los vendio este tº e vio que los yndios lo vendieron en el dho tpo e vio que no valian ny tenian mas presçio el dho ganado porque entonçes avia mucho ganado de la tierra.

Tº Pedro Guacho dixo que al tpo que la pregunta dize valia en esta çibdad un carnero e una obeja desta trra grande quatro pºs / e los corderos e pacos a dos e a tres pesos e esto es lo que sabe asi porque entonçes avia mucho //
/53r/ ganado desta trra e este tº truxo a vender ... (roto) ... vender como dho tiene.

Tº Rodrigo de Anpuero negro dixo que en el tpo que la pregunta ... (roto) ...tº vender en esta ciudad cada cabeça de obeja o carnero grande ... (roto)... pºs e lo que hera pequeño valia a mºs presçio e tanbien ... (roto) ... conprado este tº en el canpo en el dho tpo a mºs presçio de los dhos quatro pºs. Ay mas tºs como estos.

7 - Yten si saben que las dhas tierras e chacaras de Quibi valian por el dho mes de henero del año pasado de quyºs e quarenta e nueve y al presente valen mas de ocho myll pºs de buen oro porque son muchas tierras e muy buenas y en ellas muchas caxas [*sic*] de coca e de mayz e aji e de otros aprovechamiºs e mantenymiºs. Digan los tºs lo que valen e pueden valer las dhas tierras e chacaras de Quybi.

Tº Xpoval Malcachagua / dixo que este tº a visto que las dhas tierras de Quibi son buenas e le paresçe que en el tpo que la pregª dize valian tres myll e quyºs / o quatro myll pºs / porque a visto que en ellas ay muchos arboles de frutas e son buenas chacaras e tierras las de Quibi hazia la parte de abaxo son buenas e lo de arriba es andenes e pedregales y esto sabe.

Tº Diego Chunbicaxa dixo que le paresce a este tº que las dhas tierras al tpo que la pregª dize e al presente podian e podran valer tres myll e doziºs e tres myll e treziºs pºs porque la myta [*sic*] de las dhas tierras hazia abaxo es llano e donde ay arboles de frutas e coca e aji e tierra llana e del medio arriba es andenes e pedregales e por esto e por lo aver visto le paresçe que pueden las dhas tierras valer el presçio que dho tiene.

Tº don Pº Xuyo / dixo que este tº a bisto como dho tiene las dhas tierras e chacaras de Quibi en que ay chacaras de coca e mani e aji e de mayz e de arboles de fruta e son muy buenas hazia la pte de abaxo que es llano e lo otro es andenes e pedregales e que los yngas e caçiques antiguamente no acostunbravan vender en aquel tpo las trras e que agora al paresçer deste tº podran valer las dhas trras tres myll e noveçientos pºs / e que no sabe çierto si balen mas o mºs.

Tº Pº Guacho yndio natural de Ayabiri / dixo que en tpo de los yngas no se acostunbraba en esta trra vender las chacaras e que despues que los españoles entraron en ella que se bende i le paresce a este tº que baldran las dhas chacaras e tierras de Quibi quatro myll pesos por ser trras donde avia coca e muchos arboles de fruta e agi e otras comidas e que la mayor pte dellas es llana e buena trra e del medio para arryba es andenes e pedregales. Ay mas tºs que dizen las dhas tierras ser buenas y aver muchos arboles de coca y agi y otras cosas pero no dizen lo que pueden valer.

8 - Yten si saben que todo lo susodho es pu^{co} e notorio e pu^{ca} boz e fama. Los tºs se rretifican en sus dhos.

Relaçion de la provança de tºs presentados
por parte de los caçiques e yndios de Canta enco-
mendados en Niculas de Rribera en el pleito que tratan
con los yndios de Chacalla sobre las tierras del valle
de Quibi que estan juntas a las tierras e pueblos de
los yndios del dho valle de Canta.

1 - Primeramente si conosçen a las partes y a cada una dellas y tienen notiçia del valle e tierras que dizen de Quibi que estan en los //

/53v/ ...(roto)... de los dhos pueblos de Canta y de que tpo a esta pte ...(roto)... e si conosçieron al presydente Gasca e conosçieron a los yngas pasados Topa Ynga Yupangui y Guaynacapa e los demas señores naturales que an sido deste rreyno e si conosçen a Fran^{co} de Anpuero que tiene en encomyenda los yndios de Chacalla y conosçen a Nyculas de Rrybera que tiene en encomyenda los dhos yndios de Canta y si tienen notiçia de bn rrio que se llama el rrio de Quibi que parte de los termynos del dho rrepartimiº de Canta y tienen notª de una escriptura de conçierto que esta firmada del dho Rribera y del dho Anpuero y presentada en esta causa que pido se muestre a los tºs. Algºs que sy.

2 - Yten si saben que las dhas tierras que dizen de Quybi estan dentro de los termynos e pueblos del dho rrepartimiº de Canta y sus ayllos e parçialidades por quanto el dho rrio que dizen de Quibi parte los dhos termynos de Canta con los yndios que estan de la otra pte del rrio de Quibi que son Atavillos y otras parçialidades de yndios los quales dhos termynos tienen los dhos yndios de Canta desde que poblaron donde agora estan poblados e de tpo ynmemorial a esta pte sin aver avido en ella contradiçion alguna sino que son termynos conosçidos suyos. Digan lo que saben.

Tº Diego Flores de quarenta años / no le toca. Dixo que lo que sabe de la pregunta es que de ocho aºs a esta pte poco mas o mºs que aqueste tº sabe e tiene notª del dho rrepartimiº de Canta a visto rresidiendo en el dho rrepartimiº que los caçiques e yndios de Canta an tenido e poseydo las dhas tierras de Quibi e goçado dellas como suyas e que por esto sabe que estan en los trºs de Canta e porque todo aquello por alli a ellos comarcano es de los dhos yndios de Canta sin aver otros de otra pte entrellos e que esto es lo que sabe desta pregunta e que asi mismo sabe que de la otra pte del rrio son yndios atavillos de la encomyenda de M^{yn} Piçarro que confinan con los yndios de Canta.

Tº Diego de Avila de quarenta años / no le toca Dixo que lo que sabe es que este tº a estado en el dho pueblo de Quibi e Canta e que por la pte de hazia Chacalla a visto que no ay sino piedras sin planta ni se puede plantar por ser toda pedregosa e por la otra pte del rrio donde estan los yndios de Canta esta todo plantado de chacaras de coca e arboleda asi frutales como de otros arboles / donde tienen sus chacaras los yndios de Quibi e Canta e otros yndios subjetos a Canta e que estan mas çerca las dhas tierras de Quibi a Canta que no a Chacalla y en el mysmo rrio e valle de Canta e que por esto le paresçe a este tº que las dhas tierras de Quibi estan en termynos del dho rrepartimiº de Canta e por ser tan pocas que todas ellas las que estan en el mysmo pueblo de Quibi y las chacaras de coca no harian veynte hanegas de mahiz en senbradura si se senbrasen e porque lo oyo dezir a los dhos yndios de Canta e asi se da a entender viendolo por vista de ojos e que esto es lo que sabe e le paresçe desta pregª.

Tº Niculas de Rribera vsº de çinquenta años / no le toca/ salvo que los yºs de Canta son deste tº. Dixa verdad. Dixo que el dho rrio de Quibi adonde estan las dhas tierras parte limytes con los yndios de M^{yn} Piçarro e yndios de Canta e que las dhas tierras de Quibi sobre que es este pleito estan en el dho valle cabe el dho rrio metidas en medio de las tierras de los yndios de Canta por donde paresçe ser suyas de los dhos yndios de Canta e que este tº les a visto tener e poseer las dhas tierras a los dhos yndios de Canta del dho tpo de los dhos honze aºs a esta pte que a que conosçe e tiene notiçia de las dhas tierras paçificamente sin contradiçion alguna.

Tº don Fran^{co} Yayvi caçique de Cullique encomendado en Muñoz de Avila paresçe //
/54r/ de quarenta años / no le toca / dixo que lo que della sabe es que siendo este tº mochacho en tpo de los yngas antes que los españoles entrasen en esta trra vio como los yndios de Canta tenian çierta parte de las tierras de Quibi y tenian echado sus mojones y çercado lo que hera suyo y asi mysmo los yndios de Chacalla y los de Guancayo y los de Cullique heran quatro parçialidades y cada vna dellas tenia su pedaço señalado y amojonado por los yngas las quales se rriegan parte dellas por açequias y otras xagueyes y asi se las vio tener e poseer este tº en dos vezes que fue a ellas yendo a llevar la ..... de su caçique y asi mysmo se las vio tener e poseer despues que entro el marques don Fran^{co} Piçarro en esta trra e que el dho rryo de Quibi esta en las tierras de Canta e que los Atavillos estan muy apartados e que esto que dho tiene lo a visto asi despues que se acuerda e que si este tº estubiese en las dhas tierras el amostraria los termynos e mojones de cada vna de las dhas parçialidades asi como se las vio tener e poseer e que en tpo del marques tubieron çierta diferençia sobre las dhas tierras e que esto es lo que sabe desta pregª.

Tº Diego Yndio subjeto a los caçiques de Canta encomendado en Niculas de Rribera de quarenta años / dira verdad / dixo que las dhas tierras e chacaras de Quibi caen e son en los termynos de Canta lo qual parte los termynos con los yndios de Collique que son de la otra parte de Quibi los quales dhos trºs an tenido e tienen los dhos yndios de Canta desde tpo ynmemorial a esta pte sin aver tenido en ello contradiçion alguna en todo el tpo que este t^{o} se los a visto tener e poseer ni a oydo dezir que lo ayan tenido.

T^{o} Joan Capucha Yaule del pueblo de Brco del rrepartimi^{o} de Canta encomendado en Niculas de Rribera el moço paresçe de çinquenta e çinco a^{o}s; dixa verdad; dixo que dize lo que dho tiene en la primera pregunta y que es verdad que las dhas tierras y chacaras estan dentro de los termynos del dho rrepartimi^{o} de Canta como la pregunta dize y que el dho rrio de Quibi pasa por dentro de los termynos de Canta y que pasado los dhos termynos de Canta entran los Atabillos que son yndios de la encomyenda de M^{yn} Piçarro y que los dhos caçiques e yndios de Canta y sus pasados an estado poblados donde agora estan / de tpo ynmemorial a esta parte y nunca este t^{o} a visto lo en contrario.

T^{o} Capacha natural del pueblo de Atavillos encomendado en Her^{do} de Montenegro vs^{o} de mas de sesenta a^{o}s no le toca; no es xpiano; vease a /44/ no ay mas.

3 - Yten si saben que las dhas tierras e chacaras de Quibi desde vno e dos e diez e veynte e treynta e çinquenta e mas a^{o}s a esta parte que memoria de honbres no es en contrario sienpre fueron tierras e chacaras de los dhos yndios de Canta y su caçique los quales sienpre las an tenido e poseydo por suyas y como suyas labrandolas y senbrandolas y haziendo dellas como cosa suya propia e asta que pasando por las dhas tierras Guaynacapa y paresçiendoles fertiles las poblo de mytimaes para si al tpo que el dho Guaynacapa yba a las provinçias a Quito lo qual puede aver treynta e çinco o treinta e seys años. Digan lo que saben.

T^{o} don Fran^{co} Yaybi caçique de Cullique encomendado en Muñoz de Avila dixo que dize lo que dho tiene en la pregunta antes desta y que este t^{o} oyo dezir que el dho Guaynacapa tomo çierta pte de las dhas tierras de Quibi que hera de los yndios de Canta y de los naturales del dho Quibi y las dio a los dhos yndios de Chacalla que los puso alli por mitimaes y para ello enbio vn prinçipal suyo que se las quitase y las diese a los dhos yndios de Chacalla y que quando se las quitaban lloravan los dhos yndios de Canta y este t^{o} lo oyo asi dezir a sus padres y mas ançianos y era muy pu^{co} e notorio entrellos y que despues que el marques entro en esta tierra Guaman Yabre caçique prinçipal de Canta dixo a los dhos yndios de Chacalla que tenian las dhas tierras de Quibi que se fuesen dellas a sus tierras y las dexasen aquellas que heran suyas y de sus pasados y que esto es lo que sabe desta preg^{a}.

T^{o} Diego Yndio de Canta dixo que dize lo que dho tiene en la pregunta antes desta y que a visto de muy antiguo que los dhos yndios de Canta an senbrado las dhas //
/54v/ tierras y goçadolas hasta que Guaynacaba yendo a Quito y pasando por las dhas tierras puso en ellas çiertos mytimaes de los yndios yaius de anbas parçialidades y alli estubieron hasta que entraron los españoles en esta tierra lo qual puede aver el tpo que la pregunta dize poco mas o menos.

T^{o} Joan Capucha de Canta dixo que despues que este t^{o} se acuerda lo a bisto ser ansi como la pregunta lo dize en el qual tpo los caçiques yndios de Canta an tenido e poseydo las dhas tierras senbrandolas y goçando dellas y que este t^{o} vio que pasando el dho Guaynacapa por las dhas tierras de Quibi quando yba a Quito / paresçiendole que heran buenas las quito a los dhos yndios de Canta y puso en ellas mytimaes de los yndios ananyaius y urinyaius y entrellos de los dhos yndios de Chacalla y desde entonçes las tubieron hasta que los echaron de alli.

T^{o} Capacha yndio no xpiano de la encomyenda de Montenegro dixo que este t^{o} siendo mochacho vio que en tpo de Guaynacapa los yndios de Canta tenian e poseyan las dhas tierras por suyas y como suyas senbrandolas e que despues que Guaynacapa paso por las dhas tierras que no se acuerda quanto tpo puede aver yendo este t^{o} a las dhas chacaras por coca vio en ellas puestos por mytimaes los yndios yabyos de Chacalla que tenian vn pedaço de las dhas tierras de Quibi e estonçes oyo dezir como el dho Guaynacapa los puso alli por mitimaes por cocacamayos e los vio en las dhas tierras en aquella sazon e otras vezes de alli adelante. No ay mas.

4 - Yten si saben que al dho tpo que el dho Guaynacapa paso por las dhas tierras e pueblos de Canta cuyas heran / el dho Guaynacapa por paresçerle fertiles mando poner e puso çiertos yndios mytimaes que las labrasen para el y asi fueron alli por mytimaes çiertos yndios yavyos de los de Chacalla a los quales mando que las labrasen y senbrasen para el dho ynga y asi lo hizieron quatro o çinco o seys años hasta que los españoles entraron en estos rreynos. Digan lo que saben.

T^{o} Diego Yndio de Canta dixo que dize lo que dho tiene en la pregunta antes desta y que los yndios mytimaes que ansi el dho Guaynacapa dexo en las dhas tierras y mando que en ellas estubiesen algunos dellos fueron de los de Chacalla y que lo que labravan y senbravan hera para el ynga lo qual hizieron çierto tpo hasta que el dho Guaynacapa murio e que esto es lo que sabe desta preg^{a}.

T^{o} Joan Capucha de Canta dixo que dize lo que dho tiene en la pregunta antes desta y que lo que los dhos mytimaes senbravan no sabe para quien heran los quales estarian en las dhas tierras haziendo esto / el tpo que la pregunta dize poco mas o m^{o}s hasta que los españoles entraron en esta tierra.

T° Capacha yndio no xpiano dixo que lo en ella contenido lo oyo este t° dezir a los yndios yauyos mytimaes que estaban en las dhas tierras yendo a ellas a rrescatar coca que asi lo avia mandado el dho Guaynacapa como la pregunta lo dize / e que estobieron los dhos mitimaes en las dhas tierras çiertos años no sabe quantos. No ay mas.

5 - Yten si saben que despues que entraron los españoles en este rreyno que fue poco tpo despues que el dho ynga puso los dhos mytimaes en las dhas tierras de Quibi / los dhos yndios yabyos que asi puso por mytimaes dexaron despobladas las dhas tierras de Quibi e se fueron a sus pueblos y el dho caçique e yndios de Canta tornaron a tomar e poseer las dhas tierras de Quibi y asi las an tenido e poseido desde entonçes / hasta agora / senbrandolas e labrandolas y haziendo dellas como de cosa suya propia / y como lo heran de antes que pusiese los dhos mytimaes el dho Guaynacapa. Digan lo que saben.

/55r/ T° don Fran[co] Yaybi caçique de la encomyenda de Muñoz de Abila dixo que dize lo que dho tiene en las preguntas antes desta y que al tpo que el dho Guaman Yaure echo de las dhas tierras a los dhos yndios de Chacalla se quedo con ellas e se las a tenido despues aca el y sus y°s y por ellas dieron el dho Guaman Yabre y sus yndios de los dhos yndios de Chacalla dozientas cabeças de ganado por conçierto que entrellos vbo.

T° Diego Yndio de Canta dixo que lo que della sabe es que este t° bio que despues que el dho Guaynacapa murio luego los dhos mytimaes yauyos de los de Chacalla dexaron las dhas tierras y se tornaron a sus pueblos y los de Canta se bolvieron a las dhas tierras de Quibi y las an tenido y goçado dellas como de cosa suya propia como lo solian dezir antes que el dho Guaynapa [*sic*] pusiese en ellas los dhos mytimaes.

T° Joan Capucha de Canta dixo que este t° bio que despues que los españoles entraron en este rreyno los dhos mytimaes y Chacalla y yauyus dexaron las dhas tierras y las desmanpararon [*sic*] / y se fueron a sus tierras y ansi se entraron en ellas y las tornaron a tomar los caçiques e yndios de Canta como cosa suya y las an tenido e poseydo como cosa suya.

T° Capacha de la encomyenda de Montenegro dixo que lo que della sabe es que los dhos mytimaes yabyos de Chacalla dexaron las dhas trras despues que los españoles entraron en esta trra de ay a çierto tpo que no se acuerda quanto porque los dhos yndios de Canta les dieron dozientas obejas desta trra e que si no se las dieran no se las dexaran las dhas trras e que esto le dixeron los yndios de Canta e que si no les dieran las dhas dozientas obejas pelearan e rrineran sobrello los yndios de anbos rrepartimi°s / e ansi desta manera despues aca las an tenido e poseydo e tienen e poseen las dhas trras por suyas e como suyas. No ay mas.

6 - Yten si saben que teniendo e poseyendo los dhos yndios de Canta las dhas trras de Quibi como cosa suya puede aver nueve a°s poco mas o m°s / que moviendo pleito los dhos y°s de Chacalla a los dhos yndios de Canta sobre las dhas tierras diziendo que les pertenesçian por rrazon de aver sido çiertos yndios suyos mytimaes en ella y defendiendose el dho caçique e yndios de Canta ante el presydente Gasca y mandando el dho presydente Gasca que fuesen a ver las dhas tierras e procurasen de conçertar a las dhas ptes los dhos yndios de Canta se conçertaron con los dhos yndios de Chacalla por rredimyr su bejaçion / en que les darian como les dieron dozientas obejas por que se dexasen del dho pleito/segun consta por el conçierto que tengo presentado en esta causa que pido sea mostrado a los t°s. Digan lo que saben. Esta el conçierto a fojas /6/

T° Diego Flores dixo que lo que della sabe es que puede aver ocho a°s poco mas o m°s / que estando este t° en el dho rrepartimi° de Canta en el dho valle de Quibi fueron a el / el dho Fran[co] de Anpuero e Niculas de Rribera e Diego de Avila e otros españoles que no tiene memoria de sus nonbres al presente y estando en las mysmas tierras de Quibi conçertaron a los caçiques de Canta con los de Chacalla que estavan alli presentes en que por que los de Chacalla se apartasen del dr° que pretendian a las dhas tierras de Quibi e de la diferençia e pleito que sobrello tenian les diesen los caçiques e yndios de Canta dozientas obejas desta tierra y se las dieron y las rresçibieron y este ts° se las entrego por los dhos yndios de Canta e con ellas quedaron contentos los vnos y los otros e siendole mostrado el dho conosçimi° de que en esta pregunta se haze mynçion dixo que no se acuerda aberlo visto escrevir ni mas de lo que dho tiene / e que esto es lo que sabe de la preg[a].

/55v/ T° Diego de Avila dixo que lo que sabe es que puede aver el tpo contenydo en la pregunta poco mas o menos este t° salio desta ciudad con Niculas de Rribera el moço e otras personas e fueron al dho pueblo de Quibi diziendo el dho Rribera que yba a conçertar a su caçique con el de Chacalla sobre çiertas diferençias que tenian sobre bnas tierras de coca por mandado del presydente Gasca e que a lo mysmo yba Anpuero amo de los yndios de Chacalla e llegados al dho pueblo de Quibi llegaron alli el caçique de Canta que se llamava Guaman Yavre y el de Chacalla que se dezia Marcapoma e sus prinçipales e yndios antiguos y estando en el dho pueblo quatro o çinco dias vio que los dhos yndios se juntavan para tratar de las dhas tierras e vio que los dhos yndios de Chacalla dezian que vn pedaço de tierra que ellos señalavan heran suyas que se las avia dado el ynga e los de Canta dezian que heran suyas e no de los de Chacalla que si ellos labraron alguna cosa hera por mytimaes de Guaynacapa que los avia alli puesto para que le senbrasen coca y otras cosas / pero que las tierras no heran suyas e sobre esto estuvieron los dhos dias que no se conçertavan / porque el caçique de Canta no queria ningun conçierto e ynportunaçion del dho Niculas de Rribera e de otros españoles e deste t° que alli estavan / vino a conçierto con el caçique de Chacalla en que le dio dozientas obejas por el dho pedaço de tierra e unas casillas que tenian los dhos yndios de Chacalla hazia el dho pueblo de Chacalla e dello se hizo çierta escriptura e conçierto entre los dhos caçiques la qual este tº escrivio e firmo su nonbre de su letra / el qual le fue mostrado a la qual se rrefiere / e que esto es lo que sabe.

Tº Niculas de Rribera el moço. Vease a fojas quarenta e tres colinea primera.

Tº don Fran^{co} Yaybi caçique de Cullique dixo que dize lo que dho tiene en la pregunta antes desta / y que el dho Guaman Yabre caçique prinçipal de Canta dixo a este tº como los dhos yndios de Chacalla avian movido pleito a el y a los dhos sus yºs sobre las dhas trras de Quibi / y que por que se apartasen del pleito avian dado a los dhos yndios de Chacalla dozientas obejas e que esto sabe desta pregª.

Tº Diego Yndio dixo que puede aver el tpo que la pregunta dize poco mas o mºs / que este tsº vio que Niculas de Rribera el moço y Anpuero fueron a las dhas trras de Quibi y con ellos los caçiques e prinçipales y otros yndios de Chacalla y de Canta y estando en las dhas tierras de Quibi / se conçertaron en que los dhos caçiques de Canta dieron a los de Chacalla dozientas obejas por que se apartasen de lo que avian pedido sobre las dhas tierras de Quibi diziendo ser suyas y este tº vio como rresçibieron las dhas obejas y a todo ello se hallo presente.

Tº Joan Capucha dixo que lo que della sabe es que en tpo del presydente Gasca vio este tº que fueron a las dhas trras de Quibi los dhos Niculas de Rribera y Fran^{co} de Anpuero y el padre Diego de Avila y los caçiques de Chacalla y Canta y otros yndios y estando en las dhas tierras las personas suso dhas conçertaron a los dhos caçiques en la diferençia que tenian sobrellas en que los de Canta dieron a los de Chacalla por que se apartasen dellas y de no pedir las dhas tierras y que las dexasen dozientas cabeças de ganado de esta trra çien cabeças de carneros y çien cabeças de obejas y se las bio entregar y bio como las rresçibieron y hizieron vn conçierto sobre ello y este tsº lo vio escrevir que fue como lo tiene dho en esta pregunta. No mas.

7 - Yten si saben que los dhos yndios de Canta y su caçique dieron a los dhos yndios de Chacalla y a su caçique en cunplimiº del dho conçierto las dhas dozientas obejas y se las dieron y entregaron a ellos mysmos y al dho su caçique y desde entonçes aca los dhos yndios de Canta an tenido e poseydo las dhas trras sin contradiçion ni pleito alguno. Digan lo que saben.

/56r/ Tº Diego Flores dixo que dize lo que dho tiene en la pregunta antes desta e que despues aca de aver rresçivido los dhos caçiques e yndios de Chacalla las dhas ovejas por rrazon del dho conçierto a bisto que los dhos yndios de Canta an tenido e poseydo las dhas tierras de Quibi senbrandolas e cogendolas los frutos dellas sin contradiçion de los dhos yndios de Chacalla hasta agora e que esto es lo que sabe desta pregª.

Tº Diego de Avila dixo que lo que sabe es que despues de fho el dho conçierto contenido en la pregunta antes desta / este tº vio que el dho caçique de Canta ynbio çiertos yndios prinçipales e a bn español que se dize Diº Flores a vn pobleçuelo que seria dos o tres leguas de alli a entregar las dhas dozientas obejas al dho caçique de Chacalla el qual dho caçique de Chacalla enbio a vn prinçipal con çiertos yndios para que las rresçibiesen e donde a dos dias llegaron yndios de Canta e de Chacalla adonde este tº estava / con carta del dho Diego Flores / de como avia entregado las dhas dozientas obejas a los yndios de Chacalla e los yndios que en aquella sazon estavan en Quibi / dexaron las dhas tierras e se fueron dellas e el dho caçique e yndios de Canta se metieron en ellas e las an tenido e poseydo hasta agora e que despues de lo susodho este tº oyo quexarse al dho caçique de Canta Guaman Yaure para aver dado las dhas obejas porque dezia que las tierras heran suyas e que contra rrazon las avia dado e que esto es lo que sabe desta pregª.

Tº Diego Yndio dixo que dize lo que dho tiene en las preguntas antes de esta y que es verdad que este tº vio como los caçiques e yndios de Canta dieron y entregaron a los mysmos caçiques de Chacalla las dhas dozientas obejas y las rresçibieron y despues aca las an tenido e poseydo sin contradiçion alguna ny sin le mover pleito sobrello e que esto es lo que sabe desta pregª

Tº Joan Capucha de Canta / dixo que dize lo que dho tiene en la pregunta antes desta y a visto que despues de hecho el dho conçierto a esta pte an tenido los dhos yndios de Canta las dhas tierras de Quybi sin contradiçion alguna.

Tº Capacha yndio no xpiano dixo que dize lo que dho tiene en la pregunta antes desta e que despues aca de hecho el dho conçierto a visto este tº tener e poseer a los dhos caçiques e yndios de Canta las dhas trras de Quibi / e que esto es lo que sabe desta pregª. No ay mas.

8 - Yten si saben que todo lo susodho es pu^{co} e notorio e publica boz e fama. Los tºs se rretifican en sus dhos.

Pregunta añadida

1 - Yten si saben que los dhos yndios de Chacalla tienen sus tierras e terminos muy apartados de las dhas trras de Quybi e de la otra pte del dho rrio de Quibi / pasados vnos çerros que estan entre el dho rrio de Quibi e pueblos de Chacalla como es notorio por manera que nynguna continuydad ny nynguna aparençia de ser suyas ny tener drº a las dhas trras de Quibi tienen los dhos yndios de Chacalla por ser como es cosa tan distinta y apartada de las dhas tierras de Quibi los dhos pueblos de Chacalla. Digan lo que saben.

Tº Diego Flores dixo que a visto este tº que Chacalla donde los yndios tienen su asiento e pueblo es apartado de Quibi çinco leguas poco mas o mºs / y en medio esta vna sierra e que en el tpo que dho tiene que a que esta en el dho rrepartimiº de Canta no vio ni a visto en las dhas trras de Quibi ni en todo aquel valle yndios de Chacalla por lo qual e por estar //
/56v/ las dhas tierras de Quibi en las tierras e balle de los dhos yndios de Canta le paresçe a este tº que las dhas tierras de Quibi seran de los dhos yndios de Canta e que esto es lo que sabe.

Tº Diego Yndio de Canta dixo que la trra e trºs de Chacalla es muy apartado de los canta porque tienen en medio vna sierra de manera que no estan juntos sino apartados mucho lo vno de lo otro y ansi lo a visto este tº por avello andado y porque por esto lo sabe y que esto es ansi verdad.

Tº Joan Capucha dixo que los dhos yndios chacalla tienen su asiento e trra e pueblos muy apartados de las chacaras de Quybi que esta en medio un gran çerro y de la otra pte estan las dhas tierras de Canta e que dize lo que dho tiene. No ay mas.

Pu[ca] boz y fama ya se dijo
esta bien sacada del orig. y asi lo juro.
El Relator
esta byen sacada del original en lo que faze al caso y asi lo
juro en forma.

El Licen[do] de Leon El Licen[do] M[o] Luzio

Por pte de los indios de Chacalla con los de Canta sobre bnas trras llamadas Quibi
Ptes Al[o] Moreno y Fran[co] de la Torre hizo autos sin poder no estan rr[dos]. Veanse Luzio y Leon.

Don X[o]bal y don Felipe y Fran[co] por si y los demas pidieron a los de Canta que tenyendo e poseyendo ellos e sus antepasados vnas trras y chacaras de coca que dizen de Quibi desde el tpo de los ingas pasados y tenyendolas al tpo que los españoles entraron en estos rreynos / en tpo del marques Piçarro los dhos indios de Canta avian preso bn caçique suio de Chacalla llamado Vilcacapcha y lo mataron juntam[te] con 153 indios que tenia en las dhas tras porque no se querian alçar con ellos y con el inga y les tomaron las tras y no se las an querido bolver; antes dizen tener bn desistimi[o] dellas que hizo si lo hizo bn caçique de Chacalla lo qual no hizo y si lo hizo no por su voluntad y aunque lo hiçiera fueron agraviados lesos y sacrificados en el enormisimam[te] y debe ser deshecho el dho conçierto y engaño /

Pidieron que declarando las tras por suias se las mandese bolver como cosa suia y de sus pasados con los frutos y el conçierto se paresçiese rresçindido y just[a] de tr[do].

Pidieron por libres porque las tras son suias e lo fueron desde el tpo de Topa Inga Yupangui y la color que los indios de Chacalla tienen para las pedir es que en aquellas tras Guaynacapa puso çiertos mitimaes destos indios de Chacalla y señalo lo que le avian de labrar por el y al tpo que los españoles entraron en este rreyno y al de la suzeçion del Inga y no bvo para que estar alli los indios se bolvyeron a sus trras y ellos se tornaron a bolver a labrar sus tras como cosa suia propia y desde estonçes sienpre las an tenido y poseydo y tenyendolas y poseyendolas / los bxos [*sic*] las pidieron a bxo [*sic*] presydente y pusieron pleito sobre ellas y avisando mando a sus encomenderos las viesen y los conçertasen y asi los conçertaron y dieron sin dezir ninguna cosa todo lo contenido en bn conçierto que presento que pidieron por libres y just[a] a pr[a] con 6 dias.

Su Mag. da por libres los indios de Canta.

Sup[an] los de Chacalla lo aleg[do] y mas que el conçierto si su caçique lo hizo fue forçado por Fran[co] de Anpuero y que ellos no rresçibieron las ovejas sino su encom[o] Anpuero / Piden rr[on] y pr. tr[do] y alegam[o] que agora bino a su not[a] que los de Chacalla no fueron mitimaes puestos alli sino que en tpo de Guascar conforme a la costunbre que entre los indios avia se hiçiera çiertas plegarias bna por el y otra por su muger en que avia por costunbre que todo lo que se fuese con la sangre dellas en bnas ollas se ganaba de tr[no] y si alguno se lo derramaba o el que la llevava lo matavan y enterravan alli y que estonçes estos de Chacalla salieron a los trn[o]s de Quibi y por no bertelles la sangre no los echaron dellos que por esto son y entonçes an sido suyas.

/50r/ Muy p[o] s[o]r

Al[o] Moreno en nonbre del caçique e y[o]s de Chacalla en el pleito con los y[o]s de Canta sobre las tierras de Quivi / digo que a my notiçia es benido que a pedimi[o] de las ptes cont[as] esta por V. Alt. mandado dar probysion para que mys ptes no cojan ny benefiçien las chacaras que tienen en Quybi lo qual pretenden las ptes contrarias para acabar de quytar del todo vna chacarilla que a mys ptes a quedado en Quybi la qual syenpre an tenido e tienen sobre la qual no se litiga y debaxo de las dhas cavtelas pretenden acabarla de quytar a mys ptes / y por otra pte para que se puedan ap[o]vechar della / de seys dias a esta pte / les an rronpido e quebrado las açequias las ptes contrarias e quytadoselas para que no les vaya agua lo qual hazen en vengança del pleito que mys ptes les an movido y a ello no se debe dar lugar/

A V. Alt. pido y sup[co] mande se de la dha p[o]visyon syno solamente para que cada bno coja este y fize de las chacarras que tenia antes de movido este pleito fasta que se sentençie y determyne / y los dhos y[o]s no perturben a mys ptes en les quytar el agua so pena que se le detenga e que las açequyas que obieren quebrado y deshecho las tornen a poner como estaban sobre que pido just[a] e para ello etc.

El Licenc. M[o] Luzio Al[o] Moreno

/50v/ En Los Reyes a diez dias del mes de dizienbre de myll e quy[o]s e çinquenta y ocho años ante los ss e oydores en Abdiençia Rreal; la presento el dho Alonso Moreno en el dho nonbre e por los dhos ss vista mandaron que cada uno de los dhos caçiques e yndios cojan lo que oviere senbrado y no se haga ynobaçion conforme a lo mandado en la otra petiçion de los y[o]s de Canta.

Fran[co] de Carujale

/51r/ Muy pod. ss[o]r

El caçique e yndios de Canta en el pleito con los yndios de Chacalla sobre las trras / dizimos que las ptes contr[a]s an ydo a tomarnos nra coca e sementera de nras chacaras estando como esta en lites pendençia lo qual sera cabsa de que nos matemos los vnos a los otros / como ya lo an procurado otra bes / a lo qual Bra Al. no deve dar lugar pues las trras son nras y estan en nras trras / que la cabsa esta puesta en just^{a}.

A Bra Al. pedimos e suplicamos mande probeer de rremedio con just^{a} mandando so graves penas a las ptes contr^{a}s que no den ocaçion al dho escandalo ni entren en las dhas trras hasta que se determyne la cabsa / lo qual se le mande so graves penas y sobre todo pedimos just^{a}.

El Licen^{do} de Leon

/51v/ Que no se haga ynobaçion durante la causa y que el caçique de Chacalla no baia a las tras so p^{a} de pribaçion del caçicazgo ni sus indios so p^{a} de doz^{o}s açotes a cada bno. P^{o}beyose por los señores oydores en Abdiençia Rreal en diez de dizienbre de mill e qui^{o}s e çinq^{ta} e ocho a^{o}s.

Fran^{co} de Carujale

/52r/ Muy p^{o}s ss

Fran^{co} de la Torre en nombre de los yndios de Canta dixo que para que a Vra Alteza conste como los yndios yavyos daran ocasion a mys ptes para que se maten los vnos con los otros y para que conste a V. Alteza dello y que es menester poner rremedio hago presentaçion ante Vra Alteza desta carta del rreligioso que dotrina a los dhos yndios.

A Vra Alteza pido y sup^{co} mande poner rremedio en ello que no se maten sobre que pido justiçia.

El Licen^{do} de Leon

En la çibdad de Los Rreyes en doze dias del mes de dizienbre de mill e quy^{o}s e çinq^{ta} e ocho a^{o}s ante los señores oydores en Abdiençia Rreal haziendo abdiençia de rrelaçiones la presento el dho Fran^{co} de la Torre en el dho nonbre e ..... e los dhos señores mandaron que se guarde lo proveydo e que se notifique a los caçiques en persona.

Fran^{co} de Carujale

/53r/ Muy mag^{o} s^{o}r

Jesu Christo nro señor sea con V. M. / aca vinieron a Quiby obra de çinquenta yndios y prinçipales yabyos y estubieron algunos dias asta este juebes pasado / que estos de Canta los echaron que vinieron prinçipales y gente darriba y por poco se descalabraran o mataran / que los yndios yabyos dezian que venian a tomar la posesion y que esperavan ally al escrivano y a su amo a que se la fuerze a dar / y digo a V. M. que se no se rremedia que se an de matar sobre la chacra / que mas valen las dozientas obejas que dos chacras tales que ellos no sacan probecho ninguno della sino bna coquilla y yucas y mala ventura que maiz muy rruin y poco / V. Md tuvo la culpa en conçertarlos que le dieran obejas que sin obejas la obieran ellos entonçes que abia muchos testigos caçiques viejos que sabian la verdad como era de los cantas primero que con bnos sacrifiçios que traxeron los dhos yabyos de los yngas se les metieron en las tierras / V. M. ynforme desto a los señores oydores y al Lic^{do} Hidalgo que si no lo rremedia que a de aver muertos que los yabyos //

/53v/ no pretenden sino salir con la suya que tienen su punto de onrra abnque no fuera sino por que no aya muertes y rrenzillas no se las avian de dar a los yabyos sino que le dieran lo que valia por ellas / quanto mas abiendo dado dozientas obejas como dieron sin aber por que darselas sino por quitar rrenzillas y muertes conçedieron y vinieron en el conçierto. V. M. lo abise oy antes que se sentençie que los yabyos darian lo que tienen y lo que no tienen por salir con su negra onrra / y con tanto quedo.

Beso las m^{o}s de V. Md
Su capellan Hieronymo M^{yn}

Al muy mag^{o} señor
Niculas de Rribera el moço
en Los Reyes mi s^{o}r.

/54r/ En Los Rreyes diez e seys dias del mes de diz^{e} de mill e quy^{o}s e çinq^{ta} e ocho años; los señores presidente e oydores desta Rreal Audiençia en audiençia de petiçiones nonbraron por terçero en esta causa entre los yndios de Canta con los de Chaclla sobre las trras de Quivi al Licen^{do} Br^{me} Fz abogado desta Rreal Avdiençia e le mandaron lo acepte el qual juro por Dios nro señor e sobre la señal de la cruz en forma de derecho e dixo: "sy juro e amen" so cargo del qual dho juramento prometio botar en esta causa e dar su paresçer en ella e guardar secreto.

Diego Muñoz

En Los Rreyes beynte dos dias del mes de diz^{e} de myll e quy^{o}s e çinq^{ta} e ocho años / los señores presydente e oydores desta Rreal Audiençia en acuerdo de justiçia nonbraron por tr^{o} para la dha causa al dottor Colamaria de Ochoa.

Diego Muñoz

En Los Rreyes veynte e tres dias del dho mes de diz^{e} del dho año en avdi^{a} de petiçiones pu^{ca} ante los señores presidente e oydores el dho dottor Colamaria de Ochoa juro por tr^{o} en esta causa sobre la señal de la cruz en forma de derecho so cargo del qual juramento prometio botar en esta causa e dar su paresçer como alcançare por dr^{o} e guardar el secreto.

Diego Muñoz

/55r/ Los y^{o}s de Canta.

Muy p^{o}s ss

Fran^{co} de la Torre en nonbre de los yndios de Canta / digo que Vra Alteza mando hazer ynformaçion que cuyas eran las tras e coca que coxeron en las chacaras de Quibe. E yendo mys partes a dar su ynformaçion y escribiendo el çaçerdote que alli esta que aguardase para dar su ynformaçion el tsr^{o} no quyso segun consta e pareçe por esta carta de que hago presentaçion.

A Vra Alteza pido e ssup^{co} que pues consta de la maliçia de la parte contraria por esta dha carta que Vra Alteza mande al escr^{o} que buelva a hazer la dha ynformaçion por que Vra Alteza no provea con sinyestra rrelaçion lo que no sea justiçia la qual pido y el of^{o} de Vra Alteza ymploro.

El Licen^{do} de Leon — Fran^{co} de la Torre

En Los Rreyes veynte e tres de dizienbre de myll e quy^{o}s e çinquenta y ocho a^{o}s en Aud^{a} Rreal pu^{ca} ante los señores presydente e oydores / la presento en n^{e} de su parte el p^{e} e los dhos señores lo ovieron por presentado e mandaron que el dho escrivano buelva alla a hazer la dha ynformaçion pagandole su trabajo.

Diego Muñoz

/56r/ Muy mag^{o} s^{o}r

Jesu Christo nro señor sea con V. M. / ayer miercules a sol puesto supe como venia bn escrivano a ver estas chacras de Quybi y luego me parti y yo que llegaba mas aca de la puente arriba de Quibi me fue a dezir bn yndio yendo a ver si venia / como se queria partir el dho español y enbiele a dezir que me esperase y dizen estos que los yabyos le dixeron que no venia yo que se fuesen y asi se partio que quando yo llegue yba abaxo de Quivi y enbiele a dezir que nos viesemos y fuese de largo y dixeronme los yndios que dixo el español que le fuese yo a alcançar y como no traya cabalgadura que la dexe en vista por benir mas presto no le fui a alcançar. A las nuebe del dia llegue aqui a Quiby / dixeronme los yndios que les pregunto el dho como era bn pedaço de coca arriba de las casas y que quien la planto dixo que nunca los yabyos tal pie de coca senbraron ni tienen aqui casa ninguna / sino que bnos yndios de los yabyos que se llaman Pomaraqui y el otro Tanta dixeron a Guaman Yabyry //
/56v/ que se querian pasar aca a Canta que les diese el dho Guaman Yabri la dha chacra de coca; dixoles Guaman Yabry: "pasaos bosotros que yo os la dare" y supolo su caçique Marcapoma y los maltrato en Chaclla porque se querian pasar y no se pasaron ni se le dio las chacras / que no es sino bnos pedaços / el bn pedaço se planto para bn prinçipal de vista que se murio el otro dia que se llamava don Hernando Contorniza / y el otro pedaço se planto para don Antonio Malcapoma / y el otro pedaço se planto para don Filipe Astomichuy / que es muerto y su hijo lo tiene / y estos con sus obejas y su chicha lo plantaron / y don Christobal de año y medio a esta parte lo mando coger a escondidas por fuerça porque aqui en Quiby / ordinariamente no ay mas de çinco o seys yndios y quando benian a coger la coca los yabyos benian veinte y treinta yndios y se yban luego y amenazaban los yndios daqui / que por de hurto abian de pedir la coca que le llebaron / y no tienen ninguna justiçia los yabyos que su rrio y sus aguas ban a Lima //
/57r/ todas las chacras porque traen pleito si V. Md y Anpuero se ynformaran del negoçio sin dar obejas ningunas mandara el presidente que los yabyos dexaran las tierras a los de Canta / pasa desta manera / que quando Topa Ynga Yupangui / las chacras eran a los cantas y de antes que fuesen conquistados / y quien viere el rrio arriba vera y entendera que verisimamente eran suyas / y en tiempo de Guaynacapa / poseyendo los cantas sus tierras / traxeron bn sacrifiçio que llaman capacocha los yauyos desde los terminos de Xauxa que se lo entregaron los yngas que son unas conchas molidas que llaman mollo / cada yndio bn poco / y sacrificaban çiertas obejas entregadas de los dhos yngas que las trayan del Cuzco para el dho efecto onde le entregaban el sacrifiçio sacrificaban las dhas obejas y las quemaban y la sangre echaban en vnos porongos chicos y la rreçibian todos los yndios de aquel rrepartimiento todos los que eran de rrazon mançebos y hombres / no mochachos y ni biejos muy biejos y oro y plata para las guacas y rreçebido esto todos los dhos yndios como quien //
/57v/ haze chaco / casi pegado bno con otro por çerros y rriscos y por todos sus terminos dando grandes bozes con sus poroncos de sangre en sus manos y en la otra mano bn bordon porque no cayeszen / ynbocaban a todas las guacas y al sol y a la tierra diziendo que fuese el Ynga siempre moço y baliente y bibiese mucho / y su muger que fuese siempre moça y que aquel sacrifiçio se lo ofreçian en nonbre del Ynga / y algunas guacas afamadas les ofreçian en estos sacrifiçios yndios bibos los enterravan / y asi trayendo los yauyos en tiempo de Guaynacapa trayeron los dhos yavyos bn sacrifiçio destos y binieron a salir abaxo de las tierras que agora piden / y tomaron bn poquillo alli / porque yban al rrio a labar aquellos porongos despues que acabavan el dho sacrifiçios [*sic*] / y el que por mano de pecado mientras trayan la dha sangre derramaba alguna gota o alguno se la derramaba / alli lo matavan y enterravan / y de otro sacrifiçio que trayeron se entraron en parte de las tierras que piden / y quando Guascar Ynca quando entraron los españoles se entraron asta //
/58r/ junto onde esta la yglesia y por no derramar la dha sangre que trayan no osaron estos a se lo defender / y los yngas avian desterrado aqui en Quiby antes desto veinte mugeres que fueron de bn caçique de los yabyos que bibia arriba de la puente de Mama que fue traidor y lo mataron y destruyeron y aquellas mugeres las desterraron aca y como venian preñadas muchas dellas multiplicaron y hilaban y de aquello bibian / y como se entraron en las dhas tierras como digo / enbiaron yndios aca / y se hizieron muchos / y

despues que vio Guaman Yabry que no avia ya memoria de que obiese ynga alguno / echabalos que se fuesen a sus tierras que no querian dexar perder sus chacras pues que eran suyas y estando en estos debates / bino V. M. a los conçertar y diz que dieron las obejas contra su voluntad que sin dar obejas pensaban ellos de sacar sus tierras / esto pasa porque antes de agora lo e sabido yo.
/58v/ Quisera hallar aqui al español / y con tanto quedo en Quibi de camino para Cocha/ oy jueves / 22 de dezienbre.

Muy mag^{o} s^{o}r. Beso las m^{o}s a V. M.
Su capellan Hieronymo M^{yn}

/59r/ Al muy mag^{o} señor
Niculas de Rribera el moço
en Los Reyes mi s^{o}r.

/60r/ En el pleito que es entre partes. De la vna los caçiques e yndios del rrepartimientos de Chacalla y Alonso Moreno su pr. y de la otra el caçique e yndios de Canta y Fran^{co} de la Torre su pr. sobre las trras de Quivi.

Fallamos atento los avtos y meritos deste proçesso que devemos absolver y absolvemos al caçique e yndios de Canta de la demanda contra ellos puesta por parte de los dhos caçique e yndios de Chaclla e por esta nra sentençia difinytiva juzgando asi lo pronunçiamos y mandamos sin costas.

El D. Bravo de Saravia	El Licenc. Mr. de Peñalosa	El doctor Gz de Cuenca

Pronunçiose esta dha sent^{a} difynytiva por los dhos señores presydente e oydores que en ella firmaron sus nonbres estando haziendo Audi^{a} Rreal pu^{ca} en la dha çibdad de Los Rreyes en diez dias del mes de henero de myll e quy^{o}s e çinquenta y nueve a^{o}s; presentes los dhos Fran^{co} de la Torre y Alonso Moreno prs de las ptes a quien se notyffico.

Diego Muñoz

/62r/ Los y^{o}s de Chacalla.

Muy p^{o}s s^{o}r

Se suplica y pide resti^{on} con el conçierto.

Alonso Moreno en nb. de los y^{o}s de Chacalla en el pleyto con el caçique e y^{o}s de Canta sobre las chacaras y tierras de Quibi / supp^{co} de la sent^{a} por V. Alt^{a} dada en esta causa / y digo que se deve enmendar y rrevocar condenando a las ptes contr^{a}s en las dhas tierras y frutos dellas por lo siguyente:

- Lo prim^{o} por todo lo gen^{al} del dr^{o} y lo que dezir me convyene y del proçeso rresulta en favor de mys ptes que se por expreso.

- Lo otro porque por el proçeso y provanças del consta ser las dhas tierras de mys ptes y averselas dado los yngas pasados y tenerlas y poseerlas al tpo que los españoles entraron en esta trra como cosa suya / y serles despues tiranicam^{te} quitadas matando a los que en ellas estavan por servir a los xpianos y no rrebelarse contra ellos como las ptes contr^{a}s por lo qual avnque otra cosa no ovyera avyan mys ptes de ser en duda favoresçidos espeçialm^{te} tenyendo su just^{a} tan clara como la tienen pues avnque no tuvyeran provado sino solamente tener las dhas tierras al tpo de la entrada de los españoles en la tierra les bastava para conseguir su just^{a} conforme a las provisiones rreales en que se provee y manda que todas las tierras y las demas cosas esten en el estado que estavan quando los dhos españoles en su rreal nonbre entraron en la tierra.

- Lo otro porque no daña a mys ptes ni aprovecha a las contr^{a}s el sinple papel que presentaron que llaman de conçierto / porque lo es ni tal hizo ni otorgo el caçique prinçipal del rrepartimi^{o} de mis ptes / ni ya que lo hiziera podia ni puede perjudicar a los demas mys ptes ni a los y^{o}s pobres ni a los demas del dho rrepartimi^{o} pues las dhas tierras eran y son de todos en gen^{al} / y no del dho caçique solam^{te} en particular / y asi no puso perjudicar a mys ptes lo qual dho caçique quiso hazer sin poder / pues tocando a todos por todos avia de ser otorgado y aprovado / pues ninguno puede perder el dominio de su cosa sin su //
/62v/ propio hecho y consentim^{o} / espeçialm^{te} que provare en forma no aver hecho el dho caçique el dho conçierto que no rreçibio las ovejas que en el se haze mençion sino Fran^{co} de Anpuero su encomendero sin su consentim^{o} del dho caçique y contradiziendo el dho conçierto y no lo queriendo hazer antes diziendo que no queria ovejas sino las tierras que heran de sus y^{o}s / y el dho Fran^{co} de Anpuero contra la voluntad del dho caçique las rreçibio y firmo el dho conçierto / y por lo confesar y aver hecho las ptes contr^{a}s son vistos y es prueva bastante de no ser suyas las dhas tierras ni tener en ellas nada.

- Por que pido y supp^{co} a V. Alt^{a} enmyende y rrevoque la dha sent^{a} de vista condenando a las ptes contr^{a}s a rrestituçion de las dhas tierras y chacaras y frutos dellas con qt^{o} que pido y just^{a} y ser rreçibido a prueva y para ello etc.

- Otro si en n^{e} de los dhos y^{o}s y prinçipales del dho rrepartimi^{o} pido rrestituçion como de dicho mejor lugar aya de qualquyer conçierto que paresçiere perjudicar a mis ptes y pido just^{a} y para ello etc. y juro a Dios y a esta + que no la pido de maliçia.

- Otro si a V. Alt^{a} pido y supp^{co} mande que se me de en pu^{ca} forma con çitaçion de las ptes contr^{a}s la provision de Su Mg^{t} en que manda que todas las cosas esten en el estado que estavan quando los españoles entraron en esta tierra / sobre que pido just^{a}. Va sin testadura ni enmyenda.

El Licenc^{do} M^{o} Luzio Al^{o} Moreno

En los Rreyes diez y ocho de henero de myll e quy^{o}s e çinquenta y nueve años ante los señores oydores esp^{do} en Aud^{a} Rreal pu^{ca} / la presento el p^{e} en el dho nonbre / y los dhos senõres mandaron dar tr^{do} a la otra pte e que rresponda a la primera audi^{a} e que se le de cunpulsorio para lo que pide çont^{ma} la pte contraria / lo qual paso presente Ffran^{co} de la Torre pr. contrario a quien se notifico q^{to} e aperçibio para todo lo de suso cont^{do}.

Diego Muñoz

/63r/ Los y^{o}s de Canta.

Muy p^{o}s ss

Cuentan el origen como se entraron en las trras
de Quibi los de Chacalla y dizen la restituçion
porque dize que las trras aun no balian ni
balen las obejas.

Fran^{co} de la Torre en nonbre del caçique e yndios del rrepartimi^{o} de Canta en el pleito con el caçique e yndios de Chacalla sobre las tras de Quibi rrespondiendo al escripto de suplicaçion por la parte contraria presentado cuyo tenor abido aqui por rresumydo digo que Vra Alteza deve mandar confirmar la sent^{a} de vista en esta cabsa dada / sin enbargo de lo que la parte contraria dize y alega que se excluye y no deve ser considerado por lo sig^{e}:

- Lo primero por todo lo general que es aqui por espreso e por todo lo que tengo dho y alegado a que me rrefyero.

- Lo otro porque por ser tpo de verano al tienpo que se hazia la provança no prove bastantisimamente todas las preguntas del ynterrogatorio que presente y asi se viera claramente como de tpo ynmemorial a esta parte las dhas tras de Quybi eran e fueron de las dhas mys ptes y que luego como murio Guaynacapa los dhos yndios de Chacalla se bolvyeron a su tras y dexaron aquel pedaço de tras de Quybi que abian entrado lo qual provare agora dandome el termyno conpetente para ello.

- Lo otro porque agora viene de nuevo a my notiçia que los dhos yndios de Chacalla no fueron mytimaes puestos por nengun señor de los pasados en las dhas tras de Quybi sino que haziendo çiertas plegarias por el Ynga que en su lengua se llamava capacocha en las quales los yndios de vn valle salian con çiertos mates o ollas pequeñas de sangre e yban asi dando boçes e si algun yndio hazia derramar qualquyer parte de la tal sangre e los que la llevaban la derramavan le davan garrote y lo enterravan en el propio lugar y los que llevaban la dha sangre o capacocha hazian termyno hasta el lugar donde llegavan e asi vna o dos bezes los dhos yndios de Chacalla con maliçia hizieron las dhas plegarias o capa-//
/63v/ coche con la qual fueron hasta las dhas tras de Quibi antes que los dhos yndios de Canta los saliesen a rreçebir y despues por miedo de no derramar la sangre de la dha capacoche no los hecharon de las dhas tierras de Quybi avnque eran suyas con la qual color tomaron achaque los dhos yndios de Chacalla de poner mytimaes en la parte de tierra en que se avian entrado y quando Guaynacaba paso por alli no le hablaron en ello porque se abia hecho por el la dha capacoche y asi despues de muerto como dho tengo se bolvyeron a su tras los dhos partes contrarias por lo qual ningun titulo ny color pueden tener ny pretender a las dhas tras contra lo qual no enpeçe lo que la parte contraria dize que asi los hallaron los xpianos quando entraron en este rreyno e que conforme a la provysion rreal los an de dexar alli por quanto aquello se entiende y deve entender de lo que legitimam^{te} estava fho por los ss pasados deste rreyno y no por cosa tan mal hecha como las partes contrarias hizieron quanto mas que ya que aquello oviese lugar los dhos yndios que alli estuviesen en las dhas tras de Quibi avian de contribuyr y tributar con los dhos mys partes pues el trebuto que pagan y pagavan al Ynga es por las tras que tienen.

- Lo otro porque no solamente la tra que pretenden las partes contrarias pero avn todas las dhas tras de Quybi no balen tanto como las dozientas obejas que mys partes les dieron e si Vra Alteza lo manda a ver a personas beran que es berdad lo que digo porque no son abn beynte hanegas de senbradura por lo qual çesa la rrestituçion que piden las partes contrarias avnque fueran menores quanto mas siendo como son mayores y tales personas que no les conpete rrestituçion alguna.

- Por las quales rrazones e por cada vna dellas pido e ssup^{co} a Vra Alteza mande confirmar la dha sent^{a} de bista sobre que pido just^{a} e costas y el of^{o} de Vra Alteza ynploro y pido ser rreçebido a prueba por los mesmos articulos del ynterrogatorio pasado attento que no pude hazer provança por la cabsa que dha tengo y asi mesmo por las preguntas que de nuevo pusiere

- Otro si por quanto lo mas de lo susodho consiste en provança por vista de ojos asi para que se bea lo que balen las dhas//
/64r/ tras como para que se vea el lugar donde estan y asi mesmo para que se traygan los dhos e depusiçiones e e tomen de los yndios comarcanos e biejos que estan en el dho valle es neçes^{o} que se nonbren dos personas para que las bayan a ver e tasar las dhas tierras y hagan la dha provança.

- Pido e ssup^{co} a Vra Alteza mande que se nonbren dos personas para lo susodho e que bayan a lo hazer a costa de la parte cayda sobre que pido just^{a}.

El Licen^{do} de Leon Fran^{co} de la Torre

En Los Rreyes veynte e vno de dizienbre de myll e quy^{o}s e çinquenta y nueve años ante los señores oydores est^{do} en Aud^{a} Rreal; la presento el cont^{do} en n^{e} de su pte y los dhos señores ovieron la dha causa por conclusa e en los demas que pide mandaron se traygan los autos para probeer sobre ello.

Diego Muñoz

/64v/ En Los Rreyes veynte e tres dias del mes de henero de myll e quinientos e çinquenta e nueve años aviendo visto los señores presidente e oydores desta Rreal Audiençia este proçesso en lo que las partes piden se nonbren personas que vean por vista de ojos las dhas trras de Quivi sobre que es este pleito y el sitio donde estan mandaron que cada vna de las dhas ptes nonbren persona de su pte que las vayan a ver e çerca dello tomen los dhos e depusiçiones de los yndios antiguos mas çercanos que de las dhas trras tengan notiçia e sobre todo den su paresçer de lo que mas convenga proveer que sea en menor daño e perjuyzio de las dhas ptes lo qual pase e se haga ante Joan de Ençiso escr^{o} de su mag^{t} e ansy lo pronunçiaron e señalaron de sus rrubricas.

Paso ante mi

Diego Muñoz

En Los Rreyes veynte e quatro dias del mes de henero de myll e quy^{o}s e çinq^{ta} e nueve años. Notifique el avto de suso a Fran^{co} de la Torre e A^{o} Moreno prs de las dhas ptes.

Diego Muñoz

/65r/ Los y^{o}s de Chacalla

Muy p^{o}ss señores

Al^{o} Moreno en nonbre de los yn^{o}s de Chacalla en el pleyto con los de Canta sobre las tierras de Quibi. Su pr. a V. Alt^{a}. me mande dar su rreal probision para que las chacaras en que se an entrado durante este pleyto las partes contrarias se buelban a mis partes con la coca que an cojido conforme a lo probeydo por V. Alt^{a}. y pido just^{a}. y para ello etc.

El Licenc^{do}. M^{o} Luzio Al^{o} Moreno

En Los Rreyes veynte e quatro de henero de myll e quy^{o}s e çinquenta y nueve a^{o}s ante los ss presydente e oydores en Audiçia Rreal la presento el cont^{do} y los dhos señores mandaron se les de provision conforme a lo que sobre ello tienen proveydo para que las tyerras en que se ovieren entrado el dho caçique e y^{o}s de Canta se lo buelvan con lo que ovieren cogido dellas.

Diego Muñoz

/66r/ Muy p^{o}s s^{o}r

Los y^{o}s de Canta en el pleyto con los y^{o}s de Chaclla sobre las tierras de Quyvi / digo que por V. Alt. fue mandado que nombrasemos personas que fuesen a ver y a tasar las dhas trras y asy mysmo para que se haga alla la p^{o}vança.

- por tanto nosotros nombramos a Ju^{o} de Pedraza por nra pte e pedimos y suplicamos a V. Alt. lo aya por nombrado para que jure.

- otro si pedimos y suplicamos a V. Alt. m^{de} rresçebir la dha cabsa a pruva para que se hagan los articulos por donde an de declarar los t^{o}s fuera de los que tenemos presentados sobre que pidimos just^{a}. e costas y en lo neçesa al of^{o} de V. Alt.ymploramos.

El Licen^{do}. de Leon

En la çibdad de los Rreyes en veynte y çinco dias del mes de henero de mill e qui^{o}s e çinq^{ta} e nueve años ante los señores oydores en Abdiençia Rreal haziendo abdiençia de rrelaçiones; la presento Fran^{co} de la Torre en los dhos nonbres e los dhos señores dixeron que rresçibian e rresçibieron las dhas ptes a la prueva con termino de veynte dias con cargo de publicaçion concluso.

Fran^{co} de Carujale

/66v/ En Los Rreyes veynte e çinco dias del mes de hen^o del dho año de myll e quy^os e çinq^ta e nueve años. Notifyque la dha prueva a los dhos Alonso Moreno e Fran^co de la Torre prs de las ptes.

Diego Muñoz

/67r/ Los y^os de Chacalla.

Muy p^os s^or

Alonso Moreno en n^e del caçique e y^os de Chacalla en el pleyto con los y^os de Canta y su caçique / sobre las tierras de Quibi / digo que a notiçia de mys ptes es venido que V. Alt^a. ha proveydo que vaya vna persona de cada pte a ver las dhas tierras y a hazer provanças el qual es en perjuyzio de mis ptes porque tienen hecha la provança que les convyene / y las ptes contr^as han pretindido que lo susodho se haga para por yndiretas hazer provanças contradr^o por los mesmos articulos de la ynstançia pasada / y por articulos contr^os a los que tiene mi pte hechos; y porque no se le han de admitir por el camino dr^o pretenden por yndiretas que de off^o se haga; a lo qual no se deve dar lugar / y ya que se deva hazer provança por articulos nuevos y diferentes y no por los mesmos.

- por que pido y supp^co a V. Alt^a. mande que no se hagan las dhas provanças sino solam^te por articulos nuevos dados por las ptes y esamynados por V. Alt^a. y no por los pasados por la sobornaçion que en ello se presume que deve aver y se puede creer en este caso y si para ello me convyene suplicar del dho auto aquella suplicaçion ynterpongo que al dr^o de mis ptes convenga y pido cunplim^o de just^a. y costas y para ello etc.
- otro si por mi pte nonbro para que vaya por mis ptes a hazer las dhas averiguaçiones avyendose de hazer segun dho es y no de //
/67v/ otra manera / al padre fray Miguel de Orin vicario provinçial de Nra S^a. de la Md. y pido que se aya por nonbrado y just^a.
- otro si en el dho nonbre supp^co a V. Alt^a. mande que vaya otro escr^o a hazer lo que dho es que no sea el dho Ju^o de Ençiso porque lo tienen mys ptes por sospechoso y por tal lo rrecuso y juro a Dios y a esta + que no lo hago de maliçia y pido just^a. etc.
- otro si a V. Alt^a. pido y supp^co mande que cada vna de las ptes nonbre para su pte ynterprete ante quyen se aya de hazer lo que se hiziere que yo por mi pte nonbro al p^e frai Melchior Ferz. mandando que sin el no se haga nada y si se hiziere sea ninguno y pido just^a. etc.

El Licenc.^do M^o Luzio — Al^o Moreno

En Los Rreyes veynte e syete de henero de myll e quy^os e çinquenta y nueve a^os ante los señores presydente e oydores la presento el cont.^do en n^e de su ptes y los dhos señores mandaron que se lleven al semanero los ynterrogatorios que los vea / e que nonbren otra persona por terçero / e que Juan Xpoval de Frias vaya a hazer la ynform.^on e averiguaçion.

Diego Muñoz

/68r/ Muy p^os s^or

Alonso Moreno en n^e del caçique e y^os de Chacalla en el pleyto con los y^os de Canta y su caçique sobre las tierras de Quibi. Digo que yo nonbro por mys ptes al p^e frai Gaspar de Carvajal provinçial de S^to Domingo para que vea las dhas tierras y asista a las provanças y averiguaçiones.

- a V. Alt.^a pido y supp.^co lo aya por nonbrado y se cometa el juram^o del al escr^o que ha de yr a hazer la dha provança y averiguaçion sobre que pido just.^a etc.
- otro si al V. Alt.^a pido y supp.^co mande esamynar el ynterrogat^o de las ptes contr^as prim^o que se vaya a hazer la dha provança y que hasta que sea esaminado no se toman t^os ni bayan a ello sobre que pido just.^a

El Licenc.^do M^o Luzio — Al^o Moreno

En Los Rreyes veynte e ocho dias del mes de henero de myll e quy^os e çinq^ta e nueve años / ante los señores oydores en avdi.^a de rr.^on; la presento Alonso Moreno e los dhos señores mandaron que queryendo yr el dho provinçial sea por nonbrado para ello.

Diego Muñoz

/69r/ "Yo yre por servir a Nro Señor despues que pase la fiesta de Nra Señora de la Purificaçion". — Frai Gaspar de Carvajal

/70r/ Muy pod. s^or

Fran^co de la Torre / en nonbre de los yndios de Canta en el pleyto con los yndios de Chacalla / digo que a not.^a de mys ptes es benydo que Bra Al. a proveydo por rreçebtor a Ju^o Xpoval de Frias para que haga las provanças y porque mys ptes tienen por odioso y sospechoso al dho Frias y por tal lo rrecuso el qual procuro que rrecusarre a Ju^o de Ençisso por ir el sin conoçer mis ptes al dho Ju^o de Ençisso ni avello visto.

Pido y sup.^co a Bra Al. lo aya por rrecusado y mande a otro ts^o que vaya e juro a Dios en forma que esta rrecusaçion no la hago de maliçia sobre que pido just^a.

Fran^co de la Torre

En Los Rreyes veynte e ocho dias del mes de henero de myll e quy^{o}s e çinq^{ta} e nueve años ante los señores oydores en avdiençia de rrelaçion; la presento Fran^{co} de la Torre en nonbre de sus ptes e los dhos señores mandaron que vaya vn off.^{al} del secretario con el ts^{o} nonbrado.

Diego Muñoz

/7lr/ Muy pod. s^{o}r

Fran^{co} de la Torre en nonbre de los yndios de Canta en el pleyto con los yndios de Chacalla sobre las trras / digo que por Bra Al. esta proveydo e mandado que vaya vn rreçebtor a tomar las ynform^{o}s sobre lo susodho e no esta nonbrada / que persona a de yr.

A Bra Al. pido y sup.^{co} mande oy en todo el dia se nonbre la persona que oviere de yr syn sospecha sobre que pido just.

- otro sy mande Bra Al. tomar juram^{o} de las personas que estan nonbradas por terçeros para aberiguaçion de lo susodho e se de termyno para hazer las dhas p^{o}vanças / sobre que asy mesmo pido just.

Fran^{co} de la Torre

En Los Rreyes treynta dias del mes de henero de myll e quy^{o}s e çinq^{ta} e nueve años / ante los señores oydores en avdi.^{a} de rr.^{on}; la presento Fran^{co} de la Torre en n^{e} de sus ptes e los dhos señores mandaron que Fran^{co} Hrrez escr^{o} de Su Mag.^{t} vaya a entender en este negoçio con el padre Carvajal provinçial de la Orden de señor Santo Dgo.

Los terçeros vengan a jurar para ello e prorrogaron a las ptes el tr^{no} provatorio que les esta dado de veynte dias por dyez dias mas. Presente Fran^{co} de la Torre a quyen lo notifyque.

Diego Muñoz

/72r/ En Los Rreyes dos de hebr^{o} de myll e qui^{o}s e çinq^{ta} e nueve años fuera de avdiençia; la presento ante mi despues de puesto el sol e anocheçido Alonso Moreno. Diego Muñoz.

Muy P^{o}s s^{o}r

Alonso Moreno en n^{e} de los y^{o}s de Chacalla en el pleyto con los de Canta sobre las tierras de Quibi.

Supp^{co} del auto proveydo por el doctor Cuenca vro oydor y semanero por comision de V. Alt^{a} en que mando que los t^{o}s de la pte contraria se examinasen por las preguntas de la ynstançia pasada y se las admitio / y digo que se deve enmendar y rrevocar por lo gen^{al} y porque no haze al caso que las ptes contrarias sean y^{o}s / pues mis ptes lo son tanbyen; y un privilegiado contra otro igualm^{te} privilegiado no goza de su privilegio / espeçialm^{te} en este caso donde mas que en otro se presume y puede presumir sobornaçion de testigos / y es hazer alyende desto ynmortal esta causa avyendo de ser sumaria por ser entre y^{o}s / pues seria proçeder a ynfinito y dar causa a muchos perjuros y sobornos si se diese lugar a que hiziesen provanças por los mesmos articulos de la ynstançia pasada.

Por que pido y supp^{co} a V. Alt.^{a} enmyende y rrevoque el dho auto proveyendo segun esta proveydo por V. Alt.^{a} que se tiesten y borren las preguntas de la ynstançia pasada pues esta asi proveydo y es conforme a dr^{o} / sobre que pido just.^{a} y costas y para ello etc.

El Licen^{do} M^{o}. Luzio Al^{o} Moreno

/72v/ En Los Rreyes quatro dias del mes de hebr^{o} de myll e quy^{o}s e çinq^{ta} e nueve años ante los señores oydores en Avdi^{a} Rreal pu^{ca}; la presento Alonso Moreno e los dhos señores mandaron que se rresçiban los dhos ynterrogatorios e por ellos los t^{o}s que las ptes presentaren sin enbargo de la dha supp^{on}

Diego Muñoz

/73r/ Muy p^{o}s s^{o}r

Don Felipe por mi y en n^{e} de los demas y^{o}s de Chacalla en el pleyto con los de Canta sobre las tierras de Quibi / digo que V. Alt.^{a} proveyo que fuesen dos personas vna de cada pte juntam^{te} con el ters^{o} que se nonbrase a ver las dhas tierras y a hazer las provanças y que cada vna de las ptes nonbrase vn ynterprete / y por mis ptes fue nonbrado al p^{e} Carvajal / y por las ptes contrarias a Ju^{o} de Pedraza / y asi mesmo tenemos nonbrado ynterprete.

A V. Alt.^{a} pido y supp^{co} me mande dar su provision de lo susodho en que se les cometa el dho negoçio y en que se mande lo que han de hazer y el juram^{o} del dho p^{e} Caravajal se cometa al ters^{o} que va a lo susodho y pedimos just.^{a} y para ello etc.

El Licenc^{do} M^{o}. Luzio

/73v/ En la çibdad de Los Rreyes en seis dias del mes de febrero de mill e qui^{o}s e çinq^{ta} e nueve años ante los señores oydores en Abdiençia Rreal haziendo abdiençia de rrelaçiones se leyo su petiçion e por los dhos señores vista dixeron que se le de provision conforme al abto y que se comete la rrepçeçion e que vaya vna lengua del Abdiençia por todas las ptes.

Fran^{co} de Carujale

/74r/ Muy p^{o}s s^{o}r

Don Felipe por mi y en n^{e} de mi caçique e y^{o}s de Chacalla en el pleyto con los de Canta sobre las tierras de Quibi / digo que por V. Alt.^{a} fue prim^{o} proveydo que cada vna de las ptes nonbrase vn ynterprete para la provança y averiguaçiones que se han de hazer en las mesmas tierras / y despues se proveyo que fuese vno de los ynterpretes desta Rreal Audiençia / y despues se pidio por las ptes contrarias que fuese el p^{e} Gr^{mo} M^{in} por su pte para asistir a la provança con la lengua / y se proveyo asi / y nosotros no llevamos persona que asista de nra pte hasta agora.

- a V. Alt.^{a} pedimos y suplicamos mande que vaya bn ynterprete desta Rreal Audyençia solo o dos / lo que nosotros pusieremos cada pte el suyo / o todos juntos de manera que pagamos los vnos lo que los otros y no sean los bnos de mejor condiçion que los otros / sobre que pido just.^{a} y para ello etc.
- otro si a V. Al. pido y supp^{co} mande que se de al ters^{o} que va a las dhas tierras la provança que hizo en ellas Ju^{o} Xpoval de Frias sobre las tierras que pendiente esta causa nos han tomado las ptes contrarias pues esta hecha y con ella se evitaran costas y se hara mejor el neg^{o} a que el dho ters^{o} va / sobre que pido just.^{a}

El Licenc^{do} M^{o} Luzio Al^{o} Moreno

/74v/ otro sy pido y sup^{co} a Bra Alt. me mande prorrogar el termyno provatorio por otros veynte dias mas atento que mys ptes son abtores e que son pasados de los treynta con que se rresçibio a pr^{a} mas de la mytad e para ello etc.

Al^{o} Moreno

En Los Rreyes onze dias del mes de hebrero de myll e quy^{o}s e çinq^{ta} e nueve años ante los señores oydores en avdiençia de rr^{on} / la presento Alonso Moreno e los dhos señores prorrogaron el dho tr^{no} provatorio por otros veynte dias mas comunes a las dhas ptes e que solo vaya don M^{yn} Lengua a la dha ynformaçion e negoçio por anbas las dhas ptes e que se les de para el hefecto que piden la dha ynform.^{on}

Diego Muñoz

/75r/ Los y^{o}s de Chacalla.

Muy p^{o} s^{o}r

Al^{o} Moreno en nonbre de don Xpoval Vilcapoma e sus yndios sujetos con el caçique e y^{o}s de Canta sobre las chacaras e Fran^{co} de la Torre pr. en su nonbre digo que Fran^{co} Hernandez escrivano a venido a esta çibdad con las p^{o}vanças que hizo por mandado de V. Alt. a pedim^{o} de anbas ptes en esta cabsa / e por mys ptes esta cargado de los dhos que le pertenesçen.

Pido e sup^{co} a V. Alt. mande que el dho Fran^{co} Frz escr^{o} entregue luego las dhas p^{o}vanças e avtos que sobre esta cavsa trae fhos e pasaron antes al secretario Diego Muñoz ante quyen la dha cavsa pase / todo lo qual V. Alt. mande que la entregue al dho secretario oreginalmente e para ello etc.

Al^{o} Moreno

Otro si porque los dhos mys ptes tienen de presentar por t^{o}s a M^{yn} Piç^{o} y al secretario P^{o} de Avendaño e al Lic^{do} R^{o} Nyno e a Fran^{co} Velazques Talavera e a otros vezinos desta çibdad los quales son personas .... Pido e sup^{co} a V. Alt. atento que mys ptes son y^{o}s se mande cometer la rreçeçion de los dhos t^{o}s al secr^{o} desta cabsa para que juren e declaren ante el e para ello etc.

Al^{o} Moreno

/75v/ En Los Rreyes syete de março de myll e quy^{o}s e çinquenta e nueve a^{o}s en Audi^{a} Rreal ante los señores presydente e oydores en Audiençia Rreal pu^{ca} / la presento el dho pr. en n^{e} de su pte y los dhos señores mandaron que el dho Fran^{co} Hernandez entregue las dhas provanças al secretario de la causa / e que en lo demas que pide cometran la rreçepçion de los dhos t^{o}s al dho secret^{o}.

Diego Muñoz

/76r/ Los y^{o}s de Canta

Muy pod. s^{o}r

Fran^{co} de la Torre en nonbre de los yndios de Canta en el pleyto con los yndios de Chaclla sobre las trras / digo quepor mandado de Bra Al. fue Fran^{co} e otros escr^{o}s a tomar ynform^{o} sobre lo susodho / el qual la tiene en su pod.

A Bra Al. pido y sup^{co} le mande que la de horiginalm^{te} e la entregue al secr^{o} de la cabsa sobre que pido just.

Fran^{co} de la Torre

En Los Rreyes syete de março de myll e quy^{o}s e çinquenta e nueve a^{o}s en Aud^{a} Rreal pu^{ca} ante los señores presydente e oydores; la presento el dho pr. en los dhos nonbres y los dhos señores mandaron que el dho Fran^{co} Hernandez entregue la dha ynformaçion originalm^{te} al secret^{o} de la causa.

Diego Muñoz

/77r/ Los yndios de Chacalla.

Muy p^{o} s^{o}r

Al^{o} Moreno en nonbre de los caçique e y^{o}s de Chacalla en el pleito con los y^{o}s de Canta sobre las chacaras e tierras e con Fran^{co} de la Torre pr.en su nonbre digo que al dho de mys ptes convyene que Nycolas de Rribera el moço diga su dho por mys ptes en esta cabsa el qual no lo quyere dezir.

Pido e sup^{co} a V. Alt. le mande con pena que luego lo diga oy por que se cunple el tr^{no} p^{o}vatorio el qual suplico a V. Alt. me mande prorrogar por quatro dias mas e para ello etc.

Al^{o} Moreno

En Los Rreyes diez y siete de março de myll e quy^{o}s e çinquenta e nueve a^{o}s ante los señores presudente e oydores en Aud^{a} Rreal pu^{ca} de petiçiones; la presento el dho pr. y los dhos señores le conçedieron el dho termyno que pide por los dhos quatro dias e que el dho Nyculas de Rrybera oy en todo el dia declare e diga su dho so pena de diez p^{o}s.

Diego Muñoz

/78/ Muy p^{o} s^{o}r

Al^{o} Moreno en nonbre de los y^{o}s de Chacalla en el pleyto con los y^{o}s de Canta sobre las tierras e chacaras con Fran^{co} de la Torre pr.en su nonbre digo que Gomez Saravantes y Sebastian Sz de Merlo e Nyculas de Rribera el moço no quyeren dezir sus dhos en esta cabsa en que mys ptes les quyeren presentar por t^{o}s de que rreçiben daño y asy mysmo Ju^{o} de Padilla escr^{o} que sabe lo que mys ptes pretenden p^{o}bar.

A V. Alt. pido e sup^{co} mande que vn alguazil los trayga a jurar e declarar sobre que pido just. e para ello etc.

El Licenc^{do} M^{o} Luzio

Al^{o} Moreno

En la çibdad de los Rreyes en tres dias del mes de abril de myll e qui^{o}s e çinq^{ta} e nueve años ante los señores oydores en Abdiençia Rreal la presento el dho Al^{o} Moreno en el dho nonbre e los dhos señores mandaron que so pena de veynte pesos cada vno diga oy su dho.

Fran^{co} de Carujale

Notificada a Ju^{o} de Padilla en 4 de abril
t^{o}s Ju^{o} de la Peña y P^{o} Plaça

En la çiudad de Los Rreyes a quatro dias del mes de abril de myll e quy^{o}s e çinq^{ta} e nueve años yo el escrivano yuso scrito notifique por petiçion e lo en ella proveydo por los dhos señores pr^{te} e oydores al dho Ju^{o} de Padilla en su persona siendo t^{o}s Juan de la Peña e P^{o} de Plaça est^{tes} en la dha çiudad.

F^{o} de Pomareda

/78v/ E despues de lo susodho en la dha çibdad de Los Rreyes este dho dia mes y año susodho / yo el escrivano yuso scripto notifique esta petiçion y lo en ella proveydo por los dhos señores presydente e oydores al dho Gomez de Caravantes en su persona siendo t^{o}s Juan Pz de Hojeda y Al^{o} de Montalban estantes en esta dha çibdad.

F^{o} de Pomareda

E despues de lo susodho este dho dia mes y año susodhos yo el dho escr^{o} notifique esta petiçion y lo en ella proveydo por los dhos señores presydente e oydores al dho Sebastian Sanchez de Merlo en su persona syendo t^{o}s Alonso Moreno procurador estante en esta dha çibdad.

F^{o} de Pomareda

E despues de lo susodho en la dha çibdad de Los Rreyes este dho dia mes y año susodhos yo el dho escrivano notifique esta petiçion y lo en ella proveydo por los dhos señores presydente e oydores al dho Nyculas de Rribera el moço en su persona; t^{o}s Ju^{o} Monçalvo Sillero y Ju^{o} Pz de Hojeda est^{es} en esta dha çibdad.

F^{o} de Pomareda

/79r/ Los y^{o}s de Canta.

Muy pod. ss^{o}r

Fran^{co} de la Torre en nonbre de don Hr^{do} e de los demas ss e prinçipales del pueblo de Canta encomendado en Nyculas de Rribera / en el pleyto con los yndios de Chaclla encomendados en Fran^{co} de Anpuero e A^{o} Moreno en su n^{e} sobre las trras digo que el termino provatorio es pasado e dias mas.

A Bra Al. pido y supp^{co} mande hazer pu^{on} de t^{o}s en la dha cabsa sobre que pido just. etc.

Fran^{co} de la Torre

En Los Rreyes onze de abril de myll e quy^{o}s e çinquenta e nueve años en Aud^{a} Rreal ante los señores presydente e oydores; la presento el dho pr. en n^{e} de sus ptes. Y los dhos señores ovieron por fha la dha publicaçion con el termyno de la ley de consentymyento de Alonso Moreno pr. del caçique e yndios de Chacalla e les mandaron dar traslado de las p^{o}vanças e proçesado para que aleguen de su dr^{o} a la primera aud^{a} a los quales yo el dho escribano notifique.

Diego Muñoz

/80r/ Los y^{o}s de Chacalla.

Por las preguntas siguyentes se esaminen los t^{o}s que son o fueren presentados por pte del caçique e y^{o}s de Chacalla en el pleyto con el caçique e y^{o}s de Canta sobre las tierras de Quibi:

1 - Prim^{a}m^{te} sean preguntados si conosçen a las ptes y si tienen notiçia de las dhas tierras de Quibi.

2 - Yten si saben etc. que al tpo que dizen aver hecho don Fran^{co} caçique que dizen aver sido de Chacalla el conçierto de las dz^{a}s obejas de la tierra por las tierras e chacaras de Quibi / fue a ello forçado y conpelido el dho don Fran^{co} por Fran^{co} de Anpuero su encomendero que le hizo por fuerça y con temores y amenazas y malos tratamy^{o}s le hizo que se hallase presente al conçierto sin que el dho don Fran^{co} lo hiziese ni quisiese. Digan lo que saben.

3 - Yten si saben etc. que estando presente y contradiziendo el dho conçierto y no lo queriendo hazer el dho don Fran^{co} de Chacalla / lo hizo y firmo el dho Fran^{co} de Anpuero su encomendero y se tomo para si y llevo las dz^{a}s ovejas de la tierra que dizen aver dado los y^{o}s de Canta / y no las dio ni entrego al dho don Fran^{co} ni a los prinçipales ni y^{o}s de Chacalla antes las tomo y llevo y como cosa suya hizo dellas lo que quiso sin voluntad ni consentim^{o} de sus y^{o}s.

4 - Yten si saben etc. que sienpre el dho don Fran^{co} de Chacalla contradixo el dho conçierto y dezia que no queria ovejas sino las tierras de Quibi que heran de sus y^{o}s y no las podia el vender ni trocar a ovejas / y a ver que lo dezia asi el dho Fran^{co} de Anpuero rreçibio las ovejas y firmo el papel que los de Canta quisieron.

5 - Yten si saben etc. que Su Mg^{t}. tiene proveydo en todas las Yndias y en este rreyno por sus rreales provisiones y çedulas que los t^{o}s han visto originalm^{te} / que los y^{o}s y tierras y todas las demas cosas //
/80v/ esten de la manera que las hallaron los españoles quando entraron en la tal tierra sin que en nada aya ni se consienta novedad alguna. Digan los t^{o}s lo que saben y si ha sido en todas las Yndias usado y guardado lo susodho por virtud de la dha provision rreal de su Mg^{t}.

6 - Yten si saben etc. que todo lo susodho es pu^{ca} boz y fama.

El Licenc^{do} M^{o} Luzio Al^{o} Moreno

En Los Rreyes veynte y syete de henero de myll e quy^{o}s e çinquenta e nueve años ante los señores presydente y oydores en Aud^{a} Rreal pu^{ca}; la presento el dho Alonso Moreno y los dhos señores lo ovieron por presentado quanto es pertinente e mandaron que por el se examynen los t^{o}s que presentare.

Diego Muñoz

/8lr/ Los y^{o}s de Chacalla.

Por estas preguntas demas de las presentadas se esaminen los t^{o}s que son o fueren presentados por pte del caçique e y^{o}s de Chacalla en el pleyto con los de Canta sobre las tierras de Quibi:

- Yten si saben que los dhos y^{o}s de Chacalla y sus caçiques han poseydo las dhas tierras de Quibi y tenidolas como cosa suya desde el tpo de los yngas pasados que se las dieron lo qual ha tanto tpo que no ay quyen se pueda acordar dello.

- Yten si saben etc. que los dhos y^{o}s de Chacalla tenian y poseian las dhas tierras de Quibi al tpo que los españoles entraron en este rreyno y las tuvyeron mucho tpo despues de entrados. Digan lo que saben.

- Yten si saben etc. que los y^{o}s de Canta porque no se alçaron juntam^{te} con ellos los dhos y^{o}s de Chacalla y porque servian a los españoles como lo hazian al tpo que se alçaron los naturales en la tierra / mataron çiento e çinq^{ta} y^{o}s de Chacalla que estavan en las dhas tierras de Quibi y se las quitaron y tomaron y desde entonçes se las tienen tomadas y usurpadas los dhos y^{o}s de Canta. Digan lo que saben de lo dho.

- Yten si saben etc. que las dhas tierras de Quibi que los dhos y^{o}s de Canta tienen tomadas a los de Chacalla valen mas de ocho myll p^{o}s al presente y los valian de diez años a esta pte. Digan los t^{o}s lo que valian agora diez años y lo que valen agora.

El Licenc^{do} M^{o} Luzio Al^{o} Moreno

En Los Rreyes a nueve dias del mes de hebrero de myll e quy^{o}s e çinq^{ta} e nueve años ante los ss^{o}s oydores en Abdiençia Rreal de rrelaçiones la presento el dho Alonso Moreno en el dho nonbre e los dhos señores lo obieron por presentado e mandaron que los t^{o}s digan por el.

Fran^{co} de Carujale

/82r/ Los y^{o}s de Canta.

Por las preguntas siguientes sean preguntados los t^{o}s que son o fueren presentados por pte del caçique e y^{o}s de Canta encomendados en Nicolas de Rribera el moço en el pleyto que tratan con los y^{o}s de Chaclla sobre las tierras del valle de Quibi queestan juntas a las tyerras e pueblo de los y^{o}s del dho valle de Canta.

1 - Primeramente sy conosçen a las ptes y a cada vna dellas y tienen notiçia del valle o trras que dizen de Quibi / que estan en los tr^{o}s del dho valle de Canta e de que tpo a esta pte y conosçieron a los yngas pasados rreynos e señores que fueron destos rrey^{o}s espeçialmente a Topa Ynga Yupangui y a Guaynacap y tienen notiçia de vn rrio que pasa junto a las dhas tierras que se llama el rrio de Quybe.

2 - Yten si saben etc. que las dhas tierras que dizen de Quybe estan dentro de los termynos e pueblos de Canta y de sus parçialidades ny mas ny menos que esta en la figura que mostraron a los señores oydores por que los tr^{o}s parte vn çerro grande que esta entre los yabyos e y^{o}s de Chaclla y los dhos y^{o}s de Canta. Pido que los t^{o}s declaren como se parten los tr^{o}s y como caen dentro de los dhos tr^{o}s de Canta las dhas trras de Quyve. Digan lo que saben.

3 - Yten si saben etc. que las dhas tierras e chacarras de Quybe desde vno y dos y diez y veynte y treynta y çinquenta y mas años a esta parte que memorias de hombres no es en contrario siempre fueron trras e chacaras de los dhos y^{o}s de Canta y su caçique los quales siempre an tenydo y poseydo por suyas y como suyas labrandolas y sembrandolas y hasiendo dellas como de cosa p^{o}pia hasta que se metyeron en çierta p^{te} dellas çiertos y^{o}s como mytimaes de los dhos y^{o}s de Chaclla por çierto engaño que les hizieron. Digan lo que saben

4 - Yten si saben etc. que antes que vinyesen xpanos a esta trra se tenya por costumbre en ella entre los naturales de hazer çiertas rrogatybas o plegarias por los rreyes e sus señores a que llamaban ellos capacocha lo qual hera que matavan çiertas ovejas y les sacavan la sangre y la echavan en vnas ollas o mates y yban con aquello en p^{o}siçion y si alguno dellos derramava vna gota de la dha sangre o se la hazian derramar al yndio que asy la derramava o la hazia derramar lo enterravan en el p^{o}pio lugar y lo matavan para ello y esto hera ley enviolable entre ellos. Digan lo que saben.

5 - Yten si saben etc. que q^{do} hazian la dha capacocha yban los de bn pueblo hasta los terminos del otro con çiertas çerimonyas y hasta donde llegavan lo ponyan por termyno de los que llevavan la dha capacocha y lo señalavan por suyo. Digan lo que saben.

6 - Yten si saben etc. que al tpo que Guaynacap yba a Quito hizieron los dhos y^{o}s de Chaclla la dha capacocha y con la sangre susodha llevando en vnas ollas o mates fueron por la pte donde estan las dhas trras de Quybe y con las dhas çerimonyas y los dhos y^{o}s de Canta queriendolos yr a detener no osaron por no derramar la dha sangre e yncurrir en la dha pena de muerte y asy con la dha color yntentaron de se quedar y se quedaron con vna pte de las dhas tierras de Quybe / diziendo que hasta alli les pertenesçian por aver llegado con la dha capacocha hasta las dhas tierras //
/82v/ de Quybe en la pte donde llegaron y sobre que es este pleyto lo qual puede ser la tr^{a} parte de las dhas trras poco mas o menos. Digan lo que saben.

7 - Yten si saben etc. que con la dha color de la dha capacocha los dhos y^{o}s de Chaclla pusyeron mytimaes en la pte de las tierras de Quybe donde llegaron con la dha capacocha / sin tener otro dr^{o} alguno a las dhas tierras. Digan lo que saben.

8 - Yten si saben etc.que despues que el dho Guaynacap murio e antes que entrasen los xpanos en este rreyno o q^{do} entraron los dhos y^{o}s de Chaclla se salieron los dhos y^{o}s de Chaclla y los mytimaes que avian puesto por mandado de Ylla Topa Ynga y hizo de Guaynacapa y luego los dhos y^{o}s de Canta se tornaron a las dhas sus trras de Quyve ny mas ny menos que de antes las tenyan las quales an tenydo y poseydo desde entonçes hasta agora. Digan lo que saben.

9 - Yten si saben etc. que por mandado del presydente Gasca e por deferençias que tenyan sobre las dhas tierras los dhos y^{o}s de Chacalla se conçertaron con los dhos y^{o}s de Canta e a que les diesen dozientas ovejas por que se apartasen del pleyto e diferençias de las dhas trras de Quyve Digan lo que saben.

10- Yten si saben etc. que en cumplim^{o} del dho çonçierto los dhos y^{o}s de Canta dieron a los dhos y^{o}s de Chaclla dozientas cabeças de ovejas las quales les dieron y entregaron a ellos mesmos y a su caçique. Digan lo que saben.

11- Yten si saben etc. que en todas las dhas tierras de Quybe no ay veynte hanegas de sembradura q^{do} mas veynte y çinco y todas las mas dellas son andenes y tierra pedregosa y a justa e comun estimaçion cada hanega de sembradura puede valer hasta veynte p^{o}s poco mas o menos.Digan lo que saben.

12- Yten si saben etc. que las chacaras e trra sobre que es este pleyto e diferençia puede ser la terçia pte de la dha tierra o menos que fue hasta donde llegaron con la dha capacocha los dhos y^{o}s de Chaclla. Digan lo que saben.

13- Yten si saben etc. que cada cabeça de ganado de obejas de las que dieron los dhos y^{o}s de Canta a los dhos y^{o}s de Chaclla valian a seis p^{o}s y mas a justa y comun estimaçion al tpo que se las dieron. Digan lo que saben.

14- Yten si saben etc. que la cabsa por que hizieron el dho conçierto los dhos y^{o}s de Chaclla con los dhos y^{o}s de Canta valian mucho menos las trras que agora valen vn terçio y mas por darse como se dan agora mas a la labor que entonçes. Digan lo que saben.

15- Yten si saben etc. que la cabsa por que hizieron el dho conçierto los dhos y^{o}s de Canta con los dhos y^{o}s de Chaclla en las dhas tierras de Quybe hera y fue porque las dhas trras de Quybe heran y fueron de sus pasados y suyas e por quytarse de aquella molestia. Digan lo que saben.

16- Yten si saben etc. que todo lo susodho es pu^{co} e notorio e pu^{ca} boz e fama.

Otro si por quanto se enbia escrivano e personas nombradas por las partes a ver las dhas tierras pido e sup^{co} a V. Alt. m^{de} //
/83r/ que las traygan señaladas los dhos terçeros y escrivanos e den fee de como estan y asi mysmo traygan señalado el pedaço de trra sobre que es la diferençia y asy mysmo que traygan tasadas todas las dhas tierras de lo que pueden valer y las hanegas que pueden tener sobre que pido justiçia y costas y en lo nesçesario el of^{o} de V.^{a} Alt.^{a} ymploro.

El Licen^{do} de Leon — Fran^{co} de la Torre

En los Rreyes veynte e syete de henero de myll e quinyentos e çinquenta y nueve años ante los señores presydente e oydores; la presento el dho Fran^{co} de la Torre en los dhos nonbres e los dhos señores lo ovieron por present^{do} quanto es pertinente y que los t^{o}s que presentare digan por el / e se lleve este ynterrogatorio al semanero para que lo vea y en lo demas que se pide de concordia de las ptes los dhos terçeros hagan la pintura.

Diego Muñoz

En Los Rreyes veynte e syete de henero de mill e quinientos e çinquenta e nueve años. El señor doctor Gregorio Gonçalez de Cuenca oydor desta Rreal Audiençia siendo semanero aviendo visto estos ynterrogatorios en esta causa presentados por ambas partes mando que por ellos se exsaminen los t^{o}s que cada vna de las partes presentaren e que se rresçiban atento que es negoçio de yndios.

/84r/ Los y^{o}s de Canta.

Muy p^{o}s ss^{o}r

Fran^{co} de la Torre en nonbre de los yndios de Canta en el pleito con los yndios de Chacalla sobre las tras de Quybi. Digo que para que asista con la lengua que examynase los t^{o}s que se presentaren en la dha cabsa para que no se pregunte ny se declare bna cosa por otra / nonbro por terçero de my parte al padre Ger^{o}nymo M^{yn} que sabe muy bien la lengua e porque en aquello podria aver cabtela.

- Pido y ssup^{co} a Vra Alteza mande que el dho padre Gr^{mo} M^{yn} asista a la declaraçion de los t^{o}s con la dha lengua y si las partes contrarias quysieren nonbrar otro que lo nonbren para que çese qualquyer cabtela que en ello puede aver sobre que pido just^{a} e para ello etc.

- Otro si por quanto a las preguntas del ynterrogatorio que presente faltaron por hazer otras çiertas preguntas para que se entienda y sepa la verdad / a Vra Alteza pido ssup^{co} mande que los t^{o}s que se presentaren se pregunten asi mesmo por estas preg^{tas} añydidas:

- Yten si saben etc. que en tpo de Guascar Ynga que fue señor destos rreynos despues de Guaynacapa se mandaron hazer dos capacochas de las susodhas vna para su muger y otra para el y trayendo los dhos yndios de Chacalla la dha capacocha y vinyendo a lavar la sangre al dho rrio de Quybi los dhos yndios de Chacalla se entraron en lo mejor de las dhas tras de Quybi donde agora esta sembrado de coca y pacays y porque se entravan en las dhas tras los dhos yndios de Chacalla vn yndio que se dezia Chuqui Guaranca rrenpujo a vno de los dhos yndios que traya la dha capacoche diziendo que no se entrasen por su tra e porque le derramo parte de la dha sangre ahogaron el dho yndio que era de la parçialidad de los de Canta y le enterraron junto a vn molle que alli esta. Digan lo que saben.

- Yten si saben etc. que de la mesma manera mataron otros quatro yndios de los de Canta de-//
/84v/ fendiendo las dhas sus tras vno que se dezia Ynygui Chunbi y el otro Chabmecaxa y el otro Tantaçapo y el otro Guaman Llaupi.

- Yten si saben etc. que en las dhas tras de Quybi no bbo mytimays puestos por el Ynga sino çierto rresto de bn caçique que mando matar el Ynga que se llamava Collotanbo de los de Chacalla / cuya generaçion con todos los barones fue muerta por çierta culpa de trayçion / que tubo y bnas pocas de mugeres que quedaron las paso a bn corral de Quybi para que hilasen alli. Digan e declaren lo susodho como paso e por que via entraron alli las dhas mugeres y si bbo mytimaes o no y que tras le dieron en que bibieron.

- Yten si saben etc. que çiertas tras de las dhas tras de Quybi que estan cabe las casas que son las que Joan Xpoval de Frias fue a ver / son asi mesmo de los dhos yndios de Canta y la cabsa por que agora la piden los dhos yndios de Chacalla es porque despues que les dieron las ovejas y hizieron la dha transaçion / vn yndios que se llama Pomaraque e otro Tanta de los yndios de Chacalla / dixeron a Guamayabri caçique de Canta que se querian pasar a su tra a tributar con el y el dho caçique les dixo que le darian aquel pedaço de tra e otro si se pasasen como dezian / y asi con el dho conçierto le señalaron las dhas tras alos dhos yndios de Chacalla. Digan lo que saben.

- Yten si saben etc. que porque los dhos Pomaraque e Tanta dixeron que se querian pasar los hizo açotar el caçique Marcapoma y asi no se pasaron ny el dho caçique de Canta les dio las dhas tras sino se las torno a tomar para si y las senbraron de coca los yndios de Canta.

- Yten si saben etc. que puede aver tres a^{o}s poco mas o m^{o}s que bn don Joan prençipal de Canta se hizo amygo con don Xpoval prinçipal de Chacalla y sin lo pedir porque se lo rrogo el dho don Xpoval le dio liçençia para que cogese la coca del dho pedaço de tra y asi el dho don Xpoval la cojo dos o tres bezes y con el dho achaque pretenden tener dr^{o} a las dhas tras sin averselas dado nadie ny poseydolas por otra cabsa ny aviendo plantado la dha coca. Declaren los t^{o}s como y de que manera paso lo susodho.

- Yten si saben etc. que todo lo susodho es pu^{ca} boz e fama.

El Licen^{do} de Leon

/85r/ En Los Rreyes a nueve dias del mes de hebrero de myll e qui^{o}s e çinq^{ta} e nueve años ante los señores oydores en Abdiençia Rreal de rrelaçiones. La presento el dho Fran^{co} de la Torre en el dho nonbre e los dhos señores lo ovieron por presentado e mandaron que los t^{o}s digan e declaren por las dhas preguntas e que asista a la declaraçion de los dhos t^{o}s con la lengua el padre Gr^{mo} M^{yn}

Fran^{co} de Carujale

/86r/ En Los Rreyes a quatro dias del mes de março de myll e qui^{o}s e çinquenta e nueve años ante los ss^{o}r oydores en Abdiençia Rreal de rrelaçiones Alonso Moreno en nonbre de los y^{o}s de Chacalla en el pleito con los yndios de Canta presento por t^{o}a Fran^{co} de Anpuero v^{o} e rregidor desta çibdad del qual se tomo e rr^{o} juramento por Dios e por santa Maria e señal de la cruz + e palabras de los santos evangelios el qual lo hizo en forma de dr^{o} e prometio de dezir verdad.

Fran^{co} de Carujale

/87r/ Ynformaçion fha por pte del caçique de Chacalla.
El caçique e yndios de Canta ante my

Juan Xpoval de Frias — Diego Muñoz

/87v/ En el pleyto que los yndyos de Chaclla tratan con los de Canta sobre las trras de Quivi estando conclusa la causa por petiçion que en esta Rreal Audi^{a} presentaron en que los yndyos de Chaclla se quexaron de los de Canta diziendo que vna chacarilla que avia quedado a los dhos yndyos de Chaclla en las dhas trras de Quyvi sobre que no se lytiga les avian rronpido e quebrado las açequyas para que no les fuese agua / suplicaron se les mandase con pena que no se la perturben / mandoseles que cada vno cogiese lo que oviese senbrado e no oviese ynovaçion / e que el caçique de Chaclla no vaya a las trras so pena de privaçion del caçicasgo e a los yndios so p^{a} de çient açotes a cada vno / despues del qual por otra petiçion presentada por pte de los dhos yndyos de Chaclla se quexan de los de Canta diziendo que vn prinçipal dellos / avia ydo a las dhas chacaras e de la pte que a los de Chaclla avia quedado e poseyan / e le avian començado a coger la coca con lo proveydo por V. A. / suplicaron a V. A. mandase castigar al dho caçique de Canta e quytarle la chacara que les avian tomado e la coca que avian cogido e para ello fuese persona que lo averiguase / para lo qual V. A. proveyo a Joan Xpoval de Frias / el qual hizo ynform^{on}.

/88r/ Quejanse que an començado los de Canta
a cojer la coca.

Muy p^{o} s^{o}r

Alonso Moreno en nonbre del caçique e y^{o}s de Chacalla en el pleito con el caçique e y^{o}s de Canta sobre las chacaras de Quybe; digo que este miercoles pasado el dho caçique de Canta enbio a vn prençipal sujeto suyo que se dize Pavcar Vilca a las dhas chacaras de Quybe y de la pte dellas que a mys ptes an quedado e poseen echaron a los yndios que mys ptes tenian diziendo estar la cabsa sent^{da} en su favor y an començado a coger la coca que mys partes tenian e tienen çincuenta y^{o}s puestos en la dha chacara contra lo proveydo por V. Alt. de que mys ptes rresçiben daño.

A V. Alt. pido e sup^{co} mande castigar al dho caçique de Canta e quitarle la chacara que an tomado y que buelban la coca que an cogido porque de otra manera seran mys ptes defraudados / y la averiguaçion de lo susodho se cometa a quien V. Alt.^{a} fuere conseguido sobre que pedimos just^{a} e para ello etc.

El Licenc^{do} M^{o} Luzio — Al^{o} Moreno

/88v/ En Los Rreyes diez e siete dias del mes de diz^{e} de myll e quy^{o}s e çinq^{ta} e ocho años ante los señores oydores en avdi^{a} de rre^{on} pu^{ca}; la presento Alonso Moreno e los dhos señores mandaron que de ynformaçion de lo que dize e proveeran just^{a} e para la hazer vaya vn escr^{o} off^{al} del off^{o} lo qual paso estando presente Fran^{co} de la Torre pr. de la pte contraria.

Diego Muñoz

/89r/ Inf.^{on} sobre que los de Canta començaron
a coger la coca.

En el valle de Quybe juridiçion de la çibdad de Los Reyes rreynos del Peru que es de don Felipe Tavlichunbi encomendado en Rribera el moço vezino de la dha çibdad de Los Rreyes veynte e dos dias del mes de diz^{e} de myll e quynyentos e çinq^{ta} e ocho años yo Juan Xpoval de Frias escr^{o} de Sus Magestades e Su notario pu^{co} en la su Corte e en todos los sus rreynos e señorios est^{do} en el dho valle por virtud del abto proveydo e m^{do} por los señores oydores que de suso se contiene y en cunplim^{o} del fize paresçer ante mi de pedim^{o} e nonbramyento de don Felipe prinçipal del pueblo de Chacalla los yndios y t^{o}s syguientes:

T^{o} don X^{o}bal Guaquicho

E despues de lo susodho en el dho valle de Quybe en este dho dia e mes e año susodhos yo el dho escrivano tome e rresçibi juramento en forma devida de dr^{o} de vn yndio que se dixo por su nonbre don Xpoval Guaquycho prinçipal que dixo ser de los mytimaes de Canta encomendado en el dho Rribera el moço / so cargo del qual prometio de dezir verdad e siendo preguntado //
/89v/ por el tenor del dho pedim^{o} dixo que este t^{o} rreside en el dho pueblo de Quibe encomendado en el dho Nyculas de Rribera e que lo que sabe es que el myercoles pasado que se contaron catorze dias deste presente mes de dizienbre est^{do} que este t^{o} estava en el dho pueblo de Quybe vido que vino al dho pueblo don Juan de Pacavru Vilca encomendado en el dho Nyculas de Rribera que venya de la çibdad de Los Rreyes e este t^{o} vido que mando cojer de las tierras sobre que se trata p^{o} con el caçique de Chacalla a los yndios e yndias que con el traya la coca que avia en vn pedaço de las dhas trras e que estando cogendo la dha coca vido este tsgo que bn yndio que se dize Alvarado Collata yndio de el pueblo de don Xpoval Payco caçique prinçipal de Chaclla encomendado en Françisco de Anpuero dixo al dho don Juan que no cogiese la dha coca pues abia pleyto entrellos sobre ella e no estava sentençyado e este t^{o} vido que el dho don Juan dixo que todavia la abia de coger avnque le pesase porque le abia costado sus dineros senbrarla e bido que dixo a los yndios que venyan e estavan con el dho Albarado Collata //
/90r/ que se fuesen de las dhas trras e no estuviesen en ellas e que este t^{o} vido que todabia la cogio e m^{do} coger e se la llebo e dixo al dho Albarado que abia de enbiar yndios suyos e de su pueblo para que labrasen e benefiçiasen las dhas trras e cogiesen el fruto dellas y este t^{o} vido como el dho Alvarado Collata dixo al dho don Juan que se abia de quexar a los señores oydores del porque le cogia su coca e que el dho don Juan abia rrespondido que fuese que la tierra hera suya e la coca el la abia senbrado con sus yndios e que ya thenya sentençia en su fabor en que se las daban las dhas trras e que acabada de coger se fue diziendo que abia que enbiar de su trra yndios que guardasen y labrasen las dhas trras e que esto es lo que sabe desta pregunta por el juramento que fho tiene lo qual declaro por lengua de vn yndio que se dixo por su nonbre Fran^{co} Pomavilca anacona del padre Caravajal ladino en lengua de Castilla el qual juro en forma de derecho que a dho e declarado lo mysmo que el dho yndio a dho en su lengua e la verdad //
/90v/ por el juramento que fho tiene e no firmaron porque dixeron que no sabian escrebir e dixo el dho don Xpoval Guaquycho ser de hedad de treynta e çinco a^{o}s poco mas o menos.

ante mi

Juan X^{al} de Frias
escrivano de su M.^{t}

T^{o} Fran^{co} Pomavilca.

E despues de lo susodho en este dho dia e mes e año susodhos est^{do} en el dho valle yo el dho escrivano de nonbramyento del dho don Felipe hize paresçer ante mi a vn yndio que se dixo por su nonbre Françisco Pomabilca natural que dixo ser del dho pueblo de Quybe encomendado en el dho Nyculas de Rribera el moço del qual yo el dho escrivano tome e rresçibi juramento en forma devida de derecho so cargo del qual prometio de dezir verdad e syendo preguntado por el tenor del dho pedim^{o} dixo que lo que sabe es que el myercoles pasado que se contaron catorze dias deste dho mes de dizienbre e deste dho año est^{do} este t^{o} en el dho pueblo de Quybe vido este t^{o} que vino al dho pueblo //
/91r/ don Juan de Pacavru Vilca e con el çiertos yndios e yndias e este t^{o} vido que el dho don Juan mando a los dhos sus yndios que cogiesen la coca que estava en una chacra mediana de las trras de que se trata pleyto entre el dho don Juan e los yndios de Chacalla encomendados en Fran^{co} de Anpuero y este t^{o} vido que est^{do} la cogiendo llego a el dho don Juan vn yndio que se dize Alvarado Collata del caçique de Chacalla e dixo al dho don Juan que no cogiese la dha coca porque hera suya e de su caçique e que avia pleyto sobre ella e el dho don Juan le dixo que se fuese de alli porque el la abia de coger que hera suya e le abia costado sus dineros senbralla e que ya tenya sent^{a} del señor bisorrey e de los señores oydores para coger la dha coca e que este t^{o} vido que el dho Alvarado torno a rrequerir al dho don Juan que no le cogiesen la dha coca porque se yria a quexar al señor visorrey e que el dho don Juan este t^{o} vido que rrespondio que fuese norabuena que ya tenya sent^{a} en su favor sobre las dhas trras e que todabia mando coger e se cogio toda la dha coca de vna chacarilla que en ella no dexaron //
/91v/ nynguna e que acabada de coger se fue diziendo que avia de enbiar yndios de su pueblo que guardasen la dha trra e labrasen la coca e la benefiçiasen pues ya tenyan sent^{a} sobre ellas e que este t^{o} sabe e vido que de donde se cogio la dha coca es de las trras sobre que

se cae el dho pleyto e que este t° vido que los yndios e el caçique prinçipal de Chacalla estavan en ellas e las an labrado e benefiçiado e que esto sabe desta pregunta por el juramento que fho tiene e que es de hedad de quarenta años poco mas o menos e no firmo porque dixo que no sabe escrebir lo qual declaro por lengua del dho Françisco Pomavilca el qual juro en forma de derecho que a dho e declarado lo mesmo que el dho yndio a dho en su lengua e berdad por el juramento que fho tiene e no firmo porque dixo que no sabia escrebir.

ante mi Juan Xpoval de Frias
escrivano de Su M[t]

T° Alb[do] de Payco.

E despues de lo susodho en este dho //
/92r/ dia e mes e año susodhos est[do] en el dho valle de Quybe yo el dho escr° de nonbramyento del dho don Felipe hize paresçer ante mi a vn yndio que se dixo por su nonbre Albarado de Payco natural que dixo ser de la collata que es en la syerra e prinçipal que dixo ser del dho pueblo encomendado en el dho Françisco de Anpuero vezino de la dha çibdad de Los Rreyes del qual yo el dho escrivano tome e rresçibi juramento en forma devida de derecho so cargo del qual prometio de dezir verdad e syendo preguntado por el tenor del dho pedimy° dixo que lo que sabe es que puede aver dies o doze dias que este t° abaxo del dho su pueblo por mandado de don Xpoval caçique prinçipal suyo al valle de Quybe a rregar la coca que tiene en el dho valle e myrar por ella e est[do] en las dhas tierras el myercoles pasado que se contaron catorze dias deste dho mes de diziembre llego a las dhas tierras vinyendo de Lima don Juan de Pacavru Vilca con yndios e yndias suyos //
/92v/ y entraron en las tierras del dho don Xpoval que tiene en el dho valle de Quybe sobre que se trata pleytoe mando coger la coca que estava en vna chacarilla de las dhas trras y este t° como prinçipal y como yndio que abia venydo por m[do] del dho Xpoval su caçique a guardalla dixo al dho don Juan que no cogiese la dha coca pues estava en las trras del dho don Xpoval y era suia y ellos la abian senbrado e labrado e benefiçiado e abia pleyto sobre ella e el dho don Juan rrespondio que no queria syno cojella que ya hera suya que se abia dado sent[a] sobrello en su favor e m[do] a los dhos yndios que la cogiesen e asi la cogeron e llebaron syn dexar nynguna y este t° les dixo que se avia de venyr a quexar al señor bisorrey de la fuerça que le fazia y el dho don Juan rrespondio que vinyese norabuena que ya tenya sent[a] en su favor e que heran suyas e que bien podia cogellas e asy se fue e llebo toda la dha coca que en la dha chacarilla abia e diziendo que llegando a su pueblo avia de enbiar yndios de su tierra que guardasen la dha coca e labrasen las dhas //
/93r/ tierras e diziendo a este t° e a los demas yndios que alli estavan que se fuesen de alli e que esto es la verdad e lo que sabe por el juramento que fho tiene e no firmo porque dixo que no sabe escrebir e dixo ser de hedad de treynta e çinco años poco mas o menos lo qual declaro por lengua del dho Françisco Pomavilca el qual juro en forma de derecho que a dho e declarado en lengua de Castilla lo mysmo que el dho yndio a dho en su lengua e la verdad por el juramento que fho tiene e no firmo porque dixo que no sabia escrebir.

ante mi Juan Xpoval de Frias
escrivano de Su Mag[d]

T° Fr[do] Vechira.

E despues de lo susodho en este dho dia e mes e año susodhos estando en el dho valle de Quyvi yo el dho escrivano de nonbramyento del dho don Felipe Yndio tome e rresçibi juramento en forma de derecho de vn yndio que se dixo por su nonbre //
/93v/ Hernando de Vichira natural que dixo ser del pueblo de Xicamarca que es de Myn Payco encom[do] en el dho Françisco de Anpuero vezino de la dha çibdad de Los Rreyes so cargo del qual prometio de dezir verdad e siendo preguntado por el tenor del dho pedim° dixo que lo que sabe es que puede aver diez o doze dias que este t° vino al dho pueblo de Quyve en conpanya de Alvarado de Payco por mandado del dho don Xpoval caçique a rregar las tierras e coca e benefiçiar las tierras que tiene en el dho valle de Quybe sobre que se trata pleyto con el caçique e yndios de Canta e estando en el dho valle en las dhas tierras el myercoles pasado que se contaron catorze dias deste presente mes de dizienbre llego a las dhas tierras vinyendo de Lima don Juan de Pacavru Vilca con muchos yndios e yndias suyos y entraron en vna chacarilla chica senbrada de coca e la enpeço a coger e este t° vido que el dho Albarado de Payco dixo al dho don Juan que no cogiese la dha coca que hera del dho don Xpoval //
/94r/ su caçique e de sus tierras e este t° bido que el dho don Juan rrespondio que se fuese de alli que las dhas tierras heran suyas e que ya tenya sent[a] del señor bisorrey e de los señores oydores dellas e que el avia de coger la coca dellas e este tsgo vido que torno a rresponder el dho Albarado Payco que avnque avia sent[a] que todavia avia pleyto e que no cogiesen la dha coca porque se yria a quexar al señor visorrey del e el dho don Juan rrespondio que fuese donde quysiese que ya hera suya e que la avia de cojer e asi mando a los dhos yndios e yndias que consygo traya que la cogiesen e asi la cogieron toda syn dexar cosa alguna en la dha chacara e quando la ovo cogido se fue diziendo que en llegando a su trra avia de enbiar yndios de su trra que guardasen las dhas trras e las labrasen e curasen e que esto sabe desta pregunta por el juramento que fho tiene e no firmo porque dixo que no sabe escrebir e que es de hedad de treynta e çinco años poco mas o menos lo qual declaro por lengua del dho Françisco Pomavilca el qual juro en forma de derecho que a dho e declarado en su dho en lengua //
/94v/ de Castilla lo mesmo que el dho yndio a dho en su lengua e verdad por el juramento que fho tiene e no firmo porque dixo que no sabe escrebir.

ante mi Juan X[al] de Frias
escrivano de Su Mt

T^{o} Caravajal Yauyo Poma.

E despues de lo susodho en este dho dia e mes e año susodhos estando en el dho valle de Quybe de nonbram^{o} del dho don Felipe
Yndio hize paresçer ante mi a vn yndio que se dixo por su nonbre Caravajal Yavio Poma natural que dixo ser del pueblo de Cullata
que es del dho Alvarado de Payco encomendado en el dho Françisco de Anpuero del qual yo el dho escrivano tome e rresçibi
juramento en forma debida de derecho so cargo del qual prometio de dezir verdad e siendo pregunt^{do} por el tenor del dho pedim^{o}
dixo que lo que sabe es que puede aver doze o treze dias que este t^{o} vino al dho pueblo de Quybe a rregar la coca que en las dhas
trras tiene don Xpoval Payco en conpanya de Alvarado Payco e estando //
/95r/ rregando la dha coca que esta en las dhas trras llego alli don Juan caçique de Canta con muchos yndios que venya de la çibdad de
Los Rreyes y entro en las tierras del dho don Xpoval que estava guardando el dho Alvarado de Payco e en vna chacarilla que en ellas
tiene con coca e mando a los dhos yndios que cogiesen la dha coca y la enpeçaron a coger y estandola cogiendo este tsgo vido que el
dho Alvarado Payco le dixo que no cogiese la dha coca pues hera del dho don Xpoval e en su trra e el dho don Juan le rrespondio
que se fuese de alli que el la avia de coger que hera suya e de sus tierras porque ya tenyan sent^{a} sobre ello e que el señor visorrey lo
avia ya sentençiado y se las avian dado y el dho Alvarado rrespondio que todavia avia pleyto e el e su caçique avian labrado la trra
de la dha coca e que no la cogiese y todavia el dho don Juan la cogio e no dexo cosa nynguna della y se la llevo e le dixo que
llegado a su pueblo avia de enbiar yndios que la guardasen y rregasen e este t^{o} vido como el dho Alvarado Payco dixo que se avia de
venyr a quexar al señor visorrey de lo que //
/95v/ fazia e el dho don Juan rrespondio que binyese e hiziese lo que quysiese e que las trras heran suyas e thenya sentençia dello e que
esto sabe e paso e es la verdad por el juramento que fho tiene y no firmo porque dixo que no sabe escrebir e dixo que es de hedad de
quarenta años poco mas o menos lo qual declaro por lengua del dho Françisco Pomavilca el qual juro en forma de dr^{o} que a dho e
declarado en lengua castellana lo mesmo que el dho yndio a dho en la suya por el juramento que fho t^{e} e no firmo porque dixo que
no sabia escrebir.

ante mi — Juan X^{al} de Frias
escrivano de Su M^{t}

T^{o} Leonor Xuyo.

E despues de lo susodho en este dho dia e mes e año susodhos est^{do} en el dho valle de Quybe yo el dho escrivano de
nonbramyento del dho don Felipe Yndio fize paresçer ante mi a vna yndia que se dixo por su nonbre Leonor Xuyo muger que dixo
ser del dho Alvarado de Payco de la qual yo el dho escrivano//
/96r/ tome e rresçibi juramento en forma debida de dr^{o} so cargo del qual prometio de dezir verdad e siendo preguntado por el tenor del dho
pedim^{o} dixo que lo que pasa es que puede aver doze o treze dias que este t^{o} bino con el dho Alvarado Payco su marido por mandado
del dho don Xpoval caçique a curar de la coca que esta en las trras del dho valle de Quybe e est^{do} en ellas vido este t^{o} que llego alli
vinyendo de Lima don Juan caçique de Canta e muchos yndios con el y enpeçaron a sacar e coger de la coca de bna chacarilla de las
dhas tierras del dho don Xpoval que fazia el dho su marido e estava rregando Caravajal Yavrio Poma yndio e quytaron al dho
Caravajal por m^{do} del dho don Juan vna poca de coca que tenyan cogida por fuerça y el dho Alvarado Payco dixo al dho don Juan
que no cogiese la dha coca que hera del dho don Xpoval y el dho don Juan rrespondio a su marido desta tsgo que se fuese de alli de
las dhas trras que heran suyas e que el abia de coger la dha coca porque ya tenya sentençia para cogella e que la abia de coger e el
dho //
/96v/ Albarado Payco dixo que avnque tenya sentençia todavia abia pleyto e que no la cogiesen e que no rrinesen e mando a esta tsgo que
le diese dos mates de agua que bebiese e dexase la coca e el dho don Juan rrespondio que no queria su .... que la llevase e se la
bebiese el; mando coger la dha coca e entonçes el dho Alvarado Payco marido desta tsgo le dixo al dho don Juan que se avia de
quexar del al señor visorrey porque le cogia la dha coca e el dho don Juan le rrespondio que se quexase e hiziese lo que quysiese que
el abia de coger pues ya tenya sentençia para ello y asy la cogio y se la llevo syn dexar nynguna en la dha chacarilla e quando se
quyso yr dixo que en llegando a su tierra avia de enbiar yndios que guardasen e labrasen las dhas tierras e coca e que esto que dho
tiene es la verdad e lo que sabe por el juramento que fho tiene e no firmo porque dixo que no sabe escrebir lo qual declaro por
lengua del dho Françisco Pomavilca el qual juro en forma de dr^{o} que a dho e declarado en lengua de Castilla lo mysmo que la dha
yndia a dho en la suya e la verdad por el juram^{o} //
/97r/ que fho tiene e no firmo porque dixo que no sabia escrebir.

ante mi — Juan Xpoval de Frias
escrivano de Su Mag.^{d}

Vista del escr^{o}.

E despues de lo susodho en este dho dia e mes e año susodhos est^{do} en el dho valle de Quybe ante my el dho escrivano paresçio el
dho don Felipe Yndio e dixo que por que mejor conste la verdad de lo que en el caso pasa pedia e pidio a my el dho escrivano viese
por vista de ojos la chacara donde se cogio la dha coca por mandado del dho don Juan caçique de Canta e yo el dho escr^{o} fue con el
a la dha chacara que es junto al dho pueblo e casas de Quybe e lleve comygo los yndios que de suso an dho sus dhos e me señalaron
la dha chacara la qual que asi señalaron e dixeron hera la que el dho don Juan avia cogido e m^{do} coger hera de las trras sobre que se
tratava el p^{o} e abian dho sus dhos la qual yo el dho escr^{o} por my propia persona medi e de mys pasos avia de largo treynta pasos e

mas poca cosa e de ancho quynze pasos la qual paresçia estar cogida la coca della e syn nynguna coca e el dho don Felipe Yndio me pidio se lo diese //

/97v/ por escp^{o} e yo de lo vno e de lo otro di este que fha en el dho lugar dia e mes e año susodhos.

E por ende fize aquy este myo signo a tal en testym^{o} de verdad todo lo qual va en nueve fojas con esta en que va myo signo.

Juan Xpoval de Frias
escrivano de Su Mg^{t}

- Lleve de dr^{o}s de qtr^{o} dias de yda e venyda e bno de estada que son mas de diez leguas de muy mal camyno a dos p^{o}s e m^{o} cada dia.	XII p^{o}s IIII ts
- De nueve fojas de escritura.	IIII p^{o}s IIII ts
- De los juramentos de seys t^{o}s.	VII ts
- De la presentaçion del pedim^{o} e abto.	III ts
- Del signo: dos ts.	II ts
- Que son por todos: diez e ocho p^{o}s e qtr^{o} ts.	XVIII p^{o}s IIII ts

Juan X^{al} de Frias escrivano de Su M.^{t}

Auto sobre esta inf.^{on}

/98r/ En Los Rreyes quatorze dias del mes de henero de myll e quy^{o}s e çinquenta e nueve años / los señores oydores desta Rreal Audiençia aviendo visto esta ynformaçion mandaron que los yndios de Canta no se entremetan en coger la coca a los de Chacalla ny ynoven en la causa hasta que sea determinada e fenesçida por todas ynstançias y la coca que les ovieren cogido y llevado se la buelban libremente e asi lo pronunçiaron e señalaron de sus rrubricas.

Pronunçiose este avto el dia e mes e año en el contenydo.

Diego Muñoz

/98v/ Que les buelban lo que an cogido e que hasta que el pleyto se determyne no se entremetan en coger la coca.

/100r/ En Los Rreyes a ocho dias del mes de hebrero de myll e quy^{o}s e çinquenta e nueve años don Felipe prençipal de Chacalla por sy y en n^{e} de los caçiques e yndios sus ptes rrequyrio a my Fran^{co} Hernandez de Guebara escr^{o} de Su M^{t} e rrecevtor nonbrado por los senores presydente e oydores con una provision de temas.

E ansy rrequerido con la dha p^{o}bisyon el dho don Felipe pidio la guarde y cunpla e baya a sus tierras la qual yo el escr^{o} tome en mys manos e la bese e puse sobre mi cabeça con el acatam^{o} devido e dixe que estava presto de la conplir; ts^{o} Fran^{co} de la Torre.

Fran^{co} Hrrez de G^{a}
escr^{o} de Su Mag.^{t}

Secr^{o} Juan Muñoz Ryco

Fran^{co} Hrrez vro. escrivano que en el termy^{o} en que fueron rresçebidos a prueva los yndios de Chacalla y Canta en el pleito que tratan sobre las tierras de Quivi averigue en que pte dellas se an entrado los de Canta y que coca les an cogido dellas y las tierras y coca se las hagan bolver y lo ponga en estado en que estava al tpo que el pleito se començo y no ynoven durante la litis pendençia.

/100v/ Don Phelipe por la graçia de Dios Rrey de Castilla de Leon de Aragon de Ynglaterra de Françia de las dos Çeçilias de Jherusalen de Navarra de Granada de Toledo de Valençia de Galizia de Mallorcas de Sevilla de Çerdeña de Cordova de Corçega de Murçia de Jaen de los Algarves de Algezira de Gibraltar de las yslas de Canaria de las Yndias yslas e tierra firme del mar oçeano; conde de Barçelona; señor de Bizcaya e de Molina; duque de Atenyas y de neopatria; marques de Oristan de Goçiano; archiduque de Austria; duque de Borgoña y de Brabante y Milan; conde de Flandes e de Tirol etc.

a vos Fran^{co} Hernandez nro escrivano salud y graçia. Sepades que pleito esta pendiente y se trata en la nra Audiençia e Chançilleria Rreal que en la çiudad de Los Rreyes de los nros rreynos del Piru rreside; ante el nro presidente e oydores della entre el caçique e yndios de Chacalla de la bna pte y Alonso Moreno su procurador y de la otra el caçique e yndios de Canta y Fran^{co} de la Torre su procurador / sobre las tierras y chacarras de Quivi en el qual por los dhos nro presidente e oydores fue dada sent.^{a} difinitiva en que absolvieron a los dhos caçique e yndios de Canta de la demanda contra ellos puesta por parte de los de Chacalla; y por bna petiçion que el dho Alonso Moreno en su nonbre presento nos hizo rrelaçion diziendo que el caçique e yndios de Canta avia ynbiado bn prinçipal subjeto a el que se llama Paucar Bilca a las dichas tierras y en çierta pte dellas que avia quedado a los dhos sus partes y tenian e poseyan avian echado della a los yndios que los dichos sus ptes tenian puestos diziendo estar la causa sentençiada en su fabor y avian començado a coger la coca que tenian y avian puesto çinquenta yndios en la dicha chacara estando por nos probeydo y mandado que durante la causa no se hiziese ynobaçion alguna en ella por ninguna de las dichas ptes; y nos pidio y suplico mandasemos castigar al dicho caçique e yndios de Canta e quitarle la chacara que avian tomado les bolviesen la coca que avian cogido porque de otra manera los dichos sus partes serian defraudados y se ynbiase persona que lo fuese a ver y lo averiguase; e para hazer la ynformaçion e averiguaçion dello fue por nos probeydo a Juan X^{o}val de Frias nro escrivano el qual la hizo y presento ante

nos en la dicha nra Rreal Audiençia y bista por los dhos nro presidente e oydores por bn auto mandaron que los dichos yndios de Canta no se entremetan en coger la coca a los de Chacalla ni ynoben en la causa hasta que sea determinada y fenesçida por todas ynstançias y la coca que les ovieren cogido y llevado se la buelvan libremente e agora el dho Alonso Moreno en los dhos nonbres nos pidio y supp^{co} que pues estavades nonbrado para que ante vos se haga la averiguaçion; que los terçeros que las partes nonbraren an de hazer conforme a lo por nos sobre ello probeydo / vos cometiesemos la ex.^{on} y cunplim^{o} dello o como la nra mrd fuese lo qual visto. por los dichos nro presidente e oydores confiando de vos que lo hareys con toda diligençia y cuydado guardando el rr^{o} a las ptes ffue acordado que os lo deviamos cometer y que debiamos mandar dar esta nra carta para vos en la dicha rrazon; y nos tovimoslo por bien por la qual vos mandamos que en el termino que os esta dado para hazer la averiguaçion del sitio donde estan las dichas tierras de Quivi y de lo que pueden valer conforme a nra carta y provision rreal que sobre ello os esta dado veays lo susodho y os ynformays y sepais por la via que mejor pudieredes y entendieredes saber verdad como y de que manera a pasado y pasa y en que tierras y chacarras de las sobre que el dicho pleito se an entrado y an ocupado los dichos yndios de Canta durante la dicha litis pendençia tenyendolas y poseyendolas los de Chacalla antes della; o que coca les an cogido y llevado dellas . Y constandoos que durante la dicha litis pendençia se an entrado en las dichas tierras o pte dellas y las tienen las hareys bolver y rrestituyr a los dichos yndios de Chacalla con toda la coca que les ovieren cogido e llevado los de Canta en el dicho tpo sin que por esto sea visto darles ny adjudicarles en ellas mas señorio del que de derecho les pertenesçe; a los quales dichos caçique e yndios de Canta / mandareys y nos por la presente mandamos que no se entremetan en perturbar a los de Chacalla la posesyon de las dichas tierras ny en les coger la coca dellas ny ynoben en la causa cosa alguna hasta que sea determinada y fenesçida por todas ynstançias so pena de decaer de qualquier derecho que a las dichas trras tengan y pretendan tener y mas del ynteres y daño que a la otra parte por rrazon dello se le rrecresçiere e para todo ello y hazer las diligençias que en el caso convengan para mejor os ynformar y saber verdad y lo cunplir y executar bos damos poder cunplido como en tal caso se rrequiere / y por //
/10lr/ los dias que en ello vos ocuparedes ayais y lleveis de salario en cada vno dellos otro tanto como por la dicha provision nra vos esta señalado demas y allende de los derechos que oviere des de aver de la escritura que ante vos pasare y se hiziere los quales ayais y lleveys conforme al aranzel e vos sean dados y pagados por los dichos yndios de Chacalla de los quales los cobrareys e asentareys firmado de vro nonbre al fin dello / y vos aperçebimos que lo que de otra manera llevardes lo pagareys con el quatro tanto para nra camara y no fugades endeal so pena de la nra mrd y de qui^{o}s p^{o}s de oro para la nra camara.

Dada en la çiudad de Los Rreyes a veynte e syete dias del mes de henero de myll y quy^{o}s y çinquenta y nueve años.

Yo Joan Muñoz Rico sr^{o} de camara de Su Sacra Cat. Real Mag.^{d}; la fize escrebyr por su m^{do} con t.^{do} de su p^{te} e oydores.

Rrd. Sanchez Vallejo

por chançiller
Juan Muñoz Rico

T^{o} Alb. de Payco es de Chacalla.

/102r/ E despues de lo susodho este dho dia yo el dho escrivano para ynformaçion e averiguaçion de lo susodho hize paresçer ante my a Albarado de Payco natural que dixo ser del pueblo de Collata que es sujeto a don Xpoval caçique prençipal del pueblo de Chacalla encomendado en Fran^{co} de Anpuero vz^{o} de la çibdad de Los Rreyes por lengua de don M^{yn} nonbrado para esta causa por la Rreal Avdiençia se rresçibio del juramento en forma de dr^{o} por Dios e por Santa Maria e por vna señal de cruz el qual hizo el dho juramento cunplidamente e dixo: "sy juro e amen" e syendole hechas las preguntas nesçesarias rrespondio a ellas lo s^{e}:

Preguntado sy tiene notyçia de las tierras de Quybi sobre que tratan pleyto el caçique de Canta y el caçique de Chacalla dixo que las save porque las a visto y estado en ellas e entendido el pleyto que entre los dhos caçiques sobrellas se trata.

Preguntado que hedad tiene dixo que no lo sabe e que podra tener treynta años poco mas o menos e que como dho tyene es sujeto al dho ca-//
/102v/ çique de Chacalla e que dira verdad de lo que supiere y le fuere preguntado.

Preguntado sy save en que tierras de las sobre que se trata este pleyto se an entrado los yndios de Canta durante el y les an cogido la coca e que es lo que pasa zerca desto dixo que en unas chacarrillas que estan junto al pueblo de Quybi que estavan senbradas de coca que las benefiçiavan y coxian yndios yabyos de los del pueblo de Chacalla y que estas dhas tierras de donde los dhos yndios de Canta cogieron la coca a lo que este t^{o} a entendido no son de las contenidas en el conçierto e venta que hizieron entre el caçique de Chacalla y de Canta porque al tpo que se hizo las dexaron aquellas para los yndios yabyos que las tenian de antes y agora como an rreclamado y no quyeren pasar por el dho conçierto pretenden los yndios de Canta de sacar las bnas y las otras por pleyto e que no save sy en el pleyto se trata de las dhas tierras mas de que save que antes que se pusiese el dho pleyto las tenian y poseyan yndios yavyos y benefiçiavan la coca y la coxian porque se acuerda este t^{o} y bio que al tpo que se tratava del dho conçierto binyeron //
/103r/ llorando algunos yndios de los de Chacalla diziendo que no les bendyesen las dhas tierras porque ellos no tenian donde senbrar y heran pobres e conque se las dexaron y despues aca syn contradiçion de los yndios de Canta las sembraban e benefiçiavan e coxian e que quando se hizo el dho conçierto estavan presentes Fran^{co} de Anpuero y Rribera el moço y saven lo susodho e que este t^{o} dixo vn dho tanbien sobre este caso ante otro escr^{o} el qual por my el dho escr^{o} por la dha lengua le fue leydo de berbo ad berbum e torno a dezir lo mysmo e rrateficarse en el dho que dixo que paresçe; paso ante Juan Xpoval de Frias escr^{o} de Su Mag^{t}

Preguntado que tanta cantidad sera de coca la que los dhos yndios de Canta coxeron de las dhas chacarras dixo que a lo que este t^{o} le paresze aviendolo bisto dixo que seria de coca lo que los dhos yndios de Canta coxieron y que este cojieron antes que vinyese el

dho Juan Xpoval de Frias escr° a fazer la dha ynformaçion y que despues de ydo el dho escr° cojieron de las dhas chacarras cantidad de otro costal / /
/103v/ de coca antes mas que menos e que los dhos yndios decian que lo coxian porque tenian sent[a] en su fabor e que este t° cree que los dhos yndios de Canta tanbien pretenden por pleyto las dhas chacarras.

Preguntado que valdra la dha coca dixo que el primero costal que cogieron valdra quatro pesos y el costal segundo que coxieron baldra seis pesos porque era mayor lo qual todo dixo e declaro por la dha lengua y en ello se afirmo.

ante mi

Fran[co] Hrrez de G[a]
escr° de Su Mag.[t]

T° don X°bal Quichuro; es prinçipal de Quibi.

E despues de lo susodho en el dho pueblo de Quybi veynte dias del dho mes de hebrero del dho año; yo el dho escr° para ynformaçion de lo susodho hize pareszer ante mi a don Xpoval Quychiro prençipal del dho pueblo de Quybi que es del rrepartimyento de Caca [*sic*] del qual por la dha lengua fue tomado e rresçibido juramento en forma devida de dr° el qual lo hizo complidamente e rrespondio: "sy juro e amen" e syendole hechas por my el dho escr° las preguntas al caso pertenesçientes dixo lo s[e]:

/104r/ Preguntado sy sabe e tiene notiçia de las dhas tyerras de Quybi dixo que las sabe porque rresyde en ellas y es natural dellas.

Preguntado por las preguntas generales de la ley dixo que es de hedad a lo que se acuerda de mas de quarenta años e que la ba ynteres en el pleyto que se trata sobre las dhas tierras pero que por eso no dexara de dezir verdad e que querria que bençiese el pleyto los de Canta.

Preguntado que en que tyerras de las sobre que tratan pleyto los yndios de Canta con los de Chacalla se an entrado los yndios de Canta y cogido coca despues que se trata pleyto y quyen tenya e labrava las dhas tierras antes que se enpeçase este pleyto dixo que abia bn mes poco mas o menos que tenyendo e poseyendo los yndios yauyos encomendados en Fran[co] de Anpuero rrepartimyento de Chacalla çiertas tierras que estan junto al dho pueblo de Quybi de muchos años a esta parte y antes y al tpo que se enpeçase este pleyto y despues los yndios de Canta se entraron en ellas y coger çierta coca [*sic*] diziendo que //
/104v/ sobre ellas hera tanbien el pleyto e que ya avian dado sent[a] por ellos e que por eso lo coxian; e que este t° no sabe sy el dho pleyto se trata tanbien sobre estas dhas tierras en el Avdiençia Rreal e que a lo que este t° entendio al tpo que se hizo el dho conçierto entre los yndios de Chacalla y Canta de las tierras de Quybi las dhas tierras de que ansy cogeron la dha coca no entraron en el dho conçierto porque fue desta manera que estando faziendo el conçierto entre los dhos yndios de las vnas tierras e de las otras y estando presente Nycolas de Rribera el moço encomendero de los yndios de Canta y Fran[co] de Anpuero encomendero de los yndios de Chacalla llegaron çiertos yndios de los yavyos encomendados en el dho Fran[co] de Anpuero y se quexaron diziendo que para que les bendian sus tierras que ellos no lo querian porque no tenyan en que sembrar e que heran pobres e ansy vido este t° como le señalaron a los dhos yndios yavyos que eran del pueblo de Xicamarca y de Collauto çiertas tierras para ellos y los dhos caçiques y sus //
/105r/ encomenderos pusyeron mojones y las dibidieron de las demas del dho conçierto y los dhos yndios yavyos las tubieron por suyas y dende a vn año que paso lo susodho porque los dhos yndios yavyos se descuidaron de no benyr a plantallas se entraron en ellas los dhos yndios de Chacalla y las plantaron y luego vinyeron los yndios de Chacalla y benefiçiaron la dha planta y la coxieron para si y lo mysmo a visto este testygo que an fho fasta agora que despues que anda este pleyto los yndios de Canta entraron a cojer la coca diziendo que el pleyto hera sobre las dhas tierras tanbien e que ya tenyan sent[a] en su favor e syendole leydo por la dha lengua vn dho que dixo antes Juan Xpoval de Frias dixo que aquello hera tanbien la verdad y lo que avia dho y en ello se afirmo.

Preguntado que tanta cantidad de coca le paresze a este t° que podrian cojer los dhos yndios de Canta de la dha chacarras [*sic*] dixo que no hera casi nada e que valdria vn peso y no mas y lo qual dixo por la dha lengua y en ello se afirmo.

ante mi

Fran[co] Hrrez de G[a]
escr° de Su Mag.[t]

T° don Ant° Marcapoma.

E despues de lo susodho estando en el dho pueblo de Quybi a veynte e dos del dho mes de hebrero del dho año yo el dho escr° para ynformaçion de lo susodho hize paresçer ante mi a don Antonyo Marcapoma caçique que dixo ser de Canta e por lengua de el dho don M[yn] ynteprete e por my el dho escr° le fue tomado juram° en forma de dr° segun se rrequeria el qual lo hizo conplidamente e rrespondio: "sy juro e amen" so cargo del qual prometio de dezir verdad a las preguntas que por my el dho escr° se le hizieren que son las syguientes:

Preguntado sy save e tiene notiçia de las dhas tierras de Quybi dixo que las save porque son tierra de Canta.

Preguntado que años tiene dixo que no lo save.

Preguntado que en que tierras de las sobre que se trata este dho pleyto se an entrado los yndios de Canta que las toviesen los yndyos de Chacalla durante este dho pleyto e que coca les an //
/106r/ cogido dellas dixo este dho t^{o} que save çiertamente avnque no lo vido que don Juan prençipal de Canta coxio de vnas tierras en que tenyan sembrado los yndios de Chacalla vn poco de coca que no hera vn zesto e que lo coxio despues que se trata este pleyto porque la planta de la dha coca la pusieron los yndios de Canta y que los yndios de Chacalla la benefiçiavan la dha coca en las dhas tierras porque este t^{o} save çierto que despues de hecho el conçierto sobre las dhas tierras de Quybi con los yndios de Chacalla los yndios de Canta dieron a çiertos yndios de Chacalla vn pedaço de tierra porque deçian que no tenyan do sembrar e que ellos querian ser sujetos a Canta e que estas dhas tierras no heran de las del conçierto porque el conçierto fue desde las propias casas del dho pueblo de Quybi fazia Lima y las dhas tierras de donde se cojo la dha coca estan de las dhas casas fazia Canta e que es verdad que los yndios de Canta pretenden aviendo salido con su yn-//
/106v/ tençion en las dhas tierras del conçierto y quedando con ellas quedarse con estas otras en dando sent^{a} porque entienden que en todo se a de entender esta dha sentençia e pleyto. Lo qual todo dixo e declaro por la dha lengua don M^{yn} en presençia de my el dho escr^{o} y en ello dixo que se afirmava e se afirmo e no lo firmo porque no supo.

ante mi

Fran^{co} Hrrez de G^{a}
escr^{o} de Su M^{t}

T^{o} don Felipe de Chacalla.

E despues de lo susodho este dho dia yo el dho escr^{o} hize para ynformaçion de lo susodho pareszer ante mi a don Felipe prençipal que dixo ser de los yndios de Chacalla e por mi el dho escr^{o} e por lengua de don M^{yn} ynterpetre fue tomado e rresçibido juramento en forma de dr^{o} por Dios e por Santa Maria e por las palabras de los hevangelios el qual hizo el dho juram^{o} segun se rrequeria en tal caso e rrespondio: "sy juro e amen" so cargo del qual .... //
/107r/ declaro a las preguntas que le fueron fhas en la forma sig^{e}:

Preguntado sy save las dhas tierras de Quybi sobre que es este pleyto dixo que las save porque las sujeto e trae pleyto sobre ellas.

Preguntado quantos años a dixo que no lo save.

Preguntado diga e declare sy save en que tierras se an entrado los yndios de Canta que tengan y posean los yndyos de Chacalla que sean de las contenydas en el conçierto y sobre que se trata este dho pleyto e de que tantos dias a esta parte se an entrado en ellas y porque y que tenian senbrado en ellas y quanto fue lo que se les coxio dellas dixo que lo que pasa es que este t^{o} save e a entendido desde que se save acordar y de tpo ynmemorial a esta parte de avello oydo dezir a sus mayores e ançianos que todas las dhas tierras de Quybi ansy las contenydas en el conçierto como las de dr^{o} le coxieron la coca los yndios de Canta todas heran ny fue-//
/107v/ ron ny son de los yndios yavyos de Chacalla e que ansy mysmo a entendido de sus caçiques y de yndios que bieron tratar del dho conçierto que el conçierto que Fran^{co} de Anpuero hizo con los yndios de Canta por las obejas fue solamente de los buyos e casas de Quybi para abajo e que de los buyos para arriba fazia Canta lo dieron y señalaron a yndios yavyos de Chacalla que benefiçiasen e an benefiçiado en las tierras coca e que abra mas de quarenta dias que vn don Ju^{o} caçique e prençipal de Canta fue a estas dhas tierras e arranco dellas la coca que seria dos costales diziendo que ya se avia dado sent^{a} por ellas e que a entendido este dho t^{o} que por parte de los vnos yndios y de los otros salyendo con este pleyto se pretende quedar con las dhas tierras de Quybi con todas ellas e que este t^{o} ansy entiende que es el pleyto sobre todo e que se rremyte a la forma de la demenda lo qual todo dixo e declaro por estas rrazones a lo que declaro la dha lengua y en ello se afirmo y no lo firmo porque no supo.

ante mi

Fran^{co} Hrrez de G^{a}t
escr^{o} de Su Mag.

/108r/ T^{o} indio de Canta que se dize
el nonbre: *Gonçalo.*

E despues de lo susodho este dho dia yo el dho escr^{o} para ynformaçion e aberiguaçion de lo susodho hize parescer ante my a G^{o} yndio de Canta e lengua que dixo ser del qual yo el dho escr^{o} e por lengua del dho don M^{yn} ynterpetre rresçebi juramento en forma de dr^{o} y el lo hizo conplidamente segun se rrequeria en tal caso e syendo preguntado por las preguntas syguientes declaro lo sig^{e}:

Preguntado sy conosze a las dhas partes e si tiene notiçia de las dhas tierras de Quybi sobre que se trata este dho pleyto dixo que conosce a las dhas partes e tiene notyçia de las dhas tierras porque tiene pte en ellas.

Preguntado que años tiene dixo que mas de quarenta.

Preguntado en que tierras de las sobre que se trata este dho pleyto e cavsa se an entrado los yndios de Canta que tuviesen sembrado los de Chacalla e deste tpo a esta parte e que coca e semyllas les coxieron dellas porque dixo//
/108v/ que syenpre desde que se save acordar este dho t^{o} a entendido que todas las dhas tierras de Quybi ansy las que tenyan senbradas los yndios de Chacalla como las demas del conçierto heran de los yndios de Canta e que es verdad que a entendido que al tpo que se hizo el dho conçierto dexaron çiertas tierras para los yndios de los pueblos de Cullata e Aynamarca que son yndios de Chacalla para

que las senbrasen porque lloravan e se quexavan que no tenyan do senbrar e que es verdad que los dhos yndios destos pueblos an sembrado las dhas tierras e benefiçiadolas e que los yndios de Canta dando sentª por ellas pretenden quytarselas por la sentª que se dieron en su favor e ansy entienden y este tº entiende que sobre ellas es el dho pleyto e que abra quarenta dias poco mas o menos que vn prinçipal de Canta entendiendo que se avia dado sentª por los yndios de Canta avia ydo a las dhas tierras y arrancado vn poco de coca que seria vn zesto poco mas o menos e que esto save e es la verdad y en ello dixo que se afirmava e afirmo e rratefico por la dha lengua y no lo firmo porque no supo.

ante mi Fran^{co} Hrrez de Gª
escrº de Su Mag^{t}

Por bista de ojos el escrº

/109r/ E despues de lo susodho este dho dia yo el dho escrº e rrezevtor juntamente con don Antº Marcapoma prençipal de Canta y don Felipe prençipal de Chacalla e otros yndios sus sujetos fue a ver las dhas tierras do se coxio la dha coca y los dhos yndios me las mostraron y estando en ellas aviendo vysto la ynformaçion que esta tomada zerca dellas tome de la mano al dho don Felipe prençipal de Chacalla e a Juº Rroque yndio que dixeron ser de los que senbravan en ellas e les meti e apodere en la posesion que dellas tenyan de la forma y manera que en la dha provision Rreal de Su M^{t} se es leyendosela e declarandosela a los vnos e a los otros por lengua del dho don M^{yn} e dandoles a entender todo ello y las penas que en ellas se les pone los quales dixeron avello entendido e les mande a los dhos yndios de Canta diesen y entregasen luego a los yndios de Chacalla la coca que de las dhas tierras les avian cogido los quales dixeron que ansy lo farian conforme a lo que se es en la dha provision rrealsyendo trºs a lo que dho es Aº de Torregrosa e Fran^{co} Hernandez Morante.

E luego el dho don Felipe prençipal de Chacalla y el dho Juº Rroque dixeron que se davan por vien pagados y contentos de los dhos yndios de Canta de la dha coca que les coxieron por q^{to} estan satisfhos. Tºs los dhos.

E luego el dho don Antº Marcapoma pren-//
/109v/ çipal de Canta dixo que contradize el avto que se hizo en apoderalles a los yndios de Chacalla en las dhas tierras avnque aya sydo syn perjuyzio de su drº como lo dize la dha provision rreal y lo piden por tmº. Tºs los dhos.

E yo Fran^{co} Hernandez de Guebara escrº de Su Mag^{t} e rrezevtor de la cavsa presente fui a la dha aberiguaçion e avtos e lo escrebi e hize aqui myo sº tal en tmº de verdad.

Fran^{co} Hrrez de Guebara
escrº de Su Mag^{t}

Tasaçion de costas que son las sig^{es}:

- de la present^{on} e rrequerimº con la provision: quatro tomynes.		4 ts
- de çinco tºs: diez ts.	I pº	2 ts
- de ocho hojas: quatro pºs.	4 pºs	
- de la posesion: vn peso.	1 pº	
- de la contradiçion: tres ts.		3 ts
- montan: syete pºs e vn tomin.	7 pºs	I t.

rreszebilos.

Fran^{co} Hrrez de G.ª

/112r/ Don Phelipe por la graçia de Dios Rrey de Castilla de Leon de Aragon de Ynglaterra de Françia de las dos Çeçilias de Jherusalen de Navarra de Granada de Toledo de Valençia de Galizia de Mallorcas de Sevilla de Çerdeña de Cordova de Corçega de Murçia de Jaen de los Algarves de Algezira de Gibraltar de las yslas de Canaria de las Yndias yslas e tierra firme del mar oçeano; conde de Barçelona; señor de Vizcaya e de Molina; duque de Atenas y de neopatria; marques de Oristan y de Goçiano; archiduque de Austria; duque de Borgoña y de Brabante y Milan; conde de Flandes e de Tirol.

A vos el rreverendo padre Fray Gaspar de Carvajal provinçial de la Horden de Santo Domingo de los nros rreynos del Piru salud y graçia; sepades que Alonso Moreno en nonbre del caçique e yndios de Chacalla por vna petiçion que presento en la nra Audiençia y Chançilleria Rreal que rresyde en la çiudad de Los Rreyes de los dichos nros rreynos ante en nro presidente e oydores della nos hizo rrelaçion diziendo que en el pleito que en la.dicha nra Rreal Audiençia tratan los dichos sus partes con el caçique e yndios de Canta sobre las tierras de Quibi por nos estava mandado que las partes nonbrasen terçeros cada vno de su parte para que vean las dichas tyerras y el sitio dellas y lo que pueden valer y lo averiguen con testigos yndios comarcanos segun y como se contiene en la provision por nos sobre ello dada dirigida a Fran^{co} Hernandez de Guevara ante quien a de pasar lo en ella contenido y que el en nonbre de los dichos sus partes tenia nonbrado para ello por terçero a nos el dho padre provinçial y la otra parte avia nonbrado al dho Juan de Pedraza porende que nos suplicava e pedia por mrd. vos encargasemos hiçiesedes lo que por la dicha nra provision se manda pues lo teniades testado como paresçia por vra çedula en el proçeso de la dicha causa presentado y que para la examinaçion de los testigos y de lo demas que en la dicha causa se oviese de hazer fuese por anbas partes vno de los ynterpetres que rresiden en la dicha nra Rreal Audiençia o como la nra mrd fuese lo qual visto por los dichos nros presidente e oydores y lo por ellos çerca dello probeydo e mandado fue acordado que deviamos mandar dar esta nra carta en la dicha rrazon y nos tovimoslo por bien / por la qual vos encargamos que veays la dicha nra provision rreal que de suso se haze mynçion y juntamente con el dho Juan de Pedraza terçero nonbrado por parte de los dichos caçique e yndios de Canta vays al dho asiento y tierras de Quibi e alli anbos juntos e no el vno sin el otro por ante el dicho Fran^{co} Hernandez nro escrivano e vn ynterpetre de la dha nra Rreal Audiençia que a ello mandamos vaya por anbas partes hagais la averiguaçion del sitio e valor de las dichas tierras e todo lo mas en la dha nra provision rreal contenido sin exçeder della en cosa alguna para con ello poder mejor hazer en la dicha causa justiçia a las partes que en ello nos servireys. Dada en la çiudad de Los Rreyes:

A diez dias del mes de hebr^{o} de myll y qui^{o}s y nueve a^{o}s

Yo Joan Muñoz Rico sr^{o} de camara de Su Sacra Cat. Real Mag^{d}; la fize escrebyr por su m^{do} con t^{do} de su p^{te} e oydores.

R^{dor} Br^{me} Gosicano

por chançiller
Joan Muñoz Rico

En el pueblo de Quybi juresdiçion que dizen de Chacalla y Canta jueves diez y siete dias del mes de hebr^{o} de myll e quy^{o}s e çinquenta e nueve años ante my el escr^{o} e yo el escr^{o} de pedim^{o} de los caçiques e yndios de Chacalla y Canta rrequeri e de esta p^{o}vision al padre Fray Gaspar de Carabajal; ts^{o} don M^{yn} Lengua.

E luego el dho padre Carabajal obedesçio la dha probysion rreal con el acatam^{o} e rreberençia devido e dyxo que estava presto de la conplir; ts^{o}s los dhos.

Fran^{co} Hrrez de G^{a}
escr^{o} de Su Mag.^{t}

En el pueblo de Quybi a diez y ocho dias del dho mes de hebr^{o} del dho año por parte de los dhos caçiques de Chacalla y Canta se rrequyrio dara esta probisyon a Ju^{o} de Pedraza el qual dixo que la obedeçia e obedeçio e tomo e puso sobre su cabeça e que estava presto de la conplir; t^{o} el dho.

Fran^{co} Hrrez de G^{a}
escr^{o} de Su Mag.^{t}

Al provinçial de Santo Domingo que se junte con Juan de Pedraza y anbos vean el sitio de las tierras de Quivi y çerca del sitio y valor ellos cunplan lo contenido en la provision rreal por Vra Alteza çerca dello dada a pedim^{o} de los yndios de Chacalla.

secr^{o} Joan Muñoz Ryco

/113r/ Inf.^{on} de ofiçio que hazen el provinçial de los dominicos y Pedraza sobre cuyas son las trras.

E despues de lo susodho en el dho pueblo de Quybi a diez e ocho dias de myll e quynyentos e çinquenta e nueve años el muy rr^{do} padre frai Gaspar de Caravajal provinçial e Juan de Pedraça tr^{o} nonbrado por los dhos caçique e yndios de Chacalla y Canta por ante my el dho escr^{o} para averiguaçion de lo contenido en la provision rreal de Su M^{t} y por vyrtud della hizo parescer ante si a don Antonyo caçique que deudo de don Xpoval Sylasque encomendado en M^{yn} Piçarro e se rresçibio del juram^{o} en forma devida de dr^{o} segun se rrequeria e syendo preguntado por lengua de don M^{yn} Lengua ynterpetre de la Rreal Avdi^{a} y nonbrado por lo susodho se le hizieron las preguntas y el asolvio lo syguyente

Preguntado como se llana dixo que se llama don Antonyo Calpa.

Preguntado de que pueblo es caçique y que tantas leguas esta deste pueblo de Quybi dixo que su pueblo se llama Quypan encomendado en M^{yn} Piçarro e que abia seis leguas desde este pueblo de Quybi fasta alla.

Preguntado por la hedad que tiene e si por parte de los caçiques de Chacalla e de Canta le an dho que no diga verdad o si le an dado o prometido alguna cosa por que no la diga dixo que no save la hedad que tiene mas de que quando entraron los xpianos a tomar tierra hera mochacho e que no le fue da-//

/113v/ do ny prometido por las dhas partes cosa alguna para que diga el contrario de la verdad.

Preguntado sy save e a visto e tiene notiçia de las tierras de Quybi sobre que traen pleyto los caçiques de Canta y Chacalla dixo que las save porque las a visto.

Preguntado cuyas son las dhas tierras de Quybi e quyen las posee dixo que lo que este t^{o} se acuerda y a oydo y entendido muchos yndios viejos desta comarca y de su pueblo deste t^{o} tratando e platicando con ellos sobre las dhas tierras e a sus padres e aguelos deste t^{o} que las dhas tierras de Quybi sobre que se trata este pleyto al tpo que rreynava Topa Ynga Yupangue e dho Topa Ynga avia puesto por mitimaes en las dhas tierras a yndios yavyos que son del rrepartim^{o} de Chacalla y que despues de muerto el dho Topa Ynga suçedio Guaynacava el qual les avia acreszentado otros pedaços de tierras que llegavan fasta las casas del dho pueblo de Quybi y alli puso por mitimaes los mysmos yndios yavyos en otro pedaço de tierra que el dho Guaynacava avia señalado para si para que los dhos yndios yavyos se las senbrasen y labrasen para dalle trebuto de coca y otras cosas //

/114r/ e que esto como dho tiene lo a oydo ser y pasar ansy como dho tyene entre muchos yndios viejos que decian savello e avello vysto y ser dello cosa notoria e que despues desto a entendido este t^{o} honbre que fue en tpo del presydente Gasca como se conçertaron Guaman Yavri caçique prençipal de Canta y don Fran^{co} Marcapoma caçique prençipal de Chacalla en que deçia el dho caçique de Canta Guaman Yavri al dho don Fran^{co} Marcapoma: "ya saves tu que esta tierra hera mya y esta en my sytio e Ynga me lo quyto para dartela a ti y a tus yndios y oy dia yo tengo nescesidad dellas para mys yndios porque tengo falta de tierras; conçertemonos pues agora no ay Ynga e yo te dare duçientas ovejas por lo que abreis labrado y plantado y puesto andenes" y en hefeto entendio que el dho caçique de Chacalla paso por el dho conçierto y se conçerto por las dhas obejas y este t^{o} enbiava de su pueblo a chasco que lo vian y entendian e se lo decia a este t^{o} e que por rrazon de las dhas tierras antes que se hefetuase el dho conçierto andavan entrellos grandes pendençias y estavan puestos en arma para se matar los vnos y los otros y tanbien supo y entendio que los que fueron partes para que el dho conçierto se hiciese Rribera el moço con Fran^{co} de Anpuero //

114v/ los quales lo avian conçertado desta manera por metellos en paz e que tanbien entendio este t^{o} que algunos de los yndios yavyos de Chacalla les pesaba del dho conçierto y benyan llorando al caçique de Canta Guaman Yavre y le decian que eran pobres y que no tenian chacaras para coca que les dexase a ellos sus partes de chacarras que les avia dado el Inga y que como el dho Guaman Yavre caçique de Canta hera señor de lastima se les dio a los dho yndios que ansy se quexavan algunas dellas los quales es pu^{co} que las senbravan y sienbran y las poseen agora que son de las que agora se quexan que les an cogido la coca durante el pleyto.

Preguntado quales yndios estan mas çercanos a las dhas tierras de Quybe sobre que es este pleyto los de Chacalla o los de Canta dixo que los yndios yavyos de Chacalla tienen el sytio destas tierras que alinda con las dhas tierras por una parte y los de Canta alindan de la mysma manera por otra e ambas parçialidades tienen casas y rrancherias junto a las dhas tierras sobre que este pleyto.

Preguntado que fruto e ap^{o}becham^{o} tienen las dhas tierras e que se sien-//
/115r/ bra y coje en ellas dyxo que no se coxe en ellas syno coca e axi e que tiene arboledas de guabos y guayabas e que no syenbran ni cojen maiz porque es tierra hecha e apropiada para coca.

Preguntado quantos costales de coca se podran cojer en cada vn año de las dhas tierras labrandose y benefiçiandose dixo que no lo save.

Preguntado que cantidad de tierra sera la sobre que se trata este pleyto dixo que no lo save.

Preguntado diga y delcare que p^{o}s de oro podran valer las dhas tierras de Quybi dixo que no lo save.

Lo qual todo dixo e declaro en presençia de los dhos terceros nonbrados e de my el escr^{o} por lengua del dho don M^{yn} e se leyo su dho e se rratefico en ello e se le encargo e dio a entender tubyese secreto e lo firmaron los dhos terçeros.

Frai Gaspar de Carvajal J^{o} de Pedraza

ante mi Fran^{co} Hrrez de Guebara

escr^{o} de Su Mag.^{t}

/115v/ *T^{o} don Alonso Bilca.*

E despues de lo susodho este dho dia diez y ocho dias del mes de hebrero del dho año los dhos terceros para averiguaçion de lo susodho hiçieron parescer ante si a don Alonso Bilca caçique del pueblo de Puruchuca [*sic*] encomendado en Martin Piçarro e rresçibieron del juramento en forma de dr^{o} segun se rrequeria y el lo hizo conplidamente e syendoles hechas las preguntas al caso pertenescientes dixo lo syguiente:

Preguntado como se llama dixo que se llama don Alonso Bilca caçique del pueblo de Puruchuca encomendado en M^{yn} Piçarro.

Preguntado sy sabe e a bisto las tierras de Quybi sobre que este pleyto dixo que las save porque alindan con tierras de su pueblo deste t^{o}.

Preguntado por las generales de la ley dixo que no sabe la hedad que tiene mas de que se acuerda quando entraron los españoles en esta tierra porque era ya buen mochacho e que no le an dado ny prometydo cosa alguna para que diga el contrario de la verdad e que no es pariente de los dhos partes.

/116r/ Preguntado en que sytio e termyno estan las dhas tierras de Quybi sobre que es este pleyto dixo que save y a bisto que en tpo del marques don Fran^{co} Piçarro tenyan y poseyan las dhas tierras de Quybi los yndios yavyos del rrepartimy^{o} de Chacalla e antes en tpo de Guascar Ynga los vio alli e que es notiçia que este t^{o} tiene porque estavan los dhos yndios yavyos en las dhas tierras es por rrazon de que Topa Ynga Yupangue ynga se las dio y metio en ellas ponyendo los dhos mitimaes y se las avia quytado a los yndios de Canta porque ansy lo facia en su tierra deste t^{o} y en toda la tierra dese rreyno que el dho Ynga poseya los quales dhos mitimaes estobieron fasta que vino Guaynacaba a heredar el qual les acresçento e dyo otras tierras mas adelante e señalo para sy vn pedaço junto al pueblo de Quybi y puso en ellas los mysmos yavyos para que le dyesen trebuto dellas de coca que hera lo que se coxia en ellas y esto es cosa pu^{ca} entre todos los yndios viejos e ancianos que lo tratan e an tratado a la contina ofresçiendose fablar //
/116v/ sobre las dhas tierras e que los dhos yndios yavyos tubieron las dhas tierras hasta que vino el p.^{e} Gasca que estavan en este tpo para se matar los yndios de Canta y los de Chacalla sobre las dhas tierras y estando desta manera entendio este t^{o} que enbiaron a llamar el dho caçique de Canta a Rribera el moço su encomendero y el caçique de Chacalla a Fran^{co} de Anpuero su encomendero los quales vinyeron e bysto las dhas diferençias y por quytarles dellas trataron de conçertar los dhos caçiques y los conçertaron en esta manera que los yndios de Canta diesen a los de Chacalla a los de Canta [*sic*] docientas obejas de la tierra por lo que avian labrado en las dhas tierras las quales heran çiento obejas y çiento carneros e ansy se conçertaron y quedaron contentos y los de Chacalla se fueron a sus tierras y los de Canta quedaron en las dhas tierras y quedaron de no fablar mas en ello e que este t^{o} tenya yndios que yban y benian e lo vian todo lo que pasaba del dho conçierto e se lo deçian a este t^{o} e que tanbien entendio y fue pu^{co} que hiz^{o} como se hizo el dho conçierto binyeron una parçialidad de yndios yabyos //
/117r/ del rrepartim^{o} de Chacalla que heran del pueblo llamado Collata Marca los quales se quexavan diziendo que les avian vendido sus tierras e que no tenyan donde senbrar su coca y pedian e pidyeron a Guaman Yavre caçique prençipal de Canta que era el que hizo el dho conçierto que les diese alg.^{as} tierras para senbrar y el dho caçique les dio alguna parte dellas dhas tierras a los dhos yndios en

que senbrasen junto al pueblo de Quybi facia la syerra de Canta e que esto no lo a visto este tº pero que save que es cosa muy cierta y notoria entre los yndios desta comarca.

Preguntado con que tierras alindan las dhas tierras de Quybi sobre que este pleyto dyxo que alindan por una parte con tyerras de yndyos de Cuayo [*sic*] que son del hijo de contador Juan de Caçeres y por otra parte tierras de los yndios de M^{yn} Piçarro el rryo en medio e que tanbien alindan con tierras de Canta e unas syerras peladas que tras ellas estan e mojon de Chacalla e que tanbien tienen los yndios de //
/117v/ Canta y de Chacalla çiertos buyos hechos junto a las dhas tierras de Quybi las quales an estado desde el tpo de Topa Ynga Yupangue.

Preguntado que fruto es apºbechamº tienen las dhas tierras e que se coje en ellas dixo que se coje coca e axi e mani e camotes e que ay arboledas de guabos y guayabas lucumas y ortalisas e que no se coje maiz ny trigo porque no lo syenbran porque la tierra esta dyspuesta para cojerse coca.

Preguntado quanto abia de su pueblo deste tº a este asyento de Quybi dixo que fasta el pueblo donde bibe abia dos leguas e que junto al alinde de las chacarras de Quybi tiene yndios suyos y chacarras.

Preguntado que costales de coca se podran cojer en cada vn año en las dhas tierras de Quybi e que fanegadas de tierra podra tener de sementera dixo que no lo save porque nunca a senbrado en las dhas tierras.

Preguntado quantos pesos de oro //
/118r/ podran valer las dhas tierras o quantas ovejas daria el por ellas aviendolas menester dixo que no lo sabe ny lo entiende.

Lo qual todo dixo e declaro por lengua del dho don M^{yn} Lengua ynterpertre nonbrado para lo susodho en presençia de los dhos terçeros que lo firmaron de sus nonbres fuele leydo su dho por la dha lengua; se rratefico en el y encargosele que tenga secreto de lo que a dho e prometio de fazello ansy.

Frai Gaspar de Carvajal — Juº de Pedraza
ante mi — Fran^{co} Hrrez de Guebara
escrº de Su Mag.^{t}

E despues de lo susodho este dho dia los dhos terceros nonbrados para ynformaçion e averiguaçion de lo contenydo en la dha provision rreal por ante my el dho escrº hizo parescer ante sy a don Antonyo prençipal del pueblo de Çapan encomyenda del contador Juº de Caçeres //
/118v/ del qual por lengua del dho M^{yn} ynterpetre fue preguntado e se rresçibio juramento en forma de drº segun se rrequeria en tal caso e siendo preguntado dixo lo sig.^{e}:

Preguntado como se llama e de donde es natural curaca dixo que se llama don Antonyo Chunbiquybi e que su pueblo se llama Çapan que abia desde este asyento a su pueblo tres leguas.

Preguntado sy sabe las tierras de Quybi sobre que este pleyto dixo que las sabe porque las a visto.

Preguntado por las preguntas generales de la ley dixo que es de hedad de mas de quarenta años e que no es pariente de los caçiques de Canta ny de Chacalla ny le ba ynteres alguno e que no le an dado ni prometido cosa alguna para dezir el contrario de la verdad.

Preguntado en que sitio estan las dhas tierras e cuyas an sydo e son al presente dixo que las dhas tierras de Quybi a lo que este tº a oydo dezir a viejos e ançianos que oy dia //
/119r/ son bibos e a otros que son muertos e a sus padres e aguelos heran y el sytio dellas antes que ubiese yngas en el Peru de vn señor que no se acuerda de su nonbre mas de que deçian que era yndio yunga sujeto a la guaranga de Collique e que este yndio tiene oy dia vn hijo suyo bibo que rresyde en Lima e este tº le conosçe que se llama Chunbiquybe los quales las tobieron y poseyeron y sienpre tenya deferençias con los yndios yabyos de Chacalla sobre las dhas tierras fasta que entro Ynga y los dhos yndios yabyos se favoresçieron con el y binyeron a las dhas tierras de Quybi y el dho Ynga puso en ellas a los yndios yabyos y dejo en ellas mytimaes para su serviçio de los dhos yavyos y para chasques del dho Ynga para que fuesen y binyesen do les mandase e que el mysmo Inga señalo los trmºs a los de Canta y a los de Chacalla en las dhas tierras e que despues de venydos los españoles a este rreyno quando se alço Mango Ynga se alçaron los yndios de Canta y dieron sobre los dhos yndios yabyos que el Inga avia puesto por mytimaes y los mataron y los dhos yndios de Canta se //
/119v/ entraron en ellas e que asta aora los yndios yauyos de Chacalla syenpre tenyan guerra e deferençia con los yndios de Canta sobre las dhas tierras e los de Canta con los de Chacalla fasta que estando el presydente Gasca en este rreyno binyeron a las dhas tierras Rribera el moço encomendero de los yndios de Canta y Fran^{co} de Anpuero encomendero de los yndios de Chacalla y por quytallos de pasyones trataron de conçertallos y los yndios de Canta dyeron a los de Chacalla çierta cantidad de obejas por las dhas tierras e

prometian de no fablar mas en ello e que esto como dho tiene le es a este t^{o} notorio por lo que a entendido e por lo que a oydo dezir a sus mayores e mas ançianos e que sy otra cosa fuera o pasara en contrario lo ubiera oydo dezir por lo que entiende que las dhas tierras ny heran de los yndyos de Canta ny de los de Chacalla syno de los Yungas.

Preguntado con que tyerras alyndan las dhas tierras de Quybi dixo que alindan con mojones de su pueblo deste t^{o} y de M^{yn} Piçarro e yndyos yavyos de Chacalla e yndios //
/120r/ de el dho pueblo de Canta.

Preguntado que fruto e aprobechamyento se saca de las dhas tierras e que sienbran e cojen en ellas dixo que lo que este t^{o} a visto sembrar es axi e coca e pacayes e guavas e guayabas y otras cosas.

Preguntado quantos costales de coca o fanegadas de senbradura podra aver en las dhas tierras de Quybi dixo que le paresze a este t^{o} que de todas las dhas tyerras de Quybi sobre que este pleyto se podra coxer en ellas cada año de coca çinco costales de coca e otras semyllas.

Preguntado a como bale cada costal de coca dixo que baldra fasta tres pesos.

Preguntado que pesos de oro podran baler las dhas chacarras o que ovejas daria este t^{o} por ellas aviendolas menester dixo que este t^{o} daria por las dhas tierras de Quybi tresçientos pesos de oro y que no daria mas.

Lo qual todo dixo por lengua del dho don M^{yn} en presençia de los dhos ter-//
/120v/ ceros que le fyrmaron de sus nonbres; fuele encargado el secreto despues de aversele leydo.

Frai Gaspar de Carvajal — Ju^{o} de Pedraza

fho ante mi — Fran^{co} Hrrez de Guebara
escr^{o} de Su Mag.^{t}

E despues de lo susodho este dho dia los dhos Fray Gaspar de Caravajal y Ju^{o} de Pedraza terçeros nonbrados en esta cavsa para ynformaçion e aberiguaçion de lo susodho contenydo en la dha probisyon rreal por ante mi el dho escr^{o} hizo paresçer ante si a don Fran^{co} Yabyi caçique prençipal de Colli encomendado en Muñoz de Avila del qual se tomo e rresçibio juramento en forma de dr^{o} segun se rrequeria en tal caso e siendo preguntado por lengua del dho M^{yn} ynterpetre dixo lo sig.^{e}:

Preguntado como se llama dixo que se llama dixo que [*sic*] don Fran^{co} Yabyi e que es caçique de Colli que esta encomendado en Muñoz de Avila que es comarcano a este valle de Quybi.

Preguntado sy a visto y save las dhas tierras de Quybi e tyene notiçia //
/121r/ dellas dixo que las a visto y es comarcano en ellas.

Preguntado por las preguntas generales de la ley dixo que no save la hedad que tiene mas de que quando entraron los espanoles en esta tierra hera ya honbre e que no le an dado ny prometido por parte de los dhos caçiques de Canta ny por otra persona cosa alguna para que diga el contrario de la verdad ny tanpoco es pariente e que dira la verdad porque es xpiano.

Preguntado en que sytio estan las dhas tierras de Quybi sobre que es este pleyto e cuyas an sydo e son dixo que a lo que este dho t^{o} a entendido e oydo dezir a viejos de su tierra deste t^{o} y desta comarca ansi de los que agora son muertos como a otros que son bibos de cuyos nombres no tiene memoria e a sus padres e pasados deste t^{o} que las dhas tierras de Quybi antiguamente antes que ubiese Yngas heran y las tenian y poseyan vn caçique que se llamava Caxapoma que era yndio yunga que era parçialidad por sy sujeto a Inga el qual las tubo e poseyo fasta que vino a señor Topa Ynga Yupangue //
/121v/ el qual las quyto las dhas tierras al dho curaca de los yungas y las dio a los yndios yayos de la encomyenda de Chacalla que eran sus criados como los guancas y puso en ellas por mitimaes a los dhos yabyos y las dibidio y partio entre los dhos yndyos de Canta y de los yabyos y puso sus mojones y term^{o}s entre ellos y despues que murio Topa Ynga Yupangue en tiempo de Guaynacava les acresçento vn poco mas de tierra a los yavyos y tomo para si Guaynacava otro pedaço de tierra y mando a los mysmos yndios yavyos que la labrasen y beneficiasen para dalle trebuto de coca y que este t^{o} vino a las dhas tierras en tpo de Guascar Ynga y vio como estavan en las dhas tierras de Quybi todas las parçialydades de los yndios yabyos de Chacalla por mytimaes que los tenya el dho Guaynacava para chasques y para llevar cargas e que despues los yndios de Canta procuraron de sacarles de las dhas tierras a los yabyos y como mas poderosos los hecharon los quales se fueron a quexar a su caçique prençipal de Chacalla y se armo //
/122r/ el y sus yndios y binyeron a dar guerra a los de Canta e a defender las dhas tierras y estubieron los vnos caçiques y los otros a punto de se matar e anduvieron a los cabellos y en esto vinyeron a metellos en paz Rribera el moço encomendero de los dhos yndios de Canta y Fran^{co} de Anpuero encomendero de los de Chacalla y trataron de conçertallos y los conçertaron en que los yndios de Canta dieron a los yndios de Chacalla por las dhas tierras ducientas cabeças de Ganado machos y henbras por mytad e que los dhos yndios de Chacalla se avian contentado con esto y los de Canta avian quedado con las chacarras.

Preguntado con que tierras alyndan las dhas tierras de Quybi dixo que alyndan con tierras yungas de yndios de Guancayo y con tierras de los yndios de Secos de M^{in} Piçarro el rrio en medio y yndios yayos de Chacalla por las espaldas de vnas syerras y tanbien tierras de los yndios de Canta.

Preguntado que fruto se saca e dan las dhas tierras de Quybi dixo que a lo //
/122v/ que este t^{o} a entendido las dhas tierras de Quybi se coje [*sic*] coca e axi e algunas bezes mahiz e papas e guabas e otras semyllas.

Preguntado que tantas fanegadas de senbradura podra aber en las dhas tierras de Quybi o quantos costales de coca se podran cojer dixo que le paresze a este t^{o} que se podran dar en las dhas tierras fasta diez y seys fanegadas de senbradura poco mas o menos e que no sabe la coca que se podra cojer.

Preguntado que podran baler las dhas tierras de pesos de oro o en obejas o que daria este t^{o} por ellas aviendolas menester dixo que le paresze a este t^{o} que las dhas tierras podran baler hasta çien pesos e que este t^{o} no daria mas por ellas.

Lo qual todo dixo e declaro por la dha lengua e presençia de los dhos terçeros nonbrados que lo firmaron e syendole tornado a leer se rretefyco en ello e encargosele el secreto e prometiolo.

Frai Gaspar de Carvajal — Ju^{o} de Pedraza

fho ante mi — Fran^{co} Hrrez de Guebara
escr^{o} de Su Mag.^{t}

/123r/ E despues de lo susodho en el dho asyento de Quybi a veynte dias del dho mes de hebrero del dho año los dhos Frai Gaspar de Carabajal e Juan de Pedraza terceros nonbrados para lo susodho por vyrtud y en cunplimyento de la dha provisyon rreal de Su Magestad hizieron pareszer ante si a Xpoval Caxallaupi yndio del pueblo de Guarabi del rrepartimyento de Guancayo encomyenda del contador Ju^{o} de Caçeres del qual por lengua de don M^{yn} Lengua se la rresçibio juram^{o} en forma de dr^{o} segun se rrequeria en tal caso e aviendosele hecho las preguntas al caso pertenesçientes dixo lo syg.^{e}:

Preguntado como se llama dixo que don Xpoval Caxallaupi sujeto al caçique de Guancayo llamado don Juan e que es comarcano a las dhas tierras de Quybi que el mojon dellas linda con tierras de su pueblo e que fasta su pueblo abra dos leguas.

Preguntado por las preguntas generales de la ley dixo que no //
/123v/ sabe la hedad que tiene e que syrbio a Guascaringa e le trebuto e que no le an dado ny prometido por ningun yndio ny caçique diga el contrario de la verdad e que es Xpiano e dira verdad.

Preguntado sy a visto las tierras de Quybi y en que sytio estan dixo que sabe e a visto las dhas tierras de Quybi e que a sus padres e otros viejos e ancianos que son muertos les oyo dezir e tratar muchas bezes tratando e fablando de las dhas tierras de Quybi que las dhas tierras antes que vinyesen Yngas a esta tierra heran y las tenya y poseya vn señor y que este hera sujeto a otro señor que se llamava Collycapa señor de Collique desde la mar hasta este asyento de Quybi el qual la avia tenydo y poseido hasta que vino Ynga y hecho los yndios deste señor de las dhas tierras y salyendo el dho Collicapa a defender sus tierras con sus yndios le mato el Ynga e luego dyo las //
/124r/ dhas tierras a los yndyos yabyos de Chacalla de que es agora caçique el dho don Xpoval y puso en ellas de los dhos yndios yavyos mytimaes para chasques y para cargas y para que llebasen cargas y labrasen las dhas tierras para sy e para trebuto del Inga los quales syenpre estobieron en ellas hasta que entraron los españoles y despues de benydos e tanbien fasta que los dhos yndios de Canta quysieron benir e binyeron con mano armada a hechar los yavyos de las dhas tierras diziendo que estas heran sus tierras y su rrio y entrada de sus pueblos y camynos y sus yungas y que los dhos yndios de Chacalla tenian su rrio y entrada por el rrio de Mama y sus yungas y se querian matar cada dia sobre ello fasta que el Rribera el moço encomendero de los dhos yndios de Canta y Fran^{co} de Ampuero encomendero de Chacalla binyeron a las dhas tierras a metellos en paz y trataron de conçertallos //
/124v/ a los vnos e a los otros e que el que hera caçique de Canta en aquella sazon se llamava Guaman Yavre y el de Chacalla se llamava don Fran^{co} Marcapoma y el conçierto se hizo en que los yndios de Canta dyeron a los yndios de Chacalla duçientas cabeças de ganado de la tierra machos y hembras por mytad e ansy fue publico que los dhos yndios de Chacalla las rrezibieron y se contentaron y salieron de las dhas tierras y los dhos yndios de Canta se quedaron en ellas e es pu^{co} que algunos mitimaes yungas de los naturales destas tierras se quedaron en ellas y estos heran de los que heran sujetos a los yavyos e ansy zeso las pendençias fasta que agora a entendido que los yndios yabyos se arrepienten y no quyeren pasar por el conçierto y traen pleyto.

Preguntado que tierras alyndan con estas dhas de Quybi dixo que lindan con tierras de Guancayo y con tierras de los yndios de M^{yn} Piçarro que se llama Secos y que los dhos yndios yabyos estan de las //
/125r/ dhas tierras en comarca tras un zerro y los dhos yndios de Canta ansy mysmo alindan con las dhas tierras de Quybi en otras tierras que eran de los dhos yndios yungas de Collicapa que es agora rrepartim^{o} de Muñoz de Avila que es el señor que tiene dho que poseya las dhas tierras de Quybi las quales dhas tierras dyo Ynga a los dhos yndios de Canta quando dio las de Quybi a los yndios de Chacalla.

Preguntado que fruto dan las dhas tierras de Quybi dixo que dan coca e axi e yuca e mani e maiz y otras legunbres.

Preguntado que costales de coca se podran coxer de las dhas tierras de Quybi labrandose dixo que sy se labrasen todas ellas le paresze se cojerian en cada un año çien costales de coca e que cabian de senbradura quarenta fanegas y m^{o} de mahiz.

Preguntado que pesos de oro vale cada costal de coca dixo que vale tres pesos cada costal.

Preguntado que pesos de oro o obejas de la tierra podran valer las dhas tierras de Quybi e este t^{o} //
/125v/ daria por ellas aviendolas menester dixo que no lo save e que este t^{o} daria por ellas çient obejas sy las ubiese menester e si las ubiera de bender syendo suyas las diera las dhas tierras por las dhas çien obejas.

Lo qual todo dixo e declaro por lengua del dho don M^{yn} en presençia de los dhos terçeros que lo firmaron e dixo que este hera la verdad y encargosele tenga secreto dello.

Frai Gaspar de Carvajal — Ju^{o} de Pedraza

Fho ante mi — Fran^{co} hrrez de G^{a}
escr^{o} de Su Mag.^{t}

/126r/ E despues de lo susodho en el dho pueblo de Quybi este dho dia los dhos terçeros para aberiguaçion de lo que se les manda en la dha probisyon hiçieron parescer ante si a don Felipe Taulechunbi prençipal que esta en el dho valle e sytio de Quybi e por Lengua de don M^{yn} Lengua ynterprete probeido por los dhos señores oidores fue del rreçibido juramento en forma de dr^{o} segun se rrequeria en tal caso el qual lo hizo conplidamente e rrespondio: "sy juro e amen" e siendole por los dhos terçeros hechas las preguntas al caso pertenesçientes dixo lo syg.^{e}:

Preguntado como se llama dixo que don Felipe Taulechunbi.

Preguntado de que parçialydades e quyen es sujeto dixo que es yunga de los naturales que rresyden en el pueblo de Quybi sobre que es este pleyto e que antes que se conçertase el caçique de Canta con los de Chacalla sobre las dozientas obexas serbia este t^{o} con los dhos yndios yavyos a Fran^{co} de Anpuero encomendero de Chacalla porque poseyan los dhos yndios yavyos //
/126v/ sus chacarras deste t^{o} e de sus yndios e que despues que se conçertaron esta en sus tierras y serbe a Rribera el moço con los de Canta e que primero quando servia Anpuero tenyan algunas chacarras los yndios de Canta que eran tanbien de los yungas naturales y tanbien servian por cavsa de las dhas chacarras que tenyan los yndios de Canta a los de Canta.

Preguntado por las preguntas generales de la ley dixo que no sabe la hedad que tiene e que se acuerda de quando gobernaba Guaynacaba que era ya buen mochacho e que no le an dado ny prometido cosa alguna por los dhos yndios ny por alguna persona para que diga el contrario de la verdad ny tanpoco le ba ynteres en este pleyto ny se le da mas que vença una parte que otra e que dira la verdad.

Preguntado en que sytio estan las dhas tierras de Quybi sobre que es este dho pleyto dixo que sabe por cosa publica entre todos los viejos e ançianos desta comarca ansy que son ya muertos como a los que son bibos //
/127r/ e a sus padres e abuelos deste t^{o} a quyen este t^{o} lo oya dezir e tratar muchas e dibersas bezes tratando sobre las dhas tierras de Quybi sobre que es este pleyto que las dhas tierras antes que en ellas entrasen Yngas las tubo e poseyo vn señor que llamavan Collicapa que hera senor de Collyque e tenya senorio en todo este valle desde la mar hasta este asyento de Quybi el qual la tubo e poseyo çierto tiempo hasta que bino Ynga e le dio guerra a el dho Collicapa e a sus yndios e mato al dho Collicapa e se apodero en las dhas tierras de Quybi y las dio a los yndios yavyos encomyenda de Fran^{co} de Anpuero porque no quedaron yndios yungas y puso en ellas mytimaes de los yndios yabyos para chasques y para que le cojiesen coca e axi e llebasen cargas y le syrbiesen e que a los yndios cantas señalo otro çierto pedaço de tierra de los yndios yungas para que la syrbyesen en ellas como los yavyos e que vido este t^{o} que los dhos yndios yavyos estobieron en la dha posesyon que les metio Ynga de las dhas tierras fasta que entre el caçique de los //
/127v/ yndios de Canta que llamavan Guaman Yavre y don Fran^{co} Marcapoma caçique prençipal del rrepartimyento de Chacalla trataron de conçierto y enbiaron a llamar para ello sus encomenderos que fue el caçique de Chacalla a Fran^{co} de Anpuero y el caçique Canta [*sic*] a Nycolas de Rribera el moço los quales binyeron y los conzertaron en que el caçique de Canta diese a el de Chacalla dozientas obejas machos y henbras por que les dexase las dhas tierras e ansy se las dieron y los dhos caçiques de Chacalla dexaron las dhas tierras y los de Canta por rrazon del dho conçierto se entraron en ellas y las tobieron hasta que el caçique de Chacalla se arrepintio y no quyeren pasar por el conçierto y traen pleyto e que tanbien sabe e a visto este t^{o} que en çierto pedaço de chacarra que esta junto a las casas del pueblo de Quybi an senbrado y sienbran despues del conçierto çiertos yndios yabyos de los de la encomyenda de Chacalla coca y la benefiçian y la cojen por-//
/128r/ que se las dio vn hermano del dho Guaman Yavri caçique de Canta que llamaban Guaman Yauri Mychuy y tanbien el dho Guaman Yavre prençipal porque los tenyan por amygos.

Preguntado con que tierras alindan las dhas tierras de Quybi dixo que alindan con tierras de Canta y de Guancayo encomyenda de Gonçalo de Caçeres y por otra parte Secos que es rrepartimyento de M^{yn} Piçarro y vna syerra pelada que esta entre Chacalla y las dhas tierras de Quybi.

Preguntado que fruto se da y coje en las dhas tierras de Quybi dixo que se coxe coca e axi e otras cosas e plantas.

Preguntado sy la dha tyerra se senbrase toda de coca que costales se coxerian cada ano o si se senbrase de mahiz o trigo que fanegadas de senbradura daran las dhas tierras dixo que le paresze que se podrian cojer noventa costales de coca e que sy se ubiese de senbrar las dhas tierras de mahiz o trigo no lo save.

Preguntado que a como bale 1 costal de coca dixo que a dos pesos.

/128v/ Preguntado que pesos de oro podran baler las dhas tierras de Quybi sobre que este pleyto o sy les obiese de comprar que daria por ellas o sy las ubiese de vender syendo suyas por que presçio las daria dixo que no sabe lo que podran valer las dhas tierras mas de que sy fuesen suyas no las daria por nengun dinero ny otra cosa que le diesen.

Lo qual todo dyxo e declaro por la dha lengua en presençia de los dhos terceros que lo firmaron de sus nonbres e siendole tornado a leer se afirmo en ello por la dha lengua; encargosele tenga secreto y prometiolo.

Frai Gaspar de Carvajal — Ju^{o} de Pedraza

fho ante mi — Fran^{co} Hrrez de Guebara escr^{o} de Su Mag.^{t}

/129r/ E despues de lo susodho en el dho pueblo de Quybi a veynte dias del dho mes de hebrero del dho año el dho Frai Gaspar de Carabajal y Juan de Pedraza terçeros nonbrados por los yndios de Chacalla y de Canta sobre las dhas tierras de Quybi para ynformaçion del susodho contenydo en la dha provisyon rreal hizo paresçer ante sy a Gonçalo Limacho yndios que dixo ser del rrepartim^{o} de Canta y lengua dellos del qual por la dha lengua se rresçibio juramento en forma de dr^{o} segun se rrequeria en tal caso el qual hizo el dho juram^{o} e syendole hechas las preguntas al caso pertenesçientes declaro lo sig.^{e}:

Preguntado como se llama y de donde es natural dixo que se llama Gonçalo Limacho e que es natural del pueblo de Carua del rrepartimyento de Canta encomendado en Nycolas de Rribera.

Preguntado por las preguntas g^{ales} de la ley dixo que es de hedad de mas de veynte años e que se acuerda de quando entraron los xpianos en esta tierra y tanbien del Inga e que le ba ynteres en esta cavsa pero que por eso no dexara de dezir verdad e que nenguna persona le a dho ny puesto myedo ny ofreçido cosa alguna para que lo dexe de dezir.

/129v/ Preguntado en que sytio estan las dhas tierras de Quybi sobre que es este pleyto dixo que lo que este dho t^{o} a oydo y entendido de muchos yndios viejos e ancianos de Canta y desta comarca que son muertos e a otros que son bibos e lo mysmo a sus padres e abuelos deste t^{o} a quyen este t^{o} lo oya tratar e platicar muchas vezes antes que entrasen los Yngas en esta tierra las dhas tierras de Quybi heran sytio de yndios yungas que su senor es ya muerto y tiene un hijo bibo que se llama Quybi el qual la tuvo y poseyo y daba trebuto a un caçique que llamavan Collycapa senor de Collique que es agora rrepartimyento de Muñoz de Avila y tenyendola y poseyendola entro en las dhas tierras Ynga y mato los yungas que pudo y los hecho de las dhas tierras y puso en ellas yndios yabyos del rrepartimyento de Chacalla encomendados en Fran^{co} de Anpuero y puso en ellas de los dhos yndios yavyos para mytimaes para que syrbiesen al Inga e para chasques y llevar cargas y otras cosas de su serviçio los quales estobieron çiertos años contradiziendolo los dhos yndios de Canta los quales procuravan de hechallos a los dhos yndios yabyos de las dhas tierras y sobre ello //
/130r/ tenya grandes pendençias y estavan cada dia en arma para se matar sobre las dhas tierras hasta que savido lo susodho por Nycolas de Rribera el moço encomendero de los dhos yndios de Canta y por Fran^{co} de Anpuero encomendero de los yndios de Chacalla trataron de conçierto entrellos y el dho conçierto se hizo y hefetuo en que los dhos yndios de Canta dieron a los dhos yndios de Chacalla dozientas cabeças de ganado de la tyerra çiento henbras y çiento machos las quales vido este dho t^{o} que traxeron los dhos yndios de Canta a los de Chacalla para que se fuesen con ellas y Guaman Yavre caçique prençipal de Canta y los demas yndios le dixeron a este dho t^{o} como los dhos yndios de Chacalla se dieron por entregados de las dhas obejas y los dhos yndios de Canta se quedaron con las dhas tierras y las tobieron y poseyeron syn que los dhos yndios de Chacalla se quexasen ny lo contradixesen fasta que vido que le pidieron por pleyto.

Preguntado con que tierras alindan las dhas tierras de Quybi dixo que alindan con yndios de M^{yn} Piçarro e con tierras de Canta e con un çerro de Chacalla.

/130v/ // Preguntado que fruto e aprovechamyento dan las dhas chacaras dixo que dan yuca e coca e otras cosas de axi e papas.

Preguntado que pesos de oro podran baler las dhas tierras de Quybi sobre que este pleyto dixo que le paresze a este dho t^{o} que podran valer las dhas doçientas obejas que dieron los dhos yndios de Chacalla.

Preguntado sy las dhas tierras estobiesen todas senbradas de coca quanta coca se podria cojer dellas en cada un año dixo que le paresze a este t^{o} que darian las dhas tierras fasta çinquenta costales della syendo la coca bien pisada e llena en los çestos.

Preguntado que vale cada cesto de coca dixo que los cestos de coca que son de la medida valen a tres pesos y los otros a dos pesos e que no vale mas; lo qual todo dixo e declaro por lengua del dho don M^{yn} ynterpetre e se le torno a leer y afirmose en ello y los dhos terçeros en cuya presençia se hexsamyno lo firmaron de sus nonbres.

Frai Gaspar de Carvajal — Ju^{o} de Pedraza

fho ante mi Fran^{co} Hrrez de G^{a} escr^{o} de Su Mag.^{t}

/131r/ Pareszer del padre Carabajal
y de Juan de Pedraza terceros.

E despues de lo susodho en el dho pueblo de Quybi a veynte dias del mes de hebrero de myll e quinyentos e çinquenta e nueve años ante my Fran^{co} Hernandez de Guebara escr^{o} de Su Magestad y rrecevtor nombrado para la presente cavsa pareșçio el muy rr^{do} padre Frai Gaspar de Carabajal provinçial de la Horden de señor Santo Domyngo de los pedricadores [*sic*] e del monest^{o} que de la dha Horden rresyde en la çibdad de Los Rreyes e Ju^{o} de Pedraza terceros nonbrados entre los caçiques e yndios del rrepartimy^{o} de Canta y del rrepartimyento de Chacalla el dho Frai Gaspar de Carabajal nonbrado por parte de los yndios de Chacalla y el dho Juan de Pedraça nonbrado por el caçique e yndios de Canta e unanymes conformes nemyne discrepante dixeron que ellos avian rreszibido la ynformaçion que Su Magestad por su rreal provisyon les manda e avian bisto e myrado las dhas tierras e medidolas e ynformadose de los yndios e bisto e tanteado e consyderado lo que facia al caso syn fazer agrabio a laz partes a lo que avian alcançado en sus conciençias les paresce y fallan que las dhas tierras de Quybi sobre que este pleyto tyenen de sytio dos myll pasos de largo y tresçientos de ancho e tienen por linderos//
/131v/ por una pte el rrio y por la otra camyno rreal [*sic*] que viene de Lima al dho pueblo de Canta esto por lo ancho y de lo largo tyenen por linderos las tierras de Guancayo rrepartimyento de Gonçalo de Caçeres e de la otra un alinde que viene desde una quebrada que baxa de un zerro de la otra banta [*sic*] del rrio que esta en tierra de los yndios de Secos encomendados en M^{yn} Piçarro la qual linde corre al sesgo fasta llegar a la yglesia nueba del pueblo de Quybi en parejo della / las quales dhas tierras estan sytiadas conforme a los dhos de los t^{o}s por ellos tomada en termynos de yndios yungas que fueron los primeros poseedores antes que en ellas entrasen Yngas que fueron sujetos a los yndios de Collique las quales dhas tierras paresçe avellas dado el Inga a los yndios yabyos del rrepartimyento de Chacalla y las poseyeron fasta que las bendyeron por conçierto a los yndios de Canta por docientas ovejas y por virtud desta venta las poseen e que en Dios y en sus conçiençias les paresze que sy las dhas tierras estobieran en poder de españoles y las vendieran podrian valer fasta seisçientos o seteçientos p^{o}s de oro pero que entre los yndios se estiman las tierras que estan en su comarca de otra manera que las de los españoles especialmente siendo tierras de coca y por esto no podran //
/132r/ determynadamente juzgar el valor de las dhas tierras entre los yndios porque no tyenen otro mantenymy^{o} syno el que de las dhas tierras sacan labrandolas y sembrandolas y esto es su pareszer y lo firmaron de sus nonbres.

Frai Gaspar de Carvajal Ju^{o} de Pedraza
fho ante mi Fran^{co} Hernandez de Guebara
escr^{o} de Su Mag.^{t}

E yo Fran^{co} Hernandez de Guebara escr^{o} de Su Mag.^{t} e rrezebtor nonbrado para lo que dho es p^{e} fui a la dha declaraçion e t^{o}s e avtos e lo escrebi e hize aqui my s^{o} tal en tsm^{o} de verdad.

Fran^{co} Hrrez de G^{a}
escr^{o} de Su Mag.^{t}

Tasaçion de costas de cada pte las sig^{es}:

- de un rrequerim^{o} a Pedraça:quatro ts.		4 ts
- de la present^{on} de la p^{o}vision: tres ts.		3 ts
- de otro rrequerim^{o} al padre Caravajal.		4 ts
- de siete t^{o}s juram^{o}: catorze ts.	1 p^{o}	6 ts
- del parescer: seis ts.		6 ts
- de diez y ocho hojas: nueve p^{o}s.	9 p^{o}s	
- del signo: dos ts.		2 ts
Montan: treze p^{o}s e un tomyn. Rrescebilos.	13 p^{o}s	1 t.

Fran^{co} Hrrez

/133r/ En la çibdad de Los Rreyes quynze dias del mes de hebrero de myll e quy^{o}s e çinquenta e nueve años por parte de don Felipe prençipal de Chacalla y de los caçiques de Canta yo Fran^{co} Hernandez de Guevara fue rrequerido con una p^{o}bisyon rreal e rrecevtoria de Su Mag^{t} a.my cometida para que la guarde y cunpla que su tenor de la qual es el syg.^{e}

E ansy hecho el dho rrequerimy^{o} yo el dho escr^{o} tome la dha p^{o}vision rreal e la obedesci e puse sobre my cabeça como hera obligado e dixe que estava presto de la conplir e me parti oy dho dia para entender en lo susodho.
T^{o}s: Fran^{co} de la Torre e Ju^{o} Sanchez.

Fran^{co} Hrrez de G^{a}
escr^{o} de Su Mag.^{t}

secr^{o} Jhuan Muñoz Ryco

Rreçeptoria a Fran^{co} Hernandez que va por rreçeptor en el negoçio entre los yndios de Chacalla y Canta a pedim^{o} de anbas ptes con dos p^{o}s y medio de salario demas de sus derechos.
/133v/ Don Phelipe por la graçia de Dios Rrey de Castilla de Leon de Aragon de Ynglaterra de Françia de las dos Çeçilias de Jherulasen de Navarra de Granada de Toledo de Valençia de Galizia de Mallorcas de Sevilla de Çerdeña de Cordova de Corçega de Murçia de Jaen

de los Algarves de Algezira de Gibraltar de las yslas de Canaria de las Yndias yslas e tyerra firme del mar oçeano; conde de Barçelona; señor de Vizcaya e de Molina; duque de Atenas e de Neopatria; marques de Oristan y de Goçiano; archiduque de Austria; duque de Borgoña y de Brabante y Milan; conde de Flandes e de Tirol etc.

A vos Fran^{co} Hernandez nro escrivano salud y graçia. Sepades que pleito esta pendiente y se trata en la nra Audiençia y Chançilleria Rreal que rreside en la çiudad de Los Rreyes de los nros rreynos del Piru; ante el nro presidente e oydores della entre ptes de la vna el caçique e yndios de Chacalla y Alonso Moreno su procurador e de la otra el caçique e yndios de Canta y Fran^{co} de la Torre su procurador / sobre las tyerras y chacarras de Quibi / en el qual por los dichos nro presidente e oydores fue dada sent.^{a} difinitiva en que absolvieron a los dichos caçique e yndios de Canta de la demanda contra ellos puesta por los de Chacalla de la qual ffue suplicado por el dicho Alonso Moreno en los dichos nonbres y la causa fue conclusa y por los dichos nro presidente y oydores visto lo alegado por pte del dicho Fran^{co} de la Torre en los dichos nonbres por un auto probeyeron y mandaron que cada una de las dichas partes nonbren terçero de su parte que vean por vista de ojos las dichas tyerras y el sitio donde estan y se ynformen de yndios antiguos mas çercanos a ellas de lo que pueden valer y les tomen sus dichos y depusiçiones sobre ello y den en todo su paresçer de lo que mas convenga probeer que sea en menos daño y perjuyzio de las dichas partes; despues de lo qual lo rresçibieron a prueva en la dicha causa con termino de veinte dias primeros siguientes que comiençan a correr desde veinte y çinco dias del mes de henero del año de la fecha desta / y agora nos suplicaron y pidieron por mrd los procuradores de las dichas partes que a ello ynbiasemos rreçeptor de Nra Corte que hiziese las dichas probanças e ante quien pasase la dicha averiguaçion que los dichos terçeros hizieren por no aver justiçia ante quien lo poder pedir que ellos pagarian el salario que por nos les fuere señalado o como la nra mrd fuese / lo qual visto por los dichos nro presidente e oydores confiando de vos que sois tal persona qual para ello conviene fue acordado que deviamos mandar dar esta nra carta para vos en la dicha rrazon y nos tovimoslo por bien / por la qual vos mandamos que luego que con ella fueredes rrequerido vays al dho valle y tierras de Quibi y rreçibays la averiguaçion que los terçeros que las dichas partes nonbraren hizieren del sitio y valor de las dichas tierras y las probanças que çerca dello y de lo demas por ellos y cada uno dellos alegado ante vos dieren y los testigos que para ello presentaren les preguntareys que hedad han y de donde son naturales y por las demas generales / y despues por las preguntas de los ynterrogatorios que ante vos seran presentados firmados de Diego Muñoz nro escrivano de camara y al testigo que dixere que sabe la pregunta le preguntad como la sabe e al que oyo dezir lo en ella contenido que a quien e quanto tpo a e al que lo cree que como y porque lo cree por manera que cada uno dellos de rrazon sufiçiente de su dicho y depusiçion y todo ello escrito yn linpio e signado y firmado de vro nonbre çerrado y sellado en manera que haga fee lo dareys a la parte a quien tocare para que lo traya y presente en la dicha nra Rreal Audiençia para en guarda de su dr^{o} y mandamos a las personas que çerca dello fueren presentados por testigos que vengan y parescan ante vos a jurar y dezir sus dichos y depusiçiones en ello //

/134r/ a los plazos y so las penas que de nra parte les pusieredes las quales nos por la presente les ponemos y avemos por puestas y por condenados en ellas lo contrario haziendo / en lo qual todo os ocupeys los dichos veinte dias que corren y se quentan desde el dho dia veynte y çinco de henero desde presente año de la fecha desta e ayays y lleveys en cada uno dellos de salario dos p^{o}s y medio los quales vos sean dados y pagados por cada una de las dichas partes segun el tpo que por su pte os ocuparedes y los cobrareys dellos y de cada uno dellos demas y allende de los derechos que por la escritura que en ello hizieredes y ante vos pasare yobieredes de aver conforme al aranzel los quales y el dicho salario asentareys al fin de las dichas probanças para que se sepa lo que les llevardes y lo que cada una de las partes pagare porque lo que de otra manera se os diere lo pagareys con el quatro tanto para nra camara; para lo qual todo asi hazer y cunplir vos damos poder cunplido como en tal caso se rrequiere y no fugades endeal.

Dada en la çiudad de Los Rreyes a veinte e siete dias del mes de henero de myll y quinientos y çinquenta y nueve años.

Yo Joan Muñoz Rico scr^{o} de camara de Su Sacr^{a} Cath^{a} Rreal Mag.^{d}; la fize escribyr por su m^{do} con acuerdo de su p^{te} e oydores.

Rr^{or} Sanchez Vallejo — por chançiller

Juan Muñoz Rico

/134v/ En Los Rreyes seys dias del mes de hebrero de mill e quy^{o}s e çinq^{ta} e nueve años yo Diego Muñoz Ternero escr^{o} de Sus Mag^{es} e de camara en su Rreal Audi^{a} de pedim^{o} de Alonso Moreno en nonbre de los caçiques e yndios de Chaclla notifyque esta probision rreal e çite conforme a ella a Fran^{co} de la Torre pr. de los yndios de Canta siendo dello t^{o}s: G^{o} de Paredes e Joan de Segovia.

En fee de lo qual fize aqui mi signo en testim^{o} de verdad.

Diego Muñoz

/135r/ *Interr^{o} de los indios de Chacalla.*

Por las preguntas siguientes se esaminen los testigos que son o fueren presentados por parte del caçique e yndios de Chacalla en el pleyto con el caçique e yndios de Canta sobre las tierras de Quibi:

1 - Primeramente sean preguntados si conosçen a las partes y si tienen notiçia de las dhas tierras de Quibi.

2 - Yten si saben etc. que al tienpo que dizen aver el dho don Françisco caçique que dizen aver sido de Chacalla el conçierto de las dozientas ovejas de la tierra por las tierras y chacaras de Quibi fue a ello forçado y conpelido el dho don Françisco por Françisco de Anpuero su encomendero que le hizo por fuerça y con temores y amenazas y malos tratamientos le hizo que se hallase presente al conçierto sin que el dicho don Françisco lo hiziese ni quisiese. Digan lo que saben.

3 - Yten si saben etc. que estando presente y contradiziendo el dicho conçierto y no lo queriendo hazer el dicho don Françisco de Chacalla lo hizo y firmo el dicho Françisco de Anpuero su encomendero y se tomo para si y llevo las dozientas ovejas de la tierra que dizen aver dado los yndios de Canta y no las dio ni entrego al dicho don Françisco ni a los prinçipales ni yndios de Chacalla antes las tomo y llevo y como cosa suya hizo dellas lo que quiso sin voluntad ni consentimiento de sus yndios.

4 - Yten si saben etc. que sienpre el dho don Françisco de Chacalla contradixo el dicho conçierto y dezia que no queria ovejas sino las tierras de Quibi que heran de sus yndios y no las podia el vender ni trocar a ovejas y avnque lo dezia asi el dicho Françisco de Anpuero rresçibio las ovejas y firmo el papel que los de Canta quisieron.

5 - Yten si saben etc. que Su Magestad tiene proveydo en todas las Yndias y en este rreyno por sus rreales provisiones y çedulas que los testigos han visto originalmente que los yndios y tierras y todas las demas cosas //
/135v/ esten de la manera que las hallaren los españoles quando entraron en la tal tierra sin que en nada aya ni se consienta novedad alguna. Digan los testigos lo que saben y si a sido en todas las Yndias usado y guardado lo susodho por virtud de la dicha provision rreal de Su Magestad.

Yten si saben etc. que todo lo susodicho es publica voz y fama.

Diego Muñoz

/136r/ Por estas preguntas demas de las presentadas se exsaminen los testigos que son o fueren por parte del caçique e yndios de Chacalla en el pleyto con los de Canta sobre las tierras de Quibi:

1 - Yten si saben que los dhos yndios de Chacalla e sus caçiques an poseydo las dhas tyerras de Quibi e tenydolas como cosa suya desde el tienpo de los Yngas pasados que se las dieron lo qual ha tanto tpo que no ay quien se pueda acordar dello.

2 - Yten si saben etc. que los dhos yndios de Chacalla tenyan e poseyan las dhas tierras de Quibi al tpo que los españoles entraron en este rreyno y las tuvieron mucho tienpo despues de entrados. Digan lo que saben.

3 - Yten si saben etc. que los yndios de Canta porque no se alçaron juntamente con ellos los dhos yndios de Chacalla e porque servian a los españoles como lo hazian al tienpo que se alçaron los naturales en la tierra mataron çiento çinquenta yndios de Chacalla que estavan en las dhas tierras de Quybi y se las quytaron e tomaron e desde estonçes se las tienen tomadas e husurpadas los dhos yndios de Canta. Digan lo que saben de lo dho.

4 - Yten si saben etc. que las dhas tierras de Quibi que los dhos yndios de Canta tienen tomadas a los de Chacalla valen mas de ocho myll pesos al presente y los //
/136v/ valian de diez años a esta parte. Digan los testigos lo que valian agora diez años y lo que valen agora.

El Lic.^{do} Marcos Luzio — Alonso Moreno

conçertado con el original — Diego Muñoz

En el pueblo de Quybi a diez y siete dias del mes de hebr^{o} de myll e quy^{o}s e çinquenta e nueve años ante my el escr^{o} e rrezevtor parescio don Felipe prençipal por sy y en nonbre de los yndios de Chacalla e presento dos ynterrogatorios del tenor syg^{e}:

E ansy presentados yo el dho escr^{o} los rreszebi e dyxe que presentando los t^{o}s que yo estoy presto de los rreszebir e que sy algun dia estobiere por rrazon de no los presentar detenydo les cobrare de sus vienes.
T^{o}s: don M^{yn} y el padre Frai Pablo.

Fran^{co} Hrrez de G^{a}
escr^{o} de Su Mag.^{t}

En el pueblo de Quybi a veynte dias del mes de hebr^{o} del dho año ante my el dho escr^{o} p^{o} el dho don Felipe caçique de Chacalla que dixo ser por sy y en el dho nonbre presento por t^{o} a don Di^{o} Condor natural que dixo ser del pueblo de Chacalla el qual por lengua de don M^{yn} ynterpetre e por my el escr^{o} le fue tomado juram^{o} en forma por Dios e por Santa M^{a} e por una señal de cruz el qual hizo el dho juram^{o} conplidamente a lo que dixo la lengua e rrespondio: "sy juro e amen".

ante mi — Fran^{co} Hrrez de G^{a}
escr^{o} de Su Mag.^{t}

/137r/ En Quybi a veynte e tres dias del mes de hebr^{o} del dho año por pte de los yndios de Chacalla se p^{o} a Ju^{o} Chauca yndio natural que dixo ser de Guadacheri e por la dha lengua juro segun que el de suso e Anton Cachavra es de la dha encomyenda.

En Quybi a veynte e quatro dias del dho mes de hebr^{o} del dho año ante my el escr^{o} por pte del caçique de Chacalla se p^{o} por t^{o} a Alonso Paucar yndio natural de Palli del rrepartim^{o} de Chacalla y por la dha lengua juro segun que el prim^{o}.

En Quybi a veynte e seis dias del mes de hebr^{o} del dho año ante my el escr^{o} por pte de los yndios de Chacalla se pregunto a Ju^{o} Yabre yndio natural del pu^{o} de Acuchai encomyenda de Chacalla y por la dha lengua juro segun que el de suso.

Este dho dia veynte y seis de hebr^{o} del dho año por pte de los dhos yndios de Chacalla se presento por t^{o}s a Ju^{o} Yndio del pu^{o} de Cullata encomyenda de Chacalla e a M^{yn} Yusco del pu^{o} de Chicamarca de la encomyenda de Chacalla y por la dha lengua juraron segun que los de suso.

ante mi Fran^{co} Hrrez de G^{a}
escr^{o} de Su M^{t}.

/138r/ *Probança hecha por pte de los y^{o}s del rrepartimi^{o} de Chacalla.*

Don Di^{o} Condor prinçipal del valle de Chacalla / sujeto a don Xpobal / de mas de ochenta.

T^{o} el dho don Diego Condor natural de Chuya que es en el valle e rrepartimyento de Chacalla sujeto a don Xpoval Michica Mychufe [*sic*] caçique prençipal de el dho rrepartimyento de Chacalla t^{o} presentado por parte de el dho don Xpoval aviendo jurado en forma devida de dr^{o} segun segun se rrequeria [*sic*] en tal caso e syendo preguntado por el tenor de el dho ynterrogatorio por donde se pidio fuese hexsamynado por lengua de don M^{yn} Lengua ynterpetre nombrado por la Rreal Audiençia dixo e depuso lo sig.^{e}:

1 - A la primera pregunta de el dho ynterrogatorio dixo este dho t^{o} que conosze el dho don Xpoval Bilca Mychuri caçique de el rrepartimyento de Chacalla porque como dho tyene es su caçique e a los demas yndios del dho rrepartimyento e ansy mysmo conosze e a conoszido a los caçiques e gobernadores y senores que an sido y son del dho rrepartimyento de Canta por trato y conversacion que con ellos a tenydo e tiene e con sus yndios syendo como este dho t^{o} es comarcano a ellos e ansy mysmo tiene notiçia de las dhas tierras de Quibi sobre que agora se trata pleyto porque las a visto y estado en ellas y paseadolas muchas e dibersas vezes e un aguelo deste //
/138v/ testigo que es ya muerto a sembrado en ellas muchas e dibersas vezes.
Preguntado por las preguntas generales de la ley dixo este dho t^{o} que es de hedad de mas de ochenta anos poco mas o menos tiempo e como dho tiene es sujeto al dho rrepartimi^{o} e caçique de Chacalla e tiene hijos e nyetos que tanbien son sujetos e que le ba el mysmo ynteres que a los otros yndios del dho rrepartimyento pero que por eso no dexara de dezir verdad porque es xpiano e que su caçique ny otra persona no le a dado cosa alguna para que diga el contrario de la verdad.

2 - A la segunda pregunta de el dho ynterrogatorio dixo este dho t^{o} que lo que de la dha pregunta sabe es que quando vino el presydente Gasca a estos rreinos dende a ciertos dias vino a las dhas tierras de Quibi sobre que se trata este dho pleyto Fran^{co} de Ampuero encomendero del dho rrepartimyento de Chacalla y tanbien vio que bino con el Nicolas de Rrybera el moço encomendero de los yndios de Canta y llegados que fueron a las dhas tierras tierras de Quibi mandaron juntar a los yndios viejos de Chacalla los quales se juntaron y los dhos Fran^{co} de Ampuero y Nycolas de Rribera les hizieron vna pla-//
/139r/ tica en que les preguntaban que cuyas heran las dhas tierras de Quibi y los dhos yndios viejos les rrespondieron que las dhas tierras heran suyas y de sus antepasados que las avian ganado peleando por guerra y antes que entrasen Yngas en esta tyerra e que al tiempo que entraron los Yngas los fallaron en ellas e otras cosas que les platicaban e que el dho Nycolas de Rribera decia por lengua de su cacique Guaman Llavre que las dhas tierras heran de los yndios de Canta y que en estas porfias estobieron quatro o çinço dias e al cabo dellos el dho Fran^{co} de Ampuero hablo al dho Fran^{co} Marcapoma su cacique e a otros prençipales e yndios y les dixo que el tenya concertado que los yndios de Canta diesen al dho don Fran^{co} Marcapoma cacique de Chacalla duzientas obejas de la tierra machos y hembras por que dexasen las dhas tierras y les diesen al caçique e yndios de Canta y el dho don Fran^{co} Marcapoma rrespondio que no queria fazello el dho concierto porque las dhas tierras heran suyas y de sus yndios y pasados e que sy el hiciese el dho concierto y bendiese las dhas tyerras que despues del muerto lo pedirian sus hijos e que los yndios de el dho su rrepartimyento no querrian consentillo //
/139v/ y el dho Fran^{co} de Ampuero le rrenya por ello al dho al dho su cacique diziendo que tomase las dhas obejas y dexase las dhas tierras a los dhos yndios de Canta que era mejor y que ellos se fuesen a sus tierras y que era mas mejor tomar las dhas obejas que no quedar con las tyerras que eran rruines y sin provecho y tenya mucho cascaxo e piedra y el dho don Fran^{co} dezia que no queria fazer el dho concierto y el dho Fran^{co} de Ampuero se enojaba y tornaba a dezir al dho don Fran^{co} que lo hiziese y rreszibiese las dhas obejas y el dho don Fran^{co} no queria e dezia que no queria obejas synos sus tierras porque las obexas se moririan e gastarian e las tierras que nunca se acababan e que el no queria rresçibir las dhas obejas que las rresçibiese el y lo mysmo dezian muchos yndios de el dho rreparty:yento de Chacalla y se quexavan y el dho Fran^{co} de Ampuero como el dho don Fran^{co} no queria fazer el dho concierto rresçibio las dhas obejas y las fue a rresçibir a un pueblo de Canta que se llama Cullata y el dho Fran^{co} de Ampuero las hizo llebar a su casa y dio las dhas tierras a los dhos yn-//
/140r/ dios de Canta para que ellos las senbrasen y tubiesen por suyas eçepto que dejaron para los dhos yndios yabyos çierta parte de unas chacarrillas que estan junto al pueblo de Quibi y pusyeron una cruz los dhos Fran^{co} de Ampuero y Nycolas de Rribera por senal que eran desde la dha cruz façia arriba al dho pueblo de Canta y todo lo que dho tiene dixo que lo sabia este dho t^{o} como persona que a todo se hallaba presente y lo vio este dho t^{o} ser y pasar ansi por bista de ojos e que ansy mysmo vido que quando el dho Fran^{co} de Ampuero rresçibio las dhas obejas e dio las dhas tyerras de Quibi a los yndios de Canta muchos yndios de los del dho rrepartimyento de Chacalla se quexavan y lloravan diziendo que les avian quitado sus tierras y hechavan muchas

maldiciones al dho Fran^{co} de Ampuero y dezian los yndios que el dho Fran^{co} de Ampuero avia hecho lo susodho por quedarse con las obexas y tomallas para si y esto rrespondio por la dha lengua por estas propias rrazones en sustancia segun declara la lengua.

3 - A la tr^{a} pregunta de el dho ynterrogat^{o} dixo este dho t^{o} que dize lo que dho tyene en las preguntas antes desta e que sabe e bido que el dho Fran^{co} de Ampuero su encomendero tomo //
/140v/ las dhas obexas para si e las enbio a Lima a su casa con un criado suyo que se llamava Molina.

4 - A la quarta pregunta de el dho ynterrogat^{o} dixo este dho t^{o} que dize lo que dho tiene en las preguntas antes desta e que se acuerda este t^{o} que el dho don Fran^{co} Marcapoma caçique de Chacalla al tyempo que el dho Fran^{co} de Ampuero hizo el dho concierto lloro y dixo al dho Fran^{co} de Ampuero: "este conçierto yo no lo hago syno vos y Rribera español con español amygos" y esto rresponde.

E dixo ansy mysmo que vido este dho t^{o} como los caciques e yndios yavyos de Guadacheri y los de doña Maria Martel binyeron a rrenir al dho don Fran^{co} Marcapoma caçique de Chacalla y le decian que para que bendia las tierras que el Ynga le avia dado e amojonado e hazia tan gran maldad y el dho don Fran^{co} Marcapoma les dezia que el no lo avia bendido syno el dho Fran^{co} de Ampuero su encomendero y quando le fablavan desto y le rrenyan entonzes se torcia mas el dho don Fran^{co} y se quexaba y decia que lo avia de pedir y quexarse y siempre lo decia fasta que murio que abra quatro o çinco anos.

5 - A la quynta pregunta del dho ynt^{o} dixo este dho t^{o} que lo cont^{do} en la pregunta lo a //
/141r/ oydo dezir a don Xpoval cacique e que a visto y entendido que todos los mytimaes que puso el Inga en esta tyerra se estan donde los puso el Ynga porque alli los fallaron los españoles.

6 - A la sesta pregunta de el dho ynt^{o} dixo este dho t^{o} que dize lo que dho tiene y que es la berdad lo qual dixo e declaro por lengua del dho don M^{yn}.

Añadidas

1 - A la primera pregunta del dho ynterrog^{o} añadido dixo este dho t^{o} que dize lo que dho tiene en la segunda pregunta del primero ynterrogatorio e que es verdad que siempre los dhos yndios yavyos poseyeron las dhas tierras hasta que el dho Fran^{co} de Ampuero y Nycolas de Rribera hizieron el dho concierto.

2 - A la segunda pregunta de el dho ynterrogatorio dixo que dize lo que dho tyene en las preguntas antes desta e que este t^{o} bio que al tyempo que los españoles entraron en este rreino los dhos yndios yavyos tenyan e poseyan las dhas tierras de Quybi hasta que se hizo el dho concierto.

3 - A la tr^{a} pregunta de el dho ynterrogatorio dixo este dho t^{o} que este t^{o} se acuerda muy bien y lo bido desde un peñon do estava myrando lo que pasava que aviendo //
/141v/ los yndios de Canta puesto mal con los Yngas a los yndios yavyos y levantadoles un testimonyo al tpo que se alçaron los Yngas los dhos yndios de Canta llebaron por engaño çiento e çinquenta e quatro yndios cargados diziendoles que los llebaban a Canta que estaba alli su caçique y los metieron en un corral junto al pueblo de Quibi y alli los mataron a todos y este dho t^{o} lo vio de ençima de un peñal.

4 - A la quarta pregunta del dho ynterrogatorio dixo este dho t^{o} que save que las dhas tierras de Quybi son muy buenas e que valen mas de quatro myll pesos e que si fuesen suyas no las daria por ellos ni por nengun dinero porque es tierra que si syembran en ellas mahiz se coxe mucho mahiz y se syenbran coca coxen mucha coca lo qual todo que dho tiene dixo e declaro por lengua de el dho don M^{yn} Lengua ynterpetre nonbrado para esta cavsa por los senores oidores e dixo que era la verdad e aviendosele tornado a leer y declarar por la dha lengua se afirmo en ello y encargosele que tenga secreto e que no lo diga a naidie lo que a dho porque le castigara el Apo e dixo que ansy lo faria.

fho ante mi

Fran^{co} Hrrez de G^{a}
escr^{o} de Su Mag.^{t}

/142r/ Ju^{o} Chauca / no le tocan;
es de Guadacheri.

T^{o} el dho Ju^{o} Chauca yndio natural de el pueblo de Guadacheri rrepartimy^{o} de el secretario Hernan Sanchez de Merlo t^{o} presentado por parte de los caçiques prençipales yndios del rrepartimyento de Chacalla encomendado en Fran^{co} de Ampuero el qual despues de aber jurado en forma de dr^{o} segun se rrequeria e siendo preguntado por lengua de don M^{yn} yndio lengua ynterpetre de la Rreal Audiencia y nonbrado para esta presente causa por el tenor del ynterrogatorio por donde se pidio fuese hexsamynado dixo e depuso lo syguy^{e}:

1 - A la primera pregunta de el dho ynterrogatorio dixo este dho t^{o} que conosze a los caciques de Chacalla que agora son don Xpoval e conoscio a Marcapoma e a los demas prencipales que an sydo dende que este t^{o} se sabe acordar que es desde muy nyño porque en tpo de los Yngas e antes que entrasen los espanoles este t^{o} tratava con ellos porque heran todos yndios yauyos e trebutaban al Inga e que a los dhos caciques de Canta no los conosze tanto porque no a tratado con ellos e que tiene not^{a} de las dhas tierras de Quibi dende tpo de los Yngas porque a est^{do} en ellas y las //
/142v/ a vysto muchas e dibersas bezes.

Preguntado por las preguntas generales de la ley dixo este dho t^{o} que no sabe la hedad que tiene pero que se acuerda de Guaynacaba e de Guascaringa porque este t^{o} hera su criado y contador e que no es pariente amygo ny henemygo yntimo de nenguna de las dhas partes ny le ba ynteres en este dho pleyto ny le an dado ny prometido cosa alguna para que diga en este su dho el contrario de la verdad e que a jurado e dira verdad.

2 - A la segunda pregunta del dho ynterrogatorio dixo este dho t^{o} que lo que save desta pregunta sabe [*sic*] es que en tpo que estava el presydente Gasca en este rreyno y estando este dho t^{o} en la çibdad de Los Rreyes en casa de don Antonyo Bilcachava caçique prencipal del pueblo de Guadacheri y estando alli don Fran^{co} Marcapoma cacique de Chacalla vido este t^{o} vido este t^{o} como el dho don Antonyo Guadacheri [*sic*] cacique rrenya con el dho don Fran^{co} Marcapoma y le dezia: "ben aca porque as hecho tan gran maldad que vendiste por obejas las tierras de tus yndios e que avian ganado tus pasados e se las avia sustentado Inga. Que an de hazer agora tus yndios; que sera //
/143r/ dellos y de ti; de que comeran; donde yran a sembrar sus chacarras con rrazon se podran quexar tus yndios de ti y dezir que eres mal cacique e que no myras por lo que conbiene a tus yndios"; y el dho don Fran^{co} Marcapoma caçique de Chacalla vido este dho t^{o} que se le desculpava diziendo que el no avia bendido nada ny avia rresçibido tales obejas e que contra su voluntad avia tomado el dho Fran^{co} de Anpuero las dhas tierras y dadolas a los yndios de Canta y el avia rresçibido las obejas y llevadolas a su casa y se avia henojado con el dho Fran^{co} porque no queria hazer el dho concierto e que era verdad que sus yndios lloravan y se quexavan y dezian que no tenyan donde senbrar y el dho don Antonyo Guadacheri [*sic*] le tornaba a rrener e a rreprehendelle diziendo que no lo avia de consentir a su amo y el dho don Fran^{co} Marcapoma se le tornaba a desculpar y el dho don Antonyo a rrenelle y el dho don Fran^{co} callava y no sabia que rresponder de berguença y esto rrespondio a esta pregunta por la dha lengua.

3 - A la t^{a} pregunta del dho ynterrogat^{o} dixo este dho t^{o} que dize lo que dho tiene en la pregunta antes desta.
4 - A la quarta pregunta de el dho ynterro-//
/143v/ gatorio dixo este dho t^{o} que dize lo que dho tyene en la pregunta antes desta
5 - A la quynta pregunta de el dho ynterrogatorio dixo que lo que sabe desta dha pregunta es que sabe e a bysto y entendido que de la forma y manera que Ynga puso los yndios y mytimaes los an dexado estar los españoles y no los han quytado e que este t^{o} vio que Ynga saco yndios de Guadacheri y los puso por mytimaes en Xauxa e que oy dia estan alli y trebutan dende alli a su caçique de Guadacheri y esto rresponde.
6 - A la sesta pregunta del dho ynt^{o} dixo este dho t^{o} que lo que a dho es la verdad.

Añadidas

1 - A la primera pregunta de el dho ynterrogatorio dixo este dho t^{o} que lo que save desta pregunta es que desde que este t^{o} se sabe acordar que es dende Guaynacaba syempre este dho t^{o} vio por vista de ojos que los yndios yavyos del rrepartimyento de Chacalla encomendados en Fran^{co} de Anpuero tenyan e poseyan las dhas tierras de Quybi como mitimaes puestos por los Yngas pasados e tenyan alli sus casas e fazian en ellas //
/144r/ tierras [*sic*] sus sementeras desde un mojon que esta señalado oy dia fasta una quebrada que esta de la parte del rrio y por esta otra parte por la ladera del cerro do esta el mojon fasta este pueblo de Quibi e que esto oyo dezir este t^{o} a viejos e ancianos del dho pueblo de los yavyos e a otros comarcanos que los dhos yndios yavyos lo avian ganado las dhas tierras por guerras e que venydo que fue Topa Ynga Yupangue hallo alli hasta cien yndios yavyos e puso otros çinquenta yndios mas y los dio y confirmo a todos nuebamente las dhas tierras y que todo esto savia e entendia e bia este t^{o} andando entre los dhos yndios cobrando trebutos para Ynga porque era quypocamayo de todos los yndios que heran yavyos e despues que murio Topa Ynga Yupangue vino a senorear Guaynacaba y confirmo las dhas tyerras de Quibi a los dhos yndios yavyos del rrepartimyento de Chacalla y lo mysmo vido este dho t^{o} que hizo Guascar Ynga despues que suzedio en el senorio por muerte del dho Guaynacava e que esto hera cosa muy publica y not^{a} y lo es entre todos los yndios que lo saven y bieron y dello tienen not^{a} como este dho t^{o}.

3 - A la tr^{a} pregunta del dho ynte-//
/144v/ rrogatorio dixo este dho t^{o} que lo que sabe desta pregunta es que este dho t^{o} a oido dezir a muchos yndios viejos e antiguos desta comarca que al tpo que Mango Ynga se alço los yndios de Canta avian lebantado çierto testimonyo a los dhos yndios yavyos que estavan en las dhas tierras de Quybi y por engraçiarse con Ylla Topa Ynga e sus capitanes avian muerto los dhos yndios de Canta de los dhos yndios yavyos çiento e çinquenta yndios e que en esta sazon estando este t^{o} con Ylla Topa e Paucar Guaman capitanes de Mango Ynga orejones del Cuzco en vn pueblo que llaman Cocha que es de los yndios de Canta vido este dho t^{o} que llegaron ciertos yndios de Canta mensajeros de los caciques de Canta a los dhos capitanes y en presencia deste dho t^{o} les dixeron a los dhos capitanes como los dhos caciques de Canta avian muerto çiento e çinquenta yndios de los dhos yndios yavyos mytymaes porque decian que tenyan dos coraçones coraçon de yndios y de españoles e que avian enbiado cartas a dar abiso a los

españoles e que quando oyo Ylla Topa el dho mensaje estava alli con el dho Ylla Topa el cacique de los dhos yndios yavyos que se llamava Bilcapoma e el dho Ylla Topa le mando alli luego matar diziendo: "ansy que vra parçialidad tiene dos corazones que//
/145r/ avisan los españoles pues aquy morireis bos tanbien" e ansy diziendo esto le mato alli delante deste dho t^{o}.

4 - A la quarta pregunta del dho ynterrogatorio dixo este dho t^{o} que las dhas tierras de Quibi son muy buenas ansy para sembrar mahiz como para coca e que en tpo de los Yngas de las chacarras que benefiçiaban los dhos mytimaes para el Ynga le davan cada ano vnos anos çiento e çinquenta costales de coca y otros anos çiento y sesenta syn lo que le davan de caciques e yndios particulares de trebuto que le davan vnas bezes çinco costales y otras bezes quatro e que esto se coxia en las dhas tierras syn el mahiz e yuca e papas y otras cosas porque entonzes estavan las dhas tierras mejor labradas e beneficiadas e tenyan mas cuidado dellas e que a este dho t^{o} le paresze que valen muy bien los dhos ocho myll pesos que dize la pregunta y mas e que sy fueran deste t^{o} no las diera por ellos ny por nengun dinero que sy no fueran suyas y tubiera los dhos ocho myll pesos para comprallas les diera por ellas lo qual declaro por la dha lengua y en ello dixo que se afirmava e afirmo e aviendosele leydo y se le encargo el secreto y prometiolo.

fho ante mi — Fran^{co} de G^{a}
escr^{o} de Su Mag.^{t}

/145v/ Al^{o} Paucar de mas de 70 a^{o}s
sujeto a don Xpoval
es del pueblo Palli.

T^{o} el dho Alonso Paucar natural de el pueblo de Palli encomendado en Fran^{co} de Anpuero vz^{o} de la çibdad de Los Rreyes sujeto a don Xpoval caçique prençipal de el dho rrepartimyento de Chacalla t^{o} presentado por parte de el dho don Xpoval aviendo jurado en forma de dr^{o} e siendo preguntado por lengua del dho don M^{yn} Lengua ynterprete nonbrado por los senores oidores preguntado por el tenor del dho ynterrog^{o} por donde se pidio fuese hexsamynado dixo e depuso lo syg^{e}:

1 - A la primera pregunta del dho ynt^{o} dixo este dho t^{o} que conosze a las dhas partes e a cada una dellas al dho cacique don Xpoval e a los demas caciques del dho rrepartimyento de Chacalla porque como dho tiene es del dho rrepartimyento sujeto al dho don Xpoval e que al dho cacique de Chacalla e a los que an sydo antes del dende el dho tpo de Guaynacava a esta parte los a conoscido porque los a tratado y conversado muchas vezes e tiene noticia de las dhas tierras de Quybi porque las a vysto y est^{do} en ellas muchas vezes.

Preguntado por las preguntas generales de la ley dixo que sera de //
/146r/ hedad de mas de setenta años e que este dho t^{o} conoscio a Guaynacaba y le sirbio e que como dho dho tyene es sujeto al dho don Xpoval cacique del dho rrepartimyento de Canta e que tyene hijos e muger y le ba ynteres en este pleyto y desea que saque las dhas tierras el dho don Xpoval pero que no le an dado ny prometido cosa alguna para que en este dho diga el contrario de la verdad e que dira verdad e ayude Dios a quyen tobiere justicia.

2 - A la segunda pregunta de el dho ynterrogatorio dixo este dho t^{o} que lo que rresponde a lo en la dha pregunta cont^{do} es que despues que vinyeron los espanoles a esta tierra los dhos yndios de Canta sabe e byo este dho testygo que trayan diferençias con el caçique don Fran^{co} Marcapoma cacique prençipal de Chacalla e con sus prençipales e yndios sobre rrazon de que los dhos yndios de Canta queria que los de Chacalla les dexasen las dhas tierras de Quybi diziendo que eran suyas y no de los yndios yavyos y que Yga [*sic*] se las avia quitado por fuerça y sobre esto rrenyan y querian andar a las puñadas hasta que vino //
/146v/ el presydente Gasca porque en esta sazon vido este dho t^{o} que vino a las dhas tierras Fran^{co} de Anpuero encomendero de los dhos yndios de Chacalla con Nycolas de Rrybera el moço encomendero de los yndios de Canta e venydos fablaron a que a los caciques de Chacalla e a muchos yndios suyos y tanbien a los de Canta y luego los sobredhos Fran^{co} de Ampuero y Nycolas de Rribera començaron de hablar a los dhos yndios sobre el dho concierto y el dho Fran^{co} de Ampuero deçia al dho don Fran^{co} Marcapoma su cacique: "myra yo tengo ya concertado que el cacique de Canta de ducientas obejas de la tierra y que tu le dexes las chacarras y quytar os eis de pasyones porque ya te an muerto doçientos e çinquenta yndios; no te los acaben de matar todos" y el dho don Fran^{co} Marcapoma dixo al dho Fran^{co} de Ampuero su encomendero que no queria fazer el dho conçierto ny queria bender sus tierras porque heran de sus pasados y las avian ganado y que tanpoco se lo consentirian sus yndios porque eran pobres y no tenian //
/147r/ chacarras donde sembrar y que sy e las bendian que ellos no lo consentirian e las pedirian y el dho Fran^{co} de Ampuero le dezia que no syno que las avian de dar a los dhos yndios de Canta por las dhas dozientas obejas y el dho don Fran^{co} dezia que el no queria benderlas y tal haria e lo mysmo via este dho t^{o} que dezian los yndios viejos que alli estavan e ansi el dho don Fran^{co} Marcapoma dezia que para que queria el obejas que luego se le avian de morir que mas queria sus tierras e que sy las vendia que sus yndios le hecharian maldiçiones y despues las pedirian y el dho Fran^{co} de Ampuero visto que no queria el dho don Fran^{co} Marcapoma las dhas obejas las tomo el dho Fran^{co} de Ampuero y las dio a un criado suyo que llamavan Molina y se las dieron y entregaron en pueblo que llaman Ysquipanpa e ansy dixo a los dhos cacique e yndios de Canta que se entrasen en las dhas tierras y las labrasen y sembrasen para si e que quando esto bio el dho don Fran^{co} Marcapoma caçique de Chacalla començo de llorar y

los yndios e yndias pobres e viejos e otros lloravan y hechavan muchas maldiçiones al dho Fran^{co} de Ampuero e que luego como el dho Fran^{co} de Ampuero rresçibio las dhas obejas //
/147v/ los dhos yndios de Chacalla se entraron en las dhas tyerras de Quybi y las labraron y sembraron por suyas y el dho don Fran^{co} se fue llorando y desesperado a sus pueblos y lo mysmo sus yndios e que a la dha sazon entendio este dho t^{o} que los dhos Rribera el moço y Fran^{co} de Ampuero avian dado y senalado ciertos pedaços de chacrillas desde el pueblo de Quybi hazia Canta a unos yndios del pueblo de Cullata y Xicama yndios yabyos del dho rrepartimyento de Fran^{co} de Ampuero porque eran pobres y no tenyan donde sembrar las quales son de donde agora coxieron la coca los yndios de Canta sobre que se bino a hazer ynformaçion e que cree este dho t^{o} y tiene entendido que sy el dho Fran^{co} de Ampuero hiziera de voluntad del dho don Fran^{co} caçique de Chacalla el dho conçierto que rresçibiera las dhas obejas y las llebara a su casa como las llebo el dho Fran^{co} de Ampuero su encomendero y que el dho Fran^{co} de Ampuero hizo el dho conçierto por codicia de las dhas obejas e que sy el dho don Fran^{co} Marcapoma no muriera que cree y tiene por cierto este t^{o} que se fuera a quexar a la Abdiencia Rreal o al señor bisorrey e que como dejo los hijos pequeños no //
/148r/ se a podido nada fasta agora y esto rrespondio por lengua del dho don M^{in}.

3 - A la terzera pregunta del dho ynterrogatorio dixo este dho t^{o} que dizelo que dho tiene en las preguntas antes desta e que sy el dho Fran^{co} de Ampuero diera y entregara las dhas obejas al dho don Fran^{co} Marcapoma su cacique este t^{o} lo biera supiera y entendiera porque se hallo a todo presente y antes vio y entendio que el dho Fran^{co} de Ampuero sellevo las dhas ovejas a su casa a la çibdad de Los Rreyes y las enbio con el dho Molina su criado.

4 - A la quarta pregunta de el dho ynt^{o} dixo este dho t^{o} que dize lo que dho tiene en las preguntas antes desta.

5 - A la quynta pregunta de el dho ynterrogatorio dixo que lo que de la dha pregunta save es que a visto y oydo y entendido que los mytimaes que Inga puso para su serviçio se estan en la propia parte y lugar y de la manera que los puso Ynga syn que los espanoles les ayan mudado ny hecho novedad e que en Xauxa estan mitimaes de los yndios yabyos de Chacalla y Guadacheri y de otras partes desde el tpo del Inga y desde alli acuden con el trebuto a sus caciques y esto rresp.^{e} a esta //
/148v/ pregunta y lo demas en ella contenydo no lo sabe.

5 - A la quynta pregunta [*sic*] del dho ynterrogatorio dixo este dho t^{o} que dize lo que dho tiene lo qual dixo que hera la verdad y en ello dixo que se afirmava e afirmo e no lo firmo porque no supo por lengua de don M^{yn}.

Añadidas

1 - A la primera pregunta del dho ynt^{o} añadido dixo este dho t^{o} que lo que sabe desta pregunta es que en tyempo de antes que ubiese Yngas mucho tpo antes a oido dezir este dho t^{o} a sus padres e aguelos e pasados e a otros muchos yndios viejos de el dho rrepartimyento de Chacalla que los yndios yavyos de Chacalla avian hecho mucha jente de guerra y que avian llegado con ella hasta zerca de Collique y se avian buelto el rrio arriba hazia el pueblo de Quibi en el qual estava un señor que se llamava Chaumecaxa que trebutava al senor de Collique que llamavan Collicapa y que este dho Chaumecaxa les avia salido de Palli alos dhos yndios yabyos de Chacalla y le trayan camarico e chicha y otras cosas e que los dhos yndyos yavyos avian hecho su asyento en un zerro que llaman //
/149r/ Guayllayo e rresponeron e sujetaron el dho cacique Chaumecaxa el qual syrbio dende alli al cacique prençipal de Chacalla y desta manera entraron los dhos yndios yavyos de Chacalla en las dhas tierras de Quibi y despues benydo que fue Topa Ynga Yupangue a conquystar esta tierra tubo noticia de las dhas tierras de Quibi como heran tan buenas y se coxia en ellas mucha coca y enbio vn yndio orejon del Cuzco que se llamava Apar Yupangui para que senalase en ellas tierras para el dho Topa Ynga e les amojonase e pusyese yndios que las benefiçiasen y este yndio orejon tomo tierras para el Inga y puso en ellas yndios mitimaes para que las sembrasen de los yndyos yavyos y puso un mojon e termyno junto al dho pueblo de Quibi que se llama Chuquicoto y mando que dende alli para arriba fazia Canta fuese term^{o} de Canta e para yndios serranos e de alli para abaxo fazia la mar fuese de yndios yabyos e hizieron alli sus señales y casas para los yndios y despues de muerto Topa Ynga vino Guaynacava y confirmo las dhas tierras a los dhos yndyos yabyos de Chacalla y mando que se guardase en aquello lo que avia hecho su padre //
/149v/ e despues que vino Guascaringa hijo de Guaynacava a governar por la muerte de su padre vido este dho t^{o} que hizo lo mysmo e ansy vido este dho t^{o} que en esta costumbre y posesyon estobieron los dhos yndios yavyos hasta que entraron los espanoles en esta tierra y despues hasta que por rrazon del concierto que dho tiene que hizo el dho Fran^{co} de Ampuero los dhos yndios cantas se entraron en las dhas tierras de Quibi e que lo demas que dho tyene que oyo dezir a sus pasados se lo oya tratar e afirmar por cosa muy cierta e que dezian que ellos ansy lo avian bysto ser y pasar en sus tiempos e oido dezir a sus pasados e que dello hera y es y a sydo pu^{co} e notorio e publica voz y fama e comun opinyon e que nunca avian visto ny oydo dezir lo contrario lo qual dixo e declaro por lengua del dho don M^{yn}.

2 - A la segunda pregunta del dho ynterrogatorio anadido dixo este dho t^{o} que dize lo que dho tiene en la pregunta antes desta.

3 - A la tr^{a} pregunta del dho ynterrogat^{o} dixo este dho t^{o} que save que al tpo que se alço Mango Ynga contra los espanoles los dhos yndios de Canta se hizieron amygos con Ylla Topa capitan enbiado por Mango Ynga e Quyso Yupangue capitanes y por la henemystad //

/150r/ que tenyan con los yndios yavyos lebantaron testimonyo a los dhos yndios yavyos de Chacalla y dixeron al dho Ylla Topa capitan de Mango Ynga quelos dhos yndios yavyos tenyan dos coraçones uno para los espanoles y otro para los Yngas y el dho Ylla Topa dyo con una porra al cacique prençipal de Chacalla que se llama Bilcapoma en las espaldas y se fue hazia Lima y despues de ydo un yndio prençipal de Chacalla que se llama Alcos que es oy dia bibo le mato al dho cacique prençipal que se llama Bilcapoma y despues por engano tomo los dhos yndios que tenya que eran mas de çiento y çinquenta yndios y dixo que llebasen comyda a su cacique que estava en Canta y ellos lo hizieron ansy y se fueron con los caçiques de Canta los quales los enzerraron en un corral y alli les cortaron las cabezas y dixeron al dho Ylla Topa que los avian muerto porque avian enbiado mensaje a Lima a los espanoles e que querian fazer traicion e que desde entonzes se metieron yndios cantas en el dho pueblo que llaman de Quibi y amenaçavan a los yndyos yavyos que avian benydo a poseer las dhas tierras de los yndios que mataron que los avian de hechar de alli e que las tierras heran suyas e aunque les amenaçavan todavia los dhos yndios yavyos poseyan las dhas tie-//
/150v/ rras de Quybi y las labravan y senbravan y tuvieron fasta el dho conçierto y esto dixo e declaro por lengua del dho don M^{yn}.

4 - A la quarta pregunta del dho ynterrogatorio anadido dixo este dho t^{o} que las dhas tierras de Quibi son muchas e muy buenas e de mucha coca e que valen bien seis myll pesos y mas ochenta o nobenta porque se coje en ellas mahiz si senbran e papas e axi e mucha coca que cojeran cada ano en ellas de coca si las sembrasen todas mas de çient costales por ano e que si deste t^{o} fueran las dhas tierras no las diera por los dhos seis myll pesos de oro e que si las ubiera de conprar y los tuviera para dallos tanbien los dyera por ellas e que lo que dho e declarado tyene en este su dho dixo que era la verdad y lo que deste caso savia y no save otra cosa y en ello dyxo que se afirmava e afirmo e rratefico aviendoselo tornado a leer y declarar por lengua del dho don M^{yn} ynterpetre e se le encargo tenga secreto de lo que a dho e no lo diga a su cacique ny a otra persona so pena que quedara perjuro y le castigaran el qual dyxo que ansy lo faria e no lo firmo porque no supo.

fho ante mi — Fran^{co} Hrrez de G^{a}
escr^{o} de Su Mag.^{t}

/151r/ Anton Cacharure / no sabe la hedad /
acuerdase de quando entraron los españoles
en esta tierra / e no le tocan / es de Guadacheri.

T^{o} el dho Anton Cacharure natural del pueblo de Çhica caçique de Guadacheri rrepartimyento que esta encomendado en Hernan Sanchez de Merlo vz^{o} de la çibdad de los Reyes t^{o} presentado por pte del dho don Xristoval caçique prençipal del rrepartimyento de Chacalla encomendado en Fran^{co} de Ampuero aviendo jurado en forma de dr^{o} segun se rrequeria e syendo preguntado por lengua de don M^{yn} Lengua ynterpetre nombrado por los señores presydente e oidores para la presente cavsa e syendo preguntado por el dho ynterrogatorio dixo e depuso lo syg^{e}:

1 - A la primera pregunta del dho ynterrogatorio dixo este dho t^{o} que conosze a los dhos caçiques que agora son del rrepartimyento de Chacalla e a los del rrepartimyento de Canta e a los caçiques e señores que an sydo de los rrepartimy^{o}s dende que entraron los españoles en su tierra de vysta y fabla trabto y conversaçion que con ellos a tenydo e tiene e que sabe e a visto las dhas tierras que llaman de Quybi porque las a vysto y estado en ellas.

Preguntado por las preguntas generales de la ley dixo que no sabe //
/151v/ la hedad que tiene mas de que se acuerda muy bien de quando entraron los españoles en esta tierra e que no es paryente ny enemygo de nenguno de los dhos principales e yndios del dho rrepartimyento ny le toca ny enpeze nenguna de las preguntas generales de la ley e syendo por ellas pregunt^{do} e que dira verdad e ayude Dios a quyen tobiere just^{a}.

2 - A la segunda pregunta del dho ynterrogatorio dixo este dho t^{o} que lo que desta pregunta sabe es que luego como dizen que paso lo contenydo en el dho concierto que fue estando el Licenciado de La Gasca presydente que fue destos rreynos en esta tierra oyo este dho t^{o} dezir a su padre que se llama Anton Caxalluri que es ya muerto que era en aquella sazon mytimae puesto por el inga en Caranpanpa termyno de la dha p^{ovi}ncia de Chacalla que sacava alli colores le oyo dezir que al tyempo que el dho concierto de las dhas tierras se hizo por las dhas tierras de Quybi avian enbiado a llamar al dho su padre el dho Marcapoma cacique principal de Chacalla y el dho Fran^{co} de Ampuero //
/152r/ el qual avia benydo y le avian dado parte y dho le como le querian dar las dhas obejas por las dhas tierras al dho Marcapoma y que el dho su padre avia callado y no avia rrespondido nada e que avia apartado al dho Marcapoma caçique de Chacalla y dhole que no vendyese las dhas tierras ny tomase las dhas obejas y que le avia rrespondido que el faria tal y que su padre deste t^{o} le abia dho que si las vendia que le rrenyrian y querrian mal los caciques de Guadacheri y los del Arcobispo y de los yndios yavyos e que avia visto como el dho Fran^{co} de Anpuero decia al dho don Fran^{co} Marcapoma que tomase las dhas tierras y las entregase a los yndios de Canta y rrescibyese las obejas que le davan que heran mejor que no andaren pasyones e que los yndios de Canta sy no lo hazia le avian de acabar de matar los yndios que tenya sobre ellas y que el dho caçique avia dho que no queria dexar sus tierras ny tomar por ellas obejas ny otras cosas porque heran de sus yndios y que si las bendia que sus yndios se quexarian del y que las obejas se gastarian e moririan y las tierras no que syenpre estavan en pie y que nunca el dho //
/152v/ Fran^{co} de Ampuero avia podido acabar con el dho don Fran^{co} Marcapoma que hiciese el dho conçierto e rresçibiese las dhas obejas que como no avia querido que el dho Fran^{co} de Ampuero las avia rreszebido dozientas obejas y enbiadolas a Lima con un honbre e que desta manera avia pasado e los dhos yndios de Canta [*sic*] avian vysto que les tomavan las dhas tierras que avian

dado vozes y gritos y hechavan maldyçiones al dho Fran^{co} de Ampuero e a el dho su cacique e se quexaban dellos dyziendo que las tierras que tenyan para senbrar sus comydas se las quytavan y que como el dho don Fran^{co} Marcapoma avia visto llorar a los dhos yndios que llorava el tambien e que despues que paso lo susodho dende a ciertos dias vido este dho t^{o} que los caciques de Guadacheri y los de la encomyenda del Arçobispo avian enbiado a llamar al dho don Fran^{co} Marcapoma para rrenyrle por lo susodho e no quyso benyr y le enbiaron un mensajero amenaçandole e rrinendole e que el dho Marcapoma les enbio a dezir como el no avia bendydo nada que su amo lo avia fho contra su voluntad //
/153r/ e que el no avia consentydo en ello y esto dixo que sabia.

3 - A la terzera pregunta del dho ynterrogatorio dixo este dho t^{o} que no lo save e dize lo que dho tiene en la pregunta antes desta.

4 - A la quarta pregunta del dho ynterrogatorio dixo este dho t^{o} que no lo save e dize lo que dho tiene.

5 - A la quynta pregunta del dho ynterrogatorio dixo este dho t^{o} que no save mas desta pregunta de que a bisto que en muchas partes de este rreyno donde este t^{o} lo a visto por los ojos que los mytimaes que puso Inga se estan en aquel propio lugar que los puso y como los puso syn que ellos ayan mudado ny hecho novedad en ellos los españoles e que a oido dezir que es por mandado del Apo de C^{a} e que tanbien a oido dezir que estan de la mysma manera en todo este rreyno.

6 - A la sesta pregunta del dho ynterrogatorio dixo este dho t^{o} que lo que a dho es la verdad y lo que deste caso save lo qual declaro por la dha lengua.

Añadidas

1 - A la primera pregunta del dho ynterrogatorio dixo este dho testigo //
/153v/ que lo que desta pregunta sabe es que este dho t^{o} a oido dezir al dho su padre que se llamava Anton Caxalluri e a otros viejos e ancianos desta comarca que antes que binyesen los yngas a esta tierra los dhos yndios yabyos de Chacalla se avian juntado y salido de guerra a ganar tyerra y que avian ydo hasta donde agora esta puesto vn mojon y poblaron alli y mataron parte de los yndios de un senor que estava en las dhas tierras que llamavan Chunbicaxa y los demas los sujetaron y enpeçaron a hazer edefiçios y chacarras e a labrallas y cojer las sementeras e que ansy avian estado asta que vino conquystar este rreyno Topa Ynga Yupangue e sus capitanes el qual dizen que como los fallo a los dhos yndios de Chacalla mando que se estobiesen de aquella manera y les dyo chacarras y sementeras en que labrasen y les dio por termynos desde un mojon que esta junto a las casas del dho pueblo de Quybi hazia abaxo hazia la mar que fuese para yndios yungas y de alli para arriba hazia Canta fuesen serranos y despues de muerto Topa Ynga Yupangue dizen que hizo lo mysmo Guaynacava //
/154r/ y les confirmo las dhas tierras y les dio mas tierra y que despues que murio Guaynacava Guascar Ynga que sucedio en el senorio avia fho lo mysmo e que desta manera los dhos yndios de Chacalla avian tenydo las dhas tierras fasta que entraron los españoles y despues fasta que se hizo el dho concierto e que esto como dho tyene lo oyo dezir este dho t^{o} por cosa publica y notoria a los dhos viejos e ançianos los quales dezian que ellos ansy lo avian vysto ser y pasar en sus tpos e oidolo dezir a sus mayores e ancianos e que dello avia sydo y hera publico y not^{o} e publica voz y fama y esto rrespondio por la dha lengua.

3 - [*sic*] A la t^{a} pregunta del dho ynterrogatorio dixo este dho t^{o} que oyo dezir al dho su padre que quando se alço Mango Ynga los dhos yndios de Canta avian llebado por engano a su tierra a Bilcapoma caçique principal del dho rrepartimyento de Chacalla y que por bengarse dellos por la enemystad que tenyan por las tierras le avian muerto diziendole que el y sus yndios e hijos tenyan dos coraçones e que querian bien a los españoles e que heran ydos a Lima a fablar al marques e que despues que le ubieron muerto enbiaron a dezir a sus //
/154v/ yndios que su cacique estava en Canta con Inga que le llebasen camarico e creyendo los dhos yndyos que ansy hera se cargaron todos de comyda y fueron a llevar al dho su cacique y legua y media deste pueblo de Quybi en un pueblo de Canta que llamavan Acoyngo los enzerraron todos en un corral e alli los cortaron las cabeças que no quedo nenguno y despues de muertos fueron a dezille a Yllatopa e a otro capitan que andava con el que los avian muerto porque tenyan dos coraçones que avian ydo a dar avyso al marques y que todo esto los dhos yndios Cantas lo fazian de enbidia por quetalles las tierras donde senbravan coca e que avnque mataron estos yndios que syenpre avia yndios yavyos en sus asyentos y labravan las chacarras y no las dexavan perder y esto dixo por la dha lengua.

4 - A la quarta pregunta del dho ynterrogatorio dixo que a visto las dhas tierras de Quybi e que le paresce que valen ocho myll pesos e aun mas porque son muy provechosas para los yndios por amor de la coca que se coje y saca mucho della e que esto dixo que hera verdad y en ello se afirmo por lengua del dho don M^{yn} despues de averse leydo.

Fho ante my Fran^{co} Hrr^{ez} de G^{a}
esc^{o} de su M^{t}

/155r/ Juan Yaure sujeto a don Xristoval
no save la hedad que t^{e}
acuerdase de Guaynacaba e le sirvio
e fue su soldado.

T^{o} el dho Juan Yaure natural del pueblo de Acuchay de el rrepartimyento de Chacalla encomyenda de Fran^{co} de Ampuero vz^{o} de la çibdad de los Rreyes t^{o} presentado por parte de don Xristoval cacique de Chacalla en el pleyto con los caciques de Canta sobre las tierras de Quibi aviendo jurado en forma de dr^{o} segun se rrequeria e syendo preguntado por el tenor de el dho ynterrogatorio por lengua de don M^{yn} Lengua ynterpetre de la Rreal Avdiencia nombrado por los senores oidores para esta presente cavsa dixo e depuso lo syguiente:

1 - A la primera pregunta de el dho ynterrogatorio dixo este dho t^{o} que conosze a los caciques prençipales e yndios que an sydo e son del dho rrepartimyento de Chacalla porque como dho tyene es sujeto al dho rrepartimyento al cacique prençipal don Xristoval e conosze a los caçiques que an sydo e son del dho rrepartimyento de Canta desde el tpo de Guaynacava los que an sydo e son desde el dho tpo aca porque los a tratado y conversado muchas vezes //
/155v/ e que sabe e tiene notyçia de las dhas tyerras de Quybi sobre que es este pleyto porque las a vysto y est^{do} en ellas muchas e dibersas bezes.

Preguntado por las preguntas generales de la ley dixo este dho t^{o} que no sabe la hedad que tyene mas de que se acuerda de Guaynacaba y le servyo y fue su soldado e que como dho tiene es natural del rrepartimyento de Chacalla y tiene hijo e muger e le ba el mysmo ynteres que a los demas yndios del rrepartimy^{o} de Chacalla en lo de las dhas tyerras e querria que se les bolbiesen a Chacalla porque son suyas pero que por eso que no dexara de dezir verdad de lo que supiere y le fuere preguntado e que no le an dado ny prometido cosa alguna para que en este su dho diga el contrario de la verdad.

2 - A la segunda pregunta del dho ynterrogatorio dixo este dho t^{o} que lo que sabe desta pregunta es que tenyendo los dhos yndios de Chacalla las dhas tierras que llaman de Quibi desde que binyeron los yngas que se las dieron los dhos yndios de Canta al tpo que entraron los espanoles en esta tierra enpe-//
/156r/ zaron a tener pasyones con los dhos yndios de Chacalla porque tenyan las dhas tierras que dezian los de Canta que eran suyas y que se las avian dado sus pasados y ganadolas e que Ynga se las avia quitado a ellos para darselas a los de Chacalla y sobre esto se querian matar y Marcapoma cacique de Chacalla no se le davan .... por lo que decia el cacique de Canta que llamaban Guaman Yaure antes le rrespondia que el que mas pudyese abia de quedar con ellas y de otra manera no las avia de dexar y estando en esto como entendieron Nycolas de Rribera el vyejo e Fran^{co} de Ampuero que sus yndios querian rrenyr sobre las dhas tierras y estavan para matarse vinyeron al dho balle e tyerras de Quibi donde hallaron al dho Guaman Yaure caçique de Canta e a Marcapoma caçique de Chacalla y les rrineron y dezian que porque no se conformavan y heran buenos amygos y les preguntaron sobre que era la diferencia que tenyan y luego les dixeron que era sobre las dhas tierras de Quybi y los dhos Nycolas de Rribera fablaron a sus caciques para que se conzertasen y luego fablaron el dho Nycolas de Rribera y el dho Anpuero solos y des-//
/156v/ pues de aber hablado el dho Fran^{co} de Anpuero fablo con el dho Marcapoma su caçique diziendole que el tenya concerto con los dhos yndios de Canta que dyesen al dho Marcapoma duzientas obejas de la tierra machos y hembras y que ellos por ellas les dexasen las dhas tierras de Quibi y el dho Marcapoma rrespondio al dho Fran^{co} de Ampuero su encomendero que no queria hazer aquel conçierto que no queria obejas syno tenerse sus tyerras pues se las avian dado los yngas y se las avian amojonado y que sy las bendia que sus caciques syenpre le querrian mal y dirian que les avia bendido sus tierras e de sus yndios e que se moririan de hambre porque no ternyan donde senbrar e que tanbien todos los caciques de los rrepartimyentos de los Yauyos se le ternyan a mal e rrenyrian con el porque bendia las tyerras de sus yndios e que valian mas que no las obejas y que el dho Fran^{co} de Ampuero tornaba a hablar al dho caçique e a porfiar con el que lo hiziese y el dho Fran^{co} Marcapoma todavia dezia que no fazia tal venta e que sy el dho Fran^{co} de Ampuero se las vendia que sus hijos las pedirian y el dho Fran^{co} de Am-//
/157r/ puero como vido que no queria el dho don Fran^{co} Marcapoma hazer el dho concierto rreszivio las dhas obejas de los dhos caciques de Canta y las fue a rreszebir a un pueblo del dho rrepartymyento de Canta que llaman Collata linde de los pueblos de Chacalla y rreszebidas las dio y entrego a un criado suyo que llamavan Molyna que las llebase el qual las llevo a Lima por mandado del dho Fran^{co} de Ampuero y nunca el dho don Fran^{co} Marcapoma ny sus prençipales e yndios se aprobecharon de las dhas obejas y luego como las resçibio dio y entrego las dhas tierras de Quibi a los dhos yndios de Canta los quales se entraron en ellas y las labraron y sembraron como suyas y como bieron esto los dhos yndios de Chacalla enpezaron a llorar y dar bozes ansi hombres como mugeres y nynos y biejos y el dho Marcapoma cacique y hechavan maldiçiones al dho don Fran^{co} e al dho Fran^{co} de Ampuero su encomendero porque les avian bendido sus tierras y dezia el dho cacique Marcapoma: "mys hijos las pediran despues de yo muerto" porque el no hosaria por amor del dho Fran^{co} de Ampuero su encomendero e desta manera se hizo e paso lo susodho y no de otra porque este dho t^{o} //
/157v/ lo vido y se fallo presente a todo y esto dixo e declaro por lengua del dho don M^{yn} ynterpetre que dixo avello depuesto este dho t^{o} por esta rrazones [*sic*].

3 - A la terzera pregunta del dho ynterrogatorio dixo este dho t^{o} que dize lo que dho tiene en las preguntas antes desta en que se afirmava e afirmo.

4 - A la quarta pregunta de el dho ynterrogatorio dixo este dho t^{o} que dize lo que dho tiene en las preguntas antes desta.

5 - A la quynta pregunta de el dho ynterrogatorio dixo este dho t^{o} que no lo save mas de que a bisto que todos los yndios que el Inga puso para mytimaes en Xauxa y en otras partes donde este t^{o} a est^{do} a visto y entendido que se estan todavia alli de la mysma manera que los puso Ynga syn que los españoles los ayan mudado ny hecho novedad en ellos y esto rresponde.

6 - A la sesta pregunta del dho ynterrogatorio dixo este dho t^{o} que dize lo que dho tyene en las preguntas antes desta lo qual dixo que hera la verdad y en ello dyxo que se afirmava e afirmo e rratefico despues de aversele leydo por lengua del dho don M^{yn} ynterpetre.

/158r/ *Añadidas*

1 - A la primera pregunta del dho ynterrogatorio dixo este dho t^{o} que lo que sabe desta pregunta es que este dho t^{o} a oido dezir a sus padres e aguelos e a otros muchos yndios viejos e ancianos desta comarca que en tpo antes mucho que vinyesen los yngas tenyan por cost^{e} de que quando heran en una tyerra muchos yndios de juntarse e ir de guerra a conquistar otros yndios que la tubiesen sembrada e que asy los dhos yndios de Chacalla syendo muchos e tenyendo falta de tierras avian salido a los yungas a buscalla e a conquistar e que lo primero que avian fallado fue el dho pueblo que agora llaman de Juzcunga que es un mojon que esta media legua poco mas o menos del dho pueblo de Quibi e de alli fueron mas adelante fazia Lima a ver si pudieran conquistar mas y toparon con un cacique que llamavan Collicapa senor de Collique y se bolbieron a la poblacion que havian hecho en las dhas tierras de Quibi e que alli avian estado labrando y senbrando en pedaços de la tierra hasta que bino Topaynga Yupangue e les dio e anadio mas tierras de las que tenyan para que senbrasen para Ynga e para ellos e que despues que murio Topaynga Yupangue suzedio Guaynacava el qual sabe este t^{o} que //
/158v/ como supo que las dhas tierras de Quibi heran tan buenas y se coxia en ellas tan buena coca mando que sacasen del dho rrepartimyento de Chacalla a lo que oyo dezir a su padre cinquenta o sesenta yndios mas para que sembrasen para Ynga e para ellos y les vio y senalo mas tierras fazia el dho pueblo de Quybi fazia la syerra y puso un mojon junto a las dhas casas de el dho pueblo de Quybi que llaman Chuquicoto e que despues de muerto Guaynacava bino a rreynar Guascar Inga y tenyendo not^{a} de los dhos yndios aprobo y tubo por bueno lo que sus pasados avian hecho zerca de el dar de las dhas tyerras las quales tubieron los dhos yndios yavyos del dho rrepartimiento de Chacalla fasta que entraron los espanoles en esta tierra y despues fasta que los yndios de Canta se entraron en ellas por rrazon del concierto que hizo el dho Fran^{co} de Ampuero de las dhas obejas y esto rrespondio por lengua del dho don M^{yn} Lengua ynterpetre que dixo avello declarado el dho t^{o} ansy por estas rrazones.

2 - A la segunda pregunta del dho ynterrogatorio dixo este dho t^{o} que sabe la dha pregunta como en ella se cont.^{e}//
/159r/ porque este dho t^{o} lo vido ser y pasar ansy como la dha pregunta lo declara tenyendo como tiene noticia y sabe quando los espanoles entraron en esta tierra y es dello pu^{co} e notorio e publica boz y fama.

3 - A la terzera pregunta del dho ynterrogatorio dixo este dho t^{o} que sabe y es cosa publica y notoria entre los yndios viejos e moços de los dhos rrepartymyentos de Chacalla que lo saben y dello tyene noticia como este t^{o} que estando alçado Mango Ynga contra los espanoles los dhos yndios de Canta por hodio y mala voluntad que les tenyan so color que les lebantaron que los dhos yndios de Chacalla avian enbiado dos hijos de caciques a fablar al marques y por congraciarse con Ynga e con Yllatopa llebaron por engano al caçique prençipal de Chacalla que se llamaba Bilcapoma y le dixeron que tenya coraçon de españoles y de yndios y le mataron e hizieron saber a Yllatopa capitan de Ynga como le avian muerto por la dha ocasion e luego vinyeron muchos yndios a dezir a los yndios yavyos como su cacique estava en Canta e que les enbiava para que llebasen yndyos con comyda y sacaron fasta ciento e çinquenta yndios cargados y llegando a unas //
/159v/ peñas que llaman Acobingo que sera una legua deste dho pueblo de Quibi e alli los mataron a todos e que esto como dho tiene a sido y es de lo publico y notorio.

4 - A la quarta pregunta del dho ynterrogatorio dixo este dho t^{o} que las dhas tyerras de Quibi es un pedaço de tierra muy buena e que se coje en ellas coca e maiz e papapas [*sic*] y lo demas que se coje entre los yungas e que le paresze a este dho t^{o} que valen mucho e que no sabe quanto e que sy fuesen suyas que las benderia y esto rresponde.

Lo qual todo dixo que era la verdad y lo que deste caso sabia y en ello dixo que se afirmaba e afirmo e rratefico despues de abersele leydo e encargosele tenga secreto por el dho don M^{yn} Lengua ynterpetre e que lo diga a naidie [*sic*] lo que a dho so pena de perjuro e que le castigaran y ello prometio e no lo firmo porque no supo.

fho ante mi
Fran^{co} Hrrez de Guebara
escr^{o} de Su Mag^{t}.

Ju^{o} Yndio sujeto a don Xristoval
de mas de 50 / sujeto a Chacalla.

T^{o} el dho Juan Yndio que dixo ser del pueblo de Cullata del rrepartimyento de Chacalla encomendado en Fran^{co} de Am-//
/160r/ puero vz^{o} de la çibdad de los Reyes sujeto al caçique don Xristoval t^{o} presentado por parte de el dho don Xristoval cacique aviendo jurado en forma de dr^{o} e siendo preguntado por lengua de don M^{yn} yndio ynterpetre de la Rreal Avdi^{a} nombrado para esta presente cavsa e syendo preguntado por el tenor del dho ynterrogatorio por donde se pidio fuese hexsamynado dixo e depuso lo sig^{e}:

1 - A la primera pregunta de el dho ynterrogatorio dixo este dho t^{o} que conosze a los caçiques e prencipales que an sydo y son del dho rrepartimyento de Chacalla porque como dho tiene este dho t^{o} es sujeto a ellos e conosze y conoscio a los caciques e

principales que an sydo del rrepartimyento de Canta por vysta y fabla trato y conversacion syendo como estan comarcano a ellos y los conosze dende que se save acordar que es dende tpo de Guaynacava e sabe e tiene noticia de las dhas tyerras de Quibi porque las a vysto y andado y est^{do} en ellas muchas bezes.

Preguntado por las preguntas generales de la ley dixo que es de hedad de mas de cinquenta anos e que se acuerda desde el tpo de Guaynacava e que //
/160v/ como dho tiene es nasçido y criado en el dho rrepartimyento de Chacalla e a sydo casado e tiene en el un nyeto e que querria que los caciques de Chacalla saliesen con sus tierras porque a este dho t^{o} le ba ynteres pero que dira berdad de lo que supiere y le fuere preguntado e que no le an dado ny prometido cosa alguna para que en este su dho diga el contrario de la verdad.

2 - A la segunda pregunta del dho ynterrogatorio dixo que lo que sabe e a visto oydo y entendido es que tenyendo y poseyendo los dhos yndios de Chacalla las dhas tyerras de Quibi desde antes que entrasen yngas en ellas y fasta que vinyeron los espanoles al tyempo que entraron los dhos espanoles los caciques e yndios del rrepartimy^{o} de Canta y Guaman Yaure su cacique pretendieron e quisieron hechar dellas a los dhos yndios de Chacalla e Marcapoma su caçique diziendo que las dhas tierras de Quibi heran suyas y de sus pasados e que el inga se las avia quitado como mando e dadoselas a ellos e que agora que avian de tornar a dar o que sobre ello se avian de ma-//
/161r/ tar y el dho Marcapoma encomendero de los dhos yndios de Chacalla y sus yndios dezian que las dhas tierras heran suyas e que no las avian de dexar e que el que mas pudyese aquellas avia de llevar e que sobre esta rrazon estaban que se querian matar muchas bezes los dhos yndios de Chacalla defendyendo las tyerras y posesyon que tenyan y los de Canta queriendolos hechar dellas fasta que bino a noticia del licenciado de La Gasca presydente que fue destos rreinos y enbio a Nycolas de Rribera el moço encomendero de los yndios de Canta e a Fran^{co} de Anpuero encomendero de los yndios de Chacalla los quales llegaron a las dhas tierras de Quibi e fallaron que los dhos yndios estavan rrinendo y se querian dar de pedradas con hondas y dezia el dho Guaman Yaure cacique de Canta a Marcapoma cacique de Chacalla: "myra yo te quiero dar dozientas obejas y dexame las tierras pues este my rrio y my entrada y tu rrio es el de Mama" y le dezia que sy no se conzertava que avian de dar batalla y el dho Marcapoma dezia que el no queria guerra ny batalla ny menos dexar las tierras e que est^{do} en esto acabaron de llegar los dhos Rribera el moço y Fran^{co} de Ampuero y el dho Fran^{co} de //
/161v/ Ampuero dezia y rrogava al dho don Fran^{co} Marcapoma que tomase las dozientas obejas que el dho Guaman Yaure le dava por las dhas tyerras que era mejor que no las tierras y se las dexase y se fuese a Chacalla porque ya sabia que el dho Guaman Yaure y sus yndios heran muy bellacos y guerreros y que le matarian sus yndios e que el dho don Fran^{co} Marcapoma bydo este dho t^{o} que le rrespondia al dho su encomendero que no queria dar las dhas tierras ny rreszebir las dhas obejas porque las tierras no heran suyas que eran de sus yndios e que no querrian que se bendyesen porque no tenyan donde sembrar e que sy aquellas vendyese que ellos y sus hijos las avian de sacar e pedir e que le avian de bolber las obejas e que quando se las pidyesen las ternyan ya comydas y gastadas y que se quexarian del sus yndios y tanbien los yndyos caciques comarcanos de los yauyos sus amygos e parientes de rrenyrian y estarian mal con el porque bendia las tierras de sus yndios y les dexava pobres e aviendoselas dado los yngas e amojonadoselas por tierra de yndios yabyos e todavia vido este dho t^{o} que el dho Fran^{co} de Ampuero rrogava a //
/162r/ su cacique Marcapoma que hiciese el dho concierto y tomase las dhas dozientas obejas y le rrenya sobre ello y el dho don Fran^{co} Marcapoma no queria y le rrespondia lo mysmo que de antes y otras muchas cosas y el dho Fran^{co} de Anpuero dixo que pues no queria que el queria rreszebir las dhas ovejas y hazer el dho conçierto y el dho don Fran^{co} Marcapoma dixo que el no lo haria e que sy el dho Fran^{co} de Ampuero lo hiziese que el haria a sus hijos que las pidyesen las dhas tyerras e ansy el dho Fran^{co} de Ampuero hizo el dho concierto porque este dho t^{o} bydo que rresçibio las dhas obexas en vn pueblo de yndios de Canta que llaman Cullata e las avia dado y entregado a un criado suyo que llamavan Molina el qual las avia llebado el camyno de Lima a casa del dho Fran^{co} de Anpuero a lo que este t^{o} entendio e que el dho don Fran^{co} Marcapoma ny sus yndios ny rresçibieron nenguna antes decian que el dho Fran^{co} de Anpuero avia hecho el dho concierto por tomarse las dhas obexas y lleballas a su casa e que como rresçibio las dhas obejas los dhos caçiques e yndios de Canta se entraron en las dhas tierras y las labraron y los dhos yndios del dho rrepartimyento como bieron lo susodho davan bozes y hechavan muchas maldiciones al dho Fran^{co} de Anpuero y lloravan y lo mysmo el dho //
/162v/ Marcapoma lo qual dyze este dho t^{o} que vio ser y pasar por bysta de ojos e que a la dha sazon vino vn cacique que se llama Juan Mango Raqui cacique de Cullata de el rrepartimyento de Chacalla y con sus yndios que tanbyen benyan llorando y dixo al dho Fran^{co} de Ampuero e a el dho Rribera que para que les abian bendydo e quitado sus tierras que no tenyan donde senbrar e que moririan de hambre y en esto el dho Rrybera y Fran^{co} de Anpuero y el dho Guaman Yavre cacique senalaron al dho cacique e a sus yndios vnos pedacos de tierras junto a las casas del pueblo de Quibi a rraiz del cerro e pusyeron una cruz por termyno que corria el dho pedaço de tierra que ansy les davan dende alli hazia adelante fazia Canta las quales son las que don Juan principal de Canta cogio la coca sobre que vino vn esc^{o} a fazer ynformacion lo qual todo dixo e declaro por lengua del dho don M^{yn} ynterpetre el qual dijo avello declarado ansy.

3 - A la terzera pregunta del dho ynterrogatorio dixo este dho t^{o} que dize lo que dho tyene en las preguntas antes desta.

4 - A la quarta pregunta de el dho ynterrogat^{o} dixo este dho t^{o} que dize lo que dho tiene en las preguntas antes desta.

5 - A la quynta pregunta del dho ynterro-//
/163r/ gatorio dixo este dho t^{o} que lo que sabe desta pregunta es que a lo que este dho t^{o} a vysto y entendido en las partes y lugares que

a ido deste rreino despues que entraron en el los espanoles no los an mudado ny mudan e que a entendido que el Apo de Castilla lo a mandado ansy y esto rresponde.

6 - A la sesta pregunta del dho ynterrogatorio dixo este dho t^{o} que dyze lo que dho tyene lo qual dixo que era la verdad y en ello se afirmo por la dha lengua.

Anadidas

1 - A la primera pregunta del dho ynterrogatorio anadido dixo este dho t^{o} que lo que desta preg. sabe es que a oido dezir por cosa publica y notoria entre yndios viejos e ancianos del dho rrepartimyento de Chacalla e a otros desta comarca que son ya muertos de cuyos nonbres no se acuerda e lo mysmo oyo dezir a sus padres e aguelos y pasados que los dhos yndios de Chacalla quatro senores los mas prencipales del avian salido con jente a conquistar esta tierra de Quybi e que avian hallado en ella un caçique que llamavan Chaumecaxa e que este dho Chaumecaxa hera senor natural yun-//
/163v/ ga el qual se hizo amygo con ellos y les trebutaba e hizieron vn pueblo los dhos yndios yauyos en el dho balle e tierra de Quibi que no sabe su nombre y fueron todos unos e ansy estubieron en las dhas tierras de Quybi los dhos yndios yauyos e las sembraron fasta que vino Topaynga Yupangue el qual lo tubo por bueno y les confirmo las dhas tyerras diziendo que pues ellos avian sydo hombres para ganallas que se estobyesen en ellas e les mando anadir las tierras que tenyan e ansy sus capitanes de Topa Ynga avian alargado las dhas tierras de los dhos yndios yauyos fasta junto a las propias casas de Quibi do estava el dho Chaumecaxa el qual dho mojon y senal que alli se hizo llaman oy dia Chuquicoto e que en esta posesion avian est^{do} fasta que vino Guaynacava el qual se las avia confirmado y que despues de muerto el dho Guaynacava y benydo Guascar Ynga avia hecho lo mysmo e ansy las avian tenydo fasta que entraron los espanoles en esta tierra y despues fasta el concierto que hizo el dho Fran^{co} de Anpuero e aun en este dho tpo este dho t^{o} se las vido poseer por bysta de ojos a los dhos yndios yauyos e que lo demas que dho tyene este dho t^{o} como dho tyene lo oyo dezir //
/164r/ a los dhos sus padres y aguelos y pasados los quales dezian que ellos en sus tiempos ansy lo avian vysto ser y pasar en su tpos e oydo dezir a otros sus mayores e mas ancianos que ellos los quales dezian que ellos en sus tpos ansy lo avian bysto ser y pasar y oydo a los suyos en los suyos e que tal avia sydo y hera dello publico e notorio e publica boz y fama e que nunca abian bysto ny oydo dezir lo contr^{o} e que si lo contrario fuera o pasara lo obieran vysto oido dezir e no pudiera ser menos por ser yndios comarcanos e naturales muchos dellos del dho pueblo de Chacalla y esto dixo e declaro por lengua del dho don M^{yn} Lengua ynterpetre que dixo e declaro avello declarado y espaçificado e este dho t^{o} por estas rrazones aviendole sydo preguntado.

2 - A la segunda pregunta del dho ynterrogatorio dixo este dho t^{o} que dize lo que dho tiene en las preguntas antes desta e que sabe la dha pregunta como en ella se conte y declara porque lo vido ser y pasar ansi por vysta de ojos.

3 - A la terzera pregunta del dho ynterrogatorio dixo este dho t^{o} que lo que desta pregunta sabe es que sabe por cosa publica y notoria que al tpo quese alço Mango Ynga contra //
/164v/ los espanoles vino a la dha tierra de Canta y de Chacalla Ylla Topa capitan de Mango Ynga e Quiço Yupangue que venyan desbaratado [*sic*] de Lima y huyendo y estando el dho Bilcapoma cacique de Chacalla en Canta dixo uno de los caciques de Canta a Illatopa y a otro capitan y le dixo como el dho Bilcapoma cacique de Chacalla tenya coraçon de espanoles tanbyen como de yndios y que abia enbiado mensajeros al marques y el dho Yllatopa le dio con una porra y le rrino y no le mato y le dixo: "myra yo boy por my jente y luego bolbere para ir contra los espanoles" y que ansy como se fue de alli el dho Yllatopa los dhos yndios de Canta mataron al dho Bilcapoma y despues de muerto vinyeron a Chacalla a su pueblo a los yndios que rresydian en las dhas tyerras de Quibi y enganosamente les dixeron que su cacique estava con Mango Ynga esperandolos que los llevasen comyda que fuesen alla e ansy avian salido cargados ciento e cinquenta yndios creyendo que ansy hera y una legua deste pueblo de Quybi es un pueblo que llaman Acubingo avian metido entre unas peñas los dhos yndios e alli los avian muerto todos syendo como se entendio que eran maliçiosamente por bengarse dellos y esto rrespondyo por lengua del dho don M^{yn} ynterpetre.

4 - A la quarta pregunta del dho ynterrogatorio dixo este dho t^{o} que las dhas tierras de Quibi es gran pedaço de tierra y muy buenas e que a este dho t^{o} le paresze que valdran muy bien seis myll e quinyentos pesos porque se coje mucha coca en ellas sy se sienbran todas e que este dho t^{o} sy fueran suyas no las diera por ellos e que si las ubiera de conprar y tubiera el dinero diera lo mysmo y que no las dexara lo qual todo dixo e declaro ser verdad por lengua del dho don M^{yn} ynterpetre de la Rreal Audiençia nombrado para esta cavsa e se afirmo e rratefico en ello despues de abersele leido y encargosele el secreto de lo que a dho que no lo trate con persona alg^{a} so pena de perjuro e que le llevara el dyablo y prometyolo y no lo firmo porque no supo ny el dho don M^{yn} tanpoco.

Paso ante mi

Fran^{co} Herz de Guebara
escr^{o} de Su Mag^{t}.

M^{yn} Yusco sujeto a don Xristoval
desea que saque las trras don Xristoval
y bale ynteres / sujeto de Chacalla.

T^{o} el dho M^{yn} Yusco natural de el pueblo de Chicamarca que es en el rrepartimyento de Chacalla encomyenda de Fran^{co} de Ampuero sujeto a don Xristoval cacique del dho rrepar-//
/165v/ timyento t^{o} presentado por parte de el dho don Xristoval el qual avyendo jurado en forma devida de dr^{o} e syendo preguntado por lengua de don M^{yn} Lengua ynterpetre nombrado por los senores presydente e oidores e por el ynterrogatorio por donde se pidyo fuese hexsamynado dixo e depuso lo syguiente:

1 - A la primera pregunta del dho ynterrogatorio dyxo este dho t^{o} que conosze y conosçio a los caciques e prençipales que an sydo e son del dho rrepartimyento de Chacalla porque como dho tyene es su tierra y natural y sujeto al cacique e que los mysmos a conoscido e conosze a los que an sydo e son del dho rrepartimyento de Canta porque los a conoscido andando tratando e conversando con ellos syendo como son tan zercanos de su tyerra e que tyene notyçia de las dhas tyerras de Quibi porque las a vysto y estado en ellas e tiene en ellas tierras en que a sembrado desde Topa Ynga Yupangue.

Preguntado por las preguntas generales de la ley dixo que no sabe la hedad //
/166r/ que tiene e que se acuerda muy bien de Topa Ynga Yupangue e que tyene hijos e nyetos que son ya viejos e que como dho tiene le ba el mysmo ynteres que a los demas yndios del dho rrepartimyento de Chacalla en las dhas tyerras por ser natural de Chacalla y tener hijos e nyetos casados en la dha tyerra e que desea que saquen las dhas tierras los yndios de Chacalla pero que por eso no dexara de dezir verdad de lo que supiere y le fuere preguntado e que no le an dado ny prometydo cosa alguna para que en este su dho diga el contrario de la verdad.

2 - A la segunda pregunta de el dho ynterrogatorio dixo este dho t^{o} que lo que sabe desta dha pregunta es que estando los dhos yndyos de Chacalla en la posesion de las dhas tierras de Quibi e aviendolas labrado y sembrado dende el tyempo de los yngas e antes e al tiempo que entraron los espanoles en esta tierra a esta sazon vido este dho t^{o} que Guaman Yavre cacique prencipal del dho rrepartimyento de Canta vino a las dhas tierras de Quibi donde estavan los dhos yndios yauyos de Chacalla con muchos yndios de guerra armados y enplumados //
/166v/ que tomaron todos los zerros y el dho Guaman Yaure cacique de Canta enbio a dezir aldho Marcapoma cacique de Chacalla que le dexase las dhas tyerras de Quybi por bien donde no que le avia de dar batalla e que avian de pelear ellos dos solos y los demas caciques e principales con los otros prencipales e los yndios pobres con los pobres y que el que venciese avia de llevar las tierras e que desta manera y en esta quystion andubieron los vnos caciques con los otros dyferentes fasta que vido este dho t^{o} que vinyeron a las dhas tyerras de Quybi syendo avisados de las pasyones de los dhos caciques Fran^{co} de Ampuero encomendero de los yndios de Chacalla y Nycolas de Rribera el moço encomendero de los yndyos de Canta los quales luego como vinyeron enpezaron a fablar con los dhos caciques sobre que se conzertasen y el dho Fran^{co} de Ampuero dezia al dho don Fran^{co} Marcapoma su cacique que se conzertase con el dho Guaman Yare que le bendyese las dhas tierras de Quibi por obejas porque de otra manera que nunca estarian en paz e que syenpre tendrian guerra y que el dho Gua-//
/167r/ man Yaure le avia muerto muchos yndyos e que se los acabaria de matar los que le quedavan y el dho don Fran^{co} Marcapoma dixo que no queria vender las dhas tierras porque no heran suyas que eran de sus hijos e de sus yndyos e que ternyan mas probecho dellas que no de las ovexas que les pudyesen dar y que aviendo pasado tres dias el dho Fran^{co} de Ampuero torno a fablar con el dho don Fran^{co} Marcapoma su cacique y le dixo que ya tenya conzertado que el dho don Fran^{co} Guaman Yaure le diese duzientas cabecas de ganado por las dhas tierras de Quibi machos y henbras de ovejas y carneros de la tyerra que las rresçibiese y les dexase las dhas tierras porque de otra manera nunca ternyan paz y el dho don Fran^{co} Marcapoma torno a dezir al dho Fran^{co} de Ampuero que el no queria hazer el dho concierto ny queria dar por las dhas obexas las tierras que les avian dado a los yndios los yngas e que sus yndios se quexarian del y sus amygos se lo ternyan a mal vender las dhas tierrras porque sus yndios heran pobres de tierras que no tenyan donde sembrar e moririan de hambre e que no lo avia de hazer y torno a porfiar sobre //
/167v/ ello el el dho Fran^{co} de Ampuero al dho Marcapoma y no quiso y el dho Fran^{co} de Ampuero dyxo que pues no queria que el las queria vender y rreszebir las dhas obexas e ansy vido este dho t^{o} que las fue a rreszebir de los dhos yndyos de Canta a un pueblo que llaman Cullata y el dho Fran^{co} de Ampuero las dyo y entrego a vn criado suyo que llamavan Molina el qual las llevo a la dha çibdad de los Reyes y no entro en poder del dho cacique una oveja ny vn carnero que este dho t^{o} vyese ny entendyese e luego como paso lo susodho los dhos yndyos de Canta se enpeçaron a entrar e apoderarse en las dhas tyerras de Quibi y las labraron y senbraron e que los dhos yndyos de Chacalla como vieron que les quytavan las dhas tyerras todos chicos y grandes e viejos e chicos e mugeres davan bozes llorando al caçique y hechandole maldyçiones y diziendole que para que les quytava e vendya las dhas tyerras syendo de los yndyos pobres que donde se avian de yr e do avian de sembrar o que avian de hazer y que avian de comer y el dho don Fran^{co} Marcapoma vio este dho t^{o} que les decia: "myrad no digais que yo las vendo que no las bendo yo syno vro encomendero yo no puedo mas; que-//
/168r/ xaos del y no de mi que yo de mys hijos e de mandar que pidan las dhas tierras despues de yo muerto que yo no las vendo ny consyento tal"; e que luego los dhos yndyos yban a llorar al dho Fran^{co} de Ampuero e al dho Ribera y les hechavan maldiciones e que vysto esto los dhos encomenderos a unos yndios pobres del pueblo de Cullata e Xicamarca del dho rrepartimyento de Chacalla con voluntad de Guaman Yaure cacique de Canta senalaron e dyeron a los dhos yndyos de Chacalla ciertos pedacos de

tyerras en que sembrasen hazia un zerro desde junto a las casas del dho pueblo de Quybi hazia Canta y posyeron sus mojones y una cruz junto a las dhas casas que oy dia esta alli las quales dhas tierras son de donde los yndyos de Canta cojian la coca sobre lo qual vino un escr^{o} a fazer ynformaçion e que este t^{o} cree y tiene por cierto que si el dho don Fran^{co} Marcapoma vendiera de su voluntad las dhas tierras que rresçiviera las ovejas y las rrepartyera entre sus yndyos e no las llebara el dho Fran^{co} de Ampuero e que sy asta agora no les han pedydo las dhas tierras los dhos yndyos de Chacalla a los de Canta es por-//
/168v/ que se murio el dho don Fran^{co} Marcapoma e nunca hasta agora vido cacique que tenga buen entendymy^{o} e que sea honbre e que esto que dho tyene este dho t^{o} lo a vysto ser y pasar ansy por vysta de ojos fallandose a todo presente lo qual dixo e declaro por estas rrazones por lengua del dho don M^{yn} ynterpetre el qual dyxo avello declarado ansy.

3 - A la terzera pregunta del dho ynterrogatorio dyxo este dho t^{o} que dize lo que dho tiene en las preguntas antes desta.

4 - A la quarta pregunta del dho ynterrogatorio dyxo este dho t^{o} que dyze lo que dho tyene en las preguntas antes desta.

5 - A la quynta pregunta del dho ynterrogatorio dixo este dho t^{o} que a lo que este dho t^{o} a vysto y entendydo todos los yndyos mytimaes que pusyeron los yngas en muchas partes y lugares para su servyçio se estan agora ansy como Ynga los tenya syn que los espanoles los ayan mudado ny hecho novedad e que preguntado este dho t^{o} como hera ansy lo susodho le an dho que el Apo grande de Castilla lo tiene mandado y esto rrespondyo a esta dha pregunta.

/169r/ 6 - A la sesta pregunta del dho ynterrogatorio dyxo este dho t^{o} que dyze lo que dho tyene en las preguntas antes desta lo qual dyxo que era la verdad y lo que este t^{o} savia y en ello dyxo que se afirmava e afirmo por la dha lengua.

Anadidas

1 - A la primera pregunta del dho ynterrogatorio anadydo dyxo este dho t^{o} que antes que entrasen los yngas en esta tyerra los dhos yndyos de Chacalla salyeron a conquystar la tyerra de los Yauyos e las tyerras de Quybi con mucha jente de guerra y llegaron fasta junto a Collique y despues tornaron a dar la buelta y se consideraron con los yndyos que a la dha sazon avia en las dhas tierras de Quibi que heran yndios yungas sujetos a un senor que llamavan Collicapa que era senor de Collique e se concertaron con los dhos yndyos de Chacalla que no los conquistasen ny matasen syno que se mojonasen las tierras e que lo que avia de ser de cada uno e que no rrenyesen e ansy dyze que les senalaron en las dhas tierras de Quybi un mojon en un zerro que llaman Xudcunga que entonzes entre los yn-//
/169v/ dyos se llamava Chita e que alli los dhos yndyos yauyos avian hecho su poblaçion e enpeçaron a labrar las dhas tyerras de Quybi fasta otro mojon que llaman Chuquycoto que esta junto al dho pueblo e casas de Quybi e que las avyan tenydo y poseido bien dos vidas fasta que paso e vino Topa Ynga Yupangue el qual les mando que estoviesen los dhos yndyos yauyos como se estavan y el dho Topa Ynga Yupangue senalo tyerras para el y para su muger para que los dhos yndyos yavyos de Chacalla sembrasen en ellas e que esto como dho tyene lo a oydo dezir este dho t^{o} por cosa publica y notoria a sus mayores y mas ancianos que es que decian que ellos en sus tpos ansy lo avyan vysto ser y pasar e acostunbrar e oidolo dezir a sus mayores e mas ancianos los quales dezian lo mysmo e que dello avia sydo y hera publico e notorio e publica voz y fama e que nunca avian vysto ny oydo dezir lo contrario e que despues de muerto Topa Ynga Yupangue suzedio Guaynacava el qual avia mandado que los dhos yndios yauyos se estobiesen de aquella manera y senbra-//
/170r/ sen e beneficiasen las dhas tierras para el dho Guaynacava e para ellos e que esto lo save este dho t^{o} porque se acuerda bien dello e fue en su tyempo e vido benyr a lo susodho por mandado del dho Guaynacava un criado suyo orejon que no se acuerda como se llamava e que despues de muerto Guaynacava suzedyo Guascar Ynga el qual vido este dho t^{o} que hizo lo mysmo e desta manera los dhos yndyos yavyos de Chacalla tobyeron y poseyeron las dhas tyerras de Quybi y las labraron y senbraron fasta que vinyeron los espanoles a esta dha tyerra y fasta que se entregaron por el dho Fran^{co} de Anpuero al dho cacique de Canta por las dhas obejas lo qual todo dyxo e declaro este dho t^{o} por lengua del dho don M^{yn} Lengua ynterpetre que dyxo avello declarado ansy por las rrazones aviendosele preguntado.

2 - A la segunda pregunta del dho ynterrogatorio dyxo este dho t^{o} que dyze lo que dho tiene en las preguntas antes desta e que lo en esta dha pregunta contenydo lo vydo ser y pasar ansy por vysta de ojos como la dha pregunta lo //
/170v/ dize y declara y esto rresponde.

3 - A la terzera pregunta del dho ynterrogatorio dyxo este dho t^{o} que lo que save desta dha pregunta es que en tyempo de los espanoles que entraron en esta dha tyerra quando se alço Topa Ynga Yupangue desbaratado huyendo del dho desbarate y estando Bilcapoma cacique de Chacalla con Guaman Yaure cacique de Canta aguardando a Mango Ynga el dho cacique de Canta acuso al dho Bilcapoma e dyxo a Monga [*sic*] Ynga que tenia dos coracones que avia enbiado cartas al marques y el dho Mango Ynga dyzen que le rrino y le dyo con una porra y le dexo y se fue facia su asiento e que luego los dhos caçiques de Canta le avian muerto e que despues de muerto avian benydo a las dhas tierras de Chacalla e dho a los yndyos que Bilcapoma su cacique estava en Canta con Mango Ynga que les mandava que les llevasen chica [*sic*] e mahiz e otras comydas e que los dhos yndyos de Chacalla creyendo que hera ansy avian ydo çiento e cinquenta yndyos cargados destas comy-//
/171r/ das e que llegados a un pueblo que llamavan Coyngo los dhos yndyos de Canta avian metido los dhos yndyos en un corral e alli los avian muerto a todos diziendo que querian bien a los espanoles e quepor ello avian de morir e que esto que dho tyene en esta

dha pregunta le fue a este dho t^{o} publico e notorio por avello oido dezir en esta comarca entre los yndyos que dezian avello vysto e porque nunca los dhos yndyos parescieron mas muertos ny bibos.

4 - A la quarta pregunta del dho ynterrogatorio dixo este dho t^{o} que las dhas tierras de Quybi sobre que es este dho pleyto son gran golpe de tyerras e muy buenas e que a este dho t^{o} le paresze que balen mas de syete myll pesos e que si fueran suyas que no las diera por ellos e que lo que a dho era la verdad y lo que deste caso sabia para el juram^{o} que hizo y en ello dyxo que se afirmava e afirmo por lengua del dho don M^{yn} despues de aversele leydo y encargosele tenga secreto deste su dho y prometyo de fazello ansy e no lo firmo porque no supo ny tanpoco la dha lengua.

Paso ante mi
Fran^{co} Hrrez de G^{a} esc^{o} de Su M^{t}

/171v/ Fran^{co} de Ampuero
señor de la encomyenda de Chacalla
de hedad de 40.

T^{o} el dho Fran^{co} de Ampuero v^{o} de la çibdad de Los Reyes t^{o} presentado por parte de los yndyos yabyos de Chacalla aviendo jurado en forma e siendo preg^{do} dixo lo sig^{e}:

1 - A la primera pregunta dixo que conosze a los yndios e caçiques de Chacalla porque se encomyenda dellos e conosçio a Guaman Yaure cazique de Canta.

Preguntado por las generales de la ley dixo que es de hedad de quarenta años poco mas o menos tpo e que no le toca neng^{a} de las gen^{ales} e que los dhos yndios de Chacalla son de su encomyenda pero que por esto no dexara de dezir verdad.

2 - A la segunda pregunta dixo que antes y al tpo que el dho concierto se hiciese entre los yndios de Canta y de Chacalla sobre las tierras de Quybi trayan unos yndios con otros de este rrepartimy^{o} guerra y pendencias y los de Canta como favorescidos del marques cuyos heran y despues de Gonçalo Piçarro matavan y herian muchos yndios del dho rrepartimy^{o} de Chacalla e algunos los tenyan descalabrados que oy dia son bibos algunos de estos y andavan todos desasosegados y bisto esto este t^{o} y el padre Godoy enpurtunaron en esta çibdad a don Fran^{co} Marcapoma caçique de Chacalla que fuese a las dhas tierras de Quibi y este t^{o} fue alla y le llebo consygo para dar un corte en lo susodho y lle-//
/172r/ gados que fueron este t^{o} fablo con los dhos yndios para que hiciesen concierto y el dho don Fran^{co} ny sus yndios no querian fazello porque benian llorando algunos yndios de Chacalla y quexandose que las dhas tierras heran suyas que no se las vendiesen y este t^{o} contra la voluntad de los dhos yndios procuro que se hiciese y en hefeto los yndyos de Canta dieron docientas obexas pacos y guacas y el dho cacique e yndios de Chacalla no las quysieron rreszebir y los yndios de Canta las llevaron a las tierras de Chacalla y el padre Godoy le hizo saver como estavan alli a este t^{o} y este t^{o} enbio por ellas y se las traxeron a esta çibdad menos catorze poco mas o menos porque este t^{o} no se acuerda vien e que las que faltavan cree que las tomo el padre Godoy para si y las que ansy traxeron a esta çibdad dellas bendio este t^{o} y dellas paso en la carnesceria y esto rresp^{e}.

3 - A la tr^{a} pregunta dixo que dize lo que dho tiene e que es verdad que este t^{o} firmo el dho conçierto.

4 - A la quarta pregunta dyxo que dize lo que dho tiene.

/172v/ 5 - A la quynta pregunta dixo este dho t^{o} que este t^{o} se acuerda aver vysto una zedula en que Su M^{t} mandava que los yndios se estuviesen como les tomo la voz los espanoles en esta tyerra e que esta zedula se presento en un pleyto que traya Fran^{co} de Caravante y Merlo sobre los mitimaes de Xauxa e que es cosa muy publica y not^{a} que Su M^{t} tyene probeida esta zedula e se traxo a este rreyno e a esta causa cree este t^{o} que estan todos los yndios de la mysma manera que les vino la voz al tpo que el marques don Fran^{co} Piçarro entro en esta tierra porque de otra manera ubiera avido muchos pleytos.

6 - A la sesta preg^{a} dixo que dize lo que dho tiene.

Anadidas

1 - A la primera pregunta anadida dixo que este t^{o} ansy a entendido lo cont^{do} en la preg^{a} e que si alguna vez las dexasen de poseer seria huyendo de los yndios que los mataban e luego se tornarian.

2 - A la segunda pregunta dixo que sabe que los dhos yndios yavyos de Chacalla despues que estan en encomyenda de este t^{o} //
/173r/ se puede aver veynte años poco mas o menos estuvieron en las dhas tierras fasta que se hizo el concierto y oy en dia cree este t^{o} que ay algunos.

3 - A la tr^{a} pregunta dixo que en tpo de Yllatopa tenyendo este t^{o} ya encomendados los dhos yndios le fue a este t^{o} muy pu^{co} e not^{o} que los yndios de Canta mataron a Bilcapoma caçique prençipal de Chacalla con otros muchos yndios que con el estavan porque

dezian que favorescian los espanoles y en esta sazon los dhos yndios de Canta se les enpecaron a entrar en las tierras de Quibi a los yavyos y a tener con ellos guerra como contrarios que le avian muerto su cacique.

4 - A la quarta pregunta dixo que save e a entendido entre los yndios que las dhas tierras las tienen en mucho por tener agua porque en la syerra son faltos de agua y se bajan a las bertientes de los rrios y fazen sus chacarillas a mano con mucho trabajo y que el valor dellas que este t^{o} no lo save. E lo que dho tiene dixo que es la verdad y en ello se afirmo y lo firmo de su n^{e}.

Fran^{co} de Ampuero

ante my

Fran^{co} Hrrez de G^{a}
escr^{o} de Su Mag^{t}

/174r/ Ju^{o} Yauri / de 40 / no le tocan.
encom^{o} en M^{rin} Piç^{o}.

T^{o} el dho Ju^{o} Yabre del rrepartimy^{o} de M^{yn} Piçarro present^{do} por parte de los yndios de Chacalla aviendo jurado e siendo preg^{do} por lengua de Alvarado ynterpetre dyxo lo sig^{e}:

1 - A la primera pregunta dixo que tiene noticia de los yndios de los rrepartimy^{o}s de Chacalla y Canta.

Preguntado por las gen^{ale}s de la ley dixo que no sabe la hedad que tiene e paresce de quarenta anos e que no le toca nenguna de las gen^{ale}s.

2 - A la segunda pregunta dixo que este t^{o} se fallo p^{e} quando se tratava del concierto de las dhas tierras entre los yndios de Chacalla y Canta porque hera por alli su pasaje para su tierra e vido como lloravan yndios yauyos de Chacalla porque el dho Fran^{co} de Ampuero les vendia las tierras por las dhas ovejas las quales entendio que avia rr^{do} el dho Fran^{co} de Ampuero y deçia a los yndios de Chacalla que dexasen las tierras a los de Canta e que tomasen las ovejas y los yndios de Chacalla y Marcapoma su caçique decia que no queria dexallas.

Anadidas

1 - A la prim^{a} pregunta anadida dixo que este t^{o} vido desde que se save acordar algunos yndios yavyos de Chacalla pobladas las dhas tierras de Quybi y las tubieron //
/174v/ fasta que se hizo el concierto.

2 - A la segunda pregunta dixo que save e vido al tpo que entraron los espanoles en este rreyno los dhos yndios yavyos de Chacalla estavan en las dhas tierras de Quybi y esto dixo que savia e declaro por la dha lengua.

ante mi
Fran^{co} Hrrez de G^{a} escr^{o} de Su M^{t}

/175r/ Sebastian Sanchez de Merlo
de 43 e no le tocan.

T^{o} el dho Fernan [*sic*] Sanchez de Merlo v^{o} desta çibdad de Los Reyes aviendo jurado por parte de los yndios de Chacalla e siendo preg^{do} dixo lo sig^{e}:

1 - A la primera pregunta dixo que t^{e} noticia de los dhos rrepartimy^{o}s de Chacalla y Canta.

Pregunt^{do} por las gen^{les} de la ley dyxo que es de hedad de quarenta e tres anos e que no le toca nenguna de las gen^{les}.

5 - A la quynta pregunta dyxo que este t^{o} no a leydo la zedula en ella cont^{da} pero que a vysto de diez y seis o diez y siete años que esta en estas partes del Peru poco mas o menos que se a usado e guardado e usa e guarda estarse los yndios mytimaes que se hallaron quando entraron espanoles en esta tierra en la pte y lugar donde los fallaron y servirse dellos los caçiques de donde son naturales esp^{lmente} e a vysto este t^{o} y bee en los vn^{o}s desta çibdad de Los Reyes porque este t^{o} y todos los que tienen rrepartimy^{o}s de yndios yavyos tyenen yndios mytimaes en el valle de Xauxa y bee que se suben los caçiques yavyos dellos syn que se les ponga enbargo ni ynpedimi^{o} e que este t^{o} cree y tiene por cierto que mediante la zedula de Su M^{t} se guarda esta cost^{e} y que esto es lo que save y en ello se afirmo y lo firmo de su n^{e}.

Sebastian de Merlo

ante mi
Fran^{co} Hrrez de G^{a}
es^{o} de Su M^{t}

/175v/ Fran^{co} de Caravantes
de 38 / no le tocan.

T^{o} el dho Fran^{co} de Caravantes v^{o} desta çibdad de Los Reyes t^{o} presentado por pte de los yndios de Chacalla aviendo jurado en forma e siendo preg^{do} dixo lo sig^{e}:

1 - A la primera preg^{a} dixo que tiene not^{a} de los dhos rrepartimy^{o}s de Chacalla y Canta.
Pregunt^{do} por las pregs de la ley dixo que es de hedad de mas de treynta e ocho años e que no le toca nenguna de las gen^{es}.

5 - A la quynta preg^{a} dixo que este t^{o} a oydo dezir e que Su M^{t} tiene m^{do} que se guarde lo que Guaynacava hizo e no save otra cosa y lo firmo de su n^{e}.

Fr. Caravantes de Marudaz

ante mi

Fran^{co} Hrrez de G^{a} escr^{o} de Su Mag^{t}

Ju^{o} de Padilla / 28 / e no le tocan.

T^{o} el dho Ju^{o} de Padilla escr^{o} pu^{co} desta çibdad de Los Reyes t^{o} presentado por pte de los yndios de Chacalla aviendo jurado en forma e siendo preg^{do} dixo lo sig^{e}:

1 - A la primera pregunta dyxo que tiene not^{a} de los dhos rrepartimi^{o}s de Canta y Chacalla.
Preg^{do} por las gen^{les} de la ley dixo que es de hedad de mas de veynte e ocho anos e que no es p^{e} ny le toca nenguna de las gen^{les}.

/176r/ 5 - A la quynta pregunta dixo que lo que della sabe es que se quyere acordar aver oido dezir que Su Mag^{t} tiene proveydo e mandado que las cosas de los yndios se esten de la manera que los espanoles las fallaron quando entraron en este rreyno e no save otra cosa y lo firmo de su n^{e}.

Joan de Padilla
escrivano pu^{co}

ante mi
Fran^{co} Hrrez de G^{a}
es^{o} de Su Mag^{t}

Nyculas de Rribera el moço
encomendero de los yndios de Canta
de mas de 40.

T^{o} el dho Nycolas de Ribera v^{o} desta çibdad encomendero de los yndios de Canta t^{o} present^{do} por los yndios de Chacalla aviendo jurado e siendo pregunt^{do} dixo lo sig^{e}:

1 - A la primera pregunta dixo que este t^{o} se fallo presente al concierto que se hizo en Quybi de las dhas tierras y que este t^{o} vio al dho don Fran^{co} muy alegre y contento de concertarse y hizo el dho concierto a lo //
/176v/ que este t^{o} entendio de su voluntad e que nunca este t^{o} tal oyo aquella sazon ny en ningun tpo sy no es agora que lo dize la preg^{a}.

3 - A la tr^{a} pregunta dixo que los yndios de Canta vido este t^{o} que enbiaron unos principales y el cacique de Chacalla don Fran^{co} enbio otros principales y este t^{o} enbio un ombre que se dize Di^{o} Florez con ellos y los yndios de Canta entregaron las docientas ovejas a los yndios de Chacalla que enbio el dho don Fran^{co} y las llevaron a Chacalla y esto save desta preg^{a}.

4 - A la quarta preg^{a} dixo que dize lo que dho tiene.

5 - A la quynta preg^{a} dixo que lo save e a la provision e cedulas de Su Mag^{t} se rremyte.

Anadidas

1 - A la primera preg^{a} dixo que no lo save e dize lo que dho tiene e que save que se hizo el dho concierto que fue la primera vez que este t^{o} vio aquellas tierras vysto poseer a los yndios de Canta.

2 - A la segunda preg^{a} dixo que dize lo que dho tiene.

3 - A la tr^{a} preg^{a} dixo que no lo save ny otra cosa del ynterrogatorio y en ello //
/177r/ se afirmo y lo firmo de su n^{e}.

Nicolas de Ribera

ante mi
Fran^{co} Hrrez de G^{a}
es^{o} de Su Mag^{t}

E yo Fran^{co} Hrrez de Guebara escr^{o} de Su Mag^{t} e rrecevtor nonbrado para la p^{e} causa presenta presentacion de los t^{o}s juram^{o} e declaraçiones de los t^{o}s desta probança e otros autos; lo escrebi e hize aqui myo signo tal en testimy^{o} de verdad.

Fran^{co} Hrrez de G^{a} escr^{o} de Su Mag^{t}

Tasacion de costas

Del rrequerimi^{o} con la p^{o}bision		IIII ts
De present^{on} e juram^{o} de siete t^{o}s; catorze ts	Ip	VI ts
De p^{on} de dos ynterrogatorios: quatro ts		IIII ts
De un my^{o} para t^{o}s: quatro ts		IIII ts
Deste signo: dos ts		II ts
De treynta y ocho ojas: diez y nuebe p^{o}s	XIXp^{o}s	
De la ocupaçion seis dias en todo lo de Chacalla: quinze p^{o}s	VIp^{o}s	
	XXXVIIp^{o}s	IIIIts

Les rresçebi beynte e cinco p^{o}s
de besamen; lo demas

Fran^{co} Hrrez de G^{a}

/178r/ Por las preguntas siguientes sean preguntados los t^{o}s que son o fueren presentados por parte del caçique e y^{o}s de Canta encomendados en Niculas de Ribera el moço en el pleyto que tratan con los yndios de Chacalla sobre las tierras del valle de Quibi que estan junto a las tierras e pueblo de los yndios del dho valle de Canta.

1 - Primeramente si conocen a las partes e a cada vna dellas y tienen notiçia del valle e tierras que dizen de Quybi que estan en los termy^{o}s del dho valle de Canta e de que tpo a esta parte / sy conosçieron a los yngas pasados reyes e senores que fueron destos rreynos espeçialm^{te} a Topa Ynga Yupangui y a Guaynacap y tienen notiçia de un rrio que pasa junto a las dhas tierras que se llama el rrio de Quibi.

2 - Yten si saben etc. que las dhas tierras que dizen de Quibi estan dentro de los termy^{o}s e pueblos de Canta e de sus parçialidades ny mas ny menos que esta en la figura que mostraron a los señores oydores porque los tr^{o}s los parte vn çerro grande que esta entre los yabyos e yn^{o}s de Chacalla y los dhos yndios de Canta. Pido que los testigos declaren como se parten los tr^{o}s y como caen dentro de los dhos tr^{o}s de Canta las dhas tierras de Quibe. Digan lo que saven.

3 - Yten si saben etc. que las dhas tierras e chacarras de Quibe desde uno y dos y diez y veynte y treynta e çinq^{ta} e mas anos a esta parte que memorias de honbres no es en contrario siempre fueron tierras e chacaras de los dhos yndios de Canta y su caçique las quales siempre las an tenydo y poseydo por suyas y como suyas labrandolas y senbrandolas y haziendo dellas como de cosa propia hasta que se metieron en çierta parte dellas çiertos yndios como mytimaes de los dhos yndios de Chacalla por çierto engano que les hizieron.

Digan lo que saven.

4 - Yten si saben etc. que antes que vinyesen xristianos a esta tierra se tenya por costumbre en ella entre los naturales de hazer çiertas rrogativas o plegarias por los reyes e sus señores a que llamavan ellos Capacocha lo qual hera que matavan çiertas obejas y les sacavan la sangre y la hechavan en unas ollas o mates e yvan con aquello en procision y si alguno dellos derramava una gota de la dha //
/178v/ sangre o se la hazian derramar al yndio que asi la derramava o la hazia derramar lo enterravan en el propio lugar y lo matavan para ello y esto era ley ynviolable entre ellos. Digan lo que saven.

5 - Yten si saben etc. que quando hazian la dha Capacocha yban los de un pueblo hasta los termi^{o}s del otro con çiertas çerimonyas y hasta donde llegavan lo ponyan por termy^{o} de los que llevaban la dha Capacocha y lo señalavan por suyo.

Digan lo que saven.

6 - Yten si saben etc. que al tpo que Guaynacap yba a Quito hizieron los dhos y^{o}s de Chaclla la dha Capacocha y con la sangre susodha llevandola en unas ollas o mates fueron por la parte donde estan las dhas tierras de Quibe y con las dhas çerimonyas y los dhos yndios de Canta queriendolos yr a detener no osaron por no derramar la dha sangre e yncurrir en la dha pena de muerte e asi con la dha color yntentaron de se quedar y se quedaron con una parte de las dhas tierras de Quive diziendo que hasta alli les pertenesçia por aver llegado con la dha Capacocha hasta las dhas tierras de Quibe en la parte donde llegaron y sobre que es este pleyto lo qual puede ser la tierra [*sic*] parte de las dhas tierras poco mas o menos. Digan lo que saben.

7 - Yten si saben etc. que con la dha color de la dha Capacocha los dhos yndios de Chaclla pusieron mytimaes en la parte de las tierras de Quive donde llegaron con la dha Capacocha sin tener otro dr^{o} alguno a las dhas tierras. Digan lo que saben.

8 - Yten si saben etc. que despues que el dho Guaynacap murio y antes que entrasen los xristianos en este rreyno o quando entraron los dhos yndios de Chaclla se salieron los dhos yndios de Chaclla y los mytimaes que avian puesto por mandado de Ylla Topa Ynga hijo de Guaynacapa y luego los dhos yndios de Canta se //
/179r/ tornaron a las dhas sus tierras de Quive ni mas ni menos que de antes las tenyan las quales an tenydo y poseydo desde entonçes asta agora.

Digan lo que saben.

9 - Yten si saben etc. que por mandado del presidente Gasca y por deferençias que tenyan sobre las dhas tierras los dhos yndios de Chaclla se conçertaron con los dhos yndios de Canta en que les diesen dozientas obejas por que se apartasen del pleyto y diferençias de las dhas tierras de Quibe.

Digan lo que saben.

10- Yten si saben etc. que en cumplimy^{o} del dho conçierto los dhos yndios de Canta dieron a los dhos yndios de Chaclla dozientas cabeças de ovejas las quales les dieron y entregaron a ellos mesmos y a su caçique. Digan lo que saben.

11- Yten si saben etc. que en todas las dhas tierras de Quive no ay veynte hanegas de sembradura o quando mas veynte e çinco y todas las mas dellas son andenes y tierra pedregosa e a justa e comun estimaçion cada hanega de sembradura puede valer hasta veynte pesos poco mas o menos. Digan lo que saben.

12- Yten si saben etc. que las chacarras e tierra sobre que es este pleyto e diferençia puede ser la terçia parte de la dha tierra e menos que fue hasta donde llegaron con la dha Capacocha los dhos yndios de Chaclla. Digan lo que saben.

13- Yten si saben etc. que cada cabeça de ganado de obejas de las que dieron los dhos yndios de Canta a los dhos yndios de Chaclla valian seys pesos e mas a justa e comun estimaçion al tpo que se las dieron. Digan lo que saben.

14- Yten si saben etc. que al tpo que hizieron el dho conçierto los dhos yndios de Chaclla //
/179v/ con los dhos yndios de Canta valian mucho menos las tierras que agora valen vn terçio y mas por darse como se dan agora mas a la labor que entonçes. Digan lo que saben.

15- Yten si saben etc. que la causa por que hizieron el dho conçierto los dhos yndios de Canta con los dhos yndios de Chaclla en las dhas tierras de Quive hera y fue porque las dhas tierras de Quive heran y fueron de sus pasados y suyas y por quitarse de aquella molestia. Digan lo que saben.

16- Yten si saben etc. que todo lo susodho es pu^{co} y not^{o} e pu^{ca} boz y fama.

Otro si por quanto se enbia escrivano e personas nonbrados por las partes a ver las dhas tierras pido y suplico a V. Al^{a} mande que las traygan señaladas los dhos terçeros y escrivanos e den fee de como estan y ansi mysmo traygan señalado el pedaço de tierra sobre que es la diferençia y asy mysmo que traygan tasadas todas las dhas tierras de lo que pueden valer y las hanegadas que pueden tener sobre que pido justiçia y costas y en lo nesçesario el off^{o} de V. Al^{a} ynploro / El Lic^{do} de Leon / Fran^{co} de la Torre

conçertose con el original Diego Muñoz

/180r/ Muy p^{o}s señores

Ffran^{co} de la Torre en nonbre de los y^{o}s de Canta en el pleyto con los yndios de Chacalla sobre las tyerras de Quibi. Digo que para que asista con la lengua que exsamynare los testigos que se presentaren en la dha causa para que no se pregunte ny se declare vna cosa por otra nonbro por terçero de my parte al padre Geronymo M^{yn} que sabe muy bien la lengua e porque en aquello podria aver cautela.

Pido y suplico a Vra Alteza mande que el dho padre Geronymo M^{yn} asista a la declaraçion de los testigos con la dha lengua e si las partes contrarias quysieren nonbrar otro que lo nonbren para que çese qualquier cautela que en ello puede aver / sobre que pido justiçia e para ello etc.

Otro si por quanto en las preguntas del ynterrogatorio que presente faltaron por fazer otras çiertas preguntas para que se entienda y sepa la verdad. A Vra Alteza pido y supp^{co} mande que los testigos que se presentaren se pregunten asimysmo por estas preg^{tas} añadidas:

1 - Yten si saben etc. que en tpo de Guascar Ynga que fue señor destos rreynos despues de Guaynacapa se mandaron hazer dos capacochas de las susodhas una para su muger e otra para el e trayendo los dhos yndios de Chacalla la dha capacocha e vinyendo a labar la sangre al dho rrio de Quibi los dhos yndios de Chacalla se entraron en lo mejor de las dhas tyerras de Quybi donde agora esta senbrado de coca e pacays e porque se entravan en las dhas trras los dhos yndios de Chacalla un yndio que se dezia Chuqui Guaranca rrenpujo a uno de los dhos yndios //
/180v/ que traya la dha capacocha diziendo que no se entrasen por su tyerra e porque le derramo pte de la dha sangre ahogaron el dho yndio que hera de la parçialidad de los de Canta e le enterraron junto a un molle que alli esta. Digan lo que saben.

2 - Yten si saben etc. que de la mysma manera mataron otros quatro yndios de los de Canta defendiendo las dhas sus tyerras uno que se dezia Yniguy Chunbi y el otro Chaumecaxa y el otro Tanta Capo y el otro Guaman Llaupi.

3 - Yten si saben etc. que en las dhas tyerras de Quibi no huvo mytimaes puestos por el ynga syno çierto rresto de un caçique que mando matar el ynga que se llamava Collotanbo de los de Chacalla cuya generaçion con todos los varones fue muerta por çierta culpa de trayçion que tuvo e unas pocas de mugeres que quedaron las paso a un corral de Quibi para que hilasen alli. Digan e declaren lo susodho como paso e por que via entraron alli las dhas mugeres e si ovo mytimaes o no e que tierras le dieron en que bivieron.

4 - Yten si saben etc. que çiertas tyerras de las dhas tierras de Quibi que estan cabe las casas que son las que Juan Xristoval de Frias fue a ver son asimysmo de los dhos yndios de Canta / e la causa por que agora la piden los dhos yndios de Chacalla es porque despues que les dieron las ovejas e hizieron la dha transaçion vn yndio que se llama Pomaraque / e otro Tanta de los yndios de Chacalla dixeron a Guaman Yauri caçique de Canta que se querian pasar a su tierra a tributar con el y el dho caçique les dixo que le darian aquel pedaço de tierra e otros si se pasasen como dezia e asi con el dho conçierto le señalaron las dhas trras a los dhos dos yndios de Chacalla. Digan lo que saben.

5 - Yten si saben etc. que porque los dhos Pomaraque e Tanta dixeron que se querian passar los //
/181r/ hizo açotar el caçique Marcapoma e asi no se pasaron ny el dho caçique de Canta les dio las dhas tyerras syno se las torno a tomar para si y las senbraron de coca los yndios de Canta.

6 - Yten si saben etc. que puede aver tres años poco mas o menos que un don Juan prinçipal de Canta se hizo amygo con don Xristoval prinçipal de Chacalla y sin lo poder hazer por que se lo rrogo el dho don Xristoval le dio liçençia para que coxiese la coca del dho pedaço de tyerra e asi el dho don Xristoval la coxo dos o tres vezes e con el dho achaque pretenden tener drcho a las dhas tyerras sin averselas dado nadie ny poseydolas por otra causa ny aviendo plantado la dha coca. Declaren los testigos como y de que manera paso lo susodho.

7 - Yten si saben etc. que todo lo susodho es pu^{ca} boz y ffama.

El Lic^{do} de leon

conçertado con el original — Diego Muñoz

En el pueblo de Quybi a diez y siete dias del mes de hebr^{o} de myll e quy^{o}s e çinquenta e nueve años ante my el dho esc^{o} e rrecevtor por parte de los caçiques e yndios de Canta se presento dos ynterrogatorios e un testimonyo de ..... que su tenor es el syg^{e}:
e luego yo el dho rrezevtor rrequeri a los dhos yndios que me den luego t^{o}s que estoy presto de los hexsamynar donde no que sy por no me los dar me estobiere .... lo cobrare de sus bienes. T^{o}s pres^{s} Fray Pablo e don M^{yn} Lengua.

Fran^{co} Hernandez de G^{a}
escr^{o} de Su M^{t}

/181v/ E luego este dho dia los dhos yndios de Canta paresçieron ante my e dixeron que los t^{o}s de que se entendien aprobechar estan en Maca y en otras ptes e no quyeran benir que es demas para ello. Ts^{o} don M^{yn}.

E luego yo el dho escr^{o}di a los dhos yndios dos mandamy^{o}s para los t^{o}s e pueblos que me nonbraron. Ts^{o} don M^{yn}.

Fran^{co} Hrez de G^{a}
escr^{o} de Su Mag^{t}

En el pueblo de Quybi que dizen ser tierra de Canta y Chacalla a veynte dias del mes de hebrero de myll e quy^{o}s e cinquenta e nueve años ante my escriv^{o} e rrezevtor por pte del cacique e yndios de Canta.Se presento por t^{o}s a don P^{o} cacique de Guancayo encomyenda de Juan de Caceres e a don Luis Çacalla Chunbi de los quales yo el dho escr^{o} por lengua de don M^{yn} ynterpetre tome e rresçibi juram^{o} en forma de dr^{o} y ellos lo hizieron e rrespondieron: "sy ju^{o} e amen."

En Quybi a veynte e un dia del dho mes de abril del dho año ante my el escr^{o} por pte de los dhos yndios de Canta.Se presento por t^{o}s a Xristoval Caxallauxe de la encomyenda de Ju^{o} de Caceres e a Cacha yndio ynfiel natural de los Atavillos de Montenegro y juraron por la dha lengua segun su ley.

En Quybi veynte e quatro dias del mes de abril del dho año por pte de los yndios de Canta. Se presento por ts^{o}s a don Fran^{co} Arcos e Atoco yndio ynfiel e a don Ju^{o} Cachayarure e a don Fran^{co} Cacha Guaranga e Xaca yndio natural de Yacre del rrepartimy^{o} de Canta y por la dha lengua juraron segun su ley.

/182r/ En Quybi a veynte e ocho dias del dho mes de hebr^{o} ante my el escr^{o} por pte de los yndios de Canta. Se presento por t^{o}s a Cavra Bilca e A^{o} Uro e Domingo Pariana e don A^{o} Aynamare yndyos del rrepartimy^{o} de Canta los quales juraron por la dha lengua segun que los de suso.

En Los Reyes a seis de março del dho año. Paso ante my el dho escr^{o} por pte de los yndios de Canta.Se presento por t^{o}s don Fran^{co} caçique de Collique del rr^{o} de Caceres e don Diego Chunbiquybi caçique del pueblo de Quybi e Santiago yndio natural de Canares los quales juraron en forma por la dha lengua segun que los de suso.

ante mi

Fran^{co} Hrrez de G^{a}
escr^{o} de Su M^{t}

/183r/ *Probança de los y^{o}s de Canta*

Don P^{o} caçique del p^{o} de Guancayo
de Her^{do} [*sic*] de Caçeres
de mas de 30 a^{o}s / no le tocan.

T^{o} el dho don Pedro caçique del pueblo de Guancayo encomendado en Gonzalo de Cazeres hijo del contador Juan de Caceres defunto vz^{o} de la çibdad de Los Reyes t^{o} presentado por parte del caçique e yndios de el pueblo de Canta en el pleyto que tratan con los yndios de Chacalla sobre las tierras de Quibi aviendo jurado en forma de dr^{o} e siendo preguntado por el tenor del dho ynterrogatorio por donde pidio fuese hexsamynado dixo e depuso lo s^{e}:

1 - A la primera pregunta de el dho ynterrogatorio dixo este dho t^{o} que conosze a todas las dhas partes e a cada uno de los dhos pueblos de Chacalla y Canta e que a los yndios yngas a Guaynacaba no los conosçio syno de vysta que sus padres e pasados e otros ançianos le an dho que eran señores desta tierra e que tyene notiçia de las dhas tierras de Quybi porque las a vysto y est^{do} en ellas muchas vezes y su rrio alinda con las tierras deste t^{o} que es una legua de Quybi y que sabe e a bysto que el dho rrio pasa por junto a las dhas tyerras de Quybi.

Preguntado por las preguntas generales de la ley dixo que es de hedad //
/183v/ de mas de treynta anos e que no es pariente ny henemygo de neng^{a} de las dhas partes ny le an dado ny prometido cosa alguna para que en este su dho diga el contrario de la verdad e que dira verdad porque es xristiano e que ayude Dios a la pte que ubiere just^{a}.

2 - A la segunda pregunta de el dho ynterrogatorio dixo este dho t^{o} que a lo que este t^{o} a visto y entiende las dhas tierras de Quibi estan dentro de los propios termynos de Canta e mas zercanas a sus pueblos de Canta que no a los pueblos de Chacalla e que los termynos e mojones de Canta estan señalados desde un zerro que tiene un mojon viejo que a lo que este t^{o} a oydo dezir a sus antepasados antes que entrasen los yngas en esta tierra corria dende alli el termyno de Canta hacia Canta e que despues que entro Ynga en esta tierra avia quytado los mojones y señaladolos por mojones de yndios yauyos encomendados en Fran^{co} de Ampuero del rrepartimyento de Chacalla y los puso alli para mytimaes para que le syrbiesen de cargas y de chasques y de otros serviçios y andando el tpo los dhos yndios yungas del rrepartimyen-//
/184r/ to de Chacalla se fueron con el fabor de los yngas y de Guaynacaba ganando tyerra y metiendose poco a poco fasta llegar a los dhos pueblo [*sic*] de Quibi so color que yban con sus zerimonyas a adorar al demonyo y esto como dho tiene lo a entendido este dho testygo de los yndios viejos que son ya muertos y otros bibos entre los quales a sydo y es cosa notoria y esto rresponde.

3 - A la tr^{a} pregunta del dho ynterrogatorio torno este t^{o} a declarar lo mysmo que en la pregunta antes desta e mas dixo que desde que vino el presydente Gasca a estos rreinos que en aquella sazon se conçertaron los dhos yndios de Canta con los de Chacalla sobre diferençias que tenyan por las dhas tyerras syempre los dhos yndios de Canta an tenydo por costumbre de gozar y sembrar

las dhas tierras de Quibi y cojer el fruto dellas para si syn que los de Chacalla se ayan quexado fasta agora que les bee traer pleyto sobre ellas y esto rresp^{e}.

4 - A la quarta pregunta del dho ynterrogatorio dixo este dho t^{o} que lo que la pregunta dize lo a oydo dezir e platicar muchas e dibersas vezes a muchos viejos e ançianos //
/184v/ que son ya muertos e agora son bibos que en tpo de los yngas antes que entrasen xristianos en esta tierra se facia e acostunbraba fazer entre los dhos yndyos ansy desta comarca como de todo este rreyno lo que dize la pregunta de la forma y manera que en ella se contiene y declara los quales dezian que ansy lo vian e aunque tanbyen lo avyan oydo dezir a otros sus mayores y mas ançianos que dezian averlo vysto y tenello por ley como agora las leyes de los xristianos e que sy esto no fuera e pasara ansy que le paresce a este t^{o} que no se dixera ny tratara entre los viejos e antiguos e que aun mas les oyo dezir quando yban haziendo la dha zeremonya sy avia algun paso agrio que no podian pasar do estava el ofresçimyento a quyen ellos yban a buscar que tomaba el yndio que lo llebaba y lo ponya la basija en que lo llebaba en una honda como quyen tira una piedra y la arronjaba [*sic*] a el adoratorio y si por acaso no lo azertava tomaban el yndio y lo matavan luego alli syn que se mudase y lo enterravan en el mysmo paso e que esto fazian rrogando a la guaca //
/185r/ que adoravan que dyese su ciencia a Ynga e a sus hijos e para que los hiciese grandes senores e para que a ellos les diese muchas chacarras y comydas e mucha faena e que fuesen en multiplico e que desta mysma manera faziendo esta ceremonya oyo dezir que se avian yendo entrado los dhos yndios yauyos en las tierras de Quybi desde donde el inga los abia puesto por mytimaes e que Ynga se holgava dello porque los queria mucho porque eran sus criados lo qual todo a sydo y es publico y notorio a todos los yndios que lo an oydo y entendido como este dho t^{o} e nunca a visto ny oydo dezir lo contrario dello y esto rresp^{e}.

5 - A la quynta pregunta del dho ynterrogatorio dixo que lo en la dha pregunta contenydo como en ella se es y declara lo a oydo dezir este dho t^{o} a sus padres e pasados e a otros viejos e ancianos que dezian aver bisto en tpo de los yngas ser y pasar e acostumbrar lo en la dha pregunta contenydo como en ella se contiene y declara los quales lo decian tratando zerca de sus rritos e cerimonyas antiguas e que dezian que lo mysmo avian ellos oydo dezir a sus pasados y dello es cosa muy pu^{ca} y notoria entre los y^{o}s viejos desta comarca que lo //
/185v/ bieron y entre los que lo oyeron dezir como este dho t^{o} y esto rresp^{e}.

6 - A la sesta pregunta de el dho ynterrogatorio dixo que dize lo que dho tiene en las preguntas antes desta y lo demas no lo sabe.

7 - A la setima pregunta de el dho ynterrogatorio dixo este dho t^{o} que dize lo que dho tyene en las preguntas antes desta y lo demas no lo sabe.

8 - A la otaba pregunta de el dho ynterrogatorio dixo este dho t^{o} que dize lo que dho tiene en las preguntas antes desta e que tratando con sus padres e pasados e con otros viejos e ançianos desta comarca que algunos dellos son muertos e platicando sobre cosas antiguas e sobre las dhas tierras de Quybi e otras platicas que entre ellos trataban les oya e oyo dezir que despues de entrados los espanoles en este rreyno se salieron de las dhas tierras de Quybi la mytad de los yndios yauyos del rrepartimyento de Chaclla y quedaron la otra mytad pero que por eso nunca los yndios yauyos avian dexado las dhas tierras ny perdydo su posesion ny tanpoco oyo dezir ny este t^{o} entendio ny vio despues que se sabe acordar que los yndios de Can-//
/186r/ ta se entrasen en las dhas tyerras de Quybi fasta que hizieron concierto entre los dhos yndios de Canta y de Chacalla por ciertas obejas y cree este t^{o} que sy los dhos yndios de Canta se ubieran entrado en ellas de antes y tenydolas y poseydolas este t^{o} lo biera supiera y entendiera o ubiera oydo dezir y no pudiera fazer menos por ser tan comarcano a los dhos yndios de Canta y zercano a sus parçialidades que podra aver una legua de su tierra a la de las dhas tyerras de Quybi y esto rresp^{e}.

9 - A la novena pregunta del dho ynterrogatorio dixo este dho t^{o} que lo que della sabe es que supo oyo y entendio e a vibido parte dello este t^{o} que en el tpo de el presydente Gasca tenyendo los dhos yndios de Canta diferencias sobre las dhas tierras con los de Chacalla y tenyendolas y poseyendolas los dhos yndios yauyos de Chacalla trataron de conçierto en quee [*sic*] los dhos yndios de Canta prometyeron e a lo que este t^{o} oyo dezir entregaron a los dhos yndios de Chacalla duçientas cabeças de ganado çiento hembras y çiento machos por que dexasen las dhas tierras y que el dho concierto se hefetuo en presencia de Nycolas de Rribera el moço encomendero de los yndios de Canta y Fran^{co} de Ampuero encomendero de //
/186v/ los yndios de Chacalla e que lo susodho es cosa muy publica y notoria entre todos los yndios desta dha comarca porque despues que se hizo el dho concierto los dhos yndios yauyos de Chacalla dexaron las dhas tierras y los dhos yndios de Canta se entraron en ellas y las tobieron y poseyeron fasta agora que bee que se trata este dho pleyto y esto rresponde a esta dha preg^{a}.

10- A las diez preguntas de el dho ynterrogatorio dixo este dho t^{o} que dize lo que dho tiene en la pregunta antes desta e que a un hermano deste t^{o} que es ya muerto que se llamava Poma Chunbi le a oydo dezir este dho t^{o} que los dhos yndios de Canta entregaron al dho caçique de los yavyos que se llamava don Fran^{co} Marcapoma e a Guaman Yaure las dhas cabecas de ganado del dho conçierto y que ellos las avian rreszebido e llebadolas a sus pueblos y esto rresp^{e}.

11- A las honze preguntas de el dho ynterrogatorio dixo que save porque a visto las dhas tierras de Quybi que podran tener en todas ellas fasta veynte e cinco fanegas de sembradura poco mas o menos e que lo sabe porque syenbra y coje este t^{o} y lo entiende e a visto que es tierra floja e flaca //
/187r/ e mucha pedregosa e de andenes como dize la pregunta e que no sabe lo que podra valer cada fanega de sembradura y esto rresponde.

12- A las doze preguntas dixo que dize lo que dho tyene e lo demas no lo sabe.

13- A las treze preguntas de el dho ynterrogat^{o} dixo que sabe que en el tpo que se hizo el dho conçierto una obeja grande macho con hembra sy hera grande seis pesos y si hera mediana çinco pesos y si hera mas pequeña a quatro pesos esto a justa e comun estimacion porque este t^{o} a tenydo e tiene obejas que a comprado a este dho prescio a la dha sazon para combi y esto rresponde a esta dha pregunta.

14- A las catorze preguntas del dho ynterrogatorio dixo este dho t^{o} que sabe y a bysto que antes e al tpo que se hiciese el dho concierto las dhas tierras de Quybi no valian tanto como el dia de oy balen porque entonzes no tenyan dellas tanto aprobechamyento ny las abian tanto menester como agora y esto es cosa conosçida y aberiguada ansy en las dhas tierras de Quibi como en otras desta comarca que cada dia van en crescemy^{o} e valor y esto rresponde.

15- A las quynze preguntas del dho ynterrogatorio dixo que dize lo que dho //
/187v/ tyene en las preguntas antes desta y cree este t^{o} que por rrazon de aver sydo las dhas tierras de Quybi antes que vinyesen los yngas de los dhos yndios de Canta conprarian las dhas tierras porque lo mysmo hiciera este t^{o} sy ubieran sydo suyas y de sus pasados y esto rresponde.

16- A las diez y seis preguntas de el dho ynterrogatorio dixo este dho t^{o} que lo que a dho es la verdad.

Preguntas añadidas

1 - A la primera pregunta añadida dixo este dho t^{o} que dize lo que dho tiene en las preguntas del primero ynterrogatorio y demas y alyende dello dixo que oyo dezir por cosa publica e notoria a viejos e ancianos sus pasados que en tiempo de el dho Guascar Ynga se hizieron las dhas dos capacochas una para Ynga e otra para su muger e que trayendo los yndios de Canta la dha capacocha fasta su mojon la tomaron los yndios de Guancayo rrepartimyento de Gonzalo de Cazeres donde este dho t^{o} tiene su tierra y la llebaron la dha capacocha fasta la mar y su padre e aguelo deste t^{o} le dixeron que avian ydo con ella y la ofresçieron en la mar por Ynga como se solia fazer en sus plegarias y esto //
/188r/ rresponde a esta dha pregunta.

2 - A la segunda pregunta del dho ynterrogatorio dixo que no la sabe e dize lo que dho tyene en las preguntas antes desta.

3 - A la tr^{a} pregunta del dho ynterrogat^{o} dixo este dho t^{o} que no lo save e dize lo que dho tiene.

4 - A la quarta pregunta del dho ynterrogat^{o} dixo este dho t^{o} que no lo save.

5 - A la quinta pregunta de el dho ynterrogatorio dixo que no lo save.

6 - A la sesta pregunta de el dho ynterrogatorio dixo que no lo save.

7 - A la setima pregunta del dho ynterrogatorio dixo que dize que dho tiene lo qual dixo que era la verdad y en ello se afirmo syendole leydo lo qual declaro por lengua de don M^{yn} yndios ynterpetre nonbrado para lengua por el Audiencia Real de Los Reyes ante quyen pasa esta cavsa encargosele el secreto y prometiolo.

ante mi — Fran^{co} Hrrez de G^{a}
escr^{o} de Su Mag^{t}

Don Luis Çacalla Chumbi
amygo de los yndios de Canta
de hedad de 28 o 30 a^{o}s.

T^{o} el dho don Luis Zacalla Chunbi principal de el pueblo de Guarabi t^{o} presentado //
/188v/ por parte del cacique e yndios del pueblo de Canta encomendados en Nycolas de Rribera el moço vz^{o} de la çibdad de Los Reyes aviendo jurado en forma devida de dr^{o} e syendo preguntado por el tenor de el dho ynterrogatorio para que fue presentado por testigo por lengua de don M^{yn} Lengua ynterpetre nonbrado por los señores oidores dixo e depuso lo syg^{e}:

1 - A la primera pregunta de el dho ynterrogatorio dixo que conosze al cacique e yndios de Canta que agora son e conosçio a los caciques que an sydo de algunos dias a esta parte despues que se save acordar caciques de Canta e que conosze a don Xristoval cacique de los yndios de Chacalla encomendado en Fran[co] de Ampuero e algunos de sus yndios por trato e bysta e conversacion que con ellos a tenydo e tiene e que a los yngas ny a Topa Ynga Yupangue senores que dizen fueron destos rreynos no los conoszio syno de avellos oydo dezir a sus padres e pasados e lo mysmo a Guaynacava e que tiene noticia de las dhas tierras de Quybi porque las a vysto y estado en ellas algunas bezes e lo mysmo a visto el rrio que pasa por ellas que llaman el rrio de Quybi y esto rresponde.

/189r/ Preguntado por las preguntas generales de la ley dixo que es de hedad de veynte y ocho o treynta anos poco mas o menos tpo e que no es pariente amygo ny henemygo yntimo de los yndios de Chacalla ny de Canta mas de que con los yndios de Canta tyene gran amystad e son companeros porque trata con ellos este t[o] en bendelles coca y los de Canta obejas y papas pero que no le an dado ny prometido cosa alguna para que en este su dho diga el contrario de la verdad por nenguna de las dhas partes e que este t[o] es xristiano e dira verdad de lo que supiere y le fuere preguntado e que desea que venza el pleyto los yndios de Canta porque los quyere byen antes que no los de Chacalla pero que Dios de la just[a] al que la tobiere lo que declaro por sus propias rrazones por lengua del dho M[yn] ynterpetre.

2 - A la segunda pregunta del dho ynterrogatorio dyxo este dho testigo que a lo que este dho t[o] a visto las dhas tierras de Quybi estan dentro de los propios rreynos de //
/189v/ Canta y de sus parçialydades e pueblos e tan zercanos e yguales como el pueblo primero de la juresdicion [*sic*] de Chacalla y que el rrio que pasa por junto a las dhas tierras de Quibi es distinto e apartado de Chacalla e del serviçio y entrada del rrepartimyento de Chacalla porque es el propio rrio y entrada de Canta y el rrio y entrada de los de Chacalla e[s] el rrio de Mama e que este t[o] se acuerda aver oido dezir a sus mayores y mas ancianos e a sus padres e aguelos e antepasados que son ya muertos e aun a otros son bibos el dia de oy que antes que vinyesen los yngas a esta tierra heran todos unos los yndios y donde este t[o] rresyde y los de syerra e tierra de Canta y benydo que fue Hinga les puso sus mojones e dybidyolos que avia de ser de los yngas fazia la mar y lo que abia de ser fazia la sierra para Canta e que tanbien oyo dezir que el pueblo que agora esta encima del rrio de Quybi que llaman Quybi y es de Canta que antes que vinyesen los yngas hera parçialidad por si y esto rresponde a esta pregunta.

/190r/ 3 - A la tr[a] pregunta del dho ynterrogatorio dixo este dho t[o] que lo que puede declarar y rresponder a lo en ella contenydo es que a oydo dezir por cosa publica y notoria e muy çierta a yndios viejos e ancianos de la syerra y de los llanos con quyen este dho t[o] a tenydo trabto e comunycacion e a sus padres e aguelos que ya son muertos e a otros que son bibos tratando e comunycando e platicando sobre las dhas tierras de Quybi e sobre cosas pasadas que antes que ubiese yngas en estas tierras de Quybi heran las dhas tierras de un senor que no se acuerda como le dixeron que se llamava mas de que decian que era yndio yunga natural de Quybi el qual decian que trebutava como yunga a los yndios yungas e a un senor que se decia Collicapa que era senor de los yungas de la encomyenda del contador Juan de Cazeres y de Collique fasta la mar e que la juresdicion de los yungas la tenya apartada de los serranos de Canta y puesto sus termy[o]s de yungas que llegaban mas arriba del pueblo que agora llaman de //
/190v/ Quybi dos leguas mas alla fasta un pueblo que llaman Chuquycoto e que desde el dho pueblo de Chuquycoto tenya senalado por yndios yungas facia la mar y desde el dho pueblo de Chuquicoto fazia Canta hera de la syerra hazia Canta sujetos a Canta y que este dho senor lo avia tenido y poseydo mucho tpo fasta que entro Ynga en esta tierra y mato al dho Collicapa e a otros muchos yndios suyos y se entro y apodero en las dhas tierras de Quybi y puso en ellas yndios yavyos por mytimaes para su servicio para que le hiziesen ojotas e para cargas e para chasques e otros aprobechamyentos de Ynga e que estos dhos yndios yauyos heran del rrepartimy[o] de Chacalla e que entonzes el dho Ynga torno a senalar los termy[o]s de los yungas y los de los serranos facia Canta y puso sus mojones desde media legua e aun no la ay deste pueblo de Quybi facia Lima puso el dho mojon y senal para que de alli corriese el termy[o] de los yungas y se llamasen yungas y desde el dho mojon fazia Canta se llamasen serranos y tomo cierto pedaço de tierra junto al dho//
/191r/ mojon para que los dhos yndios yungas senbrasen alli para Ynga los quales dho yndios yabyos estobieron puestos en la dha parte y lugar de las dhas tierras de Quybi donde los puso Ynga fasta que murio Ynga e que ansy como moria el senor y abia otro rrey como fue Guascar Ynga y Guaynacaba acreszentavan e yban sienpre acrescentando las dhas tierras de Quibi a los yndios yauyos que ansy avia puesto por mytimaes el inga e desta manera se fueron entrando e apoderando los dhos yndios yavyos en las dhas tierras de Quybi fasta que llegaron al dho pueblo de Quybi donde agora se toman estos t[o]s que hera del dho senor que llaman Quybi que no tenya que ber con los yndios de Canta e que aunque Ynga mato a los yndios deste senor que llamavan Quybi todavia quedaron algunos yndios que se contavan y nombraban por yndios yauyos como los dhos yndios yauyos de Chacalla por no desanparar sus tierras y perdellas y que desta manera tobieron las dhas tierras los dhos yndios yavyos de Chacalla fasta que entraron los espanoles en este rreyno aunque antes que entrasen syempre //
/191v/ dizen que trayan diferencias los yndios de Canta con los de Chacalla sobre las dhas tierras y estaban en arma para se matar cada dia e que despues de entrados los dhos espanoles en este rreyno estando como estavan los dhos yndios yabyos en las dhas tierras de Quibi en el camyno de Canta por donde avian de benyr camynando los espanoles se fueron la mytad de los dhos yndios yavyos de myedo de los espanoles al pueblo de Chacalla y dexaron las dhas tierras de Quibi y quedaron la otra mytad de los dhos yndios yavyos en ellas e que despues quando vino el presydente Gasca binyeron a las dhas tierras de Quybi Nycolas de Ribera el moço encomendero de los yndios de Canta y Fran[co] de Ampuero encomendero de los dhos yndios de Chacalla y trataron de conçertar los unos yndios y los otros para que no tobiesen pasyones en que diesen los yndios de Canta a los de

Chacalla duzientas obejas de la tierra mytad machos y mytad henbras por que dexasen las dhas tierras y este t[o] entendio de yndios de los dhos rrepartimy[o]s que se las avian dado porque des-//
/192r/ de entonzes vido este dho t[o] que los dhos yndios de Chacalla yavyos se fueron a sus tierras y dexaron las dhas tierras de Quybi para los dhos yndios de Canta los quales se entraron en ellas y las an tenydo y poseido quyeta e paçificamente syn contradicion de los dhos yndios de Chacalla sy no es agora que a visto y entendido que los dhos yndios yavyos se an arrepentido y no quieren pasar por el dho conçierto e traen pleyto lo qual todo que dho tyene le ha sido a este dho t[o] publico e notorio por abello oido dezir a los dhos sus pasados los quales deçian que ansy avia sido y pasado y lo avian bisto ellos ser y pasar y oydo dezir a otros viejos e ancianos e a sus padres y pasados e que nunca avian vysto ny oydo dezir lo contrario dello e que dello avia sydo y hera pu[co] e notorio e publica voz y fama e que cree este t[o] que si lo contrario ubiera sydo y pasado tanbien se lo ubieran dho y ellos lo supieran por ser todos tan comarcanos a los dhos pueblos de Canta y lo mysmo a los de Chacalla e que tratavan unos con otros y lo mysmo con los dhos yndios de Quybi lo qual dixo e rrespondio en sustancia rrazonandolo entre el dho don M[yn] Lengua y este t[o] a lo que la lengua declaro.

/192v/ 4 - A la quarta pregunta del dho ynterrogatorio dixo este dho t[o] que lo en ella contenydo ansy e de la forma y manera que la dha pregunta lo dize y declara lo a oydo dezir este dho t[o] a sus mayores e mas ancianos e a sus padres e aguelos los quales dezian que ellos ansy lo avian bisto ser y pasar en sus tiempos e oydolo dezir a sus mayores y mas ançianos que dezian que ellos ansy lo avian visto e oido dezir a los suyos en los suyos e que nunca avian bisto ny oydo otra cosa zerca dello en contrario y este dho t[o] ansy lo cree por ser las personas que esto tratavan yndios comarcanos a los dhos pueblos de Canta y Chacalla e a las dhas tierras e rrio de Quybi e que syempre tratavan vnos con otros lo qual rrespondio por la dha lengua porque yo el dho escr[o] se lo hize rrazonar con este t[o] para que diese rrazon de su dho.

5 - A la quynta pregunta del dho ynterrogat[o] dixo que lo //
/193r/ en la dha pregunta cont[do] segun y como en ella se contiene y declara lo a oydo dezir este dho t[o] a sus mayores e mas ancianos que dezian avello visto ser y pasar ansy e acostumbrar de la forma y manera que la dha pregunta lo dize e aun dezian que ellos en sus tyempos ansy lo avian bisto e oido y entendido e oido a los suyos viejos e ancianos que dezian avian bisto ser y pasar ansy e oido a los suyos en los suyos e que dello hera pu[co] e notorio entre los yndios desta comarca y esto rrespondio por la dha lengua en buena rrazon.

6 - A la sesta pregunta del dho ynterrogatorio dixo este dho t[o] que dize lo que dho tiene en las preguntas antes desta e que a un yndio viejo natural de Guarabile pueblo deste t[o] que se llamava Tunga Caxa le oyo este dho t[o] dezir que avia bisto que en tpo del dho Guaynacava e quando yba a Quyto avian benydo los dhos yndios yavyos de Chacalla con la dha capacocha e sangre por las tierras de Quybi fazia el dho pueblo de Quybi aper-//
/193v/ çibiendo a los dhos yndios de Canta que se guardasen de la dha capacocha porque sy la derramavan avian de morir e que los dhos yndios de Canta e de Quybi que estaban en el dho pueblo de Quybi se apartavan y les dexavan entrar y no avian hecho rresystencia por que no los matasen y que desta manera avian entrado fasta el dho pueblo de Quybi con la dha capacocha e que esto no lo a oido dezir este dho t[o] a otra persona sy no es al dho yndio que dho tyene ny savia mas desta preg[a].

7 - A la setima pregunta del dho ynterrogatorio dixo este dho t[o] que dize lo que dho tiene en las preguntas antes desta e no sabe otra cosa.

8 - A la otaba pregunta del dho ynterrogaotrio dixo este dho t[o] que dize lo que dho tiene en las preguntas antes desta y lo demas no lo sabe.

9 - A la novena pregunta del dho ynterrogatorio dixo este dho t[o] que dize lo que dho tiene en las preguntas antes desta.

10- A la decima pregunta del dho ynterrogat[o] dixo este dho t[o] //
/194r/ que dize lo que dho tiene en las preguntas antes desta e que a yndios de Canta a oydo dezir este dho t[o] que las dhas obejas las dyeron y entregaron a los dhos caçique e yndios de Canta por las dhas tyerras de Quybi y las abian llebado a sus pueblos.

11- A la honzena pregunta del dho ynterrogatorio dixo este dho t[o] que a su pareszer deste t[o] las dhas tierras de Quybi podran tener de sementera de trigo e maiz fasta tres o quatro fanegas porque son todas andenes y cascajales e tierra de mucha piedra e que no sabe lo que podra valer cada fanegada de sembradura ny sabe mas desta pregunta.

12- A las doze preguntas del dho ynterrogatorio dixo que en todas las dhas tierras de Quybi podra aver de largo poco mas de media legua y que de ancho es muy angosta.

13- A las treze preguntas del del [*sic*] dho ynterrogat[o] dixo este dho t[o] que las ovejas que los yndios de Canta dyeron a los yndios de Chacalla por las dhas tierras de Quybi heran grandes que valia cada una seis pesos e si heran //
/194v/ medianas valian cinco pesos cada una y si heran mas chicas balian a quatro e a tres pesos.

14- A las catorze preguntas del dho ynterrogatorio dixo este dho tº que a lo que a entendido las dhas tierras balen cada dia mas en este terreno e que ansy le paresze que quando se hizo el dho concierto de las dhas tyerras de Quybi por las dhas obejas podian valer menos que agora aunque entre los yndios no se paresce tanto como entre españoles.

15- A las quynze preguntas del dho ynterrogatorio dixo este dho tº que dize lo que dho tiene en las preguntas antes desta y lo demas en ella contenydo no lo sabe.

16- A las diez y seis preguntas dixo que lo que a dho es verdad y en ello se afirmo por la dha lengua.

Preguntas añadidas

1 - A la primera pregunta del dho ynterrogatorio dixo que abra muchos anos que este dho tº oyo dezir tratando sobre las dhas capacochas //
/195r/ a un yndio que sy a muerto que se llamava Tunca Gaxa [*sic*] natural del pueblo de Guaravi tierra deste tº que en tpo del dho Guascar Inga se avian fho las dhas dos capacochas una para Guascar Yndia [*sic*] y otra para su muger y dezia el dho yndio que el lo abia vysto e que lo demas que dize la pregunta lo a oido dezir este dho tº a un yndio que se llamava Chunbi Quibi que es de los yndios sujetos a Canta e que lleba ynteres en este pleyto e que dize lo que dho tiene en las preguntas antes desta y esto rresponde.

2 - A la segunda pregunta del dho ynterrogatorio anadido dixo que no la sabe e dize lo que dho tiene en las preguntas antes desta.

3 - A la tr^{a} pregunta de el dho ynterrogatº dixo este dho tº que no lo save mas de lo que dho tiene en las preguntas del primero ynterrogatorio.

4 - A la quarta pregunta del dho ynterrogatorio prençipal dyxo este dho tº que no lo save e dize lo que dho tiene en las preguntas del dho primero ynterrogatº y esto rresponde a esta dha pregunta.

/195v/ 5 - A la quynta pregunta del dho ynterrogatorio dixo este dho tº que no lo sabe.

6 - A las [*sic*] sesta pregunta del dho ynterrogatorio dixo este dho tº que no lo sabe.

7 - A la setima pregunta del dho ynterrogatorio dixo este dho tº que dize lo que dho tyene en las preguntas deste su dho lo qual todo dixo que era la verdad y lo que deste caso sabia y nō otra cosa para el juramento que avia fho y en ello dixo que se afirmava e afirmo e rrateficava e rratefico aviendosele tornado a leer por lengua del dho don M^{yn} Lengua ynterpetre nonbrado para lo susodho por los dhos senores presydente e oidores el qual le encargo e dyo a entender tenga secreto deste su dho e no lo diga a nenguno de los dhos caciques e yndios de Chacalla ny de Canta so pena que quedara perjuro y le castigara el el qual prometyo de fazello ansy.

fho ante mi — Fran^{co} Hrrez de Guebara
escrº de Su Mag^{t}

/196r/ Xristoval Caxallaupe de la encomyenda
de Gº de Cazeres
no sabe la hedad que tiene e no le tocan.

Tº el dho Xristoval Caxallaupe yndio natural de Guarabi encomyenda de Gonzalo de Caçeres hijo del contador Juan de Caceres vº de la çibdad de Los Rreyes tº presentado por parte de los caciques prençipales e yndios de el pueblo de Canta encomendados en Nycolas de Rribera el moço vº de la dha çibdad de Los Rreyes aviendo jurado en forma de drº e syendo preguntado por lengua del dho don M^{yn} Lengua ynterpetre para lo susodho dixo e depuso lo sig^{e}:

1 - A la primera pregunta de el dho ynterrogatorio dixo este dho tº que save e tyene noticia de las dhas tyerras de Quibi porque las a vysto y estado en ellas desde tpo de Guascar Ynga que fue el postrero senor de los yngas e que tiene noticia de los rreyes y senores yngas que dize la pregunta por oydas porque de vysta no conoscio syno a Guascar Ynga e ansy mysmo tiene noticia del dho tpo a esta parte del dho rryo de Quybe que pasa junto a las dhas tyerras de Quybe e conosze y conoscio a los caciques que an sydo del dho tpo aca de Guascar Ynga e agora //
/196v/ son de los pueblos de Canta y de Chacalla por vista y fabla trato y conversacion que con ellos este tº tubo e a tenydo e tiene.

Preguntado por las preguntas generales de la ley dixo que no sabe la hedad que tiene mas de que se acuerda de Guascar Ynga como dho tiene e que no sabe que tenga parestenco [*sic*] ny menos le ba ynteres en esta causa ny le an dado ny prometido cosa alguna para que en este su dho diga el contrario de la verdad e que este dho tº es xristiano y dira la verdad de lo que supiere y le fuere preguntado.

2 - A la segunda pregunta del dho ynterrogatorio dixo este dho tº que lo que de la dha pregunta sabe e a entendido e oido dezir es que save que desde un mojon que esta en un zerro junto al pueblo de Quibi que no abra media legua partio Ynga los ter-//
/197r/ mynos y de aquel mojon para abajo facia la mar se contava de yndios yungas y de alli para arriba el rrio arriba fazia Chacalla e Canta se contaba la syerra e que sabe este dho tº e a bisto que dende el pueblo de Quibi e dende el dho mojon fasta el primero pueblo de Canta que se llama Quyso abra dos leguas e fasta el primero pueblo de Chacalla abra dos jornadas e que junto a las dhas tierras de Quibi esta al presente un pueblo poblado de yndios de Canta por rrazon de que conpraron el sytio y tierra los de Canta a los de Chacalla e que el rryo que pasa por junto a las dhas tierras de Quybi es termyno e rrio de drªmente de Canta e del servicio de Canta porque pasa y ba por junto al camyno y pueblos de Canta y los de Chacalla no tienen servicio ny aprobechamyento del dho rrio para sus pueblos sy no fuese rresydiendo ellos en las dhas tierras de Quibi porque su rrio y serbicio de Chacalla y de sus pueblos es el rrio de Mama que ba a Lima e que este dho tº a oido dezir e tratar entre yndios viejos e ancianos ansy de la parçialydad de donde este tº es natural como desta comarca e a sus padres e aguelos que son ya muertos los quales tratando sobre estas dhas tierras e valle e sobre cosas pasadas dezian que antes que binyesen yngas a esta tierra los yndios yungas que estan en los valles fazia la mar de la encomyenda del contador Juan de Cazeres y un cacique y senor que llamavan Chunbiquibe heran todos unos e se llamavan yungas e que este senor trebutaba a un senor de los yungas de Collyque que se llamava Collicapa el qual dho Chunbiquibe rresydia en estas dhas tierras de Quybi junto al rrio sobre que es este pleyto e tenya en ellas sus yndios y hera un senor por si que no hera sujeto a los yndios de Canta syno que solamente trebutaba al dho Collicapa senor de Collique e benydo que fue Ynga como dibidio e partio los mojones les hizo serranos los yndios que del quedaron porque Ynga mato este senor y muchos de sus yndios por //
/198r/ cierta traiçion e delyto que hizo contra el inga e que los yndios yauyos que puso Ynga en las dhas tierras de Quibi le mataron por su mandado lo qual este tº le oya dezir a sus pasados por cosa muy çierta y esto rresponde.

3 - A la trª pregunta de el dho ynterrogatorio dixo este dho tº que dize lo que dho tiene en las preguntas antes desta e que a oydo dezir este dho tº dende quanto que se sabe acordar a muchos viejos e ancianos de los quales al presente no tyene memoria porque son muertos asy de su pueblo deste dho tº como desta comarca de la syerra y de los llanos e a sus padres e abuelos tratando e platicando sobre cosas de los ygas [*sic*] pasados e otras platicas que entre ellos se ofresçian que antes que ubiese yngas en esta tierra las dhas tierras de Quebi como dho tienen heran de un senor que llamavan Chunbiquibi que hera yunga e de jeneracion de yungas y que era senor por sy el qual dava de trebuto algodon y coca y maiz y otras cosas a un cacique que llamavan Collicapa que rresydia en el valle //
/198v/ de Collique e que este senor hera senor desde esta tierra fasta la mar y tenya las orejas largas como yndios orejones del Cuzco e que el dho Chunbiquibi tenya su juresdiçion dystinta e apartada de los yndios de Canta e de sus termynos e no tenya que ber la tyerra de los serranos con su tierra e que quando entraron los yngas en esta tierra a sujetalla el primero ynga que entro que fue Topa Ynga Yupangue aguelo de Guaynacaba dio guerra al dho Collicapa cacique de Collique y el dho Chunbiquibi que estava en las dhas tierras de Quibi que trebutava al dho Cochyllaca [*sic*] fue con sus yndios a socorrer al dho Collicapa contra los yngas y los yngas en la batalla que dieron mataron a muchos yndios del dho Collicapa e al dho Collicapa e tanbien muchos yndios del dho senor Chunbiquibe y el dho Chunbiquibe se torno a sus tierras donde estubo çiertos años fasta que fue acusado que queria fazer traicion a Topa Ynga e matalle con ponçona e sabido por Topa Ynga enbio por el dho Chunbiquibi y le llebaron atado a el//
/200r/ e a sus yndios al Cuzco donde estava el dho Topa Ynga e por su mandado e alli fueron muertos por mandado del dho Topa Ynga Yupangue por la dha traicion que le querian fazer e que antes que prendiesen al dho Chunbiquibe avian puesto por Ynga Yupangue a los dhos yndios yayos del rrepartimyento de Chacalla por mytimaes en las dhas tierras de Quybi y dadoles alli un pedaço de tierra en que senbrasen para Hinga e a ellos puestos para chasques y para llebar cargas y para dar avisos e otras cosas del servicio del inga e que alli donde les puso hizo su senal y mojon que oy dia esta alli que hera media legua del pueblo de Quybi donde agora se desamynan [*sic*] los tºs y estan los dhos yndios de Canta e que desde aquel mojon para arriba façia Canta dejo para termyno de Canta e por juresdiçion de la syerra e del mojon para abajo facia Lima dejo por de los llanos e yndyos yungas e que despues de muerto Topa Ynga Yupangue suzedio en el senorio Guaynacaba / anadio a los dhos yndios yavyos que ansy avia puesto por mytimaes Ynga//
/200v/ mas tierra façia el dho pueblo de Quibi que llegaba fasta este asyento e pueblo de Quybi para que alli estobiesen por sus yndios mytimaes como los avia puesto Ynga los quales dhos yndios yauyos del rrepartimyento de Chacalla estubieron por mytimaes muchos años e tpo e que los yndios del rrepartimyº de Canta no osaban fablar de myedo que tenyan de los yngas e que despues que los yngas murieron e despues que entraron los xristianos en esta tierra luego enpeçaron los yndios de Canta a tener pasyones y diferençias sobre las dhas tyerras de Quybi y hechar dellas los yndios yauyos del rrepartimyento de Chacalla y las tobieron muchos anos que estavan para matarse fasta que vinieron Nycolas de Rribera el moço encomendero de los yndios de Canta y Fran^{co} de Ampuero encomendero de los yndios de Chacalla y trataron de concertar los unos yndios y los otros por quitalles de pasyones y el conçierto fue que los dhos yndios de Canta dyesen a los de Chacalla ducientas obejas de la tierra machos y henbras por //
/201r/ que dexasen las dhas tierras a los yndios de Canta e a este tº le es notorio por dho de los unos yndios y de los otros que los dhos yndios de Chacalla se contaron [*sic*] con el dho concierto y llebaron las dhas obejas y dejaron las dhas tierras a los de Canta y este dho tº a visto que los dhos yndios de Canta las an tenydo e poseido quyeta e paçificamente syn rrençilla ny contradiçion de los yndios de Chacalla fasta agora que a entendydo que el cacique e yndios de Canta an rreclamado del concierto e no quyeren pasar por el e piden las dhas tyerras e que lo demas que tiene dho le a sydo a este tº notorio por dho de los dhos viejos e ancianos e de sus pasados que dezian avello visto ser y pasar ansy en sus tpos en an oydolo a sus mayores e mas ancianos que

ellos e que dello avia sydo y era entre ellos cosa publica not^{a} e pu^{ca} boz y fama e comun opinyon e que nunca este t^{o} a oydo dezir cosa alguna en contrario lo qual dixo e rrespondio a esta pregunta por buenas rrazones estandolo la lengua rrazonando e preguntando.

4 - A la quarta pregunta de el dho ynterrogat^{o} dixo este dho t^{o} que sabe la //
/201v/ dha pregunta segun y como en ella se cont^{e} y declara porque este dho t^{o} lo a oydo dezir lo en ella contenydo a sus padres e aguelos e pasados e a otros biejos e ancianos que este t^{o} y ellos de cuyos nombres no se acuerda que son ya muertos tratando e platicando en las cosas a los quales les oya dezir que entre los dhos yndios se usaba e acostunbrava fazer lo contenydo en la dha pregunta por la forma que ella lo dize y declara y que hera cosa usada y guardada como las leyes de agora de los xristianos y esto rresp.^{e}

5 - A la quynta pregunta del dho ynterrogat^{o} dixo este dho t^{o} que lo en ella cont^{do} lo a oydo dezir este dho t^{o} que se usaba e acostunbraba entre los yndios en tpo de los yngas como la dha pregunta lo dyze y declara lo qual a oydo dezir e tratar e platicar a sus pasados e a otros viejos e ancianos que decian que ellos ansy lo avian visto usar e acostunbrar en sus tpos y lo avian oido dezir a sus pasados que lo avian visto fazer //
/202r/ lo mysmo en los suyos e que dello avia sydo y hera pu^{co} e not^{o} e publica boz e fama e comun opinyon y esto rresponde.

6 - A la sesta pregunta de el dho ynterrogatorio dixo este dho t^{o} que lo que desta pregunta sabe es que antes que muriese su padre deste t^{o} que se llamava Parian Quybi le oyo dezir e lo mysmo a otros muchos yndios viejos e ancianos que son muertos e de cuyos nombres no se acuerda que en tpo de Guascar Ynga que fue el postrero senor de los yndios los dhos yndios yauyos se avian entrado desde el dho mojon donde Guaynacapa les avia puesto faziendo la dha capacocha con yndios de Guascar Ynga orejones con la dha sangre fasta mas adelante deste pueblo de Quibi donde agora estan los dhos yndios de Canta e yban con la dha sangre como dize la pregunta e que los dhos yndios yauyos y orejones yban diziendo: "parta parta no derrames la sangre que moriras que el ofresçimyento de Ynga para el diablo para que le de buena vejez y le faga gran senor e le de muchos yndios e rriquezas y comydas e guarde sus yndios" e que los dhos yndios de Canta //
/202v/ de myedo de la dha capacocha e por no derramar la sangre por que no los matasen los dexavan entrar e apoderarse en la tierra y no se lo rresystian e que desta manera yban ganando tyerra e apoderandose en ella lo qual todo les oyo dezir este dho t^{o} muchas e dybersas vezes tratando sobre la dha capacocha.

7 - A la setima pregunta del dho ynterrogatorio dixo este dho t^{o} que dize lo que dho tyene en las preguntas antes desta e que nunca oyo este dho t^{o} dezir sino que los yngas pusieron los dhos mytimaes que por mandado de los yngas e como sus criados los dhos yndios yavyos llebaban la dha capacocha.

8 - A la otoba pregunta de el dho ynterrogatorio dixo este dho t^{o} que despues que entraron los espanoles en este rreyno se alço Mango Ynga y en aquella sazon vido este dho t^{o} que parte de los dhos yndios yavyos se fueron a sus pueblos de Chacalla y dexaron las tierras de Quybe e pte dellos quedaron todavia en ellas e no se acuerda este dho t^{o} aver bisto oido ny entendido que los dhos yndios de Canta entrasen en las //
/203r/ dhas tierras de Quybe sy no es despues que dieron por el dho conçierto las dhas ovejas a los de Chacalla e que sy ubieran entrado en ellas este dho t^{o} lo ubiera bisto oido y entendido y no pudiera ser menos por ser tan zercano a los dhos pueblos e tierras de Quybi e tener trato e comunycaçion con los dhos yndios de Canta e acordarse este dho t^{o} como se acuerda de quando entraron los espanoles en este rreyno y esto rresponde.

9 - A la novena pregunta de el dho ynter^{o} dixo este dho t^{o} que dize lo que dho tiene en las preguntas antes desta.

10- A la decima pregunta del dho ynterrogat^{o} dixo este dho t^{o} que dize lo que dho tyene en las preguntas antes desta e que a oydo dezir a yndios de Canta que entregaron las dhas dozientas obexas a los yndios de Chacalla por las dhas tierras y las llebaron a sus pueblos y esto rresp^{e}.

11- A las onze preguntas del dho ynterrogatorio dixo este dho t^{o} que a visto muchas vezes las dhas tierras de Quybi e que lo mas dellas es piedra y andenes e que el pareszer deste t^{o} podra aver en ellas sy se sembrasen diez fanegas y media de sembradura de mahiz o tr^{o} //
/203v/ e que no sabe lo que podra valer cada fanegada.

12- A las doze preguntas del dho ynterrogat^{o} dixo que dize lo que dho tyene en las preguntas antes desta y lo demas no lo sabe.

13- A las treze preguntas del dho ynterrogatorio dixo que valdrian las dhas obejas cada cabeça unas a seis pesos e otras a çinco e otras a quatro.

14- A las catorze preguntas del dho ynterrogatorio dixo este dho t^{o} que no lo save.

15- A las quynze preguntas del dho ynterrogatorio dixo que no lo sabe e que antes entendio del dho caçique de Canta Guaman Yavre que conprava las dhas tierras diziendo al cacique de Chacalla don Fran^{co} Marcapoma que pues no tenya aprobechamy^{o} su pueblo deste rrio se las vendyese pues su rrio hera el de Mama e Ynga le avia senalado por su tierra de Canta desde el mojon que dho tiene que esta media legua deste pueblo de Quybi e no de otra manera.

16- A las diez y seis preguntas del dho ynterrogatorio dixo este dho t° que lo que a dho es la verdad y en ello se afirmo por lengua de el dho don M^{yn}.

/204r/ *Añadidas*

1 - A la primera pregunta del dho ynterrogatorio anadido dixo este dho t° que dize lo que dho tyene en las preguntas antes desta e que ansy mysmo oyo dezir a su padre este t° en tpo de Guascar Ynga que se avian hecho las dhas dos capacochas una para Guascar Ynga y otra para su mujer.

2 - A la segunda pregunta del dho ynterrogatorio dixo este dho t° que a su padre oyo dezir que los dhos yndios de Chacalla y los orejones yendo con la dha capacocha y saliendo los dhos yndios que dize la pregunta a defender sus tierras los dhos yndios yavyos y los demas yndios que trayan las dhas capacochas los avian muerto porque lo defendian y esto $rresp^{e}$.

3 - A la t^{a} pregunta del dho ynterrogatorio dixo este dho t° que dyze lo que dho tiene en las preguntas del primero ynterrogatorio.

4 - A la quarta pregunta del dho ynterrogatorio dixo este dho t° que dize lo que dho tiene y lo demas no lo sabe y esto rresponde a esta dha pregunta.

/204v/ 5 - A la quynta pregunta del dho ynterrogatorio dixo que dize lo que dho tiene en las preguntas antes desta y lo demas no lo save.

6 - A la sesta pregunta del dho ynterrogatorio dixo este dho t° que dize lo que dho tyene en las preguntas antes desta y lo demas no lo save.

7 - A la setyma pregunta del dho ynterrogat° dixo que lo que a dho es la verdad e lo que zerca deste caso save para el juramento que a fho y en ello dixo que se afirmava e afirmo aviendosele tornado a leer y dado a entender por lengua del dho don M^{yn} Lengua ynterpetre nonbrado por los senores presydente e oidores del Avdiencia Real el qual ansy mysmo le encargo el secreto deste su dho para que no lo trate con nenguno de los dhos caçiques e yndios el qual lo prometyo.

Fho ante mi — $Fran^{co}$ Hernandez de Guebara
escr° de Su Mag^{t}

/205r/ Cacha yndio ynfiel / acuerdase de
Guaynacaba e Guascar Ynga e no le tocan.

T° el dho Capcha yndio ynfiel natural de los Atabillos rrepartimyento de Hernando de Montenegro vez° de la çibdad de Los Reyes t° presentado por parte de los caciques prencipales e yndios del pueblo de Canta encomendado en $Fran^{co}$ de Ampuero [*sic*] aviendo jurado en forma segun su ley e la $cost.^{e}$ que tiene de jurar para dezir verdad por lengua de don M^{yn} Lengua ynterpetre nonbrado por los senores presydente e oidores e syendo preguntado por el tenor del dho ynterrogatorio por donde se pidio fuese hexsamynado dixo e depuso lo syg^{e}.

1 - A la primera pregunta de el dho ynterrogat° dixo este dho t° que conosze a los caciques e yndios muchos dellos de los rrepartimyentos de Canta e Chacalla que agora son e a conosçido a los que an sydo desde el tpo que entraron los espanoles en este rreyno e que del dho tpo sabe e tiene noticia de las dhas tierras de Quibi sobre que dizen que es este pleyto porque las a visto y estado en ellas muchas vezes e que tiene //
/205v/ noticia del dho rryo de Quybi que pasa junto a las dhas tierras de Quybi e que este dho t° se acuerda de los senores yngas que fueron deste rreyno de oidas porque no los conoscio e que se acuerda este dho t° de Guascar Inga que fue el postrero senor que fue en este rreyno de los yngas.

Preguntado por las preguntas generales de la ley dixo este dho t° que no sabe la hedad que tiene mas de que se acuerda de Guaynacaba y de Guascar Ynga senores que fueron destos rreynos e que no es pariente amygo ny henemygo yntimo de nenguno de los dhos caciques prencipales yndios de los pueblos de Canta y de Chacalla que este dho t° sepa ny tanpoco le an dado ny prometido cosa alguna para que diga el contrario de la verdad e que dyra verdad de lo que supiere y le fuere preguntado.

2 - A la segunda pregunta del dho ynterrogat° dixo este dho t° que no sabe sy las dhas tierras de Quybi estan agora al presente en tierras de los yndios serranos //
/206r/ de Canta mas de que este dho t° a oydo dezir a su padre que se llamava Llopa natural del pueblo de Atabillos encomyenda de Hernando de Montenegro vz° desta dha çibdad de Los Rreyes que sera los dhos Atabillos dos jornadas destas dhas tierras de Quybi el qual le decia a este t° tratando e platicando en cosas que en antes que binyesen los yngas a esta tierra las dhas tierras de Quybi las poseya un senor que no se acuerda como se llamava e que este dho senor le dixo el dho su padre que era sujeto al rrepartimyento de Canta e que trebutaba al cacique de Canta e que quando no tenya agua por el dho rrio de Quybi que avia

sequya se juntavan los yndios de Canta y los deste senor que dho tiene y abrian unas lagunas que se fazen alla arriba en la syerra de la nyebe que cae y las hazian benyr el agua dellas por el dho rryo de Quybi y desta manera dize que el oyo dezir al dho su padre que el dho senor y sus yndios rregavan las dhas tierras de Quybi quando avia sequya con el agua de las //
/206v/ dhas lagunas e hazian sus sementeras e que tanbien oyo dezir al dho su padre que al tpo que entraron los yngas en esta tierra yendo conquistando por Chacalla se huyan yndios chequitos del dho rrepartimy^o de Chacalla y se benyan a guareszer a las dhas tierras de Quybi e a los naturales que en ellas estavan e que los dhos naturales que eran yungas los criaban e que desta manera se entravan en las dhas tierras de Quybi los dhos yndios yauyos e no save otra cosa desta pregunta.

3 - A la tr^a pregunta del dho ynterrogatorio dixo que no lo sabe e dixo que dize lo que dho tiene en las preguntas antes desta.

4 - A la quarta pregunta del dho ynterrogatorio dixo este dho t^o que lo que la dha pregunta sabe es que oyo dezir al dho su padre que en tpo de los yngas pasados salian del Cuzco yndios yngas con ciertas zeremonyas que dezian que era la pruçision del inga como agora dizen los xristianos quando ban con la pruçision rrezando e que la llamavan //
/207r/ capacocha los yndios e que yban con ella por todas las tierras del Peru rrogando por el inga e que no sabe sy yban buscando tierra ny para que otro hefeto ny se acuerda de mas y esto rresponde e dixo que no queria dezir mentira porque este dho t^o tiene mucho myedo de mentir porque el otro dia estando en Lima vio que sacavan a un xristiano los dientes porque avia mentido e que sy este t^o myntiese que le quemarian a el.

5 - A la quynta pregunta de el dho ynterrogatorio dixo este dho t^o que no la sabe e dize lo que dho tiene en las preguntas antes desta.

6 - A la sesta pregunta de el dho ynterrogatorio dixo este dho t^o que no la sabe e dize lo que dho tiene en las preguntas antes desta.

7 - A la setima pregunta del dho ynterrogatorio dixo este dho t^o que no lo sabe e dize lo que dho tiene en las preguntas antes desta.

8 - A la hotava pregunta de el dho ynterrogatorio dixo este dho t^o que no lo sabe e dize lo que dho tiene en las preguntas //
/207v/ antes desta y esto rresponde a esta dha pregunta.

9 - A la novena pregunta del dho ynterrogatorio dixo este dho t^o que oyo dezir a muchos yndios comarcanos a los dhos rrepartimyentos de Chacalla y Canta e a agunos [*sic*] yndios de Canta que despues que entraron los espanoles en este rreyno los dhos yndios de Canta Guaman Yabre cacique y don Franco Marcapoma cacique de Chacalla trayan dyferençias e andaban syenpre rrinendo sobre las dhas tierras sobre que los dhos yndyos de Canta querian hechar dellas a los yndios de Chacalla yavyos diziendo que heran suyas e que despues como el licenciado de La Gasca presydente que fue destos rreynos aviendo benydo a su noticia las dyferencias que los dhos Guaman Yavre cacique de Canta y Marcapoma caçique de Chacalla trayan sobre las dhas tyerras enbio para que los concertasen a Nycolas de Rribera el moço encomendero de los yndios de Canta e a Fran^co de Ampuero encomendero de los yndios de Chacalla los quales le dixeron a este dho t^o que avian concertado los dhos caciques en esta manera que //
/208r/ los yndios de Canta dyesen a los de Chacalla duzientas obejas por rrazon de las dhas tierras e ansy oyo dezir este dho t^o ansy mysmo que en ellas avia dado los dhos yndios de Canta a los de Chacalla y que los de Chacalla avian llebado hasta sus tierras y los de Canta se avian entrado en las dhas tierras y las avian tenydo y poseido syn que los dhos yndios de Chacalla se quexasen fasta agora que les bee andar en pleyto e a oido dezir que no quyeren pasar por el dho conçierto.

10- A la decima pregunta del dho ynterrogatorio dixo este dho t^o que dize lo que dho tyene en las preguntas antes desta.

11- A las honze preguntas de el dho ynterrogatorio dixo este dho t^o que no lo save e dyze lo que dho tyene.

12- A las doze preguntas del dho ynterrogatorio dixo este dho t^o que no lo sabe e dize lo que dho tyene.

13- A las treze preguntas del dho ynterrogatorio dixo este dho t^o que le paresze que podran valer cada cabeza de las dhas obejas henbras çinco pesos por ser henbras y los machos a quatro pesos y que sy heran pacos valdran a tres pesos y no mas y esto rresp^e.

/208v/ 14- A las catorze preguntas del dho ynterrogatorio dixo este dho t^o que no lo sabe e dize lo que dho tyene.

15- A las quinze preguntas del dho ynterrogatorio diso este dho t^o que este dho t^o a oido dezir a los dhos yndios de Canta que antes que vinyesen los yngas a esta tierra las dhas tierras de Quibi heran suyas y de sus pasados abian hecho el dho conçierto con los dhos yndios yavyos de Chacalla y dadoles las dhas obejas por las dhas tierras y por quitarse de pleytos y diferencias e que esto como dho tyene lo a oido dezir a yndios de el dho pueblo de Canta.

16- A las diez y seis preguntas del dho ynterrogatorio dixo este dho t^o que lo que a dho es la verdad.

Añadidas

1 - A la primera pregunta del dho ynterrogatorio añadido dixo este dho t° que no lo sabe e dize lo que dho tyene en las preguntas del dho ynterrogatorio primero.

2 - A la segunda pregunta del dho ynterrogat° //
/209r/ añadido dixo este dho t° que dize lo que dho tiene y lo demas no lo sabe.

3 - A la tr^a pregunta del dho ynterrogatorio añadido dixo este dho t° que dize lo que dho tiene y lo demas no lo sabe.

4 - A la quarta pregunta del dho ynterrobatorio [*sic*] anadido dixo este dho t° que no lo sabe e dize lo que dho tiene.

5 - A la quynta pregunta del dho ynterrobatorio [*sic*] anadido dixo que no lo sabe e dize lo que dho tyene.

6 - A la sesta pregunta del dho ynterrobatorio [*sic*] anadido dixo este dho t° que no la sabe e dize lo que dho tyene en la pregunta antes desta.

7 - A la setima pregunta del dho ynterrobatorio [*sic*] dixo este dho t° que dize lo que dho tiene lo qual dixo que era la verdad e lo que del caso sabia e lo dixo e declaro por lengua del dho don M^yn Lengua ynterpetre nonbrado por los dhos senores oidores para lo susodho e el qual se lo dio a entender para que se rrateficase en ello despues de avello dho e declarado e se rratefico en ello y encargosele que tenga secreto deste su dho e que no lo diga a nenguno de los dhos caçiques e yndios so pena que le castigaran el qual lo prometio.

Fho ante mi Fran^co Hrrez de G^a
escr° de Su Mag^t

/209v/ Don Fran^co Arcos / natural de Canta
y bale ynteres como a los demas.
de hedad de 50 / y querria que vençiesen sus ptes.

T° el dho don Francisco Arcos natural del pueblo de Canta encomendado en Nycolas de Rribera vz° de la çibdad de Los Rreyes sujeto a los caciques prençipales del dho rrepartimyento t° presentado por parte de los caciques del dho pueblo de Canta y gobernadores del en lugar de don Sancho e abiendo jurado en forma de dr° segun se rrequeria por lengua de don M^yn ynterpetre e syendo preguntado por el tenor del dho ynterrogatorio por donde se pidio fuese hexsamynado dixo e depuso lo sig^e:

1 - A la primera pregunta del dho ynterrogatorio dixo este dho t° que conosce y conoscio a los cacique e yndios de los dhos rrepartimyentos de Canta desde que se save acordar dende chyquito porque como dho tiene los dhos caciques de Canta son sus senyores e los de Chacalla; este dho t° a tratado y conversado con ellos y son y estan comarcanos al dho pueblo de Canta e que conoscio a Guaynacava e a Guascar Ynga e a los demas senores los conosze de oidas //
/210r/ e que no fueron en tpo deste t° gobernadores desta tierra e que tiene noticia del dho rrio e tierras de Quybi porque lo a vysto y estado en ello muchas vezes y es del rrepartimyento de Canta.

Preguntado por las preguntas generales de la ley dixo este dho t° que le paresce que seria de hedad de cinquenta años poco mas o menos tpo porque como dho tiene se acuerda de Guaynacava e del inga e que es natural del pueblo de Canta y vz° e tiene su casa y muger e hijos e que le ba ynteres en esta cavsa como a los demas caçiques e prencipales e yndios los quales querria que benciesen esta cavsa porque son suyas las tierras sobre que se trata este pleyto e que no le an dado ny prometido cosa alguna para que diga el contrario de la verdad e que dira verdad porque es xristiano.

2 - A la segunda pregunta del dho ynterrogatorio dixo este dho t° que save que las dhas tierras de Quybi estan dentro de los rreynos e juresdicion de los yndios de Canta. Preguntado como lo save dixo que antes que vinyesen //
/210v/ los yndios yngas senores deste rreyno las dhas tierras que llaman de Quibi heran de yndios yavyos de facia la mar y estava en ellas un senor por sy que no hera sujeto a los caciques de Canta el qual senor se llamava Chaumecaxa y este Chaumecaxa hera sujeto a otro senor que llamaban Collicapa senor de Collyque que es rrepartimyento de Gonzalo de Cazeres hijo del contador Juan de Caçeres al qual servia e trebutava el dho Chaumecaxa y sus yndios e que en esta dha sazon pretendiendo los dhos yndios de Canta que este dho pueblo de Quybi fuese suyo hizieron mucha jente de guerra para benir sobre el dho Chaumecaxa y sobre el dho Collicapa senor de Collique y ganalle el dho pueblo de Quibi e saviendo la gran fuerça de gente que trayan los dhos caciques de Canta tuvo temor y como estava en medio no savia a quyen acudiese y enbiava secretamente chasques a los dhos yndios de Canta secretamente y en que les dava a entender que les hera amygo y luego los dhos caçiques e yndios de Canta que//
/211r/ ansy estavan de guerra baxaron de la syerra con toda su armada y pasaron por el dho pueblo de Quybi syn fazer daño del dho Chaumecaxa ny a sus yndios y llegaron bien mas adelante del dho pueblo de Quybi facia Collique quatro leguas y savido por el dho Collicapa senor de Collique la jente e armada que llebaban los dhos yndios de Canta junto mucha jente de guerra y vyno con

ella fazia los dhos yndios de Canta los quales visto la mucha jente que traya que era mas que la dellos se rretraxeron y se hizieron fuertes en unos çerros y de alli se enbiaron mensajeros los unos a los otros y trataron de conçierto y se conzertaron en que el dho Collicapa dixo a los dhos caçiques de Canta que que era la demanda que trayan que sy lo avian por el pueblo de Quybi que el se le dava y que fuesen amygos y los dhos cacique e yndios de Canta le enbiaron a dezir que no querian otra cosa syno el pueblo de Quibi que estava en su tierra e ansy el dho Collicapa se lo dio y amojono los termy^{o}s de Collique y de Canta e puso un mojon en un zerro que se nonbrava Xudcunga e agora se llama ansi que estara media legua deste pueblo de Quybi //
/211v/ que es donde pusyeron los mytimaes los y dende alli de aquel mojon dexaron para termynos de la syerra por tierra de Canta y de alli para abajo dexaron por llanos e para los yndios yungas de Collyque e que con esto quedaron muy buenos amygos los unos y los otros e hizieron juramento entre ellos segun su ley que los yndios de Canta no tenyan mas dyferencias con ellos ny los de Collique tanpoco ny pasarian con jente los unos a las tierras de los otros para rrenyr mas sobre las dhas tierras e ansy se bolbieron los yndios de Canta a sus tierras y los de Collique tanvien y no ubo mas guerra entre ellos y se amavan y querian y se enbiavan presentes e luego dende adelante el dho Chunbicaxa senor de Quibi servia e trebutava a los yndios de Canta y ansy syrbio syenpre e fasta oy dia syrben algunos yndios que en el estan que suszedieron de aquel tpo e que esto que dho tiene le es a este dho t^{o} pu^{co} e not^{o} por avello oydo dezir a sus padres e aguelos y pasados e a otros viejos e ançianos de los dhos pueblos de Canta y de Chacalla y desta comarca que decian que ellos ansy lo avian visto ser y pasar en sus tpos //
/212r/ e oido dezir a los suyos en los suyos e que nunca a visto ny oydo dezir lo contrario y dello a sydo y es a este t^{o} pu^{co} e notorio e publica voz y fama e que ansy mysmo les a oido dezir que benido que fue Topa Ynga Yupangue a esta tierra a gobernar enbio un orejon yndio para que gobernase los yndios de Chacalla porque lo mysmo hacia en otros pueblos que ponya just^{a} en todos y este orejon traya una olla en que traya sangre e chaquira molida para ofreszer a la guaca y este se junto con los dhos yndios de Chacalla y les dixo que fuesen con el al ofresçimyento e andavan con esto por los cerros derramandolo y deziendo a la guaca que diese salud a Topa Ynga e a sus hijos e le hiciese gran senor e le diese mucha rriqueza y comydas y despues de derramada la dha capacocha andubieron a buscar agua para laballa porque tenyan por costunbre de fazello ansy y como por aquella parte donde andavan hera y es tierra seca que no ay agua binyeron de en cerro en zerro [*sic*] en busca deste rrio de Quybi y binyeron a salir alli por el dho mojon que llaman Judcunga que es el que puso Collicapa por mojon de Canta e ansy bieron al dho rrio y labaron la dha capacocha //
/212v/ y se bolbieron por los propios pasos por donde avian benydo syn poner mytimaes en las dhas tierras asta el dho pueblo de Chacalla donde fallaron que un pueblo de yndios de Chacalla yavyos y su cacique que se llamava Culluynde estava juntando yndios y diziendoles que fuesen a fablar a la guaca para que se muriese Guaynacaba porque ya avia muerto Topa Ynga por su causa porque se lo abia rrogado y que lo mysmo faria la guaca sy se lo rrogasen para que Guaynacaba muriese que avia suzedido en el senorio e que savido por el dho orejon que traya la capacocha hizo mensajero mensajeros [*sic*] al Cuzco a Guaynacava de lo susodho y avisandole dello y el dho Guaynacava enbio mandar que los matasen e ansy mataron todos los yndios que no quedaron syno las mugeres y unos nynos chequytos a los quales dhos nynos e a sus madres el dho orejon senalo Collicapa cacique de Collique por traysion de Canta y senalo tierras para Guaynacava e para los dhos yndios y desta manera se entraron y estan los dhos yn-//
/213r/ dios yavyos en las dhas tierras de Quybi e que el dho orejon y los que suzedian en su cargo yban andando el tpo con la dha capacocha y con los dhos yndios yavyos faziendo la dha zeremonya en pruzeçion y metiendose cada dia mas en las dhas tierras y tomandolas y poseyendolas e que los dhos yndios de Canta no osavan rresystirselo ansy por no fazer derramar la sangre y por que no los matasen como por que Ynga no los hiçiese matar e que tanbien savido por Guaynacava que las dhas tierras heran tan buenas y que se coxia en ella mucha e buena coca mando que se pusyesen muchos yndios yavyos en las dhas tierras y que sembrasen e beneficiasen las chacarras para el dho Guaynacava e para los dhos yndios yavyos y les senalaron chacarras y se binyeron apoderando en las dhas tierras fasta junto con el dho pueblo de Quybi e que benydo Guascar Ynga despues de muerto Guaynacava hizo lo mysmo e aumento con la capacocha hasta mas adelante del dho pueblo de Quybi y porque un yndio topo en la capacocha e hizo derramar una gota le mataron alli luego y le enterrraron el //
/213v/ qual dho yndio que ansy mataron se llamava Chuquiguaranca que oy dya tyene un hijo bibo en este pueblo de Canta e que despues que murio Guascar Inga y binyeron los espanoles a esta tierra syenpre los dhos yndios de Canta estobieron mal con los yndios yavyos de Chacalla y estavan cada dia a punto de guerra para matarse sobre las dhas tierras y Guaman Yabre cacique prencipal del pueblo de Canta dezia a los yndios de Chacalla que vien savian que las chacarras de Quibi heran suyas y que Inga se las abia quytado tiranycamente y por fuerça como tirano e que agora que avian venydo los espanoles que se las avian de dar por bien y que sy no se las davan que alli avian de morir todos y que desta manera estobieron a punto de matarse fasta que vino el presydente Gasca y enbio a los dhos Nycolas de Ribera el moço encomendero de los yndios de Canta e a Fran^{co} de Ampuero encomendero de los yndios de Chacalla y que el dho Guaman Yavre cacique de Canta no queria dar obejas nengunas diziendo que el no avia de dar nada por las tierras que heran suyas y que el dho Ribera le decia: "mira hijo mejor es que les des doçientas obejas por quytar //
/214r/ pasyones porque sy rrenys y os matais castigar os an la just^{a} e a my tanbien porque lo consyento; faz quenta hijo que tu las obejas que me las vas a dar e yo las doy por las tierras"; e ansy el dho Guaman Yuayre [*sic*] cacique por lo que el dho su encomendero le dezia e por le fazer plazer dio las dhas doçientas obejas y el dho Marcapoma caçique de Chacalla las rrescivio e se contento con ellas y del dho concierto y bio este dho t^{o} que lo tubo por bueno y se fueron los dhos yndios yabyos de las dhas tierras y las dexaron a los dhos cacique yndios de Canta y nunca mas se enojaron ni pidieron cosa alguna fasta que les pusyeron demanda a ellas diziendo que no queria pasar por el concierto e que ansy mysmo estando faziendo el dho concierto se trato que diesen a los yndios del pueblo de Cullata e Tapati yndios yavyos de Chacalla pobres les dyesen algunas chacarras para que sembrasen y estos dhos yndios fueron al dho Guaman Yavre cacique de Canta y le fablaron para que les diese tierras en que

senbrar que ellos le servirian e otras ofertas que le hizieron y el dho Guaman Yavre caçique y el dho su encomendero Rribera y el dho Fran^{co} de Ampuero le sena-//
/214v/ laron desde una cruz que se puso junto al dho pueblo de Quybi fazia arriba fazia Canta tierras para que sembrasen los dhos yndios yauyos de Chacalla e ansy las an sembrado y beneficiado fasta agora e que don Juan yndio del rrepartimyento de Canta cojo de un pedaço de las dhas tierras un poco de coca porque la avia demandado plantar a sus yndios e que era muy poca cosa y que por ello los yndios yabyos se avian ydo a quexar e avia benydo un es^{o} fazer ynformacion lo qual todo este dho t^{o} rrepondio [*sic*] por estas rrazones en sustancia por lengua del dho don M^{yn} que lo declaro ansy.

3 - A la t^{a} pregunta del dho ynterrogat^{o} dixo que dize lo que dho tiene en la preg^{a} antes desta.

4 - A la quarta pregunta del dho ynterrogatorio dixo este dho t^{o} que dize lo que dho tiene en la pregunta antes desta.

5 - A la quynta pregunta del dho ynterrogat^{o} dixo este dho t^{o} que dize lo que dho tiene en las preguntas antes desta e que ansy se fazia lo que la pregunta dize y este t^{o} lo oyo dezir e tratar a los yndios viejos.

/215r/ 6 - A la sesta pregunta dixo que dize lo que dho tiene en las preguntas antes desta e que lo en esta pregunta cont^{do} es cosa publica y notoria entre los yndios viejos desta comarca.

7 - A la setima pregunta del dho ynterrogatorio dixo este dho t^{o} que dize lo que dho tiene en las preguntas antes desta.

8 - A la otaba pregunta del dho ynterrogat^{o} dixo que dize lo que dho tyene en las preguntas antes desta.

9 - A la novena pregunta del dho ynterrogatorio dixo que dize lo que dho tiene en las preguntas antes desta.

10- A la decima pregunta del dho ynterrogatorio dixo que dize lo que dho tiene en las preguntas antes desta.

11- A las honze preguntas del dho ynterrogat^{o} dixo este dho t^{o} que a su pareszer las dhas tierras de Quybi podran fazer de senbradura fasta diez fanegadas y no mas porque es tierra muy floja e pedregosa y hecha andenes e que no save lo que podra valer cada fanegada porque en esta tierra no se arriendan las tierras.

12- A las doze preguntas del dho ynterrogatorio dixo que dize lo que dho tiene y lo demas no lo save.

/215v/ 13- A las treze preguntas del dho ynterrogatorio dixo que vien balian cada cabeza de obejas machos y hembras unas con otras de las que el dho Guaman Yavre cacique de Canta dio al dho Marcapoma caçique de Chacalla por las dhas tierras de Quybi seis y syete pesos porque eran buenas y cresçidas y este dho t^{o} las diera por ellos en aquel tpo y esto rresponde.

14- A las catorze preguntas del dho ynterrogatorio dixo este dho t^{o} que en tierras de Lima y donde ay xristianos valen agora mas las dhas tierras que no agora diez años y que entre los yndios no cree es tanto el valor a lo menos en las dhas tierras de Quybi.

15- A las quynze preguntas del dho ynt^{o} dyxo este dho t^{o} que dize lo que dho tiene en las preguntas antes desta e que el dho Guaman Yavre como dho tiene queria las dhas tierras porque eran suyas y de sus pasados.

16- A las diez y seis preguntas del dho ynterrogatorio dyxo este dho t^{o} por lengua del dho don M^{yn} ynterpetre que lo que a dho es la verdad para el juramento que hecho tiene y deste caso no save otra cosa y en ello se afirmo y no lo firmo porque no supo.

/216r/ *Preguntas añadidas*

1 - A la primera pregunta del dho ynterrogatorio anadido dixo este dho t^{o} que dize lo que dho tiene en las preguntas antes desta del primero ynterrogatorio porque alli tyene declarado a esta pregunta.

2 - A la segunda pregunta del dho ynterrogatorio dixo este dho t^{o} que lo contenydo en la dha pregunta lo a oido dezir este dho t^{o} a viejos e ancianos desta tierra que decian avellos visto matar los yndios dellos sobre defender que no se entrasen en sus tierras con las dhas capacochas.

3 - A la tr^{a} pregunta del dho ynterrogatorio dixo este dho t^{o} que dize lo que dho tyene en las preguntas del primero ynterrogatorio.

4 - A las quatro preguntas del dho ynterrogatorio dixo este dho t^{o} que dize lo que dho tyene en las preguntas antes desta.

5 - A la quynta pregunta del dho ynterrogatorio dixo este dho t^{o} que dyze lo que dho tiene en las preguntas antes desta.

6 - A la sesta pregunta del dho ynterrogat^{o} dixo este dho t^{o} que dize lo que dho tyene en las preguntas antes desta.

/216v/ 7 - A la setima pregunta del dho ynterrogatorio dixo este dho tº que lo que a dho es la verdad y en ello dixo que se afirmava e afirmo e rratefycava e rratefico aviendosele tornado a leer por lengua del dho don M^{yn} Lengua ynterpetre nonbrado por los senores presydente e oidores para esta presente causa y encargosele que tenga secreto deste su dho e que no diga a naydie [*sic*] lo que a dho so pena de perjuro y que le castigaran o sacaran los dientes como otros xristianos y prometyo de fazello ansy e no lo firmo porque no supo.

Fho ante mi

Fran^{co} Hrrez de Gª
escrº de Su Mag^{t}

Atoco yndio ynfiel subjeto a don Antº /
no sabe la hedad y acuerdase de Guaynacaba
y le syrbio y tiene muger y hijos
y le ba ynteres.

Tº el dho Atoco yndio ynfiel natural que dixo ser del pueblo de Canta encomendado en Ribera el moço vº de la çibdad de Los Reyes sujeto a don Antonyo gobernador del dho rrepartimyento tº presentado por parte de los dhos cacique e prençipales del dho rrepartimyento aviendo jurado en forma segun //
/217r/ dixo acostunbrarse segun su ley e syendo preguntado por lengua de don M^{yn} ynpertre [*sic*] de la Rreal Avdiencia nombrado para esta presente cavsa por los senores presydente e oidores e siendo preguntado por el tenor del dho ynterrogatorio dixo e depuso lo sig^{e}:

1 - A la primera pregunta del dho ynterrogatorio dixo este dho tº que conosze e a conoscido a los caciques e prencipales que an sydo del dho pueblo de Canta y del de Chacalla desde el tpo de Guaynacaba que este dho tº se sabe acordar e ansy mysmo a los caciques que del dho tpo aca an sydo e son en el dho rrepartimyento de Chacalla porque son comarcanos al rrepartimyº de Canta e a tratado con muchos dellos e que tyene notiçia de los yngas senores que an sydo destos rreynos porque este dho tº sirbio a Guaynacaba e que sabe e tyene notycia de el dho rryo e tierras de Quibe porque las a bysto y estado en ellas muchas bezes desde el dho tiempo aca.

Preguntado por las preguntas generales de la ley dixo que es de hedad //
/217v/ de muchos años pero que no los a contado e que como dho tiene se acuerda de Guaynacaba senor que fue destos rreynos e le syrbio este dho tº e fue con el a Quito e que es vzº e tiene hijos e muger e casa en el dho rrepartimyento de Canta e tiene syete hijos barones e le ba ynteres en el dho pleyto pero que por eso no dexara de dezir berdad de lo que supiere y le fuere pregunt^{do} e que no le an dado ny pºmetido cosa alguna por que deje de dezir verdad.

2 - A la segunda pregunta de el dho ynterrogatorio dixo este dho tº que sabe que las dhas tierras que llaman de Quibi son de Canta porque las dibiden y parten ciertos mojones antiguos que estan junto a las propias casas del pueblo de Quibi que llaman Chuquicoto e que deste mojon que llaman Chuquicoto para abajo a los Yauyos se quedo para yndios yavyos de fazia la mar e de alli para arriba azia la syerra de Canta fuese de Canta e que este dho tº los bydo amojonar en tpo de Guascar Inga hijo de Guaynacava e que //
/218r/ los amojono un orejon yndio que bino del Cuzco que no se acuerda como se llamaba e que otro mojon que esta mas abajo de el dho mojon de Chuquicoto sabe este dho tº e a vysto que es asiento e termyno de yndios yabyos de el rrepartimyento de Fran^{co} de Ampuero el qual dho termyno e mojon se nonbra Judcunga e que el dho mojon de Judcunga que dho tiene que esta mas abajo de el de Chuquicoto antiguamente dende muchos años antes que binyesen los yngas a conquystar en esta tierra solia ser mojon e juresdicion de los yndios de Canta puesto y senalado por un senor de Collique rrepartimyento del contador Juan de Cazeres que corria fasta alli e fasta el pueblo de Quibi en tierra e que a lo que este dho tº se acuerda aver oydo dezir muchas e diversas bezes ansy a sus padres e aguelos deste tº e a otros viejos e ancianos de el dho rrepartimyento de Chacalla y de el de Canta y de sus comarcas fue desta manera que dezian que antes muchos años que entrasen yngas en esta tierra a conquistalla hera senor del dho pueblo de Quybi e de las dhas tierras un senor que se llamaba Chaumecaxa y que este senor y sus //
/218v/ yndios trebutaban a otro curaca que era senor del pueblo de Collique hazia la mar que es el que dho tiene de la encomyenda que agora es de el contador Juan de Cazeres e que en este dho tiempo los caçiques e prencipales e yndios del dho rrepartimyento de Canta hizieron grande armada de jente de guerra porque entonzes avia muchos yndios para yr a conquistar toda la sierra del dho Collicapa cacique yunga e porque andavan en aquella sazon en temas unos caciques con otros que el dho Collicapa pretendia de ir conquystar fazia la syerra la tierra de Canta y los de Canta la tyerra de los yavyos e que ansi abian benydo e pasado por el dho pueblo de Quibi e que el senor que en el estava los avia rr^{do} muy bien e sacadoles al camyno chica [*sic*] e otras comydas porque no hosaba selles contrario ny tanpoco al dho Collicapa y tenya dos coraçones e que ansy los dhos yndios de Chacalla pasaron adelante fazia Collique e que sabido por el dho Collicapa que los dhos yndios de Canta benyan de //
/219r/ aquella manera hizo gran cantydad de jente y bino fazia la syerra donde tenya noticia que estaban los dhos yndios de Canta e que como lo supieron los dhos yndios de Canta se rretiraron por unos zerros y que est^{do} en esto avian benydo a conçierto los dhos Collicapa y los yndios de Canta en que les diese el dho pueblo de Chaumecaxa que se llamaba Quibi e le senalase los mojones de lo que avia de ser de Canta y lo que avia de ser de los llanos e que el dho Collycapa les avia senalado por mojon e termyno el dho asyento que llaman de Judcunga para que dende alli para arriba fuese de la tierra de Canta y de alli para abajo de los yngas y tyerra de Collicapa y que desta manera se avian conformado y hecho amystades e avian prometido de nunca mas fazer jente el uno contra el otro sobre los dhos termynos e que ansy se avian ydo los dhos caçiques de Collique a sus tierras y los de

Canta tanbien e que desde entonzes fasta que vino Topa Ynga Yupangue que fue muchos años e tpo syenpre los dhos yndios de
Canta avian tenydo su posecion e usado della dende el dho //
/219v/ mojon de Judcunga hazia Canta aprovechandose de el dho pueblo de Quibi e sembrando las dhas tierras de Quibi e coxiendo el
fruto dellas y tenyendo sujeçion al dho Chaumecaxa que de alli adelante por rrazon de el dho concierto les trebuto y se llamaba
suyo como de antes del dho concierto lo hera del dho Collicapa e que benydo que fue el dho Topa Ynga Yupangue a gobernar el
rreyno y a conquistar esta tierra enbio sus capitanes a Chacalla para conquistar aquella tyerra y despues que la ubieron quistado
[*sic*] e paçificado la jente çierto pueblo de yndios de los dhos yauyos de Chacalla dizen que ynbocaban e fablavan con una guaca
para que muriese Inga e que syendo el dho Inga avisado de esto enbio que matasen todos aquellos yndios y que ansy los abian
muerto todos ecepto las mugeres que estavan preñadas y nynos; por que no conosziesen mas aquella tierra ny tornasen a adorar la
guaca los avian sacado los capitanes del inga fuera de la dha tierra de Chacalla y los avian puesto en el dho termyno y mojon //
/220r/ que llaman Xudcunga que fue donde se senalo por el dho Collicapa que avia de correr el termyno de Canta e que estas dhas
mugeres avian parescido [*sic*] y cresçido sus hijos e multiplicado y se avian apoderado e aprobechado de las dhas tierras de Quibi
e que despues de muerto el dho Topa Ynga Yupangue y suzedido Guaynacaba entendiendo que las dhas tierras heran tan buenas
y que se coxia mucha coca avia mandado sacar y saco del dho rrepartimyento de Chacalla otros çien yndios mas y los puso por
sus mytimaes y les senalo tierras desde el dho asiento de Judcunga fasta el otro mojon que llaman Chuquicoto que esta junto a
las casas de el dho pueblo de Quibi e les mando sembrasen coca para ellos y para el inga y que avian est^{do} en ellas hasta que
murio Guaynacaba e aun con çiertas zeremonyas que hazian por Guaynacaba se yban entrando cautelosamente e tomaban cada
dia mas tierras e que con una capacocha que ellos llamaban avian pasado mas adelante del dho pueblo de Quibi e que benydo
que fue Guascar Ynga hijo de Guaynacaba avia //
/220v/ dado por bueno lo que el dho Su padre avia hecho e avia mandado que los dhos yndios yauyos del dho rrepartimyento de
Chacalla se estobiesen en las dhas tyerras e ansy desta manera dize que oyo dezir que las tenyan e poseyan e que los dhos
cacique e yndios de Canta no osavan fablar de myedo que tenyan a los yngas e que esto como dho tyene lo oyo dezir a los dhos
sus pasados los quales dezian que ellos ansy lo avian bisto ser y pasar en sus tiempos e oidolo dezir a otros sus mayores e
ançianos e que dello hera e avia sydo cosa publica e notoria e publica voz y fama e que nunca avian bysto ny oydo dezir lo
contrario y que ansi mysmo este t^{o} se acuerda e vido que despues de muertos los yngas y benydo los espanoles en este rreino se
les quito el myedo a los dhos yndios de Canta y luego quisieron defender las dhas sus tierras y dezian a los dhos yndios yavyos
que se salyesen dellas que vien savian que no heran suyas y que ellos y sus pasados las avian ganado e que el inga hera tirano y
se las avia quitado //
/221r/ syn podello fazer y sobre esto vido este dho t^{o} que estavan cada dia para se matar fasta que vino el presydente Gasca y en este
tiempo vinyeron a las dhas tierras Nycolas de Ribera el moço encomendero de los dhos $y^{o}s$ de Canta y $Fran^{co}$ de Ampuero
encomendero de los yndios de Chacalla y los conçertaron en esta manera que Guaman Yabre cacique de Canta dio a Marcapoma
a lo que este t^{o} entendio de los yndios en aquella sazon duzientas obejas de la tierra machos y hembras por que dexase las dhas
tierras y este dho t^{o} entendio que se hizo el dho concierto porque despues que se fueron los dhos $Fran^{co}$ de Ampuero y Nycolas
de Rribera los dhos yndios de Canta se entraron en las dhas tierras y las labraron y sembraron y los dhos yndios yauyos de
Chacalla se avian ydo a su tierra y dexadolas e que nunca se avian quexado ny contradholo fasta agora que a oido dezir que don
Xristoval caçique del rrepartimyento de Chacalla a puesto demanda por las dhas tierras y no quiere pasar por el dho conçierto lo
qual dixo e declaro por estas rrazones por lengua del dho don M^{yn}.

/221v/ 3 - A la tr^{a} pregunta del dho ynterrogatorio dixo este dho t^{o} que dize lo que dho tiene en las preguntas antes desta.

4 - En la quarta pregunta del dho $ynterrog^{o}$ dixo este dho t^{o} que dize lo que dho tiene en las preguntas antes desta e que lo en ella contenydo lo a oido dezir a sus mayores e mas ancianos que dezian avello vysto ser y pasar asy como la dha pregunta lo dize y declara e aun este dho t^{o} vio pte de lo contenydo en la dha pregunta que se fazia e acostumbrava en su tpo y esto rresponde.

5 - A la quinta pregunta de el dho ynterrogatorio dixo este dho t^{o} que dize lo que dho tiene en las preguntas antes desta e que lo contenydo en la pregunta a oido dezir que se fazia como la dha $preg^{a}$ lo dize.

6 - A la sesta pregunta del dho ynterrogatorio dixo este dho t^{o} que dize lo que dho tiene en las preguntas antes desta.

7 - A la setima pregunta del dho ynterrogatorio dixo este dho t^{o} que dize lo que dho tiene en las preguntas antes desta.

8 - A la hotava pregunta de el dho ynterrogatorio dixo este dho t^{o} que dize lo que dho tyene en las preguntas antes desta.

/222r/ 9 - A la novena pregunta del dho ynterrogatorio dixo este dho t^{o} que dize lo que dho tiene en las preguntas antes desta.

10- A la dezima pregunta de el dho ynterrogatorio dixo este dho t^{o} que dize lo que dho tyene en las preguntas antes desta en que se afirma.

11- A la honzena pregunta del dho ynterrogatorio dixo este dho t^{o} que a vysto las dhas tierras de Quibi e que son todas andenes y piedras y cascaxo y que le paresze que podra aver diez fanegadas de sembradura poco mas o menos e de el presçio que vale cada fanegada no lo save ny lo entiende.

12- A las doze preguntas de el dho ynterrogat° dixo este dho t° que dize lo que dho tiene en las preguntas antes desta.

13- A las treze preguntas del dho ynterrogatorio dixo este dho t° que le paresze que podria valer cada cabeça de obeja de las del conçierto en aquel tpo a seis pesos sy heran grandes y si heran medyanas a çinco y si heran chicas a tres pesos y a quatro.

/222v/ 14- A las catorze preguntas del dho ynterrogatorio dixo este dho t° que no lo save e dize lo que dho tiene.

15- A las quynze preguntas de el dho ynterrogatorio dixo este dho t° que le paresze e a entendido de los dhos yndios de Canta e aun por parte deste t° dyze que es ansy que sy se hizo el dho conçierto y dieron las dhas obejas fue por no dexar a los yndios de Chacalla las tierras que heran de los yndios de Canta y las avian ganado sus pasados y por quytarse de pasyones.

16- A las diez y seis preguntas dixo que lo que dho es la verdad y en ello se afirmo por la dha lengua aviendosele tornado a leer.

Añadidas

1 - A la primera pregunta de el dho ynterrogatorio dixo este dho t° que dize lo que dho tiene en las preguntas del dho primero ynterrogatorio e que este t° en aquella sazon que dize la pregunta hera pastor de Ynga pero que se fue pu^co^ y le oyo dezir a muchos yndios del dho rrepartimyento de Canta que en aquella sazon que rreynava Guascar Inga se hizieron //
/223r/ las dhas dos capacochas vna para Guascar Inga y otra para su muger e que con ellas avian entrado en las dhas tierras de Quibi los dhos yndios yauyos de el dho rrepartimyento de Chacalla e que al dho yndio que llamavan Chuqui Guaranga porque llego a defender que no entrasen con la dha capacocha y porque defendyendolo avia derramado un poco de sangre de la capacocha le avian muerto alli donde se derramo la sangre y este dho t° fue despues de pasado esto a ber el dho yndio y le bio enterrado mas adelante del dho pueblo de Quibi que fue fasta donde llegaron con la dha capacocha.

2 - A la segunda pregunta del dho ynterrogatorio anadido dixo que lo en la dha pregunta en aquella sazon lo oyo dezir a yndios de el dho pueblo de Canta que avia pasado ansy e avian muerto los yndios que dize la pregunta porque defendian la entrada con la dha capacocha en sus tierras.

3 - A las treze [*sic*] preguntas del dho ynterrogatorio dixo este dho t° que dize lo que dho tiene en las preguntas antes desta del primero ynterrogatorio.

/223v/ 4 - A la quarta pregunta del dho ynterrogatorio dixo este dho t° que no lo save e dize lo que dho tiene.

5 - A la quinta pregunta de el dho ynterrogatorio dixo este dho t° que no lo save e dize lo que dho tiene.

6 - A la sesta pregunta del dho ynterrogatorio dixo este dho t° que no lo save e dize lo que dho tiene.

7 - A la setima pregunta del dho ynterrogatorio dixo este dho t° que dize lo que dho tiene lo qual dixo que hera la verdad y en ello dixo que se afirmava e afirmo e rratefico despues de aversele leydo por lengua del dho don M^yn^ el qual le encargo tenga secreto de lo que a dho e no lo diga a nadie so pena de perjuro e que lo castigaran y prometiolo.

Fho ante mi Fran^co^ Hrrez de Guebara
escr° de Su Mag^t^

/224r/ Don Ju° Caxchayaure sujeto a don Xristoval
acuerdase de Guaynacaba y vale ynteres.

T° el dho don Juan Caxcharure natural de un pueblo que se llama Utco sujeto al cacique del dho rrepartimyento de Canta encomendado en Ribera el moço vz° de la çibdad de Los Rreyes t° presentado por parte del cacique prençipales e yndios de el dho rrepartimyento abiendo jurado en forma de dr° segun se rrequeria e syendo pregunt^do^ por lengua del dho don M^yn^ Lengua nonbrado por los senores presydente e oydores para esta presente causa e por el ynterrogatorio por donde se pidio fuese hexsamynado dixo e depuso lo syg^e^:

1 - A la primera pregunta de el dho ynterrogatorio dixo este dho t° que conosze a los caciques e prençipales e yndios que son de el dho rrepartimy° de Canta porque como dho tyene es vz° de el dho rrepartimyento e que ansy mysmo conosze a los caciques del dho rrepartimyento de Chacalla ansy los que agora son como los que an sydo de mas de quarenta años a esta parte que este dho t° se sabe acordar porque a tratado y conversado con ellos e que del dho tpo //
/224v/ aca a conosçido y conosze las dhas tierras de Quibi porque las a bysto y estado en ellas e que conoscio a Guaynacaba e tyene noticia de los senores yngas que antes del fueron de aber oido dezir e tratar dellos.

Preguntado por las preguntas generales de la ley dixo que la hedad que tyene no la sabe mas de que se acuerda de Guaynacaba porque fue en tiempo suyo e que benia donde cinquenta yndios de su pueblo en el dho rrepartimyento de Canta e tiene hijos e muger e le ba ynteres en las dhas tierras como a los demas yndios de Canta y querria que bençiesen este pleyto contra los de Chacalla pero que por eso no dexara de dezir berdad de lo que supiere e que no le an dado ny prometydo cosa alguna para que en este su dho diga el contrario de la berdad.

2 - A la segunda pregunta de el dho ynterrogatorio dixo este dho t^{o} que sabe que las dhas tierras que llaman de Quybi son de los
dhos yndios e rrepartimyento de Canta y tienen en ellas sus limytes e mojones puestos antiguamente que dibiden los termynos //
/225r/ de los yungas hazia la mar e que a lo que este dho t^{o} se acuerde aver oido dezir a sus mayores e mas ançianos naturales de el
dho rrepartimyento de Canta y de otras partes y lugares de su comarca fue desta manera que dezian que antes mucho tiempo que
binyesen los yngas a conquistar esta tierra las dhas tierras de Quibi heran juresdicion y tierras de yungas sujetas a el cacique de
Collique que llamavan Collicapa que hera senor fasta la mar e que en las dhas tierras de Quibi estava un senor en un pueblo de
yndios que llaman el pueblo de Quibi y al dho senor le llamavan Chaumecaxa e que este y sus yndios se nonbraban yungas e
serbian e trebutaban al dho Collicapa yunga senor de Collique y hera apartado de los serranos e yndios de Canta e que en este
tyempo los dhos yndios de Canta heran muy guerreros e tenyan gran fuerça de jente e querian e tenyan voluntad de benyr a
conquistar la tierra de Collique e ansy se juntaron gran cantidad de jente de guerra los dhos yndios de Canta y baxaron desta
manera de la syerra y entraron en el dho pueblo de Quibi de que era senor el dho Chaumecaxa //
/225v/ el qual dizen que no se defendyo ny hizo rresystencia alguna a la dha jente de guerra que traya por cacique e capitan Ayna
Guaranga cacique principal del dho rrepartimyento de Canta e que el dho Chaumecaxa les dio la comyda que pudo aunque no
osaba declararse a quien acuderia e que fazia dos caras e que ansy desta manera pudyeron pasar y pasaron fazia Collique los dhos
yndios de Canta y se pusyeron zerca del dho cacique Collicapa el qual como supo la nueba de la dha jente de guerra junto mucha
jente e vino con ella fazia donde estava el dho Ayna Guaranga y su jente los quales como lo supieron se retiraron y subieron por
unos cerros y se hizieron mensajeros los unos a los otros y los dhos yndios de Canta pedian al dho Collicapa que les dyese el
dho pueblo de Quibi para que fuese suyo y las dhas tierras de Quibi pues estaban junto a su pueblo e a su rrio y el dho Collicapa
les dixo que sy no lo avian por mas de por el dho pueblo de Quybi que el se lo daria que bien savia que el rrio de Quibi benya
de su tierra e del agua y nyebe que alla caya de que a ellos benya p^{o}becho e rregava sus tierras e que el le dava el dho pueblo de
Quibi e que //
/226r/ de ay adelante se tratasen y fuesen buenos amygos e tobiesen contrataçion y entrada los unos en las tierras de los otros bendiendo
y conprando como buenos amygos y que desta manera se abian conformado los unos yndyos con los otros e avian jurado e
prometydo de nunca mas darse guerra y el dho Collicapa dio y entrego las dhas tierras y pueblo de Quibi y puso mojones e
termynos que dibidian lo que avia de ser de los yungas y lo que avia de ser de la syerra e puso un mojon media legua deste
pueblo de Quibi donde llaman y nonbran oy dia Gudcunga para que dende alli para abaxo fazia la mar fuese tierra de yungas y
de alli para arriba fuese tierra de la syerra e para los dhos yndios de Canta e los dhos yndios de Canta desde entonzes se entraron
e apoderaron en las dhas tierras de Quibi y en el dho pueblo de Quibi e las labraron y sembraron quyeta e paçificamente y el dho
Chaumecaxa hera sujeto e trebutava dende alli adelante a los dhos yndios de Canta ny mas ny menos que lo fazia el dho
Collicapa senor de Collique e que en esta costumbre y posesion paçifica avian est^{do} y estobieron los dhos yndios de Canta
muchos años y tpos fasta //
/226v/ que bino a conquystar esta tierra capitanes de Topa Ynga Yupangue; enbio a conquystar a Chacalla y despues de conquistada
murio Topa Ynga Yupangue y suzedio Guyanacaba y en esta coyuntura fueron fallidos cierto caçique yndios de un pueblo que
llamavan Acupayllata de el rrepartimyento de Chacalla los quales fazian ciertas zerimonyas e ynbocaban el demonyo por que se
muriese Guainacaba e que no ubiese mas yngas e fazia entender el cacique deste pueblo que llamavan Colluinde que el avia sydo
parte y hecho que muriese Topa Ynga Yupangue con las dhas zerimonias e rritos e que sy agora se lo rrogasen que tanbien
moriria Guaynacaba e no abria mas yngas que los conquistasen e serian como de antes de lo qual dize que fue abysado el dho
Guaynacaba el qual enbio luego un capitan a aberiguallo y sabello e benydo que fue e ynformado de lo que pasaba abia muerto
el dho cacique e yndios e que no avia dexado dellos syno los nynos chequitos y las mugeres preñadas e que a estas por que
olbidasen aquella tyerra y no la conosziesen mas e no binyesen a fazer las dhas rroga-//
/227r/ tybas el dho capitan los avia sacado de el dho rrepartimyento de Chacalla y los avia puesto en el dho termyno que llaman
Judcunga donde el dho cacique Collicapa avia senalado los termynos de Canta e que alli el dho capitan les avia mandado
rresydiesen e que fuese su bibienda y les dio y senalo tyerras en que sembrasen e que despues andando el tpo sabido por
Guaynacaba que las dhas tierras donde avian puesto los dhos yndios y mugeres heran buenas y que se sembraba e coxia coca e
otras comydas mando a sus capitanes que sacasen de el dho rrepartimyento de Chacalla otros çiento yndios y los pusiesen en las
dhas tierras de Quibi y les dyesen tierras para senbrar para Ynga y para ellos y los capitanes de Guaynacaba lo hizieron ansy que
sacaron los dhos çien yndios de el dho rrepartimyento de Chacalla y los pusyeron con los demas en las dhas tierras de Quibi e
alli le senalaron tyerras en que sembrasen para Guainacaba y para ellos y les dieron desde el dho mojon de Judcunga fasta otro
mojon que hizieron que llaman Chuquicoto que esta junto a las casas de el dho pueblo de Quibi y desta manera se entraron en las
dhas tierras e benydo que //
/227v/ fue Guascar Inga a gobernar por muerte de Guaynacava su padre dixo que avia por bueno lo que su padre avia hecho y les
confirmo las dhas tyerras a los dhos yndios yauyos e cada dia se yban entrando mas so color de una zeremonya que trayan los
orejones de Guascar Inga y los dhos yndios yauyos que llamavan capacocha diziendo que era rrogativa de Ynga que pedian a la
guaca le dyese salud a el e a sus hijos e a su muger e le dyese mucha comyda e rriquezas e que era desta manera que trayan en
un baso de barro çierta sangre de obejas y chaquira molida y la derramavan en un zerro pidiendo salud para Ynga e otras cosas

que dezian e que sy algun yndio tocava en la dha capacocha que alli le matavan luego donde tocaba y derramava la sangre y le enterravan en aquel lugar y que desta manera se yban entrando en las dhas tierras e que los dhos yndios de Canta no se lo rresistian ny osaban fablar porque tenyan myedo a los yngas e que esto que dho tyene este dho t^{o} lo a oido a los dhos sus pasados los quales dezian abello bysto ser y pasar tanbien en sus tpos e oidolo dezir a sus mayores e mas ancianos que dezian que ellos ansy lo avian vysto usar y pasar y guardar y se //
/228r/ tenya por ley entre ellos como agora las leyes de los xristianos e que dello hera e avia sydo publico y notorio e publica boz y fama e que este dho t^{o} se acuerda de parte de lo susodho aver sido en su tpo e que desta manera estobieron los dhos yndios yauyos de Chacalla en las dhas tierras fasta que vinyeron los espanoles e que benydos vio este t^{o} que los dhos yndios de Canta contradezian a los dhos yndios yauyos de Chacalla la dha posesion e les dezian que les dexasen sus tierras que no heran suyas e que las avian ganado sus pasados e que los yngas se las avian quitado a ellos para darselas syn podello hazer y por esto dezian que sy no se las dexavan que agora que no avia yngas y avia espanoles que los avian de matar sobre ello y bido este dho t^{o} que sobre esta causa tenyan pasyones los dhos yndios de Canta y los de Chacalla y estaban cada dia en arma para se matar fasta que vino el Lic^{do} de La Gasca y dezian que abia mandado a $Fran^{co}$ de Anpuero encomendero de los yndios de Chacalla e a Rribera el moço encomendero de los yndios de Canta que los pusiesen en paz y los conzertasen y este dho t^{o} bido que binyeron los susodhos al dho valle e tierras de Quybe e trataron de conzertallos desta manera que el dho Guaman Yaure cacique de Canta diese a Marcapoma caçique de //
/228v/ Chacalla duzientas obejas de la tierra machos y hembras por que les dexasen las dhas tierras de Quybi e ansy bido y entendio este dho t^{o} que el dho Guaman Yaure cacique de Canta dyo las dhas obexas porque luego los dhos yndios de Chacalla se fueron a sus tierras y dexaron las dhas tierras de Quibi a los dhos yndios de Canta los quales desde entonzes se entraron en ellas y las tubieron y poseyeron syn que los dhos caciques de Chacalla fablasen mas en ello ny se quexasen hasta agora que bido que don Xpoval cacique de Chacalla pone pleyto e no quiere pasar por el dho concierto e que vido que al tyempo que se hizo el dho conçierto vnos yndios de el pueblo de Cullata y Quynoamarca yabyos del dho rrepartimyento de Chacalla fueron llorando al dho Guaman Yaure cacique de Canta y le dixeron: "pues que nro cacique Marcapoma nos a bendido las tierras que eran nras y en que sembramos y no tenemos en que sembrar dadnos vn pedaço de tierras en que sembremos y nosotros seremos vros vasallos y el dho Guaman Yaure les dio y senalo prestadas çiertas tierras junto al dho pueblo de Quibi para que senbrasen que son las de que dize que se coxo la coca este otro dia por pte delos de Canta //
/229r/ sobre que abia benydo de Lima vn escrivano a hazer ynformacion lo qual todo dixo e declaro este dho t^{o} por estas propias rrazones en sustancia por lengua del dho don M^{yn} ynterpetre.

3 - A la tr^{a} pregunta de el dho ynterrogatorio dixo este dho t^{o} que dize lo que dho tiene en las preguntas antes desta.

4 - A la quarta pregunta de el dho ynterrogatorio dixo este dho t^{o} que dize lo que dho tyene en las preguntas antes desta.

5 - A la quinta pregunta de el dho ynterrogatorio dixo este dho t^{o} que dize lo que dho tyene en las preguntas antes desta y que lo en ella $cont^{do}$ se fazia ansi como la dha $preg^{a}$ lo dize y hera pu^{co} e not^{o}.

6 - A la sesta pregunta deel dho ynterrogatorio dixo este dho t^{o} que dize lo que dho tiene en las preguntas antes desta y que lo en ella $cont^{do}$ oyo dezir por pu^{co} e notorio a yndios viejos e ancianos que dezian avello oido e visto muchos dellos.

7 - A la setima pregunta de el dho ynterrogatorio dixo este dho t^{o} que dize lo que dho tiene en las preguntas antes desta.

8 - A la otaba pregunta de el dho ynterrogatorio dixo este dho t^{o} que dize lo que dho tiene en las preguntas antes desta.

/229v/ 9 - A la novena pregunta del dho ynterrogatorio dixo este dho t^{o} que dize lo que dho tiene en las preguntas antes desta.

10- A la dezima pregunta de el dho ynterrogatorio dixo este dho t^{o} que dize lo que dho tyene en las preguntas antes desta.

11- A la honzena pregunta de el dho ynterrogatorio dixo este dho t^{o} que le paresze a este dho t^{o} que en las dhas tierras de Quibi no podra aber de diez fanegadas de tyerra arriba poco mas o menos porque este t^{o} las a andado y sabe que son las mas dellas piedra y cascaxo e andenes y tierra floxa e queno sabe a como podra baler cada fanegada y esto rresponde.

12- A las doze preguntas deel dho ynterrogatorio dixo este dho t^{o} que dizelo que dho tyene y lo demas no lo sabe.

13- A las treze preguntas de el dho ynterrogatorio dixo que le paresze a este dho t^{o} que en la sazon que dyeron las dhas obejas porlas dhas tyerras de Quibi podrian baler unas a seis y a syete $p^{o}s$ y otras a çinco y otras a quatro sy fuesen chequitas e como son las obejas ansy dize que valen.

14- A las catorze preguntas del dho ynterrogatorio dixo este dho t^{o} que no lo sabe porque entre los yndios //
/230r/ no saben que cosa es comprar ny bender tyerras e que en Lima entre los españoles a entendido y entiende este dho t^{o} que valian en aquella sazon que se hizo el dho concierto las dhas obejas menos que agora las dhas tierras y que agora balen mas porque ay mas jente.

15- A las quynze preguntas de el dho ynterrogatorio dixo este dho tº que tyene entendido que los dhos yndios de Canta si hizieron el dho conçierto y dieron a los dhos yndios yavyos las dhas obejas fue por sacar las dhas tyerras que eran suyas y las avian ganado sus pasados y por quitarse de pleito y dyferencias con los yndios yavyos.

16- A las diez y seis preguntas del dho ynterrogatorio dixo este dho tº que lo que a dho es la verdad y lo que deste caso sabia para el juramento que abia hecho y en ello se afirmo por lengua del dho don M^{yn} que lo declaro ansi.

Anadidas

1 - A la primera pregunta del dho ynterrogatorio dixo este dho tº que dizelo que dho tiene en las preguntas antes desta del ynterrogatorio prencipal e que en tpo de Guaynacaba supo y entendyo este dho tº e oyo dezir a sus pasados que se abian hecho las dhas dos capacochas //
/230v/ vna para el y otra para su muger e que en tiempo del dho Guascaringa no se hizo mas de una capacocha e que sy mas se hiziera lo supiera este dho tº porque se acuerda muy bien de aquel tyempo y tiene noticia de la que entonzes paso y esto rresponde.

E que en tpo de Guascaringa este dho tº vio que un yndio que se llamava Chuqui Guaranga quiso defender la dha capacocha de Guascaringa porque llegava a su tyerra que era mas adelante del dho pueblo de Quibi diziendo que no le entrasen sus tierras y enpeço a rrenpujar al yndio que traya la dha capacocha y derramo un poco de sangre de la dha capacocha y luego tomaron el dho yndio y le mataron y le enterraron donde cayo la dha sangre y este dho tº le byo enterrar y esto rresponde.

2 - A la segunda pregunta de el dho ynterrogatorio dixo que lo contenydo en la dha pregunta lo oyo dezir este dho tº a sus mayores e mas ancianos que abia pasado como la dha pregunta lo dize en tpo de Guaynacaba que sobre defender que no entrasen con la dha capacocha en las tierras de los dhos yndios que dize la pregunta los avian muerto los que trayan la dha capacocha y esto rrespondyo a esta dha pregunta.

/231r/ 3 - A la trª pregunta del dho ynterrogatorio dixo este dho tº que dize lo que dho tyene en las preguntas antes desta del dho primero ynterrogatorio.

4 - A la quarta pregunta de el dho ynterrogatorio dixo este dho tº que dize lo que dho tyene en las preguntas antes desta del dho primero ynterrogatorio y esto rresponde.

5 - A la quinta pregunta del dho ynterrogatorio dixo este dho tº que dize lo que dho tyene en las preguntas antes desta de el dho ynterrogatorio prençipal.

6 - A la sesta pregunta de el dho ynterrogatorio dixo este dho tº que dize lo que dho tiene en las preguntas antes desta del dho ynterrogatorio prençipal.

7 - A la setima pregunta de el dho ynterrogatorio dixo este dho tº que dize lo que dho tyene lo qual dixo que hera la verdad y en ello dixo que se afirmaba e afirmo e rrateficava e rratefico aviendosele tornado a leer por lengua del dho don M^{yn} Lengua ynterpetre nonbrado por los dhos senores presydente e oidores //
/231v/ el qual le encargo tenga secreto deste su dho e que no lo trate con nenguno de los dhos caciques so pena de perjuro e que le castigaran el qual prometio de fazello ansi e no lo firmo porque dixo que no sabia.

paso ante mi — Fran^{co} Hrrez de Guebara
escrº. de Su Mag.^{t}

/232r/ Don Fran^{co} Caxa Guaranga
sujeto a los de Canta.

Tº el dho don Francisco Caxa Guaranga natural de un pueblo que se dize Cocha que es en el rrepartimyento de Canta encomyenda del dho Nycolas de Rribera el moço vzº de la çibdad de Los Rreyes tº presentado por parte de los caciques e prençipales yndios del dho rrepartimyento de Canta aviendo jurado en forma devida de drº e syendo preguntado por lengua del dho don M^{yn} Lengua ynterpetre nonbrado por los senores presydente e oidores para esta presente causa por el tenor del ynterrogatº por donde se pidio fuese hexsamynado dixo e depuso lo syguiente:

1 - A la primera pregunta del dho ynterrogatorio dixo este dho tº que conosze e a conosçido a los caciques e prencipales que an sydo e son del dho rrepartimyento de Canta desde el tpo de Guaynacaba a esta parte porque como dho tyene es sujeto al dho rrepartimyento y lo mysmo a conoscido e conosze a los caciques prençipales e yndios que an sydo e son del rrepartimyento de Chacalla porque a tratado e conversado con ellos e que tyene noticia de las dhas tierras de Quibi porque las a vysto e estado e andado por ellas muchas vezes //

/232v/ e lo mysmo el rrio que pasa por ellas e que conoscio a Guaynacava e que tiene notycia de los demas mas Yngas [*sic*] que fueron antes del.

Preguntado por las preguntas generales de la ley dixo que podra tener cient anos porque este t° conoscio a Guaynacaba e que a tenydo e tiene hijos e nyetos e tataranyetos e que como dho tyene es natural del dho rrepartimyento de Canta e sujeto al cacique del e le ba el mysmo ynteres que a los demas yndios en esta cavsa pero que por eso no dexara de dezir verdad de lo que supiese y le fuere preg^{do} e que dize que querria que sacase las dhas tierras sus yndios e no los de Chacalla e que no le an dado cosa alguna para que no diga verdad.

2 - A la segunda pregunta del dho ynterrogatorio dixo este dho t° que sabe que las dhas tierras de Quibi son de el dho rrepartimyento de Canta e las tubieron e poseyeron antiguamente a muchos anos antes que vinyesen los Yngas a este rreyno y el rrio que por ellas pasa byene de la syerra de Canta e que lo que zerca deste caso //
/233r/ pasare que este dho t° a oido dezir muchas vezes a yndios viejos e ancianos del dho rrepartimyento de Canta e desta comarca que son ya muertos de los nonbres de los quales no se acuerda y especialmente lo oyo dezir a su padre que se llamava Gallopa e a su aguelo que se llamava Poma que dezian que muchos anos e tyempos antes que vinyesen los Yngas a fazer trebutar esta tierra los yndios de el dho rrepartimyento de Canta y su cacique que llamavan Ayna Guaranga por codicia que tenian dela tierra delos yungas e del dho pueblo de Quibi hizieron mucha jente de armada y bajaron con ella de la sierra al dho pueblo de Quibi donde avian fallado vn cacique que trebutava a otro que se llamava Chaumecaxa el qual como bido la dha jente de Canta salio de paz e a mochar los dhos yndios de Canta y con comydas el qual tenya consygo otros trescientos e que eran gran cantidad de yndios los que estos tenyan e quelos dhos caciques de Canta avian pasado adelante en busca del dho senor Collicapa para que pues hera suyo el dho //
/233v/ pueblo de Quibi pedir que se lo dyese por bien e sy no darle batalla e que llegados zerca de donde estava el dho Collicapa y sabido por el la armada que trayan los de Canta avia hecho gran cantidad de jente e avia benido con ellos azia do estava el dho Aynaguaranga cacique de Canta los quales se avian rretirado y suvido a unos zerros fuertes y que luego el dho Collicapa les avia enbiado mensajeros que supieren lo que buscavan y los de Canta les avian enbiado a dezir que querian que les dyese el dho pueblo e tierras de Quibi porque estavan zerca a su tierra y en su rrio e que el rrio que pasaba por el dho pueblo e tierras de Quibi hera suyo y bajava de su tierra [*sic*] y de las lagunas que de la nyebe se fazian y el dho Collicapa les avia enbiado a dezir que sy no lo avian por mas de por el dho pueblo e tierras de Quibi que el se lo queria dar e que fuesen buenos amygos e ansy se juntaron unos caciques e yndios con otros y se avian conformado y hechos amystades y conzertado que los unos fuesen a tratar y rrescatar a las tierras de los otros y los otros a las de los otros syn nengun //
/234r/ rrezelo y se avian abraçado y comydo y bebydo juntos e que quando el dho Aynaguaranca cacique de Canta y sus yndios se avian querido bolber a sus tierras el dho Collicapa avia ydo con ellos fasta las dhas tierras de Quibi y hecho saber al dho Chaumecaxa como su voluntad hera que fuese sujeto al rrepartimyento de Canta e alli dibidio las tyerras que avian de ser serranas y las que avian de ser de los llanos yndios yungas y pusieron un mojon y senal y le pusiexon por nonbre Judcunga que desde alli para arriba fazia la syerra fuese de la tyerra de Canta y de yndios serranos y de alli para abaxo fazia Collique fuese de yungas e ansy desde entonzes los dhos yndios de Canta se avian entrado e apoderado en las dhas tierras de Quibi y labradolas y sembradolas y lo mysmo en el dho pueblo de Quibi e que el dho Chaumecaxa caçique que dellos avia sydo sujeto e trebutava e que esto avia pasado muchos anos e tpo fasta que binyeron a gobernar este rreyno Yngas que fue el primero Topa Ynga Yupangue e que en este dho tpo el dho Topa Ynga Yupangue avia enbiado a conquistar esta tierra de Chacalla//
/234v/ e que despues que la ubieron conquistado y hecholes ser sujetos se bolbieron al Cuzco y dende a dos años avia muerto Topa Ynga Yupangue y luego avia suzedido Guaynacava e que rremando Guaynacaba un cacique de un pueblo que llaman Acopay que el cacique se llamaba Colluynde e estaba con sus yndios adorando una guaca y rrogandole que no ubiese Guaynacava por que no les hiziese serbir e trebutar e dezia a los dhos yndios suyos que Topa Ynga Yupangue avia muerto porque se lo avia rrogado el y que sy agora se lo rrogavan que tanbien moriria Guaynacava e que no servirian e que ciertos yndios que tenyan buen coraçon al Inga avian enbiado al Cuzco a fazello saber a Guaynacaba y el dho Guaynacava avia enbiado vn orejon que llamaban Apara Yupangue para que hiziese la pesquisa el qual avia benydo y se avia ynformado de lo susodho y fallo que era verdad y tomo el cacique e a sus yndios y los mato a todos que no dexo syno las mugeres que estavan prenadas e a los nynos chequitos e a estas dhas mugeres e ninos por que olvidasen la dha guaca //
/235r/ e no tornasen a caer en aquel delyto los avia sacado del dho pueblo e tierra de Chacalla y los avia puesto en las dhas tyerras de Quibi y hecho poblacion en el dho tr° que llaman Judcunga que fue donde el dho Collicapa senor de Collique avia devidydo los terminos de la syerra e de los llanos a los yndios de Canta e que les avia puesto por su corregidor vn orejon que se llamava Yungaype e que les avian dado e senalado tyerras en que senbrasen y que desta manera los dhos yndios yavyos se avian entrado en las dhas tierras de Quibi e que ansy como yba cresçiendo e multiplicando esta jente e conosciendo las dhas tierras se avian entrado en ellas cada dia mas e tanbien con cierto sacreficio que fazian por Guaynacava que llamavan capacocha e que despues que murio Guaynacava avia suzedido Guascaringa el qual como tubiese noticia de la dha jente e tierras de Quybi avia mandado que les diesen mas tierra de la que tenyan e que sembrasen para el y para ellos e ansy dize que les avian dado las dhas tierras y senaladolas e amojonadas fasta junto a las dhas casas del pueblo de Quybi que se llamava e oi dia se llama//
/235v/ Chuquicoto e que desta manera avian tenydo los dhos yndios yavyos las dhas tyerras y labradolas y sembradolas fasta que vinyeron los espanoles a esta tierra con la benyda de los espanoles los dhos yndios de Canta se alentaron e yntentaron e procuraron que los dhos yndios yavyos se las dexasen diziendo que las dhas tierras heran suyas y en sus termynos e mojones e que Ynga se las avia quitado syn rrazon e porque de cada dia se henojavan mas porque no binyese a algun dano avyan benido a

las dhas tyerras de Quibi Nycolas de Rribera el moço encomendero de los dhos yndios de Canta y Françisco de Ampuero encomendero de los de Chacalla e avian tratado con los unos yndios y con los otros que se conzertasen e ansy los conzertaron en que los dhos yndios de Canta dieron a los de Chacalla por rrazon del dho conçierto dozientas obejas de la tierra machos y henbras por que dexasen las dhas tierras e que dezian que las avia rreszebido don Fran^{co} Marcapoma caçique de Chacalla porque este dho tº no se las vio rresçebir mas de que bio este dho tº que despues que paso el dho conçierto los dhos yndios de //
/236r/ Chacalla avian dejado las dhas tierras de Quibi e ydose a Chacalla y los de Canta se avian entrado en ellas y las avia sembrado y beneficiado como suyas y las avian tenydo y poseido quyeta e paçificamente syn que por nenguno de los caciques de Canta e prençipales e yndios les fuese contradho fasta agora que es que anda este pleyto e dizen que el dho don Xpoval las pide e que lo demas que dho tyene lo a oido dezir como dho tiene a sus padres y aguelos e pasados tratando e platicando con ellos en las dhas tyerras de Quibi e yndios y en cosas pasadas que dezian que ellos en sus tpos ansy lo avian vysto ser y pasar y oido dezir a otros sus mayores que ellos que decian que ellos ansy lo avian visto a los suyos en los suyos e que tal avia sydo y hera dello publico y notorio e publica voz y fama e que nunca avian visto ny oydolo dezir al contrario ny menos este dho tº e que sy lo contrario ubiera sydo o pasado este dho tº cree lo supiera e lo obiera oido dezir y lo mysmo sus pasados por ser cosas tan senaladas y ser todos comarcanos a los dhos pueblos de Quibi e Chacalla e Canta e tierras de Quibi lo qual dyxo e declaro por estas propias rrazones abiendole sydo preguntado por la dha lengua.

/236v/ 3 - A la trª pregunta de el dho ynterrogatorio dixo este dho tº que dize lo que dho tyene en las preguntas antes desta.

4 - A la quarta pregunta del dho ynterrogatorio dixo este dho tº que lo que sabe desta dha pregunta es que a oydo dezir a sus padres e aguelos e a otros viejos e ancianos que dezian que en tpo de Topa Ynga Yupangue e aun este dho tº tyene dello notycia de lo susodho en tpo de Guaynacava que se hazia una prucesion en que benya un orejon ynga con una olla de barro en la mano e traya en ella sangre de obexas y chaquira molida y otras cosas y benyan con el muchos yndios del pueblo de donde salia el dia que salia con la dha capacocha y la derramavan por los zerros rrogando a la guaca o dymonyo que dyese mucha salud a Topa Ynga e que fuese gran senor e benciese muchas batallas e dyese salud a sus hijos y otras cosas que tenyan por costumbre de rrogalles e que yban entrando por las tierras dyziendo en su lengua: "parta parta capacocha capacocha no derrames" e que sy topava en algun yndio e derramava alguna gota de la dha capa-//
/237r/ cocha en la parte y lugar donde la derramaba alli mataban el dho yndio y le enterravan e que esto se fazia e acostumbrava e guardava como se guardan agora las leyes de los xpianos y hera dello cosa publica y notoria.

5 - A la quinta pregunta de el dho ynterrogatorio dixo este dho tº que sabe por cosa publica y notoria e costumbre e ley antiguada que se usava y guardava en tyempo de Topa Ynga Yupangue y por lo que vio y entendio en tpº de Guaynacaba de que este dho tº se acuerda que donde paravan con la dha capacocha alli senalaba por termyno todo lo que avian andado para que fuese del yndio que llebava la dha capacocha en la mano el qual ansy como asentavan dezia: "fasta aqui llego my capacocha" para que un capitan de los Yngas que alli yba le senalase el dho termyno e que como dho tiene hera cosa muy usada y acostumbrada y se tenya por mandamyº ynbiolable a lo que declaro avyendosele dado a entender a este dho tº partycularmente por la dha lengua.

6 - A la sesta pregunta del dho ynterrogatº dixo este dho tº que lo que save desta pregunta es que save por cosa por cosa [*sic*] publica y notoria e aberlo oydo dezir a yndios viejos que dezian avello vysto //
/237v/ que en tpo de Topa Ynga Yupangue que fue el primero de los Yngas los dhos yndyos de Chacalla entraron por las dhas tierras de Quibi con los dhos orejones de Ynga y que despues entraron dos bezes en tpo de Guaynacaba y llevaban la dha capacocha e yban ganando e apoderandose cada dia mas en las dhas tyerras y que despues en tpo de Guascaringa entraron con la dha capacocha los dhos yndios yavyos fasta mas adelante de las casas del dho pueblo de Quibi un tiro de arcabuz e que queriendo pasar de ally un yndio del dho pueblo de Quibi sujeto a Canta dyxo que eran aquellas sus tierras e quiso defender que no entrasen o que no pasasen de ally e derramo un poco de sangre de la dha capacocha y por esta cavsa luego ally le mataron y le enterraron en el propio lugar donde derramo la dha sangre e que tanbien a oydo dezir que lo mysmo avian fho a otros quatro yndyos del dho pueblo de Quibi que defendian que la dha capacocha no entrase en sus tyerras e que este dho tº entendyo y tiene entendydo que no osavan fablar los dhos yndyos de Canta ny rresistilles la entrada con la dha capacocha ansy por no yncurrir en la dha pena de muerte como por-//
/238r/ que los dhos yndyos yavyos heran muy queridos e pribados del Ynga y por no lo enoxar.

7 - A la setima pregunta del dho ynterrogatorio dyxo este dho tº que dyze lo que dho tyene en las preguntas antes desta.

8 - A la otava pregunta del dho ynterrogatorio dyxo este dho tº que dyze lo que dho tyene en las preguntas antes desta en que se afirma e que quando entraron los espanoles en este rreyno algunos de los dhos yndyos yavyos se yban a Chacalla de myedo de los espanoles y porque estavan en el camyno rreal por donde pasavan e que quedavan otros en las dhas tierras y las sembravan como de antes y que despues que se hiço el conçierto de las obejas los dhos yndyos de Canta entraron en las dhas tyerras de Quibi y esto rresp.^{e}

9 - A la novena pregunta del dho ynterrogatorio dyxo este dho tº que dyze lo que dho tyene en las preg.^{as} antes desta.

10- A la deçima pregunta del dho ynterrogatorio dyxo este dho t° que dyze lo que dho tyene en las preguntas antes desta y esto rresponde.

/238v/ 11- A las honze preguntas del dho ynterrogatorio dyxo este dho t° que sabe e a bysto que las dhas tyerras de Quibi son rruines e todas andenes y cascaxo e que no son buenas para sementera de trigo ny mahiz e que son buenas para coca y papas e otras legumbres y esto rresponde y lo demas en la dha pregunta contenydo dyze que no lo sabe.

12- A las doze preguntas de el dho ynterrogatorio dyxo este dho t° que dyze lo que dho tiene en las preguntas antes desta y lo demas no lo sabe.

13- A las treze preguntas del dho ynterrogatorio dixo este dho t° que le paresze que las dhas obexas que los dhos yndyos de Canta dyeron a los de Chacalla podrian baler a syete pesos grandes y si heran medyanas a seis pesos e sy mas chicas a çinco.

14- A las catorze preguntas del dho ynterrogatorio dyxo este dho t° que no lo sabe.

15- A las quynze preguntas de el dho ynterrogatorio dyxo este dho t° que dyze lo que dho tyene en las preguntas antes desta e que los dhos yndios de Canta se holgaron de sacar //
/239r/ las dhas tyerras a los de Chacalla por ser suyas y de sus pasados.

16- A las diez y seis preguntas del dho ynterrogatorio dyxo este dho t° que dize lo que dho tyene en las preguntas de el dho ynterrogatorio lo qual dyxo que era la verdad y lo que deste caso sabia e que no sabia otra cosa lo qual dyxo e declaro por la dha lengua del dho don M^{yn}.

Anadidas

1 - A la primera pregunta del dho ynterrogatorio añadido dixo este dho t° que dize lo que dho tyene en las preguntas antes desta del primero ynterrogat°.

2 - A la sesta [*sic*] pregunta de el dho ynterrogatorio dyxo este dho t° que dize lo que dho tiene en las preguntas del primero ynterrogatorio.

3 - A la terzera pregunta del dho ynterrogatorio anadido dyxo este dho t° que dize lo que dho tyene en las preguntas de el primero ynterrogatorio prençipal.

4 - A la quarta pregunta del dho ynterrogatorio dyxo este dho t° que dyze lo que dho tyene en las preguntas antes desta del dho ynterrogatorio pren-//
/239v/ çipal e que esto rresponde a esta dha pregunta.

5 - A la quynta pregunta de el dho ynterrogatorio anadydo dyxo este dho t° que dyze lo que dho tiene en las preguntas antes desta del dho primero ynterrogatorio prençipal.

6 - A la sesta pregunta de el dho ynterrogatorio dixo este dho t° que dyze lo que dho tyene en las preguntas antes desta de el dho ynterrogat° prençipal y esto rresponde.

7 - A la setyma pregunta de el dho ynterrogatorio dyxo este dho t° que dize lo que dho tiene en el dho ynterrogatorio lo qual todo que a dho dixo ser verdad y lo que deste caso sabia para el juramento que hecho avia y en ello dyxo que se afirmava e afirmo e rrateficava e rratefico aviendosele tornado a leer y dado a entender lo que ansy avia dho por lengua del dho don M^{yn} Lengua ynterpetre de la Rreal Audiencia nombrado por los senores oidores para esta causa y le encargo tenga secreto de lo que a dho e no lo diga a persona alg^{a} so pena de perjuro el qual lo prometio y no lo firmo porque no supo ny tanpoco el dho don M^{yn}.

paso ante mi — Fran^{co} Hernandez de G^{a}
escr° de Su Mag.^{t}

/240r/ Yarpa indio no chxistiano
sujeto a los de Canta.

T° el dho Yarpa yndyo natural de Yacre del rrepartimyento de Canta encomendado en Nycolas de Rrybera el moço vz° de la çibdad de Los Rreyes sujeto a sus caciques t° presentado por parte de los dhos caçiques e yndyos del dho rrepartymyento de Canta avyendo jurado en forma devyda de dr° segun dyxo que se fazia en su ley e siendo preguntado por lengua del dho don M^{yn} Lengua ynterpetre nombrado por los senores presydente e oidores para esta presente cavsa e syendo preguntado por el tenor del dho su ynterragot° dixo lo syguyente:

1 - A la primera pregunta del dho ynterrogatº dyxo este dho tº que conosce e a conoscido a los caciques prençipales e yndyos del dho rrepartymyento de Canta porque como dho tiene es sujeto a ellos y lo mysmo dyxo este dho tº que a conoscido e conosze a los dhos yndyos de Chacalla porque a tratado con ellos syendo como es tan comarcano e que tyene notycia de las dhas tierras de Quybi porque las a vysto y est^{do} en ellas muchas e dybersas vezes //
/240v/ e del dho rrio de Quybi e que tyene noticia de los Yngas senores que fueron destos rreynos especialmente se acuerda este dho tº de Guaynacava porque fue en tiempo deste dho tº.

2 - Preguntado por las preguntas generales de la ley dixo este dho tº que no save la hedad que tyene pero que se acuerda muy bien de quando Guaynacava rreynava e quando fue a Quyto que seria entonzes de fasta ocho anos poco mas o menos e que como dho tiene es natural del dho rrepartimº de Canta e sujeto a los caçiques del e que tyene yjos e muger en el dho rrepartimyento e que le ba el mysmo ynteres que a los demas prençipales e yndyos pero que por eso no dexara de dezir verdad de lo que supiere y le fue preguntado e que no le an dado ny prometydo cosa alguna para que en este su dho diga el contrario de la verdad e que desea que salga con este dho pleyto los yndyos de Canta.

2 - A la segunda pregunta del dho ynterrogatorio dyxo este dho tº que lo que save desta pregun-//
/241r/ ta es que a oido dezir este dho tº a sus padres e aguelos e pasados e a otros viejos e ancianos que son ya muertos de cuyos nonbres no se acuerda que antes que binyesen los Yngas a esta dha tyerra los dhos yndyos de Canta tenyan diferencias con los yndyos yavyos de los llanos e procuravan de matarse en byendo los unos a los otros e que ansy los dhos yndyos de Canta avian bajado con cierta cantydad de jente a los llanos a dar guerra a los yndyos yavyos e que ansy como llegaron al pueblo de Quybi que hera de yndyos yavyos donde estava un senor que llamavan Chaumecaxa le salio de paz este senor y sus yndyos y los dhos yndyos de Canta fueron fazia Collique a dar guerra al senor de Collique que llamavan Collicapa a quyen servia e trebutava el dho Chaumecaxa e que sabido por el dho Collicapa avia hecho tanbien mucha jente para benir a ver que querian los dhos yndyos de Canta el qual avia salido con ella e ydo en busca de los dhos yndyos de Canta los quales visto que el dho Collicapa traya mucha jente se avian rreti-//
/241v/ rado entre unos zerros e de alli se avian hecho mensajeros los unos a los otros y el dho caçique e yndios de Canta avian enbiado a dezir al dho Collicapa que se llanase los mojones de los yndyos serranos y los de los llanos e que le dexase el dho pueblo de Quybi e que el dho Collicapa por no rrener e bisto que el rrio que pasa por el dho pueblo de Quybi benya de Canta le avia dho que se le daria el dho pueblo de Quybi pues no queria otra cosa e que de alli adelante fuesen buenos amygos e que no se hiciesen mas guerra unos a otros en ansy el dho cacique e yndyos de Canta se juntaron con los del dho Collicapa y se dyeron por sus amygos y conzertaron que los unos entrasen a tratar en las tierras de los otros syn que se hiciesen ningun dano e prometieron de tener paz para syempre e que luego el dho Collicapa avia senalado por termynos que corryesen fazia Canta dende alli desde media legua poco menos del pueblo de Quybi en un termº que se llama Xudcunga para que dende alli para arriba fuese del termº de Canta y de alli para abajo de //
/242r/ los yndyos yungas de los llanos tyerra del dho Collicapa e que desta manera los dhos yndyos de Canta se avian buelto a sus tyerras y dende alli adelante muchos anos e tpo desde el dho mojon se avian servido de las dhas tyerras de Quybi e del dho senor que llamavan Chaumecaxa e de sus yndyos el qual les trebutava de alli adelante ny mas ny menos que al dho Collicapa e ansy estobieron fasta que entraron los Yngas en esta tyerra que tenyendo noticia de las dhas tierras las avia dado poco a poco a los yndyos yavyos de Chacalla y puestolos en ellas para que sembrasen para los Yngas e para ellos e que los primeros yndyos yavyos que entraron en las dhas tierras de Quybi a oido dezir este dho tº que fue desta manera que en tpo de Topa Ynga Yupangue se avia conquystado el dho pueblo e tierra de Chacalla e que despues de conquystada avia muerto Topa Ynga Yupangue e suzedido Guaynacava e que en esta sazon se avia fallado que un caçique del pueblo de Acopai que es en el rrepartimº de Chacalla que el dho cacique se llamava Colluyde [*sic*] façia ziertas ychecerias e rrogatybas a una guaca para que se muriese Guaynacava e deçia a sus //
/242v/ yndios que tanbien le ayudavan que aquello hera para que Guaynacava muriese por que no los hiciese serbir ny trebutar e que Topa Ynga Yupangue avia muerto porque el dho caçique se lo avia rrogado a la guaca e que lo mysmo fazia por Guaynacava sy se lo rrogavan e que desto avia sydo avisado el dho Guaynacava e avia enbyado un orejon suyo que llamavan Yungai para que supiese lo susodho el qual avia benydo e avia muerto todos los dhos yndyos e al dho cacique que no avia dexado syno solamente las mugeres que estavan prenadas y los yndyos chequytos a los quales por que otra vez no tornasen aquel delyto [*sic*] e por que olbidasen aquella tyerra los avia el dho orejon yunga sacado de su tierra y llebadolos a las dhas tyerras e valle de Quybi e que alli los avia puesto y hecho su poblacion en el dho termº e mojon que llaman de Xudcunga que hera el mojon que avia puesto el dho Collicapa senor de Collique a los yndyos de Canta para que de alli corriese su termº de Canta e que desta manera les avia puesto y dadoles tierras donde senbrasen en el dho valle e que ansy yendo multiplicando los dhos yndios avian entradose cada dia mas en las dhas tierras //
/243r/ de Quybi y las avian labrado y sembrado e que tanbien oyo dezir que en tyempo de los dhos Topa Ynga Yupangue y de Guaynacaba se hazia una plegaria y rrogativa a la guaca por los Yngas para que tubiesen salud y fuesen mui rricos y prosperos y bençiesen muchas batallas y por otras cosas que solian pedir e que a esto lo llamavan capacocha e que yban con ella los dhos yndios yavyos e que fasta donde paraban con ella alli ponian termº e dezian: "fasta aquy llega mi capacocha; fasta aquy es mi tierra" e que desta manera se yban entrando en la tierra los dhos yndios yabyos e que en la dha procesyon yban diziendo en su lengua: "parta parta no derrames la capacocha que moriras" e que los yndios de Canta no osaban defender lo susodho ansy por no henojar al Inga porque querian mucho los yndyos yavyos y heran sus criados y privados como por que no derramasen la dha capacocha y les matasen porque en aquella sazon y tpo se tenya por costunbre e ley entre los yndyos que el yndio que derramase

alguna gota de la dha capacocha se muriese por ello e ansy alli donde se derramava lo matavan y lo enterravan en el //
/243v/ mysmo lugar e tanbien oyo dezir que despues que murio Guaynacaba avyendo suzedido Guascarynga el dho Guascar Ynga avia dado a los dhos yndyos yavyos unas tierras fasta junto a las casas del dho pueblo de Quybi e que fasta alli donde llego el term^{o} avia puesto por nonbre Chuquycoto que es un mojon que esta junto a las dhas casas del pueblo de Quybi e que desta manera avian tenydo y poseido las dhas tyerras de Quybi fasta que murio Guascar Ynga y fasta que entraron los espanoles en esta tyerra e fasta que los dhos yndyos de Canta se entraron en ellas por rrazon de duzientas obejas que dieron por ellas e que esto como dho tiene lo oyo dezir este dho t^{o} a sus padres e aguelos y pasados los quales dezian que ellos en sus tpos ansy lo avian vysto ser y pasar e acostunbrar e oidolo dezir a otros mas vyejos que ellos que dezian que ellos ansy lo avian vysto de tpo ynmemorial e que dello avia sydo y hera publico e notorio e publica voz y fama e comun opynion e que nunca avian bysto ny oydo dezir lo contrario e que si lo contrario fuera o pasara lo supieran y este dho t^{o} tanbien por ser cosas tan de notar e ser tan comarcanos a los dhos pueblos de Canta y Chacalla e a las dhas tierras de Quybi e que //
/244r/ parte de lo susodho avia sydo e pasado en tpo deste t^{o} dende que se save acordar que es desde el dho tpo de Guainacava lo qual todo dyxo e declaro por lengua del dho don M^{yn} que dyxo avello declarado el dho t^{o} por estas propias rrazones en sustancia.

3 - A la terzera pregunta del dho ynterrogat^{o} dyxo este dho t^{o} que dyze lo que dho tyene en las preguntas antes desta.

4 - A la quarta pregunta del dho ynterrogatorio dyxo este dho t^{o} que dyze lo que dho tyene en las preguntas antes desta.

5 - A la quhnta pregunta del dho ynterrogatorio dixo este dho t^{o} que dyze lo que dho tiene en las preguntas antes desta.

6 - A la sesta pregunta del dho ynterrogatorio dyxo este dho t^{o} que dyze lo que dho tiene en las preguntas antes desta.

7 - A la setima pregunta del dho ynterrogatorio dyxo este dho t^{o} que dyze lo que dho tyene en las preguntas antes desta.

8 - A la otava pregunta del dho ynt^{o} dyxo este dho t^{o} que dyze lo que dho tyene en las preguntas antes desta.

9 - A la novena pregunta del dho ynt^{o} dyxo este dho t^{o} que dyze lo que dho tyene en las preguntas antes desta.

/244v/ 10- A la deçima pregunta del dho ynterrogatorio dyxo este dho t^{o} que dize lo que dho tiene en las preguntas antes desta.

11- A la honzena pregunta del dho ynterrogatorio dyxo este dho t^{o} que le paresze que las dhas tyerras sobre que es este pleyto podran tener fasta syete fanegadas de senbradura y no mas porque son todas piedra y cascaxo y hecha andenes aparejada para sembrar coca e que lo que baldra cada fanegada este dho t^{o} no lo save y esto rresponde.

12- A las doze preguntas del dho ynterrogatorio dyxo este dho t^{o} que no lo save e dyze lo que dho tiene en las preguntas antes desta.

13- A las treze preguntas del dho ynterrogatorio dixo este dho t^{o} que que le paresçe [*sic*] que las dhas ovejas y carneros que los dhos yndyos dieron a los dhos yndyos de Chacalla por las dhas tierras podrian valer aquella sazon machos y hembras unas con otras a quatro pesos porque entonzes no se bendian como agora porque ay agora mas trabto que entonzes y esto rresponde a esta preg^{a}.

14- A las catorze preguntas del dho ynt^{o} dixo este dho t^{o} que lo que desta ______//
/245r/ que entre los yndios no se estiman ny benden las tierras como entre los espanoles.

15- A las quynze preguntas del dho ynterrogatorio dixo este dho t^{o} que dyze lo que dho tyene en las preguntas antes desta e que el dho Guaman Yabre caçique del dho rrepartimyento de Canta dezia antes que diese las dhas obexas que las dhas tyerras heran suyas e que las avia de sacar syn dar obejas y que por rruego del dho Nycolas de Rribera e porque le dixo que se quitase de pleytos dyo las dhas obexas.

16- A las diez y seis preguntas del dho ynterrogatorio dixo este dho t^{o} que dyze lo que dho tyene en las preguntas antes desta lo qual todo dyxo que hera la verdad y lo que deste caso savia para el juramento que avia fho y en ello se afirmo.

Anadidas

1 - A la primera pregunta del dho ynterrogatorio anadydo dixo este dho t^{o} que dize lo que dho tiene en las preguntas antes desta del ynterrogatorio primero en que se afirma y esto rresp^{e}.

2 - A la segunda pregunta del dho ynterrogatorio dixo este dho t^{o} que se acuerda que oyo dezir a yndyos //
/245v/ del dho rrepartimyento de Canta que dizen que lo vieron y fue cosa publica y notoria que en tpo de Guascar Ynga los dhos yndyos de Chacalla yavyos avian entrado con la dha capacocha fasta mas adelante del dho pueblo de Quybi e que vn yndio que llamavan Guaranga les rrenpujo e quyso defender que no pasasen mas adelante porque tenia alli sus tierras e que derramo una gota de sangre de la dha capacocha e que luego le avian muerto por ello y le avian enterrado donde cayo la sangre porque ansy

lo acostunbravan entre ellos e que a este dho t^{o} le fue pu^{co} e not^{o} e se lo fueron a dezir al despoblado donde este t^{o} estava guardando obejas porque era obejero de Ynga y esto rresponde.

3 - A la terzera pregunta del dho ynterrogatorio dixo este dho t^{o} que dyze lo que dho tyene en las preguntas antes desta y lo demas no lo save.

4 - A la quarta pregunta del dho ynterrogatorio dixo este dho t^{o} que dyze lo que dho tyene en las preguntas antes desta y lo demas no lo save.

5 - A la quynta pregunta del dho ynterrogatorio dyxo este dho t^{o} que dize lo que dho tiene en las pregun-//
/246r/ tas antes desta y lo demas no lo sabe.

6 - A la sesta pregunta del dho ynterrogatorio dixo este dho t^{o} que no lo save e dize lo que dho tiene.

7 - A la setyma pregunta del dho ynterrogatorio dyxo este dho t^{o} que dyze lo que dho tyene lo qual dyxo que hera la verdad y en ello dyxo que se afirmava e afirmo e rrateficava e rratefico despues de averse leydo por lengua del dho don M^{yn} Lengua ynterpetre nonbrado por los dhos senores oidores para esta cavsa y encargosele tenga secreto deste su dho e no lo diga a persona alguna so pena de perjuro e que le castigaran y ello prometyo e no lo firmo porque no supo ny tanpoco la dha lengua.

paso ante mi — Franc^{co} Hrrez de G^{a}
escr^{o} de Su Mag.^{t}

Carvabilca no xristiano
sujeto a Canta.

T^{o} el dho Carvabilca natural de Ysquypanpa que es en el rrepartymyento de Canta encomendado en Nycolas de Rribera el moço vz^{o} de la çibdad de Los Rreyes t^{o} presentado por parte del cacique prencipales e yndyos del dho rrepartimyento aviendo jura-//
/246v/ do en forma de dr^{o} segun la ley e siendo preguntado por el tenor y forma del dho ynterrogatorio por donde se pidio fuese hessamynado e preguntado por lengua de don M^{yn} Lengua ynterpetre nonbrado por los senores presydente e oidores de la Rreal Audiencia para esta presente causa dixo e declaro lo syguyente:

1 - A la primera pregunta del dho ynterrogat^{o} dixo este dho t^{o} que conosze a los caçiques prinçipales e yndyos del dho rrepartim^{o} de Canta porque como dho tyene es su sujeto e ansy mysmo conosze e conoscio a los caciques que an sydo e son del dho rrepartimyento de Chacalla porque a tenydo trabto e comunycacion con ellos syendo como son tan zercanos e que se acuerda e tiene notycia de oidas de los Yngas pasados senores que fueron deste rreyno e que este dho t^{o} syrvio a Guascar Ynga e que sabe las dhas tierras que llaman de Quybi porque las a vysto y estado en ellas muchas vezes e an sydo e son de Canta.

Preguntado por las preguntas generales de la ley e que no sabe la hedad que tiene mas de que se //
/247r/ acuerda vn poco de Guaynacapa e que como dho tiene es natural de la dha tierra de Canta y sujeto a los caciques e prençipales del e tiene hijos en la dha tierra y nyetos e desea que saquen las dhas tierras de Quybi pero que por eso no dexara de dezir verdad de lo que supiere y le fuere preguntado e que no le an dado ny prometydo cosa alguna para que en este su dho diga el contrario de la verdad.

2 - A la segunda pregunta del dho ynterrogatorio dixo este dho t^{o} que save que las dhas tierras e tr^{o} de Quybi an sydo e agora son de la tierra e yndios de Canta porque en tyempo de Collicapa cacique de Collique cuyas heran las amojono e dybidio de los yungas e puso vn mojon que esta media legua deste pueblo de Quybi e le puso por nonbre Judcunga e dende alli para arriba se quedo por tierra de Canta e dende alli para abaxo por tyerra de los yungas del dho Collicapa e que este dho t^{o} a oydo dezir a su padre que se llamava Urucaxa que fue natural del dho pueblo de Canta que es ya muerto que antes que vinyesen los Yngas a esta tierra los yndios de Collique yban a rrescatar con los yndyos de Canta y los yndyos de Canta los ma-//
/247v/ tavan y lo mysmo fazian los yndyos de Collique a los de Canta quando los tomavan en sus tierras porque en aquella sazon como no avia just^{a} ny orden e todos heran senores por sy que no heran sujetos a senor no fazian syno matarse unos a otros y esto tenyan por costumbre e que en este dho tpo los dhos yndyos de Canta avian enbiado a llamar a los de Collique al dho Collicapa los quales dize este t^{o} que oyo dezir que se avian fablado los unos caciques con los otros junto al dho valle de Colique y conzertaron que los yndyos de Collique fuesen a rrescatar libremente a las dhas tyerras de Canta y los de Canta a las de Collique syn que se matasen ny hiziesen nengun dano e que se amojonasen las tierras lo que avia de ser de Canta y lo que avia de ser de Collique e que de alli adelante se tratasen como hermanos y el dho Collicapa senor de Collique avia amojonado las dhas tyerras y puesto vn mojon como dho tyene media legua del dho pueblo de Quybi fazia Collique que llaman oi dia Judcunga para que dende alli fuese la tierra e juresdicion de Canta fazia la syerra e de alli para //
/248r/ abaxo fuese juresdiçion e tierras de yungas de Collique e un senor que llamavan Chaumecaxa con todos los yndyos que tenia yungas en las dhas tierras de Quybi se los dyo el dho Collicapa a los dhos yndyos de Canta para que de alli adelante se llamasen

serranos de la tierra de Canta y los yndyos de Canta tobiesen quenta con ellos e ansy desta manera se conformaron y quedaron buenos amygos y de alli adelante entravan a rrescatar los unos yndyos en las tierras de los otros e no les fazian dano e los dhos yndyos de Canta enpezaron a labrar y sembrar en las dhas tierras de Canta juntamente con los yndyos yungas del dho Chaumecaxa que en ellas estavan y las tobieron y poseyeron fasta que Topa Ynga Yupangue vino a conquystar esta tierra y despues fasta que aviendo el dho Topa Ynga Yupangue conquystado la dha tierra de Chacalla y dexadola sujeta asy e ydose los dhos capitanes orejones de Topa Ynga al Cuzco parescio ser que vn caçique de un pueblo del dho rrepartymyento de Chacalla que llamavan Colluynde natural del pueblo de Acupay hazia el y sus yndyos çiertas //
/248v/ -tas [*sic*] hechizerias para que muriese Guaynacava para que no les hiciese serbir de lo qual dize este dho t^{o} que avia sydo avisado Guaynacava e avia enbiado un capitan suyo orejon que llamavan Yunguaype para que supiese como pasava lo susodho y lo castigase e ansy el dho Yunguaype savida la verdad de lo que pasava tomo el dho cacique Colluynde e a sus yndyos todos que adoravan en la guaca y los mato y dexo solamente los yndyos chequytos y las mugeres prenadas a los quales por que otra vez no hiciesen ellos otro trato e por que olbidasen aquella tyerra los saco della y los llebo a las dhas tierras de Quybi y los puso y poblo en el propio $term^{o}$ e mojon que llaman Xudcunga que fue el mojon en que por los dhos yndyos de Canta e Collique se avian devidido las tierras y les dio el dho orejon tyerras para que sembrasen en las dhas tierras de Quybi e que yendose multyplicando e cresciendo los dhos yndyos avian poblado e tomado cada dia mas tierras de las del dho valle de Quybi e desta manera se avian entrado e tenydo e poseido las dhas tierras y entradose en ellas e que tanbien so color de cierta zeremonya que //
/249r/ se fazia por Guaynacava y por Guascar Ynga e que rrogavan le dyese salud y le hiciese gran senor se fueron entrando los dhos yndios yavyos fasta junto a las dhas casas del pueblo de Quybi fasta un $term^{o}$ e mojon que oi dia llaman Chuquicoto las quales avian tenydo desta manera fasta que entraron los espanoles y despues fasta que por quytar a los dhos caçiques de Chacalla y de Canta de pasyones avian venydo a las dhas tyerras de Quybi $Fran^{co}$ de Ampuero encomendero de los dhos yndyos de Canta [*sic*] y Nycolas de Rribera el moço encomendero de los yndyos de Canta e avian tratado de concerto en que los dhos yndyos de Canta avian dado a los de Chacalla dzas obejas por que les dexasen las dhas tierras los quales las avian dexado por esta rrazon y los dhos yndyos de Canta las avian luego tomado y entradose en ellas y labradolas coma cosa suya syn que los yndyos de Chacalla ny otra persona se lo contradyxese fasta agora que bee que se trata este dho pleyto e que los dhos yndyos de Chacalla no quyeren pasar por la venta e que como dho tiene lo que a dho lo a oido dezir al dho su padre que dezia que ansy lo avia bysto ser y pasar en tpo suyo e oidolo dezir a otros //
/249v/ viejos e aun este dho t^{o} se acuerda e tiene noticia de parte de lo susodho desde el dho tpo de Guascar Ynga que es desde que este dho t^{o} se acuerda lo qual todo dixo por lengua del dho don M^{yn} ynterpetre que dyxo avello el dho t^{o} declarado ansi como dho tiene por las dhas rrazones.

3 - A la tr^{a} pregunta del dho ynterrogatorio dixo este dho t^{o} que dize lo que dho tyene en las preguntas antes desta en que se afirma.

4 - A la quarta pregunta del dho ynterrogatorio dixo este dho t^{o} que lo contenydo en esta dha $preg^{a}$ lo a oydo dezir este dho t^{o} al dho Vrucaxa su padre que es ya muerto que dezia que en su tyempo avia visto fazer la dha capacocha e dezian e fazian con ella las dhas zeremonias e que si algun yndio la derramava la dha capacocha defendiendo que no se entrasen en sus tierras lo matavan alli luego e que esto se usava e acostumbrava y tenya por ley como agora las de los xpianos.

5 - A la quynta pregunta del dho ynterrogatorio dixo este dho t^{o} que a oido dezir al dho su padre que en su tpo se fazia lo contenydo en la dha pregunta e aun este dho t^{o} lo vio y entendio que se fazia ansy en tpo de Guascar Ynga que es que llebava //
/250r/ la dha capacocha por las dhas tyerras donde entravan yvan diziendo en su lengua: "aparta aparta capacocha capacocha no derrames que te aran guano"; e que adonde llegavan con la dha capacocha alli dezia el yndio que la llebava: "fasta aqui llega mi capacocha fasta aquy es mi tierra" e que luego le ponyan termyno e mojon alli por tyerra del que ansy llevaba la dha capacocha e que en las dhas tyerras de Quybi syenpre llebavan la dha capacocha los dhos yndyos yavyos de Chacalla e fazian lo que dho tyene e que los dhos yndyos de Canta no osavan fablar ny perturbasen la entrada por no caer en la dha pena de muerte derramando la dha capacocha e porque entendian que a los dhos yndyos yavyos los queria mucho Ynga y heran sus criados e pribados y esto que era cosa publica y notoria.

6 - A la sesta pregunta del dho ynterrogatorio dixo este dho t^{o} que dyze lo que dho tiene en las preguntas antes desta e que lo en la dha $preg^{a}$ contenydo lo save por cosa publica y notoria e por avello oido dezir al dho su padre y esto rrespondyo a esta dha pregunta por lengua del dho don M^{yn} ynterpetre.

/250v/ 7 - A la setyma pregunta del dho ynterrogatorio dixo este dho t^{o} que dize lo que dho tyene en las preguntas antes desta e que save e vido en su tpo e oyo dezir al dho su padre que fasta donde llegavan los dhos yndyos yavyos de Chacalla con la dha capacocha la senalavan e rrayavan alli por su $term^{o}$ luego ponyan alli yndyos que sembravan aquella tyerra que acreszentavan y esto hera cosa publica y notoria.

8 - A la otava pregunta del dho ynterrogatorio dixo este dho t^{o} que dize lo que dho tyene en las preguntas antes desta e que despues de venydos los espanoles a esta tierra y poblado el marques los dhos yndyos de Canta por fuerça hecharon de las dhas tyerras de Quybi todos los dhos yndios de Chacalla yavyos y se fueron a sus tierras y los dhos yndios de Canta se entraron en las dhas tierras de Quybi e que despues queriendo los dhos yndyos de Chacalla tornar a las dhas tierras de Quybi en tiempo del presydente los dhos yndyos de Canta se le defendian y desta manera enbiaron a llamar a los dhos sus encomenderos para que los

hiziesen amygos y luego hizieron el dho concierto de las dhas obexas y los dhos yndyos de Chacalla //
/251r/ se fueron y nunca mas fablaron en las dhas tyerras fasta agora.

9 - A la novena pregunta del dho ynterrogatorio dixo este dho t^{o} que dyze lo que dho tyene en las preguntas antes desta.

10- A la decima pregunta del dho ynterrogatorio dixo este dho t^{o} que dize lo que dho tyene en las preguntas antes desta e que vido que el dho don Fran^{co} Marcapoma cacique de Chacalla no queria fazer el dho concierto ny queria tomar las dhas obejas ny las tomo y el dho Fran^{co} de Ampuero vido este dho t^{o} que las rresçibio y las mando llebar a Lima y esto rreponde. Ay mas t^{o}s que dizen aberlo oydo a y^{o}s de Canta.

11- A las onze preguntas del dho ynterrogatorio dixo este dho t^{o} que las dhas tyerras de Quybi son todo tierra de cascaxo e andenes e que le paresze a este dho t^{o} que sy se sembrasen todas de mahiz que no se senbraria en ellas de veynte fanegas arriba e que no sabe el prescio de cada fanegada e que las dhas tyerras estan mejor para coca que no para pan porque lo mas que en ellas syempre se a sembrado es coca y esto rresponde.

12- A las doze preguntas del dho ynterrogatorio dixo este dho t^{o} que dize lo que dho tyene y lo demas no lo sabe.

/251v/ 13- A las treze preguntas del dho ynterrogatorio dixo este dho t^{o} que le paresze que cada cabeza de ganado de las que los yndyos de Canta dieron por las dhas tierras de Quybi baldrian unas con otras el carnero grande de carga diez pesos y obexas tres y quatro pesos.

14- A las catorze preguntas del dho ynterrogatorio dixo este dho t^{o} que no lo save e dize lo que dho tiene.

15- A las quynze preguntas del dho ynterrogatorio dixo este dho t^{o} que al tyempo del concierto y antes el dho Guaman Yavre cacique de Canta dezia a don Fran^{co} Marcapoma caçique de Chacalla e a sus yndyos que las dhas tierras heran suyas y de sus pasados y que era su rrio y entrada este y que el de los yndyos de Chacalla hera el rrio de Mama e que cree que por esto que dezian fazian el dho concierto.

16- A las diez y seis preguntas del dho ynterrogat^{o} dixo este dho t^{o} que lo que a dho hera la verdad y en ello dyxo que se afirmava e afirmo e rratefico por lengua del dho don M^{yn}.

Añadidas

1 - A la primera pregunta del dho //
/252r/ ynterrogatorio anadido dixo este dho t^{o} que dize lo que dho tiene en las preguntas antes desta e que es verdad que oyo dezir al dho su padre e a otros viejos e ancianos que en tpo de Guascaringa se hizieron las dhas dos capacochas una para Guascar Ynga e otra para su muger e que el dho yndio que llamavan Choquy Guaranga le avian muerto porque defendiendo que no entrasen en las dhas sus tyerras avian derramado una gota de sangre de la dha capacocha e que donde llegaron y mataron el dho yndio hera mas adelante del dho pueblo de Quybi fazia Canta e que este t^{o} a vysto los huesos del dho yndyo enterrados y oy dia estan alli y esto rrespondyo por la dha lengua.

2 - A la segunda pregunta del dho ynterrogat^{o} dixo este dho t^{o} que lo contenydo en la dha pregunta lo oyo dezir este dho t^{o} al dho su padre que dezia que avia pasado ansy e que los dhos quatro yndios fueron muertos por defender la dha capacocha a los dhos yndyos yavyos que no entrasen sus tierras y se las tomasen como lo solian fazer y esto hera cosa publica entre los yndios.

3 - A la tr^{a} pregunta del dho ynterrogat^{o} anadydo dixo este dho t^{o} //
/252v/ que dyze lo que dho tyene en las preguntas del dho ynterrogat^{o} prencipal.

4 - A la quarta pregunta del dho ynterrogatorio dixo este dho t^{o} que lo que save desta pregunta es que vido que al tyempo que se fazia el dho concierto dieron para los yndios de Tapati e Allaguamarca las tierras contenydas en la dha pregunta dize para que sembrasen e que se las dyeron por rruego del dho Rribera y de Fran^{co} de Ampuero y esto rresponde a esta dha pregunta.

5 - A la quinta pregunta del dho ynterrogatorio dixo este dho t^{o} que dize lo que dho tyene y lo demas no lo sabe.

6 - A la sesta pregunta del dho ynterrogatorio dixo este dho t^{o} que dize lo que dho tyene y lo demas no lo sabe.

7 - A la setima pregunta del dho ynterrogatorio dixo este dho t^{o} que dize lo que dho tyene lo qual dyxo que era la verdad y lo que deste caso savia y en ello dyxo que se afirmava e afirmo e rrateficava y rratefico despues de averse leydo por lengua del dho don M^{yn} ynterpetre nonbrado para esta presente cavsa que le encargo el secreto y el lo prometio y no lo firmo porque no supo ny tanpoco el dho don M^{yn}.

paso ante mi

Fran^{co} Hrrez de G^{a}
escr^{o} de Su Mag.^{t}

/253r/ Alº Bro sujeto a Canta.

Tº el dho Alonso Vro natural del pueblo de Quypanpa que es en el rrepartimyento de Canta encomendado en Nycolas de Rribera el moço vzº de la çibdad de Los Rreyes sujeto al cacique prençipal del dho rrepartymyento de Canta tº presentado por parte de los caciques del dho rrepartimyento avyendo jurado en forma de drº segun se rrequeria e siendo preguntado por lengua de don M^{yn} yndio ynterpetre nombrado por los senores oidores e por el dho ynterrogatorio por donde se pidio fuese hexsamynado dyxo e depuso lo syg.^{e}:

1 - A la primera pregunta del dho ynterrogatorio dixo este dho tº que conosze a los caciques que an sydo e son del dho rrepartymyento de Canta porque como dho tyene a sydo y es su sujeto e conosze a conosçido los caciques que an sydo e son del rrepartymyento de Chacalla porque los trato e a tratado e conversado e lo mysmo tiene notiçia de las dhas tierras de Quybi porque las a vysto y estado en ellas muchas vezes e que conosçio a Guaynacava e Guascar Ynga governadores que fueron deste rreyno e que tiene notiçia de Topa Ynga Yupangue.

/253v/ Preguntado por las preguntas generales de la ley dyxo este dho tº que no save la hedad que tyene mas de que como dho tyene se acuerda de Guaynacava y le conoscio e que como dho tyene es del dho rrepartimyento de Canta sujeto a los caçiques e prençipales del dho rrepartymyento e tyene hijos en el y le ba el mysmo ynteres que a los demas yndios en que los dhos yndyos de Canta queden con las dhas tierras de Quybi pero que por eso no dexara de dezir verdad de lo que supiere y le fuere preguntado e que no le an dado ny prometido cosa alguna para que diga otra cosa de la verdad.

2 - A la segunda pregunta del dho ynterrogatorio dyxo este dho tº que save que las dhas tierras que llaman de Quybi estan dentro de los propyos termºs de Canta y lo an estado desde antes que vinyesen los Yngas porque este dho tº los a visto y su padre deste tº que se llamava Paucar que es ya muerto muchos anos a le mostro los mojones y le dixo que un mojon que esta junto al dho pueblo de Quybi que se le mostro y se nombrava Juzcunga partia los termºs de Canta y de los yn-//
/254r/ dyos yungas hazia la mar e que es alli que el mojon para arryba hazia la syerra de Canta hera y se tenya por tierra de Canta y que desde alli para abajo fazia la mar corria la juresdicion de los yungas fazia la mar de muchos anos a que la pte y esto rrespondyo a esta dha pregunta.

3 - A la terzera pregunta del dho ynterrogatorio dixo este dho tº que lo que desta pregunta save es que oyo dezir al dho su padre que se llamava Paucar que antes que ubiesen Yngas en esta tierra los pueblos de Canta y de Collique eran senores por sy hesentos e que no tenyan myedo a naidie que los avia de castigar e que en yendo yndios de la syerra de Canta a rrescatar fazia la mar entre los yndyos yavyos luego procuravan de matarlos y los matavan y que lo mysmo fazian los yndyos de Canta sy bian yndyos de los yavyos en sus tierras e que por hebitar esto los dhos yndios de Canta avian ydo a berse con el senor de Collique que llamavan Collicapa para fazerse amygos e que no se matasen los yndios e ydo el dho Collicapa les avia rresçibido bien e avian tratado de amystarse entre ellos e que los yndyos de fazia //
/254v/ la mar subiesen a la syerra a contratar y los de la syerra a los llanos syn que les hyzyesen nengun dano e que se amojonasen e senalasen entre ellos los termºs lo que avia de ser de los yungas fazia Collique y lo que avia de ser de Canta fazia la syerra e ansy el dho Collicapa se avia ydo con el caçique e yndyos de Canta hasta las tierras que llamavan de Quybi donde estavan yndios yungas que servian al dho Collicapa que tenian un senor que llamavan Chaumecaxa e avian senalado por termºs de Canta desde un poco abajo del dho pueblo de Quybi fazia Collique en un zerro que llamavan Judcunga para que de alli corriese el termyno de la syerra para que fuese de Canta y lo mysmo el dho Chaumecaxa quedava dentro de los termºs e mojones que ansy se senalavan aunque no save este dho tº sy el dho Chaumecaxa senor del dho pueblo de Quybi servia e trebutava a Canta o si era sobre si ny lo a oido dezir mas de que oyo dezir al dho su padre que despues de amojonados los dhos termºs los dhos yndyos de Canta sembravan en las dhas tierras de Quybi a bueltas con los yndyos del dho Chaumecaxa //
/255r/ caçique del dho pueblo de Quybi e que desta manera los dhos yndyos de Canta tuvieron y poseyeron las dhas tierras de Quybi con los demas yndios yungas de Collique e aunque dize e torna a dezir este dho tº sy se lo dixo ansy el dho su padre que luego los dhos yndyos de Canta avian luego sembrado las dhas tierras de Quybi mas de que decia que las tenyan por suyas despues que se puso el dho mojon e que tanbien oyo dezir al dho su padre que despues de benydos los Yngas Topa Ynga Yupangue a conquystar esta tierra que avian conquistado e sujetado para si la tyerra de Chacalla y que despues de conquistada se avian buelto al Cuzco e que en esta sazon avia muerto el dho Topa Ynga Yupangue e avia suzedido Guaynacava y en este tyempo avia fallado que un caçique del pueblo de Chacalla que se deçia Llungaype que no se acuerda vien como le dixeron que se llamava ny el pueblo tanpoco que el y sus yndyos fazian çiertas hechizerias a su guaca rrogandole que se muriese Guaynacava e que no ubiese Yngas por que no los hiziesen serbir y que syendo avisado dello el dho Guaynacava avia enbiado un orejon suyo capitan para //
/255v/ que castigase el dho caçique y los que fazian las dhas hechizerias e venydo que fue hizo pesquisa e fallo que era verdad lo que decian y luego mato al dho cacique e a sus yndios que no dexo syno los yndyos viejos y las mugeres que estavan prenadas a los quales no quiso que estobyesen en la dha tierra de Chacalla por que no se acordasen de la dha guaca y tornasen a caer en aquel yerro los saco de la dha tierra de Chacalla y los traxo al dho valle de Quybi y los avia puesto y poblado en el mysmo termyno que llaman Judcunga adonde el dho Collycapa senor de Collique avia senalado por tyerra para Canta e que estas dhas mugeres e nynos aviendo cresçido y multiplicado se avian entrado y tomado mas tierras de las que el dho orejon les avia dado quando las poblo e que desta manera e vinyendo cada dia yndyos de la dha tyerra de Chacalla y so color de cierta ceremonya que llamavan

capacocha e rrogativas que fazian por los Yngas se avian entrado en ellas y sembradolas fasta que entraron en esta tierra los espanoles e que este dho t^{o} hera obejero e panpacayo [*sic*] de los Yngas e rresydia en Bonbon al tpo que entraron los espanoles//
/256r/ andavan rrancheando los yndyos y el ganado este dho t^{o} se huyo y se vino a su tierra y nunca vio las dhas tyerras de Quibi y paso a ellas ny oyo ny entendio a lo que se acuerda otra cosa mas de lo que dho tiene lo qual le decia el dho su padre porque lo avia vysto ser y pasar en su tyempo e oidolo dezir a otros mas viejos que decian que lo avian visto e que nunca avian visto ny oydo dezir lo contrario e que entre ellos avia sydo y hera publico y not^{o} e publica voz y fama e que sy otra cosa fuera o pasara cree este dho t^{o} que el dho su padre se lo dixera lo qual declaro por estas propias rrazones en sustancia aviendole sydo pregunt^{do} partycularmente por lengua del dho don M^{yn} ynterprete que dixo avello declarado ansy. Los demas se rrefieren a lo dho en la seg.^{da} preg.^{a}.

4 - A la quarta pregunta del dho ynterrogatorio dixo este dho t^{o} que dize lo que dho tyene en las preguntas antes desta e que este dho t^{o} oyo dezir al dho su padre que en su tpo y en el de sus pasados se tenia por cost.^{e} de fazer la dha capacocha de la forma y manera que la pregunta lo dize rrogando a la guaca que dyese salud a los Yngas que entonzes rreynavan y los hiziese grandes senores y les diese muchas comydas e les hiziese venzer las bata-//
/256v/ llas e que quando yban con la dha capacocha sy algun yndio por no poder mas o de otra manera tocava en ella y la derramava alguna gota della o hera cavsa que otro la derramase lo tomavan luego ally le matavan en el propio lugar donde caya la dha sangre e que esto hera cosa usada y guardada como agora los mandamyentos del Apo e que estando este dho t^{o} en Bonbon en tpo de Guaynacava vido este dho t^{o} la capacocha que yba a Quito que la llevavan de term^{o} a term^{o} e que quando este dho t^{o} la vio la llebavan yndios guancas y esto rresponde.

5 - A la quynta pregunta del dho ynterrogatorio dixo este dho t^{o} que save la dha pregunta segun y como en ella se cont.^{e} y declara porque este dho t^{o} lo via ser y pasar ansy como la dha pregunta lo dize en tpo de Guaynacava y de Guascaringa y que dello es e a sido pu^{co} e not^{o} e publica voz y fama.

6 - A la sesta pregunta del dho ynterrogatorio dixo este dho t^{o} que dyze lo que dho tiene en las preguntas antes desta y lo demas no lo save.

7 - A la setyma pregunta del dho //
/257r/ ynterrogatorio dixo este dho t^{o} que dize lo que dho tyene en las preguntas antes desta y lo demas no lo sabe.

8 - A la otaba pregunta del dho ynterrogatorio dixo este dho t^{o} que dize lo que dho tyene en las preguntas antes desta e lo demas no lo save.

9 - A la novena pregunta del dho ynterrogatorio dixo este dho t^{o} que lo contenydo en la dha pregunta lo oyo este dho t^{o} dezir en su pueblo.

10- A la decima pregunta del dho ynterrogatorio dixo este dho t^{o} que lo contenydo en la dha pregunta lo oyo dezir a yndyos de Canta al tpo que dizen paso lo que la pregunta dize porque este dho t^{o} estaba en su pueblo y no lo bio.

11- A las onze preguntas del dho ynterrogatorio dixo este dho t^{o} que no lo sabe.

12- A las doze preguntas del dho ynterrogatorio dixo este dho t^{o} que no la sabe.

13- A las treze preguntas del dho ynterrogatorio dixo este dho t^{o} que no lo save.

14- A las catorze preguntas del dho ynterrogatorio dixo este dho t^{o} que no lo sabe.

/257v/ 15- A las quynze preguntas del dho ynterrogatorio dixo que no lo sabe e dize lo que dho tiene en las preguntas antes desta.

16- A las diez y seis preguntas del dho ynterrogatorio dixo este dho t^{o} que dize lo que dho tiene lo qual dixo que hera la verdad y lo que deste caso savia para el juramento que avia hecho y en ello se afirmo por lengua del dho don M^{yn}.

Anadidas

1 - A la primera pregunta del dho ynterrogatorio anadido dixo este dho t^{o} que dize lo que dho tiene en las preguntas antes desta en que se afirma e lo demas no lo save porque este dho t^{o} no vio syno vna capacocha y que esta la vio como dho tyene qye yba a Quito e que a oido dezir a su padre e a otros viejos e ançianos que vivyeron en tpo de los Yngas muchas capacochas.

2 - A la segunda pregunta del dho ynterrogatorio dixo este dho t^{o} que dyze lo que dho tyene y lo demas no lo save.

3 - A la tr^{a} pregunta del dho ynterrogatorio dyxo este dho t^{o} que dyze lo que dho tyene en las preguntas del dho ynterrogat^{o} prençipal y lo demas no lo save.

/258r/ 4 - A la quarta pregunta del dho ynterrogatorio dixo este dho t^{o} que no lo save.

5 - A la quynta pregunta del dho ynterrogatorio dixo este dho t^{o} que no lo save.

6 - A la sesta pregunta del dho ynterrogatorio dixo este dho t^{o} que no lo save.

7 - A la setyma pregunta del dho ynterrogatorio dixo este dho t^{o} que dize lo que dho tiene lo qual dixo que hera la verdad y en ello se afirmo por lengua del dho don M^{yn} despues de aversele leydo encargosele el secreto e prometiolo e no lo firmo porque no supo ny tanpoco el dho don M^{yn}.

paso ante mi Fran^{co} Hernandez de Guebara
escr^{o} de Su Mag.^{t}

Domingo Pariana sujeto a Canta.

T^{o} el dho Domyngo Pariana natural del pueblo de Guarua term^{o} e tierra del rrepartimyento de Canta encomyenda de Nycolas de
Rribera el moço vz^{o} de la çibdad de Los Rreyes sujeto al caçique prençipales e yndios del dho rrepartimyento t^{o} presentado por //
/258v/ parte del cacique e yndios de el dho rrepartimyento avyendo jurado en forma de dr^{o} e siendo pregunt^{do} por el tenor y forma del dho
ynterrogatorio por donde se pidio fuese hexsamynado e por lengua de don M^{yn} Lengua ynterpetre nombrado por los dhos senores
oydores para esta presente causa dyxo e depuso lo syguiente:

1 - A la primera pregunta del dho ynterrogatorio dyxo este dho t^{o} que como dho tyene conosze al caçique prencipales e yndyos del dho rrepartymyento de Chacalla que an sydo e son del dho rrepartimy^{o} de Canta desde tpo de Guascar Inga que este dho t^{o} se acuerda por ser sujeto e bibir en el dho rrepartimy^{o} e lo mysmo conosze a los caciques que an sydo e son del dho rrepartimy^{o} de Chacalla porque los a tratado e tiene noticia de aver oido dezir e tratar de los Yngas pasados senores destos rreynos e tiene notiçia de las dhas tierras y balle de Quybi porque las a vysto y estado en ellas muchas vezes.

Preguntado por las preguntas //
/259r/ generales de la ley dyxo este dho t^{o} que es de hedad de mas de cinquenta anos e que tyene muger e hijos en el dho
rrepartymyento de Canta y le ba el mysmo ynteres que a los demas caciques e yndios del dho rrepartimyento e desea que los
dhos yndyos de Canta salgan con este pleyto pero que por eso no dexara de dezir verdad de lo que supiere y le fuere preguntado
e que no le an dado ny prometido cosa alguna para que en este su dho diga el contrario de la verdad.

2 - A la segunda pregunta del dho ynterrogatorio dixo este dho t^{o} que save que las dhas tyerras de Quybi son de los yndyos de Canta
porque se las dio e amojono Collicapa caçique de Collique cuyas heran antes que vinyesen los Yngas senores que fueron destos
rreynos a estas tierras e que a lo que este dho t^{o} a oido dezir a sus padres e aguelos e a otros viejos e ançianos que an sydo en
esta dha tierra e comarca que son ya muertos que antes que entrasen los Yngas muchos anos tenyan por tema los caciques de
Canta y los de Collique de fazerse mucho dano sobre las tierras e sy yban yndyos de Collicapa a Canta los rrancheavan y ma-//
/259v/ tavan y los de Collique hazian lo mysmo a los de Canta hasta que Ayna Guaranga cacique del rrepartimy^{o} de Canta y el dho
Collicapa senor de Collique se hizieron mensajeros para que querian fablarse e que no traxesen mas jente ny armas de cada
cacique una porra e que desta manera se binyeron a juntar los dhos caçiques en las dhas tierras de Quybi e que alli avian tratado
de las diferencias que trayan sobre las dhas tierras y sobre las aguas de el rrio y el dho Collicapa dezia que el dho Ayna
Guaranga cacique de Canta tenia rrazon porque el rrio benya de su tierra e que el dho Collicapa y sus yndios rresçibian grande
aprobechamyento y rregavan sus chacarras y sementeras e que por esta rrazon el le queria dar e senalar mojones en la tierra que
avia de ser de Canta e que fuesen buenos amygos de alli adelante e no se maltratasen ny rrancheasen los yndyos e ansy el dho
Collicapa avia senalado media legua deste pueblo de Quybi en un termyno que llaman Judcunga para que de alli para arriba fazia
la syerra fuese tierras e term^{o} de Canta e que el dho Collicapa no//
/260r/ se aprobechase dellas e le avia dado los yndios yungas que servian e trebutavan al dho Collicapa para que de alli adelante les
sirbiesen e trebutasen a los dhos senores de Canta y no a Collique e que con esto el dho Ayna Guaranga cacique de Canta avia
quedado contento e abia abraçado al dho Collicapa e dadose por muy amygo suyo e que ansy se avia ydo el dho cacique de
Canta a su tierra e que dende entonzes luego los dhos yndios de Canta avian senbrado las dhas tierras de Quibi muchos anos e
servidose e aprobechadose dellas e de los yndyos que en ellas estavan e que su aguelo deste dho t^{o} tuvo tyerras en el dho valle
de Quybi senaladas y las senbro e que estando desta manera avian benydo Topa Ynga Yupangue e avia enbiado a conquystar esta
tyerra y la de Chacalla y hecholos trebutar a sy e que despues de aver apaçiguado e conquistado la dha tyerra de Chacalla se
avian ydo al Cuzco los capitanes e que en esta sazon vn cacique del dho pueblo de Chacalla que llamaban Colluynde fazia ciertas
hechizerias e sus yndyos rrogando a su guaca //
/260v/ que se muriese Guaynacava e que no ubiese Yngas porque eran bellacos e los fazian trebutar e trabajar lo qual avia sabido el dho
Guaynacava e avia enbiado un capitan suyo que llamavan Yungaype para que supiese sy hera verdad el qual avia benydo y hecho

pesquyza y fallado que era verdad lo que se le avia avia [*sic*] avisado e que el dho Yungaype orejon de Ynga avia muerto al dho Colluynde cacique e a todos los yndyos que con el fazian el dho ofrescimy^{o} e rrogatyba a la guaca que no abia dexado syno las mugeres que estavan prenadas y los yndios chequitos a los quales no quiso dexar en la dha tyerra de Chacalla por que no tornasen a adorar a la guaca e por que lo olbidasen y no se acordasen dello los llevo a las dhas tierras y valle de Quybi e los avia puesto y poblado en el dho mojon que llaman Judcunga donde encomençava el term^{o} e el term^{o} e mojon de Canta puesto y senalado por el dho caçique de Collique e que alli les avia dado y senalado tyerras en que senbrasen e que estos dhos yndios se avian crescido y multiplicado e ansy como //
/261r/ yban cresçiendo y multiplicando ansy se yban entrando cada dia mas en las dhas tierras de Quibi e que tanbien benya yndios de Chacalla de en quando en quando uno a uno dos a dos como vian que las dhas tierras heran buenas se entravan y poblavan en ellas e lo mysmo con çierta pruçicion que yban faziendo por los Yngas que llamavan capacocha e que desta manera avian entrado en las dhas tyerras los dhos yndios yavyos en tpo que rreyno Guaynacava e que despues de muerto el dho Guaynacava bino Guascar Ynga el qual confirmo lo que Guaynacava avia fho e las tierras que avia dado a los yavyos para que benefiçiasen para Guaynacava y para ellos e anadio las dhas tierras a los dhos yndios yavyos un pedaço mas para que senbrasen para Guascar Ynga e puso mytimaes yavyos el qual dho pedaço de tierra les anadyo fasta que llegaron junto al dho pueblo de Quybi donde los yndios de Canta estavan poblados e puso por mojon alli de los yndyos yavyos un nonbre que dezian Chucuycoto e oi dia se nonbra ansy e que los dhos yndios yavyos por temor //
/261v/ que tenyan a los yungas [*sic*] como bian que queria tanto a los yavyos no osavan defenderselo e que despues que avia muerto Guascar Ynga avian benydo los espanoles e que benydos los dhos yndyos de Canta alçavan aparejo para quytar las dhas tierras a los yndios yavyos e sienpre andavan en pendencias en que los dhos yndios de Canta dezian a los yndios yavyos que se saliesen destas tierras que no heran suyas agora que no hera tpo de Yngas e que tynyendo estas pasyones avian benydo conquistando estas dhas tierras los espanoles e avian muerto mucha parte de los yndyos yavyos e de los de Canta y de los yungas e que como los dhos yndyos yavyos estavan poblados junto al camyno rreal no podian defenderse tan bien como los yndios de Canta que estavan poblados en lo bajo junto al rrio en lo llano de las tierras e ansy los yabyos de temor se avian ydo muchos dellos a Chacalla e a otras partes que no avian quedado syno muy pocos e que estos que quedavan estavan en las dhas tierras e las benefiçiavan toda-//
/262r/ via y las beneficiaron y los dhos yndyos de Canta procuravan de hecharles dellas e andavan en pendencias sobrello fasta que vinyeron Rribera el moço encomendero de los dhos yndios de Canta y Fran^{co} de Anpuero encomendero de los yndios de Chacalla y trataron de conzertallos e que quando se tratava del dho conçierto le paresce a este dho t^{o} que estavan beneficiando las dhas tierras fasta diez yndios yavyos con un cacique que se llamava Caravcuri e que hecho el dho concierto se avian salido todos y entradose los yndios de Canta en las tierras que los dhos yndios beneficiavan e las avian labrado e beneficiado syn que persona alguna se lo aya contradho fasta agora e que lo que dho tiene dende que entraron los espanoles lo save este dho t^{o} e le es notorio por lo aver vysto y entendido ser y pasar ansy y lo demas como dho tiene lo oyo dezir a los dhos sus padres e pasados e a otros viejos e ancianos tratando e platicando en ello los quales dezian que ellos en sus tiempos ansy lo avian vysto ser y pasar e acostumbrar e aun lo avian oydo dezir a otros mas vie-//
/262v/ jos e ancianos que ellos los quales dezian que ellos en sus tpos ansy lo avian visto ser y pasar e avia sydo y hera dello publico e notorio e publica voz y fama e que nunca avian vysto ny oydo dezir lo contrario e que sy lo contrario fuera o pasara este dho t^{o} cree lo oviera oido dezir e supiera y esto rresp.^{e}.

3 - A la terzera pregunta del dho ynterrogatorio dixo este dho t^{o} que dize lo que dho tyene en las preguntas antes desta.

4 - A la quarta pregunta del dho ynterrogatorio dixo este dho t^{o} que lo contenydo en la dha pregunta lo a oido dezir por pu^{co} e notorio a los dhos sus padres e ancianos que en tpos de los Yngas se fazia e acostunbrava la dha capacocha de la forma y manera que la dha pregunta lo dize e que este dho t^{o} vio que en tpo de Guascar Ynga los dhos yndyos yavyos entraron con la dha capacocha rrogando a su dimonyo que diese salud a Guascar Ynga e le hizyese gran senor e le diese bitoria contra sus henemygos e otras plegarias que fazian //
/263r/ e que vido este dho t^{o} que la dha capacocha hera uno como olla en que trayan sangre de obejas e chaquira molida e que pasaron con ella los dhos yndios yavyos mas adelante de la poblacion de los de Canta e que un yndio que llamavan Chuqui Guaranga estava beneficiando su tierra e por que no pusyesen mojon en ella por tyerra de yabyos como se acostunbraba fazer detubo al yndio que llebava la capacocha y le hizo derramar una gota de sangre y luego le mataron y estando que le querian hechar un laço al pesqueço para ahorcallo dixo: "no me mateis porque todas las vezes que pare mi muger pare dos" e todavia le mataron y este t^{o} le vido enterrar en un oyo alli luego donde se derramo la sangre y esto rresponde.

5 - A la quynta pregunta del dho ynterrogatorio dixo este dho t^{o} que save la dha pregunta como en ella se cont.^{e} porque en su tyempo lo vido ser y pasar ansy como la dha pregunta lo dize y declara e lo oyo dezir a los dhos sus mayores e ancianos que dezian que lo mysmo avian ellos visto ser y //
/263v/ pasar en sus tiempos e que dello avia sydo y hera pu^{co} e not^{o} e publica voz y fama.

6 - A la sesta pregunta del dho ynterrogatorio dixo este dho t^{o} que dize lo que dho tyene en las preguntas antes desta.

7 - A la setima pregunta del dho ynterrogatorio dixo este dho t^{o} que dize lo que dho tyene.

8 - A la otava pregunta del dho ynterrogatorio dixo este dho t^{o} que dize lo que dho tyene en las preguntas antes desta.

9 - A la novena pregunta del dho ynterrogatorio dixo este dho t° que dize lo que dho tyene en las preguntas antes desta.

10- A la decima pregunta del dho ynterrogatorio dixo este dho t° que dize lo que dho tiene en las preguntas antes desta.

11- A las honze preguntas del dho ynterrogatorio dixo este dho t° que las dhas tierras de Quybi son toda cascaxo e piedra e andenes e que le paresze que podra aver fasta veynte fanegadas de senbradura e que el valor de cada fanegada que no lo save.

/264r/ 12- A las doze preguntas del dho ynterrogatorio dixo este dho t° que no lo save.

13- A las treze preguntas dixo que no lo sabe.

14- A las catorze preguntas del dho ynterrogatorio dixo que no lo save.

15- A las quynze preguntas del dho ynterrogatorio dixo este dho t° que no la sabe e dize lo que dho tiene.

16- A las diez y seis preguntas del dho ynterrogatorio dixo este dho t° que dize lo que dho tyene lo qual dixo que era la verdad y lo que deste caso savia y en ello se afirmo por la dha lengua.

Anadidas

1 - A la primera pregunta del dho ynterrogatorio anadido dixo este dho t° que no save mas de lo que dho tiene en las preguntas antes desta en que se afirma y esto rresponde.

2 - A la segunda pregunta del dho ynterrogatorio dixo este dho t° que lo en la dha pregunta contenydo lo oyo dezir este dho t° a su padre deste t° que se llamava Pariguana que dezia que avia visto matar los dhos quatro yndios en tpo de Guaynacava e que //
/264v/ los avian muerto los dhos yndyos yavyos porque les defendian la dha capacocha porque so color della dize este dho t° que les tomavan las dhas tierras a los de Canta y se entravan en ella y que los dhos yndyos de Canta no osavan fablar por cavsa de que los matase Guaynacava porque queria mucho a los yavyos porque eran sus criados.

3 - A la tr[a] preg[a] dixo que dize lo que dho tiene en las preguntas antes desta.

4 - A la quarta pregunta del dho ynterrogatorio anadido dixo este dho t° que este dho t° despues que se hizo el dho concierto a oido dezir este dho t° a yndios de Canta que avian venydo al tpo del concierto llorando yndios del pueblo de Cullata y Xicamarca yndios yavyos dando bozes que les bendian sus tierras que no tenyan donde senbrar y que el dho Rribera el moço y su cacique Guaman Yabre caçique de Canta de lastima e ynportunydad del dho Rribera les avian senalado los dhos pedaços de tierras en que sembrasen e que a lo que le dixeron y entendio fue libremente syn nengun cargo ny trebuto de lo que la preg[a] dize ni otro lo qual declaro este t° por esta rrazon.

/265r/ 5 - A la quynta pregunta del dho ynterrogatorio anadido dixo este dho t° que dize lo que dho tiene.

6 - A la sesta pregunta del dho ynterrogatorio dixo este dho t° que dize lo que dho tiene en las preguntas antes desta.

7 - A la setyma pregunta dixo que lo que a dho es la verdad para el juram° que fho tyene y en ello dixo que se afirmava e afirmo e rratefico syendole leydo afirmose en ello.

paso ante mi — Fran[co] Hrrez de G[a]
escr° de Su Mag.[t]

Don Al° Aynamare sujeto a Canta.

T° el dho don Alonso Aynamare natural del pueblo de Paro que es en el rrepartimyento de Canta encomendado en Nycolas de Rribera el moço vz° de la çibdad de Los Rreyes sujeto al caçique del dho rrepartymyento t° presentado por parte de los dhos caçiques avyendo jurado en forma devida de dr° e syendo preg[do] por el tenor y forma del dho ynterrogatorio por donde se pidio fuese hesxamynado por lengua de don M[yn] Lengua yndio ynterpetre nonbrado por los senores presydente e oidores para esta presente //
/265v/ cavsa dixo e depuso lo syguiente:

1 - A la primera pregunta del dho ynterrogatorio dixo este dho t° que conosze e a conoscido a los caciques del dho rrepartymyento de Canta porque este dho t° como dho tiene a sydo su sujeto desde el tpo que se save acordar a esta parte e lo mysmo a conoscido e conosze a los que an sydo e son del dho rrepartimyento de Chacalla porque los a tratado desde el tpo de Guaynacava a esta parte e que se acuerda de Guaynacava e de los demas senores tiene notycia e ansy mysmo sabe e tiene notiçia de las dhas tierras de Quybi porque las a bisto y estado en ellas muchas vezes.

Preguntado por las preguntas generales de la ley dixo este dho tº que es de hedad de mas de quarenta anos e aun de cinquenta porque como dho tyene se acuerda de Guaynacava e que tiene hijos e muger en el dho rrepartimº de Canta e le ba el mysmo ynteres que a los demas caciques e yndios del rrepartymyento de Canta y desea que vença el pleyto e salgan con las dhas tierras de Quybi los de Canta pero que por eso no dexara de dezir verdad de lo que supiere y le fuere preguntado e que no le an dado ny prometido cosa alguna para que diga el contrario de la verdad.

/266r/ 2 - A la segunda pregunta del dho ynterrogatorio dixo este dho tº que lo que rresponde a lo en la dha pregunta contenydo es que save que las dhas tierras de Quybi sobre que se trata este dho pleyto an sydo muchos anos antes que vinyesen los Yngas a esta tierra del dho rrepartimyº de Canta e que venydos los Yngas metyeron en ellas los yndyos yavyos del rrepartymyento de Chacalla e que las avian tenydo y estado en ellas contra su voluntad de los dhos yndios de Canta fasta que despues de entrados los espanoles en este rreyno los dhos yndyos de Canta les hecharon dellas y porque querian traer pleyto les dieron a los yndyos yavyos de Chacalla dozientas obexas.

Preguntado como lo save y de que manera pasa lo que dho tiene dixo que lo save porque lo a oido dezir a sus padres y pasados e a otros viejos e ançianos del dho rrepartymyento de Canta y de esta comarca los quales dezian que savian e avian vysto que antes que los Yngas entrasen en este rreyno las dhas tierras de Quybi heran y fueron de yndios yungas de Collique sujetos a un senor de Collique rrepartimyº //
/266v/ de Muchos [*sic*] Davila e que syendo deste senor los yndios de Canta no se llevavan bien con los dhos yndios de Collique yungas e si yban los yungas a rrescatar a la syerra los matavan o rrancheavan y si los de la syerra yban a los llanos fazian lo mysmo fasta que el cacique de Canta que llamavan Ayna Guaranga llamo sus yndios y les dixo que le parescia vyen que fuesen a juntarse con el dho Collicapa senor de Collique para fazer conpania e conçierto con el de manera que quedasen amygos e que no se matasen mas yndios e ansy baxo el dho Ayna Guaranga cacique de Canta con çiertos yndyos suyos a ver el dho Collicapa y se junto y fablo con el tratando de concierto de amystades e sobre que se amojonasen las tierras que avian de ser de Canta y las que avian de ser de los yungas y el dho Collicapa bido este dho tº que se avia conzertado con el dho Ayna Guaranga e dho le que tenia rrazon e que pues el rrio de Collique benia de su tierra de la syerra de Canta e de la nyebe e aguas que llovian e que en los llanos no llovia y el rreszevia mucho probecho del dho rrio que le queria dar las tierras de Quybi e amojonarselas para que fuesen //
/267r/ tyerras de Canta y los yndios que en ellas estavan y el dho Ayna Guaranga se lo agradesçio mucho y se abraçaron y comieron y bebieron y prometieron de nunca mas ser henemygos e ansy luego el dho Collicapa fueron al dho valle de Quibi y pusyeron mojones e pusyeron uno media legua poco menos del dho pueblo de Quibi que llaman oy dia Judcunga para que dende alli para arriba fuese de la dha tyerra e yndios de Canta e de alli para abaxo fuese tierra de Collique fazia la mar e ansy el dho Ayna Guaranga y sus yndios se fueron y luego se avian entrado en las dhas tierras y valle de Quybi e avian sembrado coca e otras comydas y de alli adelante los yndios yungas que en el dho valle de Quybi estavan avian sido serranos e sujetos a la tierra de Canta e que en esta posesyon avian estado fasta que avia venydo Topa Ynga Yupangue a conquistar esta tierra y la de Chacalla e que tenyendo la dha tierra de Chacalla conquistada un cacique que llaman Colluinde senor de el pueblo de Acupai que es en la tierra de Chacalla avia fho çiertas hechizerias e rrogativas a una guaca para que se muriese Guaynacava e para que no ubiese Yngas por-//
/267v/ que eran bellacos e los fazian trebutar e que syendo desto avisado el dho Guaynacava que estava en el Cuzco enbio un capitan suyo que llamavan Yungaipe orejon para que supiese sy hera verdad lo susodho y lo castigase e ansy el dho capitan vino al dho pueblo e hizo pesquysa de lo susodho e fallando que era verdad lo que a Ynga le avian avysado mato el dho caçique e los demas yndios que no dexo mas de las mugeres prenadas e los nynos chequitos a los quales por mandado de Guaynacava los avia desterrado y hechado de aquella tierra para que no tornasen a yncurrir en el dho caso y los avian llevado al dho pueblo de Quybi e los avian asentado y poblado en el propio termº e mojon que llaman Judcunga donde el dho Collicapa senor de Collique avia senalado por termº de la tierra de Canta e les avian dado tyerras en que sembrasen e que despues andando el tpo se avian benydo a poblar en las dhas tierras vysto que heran buenas mas yndios del rrepartymyento de Canta y las labravan y sembravan juntamente //
/268r/ con los demas yndios e que tanbien en tpo de Topa Ynga se avia hecho çierta zeremonya de una capacocha que trayan los yndios de termº a termyno e que era desta manera que llebavan en un baso de barro sangre de obejas y chaquira molida y otras cosas y que yban con ella en la mano un yndio del pueblo donde salia e aquel yndio benya declarando en su lengua e rrogando a la guaca que dyese salud a Topa Ynga e le hiziese muy rrico e poderoso e que en llegando a un termº de otros yndios asentava un mojon e dezia: "fasta aqui llego mi capacocha; fasta aqui es mi termº" e que desta manera yban ganando tierra y entrandose los unos en las tierras de los otros e que oyo dezir que en tpo de Topa Ynga avian entrado con ella en el dho rrio de Quybi y la avian labado y tornadose a salir syn tomar tierra e que despues en tpo de Guaynacava se traya la dha capacocha en el dho valle y la trayan los yndyos yavyos de Chacalla e ansy amojonavan con ella mas tierra para sy de la que tenyan e que no se lo osavan empedir //
/268v/ porque se tenya por cost.^{e} e ley entrellos que sy algun yndio rresystia la entrada con la dha capacocha y derramava alguna gota della le matavan e que desta manera defendyendo los yndyos de Canta que los dhos yndios yavyos no se le entrasen con la dha capacocha en sus tierras mataron quatro yndios de Canta y tanpoco osavan porque Guaynaca [*sic*] queria mucho los yndyos yavyos y heran sus criados y desta manera se entravan en las dhas tierras e que despues de muerto Guaynacava aviendo suzedido Guascaringa su hijo y entendiendo que las dhas tierras heran buenas e que su padre Guaynacava se las avia dado y les queria vien les dio mas tierras a los dhos yndios yavyos y les senalo el termº dellas fasta junto al pueblo de Quybi donde estavan

poblados los cantas que llaman el dho mojon Chuquicoto e que desta manera avian estado fasta que entraron en este rreyno los
espanoles e que entrados los dhos yndyos de Canta que estavan poblados junto a ellos e Guaman Yavre caçique principal les
decia e rrenia a los//
/269r/ dhos yndios yavyos que se salyesen destas tierras pues ya no hera en tpo de Yngas e que era tiempo de espanoles e andavan los
unos con los otros e diferencias fasta que vinyeron los encomenderos Rribera el moço encomendero de los yndios de Canta e
Fran^{co} de Ampuero encomendero de Chacalla y los avian concertado en que dieron los dhos yndyos de Canta dozientas obejas de
la tierra machos y henbras a los de Chacalla por que no ubiese pleyto e los yavyos dexasen las dhas tierras e que por rrazon del
dho concierto vido este dho t^{o} que los dhos yndyos yavyos se salieron de las dhas tyerras y los de Canta se entraron en ellas y
las labraron y sembraron e que los yndyos que a la sazon estavan poblados de los yavyos en el dho valle de Quybi serian diez
yndios casados y su cacique que se llamava Guaman e que algunos yndyos viejos y yndias lloravan quando se fazia el dho
concierto diziendo que les bendian sus tierras y naturales donde avian nascido e que no tenyan tierras donde sembrar e vio este
dho t^{o} que no se querian yr de las dhas tierras y su caçique de Chacalla Marcapoma los llevava por fuerça diziendo que ya esta-//
/269v/ van conzertados caçique con caçique e ansy mysmo vydo este dho t^{o} que a ciertos yndyos de Cullata e Xicamarca pobres que
davan bozes e lloravan les dieron y senalaron ciertos pedaços de tierras para que las tubiesen y rresydiesen en ellas porque los
dhos yndyos dezian que pues que su cacique les avia vendido sus tierras que los querian serbir a los yndios de Canta y contrebuir
a ellos y les dyese sus tierras en que sembrasen e que Fran^{co} de Ampuero vio este dho t^{o} que dezia que no queria que trebutasen
a Canta porque eran suyos y el dho Rribera y su cacique les dieron çiertas tierras y les pusyeron su mojon las quales son las de
donde se cojio la coca sobre que vino vn escr^{o} a fazer ynformaçion e que no save vien sy las dhas tierras se las dieron con
condicion que los yndyos fuesen sujetos a Canta mas de que este t^{o} despues aca no a vysto ny entendido que ayan trebutado e
que los dhos yndyos sembravan las dhas tierras y esto dixo e declaro por lengua del dho don M^{yn} yndyo ynterpetre que dixo
avello declarado por estas propias rrazones estandole hexsamynando y esto rresponde.

3 - A la tr^{a} pregunta del dho yn-//
/270r/ terrogatorio dixo este dho t^{o} que dize lo que dho tyene en las preguntas antes desta.

4 - A la quarta pregunta del dho ynterrogatorio dixo este dho t^{o} que dize lo que dho tyene en las preguntas antes desta.

5 - A la quynta pregunta dixo que dize lo que dho tyene en las preguntas antes desta.

6 - A la sesta pregunta del dho ynterrogat^{o} dixo este dho t^{o} que dize lo que dho tyene en las preguntas antes desta.

7 - A la setyma pregunta del dho ynterrogatorio dixo este dho t^{o} que dize lo que dho tyene en las preguntas antes desta.

8 - A la otava pregunta del dho ynterrogatorio dixo este dho t^{o} que dize lo que dho t^{e}.

9 - A la novena pregunta del dho ynterrogatorio dixo este dho t^{o} dixo este dho t^{o} [*sic*] que dize lo que dho tiene.

10- A la decima pregunta del dho ynterrogatorio dixo este dho t^{o} que dize lo que dho tiene.

11- A las onze preguntas dyxo que las dhas tierras de Quybi son lo mas dellas piedra e cascaxo e que le paresze que podra aver en todas ellas fanega y media de senbradura.

/270v/ 12- A las doze preguntas del dho ynterrogatorio dixo este dho t^{o} que no lo save e dize lo que dho tiene.

13- A las treze preguntas del dho ynterrogatorio dixo este dho t^{o} que no lo save e dize lo que dho tyene.

14- A las catorze preguntas del dho ynterrogatorio dixo este dho t^{o} que no lo sabe e dize lo que dho tiene.

15- A las quynze preguntas dixo que dize lo que dho tiene en las preguntas antes desta.

Anadidas

1 - A la primera pregunta anadida dixo este dho t^{o} que dize lo que dho tiene en las preguntas del ynterrogatorio prençipal.

2 - A la segunda pregunta dixo que dize lo que dho tiene en las preguntas del dho ynterrogatorio primero.

3 - A la t^{a} pregunta dixo que dize lo que dho tiene en las preguntas antes desta.

4 - A la quarta pregunta dixo que dize lo que dho tiene en las preguntas del ynterrogatorio primero.

5 - A la quynta pregunta dixo que dize lo que dho tyene en las preguntas del prim^{o} ynt^{o}.

/271r/ 6 - A la sesta pregunta del dho ynterrogat^{o} anadydo dixo este dho t^{o} que dize lo que dho tiene.

7 - A la setyma pregunta dyxo que lo que a dho es la verdad y en ello se afirmo e rratefico por la dha lengua despues de averse leydo.

paso ante mi Fran^{co} Hrrez de G^{a}
escr^{o} de Su M^{t}

Don Fran^{co} cacique de Collique
de Muñoz de Avila / de mas de 40.

T^{o} el dho don Fran^{co} cacique que dixo ser del pueblo de Collique encomendado en Munoz de Avila vz^{o} de la çibdad de Los Rreyes t^{o} presentado por pte del cacique prençipales e yndios del pueblo e rrepartimyento de Canta encomendado en Nycolas de Rribera el moço v^{o} de la çibdad de Los Rreyes avyendo jurado en forma de dr^{o} e siendo preguntado por lengua de don M^{yn} ynterpetre de la Rreal Avdiencia nonbrado para esta presente cavsa e por el tenor del ynterrogatorio por donde se pidio fuese hexsamynado dixo lo sig.^{e}:

1 - A la primera pregunta del dho ynterrogatorio dixo este dho t^{o} que conosçe a los caciques prençipales del //
/271v/ dho rrepartymyento de Canta e conosçio a Guaman Yavre cacique que fue del dho rrepartimyento de Canta e conosce al cacique que agora es e que tiene noticia de las dhas tierras de Quybi porque las a vysto y estado en ellas muchas vezes e que tyene notiçia de los Yngas y senores que fueron deste rreyno porque los a oido nonbrar e que no conosçio a Topa Ynga Yupangue.

Preguntado por las preguntas generales de la ley dixo este dho t^{o} que es de hedad de mas de quarenta anos e que no save ny se acuerda de los Yngas pero que se acuerda de quando entro el marques en esta tierra e que no es pariente ny henemygo de nenguna de las dhas partes ny le toca nenguna de las generales.

2 - A la segunda pregunta del dho ynterrogatorio dixo este dho t^{o} que no save mas de que a oydo dezir a sus padres e aguelos que an sydo caciques y senores del dho pueblo de Collique donde este t^{o} es cacique e a otros viejos e ançianos que son ya muertos de cuyos nonbres no se acuerda que las dhas tierras de //
/272r/ Quybi heran del dho caçique de Collique muchos anos antes que entrasen los Yngas en esta tierra e que no heran de los yndyos de Canta ny de los de Chacalla e que estando dyferentes el dho cacique de Collique que se llamava Collicapa con los caciques de Canta binyeron en concierto que el dho caçique Collicapa dio al dho caçique de Canta las dhas tierras de Quybi e que despues tenyendolas y poseyendolas los dhos yndios de Canta avian suzedido a benyr los Yngas a conquystar aquellas tierras e abian metido en ellas a los dhos yndios yavyos de Chacalla e pobladolos en ellas e que los primeros yndios que alli avian puesto hera por via de avellos desterrado por aver cometydo cierto delito contra Ynga e que ansy se avian entrado en ellas y esto rresponde.

3 - A la tr^{a} pregunta del dho ynt^{o} dixo este dho t^{o} que dize lo que dho tiene en la preg^{a} antes desta.

4 - A la quarta pregunta del dho ynterrogatorio dixo que sabe la dha pregunta como en ella se cont.^{e} porque syendo mochacho lo vido ser y pasar //
/272v/ e acostumbrar ansy como la dha pregunta lo dize y declara e ansy lo oyo dezir a sus mayores e mas ancianos los quales decian que ellos en sus tpos ansy lo avian vysto ser y pasar y acostunbrar en los suyos y oydo dezir a los suyos en los suyos e que tal avia sydo y hera dello publico e not^{o} e publica boz y fama e comun opinyon y esto rresp.^{e}

5 - A la quinta pregunta del dho ynt^{o} dixo este dho t^{o} que save la dha preg^{a} como en ella se contiene y declara porque este dho t^{o} en su tiempo lo vido ser y pasar ansy y entendio de los viejos que dezian que ansy lo avian visto fazer e que esto es cosa pu^{ca} e not.^{a}

6 - A la sesta pregunta del dho ynterrogatorio dixo este dho t^{o} que no lo save e dize lo que dho tyene en las preguntas antes desta.

7 - A la setima pregunta dixo que dize lo que dho tiene en las preguntas antes desta.

8 - A la otava pregunta del dho ynterrogatorio dixo este dho t^{o} que lo contenydo en la dha pregunta lo oyo dezir a Guaman Yavre cacique pren-//
/273r/ çipal que fue de Canta pero que este dho t^{o} no lo vio por vysta de ojos.

9 - A la novena pregunta dixo que lo contenydo en la dha pregunta lo oyo dezir a yndios de Canta al tpo que el dho presydente Gasca estava en este rreyno.

10- A la decima pregunta del dho ynterrogatorio dixo este dho t^{o} que lo en ella contenydo lo oyo dezir a yndios de Canta.

11- A las honze preguntas dixo que no lo sabe.

12- A las doze preguntas del dho ynterrogatorio dixo este dho tº que no lo save.

13- A las treze preguntas del dho ynterrogatorio dixo este dho tº que no lo save.

14- A las catorze preguntas del dho yntº dixo este dho tº que no lo save.

15- A las quynze preguntas del dho yntº dixo este dho tº que lo contenydo en la dha pregunta lo a oido dezir este dho tº a yndios de Canta.

16- A las diez y seis preguntas del dho ynterrogatorio dixo este dho tº que dize lo que dho tiene lo qual es la verdad y en ello se afirmo por la //
/273v/ dha lengua del dho don M^{yn} ynterpetre.

Anadidas

1 - A la primera pregunta del dho ynterrogatorio anadido dixo este dho tº que lo contenydo en la dha pregunta lo a oido dezir este dho tº e no lo save.

2 - A la segunda pregunta del dho ynterrogatorio dixo este dho tº que no lo save e que lo oyo dezir a yndyos de Canta.

3 - A la trª pregunta del dho ynterrogatorio dixo que no lo save.

4 - A la quarta pregunta dixo que no lo save.

5 - A la quynta pregunta dyxo que no lo save.

6 - A la sesta pregunta del dho ynterrogatorio dixo que no lo save.

7 - A la setima pregunta del dho ynterrogatorio dixo que dize lo que dho tiene lo qual dixo que era la verdad y en ello dixo que se afirmava e afirmo e rratefico por la dha lengua y no lo firmo porque no supo.

paso ante mi

$Fran^{co}$ Hrrez de Gª
escrº de Su $M.^{t}$

/274r/ Don Diº Chunbi Quibi
curaca de Quibi sujeto de Canta.

Tº el dho don Diego Chunbiquybi caçique del dho pueblo de Quybi tº presentado por pte de los caciques principales e yndios del dho rrepartimyº de Canta aviendo jurado en forma de drº e syendo preguntado por lengua del dho don M^{yn} dixo lo sig^{e}:

1 - A la primera pregunta dixo que conosze y conoscio a los caçiques que an sydo e son de los dhos pueblos de Canta y Chacalla del tpo de Guaynacava a esta parte que este tº se save acordar e tiene noticia de las dhas tierras de Quybi porque esta poblado en ellas.

Preguntado por las preguntas generales de la ley dixo este dho tº que no save la hedad que tiene mas de que se acuerda de Guaynacava e que es sujeto a Canta e le ba el mysmo ynteres que a los demas yndios del dho rrepartimyº pero que por eso no dexara de dezir verdad de lo que supiere y le fuere preguntado e que no le an dado ny prometydo cosa alguna para que en este su dho diga el contº de la verdad.

2 - A la segunda pregunta dixo que save que las dhas tierras de Quybi estan en los termºs de Canta e son tierras de Canta de muchos anos a esta parte porque este tº oyo //
/274v/ dezir a sus padres e pasados senores que an sydo del dho pueblo de Quybi e de los yndios de su parçialidad que antes que vinyesen los Yngas a esta tierra avian sydo el dho pueblo e tierras de Quybi de un senor de Collique que llamavan Collicapa e que les tenia poblado de yndyos yungas e que tenyendolas este senor las avia dado por conçierto de çierta deferencia que trayan a los yndios de Canta los quales las avian tenydo mucho tpo e sembradolas y este tº dende que nascio syenpre a estado y esta y sus padres en ellas y las an senbrado y labrado todas ellas fasta que por mandado de Ynga se pusieron çiertos yndios mytimaes que senbravan para Ynga e so color desto e de una capacocha que llamavan en aquel tpo que era cierta rrogativa se entravan los dhos yndios yavyos en las dhas tierras y se las usurpavan con favor de los Yngas y esto $rresp.^{e}$

3 - A la trª preg.ª dixo que dize lo que dho ti^{e} en las preguntas antes desta.

4 - A la quarta pregunta dixo que save lo cont[do] en la dha preg[a] porque lo vido fazer en su tpo y lo oyo dezir a sus mayores e mas ançianos.

/275r/ 5 - A la quynta pregunta dixo este dho t[o] que save la dha pregunta porque lo vido ser y pasar ansy e lo oyo dezir a sus pasados.

6 - A la sesta pregunta dixo este dho t[o] que save la dha pregunta como en ella se cont.[e] por lo aber oido dezir a yndios del dho pueblo de Canta e que en tpo de Guascar Ynga se acuerda este dho t[o] que los dhos yndios yavyos fueron con la dha capacocha como lo dize la dha preg.[a]

7 - A la setima preg[a] dixo que dize lo que dho ti[e].

8 - A la hotava pregunta dixo que save e vido que al tpo que entraron los espanoles en este rreyno se salieron algunos yndios de los de Chacalla que estavan en las dhas tierras diziendo que los espanoles los avian de hechar de alli e que los que quedaron en ellas estobieron fasta que los yndios de Canta compraron las tierras.

9 - A la novena pregunta dixo que la save como en ella se cont.[e] porque fue y paso asi y es dello pu[co] e not[o].

10- A la decima pregunta dixo que dize lo que dho ti[e].

11- A las onze preg[a] dixo que no lo save.

12- A las doze preguntas dixo que no lo save.

13- A las treze preguntas dixo que no lo save.

14- A las catorze preg[as] dixo que no lo save.

15- A las quynze preg[as] dixo que no lo save e dize lo que dho tiene y en ello se afirmo por la dha lengua.

/275v/ *Anadidas*

1 - A la primera pregunta dixo que dize lo que dho tiene e que este t[o] vido matar el dho yndio porque defendia la dha capacocha.

2 - A la segunda pregunta dixo que save la dha preg[a] por cosa pu[ca] entre los yndios.

3 - A la t[a] pregunta dixo que dize lo que dho tiene en las preguntas antes desta.

4 - A la quarta pregunta dixo que dize lo que dho tiene en las preguntas antes desta.

5 - A la quynta pregunta dixo que dize lo que dho tyene en las preguntas antes desta.

6 - A la sesta pregunta dixo que dize lo que dho tiene en las preguntas antes desta.

7 - A la setima pregunta dixo este dho t[o] que dize lo que dho tiene lo qual dixo que era la verdad y en ello se afirmo por la dha lengua y no lo firmo porque no supo.

paso ante mi Fran[co] Hrrez de G[a]
escr[o] de Su Mag[t]

/276r/ Santiago indio nral de Canares
de 50 / no le toca.

T[o] el dho Santiago yndio natural de Canares t[o] presentado por pte de los dhos cacique e yndios de Canta aviendo jurado en forma e siendo preguntado por lengua de don M[yn] Lengua de la Rreal Avdiencia e por el ynterrogatorio e pregunta segunda por donde se pidio fuese hexsamynado dixo lo syg.[e]:

1 - A la primera pregunta dixo que conoszio a Marcapoma caçique de Chacalla e a Guaman Yavre caçique de Canta por trabto y conversacion e lo mysmo a los caçiques e senores que agora son de las dhas tierras e pueblos e tiene not[a] de las dhas tierras de Quybi.

Preguntado por las preguntas generales de la ley dixo que es de hedad de mas de cinquenta anos e que no es pariente de ning^{a} de las dhas ptes ny le toca neng^{a} de las gen^{les}.

2 - A la segunda pregunta e a todas las demas del dho ynterrogat^{o} dixo este dho t^{o} que al tpo que se alço Mango Ynga e los yndios de Canta y de Chacalla contra los espanoles este t^{o} hera criado //
/276v/ del dho marques don Fran^{co} Piçarro cuyo heran los yndios de Canta y este t^{o} fue a Canta para fablar los yndios [*sic*] e apaçiguallos e vido que los yndios de Canta en aquella sazon estavan poblados y poseyan en las dhas tierras de Quybi dende el pueblo de Quybi fazia la syerra y dende alli para abaxo estavan poblados e tenyan posesyon de las dhas tierras yndios yavyos de Chacalla e que no save este t^{o} de que manera entraron alli ny cuyas son las dhas tierras mas de que el rrio que por ellas pasa viene de la syerra de Canta y es por alli la entrada e el term^{o} de Canta y los yndios de Chacalla tienen su term^{o} y entrada e rrio por Mama y esto dixo que savia y hera verdad y en ello se afirmo por la dha lengua avyendosele tornado a leer y no lo firmo porque no supo.

E otro si dixo que oyo dezir que los dhos yndios yavyos heran y^{o}s mitimaes.

ante mi — Fran^{co} Hrrez de G^{a}
escr^{o} de Su M

E yo Fran^{co} Herrez de Guebara escr^{o} de Su M^{t} p.^{e} fuy a lo que dho es e a las dhas presentaciones de t^{o}s e juram^{o} e declaraçiones e otros autos e lo escrevi e hize aqui myo s^{o} tal en tm^{no} de verdad.

Fran^{co} Hrrez de G^{a}
escr^{o} de Su Mag.^{t}

/277r/ Tasacion de costas que son las sig^{es}:

- de la present^{on} e rrequerimy^{o} con la p^{o}bision: quatro ts.		4 ts
- de la p^{on} de dos ynterrogatorios: tres ts.		3 ts
- de dos enplazamy^{o}s para t^{o}s: un p^{o}	1 p^{o}	
- de p^{on} e juram^{o} de diez y seis t^{o}s: quatro p^{o}s.	4 p^{o}s	
- de noventa y tres hojas: quarenta e seis p^{o}s.	46 p^{o}s	
- del signo: dos ts.		2 ts
- de la ynformacion de of^{o} que hizo Pedraça y el p^{e} Caravajal de la pte de Canta: seis p^{o}s y medio	6 p^{o}s	4 ts
- de doze dias de ocupacion a dos p^{o}s e m^{o}: treynta p^{o}s	*30 p^{o}s*	
Que son todos: ochenta e ocho p^{o}s e çinco ts.	*88 p^{o}s*	*5 ts*

Rresçibilos yo el escr^{o} y conforme a una carta de pago que di en Canta al cacique sy ubiere rr^{do} mas lo bolbere porque quise sacallo en limpio conforme a la p^{o}bision y despues a pedimi^{o} della atento que estava pagado se m^{do} dar oreg.^{al}

Fran^{co} Hrrez

/278r/ Los y^{o}s de Canta.

Muy pod. ss^{or}

Fran^{co} de la Torre en nonbre de don Hr^{do} e de los demas ss. e prinçipales del pueblo de Canta encomendados en Niculas de Rribera en el pleyto con los yndios de Chacalla y A^{o} Moreno en su nonbre sobre las trras / digo que visto por Vra Al. mando ver e desaminar los t^{o}s hallara que los dhos mis ptes p^{o}varon bien e bastantemente todo lo que probar debian e les conbenia para aver vitoria en esta causa; que las ptes contrarias no p^{o}varon cosa alguna que les aproveche / los t^{o}s que presentaron son syngulares e no contestes depienden de oydas e querençias vanas e no dan rrazon sufiçiente de sus dhos e depusyçiones demas de lo qual los t^{o}s que presentaron son ptes formales en esta cabsa por lo qual no hazen fee ny prueva alguna

Por tanto pido e sup^{co} a Bra Al. segun es de suso e sobre que pido just^{a} e costas el of^{o} de Bra Al. ynploro e concluyo difin^{te}.

El Licen^{do} de Leon — Fran^{co} de la Torre

En Los Rreyes catorze de abril de myll e quy^{o}s e çinquenta y nueve años ante los señores presydente e oydores en Aud^{a} Rreal pu^{ca} de petiçiones; la presento el dho Ffran^{co} de la Torre en el dho nonbre / y los dhos señores mandaron dar tr^{do} a la otra pte e que rresponda a la primera Audiençia lo qual se probeyo presente: Alonso Moreno pr. de los y^{o}s de Chacalla y se le notyfico.

Diego Muñoz

/279r/ Los y^{o}s de Canta.

Muy podrs. s^{o}r

Fran^{co} de la Torre / en nonbre de don Hr^{do} e de los demas cs e prinçipales del pueblo de Canta encomendados en Niculas de Rribera / en el pleyto con los yndios de Chacalla / e A^{o} Moreno en su n^{e} sobre las trras / digo que la dha pte contr^{a} a llebado tmny^{o} para concluyr la cabsa al qual le a sydo not^{o} e no a dho cosa alguna.

A Bra Al. pido e sup^{co} que a su rrebeldia se aya la cabsa por conclusa sobre que pido justiçia.

Fran^{co} de la Torre

En Los Rreyes diez y ocho de abril de myll e quy^{o}s e çinquenta e nueve a^{o}s en Aud^{a} Rreal pu^{ca} de petiçiones ante los senores presydente e oydores; la presento el dho pr. en n^{e} de sus ptes y los dhos señores la ovieron por presentada y la causa por conclusa.

Diego Muñoz

/279v/ En este grado de rrevista no hago tasaçion de costas porque no tengo el aranzel de España.

Diego Muñoz

/280r/ Los y^{o}s de Chacalla.

Muy p^{o}s ss

Alonso Moreno en n^{e} del caçique e y^{o}s de Chacalla en el pleyto con los y^{o}s de Canta sobre las chacaras de Quibi / digo que por V. Alt.^{a} se mando a las ptes contr^{a}s con pena de çien açotes y tresquilados y otras penas que no se entremetiesen a tomar a mys ptes tierras ni chacaras algunas y sin enbargo desto les tomaron vna chacra de coca y se la cojeron / despues de serles notificado lo susodho / y por mandado de V. Alt.^{a} se mando bolber a mys ptes y se notifico otra vez a las ptes contr^{a}s con mayores penas que no lo hiziesen otra vez / y como entonzes no fueron castigados ni se les executo la pena ni se les mando bolber la coca que cojeron / han tornado agora de nuevo con setenta y^{o}s por m^{do} de su caçique y del p^{e} Gr^{mo} M^{in} que esta en su dotrina cojer la coca de la dha chacara / y llevaron presos al dho p^{e} e y^{o}s / de que mis ptes rreçiben grande daño y fuerça de lo que se querellan ante V. Alt.^{a} en la via y forma que de dr^{o} mejor haya lugar.

A V. Alt.^{a} pido y supp^{co} mande castigar los dhos y^{o}s y a su caçique / y que buelban a mis ptes la coca que les cojeron y llevaron / sobre que pido justiçia y para ello etc.

El Licenc.^{do} M. Luzio

En Los Rreyes doze de mayo de myll e quy^{o}s e çinquenta y nueve a^{o}s ante los señores presydente e oydores est^{do} en Aud^{a} Rreal pu^{ca}; la presento el cont^{do} y los dhos señores mandaron que de ynform^{on} de lo que dize e dada probeyeron sobre ello lo que es la justiçia.

Diego Muñoz

/281r/ Relaçion de la probança
de los yndios de Chacalla
sobre las tierras de Quybi.

---

1 - Primeramente sean preguntados sy conosçen a las ptes e si tienen notiçia de las dhas tierras de Quibi.

2 - Yten si saben etc. que al tienpo que dizen aver hecho don Fran^{co} caçique que dizen aver sido de Chacalla el conçierto de las dzas ovejas de la tierra por las tierras y chacaras de Quibi fue a ello forçado y conpelido el dho don Fran^{co} por Fran^{co} de Anpuero su encomendero que le hizo por fuerça y con temores y amenazas y malos tratamy^{o}s le hizo que se hallase presente al conçierto syn que el dho don Fran^{co} lo hiziese ni quysiese.

Digan lo que saben.

T^{o} el dho Fran^{co} de Anpuero v^{o} desta çibdad señor de Chacalla de hedad de quarenta a^{o}s e no le tocan. Dixo que antes y al tpo que el dho conçierto se hiziese entre los yndios de Canta y de Chacalla sobre las tierras de Quyvi trayan unos yndios con otros de los dos rrepartimy^{o}s guerra y pendençias y los de Canta como faboresçidos del marques cuyos heran despues de G^{o} piçarro mataban y herian muchos yndios del dho rrepartimy^{o} de Chacalla e algunos le tenian descalabrados que oy dia son bivos alg^{o}s dellos y andaban todos desasosegados y visto esto este t^{o} y el padre Godoy enpurtanaron [*sic*] en esta çibdad a don Fran^{co} Marcapoma caçique de Chacalla que fuese a las dhas tierras de Qui Quibi [*sic*] y este t^{o} fue alla y le llebo consigo para dar un corte en lo susodho y llegados que fueron este t^{o} hablo con los dhos yndios para que hiziesen conçierto y el dho don Fran^{co} ny sus yndios no querian hazello porque venyan llorando alg^{o}s yndios de Chacalla y quexandose que las dhas tierras heran suyas que no se las vendyesen y este t^{o} contra la voluntad de los dhos yndios procuro que se hiziese y en hefeto los yndios de Canta dieron dozientas obejas pacos y guanacos y el dho caçique e yndios de Chacalla no las quysieron rresçebir y los yndios de Canta las llebaron a las tierras de Chacalla y el padre Godoy le hizo saber como estaban alli a este t^{o} y este t^{o} enbio por ellas y se las traxeron a esta çibdad menos catorze poco mas o menos porque este t^{o} no se acuerda bien e que las que faltaban cree que las tomo el padre Godoy para sy y las que ansy traxeron a esta çibdad dellas bendio este t^{o} y dellas paso en la carnesçeria y esto rresp.^{e}

/281v/ T^{o} Ju^{o} Chauca de Guadachiri /fol. 142 col. 2/

T^{o} Anton Cacharure de Guadachiri /fol. 151 col. 2/

T^{o} el dho Ju^{o} Yauri de la encomyenda de Anpuero de hedad de 40 años e no le tocan. Dixo que este t^{o} se hallo presente quando se trataba del conçierto de las dhas tierras entre los y^{o}s de Chacalla y Canta porque hera por alli su pasaje para su tierra e vido como lloraban yndios yauyos de Chacalla porque el dho Fran^{co} de Anpuero les vendia las tierras por las dhas ovejas las quales entendio que avia rresçebido el dho Fran^{co} de Anpuero e dezia a los yndios de Chacalla que dexasen las tierras a los de Canta e que tomasen las obejas y los yndios de Chacalla y Marcapoma su caçique dezian que no querian dexallas.

T^{o} el dho Alonso Paucar natural del pu^{o} de Palli de la encomyenda de Fran^{co} de Anpuero de hedad de mas de 70 a^{o}s rremytese /148/ col.2.

T^{o} don Diego Condor prinçipal del valle de Chacalla de hedad de mas de ochenta sujeto a don Xpoval rremytese.

Ay mas t^{o}s sujetos a Chacalla que lo dizen.

3 - Yten si saben etc.que estando presente y contradiziendo el dho conçierto y no lo queriendo hazer el dho don Fran^{co} de Chacalla lo hizo y firmo el dho Fran^{co} de Anpuero su encomendero y se tomo para si y llebo las dhas dozientas ovejas de la tierra que dizen aver dado los yndios de Canta y no las dio ny entrego al dho don Fran^{co} ny a los prinçipales ny yndios de Chacalla antes las tomo y llebo y como cosa suya hizo dellas lo que quiso syn voluntad ny consentimy^{o} de sus yndios.

T^{o} el dho don Diego Condor. Dixo este dho t^{o} que dize lo que dho tiene en las preg^{tas} antes desta e que save e vido que el dho Fran^{co} de Anpuero su encomendero tomo las dhas obejas para si e las enbio a Lima a su casa con un cryado suyo que se llamaba Molina.

T^{o} el dho Alonso Paucar. Dixo este dho t^{o} que dize lo que dho tiene en las preguntas antes desta e que sy el dho Fran^{co} de Anpuero diera y entregara las dhas obejas al dho don Fran^{co} Marcapoma que aqueste t^{o} lo biera supiera y entendiera porque se hallo a todo presente y antes bio y entendio que el dho Fran^{co} de Anpuero se llebo las dhas ovejas a su casa a la çibdad de Los Rreyes y las enbio con el dho Molina su criado.

T^{o} el dho Fran^{co} de Anpuero v^{o} desta çibdad dixo que dize lo que dho tiene e que ...(roto) ...este t^{o} formo el dho conçierto.

T^{o} el dho Nicolas de Rribera el moço v^{o} desta çibdad encomendero de los yndios de Canta de mas de 40 a^{o}s dixo que los yndios //

/282r/ de Canta vido este t^{o} que enbiaron unos prençipales y el caçique de Chacalla don Fran^{co} enbio otros prençipales y este t^{o} enbio un honbre que se dize Diego Florez con ellos y los yndios de Canta entregaron las dozientas ovejas a los yndios de Chacalla la que enbiava el dho don Fran^{co} y las llebaron a Chacalla y esto save desta preg^{ta} / Los demas se rrefyeren a lo dho.

4 - Yten sy saven etc. que syenpre el dho don Fran^{co} de Chacalla contradixo el dho conçierto y dezia que no queria obejas syno las tierras de Quibi que heran de sus yndios y no las podia el vender ny trocar a obejas y aunque lo dezia asy el dho Fran^{co} de Anpuero rresçibiollas obejas y fyrmo el papel que los de Canta quysieron.

T^{o} el dho don Diego Condor dixo que dize lo que dho tiene en las preg^{tas} antes desta e que se acuerda este t^{o} que el dho don Fran^{co} Marcapoma caçique de Chacalla al tienpo que el dho Fran^{co} de Anpuero hizo el dho conçierto lloro e dixo al dho Fran^{co} de Anpuero: "este conçierto yo no lo hago syno vos y Rribera español con español amygos" y esto rresp^{e}/ e dixo ansy mesmo que vido este dho t^{o} como los caciques e yndios yauyos de Guadacheri y los de doña Maria Martel vinyeron a rrenyr al dho don Fran^{co} Marcapoma caçique de Chacalla y le dixeron que para que vendia las tierras que el Hinga le avia dado e amojonado e hazia tan gran maldad y el dho don Fran^{co} Marcapoma les dezia que el no havia bendido syno el dho Fran^{co} de Anpuero su encomendero y q^{do} le hablaban desto y le rrenyan entonçes se corria mas el dho don Fran^{co} y se quexava y dezia que lo avia de pedir y quexarse y syenpre lo dezia hasta que murio que abra quatro años. Los demas se rrefieren a lo dho.

5 - Yten sy saven etc. que Su Mag^{t} tiene proveido en todas las Yndias y en este Rreyno por sus rreales p^{o}visiones y çedulas que los testigos han visto originalmente que los yndios y tierras y todas las demas cosas esten de la manera que las hallaron los españoles quando entraron en la tal tierra / syn que en nada aya ny se consienta novedad alguna. Digan los testigos lo que saben / y si a sydo en todas Yndias usado y guardado lo susodho por vyrtud de la dha p^{o}vision rreal de Su Mag.^{t}

T^{o} el dho Fran^{co} de Anpuero. Dixo que este dho testigo se acuerda //

/282v/ aver visto una çedula en que Su Mag.^{t} mandava que los yndios se estubiesen como les tomo la boz los españoles en esta tierra e que esta çedula se presento en un pleyto que traya Fran^{co} de Carabantes y Merlo sobre los mytimaes de Jauxa y que es cosa muy publica y not^{a} que Su Mag.^{t} tiene proveida esta çedula y se traxo a este rreyno e a esta causa cree este t^{o} que estan todos los yndios de la mysma manera que les tomo la boz al tpo que el marques don Fran^{co} Piçarro entro en esta tierra porque de otra manera ubiera savido muchos pleitos.

T^{o} el dho Fran^{co} de Carabantes v^{o} desta çibdad de hedad de treynta y ocho años e no le tocan. Dixo que este t^{o} a oydo dezir que Su Mag.^{t} tiene m^{do} que se guarde lo que Guaynacaba hizo y no save otra cosa.

T^{o} el dho Ju^{o} de Padilla escriv^{o} pu^{co} desta çibdad. Dixo que es de hedad de 28 a^{o}s e no le tocan. Dixo que lo que della save es que se quiere acordar aver oydo dezir que Su Mag.^{t} tiene proveido e m^{do} que las cosas de los yndios se esten de la manera que los espanoles las hallaron quando entraron en este rreyno e no save otra cosa.No ay mas.

6 - Yten sy saben etc. que todo lo susodho es publica boz y fama.

Añadidas

1 - Yten si saven que los dhos yndios de Chacalla e su caçique que an poseido las dhas tyerras de Quibi y tenidolas como cosa suya desde el tienpo de los Yngas pasados que se las dieron lo qual ha tanto tienpo que no ay quien se pueda acordar dello.

T^{o} el dho Juan Chauca natural de Guadacheri. No sabe la hedad que t^{e} mas de que se acuerda de Guaynacaba e de Guascaringa porque hera su criado e contador e que no le tocan. Dixo que lo que save desta preg^{ta} es que desde que este t^{o} se save acordar que es desde Guaynacaba syenpre este dho t^{o} vio por vista de ojos que los yndios yauyos del rrepartimy^{o} de Chacalla encomendados en Fran^{co} de Anpuero tenyan e poseyan las dhas tierras de Quibi como naturales puestos por los Yngas pasados e tenyan alli sus casas e hazian en ellas tierras [*sic*] sus sementeras desde un mojon que esta senalado oy dia //
/283r/ hasta una quebrada que esta de la pte del rrio y por esta otra pte por la ladera del çerro do esta el mojon hasta este pu^{o} de Quybi e alli entra lo de Canta e que esto oyo dezir este t^{o} a biejos e ançianos del dho pu^{o} de los yauyos e a otros comarcanos que los dhos yndios yauyos lo habian ganado las dhas tierras por guerras e que venydo que fue Topa Ynga Yupangui hallo alli hasta çien yndios yauyos puso otros çinquenta yndios mas y los dio y confirmo a todas nuevamente las dhas tierras y que todo esto sabia entendia e via este t^{o} andando entre los dhos yndios cobrando tributos para Ynga porque hera quypocamayo de todos los yndios que heran yauyos e despues que murio Topa Ynga Yupangui vino a senorear Guaynacaba e confyrmo las dhas tierras de Quibi a los dhos yndios yauyos del rrepartimy^{o} de Chacalla y lo mysmo vido este t^{o} que hizo Guascar Ynga despues que subçedio en el senorio por muerte de Guaynacaba e que esto hera cosa muy pu^{ca} e not^{a} y lo es entre todos los yndios que lo saben y vieron y dello tienen notiçia como este t^{o}.

T^{o} el dho Anton Cacharure natural de Guadacheri. No sabe la hedad que tiene e se acuerda de quando entraron los españoles en esta tierra e no le tocan. Dixo este dho t^{o} que lo que desta preg^{ta} save es que este t^{o} a oydo dezir al dho su padre que se llamaba Anton Caxalluri e a otros biejos e ançianos desta comarca que antes que vinyesen los Yngas a esta tierra los dhos yndios yauyos de Chacalla se avian juntado e salido de guerra a ganar tierra y que avian hido hasta donde agora esta puesto vn mojon y poblaron alli y mataron pte de los yndios de un señor que estava en las dhas tierras que llamaban Chunbecaxa y los demas los sujetaron y enpeçaron a hazer hedifiçios e chacaras e a labrarlas e coger las sementeras e que ansy habian estado hasta que vino a conquistar este rreyno Topa Ynga Yupangue e sus capitanes el qual dizen que como los hallo a los dhos yndios de Chacalla mando que se estubiesen de aquella manera y les dio chacaras y sementeras en que labrasen e les dio por termyno desde un mojon que esta junto a las casas del dho pueblo de Quybi hazia abajo hazia la mar que fuese para yndios Yngas //
/283v/ y de alli para arriva hazia Canta fuesen serranos y despues de muerto Topa Ynga Yupangue dizen que hizo lo mysmo Guaynacaba y les confyrmo las dhas tierras y que despues que murio Guaynacaba Guascar Ynga que subçedio en el señorio avia fho lo mysmo e que desta manera los dhos yndios de Chacalla avian tenido las dhas tierras hasta que entraron los espanoles y despues hasta que se hizo el dho conçierto e que esto como dho tiene lo oyo dezir este dho t^{o} por cosa publica y notoria a los dhos biejos e ançianos los quales dezian que ellos ansi lo havian visto ser e pasar en sus tienpos e oidolo dezir a sus mayores e ançianos e que dello havia sydo y hera pu^{co} y not^{o} e publica boz e fama y esto rrespondio por la dha lengua.

T^{o} el dho Juan Yauri. Dixo que este t^{o} vido desde que se save acordar alg^{o}s yndios yauyos de Chacalla poblados en las dhas tierras de Quibi e las tubieron fasta que se hizo el conçierto. Ay otros muchos t^{o}s de Chacalla que dizen lo saber.

2 - Yten si saben etc. que los dhos yndios de Chacalla tenyan e poseyan las dhas tierras de Quybi al tienpo que los españoles entraron en este rreyno y las tubieron mucho tpo despues de entrados. Digan lo que saben.

T^{o} el dho Juan Yauri. Dixo que este dho t^{o} que sabe la preg^{ta} como en ella se cont.^{e} porque este t^{o} lo vido ser y pasar ansy como la dha preg^{ta} lo declara theniendo como tiene notiçia y save quando los españoles entraron en esta tierra y dello pu^{co} e not^{o} e publica boz y fama.

T^{o} el dho Ju^{o} Yndio sujeto a Chacalla. Dixo que dize lo que dho tiene en las preg^{tas} antes desta e que save la dha pregunta como en ella se contiene e declara porque lo vido ser y pasar ansy por vista de ojos.

T^{o} el dho M^{yn} Usco dixo que dize lo que dho tiene en las preg^{tas} antes desta / e que lo en esta dha pregunta contenydo lo vido ser y pasar ansy por vista de ojos como la preg^{ta} lo dize y declara y esto rresponde.

T^{o} el dho Fran^{co} de Anpuero. Dixo que nyega que los dhos yndios yauyos de Chacalla despues que estan en encomy^{da} //
/284r/ deste t^{o} que puede aver veynte anos poco mas o menos estubieron en las dhas tierras hasta que se hizo el conçierto y oy dia cree este t^{o} que ay algunos.

T^{o} el dho Juan Yauri. Dixo que save e vido que al tienpo que entraron los espanoles en este rreyno los dhos y^{o}s yauyos de Chacalla estaban en las dhas tierras de Quybi y esto dixo que savia e declaro por la dha lengua.
Los demas se rremyten a lo dho.

3 - Yten si saven etc. que los yndios de Canta porque no se alçaron juntamente con ellos los yndios de Chacalla e porque serbian a los españoles como lo hazian al tienpo que se alçaron los naturales en la tierra mataron çiento y çinquenta y^{o}s de Chacalla que estaban en las dhas tierras de Quybi y se las quytaron e tomaron y desde entonçes se las tienen tomadas e usurpadas los dhos y^{o}s de Canta.

Digan lo que saben de lo dho.

T^{o} el dho don Diego Condor. Dixo que este dho t^{o} se acuerda muy bien e lo vido desde un peñal donde estava myrando lo que pasava que aviendo los yndios de Canta puesto mal con los Yngas a los yndios yauyos y lebantadoles un testimonyo al tpo que se alçaron los Yngas los dhos yndios de Canta llebaron por engaño çiento e çinquenta e quatro yndios cargados diziendoles que los llebaban a Canta que estava alli su caçique y los metieron en un corral junto al pueblo de Quibi y alli los mataron a todos y este dho t^{o} lo vio de ençima de un peñal.

T^{o} el dho Ju^{o} Yauri. Dixo que save y es cosa publica y notoria entre los yndios biejos e moços de los dhos rrepartimy^{o}s de Chacalla que lo saben y dello tienen notiçia como este t^{o} que estando alçado Mango Ynga contra los españoles los dhos yndios de Canta por odio y mala voluntad que se tenyan so color que les llebantaron que los dhos yndios de Chacalla avian enbiado dos hijos de caçiques a hablar al marques y por congraçiarse con Yntisa Ylla Topa llebaron por engaño al caçique prinçipal de Chacalla que se llamaba Bilcapoma y les dixeron que tenyan coraçon de españoles y de yndios y le mataron e hizieron saver a Ylla Topa cap.^{t} de Ynga como le habian muerto por la dha ocasion y luego vinyeron muchos yndios a dezir a los yndios yauyos como su caçique estava en Canta e que les enbiaba //
/284v/ que le llebasen yndios con comyda y sacaron fasta çiento y çinquenta yndios cargados y llegando a unas peñas que llaman Acobingo que sera una legua deste dho pueblo de Quibi e alli los mataron a todos / e que esto como dho tiene ha sido y es dello publico y notorio.

T^{o} el dho M^{yn} Yusco. Dixo que lo que save desta dha preg^{ta} es que en tpo de los españoles que entraron en esta dha tierra quando se alço Topa Ynga Yupangue vino el dho Topa Ynga Yupangue desbaratado huyendo del dho desbarate y estando Bilcapoma caçique de Chacalla con Guaman Yaure caçique de Canta aguardando a Mango Ynga / el dho caçique de Canta acuso al dho Bilcapoma e dixo a Mango Ynga que tenya dos coraçones que avia enbiado cartas al marques y el dho Mango Ynga dizen que le rrino y le dio con una porra y le dexo y se fue hazia Guanuco; e que luego los dhos caçiques de Canta le avian muerto e que despues de muerto avian venydo a las dhas tierras de Chacalla e dho a los yndios que Bilcapoma su caçique estava en Canta con Mango Ynga que les mandava que les llebasen chica e mayz e otras comydas e que los dhos yndios de Chacalla creyendo que hera ansy habian hido çiento y çinquenta yndios cargados destas comidas e que llegados a un pueblo que llamaban Coyngo los dhos yndios de Canta avian metido los dhos yndios en un corral e alli los avian muerto a todos diziendo que querian bien a los españoles e que por ello avian de morir e que esto que dho t^{e} en esta preg^{ta} le fue a este dho t^{o} publico y not^{o} por avello oydo dezir en esta comarca entre los yndios que dezian avello visto e porque nunca los dhos yndios paresçieron mas muertos ny bivos.

T^{o} el dho Fran^{co} de Anpuero. Dixo que en tpo de Ylla Topa tenyendo este t^{o} ya encomendados los dhos yndios le fue a este t^{o} muy pu^{co} y not^{o} que los yndios de Canta mataron a Bilcapoma caçique prinçipal de Chacalla con otros muchos yndios que con el estaban porque dezian que aborresçian los españoles y encontraron los dhos yndios de Canta que les enpeçaron a entrar en las tierras de Quybi a los yauyos y a tener con ellos guerra como contrarios que avian muerto su caçique. Ay otros muchos y^{o}s de Chacalla que la saben //

/285r/ 4 - Yten sy saven etc. que las dhas tierras de Quibi que los dhos yndios de Canta tienen tomadas a los de Chacalla balen mas de ocho mill p^{o}s al presente y los balian de diez años a esta pte. Digan los testigos lo que balian agora diez años y lo que valen agora. El Licen^{do} Marco Luzio.

T^{o} el dho Juan Chauca de Guadacheri. Este t^{o} [*sic*] que las dhas tierras de Quibi son muy buenas ansy para senbrar mayz como para coca e que en tienpo de los Yngas de las chacas [*sic*] que veneficiaban los dhos mytimaes para el Ynga le daban cada año: unos años çiento e çinquenta costales de coca y otros años çiento y sesenta / syn lo que le daban de caçiques e yndios particulares de trebuto que les daban unas vezes çinco costales y otras vezes quatro; e que esto se coxia en las dhas tierras syn el mayz e yuca e papas y otras cosas porque entonçes estaban las dhas tierras mejor labradas e benefiçiadas e tenyan mas cuidado dellas e que a este t^{o} le paresçe que valen muy bien los dhos ocho myll p^{o}s que dize la preg^{ta} e que sy fueran deste t^{o} no las diera por ellos ny por ningun dinero e que sy no fueran suyas y tubiera los dhos ocho myll p^{o}s para conprallas los diera por ellas / lo qual declaro por la dha lengua.

T^{o} el dho Anton Chaure [*sic*] de Guadacheri. Dixo que ha visto las dhas tierras de Quybi e que le paresçe que valen ocho myll p^{o}s e aun mas porque son muy provechosas para los yndios por amor de la coca que se coxe e saca mucho della y esto dixo.

T^{o} el dho Fran^{co} de Anpuero. Dixo que save e ha entendido entre los dhos yndios que las dhas tierras las tienen en mucho por tener agua porque en la syerra son faltos de agua y abaxan a las bertientes a los rrios y hazen sus chacarrillas a mano con mucho trabajo / e que el valor dellas que este t^{o} no lo sabe. Ay mas t^{o}s de los de Chacalla que la saben.

/285v/ Relaçion de la p^{o}bança de el caçique
e yndios de Canta en el pleyto con los
de Chacalla sobre las trras de Quibi.

---

1 - Primeramente sy conosçen a las ptes e a cada una dellas e tienen notiçia del valle e tierras que dizen de Quibi que estan en los trmy^{o}s del dho valle de Canta e de que tpo a esta pte y conosçieron a los Yngas pasados rreyes e señores que fueron destos rreynos esp^{al}mente a Topa Ynga y a Guainacaba y si tienen notiçia de un rrio que pasa junto a las dhas tierras que se llama el rrio de Quibi.

2 - Yten sy saben etc. que las dhas tierras que dizen de Quibe estan dentro de los trmy^{o}s e pueblos de Canta e de sus parçialidades ny mas ny menos que esta en la figura que mostraron a los señores oydores porque los trmy^{o}s pte un cerro grande que esta entre los yauyos e yndios de Chacalla y los dhos yndios de Canta. Pido que los t^{o}s declaren como se parten los trmy^{o}s y como caen dentro de los dhos trmy^{o}s de Canta las dhas tierras de Quibi. Digan lo que saben.

T^{o} el dho don P^{o} caçique del pu^{o} de Guacayo [*sic*] encom^{do} en G^{o} de Caçeres de hedad de mas de 30 e no le tocan. Dixo que lo que este dho t^{o} ha bisto y entiende las dhas tierras de Quibi estan dentro de los propios trmy^{o}s de Canta e mas çercanas a sus pueblos de Canta que no a los pueblos de Chacalla e que los trmy^{o}s y mojones de Canta estan señalados desde un çerro que tiene un mojon biejo que a lo que este t^{o} a oydo dezir a sus antepasados antes que entrasen los Yngas en esta tierra corrian desde alli el trmy^{o} de Canta hazia Canta / e que despues que entro Ynga en esta tierra avia quytado los mojones y señaladolo por mojones de yndios yauyos encomendados en Fran^{co} de Anpuero del rrepartimy^{o} de Chacalla y los puso alli para mytimaes para que le syrbiesen de cargas y de chasques y de otros serviçios y andando el tpo los dhos yndios yngas del rrepartimy^{o} de Chacalla se fueron con el fabor de los Yngas y de Guaynacaba ganando tierra y metiendose poco a poco hasta llegar a los dhos pueblo [*sic*] de Quybi so color que hiban con sus çerimonyas a adorar el demonyo y esto como dho tiene lo ha entendido este //
/286r/ dho testigo de los yndios biejos que son ya muertos y otros bibos entre los quales ha sido y es cosa notoria y esto rresponde.

T^{o} el dho Xpoval Caxallaupe encom^{do} en G^{o} de Caçeres; no supo la hedad que tiene e no le tocan. Dixo que lo que de la dha preg^{ta} save e a entendido e oydo dezir es que save que desde un mojon que esta en un çerro junto al pu^{o} de Quibi que no abra media legua partio Hinga los trmy^{o}s y de aquel mojon para lo bajo hazia la mar se contava de yndios yngas y de alli para arriva el rrio arriba hazia Chacalla e Canta se contaba de la sierra e que save este dho testigo e ha visto que dende el pu^{o} de Quibi e dende el dho mojon hasta el primero pueblo de Canta que se llama Quyso habra dos leguas hasta el primero pueblo de Chacalla habra dos jornadas e que junto a las dhas tierras de Quybi esta al presente un pu^{o} poblado de yndios de Canta / por rrazon de que conpraron el sytio e tierras los de Canta a los de Chacalla e que el rrio que pasa por junto a las dhas tierras de Quybi es trmy^{o} e rrio derechamente de Canta e del servi^{o} de Canta porque pasa y ba por junto al camyno y pueblos de Canta y los de Chacalla no tienen servi^{o} ny ap^{o}vechamy^{o} del dho rrio por sus pueblos sy no fuese rresydiendo ellos en las dhas tierras de Quibi porque su rrio y servi^{o} de Chacalla y de sus pueblos es el rrio de Mama que va / e que este dho t^{o} a oydo dezir e tratar entre yndios viejos e ançianos ansy de la parçialidad de donde este t^{o} es natural como desta comarca e a sus padres e aguelos que son ya muertos los quales tratando sobre estas dhas tierras e valle e sobre cosas pasadas dezian que antes que vinyese Ynga a esta tierra los yndios yungas que estan en los valles hazia la mar de la encomyenda del contador Ju^{o} de Caçeres y un caçique y señor que llamaban Chunbiquybe heran todos unos y se llamaban yungas e que este señor trebutaba a un señor de los yungas de Collique que se llamaba Collicapa el qual dho Tunbiquibe rresidia en estas dhas tierras de Quybi junto al rrio sobre que es este pleito y tenya en ellas sus y^{o}s y hera un señor por si que no hera subjeto a los yndios de Canta syno que solamente trebutaba al dho Collicapa señor de Collique e venydo que fue Ynga como dibidio e partio los mojones les hizo serranos los yndios que del quedaron porque //
/286v/ Ynga mato este señor y muchos de sus yndios por çierta contrataçion e delito que hizo contra el Ynga e que los yndios yauyos que puso Ynga en las dhas tierras de Quibi le mataron por su m^{do} lo qual este t^{o} les oya dezir a sus pasados por cosa muy çierta y esto rresponde.

T[o] el dho Cacha yndio ynfiel del rrepartimi[o] de Hr[do] de Montenegro; no sabe la hedad que t[e]; acuerdase de Guaynacaba e Guascar Ynga e no le tocan. Dixo que no save sy las dhas tierras de Quibi estan agora al presente en tierras de los yndios serranos de Canta mas de que este dho t[o] a oydo dezir a su padre que se llamaba Callopa natural del pu[o] de Atabillos encomyenda de Her[do] de Monten[o] v[o] desta dha çibdad de Los Rreyes que seran los dhos Atabillos a dos jornadas destas dhas tierras de Quibi el qual le dezia a este t[o] tratando e platicando en cosas que antes que vinyesen los Yngas a esta tierra las dhas tierras de Quibi las poseya un señor que no se acuerda como se llamava e que este dho señor le dixo su padre que hera subjeto al rrepartimi[o] de Canta e que trebutaba al caçique de Canta e que quando no venya agua por el dho rrio de Quibi que avia sequia se juntaban los yndios de Canta y los deste señor que dho tiene y abrian unas lagunas que se hazen alla arriba en la syerra de la nyebe que cae y las hazian venyr el agua dellas por el dho rrio de Quybi y desta manera dize que lo oyo dezir al dho su padre que el dho señor y sus yndios rregaban las dhas tierras de Quibi quando avia sequia con el agua de las dhas lagunas e hazian sus sementeras e que tanbien oyo dezir al dho su padre que al tienpo que entraron los Yngas en esta tierra yendo conquistando por Chacalla se huyan yndios chequytos del dho rrepartimy[o] de Chacalla y se venyan a guaresçer a las dhas tierras de Quibi e a los naturales que en ellas estaban e que los dhos naturales que heran yngas los criaban e que desta manera se entraban en las dhas tierras de Quibi los dhos yndios yauyos e no save otra cosa desta preg[ta].

T[o] el dho don Fran[co] Arcosna de la encomyenda de Canta vale ynteres y desea que vençan sus ptes; de hedad de mas de 50 / dixo que este dho t[o] save que las dhas tierras de Quibi estan dentro de los trmy[o]s e jurisdiçion de los yndios de Canta. Preg[do] como lo save dixo //
/287r/ que antes que vinyesen los yndios yngas senores deste rreyno las dhas tierras que llaman de Quibi heran de yndios yauyos de hazia la mar y estava en ellas un señor por sy que no hera subjeto a los caçiques de Canta el qual señor se llamaba Chaumecaxa y este Chaumecaxa hera subjeto a otro señor que llamaba [*sic*] Collicapa señor de Collique que es rrepartimy[o] de Gonçalo de Caçeres hijo del contador Ju[o] de Caçeres el qual servya e trebutava el dho Chaumecaxa e sus yndios y que en esta dha sazon pretendiendo los dhos yndios de Canta que este dho pu[o] de Quibi fuese suyo hizieron mucha gente de guerra para venyr sobre el dho Chaumecaxa y sobre el dho Collicapa señor de Collique y ganalles el dho pu[o] de Quibi y entendido esto por el dho Chaumecaxa senor de Quibi e saviendo la gran fuerça de gente que trayan los dhos caçiques de Canta tubo temores y como estaba en medio no sabia a quyen acudiese y enbiava secreptamente chasques a los dhos yndios de Canta secreptamente y en que les daba a entender que hera amygo y luego los dhos caçiques e yndios de Canta que ansy estavan de guerra bajaron de la syerra con toda su armada y pasaron por el dho pu[o] de Quibi syn fazer daño al dho Chaumecaxa ny a sus yndios y llegaron bien mas adelante del dho pueblo de Quibi hazia Collique quatro leguas y sabido por el dho Collicapa señor de Collique que la gente e armada que llebaban los dhos yndios de Canta junto mucha gente de guerra y bino con ella hazia los dhos yndios de Canta los quales visto la mucha gente que traya que hera mas que la dellos se rretraxeron y se hizieron fuertes en unos çerros y de alli se enbiaron mensajeros los unos a los otros y trataron de conçierto y se conçertaron en que el dho Collicapa dixo a los dhos caçiques de Canta que que hera la demanda que trayan que sy lo habian por el pueblo de Quybi que el se lo dava y que fuesen amygos y los dhos caçiques e yndios de Canta le enbiaron a dezir que no querian otra cosa syno el pueblo de Quibi que estaba en su tierras e ansy el dho Collicapa se lo dio y amojono los dhos trmy[o]s de Collique y de Canta y puso un mojon en un çerro que se nonbrava Judcunga y agora se llama asy que estara media legua del pueblo de Quibi que es donde pusyeron los mytimaes los Yngas y dende alli de aquel mojon dexaron para trmy[o] de la sierra por tierra de Canta //
/287v/ y de alli para abajo dexaron por llanos e para los yndios yungas de Collique e que con esto quedaron muy buenos amygos los unos y los otros e hizieron juramento entre ellos segun su ley que los yndios de Canta no ternyan mas diferençias con ellos ny los de Collique tanpoco ny pasarian con gente los unos a las tierras de los otros para rrenir mas sobre las dhas tierras e ansy se bolbieron los yndios de Canta a su tierra y los de Collique tanbien y no hubo mas guerra entre ellos y se llamaban y querian y enbiaban presentes e luego dende adelante el dho Chunbicaxa señor de Quybi servia e trebutaba a los yndios de Canta y ansy sirvio syenpre hasta oy dia syrben a los yndios que en el estan que subçedieron de aquel tpo e que esto que dho tiene es a este dho t[o] publico y notorio por avello oydo dezir a sus padres e haguelos e pasados e a otros biejos e ançianos de los dhos pueblos de Canta y de Chacalla y desta comarca que dezian que ellos ansy lo habian visto ser y pasar en sus tienpos e oydo dezir a los suyos en los suyos e que nunca a vysto ny oydo dezir lo contr[o] y dello ha sido y es a este t[o] pu[co] y notorio e publica boz y fama e que ansy mysmo les a oydo dezir que venydo que fue Topa Ynga Yupangue a esta trra a governar ynbio un horejon yndio para que gobernase los yndios de Chacalla porque lo mysmo hazia en otros pueblos que ponya just.[a] en todos y este orejon tenya una holla en que traya sangre y chaquira molida para ofresçer a la guaca y este t[o] se junto con los dhos yndios de Chacalla y le dixo que fuesen con el al ofresçimy[o] y andaban con esto por los çerros derramandolo y diziendo a la guaca que diese salud a Topa Ynga e a sus hijos e le hiziese gran señor e le diese mucha rriqueza e comydas e despues de derramada la dha capacocha andubieron a buscar agua para laballa porque tenyan por costumbre de hazello ansy y como por aquella pte donde andavan hera y es tierra seca que no ay agua vinyeron dende çerro en çerro en busca deste rrio de Quibi y vinyeron a salir a el por el dho mojon de Canta e asy fueron al dho rrio y labaron la dha capacocha y se bolbieron por los propios pasos por donde habian venydo syn poner mytimaes en las dhas tierras hasta el dho pueblo de Chacalla donde hallaron que un pueblo de yndios de Chacalla yauyos y su caçique que se llamaba Culloynde estaba juntado //
/288r/ yndios y diziendoles que fuesen a hablar a la guaca por que se muriese Guaynacaba porque ya avia muerto Topa Ynga por su causa porque se lo avia rrogado y que lo mesmo haria la guaca sy se lo rrogasen para que Guaynacaba muriese que avia subçedido en el señorio e que sabido por el orejon que traya la capacocha hizo mensajero mensajeros [*sic*] al Cuzco a Guaynacaba de lo susodho avisandole dello y el dho Guaynacaba enbio mandar que los matasen e asy mataron todos los yndios que no quedaron syno las mugeres y unos nynos chiquitos a los quales dhos nynos e a sus madres el dho orejon ynga los mando poner e sacar fuera de su tierra por que no tornasen ellos a adorar la guaca y los pusieron en el dho mojon que señalo Collicapa caçique de Collique por termyno de Canta y señalo tierras para Guaynacaba e para los dhos yndios y desta manera se entraron y estan los dhos yndios yauyos en las dhas tierras de Quybi y que el dho orejon y los que subçedieron en su cargo yban andando el tpo con la dha capacocha y con los dhos

yndios yauyos haziendo la dha çerimonya en prezesion [*sic*] y metiendose cada dia mas en las dhas tierras y tomandolas e poseyendolas e que los dhos yndios de Canta no osaban rresystirselo ansy por no hazer derramar la sangre y por que no los matasen como por que Ynga no los hiziese matar e que tanbien sabido por Guaynacaba que las dhas tierras heran tan buenas e que se coxia en ellas mucha e buena coca mando que se pusiesen muchos yndios yauyos en las dhas tierras e que senbrasen e veneficiasen las chacaras para el dho Guaynacaba e para los dhos yndios yauyos e ansy por su mandado pusieron en las dhas tierras de Quibi muchos yndios yauyos y les señalaron chacarras y se vinyeron apoderando en las dhas tierras fasta junto con el dho pueblo de Quybi / e que venydo Guascar Ynga despues de muerto Guaynacaba hizo lo mysmo e aumento con la capacocha fasta mas adelante del dho pu^{o} de Quibi y porque un yndio topo en la capacocha e hizo derramar una gota le mataron alli luego y enterraron el qual dho yndio que ansy mataron se llamaba Chuqui Guaranga que oy dia tiene un hijo bivo en este pu^{o} de Canta e que des-//
/288v/ pues que murio Pascar Ynga [*sic*] y vinyeron los espanoles a esta tierra syenpre los dhos yndios de Canta estubieron mal con los y^{o}s yauyos de Chacalla y estavan cada dia a punto de guerra para matarse sobre las dhas tierras y Guaman Yaure caçique prençipal del pu^{o} de Canta dezia a los yndios de Chacalla que bien sabian que las chacaras de Quibi heran suyas y que el Ynga se las abia quytado tiranicamente por fuerça como tirano e que agora que abian venydo los españoles que se las avian de dar por bien e que sy no se las davan que alli habian de morir todos y que desta manera estubieron a punto de matarse hasta que vino el presidente Gasca y enbio a los dhos Niculas de Rribera el moço encomendero de los yndios de Canta e a Fran^{co} de Anpuero encomendero de los yndios de Chacalla e que el dho Guaman Yaure caçique de Canta no queria dar obejas ningunas diziendo que el no havia de dar nada por las tierras que heran suyas y que el dho Rribera le dezia: "myra hijo myjor es que les des dozientas obejas por quytar pasiones porque sy rrinyes y os matais castigar os han la justiçia e a my tanbien porque lo consyento; haz quenta hijo que las ovejas que me las das a my e yo las doy por las tierras"; e ansy el dho Guaman Guare [*sic*] caçique por lo que el dho su encomendero le dezia e por le hazer plazer dio las dhas dozientas obejas al dho Marcapoma caçique de Chacalla; las rresçibio y se contento con ellas y del dho conçierto y bio este dho testigo que lo tubo por bueno y se fueron los dhos yndios yauyos de las dhas tierras y las dexaron a los dhos caçiques e yndios de Canta y nunca mas se henojaron ny pidieron cosa alguna hasta que les pusieron demanda a ellas diziendo que no queria pasar por el conçierto e que ansy mysmo estando haziendo el dho conçierto se trato que diesen a los yndios del pueblo de Cullata e Tapati yndios yauyos de Chacalla pobres les diesen algunas chacaras para que senbrasen y estos dhos yndios fueron al dho Guaman Yaure caçique de Canta y le hablaron para que les diese tierras en que senbrar que ellos le sirbirian e otras ofertas que le hizieron y el dho Guaman Yaure caçique y el dho su encomendero Rribera y el dho Fran^{co} de Anpuero //
/289r/ les señalaron desde una cruz que se puso junto al dho pueblo de Quibi hazia arriba hazia Canta tierras para que senbrasen los dhos yndios yauyos de Chacalla y ansy las han senbrado y benefiçiado hasta agora e que don Ju^{o} yndio del rrepartimy^{o} de Canta coxo de un pedaço de las dhas tierras un poco de coca porque la avia el mandado plantar a sus yndios e que hera muy poca cosa y que por ello los dhos yndios yauyos se habian hido a quexar e havia venydo un escr^{o} a hazer ynfor^{on} lo qual todo este dho t^{o} rrespondio por estas rrazones en sustançia por lengua del dho don M^{yn} que lo declaro ansy.

Ay muchos t^{o}s de Canta que declaran sobre esta.

3 - Yten si saven etc. que las dhas tierras y chacarras de Quibe desde uno e dos e diez e veynte e treynta e çinquenta e mas años a esta pte que memoria de honbres no es en contr^{o} syenpre fueron tierras de los dhos yndios de Canta y su caçique las quales syenpre las an tenydo y poseido por suyas y como suyas labrandolas y senbrandolas y haziendo dellas como de cosa propia hasta que se metieron en çierta pte dellas çiertos yndios como mytimaes de los dhos y^{o}s de Chaclla por çierto engaño que les hizieron. Digan lo que saben.

T^{o} el dho don P^{o} caçique de Guancayo encom^{do} en G^{o} de Caçeres. Dixo que torna a declarar lo mysmo que en la preg^{ta} antes desta e mas dixo que desde que vino el presidente Gasca a estos rreynos que en aquella sazon se conçertaron los dhos yndios de Canta con los de Chacalla sobre diferençias que tenyan por las dhas tierras syenpre los dhos yndios de Canta han tenydo por costunbre de gozar y senbrar las dhas tierras de Quibi y cojer fruto dellas para sy syn que los de Chacalla se ayan quexado fasta agora que los bee traer pleyto sobre ellas y esto rresponde.

T^{o} el dho Xpoval Caxallaupe de la encomyenda de G^{o} de Caçeres. Dixo que dize lo que dho tiene en las preg^{ta} [*sic*] antes desta e que a oydo dezir este dho testigo dende quanto que se sabe acordar a muchos biejos e ançianos de los quales al presente no tiene memoria porque son muertos asy de su pueblo deste dho testigo como desta comarca de la syerra y de los llanos e a sus padres e habuelos tratando e platicando //
/289v/ sobre cosas de los Yngas pasados e otras platicas que entre ellos se ofresçian que antes que vinyesen Yngas en esta trra las dhas tierras de Quibi como dho tiene heran de un señor que llamaban Chunbiquibi que hera yunga e de generaçion de yungas e que hera señor por si el qual dava de trebuto algodon y coca y mayz y otras cosas a un caçique que llamaban Collicapa que rresidia en el valle de Collique e que este señor hera señor deste esta tierra hasta la mar y tenya las orexas largas como yndios orejones del Cuzco e que el dho Chunbiquibi tenya su jurisdiçion distinta y apartada de los yndios de Canta e de sus termynos e no tenya que ver la tierra de los serranos con su tierra e que quando entraron los Yngas en esta tierra a subjetalla el primero Ynga que entro que fue Topa Ynga Yupangue aguelo de Guaynacaba dio guerra al dho Collicapa caçique de Collique y el dho Chunbiquibi que estava en las dhas tierras de Quibi que trebutaba al dho Collicapa fue con sus yndios a socorrer al dho Collicapa contra los Yngas y los Yngas en la batalla que dieron mataron a muchos yndios del dho Collicapa e al dho Collicapa e tanbien muchos yndios del dho señor Chunbiquibi y el dho Chunbiquibe se torno a sus tierras donde estuvo çiertos años fasta que fue acusado que queria hazer trayçion a Topa Ynga y matalle con ponçoña e sabido por Topa Ynga enbio por el dho Chunbequybe y le llebaron atado a el e a sus yndios al Cuzco donde estaba el dho Topa Ynga e por su mandado alli fueron muertos por mandado del dho Topa Ynga Yupangue por la dha traiçion que le querian hazer e que antes que prendiesen al dho Chunbiquybe avian puesto por Ynga Yupangue a los dhos yndios yauyos del rrepartimi^{o} de

Chacalla por mytimaes en las dhas tierras de Quibi y dadoles alli un pedaço de trra en que senbrasen para Ynga e a ellos puestos para chasques y para llebar cargas y para dar avisos e otras cosas del servi° del Ynga / e que alli donde les puso hizo su señal y mojon que oy dia esta alli que sera media legua del pueblo de Quibi donde agora se desamynan los testigos y estan los dhos yndios de Canta e que desde aquel mojon para arriva hazia Canta dexo para trmy° de Canta e por jurisdiçion de la syerra / e del mojon para abajo haçia Lima dexo por de los llanos e yndios yungas e que //

/290r/ despues de muerto Topa Ynga Yupangue subçedio en el senorio Guaynacaba; añadio a los dhos yndios yauyos que ansy abia puesto por mytimaes Ynga mas tierra hazia el dho pueblo de Quivi que llegava hasta este asyento e pu° de Quibi para que alli estubiesen por sus yndios mytimaes como los abia puesto Ynga los quales dhos y°s yauyos del rrepartimi° de Chacalla estubieron por mytimaes muchos años e tpos e que los yndios del rrepartymy° de Canta no osaban hablar de myedo que tenyan de los Yngas e que despues que los Yngas murieron y despues que entraron los xpianos en esta tierra luego enpeçaron los yndios de Canta a tener pasiones e diferençias sobre las dhas tierras de Quibi y a hechar dellas los yndios yauyos del rrepartymy° de Chacalla y las tubieron muchos años que estaban para matarse hasta que vinyeron Nycolas de Rribera el moço encomendero de los yndios de Canta y Fran^{co} de Anpuero encomendero de los yndios de Chacalla y trataron de conçertar los unos yndios y los otros por quitalles de pasiones y el conçierto fue que los dhos yndios de Canta diesen a los de Chacalla dozientas ovejas de la tierra machos y henbras por que dexasen las dhas tierras a los yndios de Canta y a este t° le es notorio por dho de los unos yndios y de los otros que los dhos yndios de Chacalla se contaron [*sic*] con el dho conçierto y llebaron las dhas obejas y dexaron las dhas tierras a los de Canta y este dho t° ha visto que los dhos yndios de Canta las han tenydo e poseydo quyeta e paçificamente syn rrençilla ny contradiçion de los yndios de Chacalla fasta agora que ha entendido que el caçique e yndios de Canta han rreclamado del conçierto e no quyeren pasar por el e piden las dhas tierras / e que lo demas que tiene dho le a sydo a este t° notorio por dho de los dhos biejos e ançianos de sus pasados que dezian avello visto ser y pasar ansy en sus tienpos e an oydolo a sus mayores e mas ançianos que ellos e que de el havia sydo y hera entre ellos cosa publica y notoria e publica boz e fama e comun opinyon e que nunca este t° a oydo dezir cosa alguna en contrario lo qual dixo e rrespondio a esta preg^{ta} por buenas rrazones estandolo la lengua rrazonando e preguntando.

/290v/ T° el dho don Luis Zacalla Chunbi de la encomyenda de Nyculas de Rriba [*sic*] el moço de hedad de 28 o 30 a°s y es amygo de los yndios de Canta porque trata con ellos. Dixo este dho testigo que lo que puede declarar y rresponder a lo en ella cont^{do} es que a oydo dezir por cosa publica y notoria y muy çierta a yndios biejos e ançianos de la syerra de los llanos con quien este dho t° a t^{do} trato e comunycaçion e a sus padres e haguelos que ya son muertos e a otros que son bibos tratando e comunycando e platicando sobre las dhas tierras de Quibi e sobre cossas pasadas que antes que ubiesen Yngas en estas tierras de Quibi heran las dhas tierras de un señor que no se acuerda como le dixeron que se llamaba mas de que dezian que hera yndio yunga natural de Quibi el qual dezian que trebutaba como yunga a los yndios yungas e a un señor que se dezia Collicapa que hera señor de los yungas de la encomyenda del contador Juan de Caçeres y de Collique hasta la mar e que la jurisdiçion de los yungas la tenya apartada de los serranos de Canta y puesto sus trmy°s de yungas que llegaban mas arriba del pueblo que agora llaman de Quybi dos leguas mas alla hasta un pueblo que llaman Chuquicoto e que desde el dho pueblo de Chuquicoto tenya señalado por yndios yungas hasta la mar y desde el dho pueblo de Chuquito [*sic*] hazia Canta hera de la syerra haçia Canta sujetos a Canta y que este dho señor lo avia tenydo y poseido mucho tpo hasta que entro Ynga en esta trra y mato al dho Collicapa e a otros muchos yndios suyos y se entro e apodero en las dhas tierras de Quybi y puso en ellas yndios yauyos por mytimaes para su servi° por que le hiziesen ojotas e para cargas e para chasques e otros aprovechamy°s de Ynga e que estos dhos yndios yauyos heran del rrepartimy° de Chacalla e que entonçes el dho Ynga torno a señalar los trmy°s de los yungas y los de los serranos hazia Canta y puso sus mojones desde media legua e aun no la ay deste pueblo de Quibi haçia Lima puso el dho mojon y senal para que de alli corriese el trmy° de los yungas y se llamasen yungas y desde el dho mojon hazia Canta se llamasen serranos y tomo çierto pedaço de tierra junto al dho //

/291r/ mojon para que los dhos yndios yungas senbrasen alli para Ynga los quales dhos yndios yauyos estubieron puestos en la dha pte y lugar de las dhas tierras de Quybi donde los puso Ynga hasta que murio Ynga e que ansy mysmo moria el señor y avia otro rrey como fue Guascar Ynga y Guaynacaba acreçentaban e yban syenpre acreçentando las dhas tierras de Quybi a los yndios yauyos que ansy avia puesto por mytimaes el Ynga e desta manera se fueron entrando e apoderando los dhos yndios yauyos en las dhas tierras de Quybi fasta hasta [*sic*] que llegaron al dho pueblo de Quibi donde agora se toman estos testigos que hera del dho señor que llamaban Quibi. Todavia quedaron alg°s yndios y estos se contaban y nonbraban por yndios yauyos como los dhos yndios yauyos de Chacalla por no desanparar sus tierras y p° dellas y que desta manera tubieron las dhas tierras los dhos yndios yauyos de Chacalla hasta que entraron los españoles en este rreyno aunque antes que entrasen syenpre dizen que trayan diferençias los yndios de Canta con los de Chacalla sobre las dhas tierras y estaban en arma para se matar cada dia e que despues de entrados los dhos españoles en este rreyno estando como estaban los dhos yndios yauyos en las dhas trras de Quibi en el camyno de Canta por donde havian de benyr camynando los españoles se fueron la mytad de los dhos yndios yauyos de miedo de los españoles al pu° de Chacalla y dexaron las dhas tierras de Quibi y quedaron la otra mytad de los dhos yndios yauyos en ellas e que despues quando vino el presidente Gasca vinyeron a las dhas tierras de Quybi Nyculas de Rribera el moço encomendero de los yndios de Canta y Fran^{co} de Anpuero encomendero de los dhos yndios de Chacalla y trataron de conçertar los unos yndios y los otros para que no tubiesen pasiones en que diesen los yndios de Canta a los de Chacalla dozientas ovejas de la tierra mytad machos y mytad henbras por que dexasen las dhas tierras y este t° entendio de yndios de los dhos rrepartimy°s que se las avian dado porque desde entonçes vido este dho testigo que los dhos yndios de Chacalla yauyos se fueron a sus trras //

/291v/ y dexaron las dhas tierras de Quybi para los dhos yndios de Canta los quales se entraron en ellas y las han tenydo y poseydo quyeta e paçificamente syn contradiçion de los dhos y°s de Chacalla sy no es agora que ha visto y entendido que los dhos yndios yauyos se han arrepentido e no quyeren passar por el dho conçierto e traen pleito lo qual todo que dho tiene le a sydo a este dho t° pu° y not° por aberlo oydo dezir a los dhos sus pasados los quales dezian que ansy abia sydo y pasado y lo avian visto ellos ser y pasar y oydo dezir a otros biejos e ançianos e a sus padres e pasados e que nunca avian visto ny oydo dezir lo contr^{o} dello e que dello havia sydo y hera pu^{co} y not^{o} y publica boz y fama e que cree este t^{o} que sy lo contrario oviera sydo y pasado tanbien se lo hubieran dho y ellos lo supieran por ser todos tan comarcanos a los dhos pueblos de Canta y lo mysmo a los de Chacalla e que trataban unos con otros y lo mysmo con los dhos yndios de Quibi lo qual dixo y rrespondio en sustançia rrazonandolo entre el dho don M^{yn} Lengua y este t^{o} a lo que la lengua declaro.

T^{o} el dho Alonso Bro sujeto a Canta / no sabe la hedad que t^{e} acuerdase de Guaynacaba y le conosçio y vale ynteres. Dixo que lo que desta preg^{ta} save es que oyo dezir al dho su padre que se llamaba Paucar que antes que vinyese Ynga en esta trra los pueblos de Canta y de Collique heran señores por si esentos e que no tenyan myedo a nadie que los avia de castigar e que en yendo yndios de la tierra de Canta a rrescatar hazia la mar entre los yndios yauyos luego procuraban de matarlos y los mataban y que lo mysmo hazian los yndios de Canta sy bian yndios de los yauyos en su tierra e que por hevitar esto los dhos yndios de Canta avian hido a verse con el señor de Collique que llamaban Collicapa para hazerse amygos e que no se matasen los yndios e hido el dho Collicapa les avia rresçebido bien e avian tratado de amystades entre ellos e que los yndios de hazia la mar subiesen a la sierra a contratar y los de la syerra a los llanos syn que les hiziesen nyngun daño e que se amojonasen y senalasen //
/292r/ entre ellos los trmy^{o}s lo que abia de ser de los yungas hazia Collique y lo que avia de ser de Canta hazia la sierra e ansy el dho Collicapa se avia hido con el caçique e yndios de Canta hasta las tierras que llamavan de Quibi donde estavan yndios yungas que serbian al dho Collicapa que tenyan un señor que llamaban Chaumecaxa e avian senalado por termy^{o}s de Canta desde un poco abajo del dho pueblo de Quybi hazia Collique en un çerro que llaban [*sic*] Judcunga para que de alli corriese el termyno de la syerra para que fuese de Canta y lo mysmo el dho Chaumecaxa quedaba dentro de los trmy^{o}s e mojones que ansy se senalaban aunque no save este dho testigo sy el dho Chaumecaxa señor del dho pueblo de Quibi servia e trebutaba a Canta o si hera sobre si ny lo ha oydo dezir mas de que oyo dezir al dho su padre que despues de amojonados los dhos trmy^{o}s los dhos yndios de Canta senbraban en las dhas tierras de Quybi a bueltas con los yndios del dho Chaumecaxa caçique del dho pueblo de Quybi e que desta manera los dhos yndios de Canta tubieron y poseyeron las dhas tierras de Quybi con los demas yndios yungas de Collique e aunque dize e torna a dezir este dho t^{o} sy se lo dixo ansy el dho su padre que luego los dhos yndios de Canta abian luego senbrado las dhas tierras de Quybi mas de que dezia que las tenyan por suyas despues que se puso el dho mojon e que tanbien oyo dezir al dho su padre que despues de venydos los Yngas Topa Ynga Yupangue a conquistar esta tierra que avyan conquistado e quystado [*sic*] para sy la tierra de Chacalla e que despues de conquistada se habian buelto al Cuzco e que en esta sazon avia muerto el dho Topa Ynga Yupangue e havia subçedido Guaynacaba y en este tienpo avian hallado que un caçique del pueblo de Chacalla que se dezia Llungaype que no se acuerda bien como le dixeron que se llamaba ny el pueblo tanpoco que el y sus yndios hazian çiertas hechizerias a su guaca rrogandole que se muriese Guaynacaba e que no hubiesen Yngas por que no les hiziesen servir y que siendo avisado dello el dho Guaynacaba avia enbiado un horejon suyo cap.^{t} para que castigase el dho caçique y los que hazian las dhas hechizerias e venydo que fue hizo pesquysa e hallo que hera verdad lo que dezian y luego mato al dho caçique //
/292v/ e a sus yndios que no dexo syno los yndios biejos y las mugeres que estaban preñadas a los quales no quiso que estubiesen en la dha tierra de Chacalla por que no se acordasen de la dha guaca y tornasen a caer en aquel yerro los saco de la dha tierra de Chacalla y los traxo al dho valle de Quibi y los abia puesto y poblado en el mysmo termy^{o} que llaman Judcunga adonde el dho Collicapa señor de Collique avia señalado por tierra para Canta / e que estas dhas mugeres e nynos aviendo creçido y multiplicado se avian entrado y tomado mas tierras de las que el dho orejon les avia dado quando las poblo e que desta manera e vinyendo cada dia yndios de la dha tierra de Chacalla y so color de çierta çerimonya que llamaban capacocha e rrogatibas que hazian por los Yngas se abian entrado en ellas e senbradolas hasta que entraron en esta tierra los espanoles e que este dho t^{o} hera obejero e panpacamayo de los Yngas e rresydia en Bonbon al tienpo que entraron los españoles andavan rrancheando los yndios y el ganado; este dho t^{o} se huyo y se vino a su tierra y nunca vio las dhas tierras de Quybi ny paso a ellas ny oyo ny entendio a lo que se acuerda otra cosa mas de lo que dho tiene lo qual le dezia el dho su padre porque lo habia visto ser y pasar en su tienpo e oidolo dezir a otros mas biejos que dezian que lo avian visto y que nunca avian visto ny oydo dezir lo contr^{o} e que entre ellos avia sydo y hera publico y notorio y publica boz y fama e que sy otra cosa fuera e pasara cree este dho t^{o} que el dho su padre se le dixera lo qual declaro por estas propias rrazones en sustançia aviendo sydo preg^{do} particularmente por lengua del dho don M^{yn} ynterpetre que dixo averlo declarado ansy.

4 - Yten sy saven etc. que antes que vinyesen xpianos a esta trra se tenya por costumbre en ella entre los naturales de hazer çiertas rrogatibas o plegarias para los rreyes e sus señores a que llamaban ellos capacocha lo qual hera que mataban çiertas obejas y les sacaban la sangre y la hechaban en unas ollas o mates e hiban con aquello en proçision y sy alg^{o} dellos derramaba una gota de la dha sangre o se la //
/293r/ hazian derramar al yndio que asy la derramaba o la hazia derramar lo enterraban en el p^{o}pio lugar y lo mataban por ello y esto hera ley ynbiolable entre ellos. Digan lo que saben.

T^{o} el dho don Luis Zacalla Chunbi. Dixo este dho t^{o} que lo en ella cont.^{do} ansy y de la forma y manera que la dha preg^{ta} lo dize y declara lo a oydo dezir este dho t^{o} a sus mayores e mas ançianos e a sus padres e aguelos los quales dezian que ellos ansy lo habian visto ser y pasar en sus tienpos e oidolo dezir a sus mayores e mas ançianos que dezian que ellos ansy lo habian visto e oidolo dezir a los suyos en los suyos e que nunca habian visto ny oydo otra cosa çerca dello en contr^{o} y este dho t^{o} ansy lo cree por ser las personas que esto trataban yndios comarcanos a los dhos pueblos de Canta y Challa e a las dhas tierras e rrio de Quybi e que syenpre trataban unos con otros lo qual rresponde por la dha lengua.

Tº el dho Xpoval Caxallaupe. Dixo que sabe la dha preg^{ta} segun y como en ella se cont.^{e} y declara porque este dho tº lo ha oydo dezir lo cont.^{do} a sus padres e aguelos e pasados e a otros biejos e ançianos que este tº y ellos de cuyos nonbres no se acuerda que son ya muertos tratando e platicando en las cosas de los Yngas y en las dhas capacochas y ofresçimyºs y en otras cosas a los quales les oyo dezir que entre los dhos yndios se usava e acostunbrava hazer lo cont.^{do} en la dha preg^{ta} por la forma que ella lo dize y declara y que hera cosa usada y guardada como las leyes de agora de los xpianos y esto rresponde.

Tº el dho Cacha yndio de la encomyenda de Her^{do} de Montenegro. Dixo que lo que de la preg^{ta} save es que oyo dezir al dho su padre que en tpo de los Yngas pasados salian del Cuzco yndios yngas con çiertas çeremonyas que dezian que hera la pruçision del Ynga como agora dizen los xpiaºs quando ban con la pruçision rresando e que la llamaban copacocha [*sic*] los yndios e que hiban con ella por todas las tierras del Peru rrogando por el Ynga e que no save sy hiban buscando tierra ny para que otro hefeto ny se acuerda de mas y esto rresponde / e dixo que no queria dezir mentira porque este dho tº tiene mucho myedo de mentir porque estotro dia estando en Lima vio que sacavan a un xpiano los dientes porque havia mentido e que si este dho tº myntiese que lo quemaran a el.
Ay otros tºs de oydas.

/293v/ 5 - Yten sy saven etc. que quando hazian la dha capacocha hiban los de un pueblo hasta los trmyºs del otro con çiertas çerimonyas y hasta donde llegavan lo ponyan por testimonyo [*sic*] de los que llevaban la dha capacocha y lo senalaban por suyo.

Digan lo que saben.

Tº el dho don Pº caçique de Guancayo. Dixo que lo en la dha preg^{ta} contenydo como en ella se cont.^{e} y declara lo ha oydo dezir este dho tº a sus padres e pasados e a otros biejos e ançianos que dezian averlo visto en tpo de los Yngas ser y pasar e acostunbrar lo en la dha pregunta contenydo como en ella se cont.^{e} y declara los quales lo dezian tratando çerca de sus rritos e çerimonyas antiguas e que dezian que lo mysmo avian ellos oydo dezir a sus pasados y dello es cosa muy publica y notoria entre los yndios biejos desta comarca que lo vieron y entre los que lo oyeron dezir como este dho tº y esto rresponde.

Tº el dho don Luis Zacalla Chunbi. Dixo que lo en la dha preg^{ta} cont.^{do} segun y como en ella se cont.^{e} y declara lo a oydo dezir este dho tº a sus mayores e mas ançianos que dezian avello visto ser y pasar ansy e acostunbrar de la dha forma y manera que la dha preg^{ta} lo dize y aun dezian que ellos en sus tpos ansy lo habian visto e oydo y entendido a los suyos biejos e ançianos que dezian abian visto ser y pasar ansi e oydo a los suyos en los suyos e que dello hera publico e notorio entre los yndios desta comarca y esto rresponde por la dha lengua en buena rrazon.

Tº el dho Xpoval Caxallaupe. Dixo que lo en ella cont.^{do} a oydo dezir este dho tº que se usava e acostunbrava entre los yndios en tienpo de los Yngas como la dha preg^{ta} lo dize y declara lo qual a oydo dezir e tratar e platicar a sus pasados e a otros biejos e ançianos que dezian que ellos ansy lo havian visto usar e acostunbrar en sus tienpos y lo habian oydo dezir a sus pasados que lo habian visto usar e acostunbrar en sus tienpos y lo avian oydo dezir a sus mayores que lo havian visto hazer lo mysmo en los suyos e que dello havia sydo y hera publico y notº e publica boz y fama e comun opinyon y esto rresponde.
Ay mas tºs que dizen que asy lo oyeron dezir.

/294r/ 6 - Yten si saven etc. que al tpo que Guaynacap hiba a Quyto hizieron los dhos yndios de Chacalla la dha capacocha y con la sangre susodha llebandola en unas ollas o mates fueron por la pte donde estan las dhas tierras de Quybe y con las dhas çirimonyas y los dhos yndios de Canta queriendolos hir a detener no osaron por no derramar la dha sangre e yncurrir en la dha pena de muerte e asy con la dha color yntentaron de se quedar y se quedaron con una pte de las dhas tierras de Quybe diziendo que hasta ally les pertenesçia por aver llegado con la dha capacocha hasta las dhas tierras de Quibe en la pte donde llegaron y sobre que es este pleyto lo qual puede ser la tierça parte de las dhas tierras poco mas o menos. Digan lo que saben.

Tº el dho don Luis Zacalla Chunbi. Dixo que dize lo que dho tiene en las preguntas antes desta e que a un yndio biejo natural de Guarabil pueblo deste tº que se llamaba Tunga Caxa le oyo este dho testigo dezir que avia visto que en tpo del dho Guaynacaba e quando hiba a Quyto havian venydo los dhos yndios yauyos de Chacalla con la dha capacocha e sangre por las tierras de Quybi hazia el dho pueblo de Quibe aperçibiendo a los dhos yndios de Canta que se guardasen de la dha capacocha porque sy la derramaban avian de morir e que los dhos yndios de Canta e de Quibi que estaban en el dho pueblo de Quibi se apartaban y les dexaban entrar y no habian fho rresistençia por que no los matasen y que desta manera avian entrado hasta el dho puº de Quibi con la dha capacocha e que esto no lo a oydo dezir este dho testigo a otra persona syno al dho yndio que dho tiene ny sabia mas desta preg.^{ta}

Tº el dho Xpoval Caxallaupe. Dixo que lo que desta preg^{ta} save es que antes que muriese su padre deste tº que se llamaba Parianquybi le oyo dezir e lo mysmo a otros muchos yndios biejos e ançianos que son muertos de cuyos nonbres no se acuerda que en tienpo de Guãsçar Ynga que fue el postrero señor //
/294/ de los yndios los dhos yndios yauyos se habian entrado desde el dho mojon de Guaynacapa e le havia puesto haziendo la dha capacocha con yndios de Guascar Ynga orejones con la dha sangre hasta mas adelante deste pueblo de Quybi donde agora estan los dhos yndios de Canta e hiban con la dha sangre como dize la preg^{ta} e que los dhos yndios yauyos e horejones hiban diziendo: "parta parta no derrames la sangre que moriras que es el ofresçimyº del Ynga para el Diablo para que de buena vejez y le haga gran señor e le de muchos yndios e rriquezas e comydas e guarde sus yndios";e que los dhos yndios de Canta de myedo de la dha capacocha e por

no derramar la sangre por que los matasen los dexaban entrar e apoderarse en la tierra y no se lo rresystian e que desta manera yban ganando tierra e apoderandose en ella lo qual todo les oyo dezir este t^{o} muchas e diversas vezes tratando sobre la dha capacocha.

T^{o} el dho don Fran^{co} Caxa Guaranga sujeto a Canta de hedad de çien años. Dixo que lo que save desta preg^{ta} es que save por cosa publica y not^{a} e averlo oydo dezir a yndios biejos que dezian avello visto que en tpo de Topa Ynga Yupangue que fue el primero de los Yngas los dhos yndios de Chacalla entraron por las dhas tierras de Quybi con los dhos orejones de Ynga y que despues entraron dos vezes en tpo de Guaynacaba y llebaban la dha capacocha e yvan ganando e apoderandose cada dia mas en las dhas tierras y que despues en tpo de Guascar Ynga entraron con la dha capacocha los dhos yndios yauyos fasta mas adelante de las casas del dho pueblo de Quybi un tiro de arcabuz e que queriendo pasar de alli un yndio del dho pueblo de Quybi subjeto a Canta dixo que heran aquellas sus tierras e quyso defender que no entrasen o que no pasasen de alli e derramo un poco de sangre de la dha capacocha e por esta causa luego alli le mataron y le enterraron en el propio lugar donde derramo la dha sangre e que tanbien a oydo dezir que lo mysmo avian fho a otros quatro yndios del dho pueblo de Quibi que defendian que la dha capacocha no entrase en sus tierras e que este dho testigo entendio y tiene entendido que no osaban hablar los dhos yndios de Canta ny rresistilles la entrada con la dha capacocha ansy por no yncurrir en la dha pena de muerte como porque los dhos yndios yauyos heran muy queridos e pribados del Ynga e por no lo henojar. Ay mas t^{o}s que lo saben por pu^{co} y not^{o} de oydas.

/295r/ 7 - Yten si saven etc. que con la dha color de la dha capacocha los dhos yndios de Chacalla pusieron mytimaes en la pte de las trras de Quibi donde llegaron con la dha capacocha syn tener otro derecho alguno a las dhas tierras. Digan lo que saben.

T^{o} el dho Carvabilca yndio ynfiel no supo la hedad que tiene mas de que se acuerda de Guaynacapa; es sujeto a Canta. Dixo que dize lo que dho t^{e} en las preg^{tas} antes desta e que save e vido en su tpo e oyo dezir al dho su padre que hasta llegaban los dhos yndios yauyos de Chacalla con la dha capacocha haseñalaban e rrayaban alli por su termyno luego ponyan alli yndios que señalasen aquella tierra que acresçentaban y esto hera cosa pu^{ca} y not^{a}.
No ay mas en esta.

8 - Yten si saven etc. que despues que el dho Guaynacapa murio y antes que entrasen los xpianos en este rreyno o quando entraron los dhos yndios de Chacalla [*sic*] se salieron los dhos yndios de Chacalla y los mytimaes que avian puesto por mandado de Ylla Topa Ynga hijo de Guaynacapa y luego los dhos yndios de Canta tornaron a las dhas tierras de Quibi ny mas ny menos que de antes las tenyan las quales han tenydo y poseido desde entonçes hasta agora. Digan lo que saben.

T^{o} el dho don P^{o} caçique de Guancayo de la encomyenda de G^{o} de Caçeres. Dixo que dize lo que dho tiene en las preg^{tas} antes desta e que tratando con sus padres e pasados e con otros biejos e ançianos desta comarca que algunos dellos son muertos e platicando sobre cosas antiguas e sobre las dhas tierras de Quybi e otras platicas que entre ellos trataban les oya e oyo dezir que despues de entrados los españoles en este rreyno se salieron de las dhas tierras de Quibi la mytad de los yndios yauyos del rrepartimy^{o} de Chacalla e quedaron la otra mytad pero que por eso nunca los yndios yauyos avian dexado las dhas tierras ny perdido su posesion ny tanpoco oyo dezir ny este t^{o} entendio ny vio despues que se save acordar que los yndios de Canta se entrasen en las dhas tierras de Quybi hasta que hizieron conçierto entre los dhos yndios de Canta y de Chacalla por çiertas obejas y cree este t^{o} que sy los dhos yndios de Canta se hubieran entrado en ellas de antes y tenydolas e poseydolas este t^{o} lo viera e supiera e entendiera e oviera oydo dezir y no pudiera ser menos por ser tan comarcano a los dhos yndios de Canta y çercano a sus parçialidades que podra aver una legua de su trra a las dhas dhas [*sic*] tierras de Quybi y esto rresponde.

/295v/ T^{o} el dho Xpoval Caxallaupe. Dixo que despues que entraron los españoles en este rreyno se alço Mango Ynga y en aquella sazon bido este dho testigo que pte de los dhos yndios yauyos se fueron a sus pueblos de Chacalla e dexaron las dhas tierras de Quibi e pte dellos quedaron todavia en ellas e no se acuerda este dho testigo aber visto oydo ny entendido que los dhos yndios de Canta entrasen en las dhas tierras de Quibi sy no es despues que dieron por el dho conçierto las dhas ovejas a los de Chacalla e que sy ubieran entrado en ellas este dho testigo lo ubiera visto oydo y entendido y no pudiera ser menos por ser tan çercanos a los dhos pueblos e tierras de Quibi e tener trato e comunycaçion con los dhos yndios de Canta e acordarsele a este dho t^{o} como se le acuerda de quando entraron los españoles en este rreyno y esto rresponde.

T^{o} el dho Fran^{co} Caxa Guaranga sujeto a Canta. Dixo que dize lo que dho t^{e} en las preguntas antes desta en que se afyrma e que quando entraron los españoles en este rreyno alg^{o}s de los dhos yndios yauyos se yban a Chacalla de myedo de los españoles y porque estaban en el camyno rreal por donde pasaban e que quedaban otros en las dhas tierras y las senbraban como de antes y que despues que se hizo el conçierto de las ovejas los dhos yndios de Canta entraron en las dhas tierras de Quibi y esto rresponde.

T^{o} el dho Carvabilca yndio ynfiel sujeto a Canta. Dixo que dize lo que dho tiene en las preguntas antes desta e que despues de venydos los espanoles a esta tierra y poblado el marques los dhos yndios de Canta por fuerça hecharon de las dhas tierras de Quybi todos los yndios de Chacalla yauyos y se fueron a sus tierras y los dhos yndios de Canta se entraron en las dhas tierras de Quibi y que despues queriendo los dhos yndios de Chacalla tornar a las dhas tierras de Quibi en tpo del presidente los dhos yndios de Canta se lo defendian y desta manera ynbiaron a llamar a los dhos sus encomenderos para que los hiziesen amygos y luego hizieron el dho conçierto de las dhas ovejas y los dhos yndios de Chacalla se fueron y nunca mas hablaron en las dhas tierras hasta agora. No ay mas.

9 - Yten si saben etc. que por m^{do} del presydente Gasca y por diferençias que tenyan sobre las dhas tierras los dhos yndios de Chacalla se conçertaron con los dhos yndios de //
/296r/ Canta en que les diesen dozientas obejas por que se apartasen del pleyto y diferençia de las dhas tierras de Quybi. Digan lo que saben.

T^{o} el dho don P^{o} caçique de la encomyenda de G^{o} de Caçeres Dixo que lo que save della es que supo oyo y entendio e aun vido pte dello este t^{o} que en el tpo del presidente Gasca tenyendo los dhos yndios de Canta diferençias sobre las dhas tierras con los de Chacalla tenyendolas y poseyendolas los dhos yndios yauyos de Chacalla trataron de conçierto en que los dhos yndios de Canta prometieron e a lo que este t^{o} oyo dezir entregaron a los dhos yndios de Chacalla dozientas cabezas de ganado çiento henbras y çiento machos por que dexasen las dhas tierras y que el dho conçierto se hefetuo en presençia de Nyculas de Rribera el moço encomendero de los yndios de Canta y Fran^{co} de Anpuero encomendero de los yndios de Chacalla e que lo susodho es cosa muy publica y notoria entre todos los yndios desta dha comarca porque despues que se hizo el dho conçierto los dhos yndios yauyos de Chacalla dexaron las dhas tierras y los dhos yndios de Canta se entraron en ellas y las tobieron y poseyeron hasta agora que bee que se trata este dho pleyto y esto rresponde desta dha preg.^{ta}

T^{o} el dho Cacha yndio del rrepartimy^{o} de Montenegro. Dixo que oyo dezir a muchos yndios comarcanos a los dhos rrepartimy^{o}s de Chacalla y de Canta e a alg^{o}s yndios de Canta que despues que entraron los españoles en este rreyno los dhos yndios de Canta Guaman Yaure caçique y don Fran^{co} Marcapoma caçique de Chacalla trayan diferençias e andavan syenpre rrinendo sobre las dhas tierras e sobre que los dhos yndios de Canta querian hechar dellas a los yndios de Chacalla yauyos diziendo que heran suyas e que despues que vino el Licen.^{do} de La Gasca presidente que fue destos rreynos aviendo venydo a su notiçia las diferençias que los dhos Guaman Yaure caçique de Canta y Marcapoma caçique de Chacalla trayan sobre las dhas tierras enbio para que los conçertasen a Nyculas de Rribera el moço encomendero de los yndios de Canta e a Fran^{co} de Anpuero encomendero de los yndios de Chacalla los quales le dixeron a este dho t^{o} //
/296v/ que avian conçertado los dhos caçiques en esta manera que los yndios de Canta diesen a los de Chacalla duzientas obejas por rrazon de las dhas tierras e ansy oyo dezir este dho testigo ansy mysmo que se las havian dado los dhos yndios de Canta a los de Chacalla y que los de Chacalla avian llebadolas a sus trras y los de Canta se avian entrado en las dhas tierras y las abian tenydo y poseido syn que los dhos yndios de Chacalla se quexasen hasta agora que les vee andar en pleito e a oydo dezir que no quieren pasar por el dho conçierto. No ay mas.

10- Yten si saven etc. que en cunplimy^{o} del dho conçierto los dhos yndios de Canta dieron a los dhos yndios de Chacalla dozientas cabezas de ovejas las quales les dieron y entregaron a ellos mesmos e a su caçique. Digan lo que saben.

T^{o} el dho don P^{o} caçique de Guancayo de la encomyenda de G^{o} de Caçeres. Dixo que dize lo que dho tiene en las preguntas antes desta e que a un hermano deste t^{o} que es ya muerto que se llamaba Poma Chunbi le a oydo dezir este dho t^{o} que los dhos yndios de Canta entregaron al dho caçique de los yauyos que se llamaba don Fran^{co} Marcapoma e a Guaman Yabre las dhas cabezas de ganado del dho conçierto y que ellos las avian rresçebido e llebadolas a sus pueblos y esto rresponde.

T^{o} el dho Carvabilca no xpiano subjeto a Canta. Dixo que dize lo que dho tiene en las preg^{tas} antes desta e que vido que el dho don Fran^{co} Marcapoma caçique de Chacalla no queria hazer el dho conçierto ny queria tomar las dhas ovejas ny las tomo y el dho Fran^{co} de Anpuero vido este dho testigo que las rresçibio y las mando llevar a Lima y esto rresponde. Ay mas t^{o}s que dizen abello oydo a yndios de Canta.

11- Yten si saven etc. que en todas las dhas tierras de Quivi no ay beynte anegadas de senbradura o quando mas veynte e çinco y todas las mas dellas son andenes y tierra pedregosa e a justa e comun estimaçion cada hanega de senbradura puede valer hasta veynte p^{o}s. Digan lo que saben.

T^{o} el dho don P^{o} caçique de Guancayo de la encomyenda de G^{o} de Caçeres. Dixo que save porque ha visto las dhas tierras de Quivi que podran tener en todas ellas hasta veynte e çinco anegas de senbradura poco //
/297r/ mas o menos y que lo sabe porque sienbra y coxe este t^{o} y lo entiende y ha visto que es tierra floxa e flaca e mucha pedregosa e de andenes como dize la preg.^{ta} e que no save lo que podra valer cada anega de senbradura y esto rresponde.

T^{o} el dho don Luis Çacalla Chunbi. Dixo que a su paresçer deste t^{o} las dhas tierras de Quybi podran tener de sementera de trigo e mayz hasta tres o quatro hanegas porque son todas andenes e cascaxales e tierra de mucha piedra e que no sabe lo que podra valer cada anegada de senbradura ny sabe mas desta preg.^{ta}

T^{o} el dho Xpoval Caxallaupe. Dixo que a visto muchas vezes las dhas tierras de Quybi e que lo mas dellas es piedra e andenes e que al paresçer deste t^{o} podra aver en ellas sy se senbrasen diez anegas y media de senbradura de mayz o trigo e que no sabe lo que podra valer cada hanegada.

T^{o} el dho don Fran^{co} Arcosna de la encomyenda de Canta Dixo que a su paresçer las dhas tierras de Quybi podran hazer de senbradura hasta diez anegadas y no mas porque es tierra muy floxa y pedregosa y hecha andenes e que no save lo que podra valer cada hanegada porque en esta tierra no se arriendan las tierras.

Tº el dho Atoco yndio sujeto a Canta. Dixo que ha visto las dhas tierras de Quibi e que son todas andenes e piedras y cascajo y que le paresçe que podra aver diez hanegadas de senbradura poco mas o menos e que el preçio que vale cada hanegada no lo save ny lo entiende.

Tº el dho don Juan Caxayaure del rrepartimyº de Canta. Dixo que le paresçe a este dho testigo que en las dhas tierras de Quibi no podra aver de diez anegadas de tierra arriba poco mas o menos porque este tº las ha andado y save que son la mas dellas [*sic*] piedra y cascaxo y andenes y tierra floxa y que no save a como podra baler cada hanegada y esto rresponde.

Tº el dho don Fran^{co} Caxa Guaranga de la encomyenda de Canta. Dixo que save e a visto que las dhas tierras de Quibi son rruynes e todas andenes y cascaxo e que no son buenas para sementera de trigo ni mayz e que son buenas para coca e papas e otras legunbres y esto rresponde y lo demas en la dha preg^{ta} contenydo dize que no lo sabe. Ay mas tºs de Canta que dizen son rruynes y pocas.

12- Yten si saben etc. que las chacarras e tierras sobre que es este pleyto e diferençia puede ser la terçya pte de la dha tierra o menos que fue hasta donde llegaron con la dha //
/297v/ capacocha los dhos yndios de Chacalla. Digan lo que saben.

Tº el dho don Luis Zacalla Chunbi. Dixo que en todas las dhas tierras de Quibi podra aver de largo poco mas de media legua y que de anplo es muy angosta. No ay mas en esta.

13- Yten si saven etc. que cada cabeza de ganado de obejas de las que dieron los dhos yndios de Canta a los dhos yndios de Chacalla valian seys pºs e mas a justa e comun estimaçion al tpo que se las dieron. Digan lo que saben.

Tº el dho don Pº caçique de Guancayo de la encomyenda de Gº Caçeres. Dixo que save que en el tpo que se hizo el dho conçierto una obeja grande macho con henbra sy hera grande seys pºs y si hera mediana çinco pºs y sy hera mas pequena quatro pºs esto a justa e comun estimaçion porque este tº ha tenydo e tiene ovejas que ha conprado a este dho preçio a la dha sazon para comer y esto rresponde a esta dha preg.^{ta}

Tº el dho don Luis Zacalla Chunbi. Dixo que sy las ovejas que los yndios de Canta dieron a los yndios de Chacalla por las dhas tierras de Quibi heran grandes que valia cada una seys pºs e si heran medianas valian çinco pºs cada una e sy heran mas chicas valian a quatro e a tres pºs.

Tº el dho Xpoval Caxallaupe. Dixo que baldrian las dhas obejas cada cabeza unas a seys pºs e otras a çinco e otras a quatro.

Tº el dho Cacha yndio. Dixo que le paresçe que podrian valer cada cabeza de las dhas ovejas henbras çinco pºs por ser henbras y los machos a quatro pºs y que si heran pacos baldrian a tres pºs y no mas y esto rresponde.

Tº el dho don Fran^{co} Arcosna natural de Canta. Dixo que bien balia cada cabeza de obeja machos y henbras unas con otras de las que el dho Guaman Yaure caçique de Canta dio al dho Marcapoma caçique de Chacalla por las dhas tierras de Quibi seys y siete pºs porque heran buenas y creçidas y este dho tº los diera por ellas en aquel tpo y esto rresponde. Ay mas tºs de Canta deste tenor/

14- Yten si saven etc. que que al tienpo [*sic*] que hizieron el dho conçierto los dhos yndios de Chacalla con los dhos yndios de Canta valian mucho menos las tierras que agora valen //
/298r/ un terçio y mas por darse como se dan agora mas a la labor que entonçes. Digan lo que saben.

Tº el dho don Pº caçique de Guancayo. Dixo que save e ha visto que antes e al tienpo que se hiziese el dho conçierto las dhas tierras de Quybi valian tanto como el dia de oy valen porque entonçes no tenyan dellas tanto aprovechamyº ny las abian tanto menester como agora y esto es cosa conosçida e aberiguada ansy en las dhas tierras de Quibi como en otras desta comarca que cada dia van en cresçimyº e valor y esto rresponde.

Tº el dho don Luis Zacalla Chunbi. Dixo que a lo que ha entendido las tierras valen cada dia mas en este rreyno e que ansy le paresçe que quando se hizo el dho conçierto de las dhas tierras de Quibi por las dhas obejas podian valer menos que agora abnque entre los yndios no se paresçe tanto como entre españoles.

Tº el dho don Fran^{co} Arcosna. Dixo que en tierra de Lima y donde ay xpianos valen agora mas las tierras que no agora ha diez años y que entre los yndios no creçe tanto el valor a lo menos en las dhas tierras de Quybi. Ay mas tºs que dizen valen agora mas por aver mas gente que al tpo del conçierto.

15- Yten sy saben etc. que la causa por que hizieron el dho conçierto los dhos yndios de Canta con los dhos yndios de Chacalla en las dhas tierras de Quibi hera y fue porque las dhas tierras de Quibi heran y fueron de sus pasados y suyas y por quytarse de aquella molestia. Digan lo que saben.

Tº el dho don Pedro de la encomyenda de Gº de Caçeres Dixo que dize lo que dho tiene en las preguntas antes desta y cree este dho tº que por rrazon de aber sido las dhas tierras de Quibi antes que vinyesen los Yngas de los dhos yndios de Canta conprarian las dhas tierras porque lo mysmo hiziera este tº sy ubieran sydo suyas y de sus pasados y esto rresponde.e

Tº el dho Cacha yndio de la encomyenda de Herdo de Montenegro. Dixo que este dho tº a oydo dezir a los dhos yndios de Canta que antes que vinyesen los Yngas a esta tierra las dhas tierras de Quibi heran suyas y que por esta rrazon de ser suyas y de sus pasados avian hecho el dho conçierto con los dhos yndios yauyos de Chacalla y dadoles las dhas obejas por las dhas tierras y por quytarse de pleytos e diferençias e que esto como dho tiene lo ha oydo dezir a yndios del dho puº de Canta.

/298v/ Tº el dho Atoco yndio de Canta. Dixo que le paresçe e a entendido de los dhos yndios de Canta / e aun por pte deste tº dize que save que sy se hizo el dho conçierto y dieron las dhas obejas fue por no dexar a los yndios de Chacalla las tierras que heran de los yndios de Canta y las avian ganado sus pasados y por quytarse de pasiones.

Tº el dho don Juº Caxcha Yaure de la parçialidad de Canta. Dixo que este tº tiene entendido que los dhos yndios de Canta sy hizieron el dho conçierto y dieron a los dhos yndios yauyos las dhas ovejas fue por sacar las dhas tierras que heran suyas y las avian ganado sus pasados y por quytarse de pleyto y diferençias con los yndios yavyos.
Ay mas tºs de Canta que dizen lo mysmo.

16- Yten si saven etc. que todo lo susodho es publico y notorio y publica boz y fama.

El Licendo de Leon Franco de la Torre
Rratificanse los testigos.

Relaçion de mas probança por las preguntas
añadidas por la pte de los yndios de Canta.

---

1 - Yten si saven etc. que en tpo de Guascar Ynga que fue señor destos rreynos despues de Guaynacapa se mandaron hazer dos capacochas de las susodhas una para su muger e otra para el e trayendo los dhos yndios de Chacalla la dha capacocha e vinyendo a labar la sangre al dho rrio de Quybi los dhos yndios de Chacalla se entraron en lo mejor de las dhas tierras de Quibi donde agora esta senbrado de coca e pacays e porque se entraban en las dhas tierras los dhos yndios de Chacalla un yndio que se dezia Chuqui Guaranga rrenpujo a bno de los dhos yndios que traya la dha capacocha diziendo que no se entrasen por su tierra e porque le derramo parte de la dha sangre ahogaron el dho yndio que hera de la parçialidad de los de Canta e le enterraron junto a un molle que alli esta. Digan lo que saben.

Tº el dho don Pº caçique de Guancayo. Dixo que dize lo que dho tiene en las pregtas del primer ynterrogº / y demas y aliende dello dixo que oyo dezir por cosa publica y notoria a biejos e ançianos sus pasados que en tienpo del dho Guascar Ynga se hizieron las dhas //
/299r/ dos capacochas una para Ynga e otra para su muger e que trayendo los yndios de Canta la dha capacocha hasta su mojon la tomaron los yndios de Guancayo rrepartimyº de Gº de Caçeres donde este dho tº tiene su tierra y la llebaron la dha capacocha hasta la mar y su padre e aguelo deste tº le dixeron que avian hido con ella y la ofresçieron en la mar por Ynga como se solia hazer en sus plegarias y esto rresponde a esta dha pregunta.

Tº el dho don Luis Çacalla Chunbi. Dixo que abra muchos años que este tº oyo dezir tratando sobre las dhas capacochas a un yndio que es ya muerto que se llamaba Tunga Caxa natural del pueblo de Guaravi tierra deste tº que en tienpo del dho Guascar Ynga se avian hecho las dhas dos capacochas una para Guascar Ynga y otra para su muger y dezia el dho yndio que el lo havia visto e que lo demas que dize la preg.ta lo a oydo dezir este dho tº a un yndio que se llama Chunbiquibi que es de los yºs sujetos a Canta e que le ba ynteres en este pleito e que dize lo que dho tiene en las preg.tas antes desta y esto rresp.e

Tº el dho Atoco. Dixo que se rretifica en lo dho en el primer ynterrogº e que este tº en aquella sazon que dize la preg.ta hera pastor del Ynga pero que le fue puco y lo oyo dezir a muchos yndios del dho rrepartimyento de Canta que en aquella sazon que rreynaba Guascar Ynga se hizieron las dhas dos capacochas una para Guascar Ynga y otra para su muger e que con ellas avian entrado en las dhas tierras de Quybi los dhos yndios yauyos del dho rrepartimyº de Chacalla e que al dho yndio que llamaban Chuqui Guaranga porque llego a defender que no entrasen con la dha capacocha y porque defendiendolo avia derramado un poco de sangre de la capacocha le avian muerto alli donde se derramo la sangre y este dho testigo fue despues de pasado esto a ver el dho yndio y le bio enterrado mas adelante del dho pueblo de Quybi que fue hasta donde llegaron con la dha capacocha.

Tº el dho don Juº Caxcha Yaure de la encomyenda de Canta Dixo que dize lo que dho tiene en el ynterrogº prinçipal e que en tpo de Guaynacaba supo y entendio este dho tº e oyo dezir a sus pasados que se avian hecho las dhas dos capacochas una para el Ynga y otra para su muger e que en tpo del dho Guascar Ynga no se hizo mas de una capacocha e que sy mas se hiziera lo supiera este dho

t° porque se acuerda muy bien de aquel tpo y tiene notiçia de lo que entonçes paso / y esto rresponde e que en tpo de Guascar //
/299v/ Ynga este dho testigo vio que un yndio que se llama Chuqui Guaranga quyso defender la dha capacocha de Guascar Ynga porque llegaba a su tierra que hera mas adelante del dho pu° de Quybi diziendo que no le entrasen sus tierras y enpeço a rrenpujar al yndio que traya la dha capacocha y derramo un poco de sangre de la dha capacocha y luego tomaron el dho yndio y lo mataron y lo enterraron donde cayo la dha sangre y este dho testigo le vio enterrar y esto rresponde. Ay mas t°s que se afirman en lo dho en el ynterrog° primero.

2 - Yten si saven etc. que de la mysma manera mataron otros quatro yndios de los de Canta defendiendo las dhas sus tierras: uno que se dezia Ynigui Chunbi y el otro Chaumecaxa y el otro Tanta Çapo y el otro Guaman Llaupe.

Los testigos dizen que lo oyeron en aquel tpo y se rrefyeren a lo dho.

3 - Yten si saven etc. que en las dhas tierras de Quibi no hubo mytimaes puestos por el Ynga syno çierto rresto de un caçique que mando matar el Ynga que se llamava Collotanbo de los chacallas cuya generaçion con todos los baronos fue muerta por çierta culpa de trayçion que tubo e unas pocas de mugeres que quedaron las paso a un corral de Quibi para que hilasen alli. Digan y declaren lo susodho como paso y por que via entraron alli las dhas mugeres e si ovo mytimaes o no e que tierras le dieron en que bibieron. No ay t°s.

4 - Yten si saven etc. que çiertas tierras de las dhas tierras de Quybi que estan cabe las casas que son las que Ju° Xpoval de Frias fue a ver son ansy mysmo de los dhos yndios de Canta e la causa por que agora la piden los dhos yndios de Chacalla es porque despues que les dieron las ovejas e hizieron la dha transaçion un yndio que se llama Pomaraque e otro Tanta de los yndios de Chacalla dixeron a Guaman Yaure caçique de Canta que se querian pasar a su tierra a tributar con el y el dho caçique les dixo que les darian aquel pedaço de tierra o otros sy se pasen como dezia e asy con el dho conçierto le sanalaron las dhas tierras a los dhos dos yndios de Chacalla Digan lo que saben.

/300r/ T° el dho Carvabilca no xpi° sujeto a Canta. Dixo que lo que sabe desta pregunta es que vido que al tienpo que se hazia el dho conçierto dieron para los yndios de Tapati e Allaguamarca las trras contenydas en la dha pregunta dize para que senbrasen e que se las dieron por rruego del dho Rribera y de Fran^co^ de Anpuero y esto rresponde a esta dha preg.^ta^

T° el dho Domyngo Pariana subjeto a Canta. Dixo que este dho testigo despues que se hizo el dho conçierto a oydo dezir este dho testigo a yndios de Canta que avian venydo al tienpo de conçierto llorando yndios del pueblo de Collata y Xicamarca yndios yauyos dando bozes que les vendian sus tierras que no tenyan donde senbrar y que el dho Rribera el moço y su caçique Guaman Yaure caçique de Canta de lastima e ynportunydad del dho Rribera les avian senalado los dhos pedaços de trras en que senbrasen e que a lo que le dixeron y entendio fue libremente syn ningun cargo ny tributo de lo que la preg.^ta^ dize ny otro lo qual declaro este dho testigo por esta rrazon. No ay mas.

5 - Yten si saven etc. que porque los dhos Pomaraque e Tanta dixeron que se querian pasar los hizo açotar el caçique Marcapoma e asy no se pasaron ny el dho caçique de Canta les dio las dhas tierras syno se las torno a tomar para sy y las senbraron de coca los yndios de Canta. No ay mas.

6 - Yten si saven etc. que puede aver tres años poco mas o menos que un don Juan prinçipal de Canta se hizo amygo con don Xpoval prinçipal de Chacalla y syn lo poder hazer porque se lo rrogo el dho don Xpoval le dio liçençia para que cogiese la coca del dho pedaço de tierra e asy el dho don Xpoval la coxo dos o tres vezes e con el dho achaque pretenden tener derecho a las dhas tierras syn averselas dado nadie ny poseidolas por otra causa ny aviendo plantado la dha coca.

Declaren los testigos como y de que manera paso lo susodho. No ay mas t°s.

7 - Yten sy saben etc. que todo lo susodho es publica boz y fama. El Liçen^do^ de Leon

Rratificanse los t°s.

Esta bien sacada del original y asi lo juro. El Relator

Digo lo mesmo. El Licen^do^ de Leon

El Licenc.^do^ M° Luzio

/301r/ En el pleito entre partes:
de la una los yndios de Chaclla
y de la otra los yndios de Canta
y Alonso Moreno y Fran° de la Torre sus pr°s.

Fallamos atento lo nuevamente alegado y provado que devemos rrevocar y rrevocamos la sent.ª en esta causa dada por nos el presidente e oydores desta Rreal Audiençia de que por pte de los dhos yndios de Chaclla fue suplicado y haziendo en el caso lo que de justiçia deve ser hecho condenamos a los dhos yndios de Canta a que buelban e rrestituyan las dhas trras de Quivi sobre que es este pleyto a los dhos yndios de Chaclla; con que primeramente los dhos yndios de Chaclla buelban e rrestituyan a los dhos yndios de Canta las dozientas ovejas que por el conçierto en esta causa presenta aparesze que rresçibieron dellos y por esta nra sent.ª juzgando asi lo pronunçiamos y mandamos en grado de rrevista.

El D. Bravo de Saravia — El Licen. Mr. de Peñalosa — El Doctor Gz de Cuenca

Dada e pronunçiada fue esta sent.ª por los dhos señores presydente e oidores desta Rreal Audiençia haziendo audiençia publica en Los Rreyes a honze dias del mes de Jullio de mill e quinyentos e çinquenta e nueve años estando presentes los procuradores de las dhas partes a los quales lo notifique.

Diego Muñoz

/302r/ Los yºs de Chacalla. — pide executoria.

Muy pº sºr

Alº Moreno en nonbre del caçique e yºs de Chacalla en el pleyto con los yºs de Canta y con Franᶜᵒ de la Torre pr. en su nonbre digo que por V. Alt. fue pºnunᵈᵃ sent.ª de rrevista en la dha cabsa.

Pido e sup.ᶜᵒ a V. Alt. me mande dar su rreal executoria della e para ello etc.

Alº Moreno

En Los Rreyes catorze de jullio de myll e quyºs e çinquenta y nueve aºs en Aud. Rreal ante los señores presydente e oydores / la presento el cont.ᵈᵒ en el dho nonbre y los dhos señores mandaron que se le de la dha ex.ª que pide dando ante todas cosas la s.ª

Diego Muñoz

/303r/ Los yºs de Canta. — Muy pºs ssºr

Suplican de la sentençia para la pers.ª rreal.

El caçique e yndios de Canta en el pleito con los yndios de Chacalla sobre la trra de Quivi y nras juridiçion y termyºs suplicamos de la sentençya dada y pºnunçiada por vro presydente e oydores desta Rreal Audiª para ante Vra Persona Rreal y los de Vro Consejo Rreal de Yndias / y hablando con el acatamyº que devo digo que la dha sentençya de rrevista se deve y a de enmendar mandando confirmar la sentençia de vista de vro presydente e oydores desta Rreal Audiençia.

A V. Al.ª pido y sup.ᶜᵒ que mandandome admytir la dha suplicaçion mande enmendar y rrevocar la dha sentençya de rrevista y mande confirmar la sentençia de bista sobre que pido just.ª.

El Licenᵈᵒ de Leon

En Los Rreyes catorze dias del mes de Jullio de myll e quinientos e çinquenta y nueve aºs ante los señores presydente e oydores estando haziendo Aud.ª Rreal publica / presento esta pet.ᵒⁿ Ffranᶜᵒ de la Torre pr. de los dhos yºs; e los dhos señores la ovieron por presentada y le conçedieron la dha segunda suplicaçion para que la pueda proseguir ante Su Mag.ᵗ y señores de Su Rreal Consejo de Yndias donde mandaron se presente con el proçeso //
/303v/ de la dha causa originalmente dentro del termyno de la hordenança y so las penas della e aperçibieron a las partes que si tienen algo que dezir e alegar en la dha causa lo digan y aleguen en esta Rreal Avdiençia antes de sacar della el dho proçeso porque ante Su Mag.ᵗ ny Su Rreal Consº no an de alegar ny probar otra cosa alguna y por los mysmos autos se a de determynar la causa y los mandaron çitar para que vayan o enbien en seguimyento della en forma acostunbrada y en su rrebeldia les senalaron los estrados del Consº Rreal de Su Mag.ᵗ

Diego Muñoz

Este dho dia notifyque el dho auto a los procuradores de las dhas ptes Franᶜᵒ de la Torre e Alonso Moreno que presentes estavan.

Diego Muñoz

/304r/ En Los Rreyes diez e siete dias del mes de Jullio de myll e quyºs e çinquenta e nueve años / fuera de avdiençia; la presento Alonso Moreno con protestaçion de la rretificar para la prov.ª

Muy pºs señores

Suplican del auto de aver otorg.^{do} la sup.^{on} la r^{l} p^{a}.

Alonso Moreno en nonbre del caçique e yndios de Chacalla en el pleito con el caçique e yndios de Canta sobre las tierras de Quibi e con Fran^{co} de la Torre en su nonbre supp^{co} del auto probeido por vro presidente y oydores en que otorgaron a las ptes contrarias la suplicaçion para España y mandaron a mys ptes dar fianças y digo que en lo susodho se deve el dho auto enmendar y rrebocar por lo siguiente:

- lo prim^{o} por todo lo gen.^{al} del dr^{o} e lo que dezir y alegar me conviene que es por expreso.

- lo otro porque esta causa es entre yndios y que se devia hazer y aver hecho sumariamente y sola la verdad sabida y no se avian de admytir las largas y dilaçiones que de contrario se p^{o}curan ynjustamente y sin tener causa de litigar y lo hazen al feto de estarse en las tierras entendiendo que mys ptes no tienen fiança alg.^{a} que dar y no se ha de dar lugar a semejante calunya espeçialmente no valiendo como no valen las tierras la cantidad de la hordenança como pareçe por el pareçer del padre Carvajal y Juan de Pedraça que sin pasion vieron e tasaron las dhas tierras y sin la afiçion que mys ptes les tienen y ya que por alguna via se deviese otorgar la dha suplicaçion avia de ser sin fianças pues es notorio ser las tierras de my pte y no tener ning.^{a} defensa justa las ptes contrarias y estandose como se estan ay las tierras sin que en ellas pueda aver diminuçion ny fraude en que puedan ser perjudicadas las ptes contrarias porque mys ptes no tienen fianças que dar y que asi lo juro a Dios y a esta + en anymas de mys ptes en forma.

Por que pido e supp.^{co} a V. Al.^{a} enmyende los dhos autos denegando la supp.^{on} de las ptes contrarias y declarando no ser mys partes obligados a dar fiança alguna sobre que pido just.^{a} y costas e para ello etc.

El Licenc.^{do} M^{o} Luzio — Al^{o} Moreno

/304v/ En Los Rreyes diez y ocho de Jullio de myll e quy^{o}s e çinquenta e nueve a^{o}s en Aud.^{a} Rreal pu^{ca} ante los señores presydente e oydores en Audiençia Rreal pu^{ca} / la presento el cont.^{do} y los dhos señores mandaron traer los autos de la causa para proveer sobre ello.
Diego Muñoz

Va en provision sobre la supp.^{on} para la ps.^{a} rreal si a lugar o no de conçederse o no [*sic*]. No se tasa a los yndios dr^{o}s porque no ay arranzel.

Diego Muñoz

/305r/ Los caçiques de Chacalla.

Muy p^{o}s ss^{or}

pide se tase lo que an de pagar de las obejas.

Alonso Moreno en nonbre de los caçiques de Chacalla en el pleito con los de Canta sobre las chacaras de Quibi / digo que por V. Al.^{a} fue esta causa sentençiada en rrevista en favor de mis partes con que buelvan a las partes contrarias las dozientas ovejas que dizen que dieron a mys ptes y porque querrian pagar las dhas ovejas a los presçios que valian quando las rreçibieron pues no son obligados a mas y por el proçeso pareçe el valor que entonzes valian.
A V. Al.^{a} pido y supp.^{co} lo mande ver y tasar lo que mys ptes han de pagar sobre que pido just.^{a} e para ello etc.

El Licenc.^{do} M^{o} Luzio — Al^{o} Moreno

En Los Rreyes en veinte e ocho dias del mes de Jullio de mill e quinientos e çinquenta e nueve anos ante los senores presidente e oidores desta Rreal Avdiençia est.^{do} haziendo avdiençia pu^{ca} / la presento el cont.^{do} e los dhos senores presidente e oidores mandaron que se lleve al acuerdo.

Diego Muñoz

/306r/ Piden los partos y pospartos y mejoramy^{o}s
y que pagaran los frutos.

Los caçiques de Canta por nosotros y en n^{e} de los demas yndios de nro rrepartimi^{o} en el pleyto con los yndios de Chacalla sobre las tierras de Quibe no apartandonos de la suplicaçion que tenemos hecha por ante la Persona Rreal / dezimos que a nra notiçia es benido que las partes contrarias an pedido que quieren pagarnos las dozientas ovejas en dineros y como balian quando nosotros las dimos y porque ya que se deshaze la dha transaçion que entre nosotros se hizo conforme a dr^{o} se a de bolver en el punto y estado en que estaba quando se las dimos bolbiendo a cada uno lo suyo ni mas ni menos que lo dio y ansi los dhos y^{o}s de Chacalla son obligados a nos bolber las mismas ovejas que les dimos en ovejas y con los partos y pospartos que an rrentado e multiplicado desde que se las dimos hasta agora y ansi mesmo los mejoramientos que emos hecho en las dhas trras en coca y arboles mandandolo apreçiar en lo que bale que nosotros estamos prestos por la conpulsion que se nos haze por la dha sent.[a] de dar a las partes contrarias los frutos y rrentas que las dhas tierras an rrentado desde el dho conçierto y transaçion aca.

E por tanto a V. Al. pedimos y suplicamos mande a la parte contraria que para cunplir con la sent.[a] que en su fabor se dio que nos buelban las dhas obejas que les dimos en obejas y con partos y pospartos y los mejorami[o]s que emos hecho en las dhas tierras que nosotros estamos prestos de descontar o pagar lo que paresçiere aver cojido de los dhos frutos y rrentas sobre que pedimos just.[a] y costas y bro rreal of[o] ynploramos.

El Licen[do] de Leon

/306v/ En Los Rreyes veynte e nueve dias del mes de Jullyo de myll e quy[o]s e çinquenta e nueve años ante los señores oydores en avdiençia de rr.[on] / se presento esta pet.[on] e los dhos señores mandaron que los dhos yndios de Chaclla buelban e paguen a los de Canta las ovejas de la condenaçion en ovejas e no en dinero asy como las rresçibieron.

Diego Muñoz

/307r/ El caçique de Chacalla.

Muy p[o]s s[o]r

Don Xpoval Vilcapoma caçique de Chacalla encomendado en Fran[co] de Ampuero vez[o] desta çiudad por mi y en nonbre de los demas yndios del dho rrepartimi[o] demando ante V. Al. al dho Fran[co] de Ampuero / e digo que por el mes de henero del año pasado e qui[o]s e quarenta e nuebe el dho Fran[co] de Ampuero e Niculas de Rribera en quien estaban y estan encomendados los yndios de Canta hizieron çierto conçierto por el qual dixeron que don Fran[co] caçique de Chacla mi anteçesor se abia de los dhos yndios de Canta dozientas ovejas de la tierra por rrazon de las diferençias que entre mis yndios e los de Canta avia sobre las tierras de Quive siendo lo susodho al contrario de la verdad porque ni el dho don Fran[co] ni los dhos sus ynd[o]s tal conçierto hizieron ni ellos rreçibieron las dhas dozientas ovejas syno que el dho Fran[co] de Ampuero a fin de rreçibir como rreçibio para sy las dhas dozientas obejas hizo y formo el dho conçierto como a el le pareçio / y aora en el pleito que yo e los dhos mis yndios hemos puesto a los dhos yndios de Canta sobre las dhas trras de Quive por sent.[a] de rrevista de V. Al. nos estan mandadas bolver las dhas trras con que ante todas cosas bolvamos y rrestituyamos a los dhos yndios de Canta las dhas dozientas ovejas que por el dho conçierto dieron y asi esta obligado el dho Fran[co] de Ampuero como persona en cuyo poder entraron las dhas ovejas a nos las dar bolver y rrestituyr o pagar con todos los partos y pospartos que dellas ovieren proçedido y pudieron proçeder y proçedieren hasta la rreal rrestituçion pues conforme a la dha sent.[a] nosotros estamos obligados a las bolver a los dhos yndios de Canta e no al dho Fran[co] de Ampuero.

A V. Al. pido y suplico que en el caso me haga breve y entero cumplimi[o] de justiçia y mande pareçer ante sy al dho Fran[co] de Ampuero y que con juram[o] declare si es verdad que rreçibio el para sy las dhas dozientas ovejas y se aprovecho dellas y pareçiendo ser ansy vale por la via que mas me conbenga conpela y apremie al dho Fran[co] de Ampuero a que me de y pague las dhas dozientas ovejas con todos los partos y pospartos hasta la rreal rrestituçion e pido en todo cumplimi[o] de Just.[a] y costas y para ello etc.

El Liçenz[do] Hier.[mo] Lopez Sanchez Vallejo

/307v/ En Los Rreyes en veinte e ocho dias del mes de Jullio de myll e quinientos e çinquenta e nueve años ante los señores presidente e oidores desta Rreal Audiençia estando haziendo audiençia pu[ca] / la presento el cont.[do] e los dhos señores mandaron que el dho Fran[co] de Ampuero parezca ante el semanero a declarar sobre lo susodho.

Diego Muñoz

En Los Rreyes dos dias del mes de agosto de myll e quy[o]s e çinquenta e nueve años el dho Fran[co] de Ampuero vz[o] e rregidor desta çiudad paresçio ante el señor dottor Saravia oydor desta Rreal Audiençia e juro el dho Fran[co] de Anpuero por Dyos nro señor e sobre la señal de la cruz en forma de derecho e dixo: "sy juro" e amen e prometio dezir verdad e aviendo jurado e siendo preguntado por el tenor de la dha demanda dixo que este declarante tiene dho su dho sobre lo contenydo en la dha demanda al qual se rrefiere en el proçesso de pleyto de que en ella se haze mynçion siendo presnetado por testigo en la dha causa e que siendole mostrado si otra cosa toviere que declarar la declarara / e que esto es la verdad para el juram[o] que hizo e lo firmo de su nonbre.

Fran[co] de Ampuero paso ante mi
Diego Muñoz

/308r/ Los caziques e y[o]s de Chacalla.

Muy p[o]s ss[o]r

Pide se les m[de] pagar en din[o] lo que avian de dar en oyejas.

Don Felipe por my y en nonbre de los caçiques e yndios de Chacalla en el pleito con los de Canta sobre las tierras de Quibi / digo que al tiempo que nosotros rreçibimos las ovejas que se dieron a Fran[co] de Anpuero que agora se nos mandan bolver no valian a mas de a dos pesos cada vna y a menos porque la mytad heran pacos y entre ellos no avia hembras y no seria justo que pues se las llevo nro encomendero y luego se constituyeron se nos mandasen bolver agora en obejas queriendo como queremos nosotros pagar lo que entonzes valian o dar por ellas obejas de Castilla que al presente valen mas que entonzes valian las otras.

A V. Al.[a] suplicamos nos haga mrd de mandar que paguemos en dinero lo que valian las obejas quando se dieron o se rreçiban por ellas otras tantas obejas de Castilla pues las tenemos y ninguna de esotras y en ello se nos hara just.[a] y mrd.

Don Felipe

En Los Rreyes en veinte e dos dias del mes de agosto de myll e quinientos e çinquenta e nueve anos ante los senores presidente e oidores desta Rreal Aud.[a] haziendo audiençia pu[ca] y rreal / la presento el cont.[do] e los dhos señores mandaron que se guarde la sent.[a] en la causa dada.

Diego Muñoz

/309r/ Los y[o]s de Chacalla.

Muy p[o]s señores

Pide se m[de] tasar lo que podian baler las obejas.

Don Felipe por my y en nonbre de los demas caçiques e yndios de Chacalla en el pleito con los de Canta sobre las tyerras / digo que por V. Al.[a] esta man[do] que nosotros demos a las partes contrarias las dozientas obejas que dizen avernos dado y porque nosotros no tenemos obejas que dar y queremos pagar el valor de las dichas obejas a como agora valen teniendo consideraçion a que heran todos machos y pacos y que tenyan carache y ay personas en este pueblo que lo vieron entonzes y lo podran tassar conforme al presçio que agora tuviera syendo bivo.

A V. Al.[a] pido y supp.[co] mande de off[o] o como de dr[o] mejor oviere lugar que las tassen dos personas de las que las vieron como son el padre Godoy y los demas que han declarado como t[o]s en esta causa para que nosotros las paguemos y se nos entreguen nras tierras y pido just.[a] etc.

Don Phelipe

En Los Rreyes veynte e seys de agosto de myll e quy[o]s e çinquenta y nueve años ante los senores oydores en audiençia de rrelaçion se leyo esta pet[on]. e los dhos señores mandaron que cunplan la ex.[a] de las sent.[as] en la causa dada.

Diego Muñoz

/310r/ Muy p[o] señor

Pide exsecutoria de la sent.[a] de rrevista.

Diego de Ocanpo en nombre de don Juan Baptista Guaraca caçique prinçipal de Chacalla y de los prinçipales e yndios a el sujetos encomendados en Fran[co] de Anpuero digo que las dhas mis ptes ha muchos años que trataron pleyto con los yndios de Canta encomendados en Nyculas de Rribera sobre las trras e chacaras de Quivi en la qual cabsa se sentençio por ellas en rrevista.

Pido y sup.[co] a Vra Al. mande dar exsecutoria de las dhas sentençias para que se guarden y cunplan para lo qual.

Di[o] de Ocanpo

En Los Rreyes a veynte de novienbre de myll e quy[o]s e sesenta y çinco años ante los ss[o]r presydente e oydores estando en abdiençia pu[ca] / la presento el cont.[do] y los dhos ss[o]r mandaron que se le de la executoria que pide estando sentençiado en rrevista.

Fran[co] Lopez

/311r/ Muy p[o] señor

Piden ex.[a] de las sent.[as] de vista e Rrebista.

Diego Ocanpo en n[e] del caçique e yndios de Chacalla encomendados en Fran[co] de Anpuero vz[o] desta çibdad en la cabsa con los yndios de Canta sobre las chacaras de Quivi / digo que V. Al. dio sentençya de rrevista en esta cavsa syete años ha en que mando rrestituyr a mys partes las dhas trras y aunque las partes contrarias suplicaron segunda vez para ante Vra Persona el proçeso lo qual paresçe pues esta

A Vra Al. pido y sup. mande dar a mys ptes executoria de las dhas sentençias syn fianças algunas para lo qual etc.

El Licen[do] de Leon   Diego de Ocanpo

En Los Rreyes en veynte y quatro del mes de mayo de mill e quinientos y sesenta y seis años ante los señores presidente e oydores en audiençia pu[ca] / la presento el contenido en nombre de su parte.

Los dhos señores mandaron que se traygan los autos para proveer en ello.

Fran[co] Lopez

/311v/ En Los Rreyes en primero de junyo de myll e quy^{o}s e sesenta e siete años los señores presidente e oydores desta Rreal Audiençia aviendo visto esta causa mandaron que a los yndios de Chacalla se les de la exsecutoria que piden sin fianças.

Avendaño

Se les de a los y^{o}s de Chacalla la exsecutoria syn fianças en primero de junyo 1567.

/312r/ Y^{o}s de Chacalla.

Muy p^{o} señor

Diego de Ocanpo en n^{e} del caçique e yndios de Chacalla en la cabsa con los yndios de Canta sobre las chacaras de Quivi / digo que estando en esta cabsa mandaba dar exsecutoria de los abtos llevo Juan de Arrandolaça el proçeso y no lo a buelto y diz que quyere suplicar por molestar a mis partes.

A V. Al. pido y sup.^{co} mande lo buelva luego con pena y que el secretario me de la dha causa exsecutoria syn enbargo de su suplicaçion para lo qual etc.

Diego de Ocanpo

En Los Rreyes a diez e siete dias del mes de junyo de myll e qui^{o}s e sesenta y siete a^{o}s ante los ss. pres.^{e} e oid^{o}s est.^{do} en audiençia publica / la presento Diego de Ocanpo en n^{e} de su pte. y los dhos señores mandaron que oy en todo el dia lo buelva so pena de seys p^{o}s en contr^{o} de Ju^{o} de Arrandolaça a quien se not^{o}.

Avendaño

/313r/ Don Felipe Tauri.

Muy p^{o}s s^{o}r

Piden pr.^{on} de amparo/

Don Felipe Tabri Chunbi y Alonso Cahuaca prençipales de el pueblo de Quibe por nos y en nonbre de los demas prençipales e yndios de los rrepartimientos dezimos que nosotros tenemos çiertas tierras e chacarras en el dho pueblo de Quibe e su comarca e somos yungas y los yndios yavyos de Fran^{co} de Anpuero nos ynquietan y dizen que ellos tienen alli tierras siendo serranos e no teniendo en ellas cosa alguna syno nosotros desde la fundaçion del pueblo y de tpo de los Yngas pasados asy nosotros como nros antepasados / y dello rresçebimos molestia y daño a que V. Al. no debe lugar.

A V. Al. pedimos y suplicamos nos mande dar probysion de anparo pues es cosa nra e la emos tenido e tenemos asy nosotros como nros pasados de tpo ynmemorial a esta pte mandando que los dhos yndios yavyos del dho Fran^{co} de Anpuero no se entremetan en ello con pena que para ello se les persiga sobre que pedimos just.^{a} e para ello etc.

Don Felipe

En Los Rreyes a diez e nueve dias del mes de junyo de myll e qui^{o}s e sesenta e siete a^{o}s ante los ss. pres.^{e} e oidores est^{do} en audiençia pu^{ca} / la presento el cont.^{do} y los dhos señores mandaron dar tr.^{do} a la otra pte e que rresponda a la primera aud.^{a} / este traslado ha de ser a los yndios de Chacalla de Fran^{co} de Anpuero.

Avendaño

/314r/ Don Phelipe Tauli Chunbi.

Muy p^{o} s^{o}r

Don Felipe Taule Chunbi y Al^{o} Caguaca caçiques de Quybe por nosotros y en nonbre de nuestros y^{o}s yungas como terçeros por nuestro ynteres dezimos que a nuestra notiçia es benydo que entre los caçiques de Canta y Chacalla a mas de seis años que se trajo pleyto sobre nuestras propias tierras de Quybe donde estamos poblados y son nuestras de tpo ymemorial a esta pte estando como estamos en posesion dellas syn saver que entre los susodhos se trataba el dho pleyto sobre nuestra haçienda y ansi porque el caçique de Canta no probo señorio en las dhas tierras se adjudicaron las dhas nuestras tierras al caçique de Chacalla en çierta fforma y por quanto como dho tenemos de tpo ymemorial a esta pte nosotros y nuestros pasados hemos estado sienpre en posesion de las dhas tierras y nos pertenesçen como haçienda nuestra y como cosa que sienpre hemos labrado y desde el tpo de los Yngas y antes que nos conquystasen y despues quando entraron los españoles en este rreyno sienpre las hemos poseydo desfrutado y labrado fasta aora y si alguna vez se an entrado algunos y^{o}s en ellas a sido por ser nosotros yungas y pocos y no podernos defender y nos las an tomado tiranycamente a ellas y a nosotros p^{o}s sienpre las hemos tenydo y nos es justo que por el pleyto que entre los susodhos paso nosotros perdamos nuestras tierras y haçiendas.

Por tanto como mejor de dr^{o} podemos y debemos suplicamos de qualquyer sentençia que en lo susodho se aya dado en que se declare no pertenesçernos a nosotros las dhas tierras de Quybe.

/314v/ Pedimos y suplicamos a V. Al.^{a} nos haga just.^{a} por la bia e forma que mas nos conbenga y si mas pedimy^{o} es nesçesario V. Al.^{a} enmyende y rrevoque la dha s.^{a} y nos mande amparar en la posesyon que hemos tenydo e tenemos de las dhas tierras pues nos pertenesçen sobre que pedimos just^{a} y costas y el of^{o} de V. Al.^{a} ynploramos.

El Licen^{do} de Leon

En la çiudad de Los Rreyes a diez y ocho dias del mes de junyo de myll e qui^{o}s y sesenta y siete años ante los señores presidente e oydores estando en audiençia de rrelaçiones se leyo esta petiçion y los dhos señores mandaron traer todos los autos para los ver y proveer.

Alvaro Ruiz de Nabamuel

En Los Rreyes en veynte de junyo de myll e qui^{o}s e sesenta e siete años los señores presidente e oydores desta Rreal Audiençia aviendo visto esta causa mandaron que a los dhos yndios de Chacalla se le de la exsecutoria que les esta mandada dar sin enbargo de lo alegado por la parte contraria sin perjuyzio del derecho de las partes y mandaron dar traslado destas dos petiçiones a los yndios de Chacalla / presente Diego de Ocampo pr. contrario a quien se notifico.

Avendaño

/315r/ Muy p^{o} s^{o}r

Pide se haga infor^{on} a quien pertenesçen.

Joan de Arrandolaça en n^{e} de los yndios yungas del asiento de Quybe sobre las trras que alli piden los yndios de Chacalla / digo que mys ptes como posehedores que son al presente y lo han sido desde antes que los españoles entrasen en este rreyno y como terçeros opositores por su ynteres contradixeron el sacar de la ex.^{a} de los dhos yndios de Chacalla contra los indios de Canta porque las dhas trras de Quybe no son de alguno ny nynguno dellos sino de mys ptes que las tienen y poseen desde el tpo ynmemorial a esta pte e al presente estan en ellas y lo estaban al tpo que los españoles entraron en este rreyno y ansi suplicaron mis ptes de la dha sentençia de rrebista aunque no avia neçesidad por ser suyas las dhas trras y estar en posesion dellos de lo qual estoy presto de dar luego ynformaçion o que se tome de of^{o} para que conste de lo que digo pues estan aquy yndios comarcanos que lo pueden saber y es ansi que los dhos yndios de Chacalla han sacado ex.^{a} e por virtud della han ynobado o quyeren ynobar despojando a mys ptes de su posesion lo qual es contrato de dr^{o}.

A V. Al. pido e sup.^{co} que pues agora esta ^{a}qu^{i} el corr.^{or} Luys de Saravia y estan aquy todos los yndios comarcanos V. Al. m^{de} de of^{o} que se haga ynformaçion por los articulos que cada una de las ptes diere y constando de lo que digo se haga lo que tengo pedido y V. Al. anpare en su posesion a mys ptes e se rreboque ante todas //
/315v/ cosas el atentado por las ptes contrarias pues no se puede ynobar pendiente la dha lid e suplicaçion sobre que pido just.^{a} e para ello etc.

El Licen.^{do} de Leon J. Arrandolaça

Que se traya esta petiçyon y el proçeso; proveyose en aud.^{a} de rrelaçiones por los señores presidente e oidores desta Rreal Audiençia en Los Rreyes en treinta dias del mes de junyo de myll e quiny^{o}s y sesenta y siete años.

Alvaro Ruiz de Nabamuel

/316r/ Don Ju^{o} Baptista caçique de los Chupachos.

Muy p^{o} señor

Don Juan Bap^{ta} Caxaguaraca caçique prinçipal del rrepartimy^{o} de Chacalla encomendado en Fran^{co} de Anpuero vz^{o} desta çibdad ante V. Al. me querello y acuso criminalmente a don Pedro Guari y a don Christoval Carvabilca y a don Gonçalo Carvabilca y a los demas que paresçieren culpados y contando el caso premisas las solenidades del dr^{o} / digo que es asy que tenyendo e poseyendo los caçiques mys anteçesores de Chacalla y sus yndios quieta y paçificamente unas trras en el valle de Quivi por nos las aver dado los Yngas señores que fueron deste rreyno / el caçique de Canta y sus y^{o}s porque los dhos mys anteçesores y sus yndios servian a V. Al. y no se quysieron alçar e rrebelar contra vro serv^{o} como los dhos caçiques e yndios de Canta lo hizieron entraron en las trras del dho valle de Quivi / y mataron a un prinçipal e çiento e çinquenta y tantos yndios que alli tenyamos / y les tomaron sus haziendas e trras de lo qual nos querellamos ante vra just.^{a} y aunque sobre ello se hizo algun castigo y se mandaron bolber las dhas trras todavia las tornaron a tomar y se quedaron con ello hasta que despues ocurrimos a esta Rreal Abdiençia en la qual por sentençia de rrevista y executoria se nos mandaron bolver las dhas trras y aviendo tomado posesyon dellas por la dha vra exsecutoria rreal los susodhos por mi acusados con poco temor de Dios y en gran menospreçio de vras justiçias y desacato desta Rreal Abdiençia se tornaron a entrar en las dhas trras y derribaron los mojones que por mandado de vras just.^{as} se avian puesto en ellas juntando para ello mucha gente armada y a punto de guerra en lo qual cometieron grandes y graves delictos //
/316v/ dignos de grave puniçion y castigo exenplar.

Por que pido y sup.^{co} a V. Al. condene a los susodhos por my acusados en las mayores y mas graves penas que por fuero y dr^{o} hallare estableçidos contra los que cometen semejantes delictos y las mande exsecutar en sus personas e bienes por que a ellos sea castigo y a otros en exenplo sobre que pido just.^{a} y las costas y para ello etc. e juro a Dios en forma que esta querella no la pongo de maliçia.

El Lic^{do} Falcon

En Los Rreyes primero de julio de myll e quiny^{o}s e sesenta e siete a^{o}s ante los señores pres.^{te} e oydores desta Rreal Aud.^{a} en aud.^{a} publica se presento esta pet.^{on} y los dhos señores mandaron que se trayga la ynformaçion que sobre ello esta hecha y se traya.

Avendaño

/317r/ El caçique e y^{o}s de Chacalla.

Muy p^{o} señor

Diego de Ocanpo en nombre del caçique e yndios de Chacalla de la encomyenda de Fran^{co} de Anpuero vz^{o} desta çibdad / rrespondyendo a un pedimyento fho por don Felipe Taule Chunbi y alonso Caguaca caçiques que dizen ser del valle y pueblo de Quivi en que en efeto dizen que las trras del dho valle de Quivi que por sentençia de rrevista y exsecutoria desta Rreal Abdiençia estan adjudicadas a my pte y mandadas entregar y entregadas son suyas y les pertenesçen y que la dha sentençia y exsecutoria no les perjudica por ser dadas entre otras partes piden que asy se declare / digo que V. Al. ha de dar por nynguno el dho pedimyento porque por el dho proçeso consta que las dhas trras pertenesçen a mys partes y a sus yndios y por tales suyas se las mandaron entregar y no faze al caso dezir que no se litiga con ellos porque es notorio que en este rreyno dende que se descubrio todos los proçesos y pleytos se hazen y an fho con el caçique prinçipal en nombre de sus yndios syn otro poder ny lo pueden dar los yndios porque no tienen conçejos para poder dar en nombre de todos poder y asy se ha admytido y admyte el caçique prinçipal para poder pleytear y dar poder en nombre de todos en espeçial que estos son yndios ya puestos por los serranos para dezir lo que dizen y es cosa evidente que an sabido del pleyto que se tratava syendo cosa tan publica y aun es de creer que an contribuydo para las costas del pleyto y sy se diese lugar a que se bolviesen los pleytos que se an fho los caçiques seria destruyr e ynquietar toda la trra.

/317v/ Por que pido y sup.^{co} a V. Al. declare no aver lugar el dho pedimyento y mande que no se syga mas pleyto sobre esta cabsa pues esta fenesçida por sentençia de rrevista sobre que pido just.^{a} y las costas y para ello etc.

El Lic.^{do} Falcon Ocanpo

En Los Rreyes primero de julio de myll e qui^{o}s y sesenta y siete a^{o}s ante los señores pres.^{te} e oydores desta Rreal Audi.^{a} estando en audi^{a} pu^{ca} se presento esta pet.^{on} y los dhos señores la ovieron por conclusa y mandaron que se trayga todo.

Avendaño

/318r/ A pedimi^{o} del caçique prinçipales e yn^{o}s del rrepartimi^{o} de Chacalla contra el caçique prençipales e yn^{o}s del rrepartimi^{o} de Canta sobre çiertas tierras e ovejas.

/318v/ Don Phelipe por la graçia de Dios rrey de Castilla de Leon de Aragon de las dos seçilias de Jerusalen de Navarra de Granada de Toledo de Valençia de Galiçia de Mallorcas de Sevilla de Çerdeña de Cordova de Corçega de Murçia de Jaen de los Algarves de Algezira de Gibraltar de las yslas de Canaria de las Yndias yslas e tierra firme del mar oçeano; conde de Flandes y de Tirol etc.

Al nro presidente e oidores de la nra Audiençia y Chançilleria Rreal que por nro mandado rreside en la çiudad de Los Rreyes de los nuestros rreinos y provinçias del Peru y a todos los nuestros corregidores y juezes de rresidençia y alcaldes ordinarios y otros qualesquier nuestros juezes e justiçias della e a cada vno y qualquier de vos en vra jurisdiçion salud y graçia.

Sabed que pleito se trato en la dha nra Rreal Audiençia ante el presidente e oidores della entre partes de la vna don Xpoval Vilcapoma caçique prençipal de Chacalla y don Phelipe prençipal del dho rrepartimyento y Fran^{co} Yndio / por si y en nonbre de los demas yndios e prençipales del dho rrepartimyento y de la otra don Hernando caçique prençipal del rrepartimyento de Canta y sus yndios e prençipales y Alonso Moreno y Françisco de la Torre y Diego de Ocanpo procuradores en sus nonbres / sobre rrazon que pareçe que en la dha çiudad de Los Rreyes ante los dhos presidente e oidores en çinco dias del mes de otubre de mill e quinientos e çinquenta e ocho años paresçio el dho don Xpoval Vilcapoma caçique prençipal del dho rrepartimi^{o} de Chacalla y don Phelipe prençipal y Fran^{co} Yndio por si y en nonbre de los demas yndios //

/319r/ y prençipales del dho rrepartimi^{o} y por un pedimi^{o} que presentaron dixeron que teniendo y poseyendo ellos y sus antepasados vnas tierras y chacarras de coca que se dizen de Quibi desde el tienpo de los Yngas pasados de tienpo ynmemorial a esta parte y que las tenian al tienpo que los españoles entraron en esta tierra los caçiques e yndios del rrepartimi^{o} de Canta en tienpo del marques don Françisco Piçarro por engaño prendieron un caçique que entonçes era suyo que se dezia Vilcacapcha y lo mataron y juntamente con el çiento e çinquenta e tres yndios que tenian las dhas tierras y muerto que los ovieron les tomaron las dhas tierras lo qual hizieron porque el dho su caçique e yndios servian a los españoles y estavan bien con ellos y no se alçaban con el Ynga quando se alço en este rreino y aunque les han pedido muchas vezes las dhas tierras no las han querido bolver antes dizen tener un conçierto y desistimi^{o} dellas que dizen avelles hecho vn caçique que dizen aver sido suyo el qual nunca tal hizo a lo menos por su voluntad y aunque lo hiziera en el fueron todos los caçiques e prençipales del dho rrepartimi^{o} de Chacalla engañados enormysimamente y les deve ser desecho el dho engaño de manera que ellos ayan las dhas tierras y queden con ellas como cosa que es suya e nos pidieron y

suplicaron que declarando las dhas tierras por suyas se las mandemos bolver y entregar libremente como //
/319v/ cosa que es suya de sus antepasados con los frutos y rrentas que han llevado hasta la rreal rrestituçion e que si conçierto alguno pareçiere que les pueda perjudicar declarando ser en el engañados enormisimamente lo rreçindamos e anulemos conforme a derecho de lo qual se mando dar traslado a la otra parte para que rrespondiese a la primera y siendo notificado a Fran[co] de la Torre procurador de la otra parte en rrespuesta de la dha demanda dixo que la negava en todo y por todo segun y como se ponia y que deviamos absolver y dar por libre al dho su parte de lo contra el pedido porque la dha demanda no avia sido puesta por parte ni contraparte obligada y por todo lo demas general que avia por espresado y por no ser puesta en tienpo ni en forma y porque las dhas tierras son de sus partes que las tubieron e poseyeron desde antes de Topa Ynga Yupangui y la color que las dhas partes contrarias tienen para pedir fue porque en aquella parte y lugar Guaynacaba puso çiertos mitimaes de los yndios de los pueblos de la parte contraria por çiertas causas que le movieron y en la dha tierra le señalo lo que avian de labrar para el dicho Guaynacaba y despues que los españoles entraron en este rreino y çeso la sujeçion del Ynga y no avia para que estar alli los dhos mitimaes se bolvieron a sus tierras y asi los dhos sus partes lo tornaron a labrar como cosa //
/320r/ suya propia que era y es y desde entonçes aca sienpre lo han tenido e poseydo y teniendolo y poseyendolo las partes contrarias pidieron las dhas tierras a nro presidente el Liçençiado Gasca y pusieron a pleito a sus partes sobre las dhas tierras y andando en el dho pleito el dho nro presidente mando a los encomenderos de su parte y de la pte contraria que fuesen a ver las dhas tierras y procurasen de los conçertar y asi las partes contrarias y su parte fueron a ver las dhas tierras y estando en la dha diferençia y debate se vinieron a conçertar y hazer transaçion en el dho pleito y no deviendo sus partes cosa alguna por se quitar del dho debate dieron a las partes contrarias todo lo contenido en esta transaçion y conçierto de que hizo presentaçion por que no les mobiesen mas pleitos sobre las dhas tierras por lo qual aunque las dhas tierras fueran de las dhas partes contrarias y tubieran derecho alguno a ellas aquel avian perdido e no podian pedir mas las dhas tierras ni poner la demanda que ponen quanto mas siendo como son de sus partes y el titulo y transaçion que presento es del tenor siguiente:

"Digo yo don Françisco señor del pueblo de Mama y Chacra que sirvo a Françisco de Anpuero y lo tengo por señor y amo y le sirvo como su caçique que es verdad que por quanto entre vos Guaman Yare señor del pueblo //
/320v/ de Canta y pueblos y prençipales a el subjetos y teneis por amo y le servis a Niculas de Rribera y por quanto entre vos el dho Guaman Yaure y mi don Fran[co] teniamos diferençia sobre çiertas tierras mucho tienpo a que se dizen Quibi que confinan con los yndios de Guancayo en el mesmo rrio y por arriba hazia la sierra con yndios de Canta y los dhos nros amos Fran[co] de Anpuero y Niculas de Rribera por nos hazer bien y quitarnos de diferençia y pasion vinieron anbos a dos juntamente a las tierras del dho Quibi y todos juntos ellos y nosotros nos conçertamos en esta manera que me diesedes vos el dho Guaman Yaure dozientas ovejas y que yo el dho don Françisco os doy las dhas tierras de Quibi para vos y para vros deçendientes libres y desenbaraçadas / y digo que agora ni en ningun tpo por mi ni yndio mio ni mi amo os sera puesto entreval alguno y porque es verdad que rreçevi las dhas dozientas ovejas de que me doy por contento y pagado a toda mi voluntad / os doy esta carta firmada de mi nonbre y de mi amo Fran[co] de Anpuero y porque yo no se firmar rrogue al padre Garçia de Godoy firmase por mi; testigos que fueron presentes a todo lo sobredho: nros amos Fran[co] de Anpuero y Niculas de Rrivera y el padre Lorenço de Godoy y Di[o] de Avila y Pedro Ruiz de Barahona y Diego de //
/321r/ Flores y en lo que toca a las tierras fuera del rrio hazia la parte de Chacalla se partieron desta manera que la tierra que cae de los yndios de Canta ado esta una + para abaxo hazia Guancayo son del dho Guaman Yaure y de la + para arriba del dho don Fran[co] y porque es verdad lo firmaron de sus nonbres.

Fecho sabado a veinte e seis dias del mes de henero de mill e qui[o]s y quarenta e nueve años. Françisco de Anpuero; Niculas de Rrivera; Garçia de Godoy;por testigo Pedro Rruiz; por t[o] Diego de Avila." de lo qual fue mandado dar traslado a la otra parte y fue notificado al dho Alonso Moreno procurador contrario y rresçevido a prueba con termyno de seis dias y prorrogado por otros veinte mas comun a las ptes las quales hizieron sus provanças con t[o]s y escrituras y alegaron de bien provado y se concluyo la dha causa e vista por los dhos presidente e oidores dieron e pronunçiaron en ella sent.[a] de vista del tenor siguiente:

"En en pleito que es entre partes de la vna los caçiques e yndios del rrepartimi[o] de Chaclla y Alonso Moreno su procurador y de la otra el caçique e yndios de Canta y Françisco de la Torre su procurador sobre las tierras de Quibi / ffallamos atento los autos y meritos deste proçeso que devemos absolver y absolvemos al caçique e y[o]s //
/321v/ de Canta de la demanda contra ellos puesta por parte de los dhos caçique e yndios de Chaclla e por esta nra sentençia difinitiba juzgando ansi lo pronunçiamos y mandamos sin costas/el marques / el dotor Bravo de Saravia el Liçençiado Mercado de Peñalossa / el dotor Gonçalez de Cuenca."

La qual fue pronunçiada por los dhos presidente e oidores en la dha çiudad de Los Rreyes en diez dias del mes de henero de mill e qui[o]s y çinquenta y nueve años y fue notificada a los procuradores de las dhas partes y por pte del caçique e yndios del dho rrepartimi[o] de Chacalla fue suplicado della y espresando contra ella agravios dixeron que se devia emendar y rrevocar condenando a las ptes contrarias en las dhas tierras y frutos dellas por todo lo general del derecho y que dezir le conviene y del proçeso rresulta en ffavor de sus partes que ha por espreso y porque por el proçeso e provanças del consta ser las dhas tierras de sus partes y averselas dado los Yngas pasados y tenerlas e poseerlas al tienpo que los españoles entraron en esta tierra como cosa suya y serles despues tiranycamente quitadas matando a los que en ellas estavan por servir a los cristianos y no rrevelarse contra ellos como las ptes contrarias. //
/322r/ Por lo qual avnque otra cosa no obiera avian sus partes de ser en duda faboresçidos espeçialmente teniendo justiçia tan clara como la tiene pues aunque no tubieran provado sino solamente tener las dhas tierras al tienpo de la entrada de los españoles en la tierra bastava para conseguir su justiçia conforme a las provisiones rreales en que se provee y manda que todas las tierras y las demas cosas

esten en el estado que estavan quando los dhos españoles en su rreal nonbre entraron en la tierra e porque no daña a sus ptes ni aprovecha a las contrarias el sinple papel que presentaron que llaman de conçierto porque ni lo es ni tal hizo ni otorgo el caçique prinçipal del dho rrepartimyento ni ya que lo hiziera podia ni puede perjudicar a los dhos sus partes ni a los y^{o}s pobres ni a los demas del dho rrepartimyento pues las dhas tierras eran y son de todos en general e no del dho caçique solamente en particular e asi no pudo perjudicar a sus ptes lo que el dho caçique quiso hazer sin poder pues tocando a todos por todos avia de ser otorgado y aprovado pues ning^{o} puede perder el dominyo de su cosa sin su propio hecho y consentimyento espeçialmente que provara en forma//
/322v/ no aver hecho el dho caçique el dho conçierto y que no rreçibio las obejas que en el se haze minçion sino Fran^{co} de Anpuero su encomendero sin su consentimyento del dho caçique y contradiziendo el dho conçierto y no lo queriendo hazer antes diziendo que no queria ovejas sino las tierras que heran de sus yndios y el dho Françisco de Anpuero contra la voluntad del dho caçique las rreçivio y firmo el dho conçierto y por lo confesado y aver hecho las partes contrarias son vistos y es prueva bastante de no ser suyas las dhas tierras ni tener en ellas nada e pido e suplico se emendase y rrevocase la dha sentençia de vista condenando a las partes contrarias a rrestituçion de las dhas tierras e chacarras y frutos dellas con costas y asimysmo pidio rrestituçion como mejor lugar obiere de derecho de qualquier conçierto que paresçiere perjudicar a sus ptes y juro que no lo pedia de maliçia y asimismo pidio e sup.^{co} se le mandase dar en pu^{ca} forma con çitaçion de la pte nra rreal provision en que se manda que todas las cosas esten en el estado en que estavan quando los españoles entraron en esta tierra de lo qual se mando dar treslado a la otra pte e fue notificado a Fran^{co} de la Torre pr. contrario el qual por otra petiçion que presento en nonbre de los dhos caçique e prençipales e y^{o}s de Canta //
/323r/ rrespondiendo a la dha suplicaçion dixo que deviamos demandar confirmar la sentençia de vista en esta causa dada sin enbargo de lo que la otra pte dize alega que se escluye por todo lo general que avia por espresado y por todo lo que tiene dho y alegado e porque por ser tpo de verano al tpo que se hazia la provança no probo bastantisimamente todas las preguntas del ynterrogatorio que presento e asi se viera claramente como de tienpo ynmemorial a esta parte las dhas tierras de Quibi eran e fueron de las dhas sus ptes y que luego como murio Guaynacaba los dhos yn^{o}s de Chacalla se bolvieron a sus tierras y dexaron aquel pedaço de trras de Quibi que avian entrado lo qual provara agora dandole tienpo conpetente para ello y porque agora biene de nuevo a su notiçia que los dhos y^{o}s de Chacalla no fueron mytimaes puestos por ningun señor de los pasados en las dhas tierras de Quibi sino que haziendo çiertas plegarias por el Ynga que en su lengua se llamava capacoche en las quales los y^{o}s de un valle salian con çiertos mates o ollas pequeñas de sangre yvan ansi dando bozes y si algun yndio hazia derramar qualquier pte de la tal sangre o los que la llevaban la derramavan le davan garrote y enterraban en el propio lugar e los que llevavan //
/323v/ la dha sangre o capacoche hazian termino hasta el lugar donde llegavan e asi vna o dos vezes los dhos yndios de Chacalla con maliçia hizieron las dhas plegarias o capacoche con la qual fueron hasta las dhas tierras de Quibi antes que los dhos yndios de Canta le saliesen a rresçebir y despues por miedo de no derramar la dha sangre de la dha capacoche no los hecharon de las dhas tierras de Quibi aunque heran suyas con la qual color tomaron achaque los dhos y^{o}s de Chacalla de poner mitimaes en la pte de tierra en que se avian entrado quando Guaynacava paso por alli no le hablaron en ello porque se avia hecho por el la dha capacoche y asi despues de muerto como dho tiene se bolvieron a sus tierras las dhas ptes contrarias por lo qual ningun titulo ni color pueden tener ni pretender a las dhas tierras contra lo qual no enpeçe lo que la pte contraria dize que asi lo hallaron los christianos quando entraron en este rreino e que conforme a nra rreal provision los an de dexar alli por quanto aquello se entiende y deve entender de lo que ligitimamente estava hecho por los señores pasados de este rreino e no por cosa tan mal hecha como las ptes contrarias hizieron quanto mas que ya que aquello oviese lugar los yn^{o}s que alli estubiesen en las //
/324r/ dhas tierras de Quibi avian de contribuir e tributar con los dhos sus partes pues el tributo que pagan e pagavan al Ynga es por las tierras que tenian e porque no solamente la tierra que pretenden las partes contrarias pero aun todas las dhas tierras de Quibi no valen tanto como las dozientas ovejas que sus ptes les dieron y que si por nos se provee que lo vayan a ver personas hallaran ser verdad lo que dize porque aun no son veinte fanegas de senbradura por lo qual çesa la rrestituçion que pide las ptes contrarias aunque fueran menores quanto mas siendo como son mayores e tales personas que no les conpete rrestituçion alguna por las quales rrazones nos suplico mandasemos confirmar la dha sentençia e pidio ser rreçivido a prueba por quanto lo mas de lo susodho consiste en provança por vista de ojos para que se vea el valor de las dhas tierras y el lugar donde estan e para que se traigan los dhos e depusiçiones e se tome de los yndios comarcanos e viejos que estan en el dho valle e para ello es neçesario que se nonbren dos personas que las vayan a ver e tasar e hazer las dhas provanças e con esto se ovo la causa por conclusa y para lo demas se pronunçio un auto del tenor sig.^{te}:

/324v/ "En Los Rreyes veinte e tres dias del mes de henero de mill e qui^{o}s e çinquenta e nuebe años aviendo visto los señores presidente e oidores desta Rreal Audiençia este proçesso en lo que las partes piden se nonbren personas que vean por vista de ojos las dhas tierras de Quibi sobre que es este pleito y el sitio donde esta mandaron que cada una de las dhas partes nonbren persona de su pte que las vayan a ver e çerca dello tomen los dhos e depusiçiones de los yndios antiguos mas çercanos que de las dhas tierras tengan notiçia e sobre todo den su paresçer de lo que mas convenga proveer que sea en menos daño e perjuizio de las dhas partes lo qual pase y se haga ante Juan de Ençiso escrivano de Su Mag.^{t} y asi lo pronunçiaron y señalaron de sus rrubricas" el qual dho auto fue notificado a los procuradores de las partes y las dhas partes rreçevidas a prueva con termino de veinte dias con cargo de publicaçion y conclusion y en cunmlimyento del dho auto fueron nonbrados al provinçial de Santo Domingo e Juan de Pedraça para que fuesen a ver el dho sitio de las dhas tierras de Quibi y çerca del sitio y valor de las dhas tierras y para que hiziesen çiertas provanças para averiguar la verdad y se les dio para ello una nra rreal comysion del tenor sig.^{te}:

/325r/ "Don Phelipe por la graçia de Dios rrey de Castilla de Leon de Aragon del Ynglaterra de França de las dos Seçilias de Jerusalen de Navarra de Granada de Toledo de Valençia de Galiçia de Mallorcas de Sevilla de Çerdeña de Cordova de Corçega de Murçia de Jaen de los Algarves de Algezira de Gibraltar de las yslas de Canaria de las Yndias yslas e tierra firme del mar oçeano; conde de

Barçelona; señor de Vizcaya e de Molina; duque de Atenas y de neopatria; marques de Oristan e de Goçiano; archiduque de Austria; duque de Borgoña e de Bravante y Milan; conde de Flandes y de Tirol etc.

A vos el rreverendo padre Frai Gaspar de Carvajal provinçial de la Horden de Santo Domingo de los nros rreinos del Peru salud y graçia.

Sepades que Alonso Moreno en nonbre del caçique e yn^{o}s de Chacalla por una petiçion que presento en la nra Audiençia e Chançilleria Rreal que rreside en la çiudad de Los Rreyes de los nros rreinos antes el dho nro presidente e oidores della nos hizo rrelaçion diziendo que en el pleito que en la dha nra Rreal Audiençia tratan los dhos sus partes con el caçique e yn^{o}s de Canta sobre las tierras de Quibe / por nos estava mandado que las partes nonbrasen terçeros cada vno de su parte para que vean la [*sic*] dichas tierras y el sitio dellas y lo que pueden valer y lo averiguen//
/325v/ con testigos yndios comarcanos segun y como se cont.^{e} en la provision por nos sobre ello dada dirigida a Fran^{co} Hernandez de Guevara ante quien a de pasar lo en ella contenido y que el en nonbre de los dhos sus ptes tenia nonbrado para ello por terçero a vos el dho padre provinçial y la otra parte avia nonbrado al dho Juan de Pedraza porende que nos suplicava e pedia por merçed vos encargasemos hiziesedes lo que por la dha nra provision se manda pues lo teniades açeptado como pareçia por vra çedula en el proçeso de la dha causa presentado y que para la exsaminaçion de los t^{o}s y de lo demas que en la dha causa se oviese de hazer fuese por anbas partes vno de los ynterpretes que rresiden en la dha nra Rreal Audiençia e como la nra merçed fuese lo qual visto por los dhos nro presidente e oidores y lo por ellos çerca dello proveido e mandado fue acordado que deviamos mandar dar esta nra carta en la dha rrazon y nos tuvimoslo por bien por la qual vos encargamos que veais la dha nra provision rreal que de suso se haze minçion y juntamente con el dho Juan de Pedraza terçero nonbrado por parte de los dhos caçique e yn^{o}s de Canta vais al dho asiento y trras de Quibi e alli anbos juntos e no el uno sin el otro por //
/326r/ ante el dho Françisco Hernandez nro escr^{o} e un ynterprete de la dha nra Rreal Audiençia que a ello mandamos vaya por anbas ptes hagais la averiguaçion del sitio e valor de las dhas tierras e todo lo demas en la dha nra provision rreal contenido sin exçeder della en cosa alg.^{a} para con ello poder mejor hazer en la dha causa justiçia a las partes que en ello nos servireis. Dada en la çiudad de Los Rreyes a diez dias del mes de hebrero de mill e qui^{o}s y çinq.^{ta} e nueve años.

El marques / el dotor Bravo de Saravia / el Liçençiado Mercado de Peñalossa / el dotor Gonçalez de Cuenca.

Yo Juan Muñoz Rrico escriv^{o} de camara de Su Sacra Catolica Rreal Mag.^{t} la fis escrevir por su mandado con acuerdo de su presidente e oidores.

Rregistrada Bartolome Gascon.
Por Chançiller: Juan Muñoz Rrico".

Los quales hizieron çiertas probanças y aberiguaçiones y al fin dellas dieron su pareçer como personas que lo avian visto y entendido del tenor sigui^{te}:

"E despues de lo susodho en el dho pueblo de Quibi a veinte dias del mes de hebrero de mill e qui^{o}s y çinq.^{ta} y nueve años ante my Françisco Hernandez de Guebara escrivano de Su Mag.^{t} y rreçevtor nonbrado para la presente causa pareçio el muy rreverendo padre Fray Gaspar de Carvajal provinçial de la Horden de Señor Santo Domingo de los predicadores del monesterio //
/326v/ que de la dha Horden rreside en la çiudad de Los Rreyes e Ju^{o} de Pedraza terçeros nonbrados entre los caçique e yn^{o}s del rrepartimi^{o} de Canta y del rrepartimiento de Chacalla el dho Fay [*sic*] Gaspar de Carvajal nonbrado por parte de los y^{o}s de Chacalla y el dho Juan de Pedraza nonbrado por el caçique e yndios de Canta e unanymes conforme nemine discrepanto dixeron que ellos avian rreçevido la ynformaçion que Su Mag.^{t} por su rreal provision les manda e avian visto e mirado las dhas tierras e medidolas e ynformadose de los yn^{o}s e visto e tanteado e considerado lo que fara al caso sin hazer agravio a las partes a lo que avian alcançado en sus conçiençias le pareçe y fallan que las dhas tierras de Quibi sobre que es este pleito tienen de sitio dos mill pasos de largo y trezientos de ancho e tienen por linderos por una parte el rrio y por la otra camino rreal que viene de Lima al dho pueblo de Canta esto por lo ancho y de lo largo tienen por linderos las tierras de Guancayo rrepartimy^{o} de Gonçalo de Caçeres e de la otra una linde que viene desde una quebrada que baxa de un çerro de la otra banta [*sic*] del rrio que esta en tierra de los y^{o}s de Secos encomendados en Martin Piçarro la qual linde corre al sesgo hasta llegar a la yglesia nueva del pueblo de //
/327r/ Quibi en parejo della las quales tierras estan situadas conforme a los dhos de los testigos por ellos tomada en terminos de yn^{o}s yungas que fueron los primeros poseedores antes que en ellas entrasen Yngas que fueron quietos a los yndios de Collique las quales dhas tierras pareçe avellas dado el Ynga a los yndios yauyos del rrepartimyento de Chacalla y las poseyeron hasta que las vendieron por conçierto a los yn^{o}s de Canta por doçientas ovejas y por vyrtud desta venta las poseen e que en Dios y en sus conçiençias les paresçe que si las dhas tierras estovieran en poder de los españoles y las vendieran podrian valer hasta seisçientos o seteçientos pesos de oro pero que entre los yndios se estiman las tierras que estan en su comarca de otra manera que las de los españoles espeçialmente siendo tierras de coca y por esto no podran determinadamente juzgar el valor de las dhas tierras entre los yn^{o}s porque no tienen otro mantenimy^{o} syno el que de las dhas tierras sacan labrandolas y senbrandolas y esto es su paresçer y lo firmaron de sus nonbres: Fray Gaspar de Carvajal / Juan de Pedraza.

Paso ante mi Françisco Hernandez de Guevara escrivano de Su Mag.^{t}"

Y asimismo las dhas ptes hizieron otras provanças y cada vna dellas por si y fue con ellas avida la dha causa por conclusa e por los dhos presidente //
/327v/ e oidores dada e pronunçiada en ella sentençia del tenor sig.^{te}:

"En el pleito entre partes de la una los yn^{o}s de Chaclla y de la otra los yn^{o}s de Canta y Al^{o} Moreno y Fran^{co} de la Torre sus procuradores / hallamos atento lo nuevamente alegado y provado que devemos rrebocar y rrevocamos la sentençia en esta causa dada por nos el presidente e oidores desta Rreal Audiençia de que por parte de los yn^{o}s de Chacalla fue suplicado e haziendo en el caso lo que de justiçia deve ser hecho condenamos a los dhos yn^{o}s de Canta a que buelvan y rrestituyan las dhas tierras de Quibe sobre que es este pleito a los dhos yn^{o}s de Chacalla con que primeramente los dhos yn^{o}s de Chacalla buelvan y rrestituyan a los dhos yn^{o}s de Canta las dozientas ovejas que por el conçierto en esta causa presentado paresçe que rresçibieron dellos y por esta nra sentençia juzgando asi lo pronunçiamos y mandamos en grado de rrevista El marques / el dotor Bravo de Saravia / el lic.^{do} Mercado de Peñalosa / el dotor Gonçalez de Cuenca."

La qual fue dada e pronunçiada por los dhos presidente e oidores en la dha çiudad de Los Rreyes en onze dias del mes de jullio de mill e qui^{o}s y çinquenta y nueve años y fue notificado a Fran^{co} de la Torre y Al^{o} Moreno procuradores de las partes que se hallaron presentes al pronunçiamy^{o} de la dha sent.^{a} y despues por parte del caçique prençipales e yn^{o}s //
/328r/ del dho rrepartimy^{o} de Chacalla nos fue pedido y suplicado por una petiçion que ante nos presento en la dha Rreal Audiençia que le mandasemos dar e diesemos exsecutoria de las dhas sent.^{as} de vista e rrevista en esta causa dadas e pronunçiadas y por el dho presidente e oidores fue proveido que se le diese dando ante todas cosas las fianças que es obligado a dar conforme a nra rreal hordenança de la qual sent.^{a} por parte de los dhos caçique prinçipales e yn^{o}s del dho rrepartimyento de Canta por una petiçion que presento ante nos en la dha nra Rreal Audiençia fue suplicado de la dha sent.^{a} de rrevista para ante nra Persona Rreal y nos suplico le admitiesemos la dha suplicaçion mandando emendar y rrebocar la dha sentençia de rrevista y confyrmar la de vista lo qual visto por los dhos presidente e oidores la obieron por presentada y le conçedieron la dha segunda suplicaçion para que la pueda proseguir ante nos e los del nro Consejo Rreal de las Yndias y mandaron que se presentase con el proçeso de la dha causa originalmente dentro del termino de la hordenança e so las penas della e aperçibieron a las partes que si tubieren algo que dezir e alegar en la dha causa lo digan y aleguen //
/328v/ en la dha nra Rreal Audiençia antes de sacar el dho proçeso porque ante nos ni en el dho nro Consejo Rreal de las Yndias no an de alegar ni provar otra cosa alguna e que por los mesmos autos y meritos del dho proçeso se a de determinar la dha causa e para ello mandaron çitar a las ptes para que vayan o enbien en seguimy^{o} del dho pleito en la forma acostunbrada y en su rrebeldia le señalaron los Estrados Rreales del dho nro Consejo Rreal de las Yndias lo qual ffue notificado a Françisco de la Torre y Alonso Moreno procuradores de las ptes e por parte del caçique e yndios del dho rrepartimyento de Chacalla ffue suplicado de aver otorgado la suplicaçion para ante nra Persona Rreal y pedido mandasemos enmendar y rrebocar el dho auto por todo lo general del dr^{o} e dezir e alegar le conviene que avia por expreso e porque esta causa es entre yndios e que se devia hazer e aver hecho sumariamente y sola la verdad sabida y que no se avian de admitir las dilaçiones que de contrario se procuran ynjustamente y sin tener causa de litigar que lo hazen a hefeto de estarse en las tierras entendiendo que sus ptes no tienen fiança alguna que dar e no se a de dar lugar a semejante calunia //
/329r/ espeçialmente no valiendo como no valen las tierras la cantidad de la hordenança como pareçe por el pareçer del padre Carvajal e Juan de Pedraza que syn pasion bieron y tasaron las dhas tierras e sin la afiçion que sus partes les tienen e ya que por alguna via se deviese otorgar la dha suplicaçion avia de ser sin fianças pues es notorio ser las tierras del dho su pte y no tener alguna defensa justa las ptes contrarias y estandose como se estan ay las tierras sin que en ellas pueda aver diminuçion ny fraude en que puedan ser perjudicadas las partes contrarias e porque las dhas sus partes no tienen fianças que dar e asi lo juraron en forma e nos suplicaron mandasemos enmendar los dhos autos denegando la dha suplicaçion a las ptes contrarias e declarando no ser sus partes obligados a dar fiança alguna despues de lo qual por una petiçion que el dho caçique del rrepartimyento de Canta presento por si y en nonbre de los demas yndios del dho rrepartimyento no se apartando de la suplicaçion que tienen hecha para ante nuestra Persona Rreal dixeron que a su notiçia era benido que las ptes //
/329v/ contrarias avian pedido que quieren pagarles las dozientas ovejas en dineros y como valian quando ellos las dieron e porque ya que se deshaze la dha transaçion que entre ellos se hizo conforme a derecho se ha de bolver en el punto y estado en que estava quando se las dimos bolviendo a cada vno lo suyo ni mas ni menos que lo dio y asi los dhos y^{o}s de Chacalla son obligados a les bolver las mismas ovejas que les dieron en ovejas y con los partos y pospartos que han rrentado e multiplicado desde que se las dieron hasta agora e asimismo los mejoramy^{o}s que ha hecho en las dhas tierras en coca y arboles mandandolo apreçiar en lo que vale que ellos estan prestos por la conpulsion que se les haze por la dha sent.^{a} de dar a las ptes contrarias los frutos y rrentas que las dhas tierras han rrentado desde el dho conçierto y transaçion aca; e nos suplicaron mandasemos a la pte contraria que para cunplir con la sent.^{a} que en su ffavor se dio que les buelvan las dhas ovejas que les dieron en obejas y con partos y pospartos y los mejoramyentos que ha hecho en las dhas tierras que ellos estan prestos de descontar o pagar lo que pareçiere aver cogido //
/330r/ de los dhos frutos y rrentas / lo qual visto por los dhos presidente e oidores mandaron que los dhos yndios del rrepartimy^{o} de Chacalla buelvan e paguen a los del rrepartimy^{o} de Canta las ovejas de la condenaçion en ovejas e no en dinero como las rreçivieron despues de lo qual el dho don Xpoval Vilcapoma caçique del dho rrepartimy^{o} de Chacalla encomendado en Françisco de Anpuero por si y en nonbre de los demas yndios del dho rrepartimyento puso demanda al dho Fran^{co} de Anpuero diziendo que por el mes de henero del año pasado de quinientos e quarenta e nueve el dho Fran^{co} de Anpuero e Niculas de Rrivera en quien estavan y estan encomendados los y^{o}s del dho Rrepartimyento de Canta hizieron çierto conçierto por el qual dixeron que don Fran^{co} caçique de Chacalla su anteçesor rreçevia de los dhos y^{o}s de Canta dozientas ovejas de la tierra por rrazon de las diferençias que entre sus yndios y los del rrepartimy^{o} de Canta avia sobre las tierras de Quibe siendo lo susodho al contrario de la verdad porque ni el dho don Fran^{co} ni los dhos sus yn^{o}s tal conçierto hizieron ni ellos rresçibieron las dhas dozientas ovejas sino que el dho Fran^{co} de Anpuero a fyn de rreçevir como rresçibio para si las dhas dozientas //
/330v/ ovejas hizo e fyrmo el dho conçierto como a el le paresçio y agora en el pleito que el y sus yn^{o}s an puesto a los dhos yn^{o}s de Canta

sobre las dhas tierras de Quibe por sent.^{a} de rrevista les estan mandadas bolver las dhas tierras con que ante todas cosas buelban y rrestituyan a los dhos yn^{o}s de Canta las dhas dozientas ovejas que por el dho conçierto dieron y que asi esta obligado el dho Françisco de Anpuero como persona en cuyo poder entraron las dhas ovejas a se las dar bolver y rrestituir o pagar con todos los partos y pospartos que dellas ovieren proçedido o pudieren proçeder y proçedieren hasta la rreal rrestituçion pues conforme a la dha sent.^{a} ellos estan obligados a las bolver a los dhos yndios de Canta e no al dho Fran^{co} de Anpuero y nos pidio y suplico le hiziesemos breve y entero cunplimy^{o} de justiçia mandando pareçer al dho Fran^{co} de Anpuero y que con juramento declare si es verdad que rresçibio el para si las dhas dozientas ovejas e se aprovecho dellas y paresçiendo ser asi fuese conpelido e apremiado el dho Françisco de Anpuero a que les de y pague las dhas dozientas ovejas con todos los partos e pospartos hasta la rreal rresti-//

/331r/ tuçion y vista se proveyo que el dho Françisco de Anpuero paresçiese e asi paresçio y hizo la dha declaraçion despues de lo qual la pte de los caçiques e yndios del dho rrepartimyento de Chacalla por petiçion que presentaron dixeron que al tienpo que ellos rresçibieron las ovejas que se dieron al dho Fran^{co} de Anpuero que agora se les mandan bolver no valian mas de dos pesos cada vna y a menos porque la mytad heran pacos y entre ellos no avia henbras y no seria justo que pues las llevo su encomendero y se consumieron que se les mandasen bolver en ovejas queriendo como ellos quieren pagar lo que entonçes valian o dar por ellas ovejas de Castilla que al presente valen mas que entonçes valian las otras e nos suplicaron mandasemos que pagasen en dinero lo que valian las ovejas quando se dieron o se rresçiban por ellas otras tantas ovejas de Castilla pues las tienen y ffue proveido que se guardase la dha sentençia e despues por parte de los dhos yndios de Chacalla fue tornado a suplicar le mandasemos dar la dha exsecutoria segun que la avian pedido //

/331v/ y se mandaron traer los autos e vistos por los presidente e oidores fue pronunçiado un auto del thenor siguiente:

"En Los Rreies en primero de junio de mill e quinientos y sesenta y siete años los señores presidente e oidores desta Rreal Audiençia aviendo visto esta causa mandaron que a los yn^{o}s de Chacalla se les de la exsecutoria que piden sin fianças."

E agora la parte de los dhos caçique prençipales del dho rrepartimyento de Chacalla nos pidio e suplico le mandasemos dar nra carta Executoria Rreal de las dhas sentençias para que lo en ello contenido oviese hefeto o que sobre ello proveyesemos como la nra merçed fuese lo qual visto por los dhos presidente e oidores fue acordado que deviamos mandar dar esta nra carta para vos en la dha rrason y nos tovimoslo por bien por la qual vos mandamos que veais las dichas sentençias de vista y rrevista en la dha causa dadas e pronunçiadas por los dhos presidente e oidores que de suso van yncorporadas e las guardeis y cunplais y executeis y hagais guardar cunplir y executar //

/332r/ en todo e por todo como en ellas se contiene e contra el tenor y forma dellas no vais ni paseis ni consintais yr ni pasar en manera alguna so pena de la nra merçed e de cada qui^{o}s p^{o}s de oro para la nra camara.

Dada en Los Rreyes a diez y nueve dias del mes de junio de mill e qui^{o}s y sesenta y siete años.

Yo P^{o} de Avendaño escrivano de camara de la Mag.^{tt} Rreal; la fiz escrevir por su mandado con acuerdo del su presidente e oidores.

Rr.^{da} Juan de Murga
Chanciller Gaspar de Solis

/332v/ En la çiudad de Los Rreyes de sus rreinos del Peru veinte e çinco dias del mes de junio año del Señor de mill e qui^{o}s e sesenta e siete años ante el muy mag^{o} señor Gomes Caravante de Maçuelas alcalde ordinario en esta çiudad por Su M.^{t} pareçio Di^{o} de Ocanpo pr^{o} de Corte en nonbre de los caçiques e prinçipales del rrepartimi^{o} de Chacalla encomendado en Fran^{co} de Anpuero e presento esta carta executoria rreal de Su M.^{t} e pidio que la obedesca e cunpla e en cunplimi^{o} della le mande dar e de su mandami^{o} para que sus y^{o}s sean metidos e anparado en la tenençia e posesion de las tierras e de lo demas cont.^{do} en esta dha prov.^{on} e pidio just. e luego el dho señor alcalde abiendo bisto esta prov.^{on} rreal de Su M.^{t} la tomo e beso e puso sobre su cabeça e dez que esta presto de la cunplir como en ella se contiene como carta e mandado de su rrey e s^{o}r nat.^{al} caçique en Dios nro s^{o}r dexe bivyr e rreinar por largos tpos / t^{o}s: Al^{o} de Valençia e Ant^{o} Lopes.

Pedro de Valverdeescr^{o} pu^{co}

/333r/ Alguazil mayor desta çiudad de Los Rreyes o vro lugarteni.^{e} y Ant^{o} Lopes alguazil del Canpo o qualquier de vos; yo os mando que vais a unas tierras e chacarras de coca que se dizen de Quibi que poseen los yndios del pueblo de Canta e dando e entragando los caçiques e prinçipales de Chacalla a los caçiques e yndios de Canta dozientas ovejas / meted e anparad en la ten.^{a} e posesion de las dhas tierras e chacaras de Quibi a los caçiques e p.^{les} del dho rreptimi^{o} de Chacalla o a qualquier dellos para que las tengan e posean conforme a las sentençias dadas en la causa que tratan en esta probision rreal e conforme a ella e asi lo cunplid so pena de çien p^{o}s para la camara de Su Mag.^{t} Fho en los Rreyes a veinte e çinco dias del mes de junio de mill e qui^{o}s e sesenta e siete años.

Gomez Caravantes
de Macuela

m^{do} del s^{o}r Alcalde
P^{o} de Valverde escr^{o} pu^{co}

/334r/ Entregan las ovejas conforme a la sentençia.

En el asiento e pueblo de Quybi termyno e jurisdiçion que dizen ser del pueblo de Canta encomendado en Nyculas de Rribera vezino de la çibdad de Los Rreyes en sabado veynte e ocho dias del mes de junyo año del Señor de myll e qui^{o}s e sesenta e siete años estando dentro de unas tierras que se dizen de Quybi contenidas en bna Executoria Rreal de Su Majestad sobre que se a tratado pleito en la Rreal Avdiençia de la dha çibdad de Los Rreyes entre el caçique e yndios del pueblo de Chacalla encomendado en Fran^{co} de Anpuero de una parte e de la otra el caçique e yndios del dho pueblo de Canta y estando presente Antonyo Lopez alguazil del Canpo por ante my Alonso Diaz de Gibraleon escrivano de Su Mag.^{t} e de los testigos de suso exp.^{o}s paresçio presente don Juan Bautista Caxa Guaraca caçique prinçipal del dho rrepartimy^{o} de Chacalla por si e en nonbre de los prinçipales e yndios del dho rrepartimyento e presento una executoria rreal de Su Magestad hemanada de los señores presidente e oydores de la dha Rreal Audiençia desta dha çibdad e sellada con su rreal sello; con un mandamy^{o} encorporado al cabo della de Caravantes de Maçuelas alcalde hordinario de la dha çibdad de Los Rreyes e firmado de Pedro de Balverde escrivano pu^{co} della e presentada le pidio e rrequyrio que le metiese e diese la posesion de las dhas tierras de Quybi de que en ella se hazia mynçion como por el dho mandamy^{o} el dho alcalde se lo mandava e porque hera obligado a entregar primero dozientas cabeças de ovejas que el cunplimyento dello fazia presentaçion e muestra dellas para que se cunpliese lo contenydo en la dha executoria e mandamy^{o} e para ello pidio que se buscase al caçique e prinçipales de Canta sy estavan en el dho pueblo de Quybi para que las rresçibiesen y el dho Antonyo Lopez alguazil por ante my el dho escr^{o} busco //

/334v/ en el dho pueblo de Quybi para ver sy estava en el el caçique prinçipal del dho pueblo de Canta para que rresçibiese las dhas dozientas cabeças de ovejas e en cunplimy^{o} dello se pregunto a çiertos prinçipales del dho pueblo de Canta e yndios del que estavan en el dho pueblo de Quybi los quales dixeron que el dho caçique prinçipal de Canta estava en la dha çibdad de Los Rreyes e que no estava en Canta ny alli e el dho Antonyo Lopez alguazil por ante my el dho escrivano de pedimy^{o} del dho don Juan Bautista caçique del dho pueblo de Chacalla fizo paresçer ante si a don Fran^{co} Poma prençipal del dho pueblo de Canta y hermano de don Fran^{co} Guaman Guaranga caçique prinçipal del dho pueblo de Canta e don Pedro prinçipal del pueblo de Llaque e don Xpoval Carvavilca e Fran^{co} Pomacaxa todos prinçipales del dho pueblo de Canta e otros muchos prinçipales e yndios del dho pueblo de Canta que presentes estavan e asy paresçidos de pedimy^{o} del dho don Joan Bautista por si e en nonbre de los prinçipales e yndios del dho pueblo de Chacalla por ante my el dho escr^{o} rrequyrio a los dhos prinçipales atento que el dho su caçique no estava alli syno en la çibdad de Los Rreyes como ellos lo avian confesado que rresçibiesen luego las dhas dozientas cabeças de ovejas que por la dha rreal executoria se le mandavan dar las que dioles en la plaça del tanbo del dho pueblo de Quybi; conte e ovo en las que asi traxeron dozientas cabeças de ganado ovejuño de la tierra e asy contadas los dhos prinçipales y espeçialmente el dho don Fran^{co} Poma dixeron que el no hera prinçipal del dho pueblo de Canta e que el caçique prinçipal del dho pueblo de Canta estava en la çibdad de Los Rreyes e que con el se entienda lo susodho porque no quiere rrresçibir oveja nynguna ny los demas prinçipales ny entremeterse en ello.

A todo lo gen.^{al} fueron presentes por testigos: Pedro de Niça ynterpetre de la dha Rreal Avdiençia por cuya lengua paso e se hablo todo lo susodho e el dho Antonyo Lopez alguazil e otros muchos prinçipales.

/335r/ E luego el dho don Joan Bautista caçique del dho pueblo de Chacalla por si e en nonbre de los dhos sus prinçipales e yndios por la dha lengua del dho Pedro de Niça por ante my el dho escr^{o} torno a rrequyrir al dho don Fran^{co} Poma que pues el es prinçipal y hermano del dho caçique de Canta que rresçiba las dhas dozientas cabeças de ovejas que presentes estavan e yo el dho escr^{o} avia contado que por la dha executoria rreal se le mandavan dar pues el dho caçique no estava alli para las rresçibir e no queriendolas rresçibir ny entregarse en ellas / que el cunplia lo que por su parte hera obligado e desde luego hazia dexaçion dellas e se desistia e apartava del señorio que en ellas tenya e las dexava alli en la dha plaça del dho pueblo de Quybi e protesto que si se perdiesen o muriesen o otro algun rriesgo por ellas vinyese que no fuese a su culpa e cargo syno al de los dhos caçiques e prinçipales del dho pueblo de Canta porque el con lo susodho tenya cunplido e cunplia con lo que por la dha executoria rreal e mandamy^{o} del dho alcalde se le mandava e pues el avia cunplido con ello e traydo las dhas ovejas e dexadolas alli contadas en su tierra dellos e no las avian querido rresçibir que debaxo de la dha protestaçion que tenya hecha atento que el dho caçique de Canta estava en la dha çibdad de Los Rreyes rrequyrio por ante my el dho escrivano al dho Antonyo Lopez alguazil por la dha lengua del dho P^{o} de Niça estando presentes los dhos prinçipales e yndios del dho pueblo de Canta que le de y meta en la posesion de las dhas trras de Quybi como por la dha executoria e mandamy^{o} se le manda e dada le anpare e defienda en ella dexandole quyeto e paçifico en la dha posesion de las dhas trras de Quybi para que las tenga el e sus prinçipales e yndios e posean como cosa suya que son e lo pidio por testimy^{o} para guardar de su derecho. T^{o}s los dhos.

/335v/ E luego el dho don Fran^{co} Poma herm^{o} que dixo ser del dho caçique de Canta en presençia de los dhos prençipales dixo que el no tiene que dezir cosa ninguna a lo susodho mas de lo que dho tiene e que el dho don Juan Bautista ponga en el dho pueblo de Quybi los yndios yungas sujetos a el e que ellos tienen en el dho pueblo de Quybi. T^{o}s los dhos e el dho P^{o} de Niça por cuya lengua paso lo susodho.

E luego el dho Antonyo Lopez alguazil en cunplimyento de la dha executoria rreal e por virtud del dho mandamyento del dho alcalde por ante my el dho escrivano atento a que el dho don Joan caçique del dho pueblo de Chacalla avia traydo las dhas dozientas cabeças de ganado para las dar como hera obligado e visto que no las quisieron rresçibir los dhos prinçipales de Canta e que las dexo alli en la plaça del dho pueblo de Quybi e se desistio dellas tomo por la mano al dho don Joan Bautista caçique prinçipal del dho pueblo de Chacalla e asy tomado le metio dentro de las tierras de Quybi de que en la dha executoria se faze mynçion que son en el

valle del dho pueblo de Quybi las quales enpieçan por lo alto de la barranca del valle donde esta el dho pueblo de Quybi desde un çerrillo de piedras que esta junto e frontero de la yglesia nueva del dho pueblo de Quybi donde solia estar una crus e desde alli hasta dar en la otra banda del rrio por donde se haze una quebrada que viene a dar a la dha yglesia nueva del dho pueblo de Quybi e de alli para abaxo corriendo por el dho valle hasta confinar con las tierras de Guancayo que agora dizen que son de Lorenço Estupinan e asy metido en ellas dixo que le dava e dio la posesion de todas ellas como por el dho mandamyº se le manda e conforme a la dha //
/336r/ executoria rreal e el dho don Joan por si e en nonbre de los dhos prinçipales e yndios del dho pueblo de Chacalla aprehendiendo la dha posesion de las dhas tierras se paseo por ellas e cortando de las rramas de los arboles que por alli avia y echandolas de una pte a otra e de otra a otra e tirando asimysmo de las piedras e terrones que por alli avia todo lo qual dixo que fazia e fizo en señal de verdadera e paçifica posesion que tomava e aprehendia e tomo e aprehendio de las dhas tierras rreal avto en el corpora evelas y la qual dixo que tomava e aprehendia syn que por persona alguna le fuese contradho ny puesto nyngun enbargo ny ynpadimyº en ella e de como la tomo e aprehendio en la manera susodha e de como quedo en la dha su posesion paçifica por ante my el dho escrº e en presençia del dho alguazil e del dho Pedro de Niça ynterpetre mando hazer e hizo un mojon grande de piedra e lodo en el dho çerrillo de piedras donde solia estar la dha crus frente de la dha yglesia del dho pueblo de Quybi e otro de la otra vanda del rrio en medio de la quebrada que deçiende de la sierra e en medio destos dos mojones que es en medio del dho valle estava hecho otro mojon grande de piedra que paresçia ser hecho de tienpo antiguo e con estos tres mojones quedaron amojonadas las dhas trras e de alli para abaxo hazia Guancayo corriendo por el dho valle abaxo hasta confinar con las dhas trras de Guancayo el dho don Joan asimysmo tomo e aprehendio la posesion de todas ellas paseandose por ellas e cortando de las rramas de los arboles e de las piedras que alli avia y echandolas e tirandolas de una pte a otra e de otra a otra e en señal de la //
/336v/ dha su posesion para que se entendiese e supiese hasta donde llegavan las dhas sus tierras subio a lo alto del valle adonde estava una crus alta puesta junto al camyno rreal que viene a Lima e alli enfrente de la dha crus el dho don Joan mando hazer e fizo vn mojon de piedra el qual quedo hecho e acabado de hazer pidio a my el dho escrº le diese por testimyº como el dho alg.zil le avia dado la posesion de las dhas tierras de suso declaradas e de como el la avia [*sic*] tomado quyeta e paçificamente e sin contradiçion de persona alguna que ay se estoviese ny se lo contradixese e de como el dho alguazil le dexo en ella quyeta e paçificamente en la manera que de suso esta dho e declarado e yo el dho escrº de su pedimyº le di ende este que es fecho en el dho dia mes e año susodhos. Tºs los dhos.

E luego el dho Antonyo Lopez alguazil en continuaçion de la dha posesion por ante my el dho escrivano subio a lo alto del dho valle donde esta el dho pueblo de Quybi e desde el mojon primero que se fizo donde solia estar vna crus junto a la yglesia nueva del dho pueblo bajando de alli yendo hazia Canta donde estan las casas del dho pueblo de Quybi desde alli la barranca del valle en la mano hasta donde se parten las tierras de Canta e Chacalla hasta un arroyo que se dize Cacalarca / tomo por la mano al dho don Joan Bautista e le metio dentro de las tierras que por alli avia hasta confinar con la syerra e de alli para abaxo hasta el postrer mojon de piedra que se fizo donde estava la dha crus e donde se acabavan las dhas tierras de Quybi de que se le dio la dha posesion e asy tomado le metio dentro dellas e dixo que le dava e dio la posesion de todas ellas segun e como le avia dado de las trras del dho valle e el dho don Joan por si e en nonbre de sus prinçipales e yndios se paseo por ello e dixo que tomava e aprehendia la pose-//
/337r/ sion de todo ello quyeta e paçificamente e sin contradiçion de persona alguna atento que lo susodho entra e se yncluye en las tierras de Quybi que por la dha executoria se le mandan dar e en señal de la dha su posesion que tomo e aprehendio de lo susodho mando hazer e se fizieron desde donde estan fundadas las casas del dho pueblo de Quybi junto a la barranca del valle e de la vanda de un arroyo de agua que por alli corre e el dho arroyo arriba hazia Canta por la orilla de la dha barranca hasta el dho arroyo Cacalarca se hizieron nueve mojones de piedra e lodo con los quales quedaron amojonadas e señaladas las tierras de Quybi que por la dha executoria e mandamyento se le mandan dar; todo lo qual el dho don Joan dixo que hazia e hizo en señal de la dha su posesion que tomava e aprehendia e tomo e aprehendio segun e como de las primeras tenya tomada e pidio a my el dho escrº le diese por testimyº de como el dho alguazil le dava la dha posesion quyeta e paçificamente e sin contradiçion de persona alguna e de como el quedava en ella en la manera susodha e yo el dho escrº de su pedimyº le di ende este que es fecho en el dho dia mes e año susodhos.
Tºs los dhos.

E luego en este dho dia mes e año susodhos acabado de tomar la posesion de las dhas tierras de Quybi el dho don Joan e dadosela el dho alguazil e hecho los mojones de que en ellas se haze mynçion el dho don Joan Bautista por sy e en n.e de los dhos sus prinçipales e yndios del dho pueblo de Canta por ante my el dho escrº por lengua del dho Pedro de Niça rrequyrio al dho don Franco Poma e a todos los demas prinçipales del dho pueblo de Canta que son los dhos don Pedro prinçipal del pueblo de Llaque e don Xpoval Carvabilca e a Franco Pomacaxa e a don Pedro Yavre e a don Gº Carvabilca e a otros muchos prinçipales e yndios del pueblo de Canta que presentes estavan que no toquen //
/337v/ en los mojones que tiene hechos por donde tiene amojonadas e señaladas sus tierras de Quibi de que oy dho dia se le a dado posesion y el la a tomado ny se los derriben con aperçibimyº que les haze e si a ellos le tocaren o se los derribaren o mandaren derribar que se quexara dellos como de personas que van contra el mandamyento rreal e quebrantan sus rreales provisiones demas que cobrara dellos todas las perdidas costas e daños que por rrazon dello le vinyere y el dho Antonyo Lopez alguazil de pedimyº del dho don Joan por lengua del dho Pº de Niça e por la suya que les dio a entender que asi lo hiziesen e cunpliesen e no lo quebrantasen lo que por el dho don Joan les hera rrequerido so pena de quyºs pºs para la camara e fisco de Su Mag.t demas que se proçederia contraellos por todo rrigor de justiçia los quales dhos prinçipales no rrespondieron cosa ninguna Tºs los dhos Pedro de Niça e Antonyo Lopez.

E luego el dho don Joan asymysmo por ante my el dho escrº por la dha lengua del dho Pº de Niça rrequyrio a los dhos prinçipales e yndios del dho pueblo de Canta de suso declarados que las aguas que las dhas tierras tienen de Quybi de que tomo e se le dio la

dha posesion e pasan por ellas que no se las quyten ny le perturben el rregar de las dhas tierras con aperçibimyento que les haze que asymysmo se quexara dellos como de personas que procuran ynquietar a los dhos naturales e quebrantar el mandamy^{o} de Su Magestad demas de cobrar dellos todos los daños perdidas e menoscabos que por rrazon dello le vinyeren y el dho Antonyo Lopez alguazil de su pedimyento se lo rrequyrio e mando asi lo cunplan e guarden e no lo quebranten so pena de quy^{o}s p^{o}s para la camara e fisco de Su Magestad los quales dhos prinçipales no rrespondieron cosa ninguna. Testigos los dhos P^{o} de Niça e Ant^{o} Lopez.
/338r/ E luego el dho don Joan Bautista caçique del dho pueblo de Chacalla por ante my el dho escr^{o} por la dha lengua del dho P^{o} de Niça rrequyrio al dho don Fran^{co} Poma e a los demas prinçipales e yndios del dho pueblo de Canta que presentes estavan que le dexasen libres e desenbargadas e se las desocupe e se salga dellas a hazer casas a otra pte porque el por si e en nonbre de los dhos sus prinçipales e yndios las quyere poblar e hazer casas en ellas con aperçibimy^{o} que no lo haziendo asi si alg.^{a} cosa sobre ello suçediere que no sea a su culpa e cargo syno al dellos porque el se lo rrequyere por bien de paz e concordia / lo qual asymysmo por my el dho escr^{o} fue notificado a Antonyo Mychuay e M^{yn} Monça e Alonso Cavac yndios naturales del dho pueblo de Quybi que le dexen libres e desenbargadas sus casa e tierras e se vayan a bivir a otra pte todos los quales juntamente con Domyngo Çaguaque yndio del dho pueblo de Quybi dixeron que lo oyan e que en Lima estava su caçique; a todo lo qual fueron presentes por testigos los dhos P^{o} de Niça por cuya lengua paso lo susodho e el dho Antonyo Lopez e M^{yn} Piçarro prinçipal del rreptmy^{o} de Chacalla e Joan Yapa prinçipal del dho rreptmi^{o}.

E despues de lo susodho en el dho dia mes e año susodhos desde a obra de una ora poco mas o menos despues de aver dado la posesion de las dhas tierras de Quybi de suso declaradas al dho don Joan caçique de Chacalla e quedado en ellas quyeta e paçificamente como esta dho e declarado en la dha posesion los prinçipales e yndios del dho pueblo de Chacalla vinyeron ante my el dho escr^{o} en Antonyo Lopez alguazil dando bozes que los prinçipales e yndios de Canta a quyen se avia notificado todo lo de suso contenido e con quyen avian pasado todos los autos desta dha posesion //
/338v/ le avian quytado e derribado los dhos mojones e que asi en presençia de my el dho escr^{o} avia puesto en las dhas trras que pedian e rrequerian lo saliese a ver para que lo viese por vista de ojos e como tenyan en poco los mandatos de la Rreal Justiçia e luego yo el dho escr^{o} en cunplimy^{o} de lo susodho sali del tanbo del dho pueblo de Quybi donde estava y el dho Antonyo Lopez alguazil comygo y el dho P^{o} de Niça ynterpetre e vide como don P^{o} Guari e don Xpoval Carvavilca e don G^{o} Carvavilca estavan en el çerro donde el dho mojon primero se avia hecho junto a la yglesia del dho pueblo de Quybi que ellos e los demas prinçipales e yndios de Canta acabavan de deshazer el dho mojon e se deçendian del dho çerro despues de aver acabado de deshazer el dho mojon e deçendidos abaxo llego a ellos por ante my el dho escr^{o} el dho Antonyo Lopez alguazil e les pregunto que por que avian derribado el dho mojon los quales dhos prinçipales de suso declarados dixeron que avian derribado el dho mojon porque les tomavan sus tierras y el dho alguazil por aver delinquydo en lo susodho prendio a los dhos prinçipales e asy presos los tuvo para los traer a la çibdad de Los Rreyes e los entregar por presos al alcayde de la Carçel Rreal de Corte desta dha çibdad. T^{o}s los dhos.

E luego el dho don Joan Bautista caçique del dho pueblo de Chacalla por sy e en nonbre de los dhos sus prinçipales e yndios por ante my el dho escr^{o} dixo que se querellava e querello de los dhos prinçipales de suso declarados e de todos los demas prinçipales e yndios del dho pueblo de Canta que fueron en derribar los dhos mojones e paresçieron averlos derribado e pidio al dho alguazil //
/339r/ los trayga presos a la dha çibdad de Los Rreyes para que en ella por los señores presidente e oydores sean castigados como personas que cometieron delito en quebrantar lo proveydo e mandado por Su Magestad e protesto de dar la dha querella mas en forma ante los dhos señores presidente e oydores de los dhos prinçipales e yndios de Canta la qual juro en forma.

E despues de lo susodho en beynte e nueve dias del dho mes de junyo del dho año de myll e quy^{o}s e sesenta e siete años el dho Antonyo Lopez alguazil de pedimyento del dho don Joan Bautista despues de aver derribado los prinçipales e yndios de Canta los dhos mojones por ante my el dho escr^{o} torno a dar la posesion de las dhas trras al dho don Joan e le anparo en ellas e el dho don Joan la tomo segun e como primero la avia tomado e lo pidio por testimy^{o} e yo el dho escr^{o} le di ende este que es fho en el dho dia mes e año susodhos. T^{o}s los dhos Antonyo Lopes alguazil e el dho P^{o} de Niça e los dhos M^{yn} Piçarro e Joan Yapa prinçipales del dho pueblo de Chacalla.

E yo el dho Alonso Diaz de Gibraleon escr^{o} de la Mag.^{d} Rreal en la su Corte rreynos e señorios presente fuy en uno con los dhos testigos e a todo lo que dho es e en esta dha posesion e autos se contiene //
/339v/ e de pedimy^{o} del dho don Joan Bautista caçique del pueblo de Chacalla lo escrevi e por ende fis aquy este myo signo a tal en testimy^{o} de verdad.

Alonso Diaz de
Gibraleon escrivano

Dr^{o}s desta posesion e auto de que en ella se haze mynçion: ocho p^{o}s de plata corriente.

E despues de lo susodho en el dho dia domyngo veynte e nueve de junyo del dho año de myll e quy^{o}s e sesenta e siete a^{o}s el dho don Joan Bautista caçique del dho pueblo de Chacalla paresçio ante my el dho escr^{o} e me presento vna provision rreal hemanada de los señores presidente e oydores del Avdiençia e Chançilleria Rreal que rreside en la dha çibdad de Los Rreyes el thenor de la qual es este que es el syguiente:

/340r/ Don Phelipe por la graçia de Dios rrey de Castilla de Leon de Aragon de las dos Seçilias de Jerusalen de Navarra de Granada de Toledo de Valençia de Galizia de Mallorcas de Sevilla de Cerdeña de Cordova de Corçega de Murçia de Jaen de los Algarves de Algezira de Gibraltar de las yslas de Canaria de las Yndias yslas y tierra firme del mar oçeano; conde de Flandes e de Tirol etc.

por quanto Diego de Ocanpo en nonbre de don Juan Baptista caçique del rrepartimiento de Chacalla prinçipales e y[o]s del por una petiçion que presentaron en la nuestra Rreal Audiençia y Chançilleria Rreal que por nuestro mandado rreside en la çiudad de Los Rreyes ante el presidente e oydores della nos fue hecha rrelaçion diziendo en la causa que an tratado con el caçique e yn[o]s de Canta sobre las tierras de Quibi que los dhos sus partes an ydo a entregar a los susodhos las dozientas ovejas conforme a la sentençia dada en la dha causa y que el dho caçique de Canta anda en cubierto por Maranga y otras partes desta dha çiudad por no las yr a rreçebir y molestar a los dhos sus partes y nos fue pedido y suplicado mandasemos proveer de manera que entregandolas a qualesquier yndios o prinçipales de Canta cunplan los dhos sus partes porque de otra manera nunca se acabara la dha causa o como la nuestra merçed fuese lo qual visto por el dho presidente e oydores fue acordado que deviamos mandar dar esta nuestra carta para vos en la dha rrazon e nos tovimoslo por bien / por la qual mandamos que buscando en su casa al dho caçique de Canta la parte del dho caçique de Chacalla para le entregar las dhas dozientas ovejas y haziendolo saber en ella y no lo hallando para el dho entrego las entreguen a sus prençipales del dho rrepartimiento de Canta y entregadas con su carta de rreçibo sea visto aver cunplido el dho entrego con el el dho caçique de Chacalla.

Dada en Los Rreyes a veinte y ocho de junyo de mill e qui[o]s y sesenta y siete años.

Yo P[o] de Avendaño escrivano de Camara de la Mag.[t] e la fiz escrevir por su mandado con acuerdo del presidente e oydores.

/340v/ El Licen[do] Castro — El D. Bravo de Saravia — El Lic.[do] don Alv[o] Ponçe de Leon — El Lic[do] Ser[o] de Vigil

Para que si la parte del caçique de Chacalla buscare al caçique de Canta en su casa y haz.[do] lo saber en ella no le hallaren para le entregar las dozientas ovejas se entreguen a los prinçipales del dho rrepartimiento de Canta y con esto cunpla con el dho entrego.

/342r/ Para que si la parte del caçique de Chacalla buscare al caçique de Canta en su casa y haziendolo saber en ella no le hallaren para le entregar las dozientas ovejas se entreguen a los prinçipales del dho rrepartimy[o] de Canta y con esto cunpla con el dho entrego.

Don Felipe por quanto Diego de Ocanpo en nonbre de don Joan Bautista caçique del rrepartimy[o] de Chacalla prinçipales e yndios del por bna peti[on] que presentaron en la nra Rreal Audiençia e Chançilleria que por nro mandado rreside en la çibdad de Los Rreyes ante el presidente e oydores della nos fue hecha rrelacion diziendo en la causa que han tratado con el caçique e yndios de Canta sobre las tierras de Quyvi que los dhos sus partes han ydo a entregar a los susodhos las dozientas ovejas conforme a la sentençia dada en la dha causa y que el dho caçique de Canta anda en cubierto por Maranga y otras partes desta dha çibdad por no las yr a rreçevyr y molestar a los dhos sus partes e nos fue pedido e suplicado mandasemos probeer de manera que entregandolas a qualesquier yndios o prinçipales de Canta cunplan los dhos sus partes porque de otra manera nunca se acabara la dha causa e como la nra md fuese lo qual visto por el dho presidente e oydores fue acordado que debiamos mandar dar esta nra carta para vos en la dha rrazon e nos tobimoslo por bien por la qual mandamos que buscando en su casa al dho caçique de Canta la parte del dho caçique de Chacalla para le entregar las dhas dozientas ovejas y haziendolo saber en ella y no lo hallando para el dho entrego las entreguen a sus prinçipales del dho rrepartimy[o] de Canta y entregadas con su carta de rreçibo sea visto aver cunplido el dho entrego con el el dho caçique de Chacalla. Dada en Los Rreyes a veinte e ocho de junyo de myll e quy[o]s e sesenta e siete años.

/343r/ Muy pr[o] s[o]r
Pide que pues no los pude prender el aguazil m[de] soltar a los dhos.

Joan de Arrandolaça en n[e] de don P[o] y don Gonçalo yndios prinçipales de Canta y don Xpoval e don Fran[co] asimysmo yndios prinçipales de Canta / digo que a los dhos mis partes traxo presos un alguazil llamado Ant[o] Lopez porque no quysieron rresçibir çiertas ovejas de los yndios de Chacalla y tomaron color para ello porque dixieron que no pusiesen en sus chacaras çiertos paredones que ponyan diziendo que yban contra la ex.[a] emanada desta Rreal Aud.[a] siendo como es ansi que los dhos prinçipales no supieron de la dha ex.[a] cosa alguna ny les fue notificada mas de verse yr a despojar de sus trras lo qual no se puede hazer conforme a dr[o] hasta ser oydos los posehedores dellas.

A V. Al. pido e sup.[co] m[de] soltar librem[te] a los dhos mis ptes pues no los pudo prender el dho alguazil ny tubo juridiçion para ello e sobre todo pido just.[a] e para ello etc.

El Licen.[do] de Leon — J. Arrandolaça

/343v/ En Los Rreyes prim^{o} dia del mes de jullio de mill e qui^{o}s y sesenta y siete años ante los señores presidente e oydores en audiençia publica se leyo esta petiçion y los dichos señores la mandaron llevar a la carçel con la ynformaçion.

Ju^{o} G^{o}s Rincon

/344r/ Confision de don P^{o} Yavre.

E despues de lo susodho en la dha çibdad de Los Rreyes en treynta dias del mes de junyo de myll e quy^{o}s e sesenta e siete años est.^{do} preso en la carçel rreal de Corte don P^{o} Yavre prinçipal del pueblo de Canta que por otro nonbre se dixo don P^{o} Guari del qual fue tomado e rresçebido juram^{o} en forma devida de dr^{o} el qual lo fizo e so cargo del prometio de dezir verdad e le fueron fhas las preguntas syguientes:

- Preguntado como se llama e sy es prinçipal del pueblo de Canta dixo que se llama P^{o} Yavre e que es prinçipal del pueblo de Canta.

- Preguntado sy el viernes que agora paso que se contaron veynte e siete deste presente mes de junyo si vido que fueron al pueblo de Quybi un alguazil e un escr^{o} a dar la posesion de las trras de Quybi a don Joan Bautista caçique de Chacalla dixo que es verdad que los vido en el dho pueblo de Quybi e que llegaron el viernes el dho alguazil e escr^{o} e que yvan a dar la dha posesion de las dhas trras de Quybi.

- Preguntado sy el sabado luego syguyente rrequyrieron a este confesante e a don Fran^{co} Poma e a otros prinçipales de Canta que estavan alli el dho alguazil e escr^{o} de parte del dho don Joan Bautista que rresçibiesen las dozientas ovejas que se le mandavan dar de la trra e que no las quysieron rresçibir e se quedaron alli / dixo que es verdad que vido las dhas ovejas e que este confesante ny los demas prinçipales de Canta no las quisieron rresçibir diziendo que en los tienpos pasados los caçiques e yndios por el amystad que tenyan les avian dado las dhas ovejas e que no estava //
/344v/ alli su caçique prinçipal e que asy no las rresçibieron el ny don Fran^{co} ny otros prinçipales que estavan alli.

- Preguntado sy es verdad que despues de aver contado las dhas ovejas e no aviendolas querido rresçibir sy Antonyo Lopez alguazil dio la posesion de las dhas trras de Quybi al dho don Joan Bautista caçique del dho pueblo de Chacalla / dixo que oyo dezir al dho alguazil e escr^{o} como se tratava de dar la posesion de las dhas trras al dho don Joan pero que quando le metieron en la posesion dellas no lo vido este confesante.

- Preguntado sy es verdad que despues de aver tomado la posesion de las dhas trras de Quibi el dho don Joan Bautista se mando hazer e se fizo un mojon grande de piedra e lodo donde solia estar una crus enfrente de la yglesia e despues de hecho sy este confesante e los demas yndios e prinçipales de Canta que alli estavan sy lo derrocaron e deshizieron / dixo que es verdad que la crus que esta frente de la yglesia e junto della que esta alli la dha crus e que el mojon que agora nuevamente se puso endonde la pregunta dize lo pusieron e metieron treynta e siete braças en la trra adentro de los yndios de Canta e que junto alli va una açequia donde los dhos yndios de Canta sacan una açequya para rregar sus sementeras e que como estava el dho mojon puesto en su //
/345r/ perjuyzio este confesante e los demas prinçipales que estan presos e otros muchos yndios e yndios [*sic*] que le ayudaron asi del pueblo de Quybi como de Canta e desfizieron e derrocaron e quytaron de alli el dho mojon que estava rrezien hecho e puesto por el dho alguazil e escr^{o} e asi derribado el dho mojon llego luego yncontinente el dho Antonyo Lopes alguazil e prendio a este confesante e a otros yndios e ato las manos atras a este confesante e a otro yndio con el e que luego desde a un rrato le desato e lo traxo preso a esta carçel donde agora esta al qual traxo porque avia derribado el dho mojon que el avia mandado poner e que esta es la verdad de lo que pasa e sabe para el juram^{o} que fizo e no lo firmo porque no sabia lo qual declaro por lengua de Juan de Alvarado ynterpetre desta Rreal Avdiençia el qual no lo firmo porque no sabia escrevyr.

ante mi A^{o} Diaz de Gibraleon
escrivano

/346r/ Confision de don G^{o} Carvavilca.

E despues de lo susodho en la dha çibdad de los Rreyes en primero dia del mes de jullio de myll e qui^{o}s e sesenta y siete años estando preso en la carçel rreal desta Corte don Gonçalo Carvavilca prinçipal del pueblo de Canta del qual fue tomado e rresçibido juramento en forma devida de dr^{o} el qual lo fizo e so cargo del prometio de dezir verdad e siendo preguntado por lengua de Joan de Alvarado ynterpetre desta Rreal Audiençia le fueron fhas las pgtas syg.^{es}:

- Preguntado como se llama dixo que don Gonçalo Carvavilca e que es prinçipal del pueblo de Canta.

- Preguntado sy vio que fueron al pueblo de Quybi el viernes que agora paso un alguazil e un escr^{o} a dar la posesion de las trras de Quybi a don Joan Bautista caçique de Canta [*sic*] por virtud de una executoria rreal de Su Mag.^{t} Dixo que es verdad todo lo contenido en la dha preg.^{a} como en ella se contiene porque asi lo vido.

- Preguntado sy vio que luego el sabado syguyente traxeron alli al dho pueblo de Quybi el dho Joan caçique de Chacalla dozientas ovejas para las entregar al caçique e prinçipales de Canta e que por no las rresçibir le dieron la posesion de las dhas trras de Quybi / dixo que no vido las dhas ovejas ny menos dar la dha posesion porque el estava metido en su casa e no lo vido.

- Preguntado sy despues que se le dio la posesion de las dhas trras al dho don Joan caçique de Chacalla sy mando fazer e fizo çiertos mojones por termy^{o}s e se fizo //
/346v/ uno frente de la yglesia ençima de un çerrillo de piedras que estava alli / dixo que no lo vido porque estuvo en su casa.

- Preguntado que si despues de hecho el dho mojon este confesante e don P^{o} Yavre e otros prinçipales e muchos yndios de Canta sy derrocaron e quytaron el dho mojon que estava hecho / dixo que es verdad que frontero de la yglesia que esta en el dho pueblo de Quybi junto a vnas peñas por donde yva vna açequya de agua de los yndios de Quybi pusieron alli un mojon los yndios de Chacalla e que con el dho mojon quebraron el açequia e que como vieron esto los yndios de Quybi lloraron e dieron gritos diziendo que por que les quytavan sus trras e asy fue este prinçipal e otros yndios de Canta a vello e quando llego junto ally a el dho mojon estava deshecho que lo avian deshecho los yndios de Quybi e porque el alguazil fallo a este confesante junto al dho mojon quando lo deshazian lo prendio a este confesante e ato las manos atras e lo traxo preso a esta carçel.

- Preguntado sy se fallaron en defazer el dho mojon los demas prinçipales que estan presos / dixo que los vido alli quando desfizieron el dho mojon e que este confesante yva con todos ellos a faboresçellos para que el dho mojon se desfiziese como se desfizo e quyto e que esta es la verdad //
/347r/ para el juramento que fizo de lo que pasa e vio lo qual declaro por lengua del dho Joan de Alvarado el qual no lo firmo porque no sabia escrevyr.

ante mi — A^{o} Diaz de Gibraleon
escrivano

/348r/ Confision de don Fran^{co} Pomacaxa.

E despues de lo susodho en la dha çibdad de Los Rreyes en prim^{o} dia del mes de jullio de myll e quy^{o}s e sesenta e siete a^{o}s estando preso en la carçel rreal de Corte don Fran^{co} Pomacaxa prinçipal del pueblo de Canta le fue tomado e rresçibido juramento en forma devida de dr^{o} el qual lo fizo e so cargo del le fueron fhas las preguntas syg.^{es}:

- Preguntado como se llama dixo que don Fran^{co} Pomacaxa e que es prinçipal del pueblo de Canta.

- Preguntado sy este confesante se hallo presente quando los prinçipales e yndios de Canta deshizieron un mojon que avian hecho los yndios de Chacalla junto a la yglesia ençima de vn çerrillo de piedras el sabado que agora paso dixo que es verdad que quando este confesante llego al pueblo de Quybi que fue el sabado por la tarde ya el dho mojon estava deshecho e que es verdad que vio estar alli junto al dho mojon a todos los prinçipales que agora estan presos e a otros muchos yndios e que este confesante no los ayudo a deshazer e que es verdad que porque los yndios e yndias de Quybi lloravan mucho fue este dho confesante e los demas prinçipales que estan presos a ver el dho mojon que estava deshecho e alli //
/348v/ hallo al dho alguazil e lo prendio como los demas e lo traxo a la carçel rreal desta Corte donde agora esta e que esta es la verdad para el juram^{o} que fizo e no lo firmo lo qual declaro por la dha lengua.

ante mi — A^{o} Diaz de Gibraleon
escrivano

Confision de don X^{o}val Carvavilca.

E despues de lo susodho en este dho dia mes e año susodhos estando preso en la carçel rreal desta Corte don Xpoval Carvavilca yndio de Canta del qual fue tomado e rresçebido juramento en forma devida de dr^{o} el qual lo fizo e siendo preguntado por la dha lengua le fueron fhas las preguntas syg.^{es}:

- Preguntado como se llama dixo que don Xpoval Carvavilca e que es yndio de Canta.

- Preguntado sy ayudo a deshazer el mojon que estava hecho por la just.^{a} de pedimy^{o} de Joan caçique de Chacalla junto a la ygle-//
/349r/ sia del pueblo de Quybi o si vio quyen lo desfizo / dixo que es verdad que est.^{do} este confesante en su casa oyo llorar e dar gritos a muchos yndios e yndias del pueblo de Quybi e quando ya el dho mojon estava deshecho lo que avian derribado los yndios de Quybi este confesante fue alla e vido como avian desfho el dho mojon que aquel dia sabado la dha just.^{a} avia fho de pedimy^{o} del caçique de Chacalla e este confesante vio como deshazia el açequya para poner el dho mojon e que por esto lo avian deshecho e que el dho alguazil prendio a este confesante aquy junto a su casa en un patio e lo traxo adonde los otros estavan presos atadas las

manos atras e asy lo traxo a la carçel rreal desta Corte donde agora esta e que esta es la verdad para el juram^o que fizo lo qual declaro por la dha lengua e no lo firmo porque no sabia.

ante mi A^o Diaz de Gibraleon
escrivano

/350r/ Don Juan Baptista Caxaguaraca

Muy p^o señor

Pide se tomen sus confisiones a los prinçipales para que declaren donde estan dos prinçipales que faltan.

Don Juan Baptista Caxaguaraca caçique prinçipal de Chacalla en la querella caçiquil que tengo dada de don Pedro Guari y don Christoval y don G^o Carvavilca prinçipales de Canta / digo que en las trras y chacaras de Quivi estavan dos prinçipales que se dezian Juan Chancha Guaman y Fran^co Michuay con otros yndios a ellos subjetos los quales an sydo y son subjetos a my y a mys antepasados en el dho caçicasgo / y de un mes a esta parte que por esta Rreal Abdiençia se me mando dar executoria de las dhas trras los dhos prinçipales los an escondido o muerto que no paresçen / como an fho a otros muchos.

A Vra Al. pido y sup.^co mande se tomen sus confisyones a los dhos prinçipales que estan presos para que digan y declaren donde estan o que se an fho los dhos yndios e yndias que asy estavan en las dhas trras de Quivi sobre que pido just.^a y para ello etc.

Di^o de Ocanpo

En Los Rreyes quatro de jullio de myll y quinientos y sesenta y siete años ante los señores presidente y oidores desta Rreal Audiençia est^do en audiençia publica se presento esta pet.^on //
/350v/ y los dhos señores mandaron que declaren adonde estan o que se an fho e asta que declaren no se muestre esta petiçion.

Avendaño

/351r/ Confision de don P^o Yavre.

E despues de lo susodho en quatro dias del dho mes de jullio del dho año de mill e qui^os e sesenta e siete años en cunplimyento de lo proveydo e mandado por los señores presidente e oydores fue tomado e rresçibido juramento e forma devida de dr^o de don P^o Yavre prinçipal del pueblo de Canta el qual lo fiz e so cargo del le fueron fhas las preguntas syg.^es:

- Preguntado si conosçio a don Joan Chancha Guaman e a Fran^co Mychuay e si conosçio a otros yndios a ellos sujetos los quales heran prinçipales en las trras de Quybi sobre que es este pleito / dixo que es verdad que este que declara conosçio al dho Joan Chancha Guaman el qual no hera prinçipal syno yndio particular el qual hera quyta que quyere dezir çimarron que puede aver un año poco mas o menos que no lo a visto e que al Fran^co Mychuay que no lo conosçe e que no heran naturales de Quibi syno que unas vezes que estavan en Guancayo e otras vezes en Quybi e otras vezes en Chacalla los quales dhos dos prinçipales el uno dellos que se dize Joan Chancha Guaman es natural de Chacalla e que al otro no le conosçe como dho t.^e //
/351v/ e que no sabe este que declara que tuviesen indios sujebtos a ellos nyngunos porque no heran prinçipales.

- Preguntado sy los dhos dos prinçipales Joan Chancha Guaman e Fran^co Mychuay si heran sujebtos a don Joan Bautista caçique de Chacalla e a sus antepasados / dixo que no sabe sy heran sujebtos del dho caçique de Chacalla ny syno porque heran yndios que se andavan de pueblo a pueblo.

- Preguntado que avito trayan los dhos yndios prinçipales sy los trayan de Quybi o de Chacalla o de Canta dixo que no trayan abito de serranos ny de yungas ny de Canta ny de Chacalla syno que vnas vezes trayan motas de algodon e otras vezes de lana e una guaraca de cabuya como los yndios de Guaylas.

- Preguntado que si de un mes a esta pte poco mas o menos los dhos dos prinçipales faltan porque los an escondido o muerto e no paresçen / dixo que de un año a esta pte no a visto al dho Juan Chancha Guaman ny sabe lo que del se a hecho ny sy es bivo o muerto porque el otro no le conosçe e que es verdad que el dho //
/352r/ Joan Chancha Guaman tiene muger e hijos oy dia en el pueblo de Quybi.

- Preguntado que que se an fho los yndios e yndias que solian estar en las trras de Quybi del pueblo de Chacalla / dixo que estaran en Quibi e en Guancayo adonde solian estar e que esta es la verdad de lo que pasa e sabe para el juramento que fizo e no lo firmo porque no sabia lo qual declaro por lengua de Joan de Alvarado ynterpetre desta Rreal Avdiençia.

ante mi A^o Diaz de Gibraleon
escrivano

Confision de don G° Carvavilca.

E despues de lo susodho en la dha çibdad de Los Rreyes en el dho dia mes e año susodhos en cunplimy° de lo proveydo e mandado por los señores presidente e oydores fue tomado e rresçibido juramento en forma devida de dr° de don Gonçalo Carvavilca Prinçipal del pueblo de Canta el qual lo fizo e so cargo del prometio de dezir verdad e le fueron fhas las preg.[as] syg.[es]:

/352v/ - Preguntado si conosçe a Joan Chancha Guaman e a Fran[co] Mychuay prinçipales que solian estar en las trras de Quybi con otros yndios e yndias a ellos sujebtos / dixo que es verdad que este confesante conosçio a los dhos Joan Chancha Guaman e Fran[co] Mychuay de seys meses a esta pte los quales conosçio estar en el pueblo de Quybi vinyendo este confesante del pueblo de Canta a esta çibdad e que no heran prinçipales syno yndios comunes e que antes de agora este confesante los conosçio rresidir en el pueblo de Aguarny rrepartimyento de Guancayo encomendado en G° de Caçeres e otras vezes yvan al pueblo de Chacalla e que no sabe este confesante que tuviesen yndios ny yndias nyngunos a ellos sujebtos porque no sabe sy heran prinçipales e que este confesante oyo dezir a muchos yndios que porque el caçique de Chacalla los maltratava andavan de pueblo en pueblo los dhos prinçipales.

- Preguntado sy los dhos dos prinçipales de suso declarados sy an sydo e son sujebtos a don Joan Bautista caçique de Chacalla e a sus antepasados / dixo que a este confesante le paresçe que los dhos dos prinçi-//
/353r/ pales heran sujebtos al dho caçique de Chacalla porque yvan alla algunas vezes.

- Preguntado sy de un mes a esta parte poco mas o menos los dhos prinçipales faltan del dho pueblo de Quybi e el caçique e prinçipales del pueblo de Canta los an muerto e escondido e no paresçen / dixo que no sabe donde estan los dhos dos prinçipales e sy estan en Quybi o en Chacalla o en otra parte ny sabe sy son muertos ny bivos porque de çinco o seys meses a esta parte no los a visto.

- Preguntado sy tienen mugeres e hijos en el dho pueblo de Quybi dixo que en el pueblo de Quybi a visto este que declara que tienen alli sus rranchos e bahareques e no sabe sy tienen alli sus mugeres e hijos o en otra pte.

- Preguntado que abito trayan los dhos dos prinçipales sy heran de yungas o serranos dixo que no trayan abito de Canta ny de Chacalla syno que unas vezes trayan abito de algodon e otras vezes de lana e como tenyan la posibilidad asi andavan e que esta es la verdad para el juramento que fizo lo qual declaro por la dha lengua.

ante mi A° Diaz de Gibraleon
escrivano

/353v/ Confision de don Fran[co] Pomacaxa.

E despues de lo susodho en la dha çibdad de Los Rreyes en el dho dia quatro dias del dho mes de jullio del dho año de myll e quy°s e sesenta e siete años en cunplimy° de lo proveydo e mandado por los señores presidente e oydores fue tomado e rresçibido juramento en forma devida de dr° de don Fran[co] Pomacaxa prinçipal del pueblo de Canta el qual lo fizo e so cargo del prometio de dezir verdad e siendo preguntado por el thenor del dho pedimy° dixo lo syguiente:

- Preguntado si conosçe a Joan Chancha Guaman e a Fran[co] Mychuay e sy sabe que son prinçipales del pueblo de Quybi e que tienen otros yndios e yndias a ellos sujebtos / dixo que puede aver dos meses poco mas o menos que este confesante vido a los dhos dos prinçipales en el pueblo de Quybi e que sabe que no son prinçipales syno yndios comunes porque no tenyan yndios a ellos sujebtos ny heran mandones e que algunas vezes estavan en el pueblo //
/354r/ de Guarny rrepartimy° de G° de Caçeres e otras vezes en Quibi e otras vezes en Chacalla.

- Preguntado sy los dhos Joan Chancha Guaman e Fran[co] Mychuay sy son yndios prinçipales sujebtos a don Joan Bautista caçique de Chacalla e a sus antepasados / dixo que no sabe sy son sujebtos al dho caçique de Chacalla o no mas de que a oydo dezir que son del rrepartimy° de Chacalla.

- Preguntado que abito trayan sy heran de serranos o de yungas / dixo que algunas vezes los via este confesante traer abitos de serranos e otra vezes de yungas vnas vezes de lana e otras vezes de algodon e guaracas de cabuya.

- Preguntado sy son casados en el pueblo de Quybi e sy tienen alli sus mugeres e hijos e casas / dixo que no sabe donde tienen sus mugeres e hijos ny donde tienen casas porque son yndios que andan de pueblo en pueblo e no tienen asyento en nynguna parte.

- Preguntado de donde son naturales dixo que no lo sabe.

/354v/ - Preguntado que diga e declare que se an fecho los dhos dos prinçipales porque de un mes a esta parte faltan sy los an muerto o escondido el caçique e prinçipales e yndios de Canta e que se a fecho dellos / dixo que no sabe lo que se a fecho dellos pero que a lo que a este que declara le paresçe que podran estar en el pueblo de Guarny rrepartimy^{o} de G^{o} de Caçeres e que por aver ydo estotro dia el caçique de Chacalla a Quybi que por myedo del podria ser que se oviesen ydo alli e que estoviesen en el dho pueblo pero que no sabe sy son bivos o muertos o faydos e que esta es la verdad de lo que sabe para el juramento que fizo e no lo firmo porque no sabia lo qual declaro por la dha lengua.

ante mi A^{o} Diaz de Gibraleon
escrivano

/355r/ Confision de don Xpoval Carvavilca.

E despues de lo susodho en la dha çibdad de Los Rreyes en quatro dias del dho mes de jullio del dho año de myll e quy^{o}s e sesenta e siete años en cunplimyento de lo proveydo e mandado por los señores presydente e oydores fue tomado e rresçibido juramento en forma devida de dr^{o} de don Xpoval Carnavilca prinçipal del pueblo de Canta el qual lo fizo e so cargo del prometio de dezir verdad e siendo preguntado por el thenor del dho pedimy^{o} dixo lo syguyente:

- Preguntado si conosçe a Joan Chancha Guaman e a Fran^{co} Mychuay e de que tienpo a esta parte e si sabe que son prinçipales del pueblo de Quybi dixo que es verdad que conosçe a los dhos Joan Chancha Guaman e Fran^{co} Mychuay de un año a esta parte poco mas o menos e que de un mes a esta pte no los a visto e que no son prinçipales syno yndios pobres particulares e que no tienen yndios ny yndias a ellos sujebtos porque son çimarrones e porque andan de pueblo en pueblo.

- Preguntado sy son casados e tienen casas e mugeres e hijos en el pueblo de Quybi dixo que es verdad que este que declara vido tener sus //
/355v/ rranchos e bahareques a los dhos dos yndios algunas vezes en el pueblo de Quybi e sus mugeres alli e otras vezes en Guarny e otras vezes en Chacalla.

- Preguntado sy los dhos Joan Chancha Guaman e Fran^{co} Mychuay si son sujebtos a don Joan Bautista caçique de Chacalla e a sus antepasados o a quyen son sujebtos / dixo que es verdad que este confesante oyo dezir a los susodhos que heran sujebtos de los yndios de Chacalla a este confesante veya algunas vezes que yvan a tener algunas pasquas al pueblo de Chacalla.

- Preguntado que diga e declare que se an fho los dhos Joan Chancha Guaman e Fran^{co} Mychuay si son muertos o bivos o estan escondidos porque de un mes a esta parte faltan e sy los an muerto o escondido los caçiques e prinçipales del dho pueblo de Canta / dixo que no sabe este confesante que se an muertos ny que se a hecho dellos e que a el que deven de andar por los pueblos donde solian estar e que no sabe otra cosa.

- Preguntado que abito trayan sy heran de serranos o de //
/356r/ yungas dixo que es verdad que los dhos dos yndios son pobres e que quando tenyan rropa de algodon o de lana que aquello trayan e que las guaracas trayan de cabuya e que esta es la verdad de lo que pasa e sabe para el juram^{o} que fizo e no lo firmo porque dixo que no sabia lo qual declaro por la dha lengua del dho Joan de Alvarado.

ante mi A^{o} Diaz de Gibraleon
escrivano

/357r/ Don Felipe Taulichunbi.

Muy pr^{o} s^{o}r

El don Felipe salio a la causa despues de aver avido sentençia de vista y rrebista en favor de los yndios de Chacalla, como yndios sujetos a Canta.

Joan de Arrandolaça en n^{e} de don Felipe Taulichunbi Alonso Caguaca prinçipales yungas del asiento de Quybi y el pueblo que alli esta en el pleito con el caçique e yndios de Chacalla sobre las dhas trras de Quybi / digo que por V. Al. fue m^{do} dar tr^{do} del pedimy^{o} e suplicaçion que mys ptes hizieron al dho caçique e yndios de Chacalla y que por aquello no se ynpidiese la execut.^{a} segun mas largo consta e paresçe por los autos del proçeso cuyo tenor avido aquy por rresumydo / digo que las ptes contr^{as} pretenden sacar la dha executoria y con ella ynobar e despojar a mys ptes de las dhas sus trras no pudiendose hazer ny ynobar en la dha causa hasta que mis ptes sean oydas por no averse hecho con ellos la causa sino entre otros terçeros que no tenyan el dr^{o} que mys ptes tienen a las dhas trras y en todas vras Avdiençias Rreales e Consejos aviendo terçero que salga a la causa aunque sea despues de la sentençia de rrebista que pretenda dr^{o} a la cosa sobre que se ha tratado pleito sienpre le oyen e ynpiden la ex.^{a} mayorm.^{te} estando en posesion como mys ptes estan de las dhas trras de tpo ynmemorial a esta pte antes y despues de los Yngas y demas desto mostraran como jamas las dhas trras nunca fueron quytadas las dhas trras [*sic*] a mys ptes asi por los Yngas ny por otra persona alguna.

Por tanto a V. Al. pido e sup.^{co} m^{de} que no se ynobe en la dha causa ny se les de la dha ex.^{a} hasta que mys ptes sean oydas y pues V. Al. mando que si se le diese fuese sin perjuyzio del dr^{o} de mys ptes se entiende tanbien sin perjuyzio de su posesion sobre que pido just.^{a} e para ello que sy es neçesario sup.^{co} del dho auto en que se manda dar la dha ex.^{a}

El Licen.^{do} de Leon J. Arrandolaça

/357v/ En Los Rreyes veinte e siete de junyo de myll e qui^{o}s e sesenta e siete a^{o}s ante los señores pres.^{te} e oydores desta Rreal Audiençia en audiençia pu^{ca} le presento esta pet.^{on} y los dhos señores probeyeron lo proveydo.

Avendaño

/358r/ Los yn^{o}s yungas del asiento de Quibe.

Muy pr^{o} s^{o}r

Joan de Arrandolaça en n^{e} de los yndios yungas del asiento de Quybe en el pleito con los de Chacalla sobre las trras / digo que Di^{o} de Ocanpo pr. contr^{o} llebo tr^{no} para rresponder a my demanda e suplicaçion e no rresponde cosa alguna.

Pido e sup.^{co} a V. Al. m^{de} aver por conclusa la causa sobre este articulo e se manden ver los autos y fazer en la causa segun tengo pedido e suplicado e para ello etc.

J. Arrandolaça

En Los Rreyes veinte e siete dias del mes de junio de myll e quiny^{o}s e sesenta y siete a^{o}s ante los señores presidente e oydores desta Rreal Audiençia en aud.^{a} pu^{ca} se presento esta pet.^{on} y los dhos señores mandaron dar traslado a la otra parte e que a la primera audiençia rresponda donde no la ovieron por conclusa. Presente Diego de Ocanpo a quien se notifico.

Avendaño

/359r/ Los y^{o}s de Chacalla.

Muy p^{o} señor

Diego de Ocanpo en n^{e} del caçique don Ju^{o} Bautista e yndios de Chacalla en la cabsa con el caçique e yndios de Canta sobre las trras de Quivi / digo que mys partes an ydo a entregar a los susodhos las dozientas ovejas conforme a la sentençia y el dho caçique de Canta dizen que anda encubierto por Maranga y en otras ptes desta çibdad por no las yr a rresçibir e molestar a mys partes.

A V. Al. pido y sup.^{co} mande probeer que entregandolas a qualesquyer yndios e prinçipales de Canta cunplan mys ptes porque de otra manera nunca se acabara esta cabsa para lo qual etc.

Diego de Ocanpo

En Los Rreyes veinte e siete de junio de mill e qui^{o}s y sesenta y siete años ante los señores presydente e oydores desta Rreal Audiençia, y los dhos señores mandaron que buscandole en su casa al caçique prençipal y no le hallando se le entreguen a los prençipales del dho rrepartimi^{o} y con esto cunplan.

Avendaño

/360r/ Don Ju^{o} Bautista.

Muy p^{o} señor

Don Juan Baptista Caxaguaraca caçique prinçipal de Chacalla en la querella que tengo dada de los caçiques e yndios de Canta por aver derribado los mojones que por exsecutoria desta Rreal Abdiençia se pusyeron en las trras de Quivi / digo que aviendo vras justiçias tornado a hazer los dhos mojones y anparado a mi y a mys yndios en la dha posesyon y traydo presos algunos prinçipales que fizieron la dha fuerça contra vra Rreal Exsecutoria / aora de nuevo los dhos caçiques e yndios de Canta añadiendo delicto a delicto e fuerça a fuerça an tornado a derribar los dhos mojones en gran desacato de vra just.^{a} y desta Rreal Abdiençia en lo qual an cometido mayores delictos.

Pido y sup.^{co} a V. Al. mande que ante todas cosas vaya un alguazil desta Corte a hazer ynformaçion de lo susodho y prender los culpados y poner los dhos mojones que estavan puestos por exsecutoria desta Rreal Abdiençia sobre que pido just.^{a} y para ello etc. e juro a Dios en forma que no lo pido de maliçia.

El Lic.^{do} Falcon

/360v/ En la çiudad de Los Rreyes a tres dias del mes de jullio de myll y qui^{o}s y sesenta y siete a^{o}s ante los señores presidente e oydores estando haziendo audiençia de rrelaçiones se leyo esta petiçion y los dhos señores mandaron que se trayga todo.

Alvaro Ruiz de Nabamuel

/361r/ Don Felipe Taulichunbi.

Muy ps^{o} s^{o}r

Don Felipe Tablichunbi e Alonso Caguac prinçipales de Quibi / dezimos que por el pleyto que traen los caçiques de Chacalla con los de Canta y por aver sacado la executoria que sacaron desta Real Abdiençia no solamente nos quyeren tomar las dhas nras trras pero aun nos quyeren arrancar y arrancan todas las sementeras de coca e agi e mayz e otras sementeras que alli tenemos syendo como es contra just.[a] aunque se nos ayan de quitar las dhas tierras que nos quiten lo que emos senbrado syno que se coja mayormente syendo de nosotros las dhas trras y que tenemos suplicado y nos emos ofresçido a provar.

Por tanto a V. Al. pedimos suplicamos se les mande con graves penas no arranquen cosa alguna de lo que esta senbrado ny se haga ynovaçion fasta que la cabsa se fenezca sobre que pido just.[a] e para ello etc.

El Licen.[do] de Leon

E asimysmo una huerta que tenemos con naranjos e otras frutas en el dho asiento de Quybi los yndios de Chacalla an entrado e destruydo totalmente ellos y con sus ganados e para ello etc.

/361v/ En Los Rreyes a tres dias del mes de jullio de mill y qui[o]s y sesenta y siete a[o]s ante los senores presidente y oidores en audiençia de rrelaçiones se leyo esta petiçion y los dhos senores mandaron que por esta vez

Alvaro Ruiz de Nabamuel

/362r/ Don Phelipe caçique e los y[o]s de Quibe.

Muy ps[o] s[o]r

Don Felipe caçique de los yn[o]s de Quibi en el pleyto con los yndios de Chacalla / digo que demas de la ynobaçion y fuerça que nos an fho los dhos yndios bino ayer bn yndio por la costa y me a çertificado que los dhos yndios de Chacalla nos an destruydo las parras arboles y verdura que tenyamos en una huerta e nos an arrancado la coca como paresçe por las muestras que dello traen que son estas de que hago demostraçion lo qual V. Al. deve rremediar hasta tanto que se fenezca la dha cabsa.

Por tanto a V. Al. pido y sup.[co] mande poner rremedio en lo susodho e mandar con pena a los dhos yndios que salgan de las dhas tierras y saquen dellas el ganado que an metido alli conque destruyen todas nras chacarras sobre que pedimos just.[a] e para ello etc.

El Licen.[do] de Leon J. Arrandolaça

En Los Rreyes quatro de julio de myll e quiny[o]s e sesenta e siete a[o]s ante los senores presydente e oydores desta Rreal Aud.[a] en aud.[a] pu[ca] presento esta pet.[on] y los dhos senores mandaron que se traygan los autos para mañana.

Avendaño

/363r/ Don Felipe Taulichunbi e Alonso Caguac prinçipales de Quibi.

Muy p[o]s s[o]r

Don Felipe Tablichunbi y Alonso Caguac prençipales del pueblo de Quibi en el pleyto con los yn[o]s de Chacalla dezimos que nosotros truximos el modelo en forma gruesa y demostrativa por donde claro paresçe que nuestro pueblo esta en las mismas trras que los dhos y[o]s de Chacalla nos quieren quytar y quitandonoslas quedamos syn nyngun rremedio de poder bibir ny sustentarnos por lo qual aunque fuera verdad que el Ynga nos las oviera quytado no lo pudo hazer en nro perjuyzio aunque lo hiziera de poderio absoluto por ser como hera tiranya notoria y contra dr[o] natural mayormente que aunque se diga que las dio a çiertos mytimaes y aquellos se quisieron aprovechar de la tal md no consta por todo el proçeso quyen son los tales mytimaes para que puedan pedir las dhas trras ny se le puedan dar y nro derecho es çierto pues estamos no solamente en posesion pero aun no emos decaydo del señorio que dellas tenemos demas de que provaremos otras muchas cabsas por donde los dhos yndios de Chacalla no tienen derecho algu[o] a las dhas tierras en la prosecuçion desta cabsa y porque para la justificaçion della conbiene que ante todas cosas se berifique el modelo que ante vro presydente e oydores presentamos.

Pedimos y suplicamos a V. Al. vaya una persona syn sospecha que averigue sy es verdadero el dho //
/363v/ modelo por que sobre el se a de fundar todo el dho pleyto y constara por ello de nra justiçia y de la ynjustiçia de la pte contraria para lo qual etc.

El Licen[do] de Leon J. Arrandolaça

En Los Rreyes a quatro dias del mes de julio de mill e qui[o]s e sesenta e siete a[o]s ante los senores presidente e oydores desta Rreal Audiençia estando en audiençia publica se presento esta petiçion e los dhos señores proveyeron que para el primer acuerdo tengan ynformado. Presentes el Lic[do] Falcon y Di[o] de Ocanpo a quyen se notifico.

Abendaño

/364r/ Don Phelipe Taulichunbi.

Muy p[o]s s[o]r

Don Felipe Tablichunbi caçique yunga del pueblo de Quibi y Al[o] Çaguaca prinçipal en el pleyto con los yndios de Chacalla sobre las trras / digo que mi abogado quiere ynformaçion de dr[o] en la dha cabsa y pues es negoçio que nos ynporta toda nra hazienda.

A V. Al. pedimos e suplicamos no se determine la dha cabsa hasta que ynforme el dho nro letrado sobre que pedimos just.ª e para ello etc.

El Licen^{do} de Leon J. Arrandolaça

Otro si pedimos e suplicamos a V. Al. mande soltar a los dhos prençipales que estan presos porque todos ellos estan enfermos que sy es nesçesario daremos fianças de paga e lo juzgan dar sentençiado en la dha cabsa sobre que pedimos just.ª e para ello etc.

En Los Rreyes ocho de julio de myll e quinyºs e sesenta y siete aºs ante los senores presidente e oydores desta Rreal Audiençia en audiençia publica / sepresento esta pet.^{on} y los dhos señores mandaron que ynforme para el primer acuerdo y en lo demas se oye.

Avendaño

/365r/ Don Felipe Taulichunbi.

Muy pº sºr

Don Felipe Taulechunbi y Alº Caguac prençipales de Quybe por nosotros y en nonbre de los demas yºs del dho pueblo / dezimos que a nra notiçia es benydo que los yºs de Chacalla an traydo un modelo de las dhas tierras diferente del que nosotros dimos poniendo falla en ellos pueblos [*sic*] donde no estan a fin de tomarnos nras tierras y enfraudar el negoçio para que no se entienda y pues toda esta causa esta en las discrepçion [*sic*] de la tierra para que se bea como quedamos despojados totalmente y sin rremedio de poder senbrar ny sustentarnos a nosotros ny a nros hijos por donde seremos conpelidos a yr a buscar tierras extrañas donde nos podamos sustentar y a nuestras mugeres y hijos abiendo naçido todos nros antepasados alli y nosotros y que aquellas tierras son nuestras desde tienpo ynmemorial a esta pte y sienpre hemos sido besitados y una cordillera de grandisimos çerros nos pte las tierras y limites demas de que como tenemos alegado aunque el Inga nos las obiera quitado para si o para otra no lo pudiera aver hecho ny lo pudo hazer pues fuera notoria tirania querernoslas quytar pues hera quytarnos las bidas con ello mayormente que si algo quyto seria y fue para si e ya aquello esta quitado pues ny parezen los mytimaes que por si puso ny otra cosa alguna y pues todo depende del dho modelo.

Pedimos y suplicamos a Vra Al.ª que se enbiese una persona de confiança a costa de entranbas ptes //
/365v/ con un escribano para que aberigue la verdad del dho nuestro modelo y aberigue e se aberigue [*sic*] el pleyto sobre que tratamos para que pedimos just.ª para lo qual etc.

El Licen^{do} de Leon

En Los Rreyes onze de julio de myll e quiºs y sesenta y siete aºs ante los señores pres.^{te} e oydores desta Rreal Audiª estando en audiª rreal pu^{ca} / se presento esta petiçion y los dhos señores mandaron que se juntan anbas partes y hagan un modelo de conformydad para proveer al primer acuerdo lo qual se proveyo presentes los dhos Juan de Arrandolaça e Diº de Ocanpo procuradores de las partes a quien se notifyco.

Avendaño

/366r/ Los yºs de Chacalla.

Muy pº señor

Diego de Ocanpo en nonbre de los yndios de Chacalla en la cabsa con los yndios de Canta sobre las chacaras de Quivi / digo que las partes contrarias a fin de dilatar esta cabsa no se quyeren juntar con nosotros a fazer el modelo de las dhas trras y mojones y pues en esta çibdad esta el padre Fray Gaspar de Carvajal persona que fizo la averiguaçion que esta puesta en el proçeso y entendio los linderos que tienen las dhas trras.

A V. Al. pido y sup.^{co} mande determynar la dha cabsa y que el dho padre Fray Gaspar de Carvajal vaya a deslindar las dhas trras sobre que pido just.ª y para ello etc.

Diego de Ocanpo

/366v/ En Los Rreyes a quinze de julio de myll e quiºs y sesenta y siete aºs ante los señores presidente e oydores desta Rreal Audiençia estando en audiençia publica se presento esta pet.^{on} y los dhos senores mandaron que para el primer acuerdo las partes tengan ynformado lo qual se proveyo presente Joan de Arrandolaça a quien se notifico.

Avendaño

/367r/ Don Phelipe Tablichunbi.

Muy pºs sºr

Don Felipe Tablichunbi caçique de Quibi por mi y en nonbre de los demas yndios del dho pueblo digo que esotro dia truximos ante vro presidente e oydores el modelo y pintura de nras tierras e pueblo de Quibi y de Canta y Chacalla para que se biese como nos quyeren tomar todas nras tierras e hazienda contra justiçia y porque los ynºs de Chacalla truxeron otra pintura Vra Al. mando por ser diferente y contraria que se hiziese de comun consentimiº una pintura o modelo en casa de Alvarado ynterpetre adonde nosotros fuimos y el dho Alvarado pidio que cada uno truxese su modelo que havian hecho y los dhos yn °s de Chacalla dixeron que el suyo ya lo avian hecho pedaços y porque llevaron ynºs biejos de la comarca para lo hazer se binyeron a conformar con mi modelo e pintura y dezir que aquel hera verdadero y asy de comun consentimiº hizimos otro ny mas ny menos que el que yo avia traydo para que V. Al. vea con las cabtelas que andan los dhos ynºs de Chacalla y pues esta hecho de conformidad.

Pedimos e suplicamos a V. Al. que el dho modelo se trayga y se castigue a los dhos yºs de Chacalla que traxeron el otro modelo pues pretendieron engañar a vro presydente e oydores sobre que pido just.ª e para ello etc. porque en tienpo del Ynga quyen hazia otro tanto aunque fuesen diez pueblos los asolava todos.

El Licen[do] de Leon J. Arrandolaça

/367v/ En Los Rreies quinze de julio de mill e quinientos e sesenta y siete años ante los señores presidente e oidores desta Rreal Audiençia / se presento esta pet.[on] y los dhos señores mandaron lo proveydo.

Avendaño

/368r/ En la causa que es entre partes de la
una los yndios de Chacalla y de la otra
los de Quyvy sobre las tierras de Quyvy.

En la çiudad de Los Rreyes en diez e ochos dias del mes de julio de myll e quyºs e sesenta e siete años / los señores presidente e oidores desta Rreal Audiençia aviendo visto la dha causa mandaron que vaya una persona desta Corte a las dhas tierras de Quyvy sobre que se ha tratado este pleyto para que conforme a la exsecutoria sobre ellas dada de la posesion dellas a los dhos yndios de Chacalla y ponga los mojones en ellas. Y asi lo proveyeron e mandaron e señalaron de sus rrubricas.

El qual dho auto fue pron.[do] por los dhos señores pres.[te] e oydores el dho dia mes e año en el cont.[do] est.[do] presentes: Juan de Arrandolaça e Diº de Ocanpo procuradores de las ptes a quien se notifyco.

Avendaño

/369r/ Los yºs de Quibi.

En Los Rreyes beynte e un dias de jullio
de myll e quyºs e sesenta e syete aºs / Juº
de Arrandolaça prd de su pte presento esta petiçion
con protestaçion de la rratificar el primero dia
de abdiençia.
Luys de Billarreal escrº de Su Mag[t]

El don Felipe y los demas de Quibi sup.[an] de la sentençia; no parece ser sus sujetos.

Muy prº sºr

Don Felipe por mi y en n[e] de los demas yndios del asiento de Quybi en el pleito con los yndios de Chacalla sobre las trras del dho nro pueblo / sup[co] del auto e sent[a] dada e pronunçiada por el pr[te] y oydores en que en efeto mandaron se ynbiase persona a entregar la posesion e amojonar las dhas trras de Quybi y executar la ex.ª que esta dada contra los yndios de Canta segun mas largo en la dha sentençia se cont.[e] cuyo tenor avido aquy por rresumydo Digo que el dho auto es de enmendar y rrebocar por lo general y lo que rresulta del proçeso en su fabor que es aquy por espreso e para lo que dho y alegado tengo a que me rrefiero / y porque poseyendo yo como poseo juntamente con mys yndios las dhas trras de tpo ynmemorial a esta pte y tenyendolas y poseyendolas como las poseo siendo sºr dellas yo e mys yndios nyng.ª cosa me perjudica del pleito que se trata entre los yndios de Canta e Chacalla para ser oydo e ynpedir la ex.ª que por pte de los dhos yndios de Chacalla se pide mayormente aviendo como he suplicado de la dha ex.ª y sent.ª de rrebista que en su Rreal Aud.ª se dio.

Lo otro porque aunque no ubiera otra cosa sino la transaçion e conçierto que en n[e] suyo se hizo por los yndios de Canta conçertar obejas con los dhos yndios de Chacalla para que no entrasen en las dhas trras los dhos yndios de Chacalla por prometer como prometieron los dhos yndios de Chacalla de no entrar mas en ellas por las //
/369v/ dhas dozientas ovejas que de los dhos yndios de Canta rreçibieron no me pueden ynquyetar ny pedir las dhas trras aunque algun drº tubieran a ellas quanto mas no tenyendolo / contra lo qual no conpete dezir que los Yngas pasados dieron a los dhos yndios de Chacalla çierta pte de las dhas trras porque demas de no aver hecho lo susodho los dhos Yngas ny quytadonos a nosotros las dhas trras si alguna pte tomo dellas seria e fue para si ponyendo en ellas mytimaes para que las senbrasen para el y no para los dhos yndios de Chacalla e aquello esta ya acabado por no aver Ynga y consta y como consta notoriamente no pertenesçer las dhas trras a los dhos yndios de Chacalla ny de los mytimaes que alli puso el Ynga a quyen dizen que dio las dhas trras y quando consta de non jure adjuntis no deve ser admytido el que pide y deve ser oydo qualquier terçero posehedor / quanto mas que siendo como heran nras las dhas trras aunque constara para si probar que el dho Ynga nos las quyto nos avian de ser rrestituydas aunque no estubieramos en

posesion dellas pues nyngun rrey puede quytar a su basallo su hazienda si no es dandole otra tal cosa en rreconpensa e de poderio absoluto e por causa de la rrepu^{a} todo lo qual çeso aquy.

- lo otro porque nosotros no tenemos otra cosa alguna donde poder senbrar ni de que nos poder sustentar si no es las dhas trras que nos quyeren quytar e los dhos yndios de Chacalla tienen muy gran abundançia de trras y estan en trra muy mas fertil que nosotros y de las dhas trras no se pueden aprobechar si no es para cosas de vino y nosotros las queremos para sustentar a nosotros y a nros hijos como lo hizieron nros pasados y como nasçidos y criados en ellas y solo tratamos de defender la vida y lo que a nosotros nos haze tanto daño e a los dhos yndios de Chacalla no aprobecha mas de por salir con la thema que han tomado por nyng.^{a} via se deve hazer.

- lo otro que ya que todo lo susodho çesara que no çesa para poder pedir la dha ex.^{a} e tomar las dhas nras trras avian de paresçer los mysmos //
/370r/ yndios mytimaes o yndios a quyen dizen los dhos yndios de Chacalla que dieron las dhas trras.

Por tanto a V. Al. pido e sup.^{co} enmyenda y rreboque el dho auto mandandonos anparar en la dha nra posesion y que seamos oydos de nuebo sobre que pedimos justiçia e costas e para ello etc. e pido ser rr^{do} a pr.

El Licen^{do} de Leon J. Arrandolaça

En Los Rreyes veinte e tres de jullio de myll e qui^{o}s e sesenta y siete a^{o}s ante los senores presidente e oydores desta Rreal Audi^{a} estan en audiençia pu^{ca} / se presento esta petiçion y los dhos señores proveyeron lo proveydo sin enbargo y se rremyte a Su Magestad con la causa prinçipal lo qual se proveyo presente Diego de Ocanpo y Ju^{o} de Arrandolaça procuradores de las partes a quien se notifico.

Avendaño

/371r/ El caçique de Canta.

Muy p^{o} señor

El caçique de Canta por my y en nonbre de mys y^{o}s / me querello ante V. Al.^{a} de un español que esta en el balle de Tacotay guardando bacas y premysas las solenydades del dr^{o} digo que el dho español con poco temor de Dios y en menospreçio de Vra Justiçia Rreal bino a mys pueblos e diçiendo a mis prençipales que porque no querian rresçebir çiertas obexas de la tierra que trayan los y^{o}s de Chacalla les dio muchos espaldarazos y a un prençipal llamado don Pedro le dio muchos pomazos con la mançana del espada en los pechos y cabeza de tal manera que el dho don Pedro esta a la muerte y ansimismo quebro todas las baras de mys alguaçiles de mys pueblos haçiendoles muchos malos tratamy^{o}s dandoles de golpes en lo qual el dho español cometio delito de fuerza e ynjuria e por ello debe de ser castigado conforme a dr^{o}.

Por tanto a V. Al.^{a} pido y supl.^{co} me haga just.^{a} por la bia e forma que mas me conbenga y constando de las dhas ynjuria y fuerza condene al dho español en las mayores e mas graves penas que por fuero y por dr^{o} hallarse estableçidas contra los que semejantes delitos cometen las quales mande executar en sus personas e bienes//
/371v/ por que a el sea castigo y a otros en exenplo sobre que pido just.^{a} y el of^{o} de Bra Al.^{a} ynploro.

El Licen^{do} de Leon

En Los Rreyes veinte e tres de julio de myll e qui^{o}s de sesenta y siete a^{o}s ante los señores presidente e oydores desta Rreal Audiençia estando en audiençia publica / se presento esta petiçion y los dhos señores mandaron que den ynformaçion de lo que dizen en esta petiçion.

Avendaño

/372r/ En Los Rreyes veinte e tres dias del mes de jullyo de myll e qui^{o}s e sesenta e syete anos don Françisco caçique prençipal de Canta por sy y en n^{e} de los demas caçiques e prençipales del dho rrepartimyento presento por testigo un yndio que mediante Juan de Alvarado ynterpetre desta Rreal Abdiençia dixo llamarse Alonso Carvabilca natural del pueblo de Yscaypanpa rreptimy^{o} de Canta del qual fue rresçibido juramento en forma de derecho so cargo del qual prometio de dezir verdad e syendo preguntado por la dha querella dixo que lo que dello sabe e pasa es que abra seys dias poco mas o menos que un español que guarda çiertas bacas e rreside en el balle de Tocota fue al pueblo de Ayxa en el dho rreptimyento de Canta e un sabado a mediodia poco mas con muchas cozes e amenazas diziendo: "perrazos que hazeys ay" a çiertos yn^{o}s que en el dho pueblo estavan; "por que no rresçibis estas dozientas obejas que los yn^{o}s de Chacalla os dan" e arremetio a un prençipal hermano del dho don Fran^{co} caçique prençipal que se llama don Pedro e syn cabsa alguna le asyo de los cabellos e le dio //
/372v/ muchos golpes e coçes e luego tomo la espada e le dio de espaldarazos con ella syn sacalla de la bayna e con el pomo le dio de pomazos e le arrastro por el suelo e porque este t^{o} llego al dho español como alguazil que es e le dixo que no le hiziese mas mal al dho don Pedro el dho español arremetio a este t^{o} e le dio de golpes e le tomo la vara de alguazil que llevava e se la quebro e que esto

es la verdad para el jur^{o} que hizo e que el dho don P^{o} esta muy malo de los dhos malos tratamyentos e que el dho español no sabe como se llama mas de que si le ve le conosçera.

ante mi Avendaño

E para ynformaçion de los dhos el dho don Fran^{co} caçique presento por testigo un yndio que mediante el dho Juan de Alvarado ynterpetre dixo llamarse Di^{o} Tantas natural del pueblo de Bisca rrepartimyento de Canta del qual se rresçibio juram^{o} en forma de derecho so cargo del qual prometio de dezir verdad //

/373r/ e syendo preguntado por la dha querella dixo que lo que sabe e pasa es que el sabado proximo que paso que se contaron diez e syete del presente estando este testigo en el pueblo de Ayxa con otros yndios oyo rruydo e bozes de yndio que se quexava en la plaça o panpa del dho pueblo donde se suelen juntar los yndios y este t^{o} fue a ver que hera e vido que un español que rreside en el balle de Tocota y guarda çiertas bacas estava en un caballo e con una espada en la mano e vido que don Pedro prençipal hermano de don Fran^{co} caçique prençipal del dho rrepartimi^{o} de Canta se estava quexando e dando bozes diziendo que aquel español le avia maltratado e dado de pomazos con la mançana del espada e arrastrado de los cabellos e los y^{o}s que ally estavan dezian que aquel español a mandado al dho don P^{o} y el dho español dezia al dho don P^{o} que por que no tomava las dozientas obejas y luego este t^{o} como vido al dho don P^{o} tan maltratado e la cara llena de cardenales bino despues a dar mandado dello al dho don Françisco caçique prençipal del dho rrepartimi^{o} de Canta e que el dho español no sabe como se llama mas de que si vele le conosçera e que esta es la verdad para el //

/373v/ juramyento que hizo e no firmo ny el dho ynterpetre porque dixo que no sabia e que es de hedad de mas de treynta años e que aunque es del rrepartimi^{o} de Canta a dho verdad e que no le enpegan ny tocan las demas preguntas g^{les} de la ley que le fueron fhas.

ante mi Avendaño

E este dho dia mes e año susodhos el dho don Fran^{co} presento por t^{o} a un yn^{o} que mediante el dho ynterpetre dixo llamarse D^{o} Guaman Laquila que es natural del pueblo de Capa rrepartimi^{o} de Canta del qual se rresçibio juramento en forma de derecho so cargo del qual prometio de dezir verdad e syendo preguntado por la dha querella dixo que lo que della sabe es que el sabado proximo que paso que se contaron diez e syete del presente estando este t^{o} en el pueblo de Ayxa con don P^{o} prençipal e otros yndios con ellos //

/374r/ llego al dho pueblo de Ayxa un español a caballo que es el que rresyde en el balle de Tocota e guarda çiertas bacas en el que no sabe como se llama mas de que si le ve le conosçera e hasta treynta yn^{o}s de Chacalla con el; como llego dixo al dho don Pedro que rresçibiese sus doz.^{as} obejas que alli las tenian los yndios y el dho don P^{o} dixo que no estava alli su caçique que el se lo ynbiaria a dezir e que las obejas que ellos avian dado no heran de aquella manera porque aquellas que les querian dar estavan flacas e llenas de carache e no heran de rresçebir porque no valian nada / e luego el dho español arremetio con el caballo al dho don P^{o} e le asyo de los cabellos e le truxo arrastrando e luego le dio de coçes e golpes e con la espada dio de espaldarazos e golpes e pomazos con el pomo del espada de que el dho don P^{o} quedo muy maltratado e lleno de cardenales e esta muy enfermo y en cama del dho mal tratamy^{o} e porque un alguazil llamado Al^{o} Carvavilca dixo al dho español que no maltratase de aquella manera al dho don P^{o} que hera su caçique el dho español arremetio a el e le dio de golpes e le arrastro de los cabellos e le quebro la bara que de alguazil tenia en la mano e que esto que dho tiene es la verdad e lo que //

/374v/ que [*sic*] save deste caso para el jur^{o} que hizo e que aunque este t^{o} es natural de Canta a dho la verdad e no firmo porque no supo.

ante mi Avendaño

/375r/ Don Ju^{o} Batista Caxaguaraca.

Muy ps^{o} señor

Don Juan Bap.^{ta} Caxaguaraca caçique prinçipal de Chacalla en la cabsa con don Felipe Taulichunbi y los demas sobre las trras y chacaras de Quibi / digo que esta cabsa esta determynada en rrevista.

A V. Al. pido y supl.^{co} pues por el proçeso de la dha cabsa consta que el padre Fray Gaspar de Carvajal de la Orden de Señor Santo Domingo faze averiguaçion quales heran las dhas trras y los linderos y mojones que tienen V. Al. probea y mande que el dho padre juntamente con un alguazil desta Corte nos vaya a anparar y amojonar las dhas trras y heche dellas a las partes contrarias sobre que pido just.^{a} y para ello etc.

El Licen^{do} de Leon Diego de Ocanpo

En Los Rreyes en veynte e quatro de jullio de myll e qui^{o}s e sesenta e siete años ante los señores presidente e oydores desta Rreal Aud.^{a} / se presento esta pet.^{on} e proveyeron que queriendo yr el dho Fray Gaspar para el dho efeto vaya quando fuere el dho alguazil.

Avendaño

/376r/ Don Fran^{co} Guaman Guaranga.

Muy p^{o} s^{o}r

Don Fran^{co} Guaman Guaranga caçique de Canta por my y en n^{e} de mys yn^{o}s no apartandome de la suplicaçion que se a fha por ante V. Persona Rreal / digo que por V. Al. a sydo mandado que V. Persona baya a cunplir la executoria que esta dada en la Rreal Audiençia e aparta las trras de Quybe y por quanto las obejas por las quales pretenden pagar son obejas pacos e viejas e llenas de carache e gusanos e las que nosotros les dimos fueron obejas grandes henbras escogidas por dos españoles que agora estan aquy e pues se manda deshazer el conçierto que nros encomenderos hizieron entre nosotros no es justo que nos llenen los partos e pospartos de las dhas obejas e sobre todo que nos las den con carache e pacos.

Por tanto a V. Al. pido e sup^{co} que las obejas que nos dieren sea de la manera mysma que nosotros las dimos con los partos e pospartos que si nosotros fueremos obligados a bolver los frutos de las dhas trras nosotros lo bolveremos entretanto que la cavsa se determyna en Vro Consejo y por Vra Persona Rreal en grado de segunda suplicaçion sobre que pedimos just.^{a} e para ello etc.

El Licen^{do} de Leon J. Arrandolaça

/376v/ En Los Rreyes veinte y nueve de jullio de mill y qui^{o}s y sesenta y siete años ante los señores presidente y oidores desta Rreal Audiençia / se presento esta pet.^{on} y los dhos señores mandaron que la persona que fuere a las tierras vea si son buenas de dar y rresçebir las obejas que se les dan y si no fueren tales se las haga dar.

Avendaño

/377r/ El caçique e y^{o}s de Chacalla.

Muy ps^{o} señor

Diego de Ocanpo en n^{e} del caçique e yndios de Chacalla en la cabsa con los yndios de Canta e don Felipe e los demas yndios yungas sobre las trras e chacaras de Quivi digo que Vra Al. por abto de vista e rrevista tiene mandado que una persona desta Rreal Abdiençia por ante vro escrivano anpare mys partes en la posesyon de las dhas trras y se las amojone y porque las ptes contrarias por la enemystad que a mys partes tienen les querran quitar las aguas con que se rriegan las dhas trras y seria como sy no se las diesen.

Pido y sup^{co} a V. Al. probea la persona y el escrivano que an de yr a lo susodho y que asymysmo vaya a ello el padre Carvajal y se entregue al escrivano que a ello fuere el proçeso para que por el se pongan los mojones y se les cometa y mande que asymysmo averiguen las aguas con que las dhas trras se an rregado y se pongan por espreso y se mande a las ptes contrarias que no se las quyten ny pte dellas ny los mojones que se pusyeren so graves penas para lo qual etc.

El Lic^{do} de Leon Diego de Ocanpo

Otro sy pido y sup.^{co} a V. Al. que pues consta que los prinçipales de Canta derribaron e hizieron derribar los mojones e ympidieron el cunplim^{o} de la carta executoria V. Al. probea y mande que el escrivano y alguazil que fueren a anparar mys partes en la posesyon de las trras vayan a costa de las ptes contrarias sobre que pido just.^{a} y para ello etc.

Diego de Ocanpo

/377v/ En Los Rreyes veinte e nueve de jullio de mill y qui^{o}s y sesenta y siete años ante los señores pres.^{te} e oydores desta Rreal Aud.^{a} estando en aud.^{a} pu^{ca} / se presento esta petiçion y los dhos señores mandaron en lo primero que se haga como lo pide y se ponga en la comysion que se diere y en lo segundo que vaya a costa de anbas partes lo qual se proueyo presente Juan de Arrandolaça procurador contrario a quien se notifico.

Avendaño

/378r/ Caçique de Chacalla. En prim^{o} de agosto 1567

Muy p^{o} señor

Diego de Ocanpo en n^{e} del caçique prinçipales e yndios de Chacalla en la cabsa con el caçique e yndios de Canta sobre las trras e chacaras de Quivi / digo que ha muchos dias que V. Al. tiene probeydo que vaya a las dhas chacaras e trras una persona rreçeptor y alguazil juntamente con el padre Fray Gaspar de Carvajal a anparar a mys partes en la posesyon que tienen tomada de las dhas trras y aunque mis partes an ydo muchas vezes por el m^{o} para que se haga el secretario dize que no esta señalada la persona que lo a de hazer.

A Vra Al. pido y sup.^{co} mande señalar la dha persona y que parta luego a hazerlo / y para ello se le de el proçeso original porque en el esta la averiguaçion que el dho padre Fray Gaspar de Carvajal fizo de quienes heran las dhas trras para lo qual etc.

Diego de Ocanpo

En Los Rreyes en primero de agosto de mill e qui^{o}s y sesenta y siete años ante los señores presidente y oidores desta Rreal Audiençia / se presento esta pet.^{on} y los dhos señores lo rremitieron a Su Señoria para que nonbre persona.

/379r/ Don Phelipe caçique de Quivi.

Muy p^{o} s^{o}r

El don Felipe y los demas y^{o}s de Quibe se quexan de los de Chacalla diziendo les destruyen quanto siembran.

Don Felipe caçique de Quibi / digo que los y^{o}s de Chacalla con el color de la sentençia que en esta Rreal Audiençia se a dado van cada dia a nras chacaras y como negros çimarrones o gente de guerra nos destruyen y asuelan nras chacaras y el mayz y la coca y todo lo demas que tenemos senbrado sin ozalles rresistir ni rresistir [*sic*] ny defendernos porque luego nos amenazan que nos an de hazer prender como prendieron a los demas prinçipales que hizieron prender por lo de los mojones en lo qual V. Al. deve poner rremedio mandando que vaya persona a partir las dhas tierras y que se concluya por esta causa castigar a los dhos y^{o}s que nos rroban.

A V. Al.^{a} pido y sup.^{co} m^{de} que sin perjuizio de la suplicaçion que esta ynterpuesta para Vra Persona Rreal vaya la persona que a de yr a partir las dhas tierras y ver lo que nos ha de dejar para que ansi podamos sustentar los yungas que ally estamos pues ha sido aquella tierra nra desde el tpo ynmemorial a esta parte sobre que pido just.^{a} y para ello etc.

El Licen^{do} de Leon

En Los Rreyes ocho de ag^{o} de mill e qui^{o}s y sesenta y siete años ante los señores presidente y oidores desta Rreal Audiençia / se presento esta petiçion y los dhos señores mandaron lo proveido.

Avendaño

/380r/ Los y^{o}s de Chacalla.

Muy p^{o} señor

Diego de Ocanpo en n^{e} de los yndios de Chacalla en la cabsa con los yndios de Canta y los yungas sobre las trras de Quivi / digo que V. Al. mando que una persona desta Corte fuese a anparar a mys partes en la posesyon que tomaron de las dhas trras por executoria desta Rreal Abdiençia y a poner los mojones y agora las ptes contrarias an pedido que mys partes les entreguen las dozientas ovejas que por la dha executoria les estan mandadas dar buenas diziendo que no lo son las que les an entregado y porque mys partes an cunplido con la dha carta executoria y por los abtos del proçeso paresçio que an entregado las dhas dozientas ovejas y lo que en contrario se pide el de maliçia porque se les an muerto algunas de las dhas ovejas y quyeren que mys partes se las tornen a dar bivas.

A V. Al. pido y sup.^{co} probea y mande que los dhos yndios de Canta traygan ante la persona que asy fuere las ovejas que mys ptes les entregaron e que sy paresçiere no ser de dar e tomar algunas dellas mys partes les entreguan otras que sean buenas en lugar de las que les bolvieron y que con esto ayan cunplido / sobre que pido just.^{a} y para ello etc.

El Lic.^{do} Falcon D. Ocanpo

/380v/ En Los Rreyes a ocho de agosto de myll e qui^{o}s e sesenta e siete a^{o}s ante el presydente e oydores desta Rreal Audiençia estando en audi.^{a} pu^{ca} / se presento esta petiçion y los dhos señores mandaron que las ovejas las traygan las personas a quien las entregaron para que la persona a quien esta cometido las vea si son de dar y tomar y no lo siendo les haga dar otras que lo sean lo qual se proueyo presente Juan de Arrandolaça procur^{or} contr^{o} a quien se notifico.

Avendaño

/381r/ Don Joan Bautista.

Muy p^{o} señor

Don Juan Bap.^{ta} caçique prinçipal de Chacalla en la cabsa con los yndios de Canta sobre la posesyon de las trras de Quivi / digo que V. Al. mando que una persona fuese a anparar a my y a mys yndios en la posesyon de las dhas trras / y amojonarlas y aora a venido a my not.^{a} que Luis de Villarreal rreçeptor desta Rreal Abdiençia procuro yr con la dha comysion visto que no se la an dado / pretende yr con la persona que va a executar lo por V. Al. mandado para favoresçer a las ptes contrarias y meternos en mas pleytos como honbre que a soliçitado y soliçito el dho pleyto / a que no se ha de dar lugar.

A V. Al. pido y sup.^{co} probea y mande que el dho Luis de Villarreal no vaya a las dhas trras y que solo se vayan la persona que V. Al. ha señalado y el padre Fray Gaspar de Carvajal como esta mandado sobre que pido just.^{a} y para ello etc.

El Lic^{do} Falcon

En Los Rreyes diez y seis de ag.^{to} de mill e qui^{o}s y sesenta y siete a^{o}s ante los señores pres.^{te} y oidores desta Rreal Audiençia / se presento esta pet.^{on} y los dhos señores proveyeron que queriendo yr Ffray Gaspar vaya como esta proveido y lleve el escrivano que le paresçiere.

Avendaño

/382r/ En la çibdad de Los Rreyes de los rreynos e probinçias del Peru a catorze dias del mes de ag^{o} de myll e qui^{o}s e sesenta e siete años paresçio presente Juan Bautista Caxaguaraca caçique prinçipal del rrepartimi^{o} de Chacalla encomendado en Fran^{co} de Anpuero vz^{o} desta dha çibdad y me rrequerio con una provysion rreal de Su Magestad a my dirigida su thenor de la qual es este que se sigue:

Don Phelipe por la gracia de Dios rrey de Castilla de Leon de Aragon de las dos Seçilias de Jerusalen de Nabarra de Granada de Toledo de Valençia de Galizia de Mallorcas de Sevilla de Çerdeña de Cordoba de Corçega de Murçia de Jaen de los Algarves de Algeçira de Gibraltar de las yslas de Canaria de las Yndias yslas e tierra fyrme del mar oçeano; conde de Flandes e Tirol etc.

a bos Br^{me} de Prol nuestro rresçetor en la nra Avdi^{a} e Chançelleria Rreal que por nro man^{do} rresyde en la çibdad de Los Rreyes a quyen esta sometido el dar la posesyion de las tierras de Quybe e poner los mojones conforme a nra carta ex.^{a} rreal por nos mandada dar Salud e graçia / saved que pleyto se a tratado y esta pendiente en grado de segunda suplicaçion en la dha nra Rreal Audiençia entre partes: de la una el caçique e y^{o}s de Chacalla e de la otra el caçique e yn^{o}s de Canta sobre rrazon de unas tierras de Quybe en la qual aviendose dado sent.^{a} de vista e rrevista fue mandada dar nra carta ex.^{a} rreal a los dhos yndios de Chacalla para que les fuesen bueltas y rrestituydas las dhas //
/382v/ tierras de Quybe sobre que a sydo y es este pleyto con que los dhos yndios de Chacalla volviesen e rrestituyesen a los dhos yndios de Canta las dozientas obejas que rresçibieron por un conçierto presentado en la dha causa segun mas largamente consta e paresçe por la dha ex.^{a} despues de lo qual paresçio presente en la dha nra Rreal Avdi^{a} don Felipe Taurichunbi e Al^{o} Saoac prinçipales del pueblo de Quybe por sy e en n^{e} de los demas prinçipales e yndios del dho rrepartimi^{o} e por una pet.^{on} que presentaron nos hizieron rrelaçion diziendo que ellos tenyan çiertas tierras e chacaras en el dho pueblo de Quybi e su comarca y que son yungas e los yndios yauyos de Fran^{co} de Anpuero los ynquyetan e dizen que ellos tienen alla tierras seyendo serranos no teniendo en ellas cosa alguna sino ellos desde la fundaçion del pueblo e de tienpo de los Yngas pasados e porqûe dello rresçiben molestia nos suplicamos le mandasemos dar nra carta e probisyon rreal de anparo pues es cosa suya e lo an tenydo e tienen ansy ellos como sus antepasados de tienpo ynmemorial a esta parte mandando que los dhos yndios yauyos no se entremetan en ello de que fue man.^{do} dar traslado a la otra pte de los yn^{o}s de Chacalla e Diego de Ocanpo en su n^{e} por una pet.^{on} que presento en rrespuesta del dho pedimi^{o} dixo que se devia dar por nyng^{o} el dho pedimi^{o} porque por el proçeso consta que las dhas tierras pertenesçen a los dhos sus partes e a sus yn^{o}s e por tales suyas se las mandaron entregar e que no hazia al caso dezir que no se litigo con ellos pues hera not^{o} en los dhos nros rreynos del Peru desde que se descubrieron que todos los dhos pleytos se an hecho e hazen con el caçique prinçipal en n^{e} de sus yndios syn otro poder ny lo //
/383r/ pueden dar los yndios porque no tienen conçejos para poder dar en n^{e} de todos poder e ansy se a admytido e admyte el caçique prinçipal para poder pleytear e dar poder en n^{e} de todos en el porque estos son yndios ynpuestos por los serranos para dezir lo que dizen e que hera cosa evidente que an sabido del pleyto que se trataba seyendo cosa tan pu^{ca} y aun es de creer que an contribuydo para las costas del pleyto de que si se diese lugar a que se vos biesen los pleytos que se an hecho con los caçiques seria destruyr e ynquyetar toda la tierra e nos fue pedido e suplicado declarasemos no aver lugar el dho pedimi^{o} e que mandasemos que no se syguiese mas pleyto sobre esta causa pues esta fenesçida por sent.^{a} de rrebista e por otra pet.^{on} que presentaron en la dha nra Rreal Aud.^{a} don Felipe Taulichunbi e Al^{o} Caguaca caçiques de Quybe por sy e por los demas yndios yungas como terçeros por su ynteres dixeron que a su not.^{a} hera benido que entre los caçiques de Canta e Chacalla a mas de seis años que se traxo pleyto sobre sus propias tierras de Quybe donde estan poblados e son suyas de tpo ymemorial a esta parte y est.^{do} como estan en posesyon dellas syn saber que entre los susodhos se trataba el dho pleyto sobre su hazienda e ansy porque el caçique de Canta no probo señorio en las dhas tierras se adjudicaron al caçique de Chacalla en çierta forma e por q^{to} como dho tiene de tpo ymemorial a esta parte ellos e sus antepasados an estado en posesyon sienpre de las dhas tierras e les pertenesçen como hazienda suya e como cosa que sienpre an labrado desde el tpo de los Yngas antes que los conquystasen y despues quando entraron los españoles en los dhos nuestros rreynos syenpre las an posydo desfrutado e labrado hasta agora e sy alguna bez se an entrado algunos yndios en ellas a sydo por ser ellos yungas e pocos y no se poder defender y no se las an tomado tiranycamente y sienpre las an tenido //
/383v/ e que no es justo que por el pleyto que entre los susodhos paso ellos pierdan sus tierras e haziendas por las causas como mejor de dr^{o} podian e devian suplicaron de qualesquier sent.^{as} que en lo susodho se aya dado e que se declare no pertenesçerles a ellos las dhas tierras de Quibi e nos suplicaron les mandasemos hazer justiçia por la via e forma que mas les conbenga e syon e que si hera nesçes^{o} suplicaba del dho auto en que se mandaba dar la dha ex.^{a} lo qual visto por el dho presydente e oydores fue mandado guardar lo probeydo sobre lo qual por las dhas partes fueron alegadas çiertas rrazones y traydos çiertos modelos e se obo la causa por conclusa la qual vista por el dho presydente e oydores probeyeron un auto del tenor seguyente:

"En la causa que es entre partes: de la una los yndios de Chacalla e de la otra los de Quybi sobre las tierras de Quybi en la çibdad de Los Rreyes en diez e ocho dias del mes de jullio de myll e qui^{o}s e sesenta e siete años los señores presidente e oydores desta Rreal Aud.^{a} aviendo visto la dha causa mandaron que vaya una persona desta Corte a las dhas tierras de Quybi sobre que se a tratado este pleyto para que conforme a la ex.^{a} sobre ellas dada de la posesyon dellas a los dhos yndios de Chacalla e ponga los mojones en ellas e ansy lo probeyeron y mandaron e señalaron de sus rrubricas."

"El qual dho auto fue pronunçiado por los dhos señores presydente e oydores el dho dia mes e año en el cont.^{do} estando presentes: Juan de Arrandolaça e Di^{o} de Ocanpo procuradores de las partes a quyen se notifico.

Avendaño"

Del qual dho auto por don Felipe por sy e en n^{e} de los demas yndios del asyento de Quybe fue suplicado por bna pet.^{on} que presento en veinte e tres de jullio de myll e qui^{o}s e sesenta e siete años / dixo que se avia de enmendar e rrevocar el dho auto por lo

general e lo que rresulta del proçeso en su fabor y por lo que tiene dho y alegado e porque poseyendo como posee juntamente con sus yºs las dhas tierras de tpo ynmemorial a esta parte y seyendo señor dellas el y sus yndios nyng.ª cosa le perjudica el dho pleyto que se trato entre los yndios de Canta y Chacalla para ser oydo e ynpedir la ex.on que por los de Chacalla se pide mayormente aviendo suplicado de la dha ex.ª e sent.ª de rrebista e porque aunque no obiera otra cosa syno la transaçion e conçierto que en ne suyo se hizo por los yndios de Canta conçertandose con los dhos yndios de Chacalla porque no entrasen en las dhas tierras los dhos yndios de Chacalla prometer como prometieron de no entrar en ellas por las dhas dozientas obejas que rresçibieron de los dhos yndios de Canta no les podian pedir ny ynquyetar //
/384v/ las dhas tierras abnque algun drº tobieran a ellas quanto mas no tenyendolo contra lo qual no enpeçe dezir que los Yngas pasados dieron a los dhos yndios de Chacalla çierta parte de las dhas tierras porque demas de no aver hecho lo susodho los dhos Yngas ny quytadoles a ellos las dhas tierras sy alguna parte tomo dellas seria e fue para sy ponyendo en ellas mytimaes que las senbrasen para el y no para los dhos yndios de Chacalla e alegaron otra rrazones en sus petiçiones contenydas e nos suplicaron mandasemos enmendar e rrebocar el dho auto y anparar en la dha su posesyon e que sean oydos de nuebo lo qual visto por el dho presydente e oydores probeyeron lo probeydo syn enbargo e que se rremyta a la nra Persona Rreal con la causa prinçipal lo qual fue notificado a Juº de Arrandolaça e Diego de Ocanpo prºs de las partes e agora por parte de los dhos yndios de Chacalla nos fue pedido e suplicado que pues por auto de bista e rrebista estaba proveydo que fuese una persona a les anparar en la posesyon de las dhas tierras y a que se las amojonasen que sy mas pedimiº hera nesçesario mandasemos enmendar e rrebocar la dha sª e le mandasemos anparar en la posesyon que a tenydo e tiene de las dhas tierras lo qual visto por el dho presiedente e oydores probeyeron que a los dhos yndios de Chacalla se les de la exª que les esta mandada dar syn enbargo de lo alegado por la parte contraria syn perjuizio del drº de las partes e mandaron dar traslado de las dhas petiçiones a los yndios de Chacalla lo qual se notifico a Diº de Ocanpo su pror despues de lo qual en veinte e siete dias del mes de junyo deste presente año de myll e quiºs e sesenta e siete por una peton que presento Juan de Arrandolaça en ne de don Felipe Taulichunbi e Alº Çaguac prinçipales yungas del dho pueblo de Quibe y el pueblo que alli esta en el pleyto con el caçique e yndios de Chacalla sobre las dhas tierras de Quibi dixo que las partes contrarias pretendian sacar la dha exª e con ella ynobar e despojar a sus partes de las dhas sus tierras no pudiendo hazer ny ynobar en la dha causa hasta que sus partes sean oydas por no averse hecho con ellos la causa syno entre otros terçeros que no tenian el drº que sus partes tienen a las dhas tierras e que en todas las nras Audas Rreales e Consejos aviendo terçero que salga a la causa aunque sea despues de la sentª de rrebista que pretenda drº a la costa sobre que se a tratado pleyto syenpre le oyen e ynpiden la exon mayormente estando en posesyon como sus partes estan de las dhas tierras de tpo ynmemorial a esta parte antes e despues de los Yngas y demas desto mostraran como jamas las dhas tierras nunca fueron quytadas a su parte por los Yngas ny por otra persona algª y nos fue pedido e suplicado mandasemos que no se ynobase en la dha causa ny se les diese la dha exª hasta que sus partes fuesen oydas e que pues estava mandado que se les diese syn perjuizio del drº de sus partes que se entendiese tanvien syn //
/384r/ perjuizio de su pose para que tobiese efeto se mandase nonbrar e que fuese a costa de las partes contrarias y se le entregue el proçeso original para que por el se pongan los dhos mojones e visto por el dho presidente e oydores proveyeron que la persona que fuere a lo susodho sea a costa de anvas partes e que al que se señalare se entregue el proçeso e que el nonbramiº de la tal persona se rremetia al nro presidente el qual nonbro a vos el dho Brme de Prol nro escrº Rreal e rreçetor de la dha nra Rreal Aud.ª e confiando de vos que sois tal persona que guardareis nro serviº e que vien e diligentemente hareis lo que por nos vos fuere mando e que entendereis en lo susodho como convenga / fue acordado que deviamos mandar dar esta nra carta para vos en la dha rrazon e nos tobimoslo por bien por la qual vos mandamos que luego que con esta nra carta e provisyon rreal fueredes rrequerido hos partais en cunplimº del auto que de suso va yncorporado llevando vara de nra Justicia Rreal e vais al dho rrepartimiº de Canta e tierras de Quibi sobre que a sydo //
/385r/ y es el dho pleyto y puesto en ellas beays nra carta ex.ª rreal que esta dada a los dhos yndios de Chacalla sobre las dhas tierras e al tenor e forma della metais en la posesyon de las dhas tierras de Quybi a los dhos yndios de Chacalla e metidos que los ayais en la dha posesyon en forma vereys por vista de ojos e averiguareys presentes las partes a quien toca e sabreis por ynformaçion e por los autos del dho proçeso las partes e lugares de las dhas tierras de Quybi donde estaban puestos mojones en tpo del Ynga por donde se viese estar dibidydas las dhas tierras con sus comarcanos e por aquellas partes por donde hallaredes que estavan los dhos mojones en tpo antiguo los amojonareis e hareis amojonar de nuevo o rrenovareis en los dhos lugares e limytes ponyendo mojones tan altos como un estado de honbre a costa de las partes entre quyen se partiere los dhos limytes para que por el dho amojonamiº se conozcan las dhas tierras que se yncluyen en las dhas tierras de Quibi para que de aquellas pueda usar e use los dhos yndios de Chacalla y no mas e mandamos a los dhos ynºs de Chacalla e a los comarcanos de las dhas tierras de Quibi que guarden el dho amojonamiº que ansy por bos fuere hecho e que no quyten ny muden los dhos mojones ny alguno dellos de la parte e lugar donde los pusyeredes so las penas que vos de nra parte les pusyeredes las quales les ponemos e avemos por puestas y es nra mçd que esteis en hazer e cunplir lo susodho en yda buelta y estada quynze dias y que ayais e lleveis de selario para vra costa e mantenimiº cada uno de los dhos quinze dias quatro pesos conforme a la hordenança de la dha nra Avd.ª a costa de anvas partes syn llevar ni cobrar drºs algºs de las ynformaçiones e autos ny de otra cosa que ante vos pasaren y se hiziere como juez de comysion ny como escrº e por escusar de costas a los dhos naturales mandamos que no lleveis escrº e que ante vos como nro Consejo Rreal pase todo e que vos hordeneis e agais todo lo que se ofresçiere aver de hazer en rrazon de lo susodho syn les pedir ni //
/385v/ llevar costa alguna por ello el qual dho selario mandamos a las dhas partes que os den e paguen cada una dellas luego que por bos fueren rrequeridos e que no hos lo dando e pagando lo podais aver e cobrar dellas e de sus bienes conpeliendoles a ello por todo rrigor de drº que para entender en todo lo susodho e cada una cosa e parte dello e cobrar el dho selario vos damos poder cunplido tal qual para en tal caso se rrequyere e no fagades endeal por alguna manera so pena de la nra mçd e de quiºs pesos para la nra camara. Dada en Los Rreyes a ocho de agosto de myll e quiºs e sesenta e syete años / yo el escrº Pedro de Avendaño escbº de camara de la Mag.t Rreal la fyz escrevir por su man.do con acuerdo del su presydente e oydores.

Rregistrada Juan de Murga
Chançiller Gaspar de Solis

El Lic.^{do} Castro / Dotor Brabo de Sarabia / El Lic.^{do} don Alvaro Ponçe de Leon / El Liçençiado Serrano de Vigil.

En la qual dha probisyon que de suso ba ynserta e yncorporada de Su Mag.^{t} e sellada con su rreal sello dixo me rrequeria e rrequerio para que conforme a ella e a la ex.^{a} de que en ella se haze mençion fuese con el al dho pueblo de Quybi e a las demas partes e lugares que fuese nesçesario a le dar y entregar la posesyon de las tierras de Quybi de que en ellas se haze mençion y las amojonar dibidir e apartar de las tierras comarcanas a ellas que ellos estaban prestos de guardar e cunplir desta parte lo que por ellas les es mandado e que haziendolo ansy aria lo que no hera cometido e mandado e protesto lo en contrario hecho e luego yo el dho escrivano e rresçetor susodho tome la dha probisyon rreal en mis manos e la ovedesçi con el acatamy^{o} debido e dixe e rrespondi que estaba presto de la cunplir en todo e por todo segun e como en ella es contenido e se contiene y de me partir luego a entender en lo susodho. T^{o}s Antonyo Pz de Villareçes e P^{o} de Prado e Al^{o} de Balle.

/386r/ E despues de lo susodho en la dha çibdad de Los Rreyes a diez e seis dias del mes de ag^{o} e del dho año paresçieron presentes los dhos Ju^{o} Bautista caçique prinçipal de Chacalla e otros yndios yauyos del dho rrepartimi^{o} e me pidieron que por q^{to} estaba mandado que el padre Fray Gaspar de Carvajal de la Horden de Santo Domyngo como persona que savia las dhas tierras de Quybi e por donde se amojonaban partian e dibidian fuese a entender en ello e que yo el dho escrivano e persona nonbrada asystiese con el en todo lo susodho si lo quysiese hazer el dho padre Fray Gaspar que yo el dho escrivano lo fuese a tratar con el para que juntos fuesemos a entender en ello como los señores presydente e oydores lo mandavan e yo dixe que ansy lo cunpliria. T^{o}s Esteban Pz ... e Fran^{co} de Anpuero.

B.^{me} de Prol escr^{o}

E despues de lo susodho en los dhos dias del dho mes e año yo el presente escrivano fuy al monesterio de Santo Domingo esta dha çibdad de Los Rreyes adonde halle al dho padre Fray Gaspar de Carbajal e le hize saver como los dhos señores presydente e oydores desta dha Rreal Avd.^{a} le rrogavan y encargavan que fuese a entender en este dho negoçio de las tierras de Quybi a pedimi^{o} de los yndios de Chacalla con los de Canta e sobre todo lo demas en la dha provisyon e carta ex.^{a} contenydo / el qual dixo que obedesçia lo que le hera man.^{do} e que estaba presto de luego yr e partirse mañana a entender en ello juntamente con yo el dho escrivano e persona a quyen estaba cometido e lo firmo de su n.^{re}

B.^{me} de Prol escr^{o}

/386v/ E despues de lo susodho estando en el pueblo de Quybi a veynte dias del dho mes de ag.^{to} del dho año / yo el presente escr^{o} yntime e notyfique a los dhos don Ju^{o} Bautista caçique prinçipal de Chacalla encomendado en Fran^{co} de Anpuero e a otros prinçipales e yndios del dho rrepartimy^{o} e les hize saber como hera venydo a este dho pueblo a entender juntamente con el padre Fray Gaspar de Carbajal en les entregar las tierras de Quybi sobre que avian traydo pleyto con los yndios de Canta que para que se cunpliese y efetuase ellos desta parte traxesen las dosientas obejas de la tierra que avian de entregar a los dhos yndios de Canta que fuesen de dar e rresçibir lo qual les dixeron que ellos no las tenyan porque las tenyan los yndios de Canta e que tenyan probisyon para que ellos las traxesen e que a ellos se le mandase las traxesen porque las que no fuesen de dar e rresçibir ellos darian darian [*sic*] otras como los señores presidente e oydores lo mandavan lo qual dixeron e declararon ansy por lengua de Ju^{o} de Albarado ynterpetre.

B.^{me} de Prol escr^{o}

E despues de lo susodho en beynte e bn dias del dho mes e año el dho padre Fray Gaspar de Carbajal e yo el dho escrivano y personas nonbradas para lo susodho hizimos paresçer ante nos a los caçiques e prinçipales de Chacalla e a los prinçipales e yndios de Canta a los quales por lengua de el dho Ju^{o} de Albarado les fue dado a entender como venyamos a entender en este negoçio y que para lo poner en hefeto los yndios de Chacalla avian de dar y entregar a los de Canta que estaban //
/387r/ presentes dozientas obejas de la tierra las quales dezian aver ya rresçibido los dhos yndios de Canta que sy hera ansy las traxesen para que se biesen sy heran de dar e rresçibir e las que no lo fuesen que les hariamos dar otras que lo fuesen/que luego los dhos yndios e prinçipales de Canta rrespondieron por la dha lengua que nunca tales obejas avian rresçibido ny tenyan en su poder ny savian dellas que porque heran todas ellas pacos e no obejas e llenas de carache que se querian morir no las avian rresçibido que adonde las dexaron los yndios de Chacalla las buscasen e traxesen para ellos las rresçibir como por la dha ex.^{a} e probisyon rreal se le manda / e luego los dhos yndios de Chacalla por la dha lengua dixeron que sienpre avian tenydo e tenyan los yndios de Canta las dhas obejas en su poder e que ellos las avian de traer e no los yndios de Chacalla e que se les avian muerto algunas e que las avian de rresçibir en q.^{ta} e tobieron otras rrazones los unos con los otros.

E visto lo susodho e que los unos por los otros dexaban de yr a buscar y traer las dhas obejas e que hasta que se entregasen no se podia entender en el dho neg^{o} mandamos que los yndios de Chacalla e los yndios de Canta enviasen de cada parçialidad diez yndios e que estos fuesen a buscar las dhas obejas e las traxesen para el dho efeto con pena de dozientos pesos al que lo dexase de hazer a los quales se le dixo y declaro ansy por el dho Ju^{o} de Albarado ynterpetre y dixeron que ansy lo cunplirian.

B.^{me} de Prol escr^{o}

/387v/ E despues de lo susodho en beynte e dos dias del dho mes e año por la dha lengua de Juan de Albarado ynterpetre les fue mandado a los dhos yndios de Canta e de Chacalla a entranbas las dhas parçialidades que traxesen las dhas obejas como les estaba mandado so la dha pena y estando presentes la parçialidad de Canta dixeron que ellos no las avian rresçibido e que los yndios que avian enbiado de su pte para las ayudar a traer se avian buelto parte dellos porque las tenyan en su poder los de Chacalla / e luego la parçialidad de Chacalla dixeron que los de Canta tenyan en guarda de las dhas obejas muchos yndios que no las querian dexar traer por que no biese hefeto lo cont.^{do} en la dha ex.^{a} que nos pedian e rrequerian le diesemos y entregasemos las dhas tierras e posesyon dellas. E luego les fue man^{do} a entranbas las dhas parçialidades que entranbas trayan las dhas ovejas como les tenemos mandado e so las dhas penas e otros dozientos pesos de oro lo qual se le dio a entender por el dho Ju^{o} de Albarado ynterpetre de la dha Rreal Abdiençia los quales exsecutaran por ellas para la camara de Su Magestad.

B.^{me} de Prol escr^{o}

E despues de lo susodho en el dho pueblo de Quibi a veynte e tres dias del dho mes e año el dho Ju^{o} Bautista caçique prinçipal de Chacalla por sy e los demas yndios de su parçialidad e ayllo dixeron que por q^{to} las obejas rresçiben gran trabaxo en vaxar al dho pueblo de Quibe e se podrian morir muchas dellas por estar flacas que rrequyeren una dos e tres bezes a los prinçipales e yndios de Canta que estaban presentes que las fuesen a rresçibir a la syerra donde las tenian e que el padre //
/388r/ Fray Gaspar de Carvajal e yo el dho escrivano las fuesemos halla a ver si eran de dar e rresçibir lo qual luego yo el dho escr^{o} por la dha lengua les hizimos saver a los dhos yndios de Canta e dixeron que ellos no querian yr a rresçibir las dhas obejas a la syerra syno en el pueblo donde les quytaban las tierras como se mandaba por la dha rreal provisyon e que demas desto no se podia yr a pie ny a caballo donde estaban las dhas obejas e que aunque fuesen no las tenyan hally todas dozientas cabezas porque faltaban muchas e de las que avia estaban llenas de carache e heran pacos e no de rresçibir e como faltase alguna de las dhas dozientas cavezas no las avian de rresçibir que prim^{o} que les quytasen sus tierras le diesen las dhas dozientas obejas e los unos e los otros lo pidieron ansy por testimonyo por la dha lengua de Ju^{o} de Albarado ynterpetre.

E visto lo susodho se les mando traxesen las dhas ovejas entranvas parçialidades como les estaba mandado e so la dha pena lo qual les declaro ansy el dho Ju^{o} de Albarado atento que no las querian yr a rresçibir adonde ellas estan e so las dhas penas. Presentes los prinçipales del [*sic*] Chacalla e Canta a quyen se notifico.

E despues de lo susodho en el dho pueblo de Quybe veynte e siete dias del dho mes e año los yndios de Canta e Chacalla traxeron vn hato de obejas e pacos los quales yo el dho escrivano myre y conte una a una en presençia del dho padre Fray Gaspar de Carbajal y contadas halle que avia çiento e treze obejas machos e henbras e dos corderos chiquytos por todas entre las quales avia diez e ocho que no heran de rresçibir porque estaban llenas de carache apartes las quales luego di y entregue //
/388v/ a los prinçipales e yndios de Canta para e quenta [*sic*] de las dozientas obejas que le avian de entregar por manera que les quedan hobejas buenas de dar e rresçibir nobenta e çinco cabezas los quales dixeron que no las querian rresçibir porque demas de ser pacos e chiquytos no se le cunplian las dozientas cabezas que le avian de entregar prim^{o} que les diesen la posesyon de las tierras sobre que an traydo pleyto e que me piden e rrequyeren no les de la posesyon de las dhas tierras hasta que le entreguen las dhas dozientas obejas y protestaron lo que en cont^{o} se hiziese.

E luego los dhos prinçipales e yndios de Chacalla aviendoles preguntado por la dha lengua que son de las demas obejas a cunplimi^{o} de las dhas dozientas que an de dar a los yndios de Canta dyxeron que todas las dhas dozientas obejas ellos las avian ya entregado por ante Gibraleon a los dhos yndios de Canta e que syenpre ellos las avian tenydo despues aca en su poder e las avian guardado e que ellos avian de dar cuenta dellas y traerlas para que ellos se las entregasen e que si se an muerto o ellos las an comydo que no son obligados a entregarlas agora otra vez que piden e rrequyeren a my el dho escrivano le entregue la posesyon de las dhas tierras que las obejas que faltan que las pida a Fran^{co} de Anpuero que esta obligado a pagar las dhas dozientas obejas e que ellos tienen ya cunplido de su parte todo lo que heran obligados porque Fran^{co} de Anpuero su encomendero avia conprado las dhas dozientas obejas y las traxo del rrepartimi^{o} de Montenegro a este asyento de Quybe e las //
/389r/ entregaron todas dozientas a los dhos yndios de Canta e a los prinçipales del dho rrepartimi^{o} que si algo quysieren e faltaren las pidan a quyen con dr^{o} puedan porque ellos ya an cunplido e que sy quysieren lo pidan a Anpuero e que les den la posesyon de sus tierras e que sy no quysieren rresçibir las dhas ovejas que sea a su culpa e cargo e yo el dho escr^{o} rrespondi que hasta tanto que se entregasen los de Canta en las dhas dozientas obejas no les podra dar la posesyon de las dhas tierras e que sobre ello ocurriesen a los señores presydente e oydores que sobre ello probeyesen lo que fuesen servidos / lo qual se les dio a entender por la dha lengua.

B.^{me} de Prol escr^{o}

E luego yo el dho escrivano y el dho padre Fray Gaspar de Carvajal mandaron a entranbas las dhas parçialidades que atento que los unos ny los otros no quyeren rresçibir las dhas obejas ny guardarlas por que no se pierdan e se mueran hasta que por la Rreal Aud.^{a} se probea syn perjuizio del dr^{o} de cada una de las partes que cada una nonbre seis yn^{o}s e se guarden e apaçenten las dhas obejas y las curen de manera que no se pierdan e tengan q^{ta} con lo que dellas nasçieren so pena que si se murieren o se perdieren sea a costa de entranbas partes e ansy se lo notifique en presençia dellos y por lengua del dho Ju^{o} de Albarado ynterpetre.

B.^{me} de Prol escr^{o}

E luego la parçialidad de Canta dixo que no queria dar los seis yn^{o}s que le fue man^{do} para guardar las dhas obejas como le estaba man^{do} / porque no se le acababan de entregar todas dozientas como se le avian //
/389v/ de dar y entregar antes e primero que se le diese la posesyon de las tierras al caçique e yndios de Cachalla [*sic*] e que de mandarselo ansy apelavan para que ante los señores presydente e oydores de la Rreal Aud.^{a} de Los Rreyes e lo pidieron por testimonyo. T^{o}s a

todo lo que dho es: el padre Fray Gaspar de Carbajal e Fray Graviel de Obiedo e Fray Pedro de Arreal e porende lo fyrme de my nonbre.

B.[me] de Prol scr[o]

E luego yncontinente traxeron e presentaron los yndios de Chacalla veynte e çinco pellejos de obejas cabrunos e uvillo que dixeron se avian muerto de las obejas que avian entregado a los yndios de Canta e que aquellas e todas las demas que faltavan a cunplimi[o] de las dhas dozientas obejas no heran obligados a las entregar otra bez a los de Canta porque se avian muerto por suyas y en sus propias tierras y que me pedian a my el dho escr[o] ansy lo mandase e declarase e lo pidieron por testimonyo.

B.[me] de Prol scr[o]

E luego est.[do] presentes los prinçipales de Canta dixeron que ellos no avian rresçibido las dhas obejas ny pte dellas ny heran para rresçibir por ser todas llenas de carache y enfermas e que por ser tales no las avian querido rresçibir e que dandoselas agora todas dozientas de dar y rresçibir las tomarian e si faltase algunas que no las querian e que no hiziese costa de lo que pedian ny les diese posesyon hasta que cunpliesen la carta ex.[a] e lo pidieron por testi[o].

/390r/	El Liçen[do] Castro	El D. Bravo de Saravia	El Liç[do] don Alv[o] Ponçe de Leon	El Liç[do] Ser[o] de Vigil

S[o] Avendaño

Para que B.[me] de Prol vea las ovejas que estan juntas para el entrego de los yn[o]s de Canta y si son de dar y rresçebir e las que no lo fuere conpela a los yn[o]s de Chacalla que luego den otras en su lugar que sean tales con toda brevedad y concluido esto les de la posesion de las tierras de Quibi conforme a la executoria rreal de manera que en lo uno y en lo otro no aya dilaçion.

por dupp.[da]

/390v/ Don Phelipe por la graçia de Dios rrey de Castilla de Leon de Aragon de las dos Seçilias de Jerusalen de Navarra de Granada de Toledo de Valençia de Galizia de Mallorcas de Sevilla de Çerdeña de Cordova de Corçega de Murçia de Jaen de los Algarves de Algezira de Gibraltar de las yslas de Canaria de las Yndias yslas y tierra firme del mar oçeano; conde de Flandes y de Tirol etc.

A vos Bartolome de Prol nro rreçeptor a quien esta cometida el dar de la posesion de las tierras de Quibi a los yn[o]s de Chacalla / salud y graçia /

Sabed que Diego de Ocanpo en nombre de los caçiques e yn[o]s del dho rrepartimiento de Chacalla por una petiçion que ante nos presento en la nuestra Rreal Audiençia y Chançilleria que por nuestro mandado rreside en la çiudad de Los Rreyes ante el presidente e oydores della nos fue hecha rrelaçion diziendo que nos mandamos dar una nuestra carta y provision rreal para que los yn[o]s de Canta traxesen las ovejas que rreçibieron para que las viesedes si eran de rreçebir y no lo syendo proveyesedes que le diesen otras buenas pues por los autos del proçeso avia constado que los dhos sus partes se las avian entregado y que siendo rrequeridos con la dha nuestra provision rreal los dhos yn[o]s de Canta no lo avian querido cunplir diziendo que no avian rreçebido las dhas ovejas y que entre anbos rrepartimientos avia pleitos y diferençias sobre aver rreçebido las dhas ovejas o no y quien las avia de traer / y que por esta causa se dilata la execuçion de lo por nos proveydo y que pues sus partes tomaron e tienen posesion de las dhas tierras por executoria de la dha nuestra Rreal Audiençia y estan en posesion dellas que no era justo que con maliçias os detubiesedes en poner los mojones dellas ni se hiziesen tantas costas espeçialmente aviendose como se an offreçido los dhos sus partes a trocar las ovejas que les bolvieren si no pareçieren de dar y tomar y nos fue pedido y suplicado mandasemos dar nuestra provision rreal para que ante todas cossas se pongan los dhos mojones y que los dhos sus partes sean anparados en la dha su posesion dando fianças de trocar las ovejas que pareçieren no ser de dar ni tomar / o como la nuestra merçed ffuese / lo qual visto por el dho presidente e oydores fue acordado que deviamos mandar dar esta nuestra carta para vos en la dha rrazon e nos tobimoslo por bien por la qual bos mandamos que luego que con ella fueredes rrequerido hagais traer ante vos las ovejas que estan juntas para el dho entrego y veais si son de dar y rreçebir e las que no lo fueren conpelereys a los dhos yn[o]s de Chacalla a que luego den otras en su lugar que sean tales con toda brevedad y concluido esto los anpareys en la posesion que de las dhas tierras de Quivi tienen / o les dareys otras de nuevo conforme a nuestra Executoria Rreal que sobre ello esta dada no consintiendo que en lo uno ni en lo otro aya dilaçion alguna / lo qual ansy //
/391r/ hazed y cunplid so pena de quinientos pesos para la nuestra camara.

Dada en Los Rreyes a veynte e siete dias de agosto año de mill y quinyentos y sesenta y siete años.

Yo el scr[o] P[o] de Avendaño secrivano de camara de la Mag.[t] Rreal la fiz escrevyr por su mandado con acuerdo del presidente e oydores.

Rr.[da] Juan de Murga

Chançiller
Gaspar de Solis

/392r/ El Liçen[do] Castro — El D. Bravo de Saravia — El Lic[do] don Alv[o] Ponçe de Leon — El Lic[do] Ser[o] de Vigil

Secr[o] Avendaño

Que Bartolome de Prol rreçeptor quando diere la posesion de las tierras de Quivi a los yndios de Chacalla y las amojonare vea ansimismo las aguas con que hasta agora se an rregado y las ponga por escrito y las que ansi averiguare con que se an rregado ninguna persona se las quite so çierta pena.

/392v/ Don Phelipe por la graçia de Dios rrei de Castilla de Leon de Aragon de las dos Seçilias de Jerusalen de Navarra de Granada de Toledo de Valençia de Galiçia de Mallorcas de Sevilla de Cordova de Corçega de Murçia de Jaen de los Algarves de Algeçira de Gibraltar de las yslas de Canaria de las Yndias yslas y tierra firme del mar oçeano; conde de Flandes y de Tirol etc.

A vos Bartolome de Prol nuestro rreçeptor a quien esta cometido el dar de las posesiones de las tierras de Quibi a los yndios de Chacalla y poner los mojones en ellas / salud y graçia / Saved que Diego de Ocanpo en n[e] del caçique e yndios de Chacalla en la causa con los yndios de Canta sobre las tierras de Quibi presento ante nos una petiçion en la nra Rreal Audiençia y Chan.[a] que por nro mandado rreside en la çiudad de Los Rreies de los nuestros rreinos del Piru / ante el presidente y oidores della por la qual nos hizo rrelaçion diziendo que por autos de vista y rrevista esta por nos mandado que una persona vaya a anparar a los dhos sus partes en la posesion de las dhas tierras y se las amojonar / y porque las partes contrarias por la enemistad que les tienen les querran quitar las aguas con que se rriegan las dhas trras de Quibi y si se hiziese seria como no me las diesen [*sic*] nos pidio y suplico mandasemos a la persona y escrivano que an de yr a lo susodho que averiguen las aguas con que las dhas tierras se an rregado y se pongan por escrito en el dho proçeso de la causa y se mande a las partes contrarias que no se las quiten ny parte dellas agora ni en ningun tienpo so graves penas / o como la nuestra merçed fuese / lo qual visto por el dho presidente y oidores ffue acordado que deviamos mandar dar esta nra carta para vos en la dha rrazon e nos tovimoslo por vien por la qual vos mandamos que al tienpo que diesedes la posesion de las dhas tierras de Quibi a los dhos yndios de Chacalla y pusieredes en ellas los mojones en cunplimi[o] de la nra Rreal Ex.[a] que ante vos sera presentada veais e averigueis ansimismo con çitaçion de las ptes las aguas con que las dhas tierras de Quibi se an rregado //
/393r/ hasta agora y de las syerras y lugares de donde biene y por donde an entrado y entran en el dho rregadio y en que cantidad y en que tantos topos de tierra se tenya por costunbre de rregar con ella y lo asentareys todo particularmente por escrito para que se entienda lo que en esto a pasado y las aguas con que ansi averiguaredes averse rregado las dhas trras de Quivi mandamos que ningun caçique ny prençipal ny yndio ni otra ninguna persona se las quite ni parte dellas agora ni en ningun tienpo y las dexen yr libremente como hasta aqui lo an fho y pasar el dho rregadio so pena de pagar el daño que siguiere al dueño o dueños de las dhas tierras y mas mill pesos de oro dar a nra camara y no fagades endeal por alguna manera so pena de la nuestra mrd y de qui[o]s p[o]s de oro para nuestra camara.

Dada en Los Rreyes a treynta dias del mes de agosto de mill e qui[o]s y sesenta y siete a[o]s.

Yo el secr[o] P[o] de Avendaño escrivano de camara de la Mag.[t] Rreal la fiz escrevyr por su mandado con acuerdo del presidente e oydores.

Rr.[da] Juan de Murga — Chançiller Gaspar de Solis

/394r/ El Liçen[do] Castro — El D. Bravo de Saravia — El Liç[do] don Alv[o] Ponçe de Leon — El Liç[do] Ser[o] de Vigil

Secr[o] Avendaño

Para que Bartolome de Prol vea las ovejas que estan juntas para el entrego de los yn[o]s de Canta para el entrego de los yndios de Canta [*sic*] e si son de dar y rresçebir y las que no lo fuere conpela a los yn[o]s de Chacalla que luego den otras en su lugar que sean tales con toda brevedad y concluido esto les de la posesion de las tierras de Quivi conforme a la Ex.[a] Rreal de man.[a] que en lo uno y en lo otro no aya dilaçion.

/394v/ Don Phelipe por la graçia de Dios rrey de Castilla de Leon de Aragon de las dos Seçilias de Jherusalen de Navarra de Granada de Toledo de Valençia de Galizia de Mallorcas de Sevilla de Çerdeña de Cordova de Corçega de Murçia de Jaen de los Algarves de Algezira de Gibraltar de las yslas de Canaria de las Yndias yslas y tierra firme del mar oçeano; conde de Flandes y de Tirol etc.

A vos Bartolome de Prol nuestro rreçeutor a quien esta cometido el dar de la posesion de las tierras de Quibi a los yndios de Chacalla / Salud y graçia /

Sabed que Diego de Ocanpo en nombre de los caçiques e yndios del dho rrepartimiento de Chacalla por una petiçion que ante nos presento en la nuestra Rreal Audiençia y Chançilleria que por nuestro mandado rreside en la çiudad de Los Rreyes ante el presidente e oydores della nos fue hecha rrelaçion diz.^{do} que nos mandamos dar una nuestra carta y provision rreal para que los yndios de Canta traxesen las ovejas que rreçibieron para que las viesedes si eran de rreçebir y no lo siendo proveyesedes que le dieredes otras buenas pues por los autos del proçeso avia constado que los dhos sus partes se las avian entregado y que siendo rrequeridos con la dha nuestra provi^{on} rreal los dhos yn^{o}s de Canta no la avian querido cunplir diziendo que no avian rreçebido las dhas ovejas y que entre anbos rrepartimientos avia pleito y diferençias sobre aver rreçebido las dhas ovejas o no y quien las avia de traer y que por esta causa se dilata la execuçion de lo por nos proveydo y que pues sus partes tomaron y tienen posesion de las dhas tierras por executoria de la dha nra Rreal Audiençia y estan en posesion dellas / que no era justo que con maliçias os detubiesedes en poner los mojones dellas ni se hiziesen tantas costas espeçialmente aviendose como se an ofreçido los dhos sus partes a trocar las ovejas que les bolvieron si no pareçieren de dar y tomar y nos fue pedido y supp.^{do} mandasemos dar nuestra prov.^{on} rreal para que ante todas cosas se pongan los dhos mojones y que los dhos sus partes sean anparados en la dha su posesion dando fianças de trocar las ovejas que pareçieren no ser de dar ni tomar / o como la nuestra merçed fuese / lo qual visto por el dho presidente e oydores ffue acordado que deviamos mandar dar esta nra carta para vos en la dha rrazon y nos tovimoslo por bien por la qual vos mandamos que luego que con ella ffuesedes rrequerido hagais traer ante vos las ovejas que estan juntas para el dho entrego y veais si son de dar y rreçebir y las que no lo fueren conpelereys a los dhos yn^{o}s de Chacalla a que luego den otras en su lugar //

/395r/ con toda brevedad y concluido esto los anparareys en la posesion que de las dhas tierras de Quibi tienen o les dareys otras de nuevo conforme a nuestra Executoria Rreal que sobre ello esta dada no consintiendo que en lo uno ni en lo otro aya dilaçion alguna lo qual ansi hazed y cunplid so pena de quinientos p^{o}s para la nra camara.

Dada en Los Rreyes a veynte e siete de agosto de mill y quiny^{o}s y sesenta y siete a^{o}s.

Yo el scr^{o} P^{o} de Avendaño escrivano de camara de la Mag.^{tt} Rreal la fiz escrevyr por su mandado con acuerdo del presidente e oydores.

Rr^{da} Juan de Murga

Chançiller
Gaspar de Solis

En Los Rreyes treynta dias del mes de ag^{to} de myll e qui^{o}s e sesenta e siete años me rrequerieron con esta probisyon de Su Mag.^{tt} e con otra antes desta para que las cunplays en su cunplim^{o} les de la posesyon de las dhas tierras contenydas en la dha carta ex.^{a} y de las aguas con se rregan [*sic*] e an solido rregar e se las amojonen como por la dha rreal provisyon se me mandava que ellos estan prestos de entregar las diez e ocho obejas que de las çiento e treze yo el presente escr^{o} desheche por estar malas e llenas de carache y no de dar e rresçebir so las penas contenydas en la dha probisyon juntamente con el padre Fray Gaspar de Carbajal //

/395v/ so las penas en las dhas probisyones contenydas dixe e rrespondi que las obedesçia e que estava presto de las cunplir en todo e por todo.

B.^{me} de Prol scr^{o}

E luego yncontinente los prinçipales e yndios de Chacalla traxeron e presentaron ante my el dho escrivano y ante el dho padre Fray Gaspar de Carbajal las dhas diez e ocho ovejas que avia yo sacado del dho monton por no ser de rresçibir que con ellas son çiento e treze obejas de dar e rresçibir e contadas e myradas todas ellas por paresçer que heran de dar e tomar mande a los yndios de Canta que ally estaban en el propio pueblo de Quybe que se disen don Her.^{do} ..ynac Guaranga e don Fran^{co} Saroa Poma hermano de don Fran^{co} caçique prinçipal de Canta e don Fran^{co} Pilco e don Fran^{co} Arcos e otros prinçipales e yn^{o}s que rresçiban las dhas obejas e las tengan por suyas e como suyas e las guarden e benefyçien como tales con apresçibimi^{o} [*sic*] que les hize que sy se murieren algunas o se perdieren sea a su culpa e cargo y no de los yndios de Chacalla los quales dixeron que el caçique prinçipal estaba en Lima e que ellos no heran caçiques e que no las querian rresçibir quanto mas que no se las davan todas juntas como Su Magestad lo mandava que ellos protestaban de no las rresçibir ny guardar e que si se perdiesen o muriesen fuese a culpa e cargo de los dhos yndios de Chacalla e no dellos por no se las dar ny entregar todas juntas como se mandava e que si hera nesçesario apelavan de lo por my probeydo e mandado para ante los señores presydente e oydores e lo pedio por testimonio. T^{o}s: el padre Fray Grabiel y Fray P^{o} de Albarado.

B.^{me} de Prol scr^{o}

/396r/ E luego yncontinente los dhos yndios de Chacalla dixeron que se disystian e apartavan de todas las dhas obejas e del dr^{o} que a ellas tenian e poseyan e les pertenesçia en qualquier manera y lo çedian e traspasaban en los dhos yndios de Canta que estaban presentes para que agan dellas como de cosa suya propia porque las que faltaban a cunplimi^{o} de las dhas dozientas ovejas si se avian perdido e muerto hera a su culpa e cargo e con esto abian cunplido conforme a la provisyon de Su Magestad atento que ya ellos abian traydo y entregado a los dhos yndios todas las dhas dozientas ovejas en presençia de Alonso Diaz de Gibraleon escrivano e ansy lo pidio por testimonyo. T^{o}s : los dhos y el dho Ju^{o} de Albarado ynterpetre que lo declaro ansy a las partes.

B.^{me} de Prol scr^{o}

E luego yncontinente el dho dia mes e año susodhos el dho Fray Gaspar de Carbajal como persona que dixo que sabia quales heran las tierras de Quybe sobre que se abia tratado este pleyto e las avia medido e tasado e ynformadose vien quales heran en presençia de my el dho B.^{me} de Prol escr^{o} e de los t^{o}s de yuso escritos fue a las dhas tierras de Quybe y dende la otra vanda del rrio grande por donde se haze vna pequena quebrada que baxa de lo alto del çerro de la parçialidad //
/396v/ de los yndios de Xecas de Ju^{o} Piçarro e viene derecho por una azequya arriba y paredones biejos hasta llegar a lo alto del çerro donde esta la yglia [*sic*] una cruz e el dho çerro dizen se llama Chuquicoto e den dende [*sic*] ally hasta la dha cruz se pusyeron veynte e siete mojones de a estado de un honbre de doze a doze braças del uno al otro e dende la la dha cruz al sesgo por detras del pueblo de Quybe e junto a el se pusyeron çinq.^{ta} mojones de a estado de honbre cada uno de doze braças a doze braças que ba por el arroyo arriba en lo alto de la syerra hasta llegar al çerro de Guavyarinca e se rriegan las dhas tierras de lo alto con el dho arroyo que dizen Cacalarca por donde viene un arroyo e azequya de agua al pueblo de Quybe e con los dhos mojones quedaron amojonadas todas las dhas tierras que el dho padre Carbajal dixo e senalo ser de los yndios de Chacalla conforme a la dha Rreal Ex.ª que lindan las dhas tierras con tierras de los yndios de Canta por el rrio e llanada de las tierras e por la otra parte las syerras de los yndios yauyos e por la otra parte por el rrio abaxo con tierras e chacras de los yndios de Guancayo de G^{o} de Cazeres las quales dhas tierras de suso declaradas que el dho padre Fray Gaspar de Carbajal dixo e declaro ser las sobre que se trayan los dhos pleytos e amojonarse por las partes e lugares que ansy quedan amojonadas pertenesçer a los yndios de Chacalla. T^{o}s que fueron presentes a todo lo que dho es: Ju^{o} de Albarado y el padre Fray Grabiel de Obiedo y Fray Pedro de Hrreal //
/397r/ de la Horden de Señor Santo Domyngo y en presençia de las dhas partes los yndios de Canta e Chacalla.

B.^{me} de Prol scr^{o}

E luego yncontinente el dho padre Fray Gaspar de Carbajal en presençia de my el dho escrivano y entranbos a dos dymos la tenençia e posesyon de las dhas tierras de suso deslindadas e declaradas e amojonadas a los prinçipales e indios de Chacalla y le meti en la tenençia e posesyon de todas ellas e le anpare e defendi en ellas e si hera nesçesario se la daba de nuebo e ellos se entraron en ellas de una parte a otra e tomaron tierra piedra e arboles e hecharon a una parte e a otra e dixeron por lengua del dho Ju^{o} de Albarado que se davan por metidos e anparados en la tenençia e posesyon de las dhas tierras de suso deslindadas e declaradas como Su Magestad por sus rreales probisyones lo manda e yo el dho escrivano le di la dha posesyon corporal rreal virtual … de todas ellas e le anpare e defendi en la que tenya tomada e luego paresçieron presente don Felipe Taulichunbi Al^{o} Çagua prinçipales yungas del dho pueblo de Quybe e dixeron que por sy e en nonbre de los demas yndios yungas del dho pueblo de Quybi contradizian el amojonamy^{o} de las dhas tierras e daçion de posesyon e anparo dellas a los dhos yndios de Chacalla porque las dhas tierras heran suyas e las tenyan senbradas y que protestaban de rreclamar con todo ello e lo pedian por //
/397v/ testimonyo e a los presentes le fuesen dello / todo lo qual pidieron por lengua del dho Ju^{o} de Albarado ynterpetre.

B.^{me} de Prol scr^{o}

E luego yncontinente les di y entregue la posesyon de las aguas con que las dhas tierras se an solido e suelen rregar para que de todo ello se aprobechen como de cosa suya e los dhos yndios de Chacalla hizieron las açequyas e tomaron agua e començaron a rregar algunas de las dhas tierras por birtud de lo qual tomaron e yo les di la posesyon corporal rreal virtual de las dhas tierras con todo lo a ellas anexo e pertenesçiente e aguas mamantes e corrientes e con que se rriegan e suelen rregar e los dhos yndios se dieron por metidos e anparados en ellas y los dhos yndios yungas se lo contradixeron como dho es e luego los dhos prinçipales e yndios de Chacalla de como quedan en la dha posesyon lo pidieron por testimonyo. T^{o}s que fueron presentes a lo que dho es: los susodhos.

B.^{me} de Prol scr^{o}

E luego yncontinente yo el dho escrivano e persona nonbrada para lo que dho es con asystençia del padre Fray Gaspar de Carbajal les defendi e anpare a los dhos prinçipales e yndios de Chacalla en la su posesyon de las dhas tierras e aguas con que se rriegan e mande que nyng.^{a} persona les quytase ny rremoviese de la dha su posesyon ny derribese los dhos mojones ny alguno dellos so pena de qui^{o}s pesos de oro para la camara y fysco de Su Magestad en las quales dhas penas cada una dellas les dava por condenado lo contrario haziendo lo qual les dio a entender la dha lengua ansy a las partes. T^{o}s los susodhos.

B.^{me} de Prol

/398r/ E luego yncontinente el dho dia mes e año susodhos estando en el dho pueblo de Quybe los dhos prinçipales e yndios de Chacalla dixeron que pues estaban metidos en la tenençia e posesyon de las dhas tierras que ellos querian dexar en ellas en su nonbre treynta yndios de su parçialidad e ayllo para que las labrasen e benefiçiasen que se lo diese ansy por fee y testimonyo e que nyng.^{a} persona les perturbase en la dha su posesyon con pena que para ello se les pusyese lo qual yo el dho escr^{o} ansy mande e que nadie les perturbase en todo ello ny en parte dello so la dha pena lo qual les dio a entender la dha lengua e fueron t^{o}s los susodhos.

B.^{me} de Prol scr^{o}

E luego por my el dho escrivano fue mandado notificar por la dha lengua a los dhos yndios de Chacalla e de Canta todo lo probeydo de suso en sus personas para que lo guarden e cunplen so las dhas penas y los dhos yndios de Canta e los dhos yndios yungas lo contradixeron como dho tenyan e que sy hera nesçesario apelavan de todo ello para ante los señores presydente e oydores e lo pidieron por test^{o}. T^{o}s los dhos.

B.^{me} de Prol scr^{o}

E luego por la dha lengua les mande e declare la dha lengua [*sic*] que el fruto que estaba senbrado lo dexasen gozar a los yndios de Canta e yungas que lo tenyan senbrado hasta alçar el fruto de todo ello so las dhas penas e que en todo lo demas guardasen y cunpliesen lo //

/398v/ de suso probeydo e mandado lo qual les dixe e declare a las dhas partes por la dha lengua del dho Ju^{o} de Albarado e los dhos yndios de Chacalla dixeron que ansy lo cunplirian con que no labrasen mas dende agora en adelante los dhos yndios de Canta e yungas en las dhas tierras agora de nuebo y ansy lo pedieron por testim^{o} e fueron a todo ello presentes por t^{o}s los susodhos.

B.^{me} de Prol scr^{o}

E luego los dhos yndios de Canta dixeron que de todo lo por my el dho rresçetor fecho se sentian por agraviados e lo contradizian todo ello e no querian rresçibir las dhas obejas como dho tienen e apelavan de todo ello como apelado tenyan e lo pedian por test^{o}. T^{o}s los susodhos.

ante my B.^{me} de Prol scr^{o}

Dr^{o}s: el selario de quynze dias; los de Chacalla me pagaron e los otros la mytad y me deben la rresta.

/399r/ Don Felipe caçique de Quibi.

Muy p^{o}s. s^{o}r
El don Felipe muestra aqui no ser
sujeto a los de Chacalla.

Don Felipe caçique del pueblo de Quibi por mi y en n^{e} de mis yndios / digo que el caçique e yndios de Chacalla nos an destruido todos los cocales y maizales y nos an quemado una chacara de frisoles en que avia tres anegas y media de senbradura con color de la sent.^{a} que se a dado en su favor e asimismo nos quieren echar del pueblo que alli tenemos poblado e nos an echo otros muchos malos tratamientos e pues agora ba el rreçevtor a partir las dhas tierras y que los yndios de Canta rresçiban el ganado que les avian dado.

A V. Al. pido y suplico se rresçiva ynform.^{on} de los dhos dayños e ynjurias que nos an echo para que se nos probea de just.^{a} y nos paguen sobre que pedimos just.^{a} y costas e para ello etc.

El Licen^{do} de Leon

En Los Rreyes doze de agosto de myll e qui^{o}s e sesenta e siete a^{o}s ante los señores presydente e oydores desta Rreal Audiençia estando en audiençia pu^{ca} se presento esta pet.^{on} y los dhos señores lo cometieron a Bartolome de Prol que tome ynformaçion sobre todo.

Avendaño

/400r/ Don Antonio y don Her.^{do}

Muy mag^{o} y muy Rr^{do} s^{o}r

Don Antonio y don Hern^{do} prençipales de Canta por nosotros y en n^{e} de los demas prençipales e yndios del dho rrepartimi^{o} dezimos que a nra ynportunaçion e rruego bino a este pueblo de Quivi Luys de Villa Rreal rresçeptor de la Rreal Avdiençia a pedir por nosotros lo que nos conbinyese y aviendolo de hazer asy se a ydo a Lima syn fenesçer a lo que emos benido y pues nosotros no tenemos quyen buelba por nosotros.

Pedimos a V. Al. y rrequerimos mande que Gibraleon no buelva por los yndios de Chacalla ny este aqui / con protestaçion que hazemos que si V. Al. no lo probee ny manda asy nos yremos a quexar a Su Mag.^{t} e a su Rreal Abdi^{a} sobre que pedimos just.^{a} e testimonio de lo que V. Al. sobre ello probeyere.

Don Ant^{o} y don Her.^{do}

En Los Rreyes veinte y seis de ag^{to} de myll y qui^{o}s y sesenta y siete años ante los señores presidente y oidores desta Rreal audiençia en audiençia pu^{ca} se pres^{to} esta pet.^{on} y los dhos señores mandaron.

Avendaño

/400v/ En el pueblo de Quybe veynte y tres dias del mes de ag.^{to} de myll e qui^{o}s e sesenta y siete años el muy rr^{do} padre Fray Gaspar de Carbajal y en my presençia como juez de comysion se presento esta petiçion e visto se mando al dho Gibraleon que luego salga del dho asyento porque no conviene que trayga desasegados [*sic*] los yndios de la una parçialidad con la otra so pena de çien p^{o}s para la camara de Su Mag.^{t} e que a su costa le ynbiare persona a la çibdad de Los Rreyes el qual dixo que yo el dho rresçetor no hera parte para se lo mandar que no traya comysion para lo hechar del valle e que quando la tubiese lo haria e no de otra manera y por no hazerlo el dho padre Carbajal se lo torno a mandar y aun a rrogar e dixo que traxesen provisyon de la Rreal Avdiençia para ello que no lo haria de otra manera lo qual dixo con grandes bozes dando ocasion a que se atrabesasen con el. T^{o}s; el padre Obiedo / el padre Perez / en fee de lo qual lo firme de my n.^{e}

B.^{me} de Prol scr^{o}

/401r/ Don Phelipe caçique de Quibe.

Muy p^{o}s. s^{o}r

Agravianse del amojonamy^{o} echo.
No pareçe ser estos y^{o}s sujetos a
Chacalla mas que a Canta.

Don Felipe caçique de Quibi por mi y en nonbre de los demas yndios yungas del dho pueblo sin perjuyzio de mi dr^{o} digo que V. Al. mando yr al padre Fray Gaspar de Carabajal para que señalase las tierras que avian de aver los yndios de Chacalla y llevo consygo a Prol rresçetor desta Rreal Abdiençia y por soliçitador de los dhos yn^{o}s de Chacalla Gibraleon que tantas vezes se le mande que no lo llevase y aun que se le mando estando alli que se biniese no lo quiso fazer antes nos hizo malos tratamy^{o}s y el dho padre Carabajal demas de lo que declaro la otra vez tan contra justiçia agora de nuevo por dar contento a los dhos y^{o}s de Chacalla nos toma todo nro pueblo de Quibi y nras casas y çiertas chacarrillas de coca que estan detras de nras casas syn pedillo los dhos y^{o}s de Chacalla syno que se contentavan con lo que primero estava mandado que es hasta donde estava la dha yglesia y porque Hernando Astobilca de nra parçialidad que es un yndio biejo que sabia lo que prim^{o} pedian los dhos yndios de Chacalla y dezia los limites de la dha tierra el dho padre Carabajal le tomo por los cabellos y le dio muchas coçes por donde se ve su pasyon y que es verdadera la rrecusaçion que sienpre le emos fho y la afiçion que tiene a los dhos y^{o}s de Chacalla y pues deçidio de los termi^{o}s de la comisyon que llevava y es nyngu^{o} todo lo que hizo por la rrecusaçion que sienpre le pusymos.

A V. Al. pedimos e suplicamos se de todo por nyngu^{o} y se mande rrebocar todos los dhos mojones y que no se nos quite nro pueblo pues nunca a avido pleyto sobre el ny sobre las chacarrillas particulares que estan detras del //
/401v/ y sy es nesçesario apelo del modo que no uvo en poner los dhos mojones e pido just.^{a} e para ello etc.

Otro si por q^{to} aunque V. Al. mando que nos entregasen las obejas de dar y rresçibir antes que le diesen la posesyon de las dhas tierras y no se a hecho segun consta e paresçe por esta petiçion e abtos de que fizemos presentaçion a V. Al. pedimos e suplicamos mande que la posesyon de las dhas tierras no se de y si estuviere dada se de por nynguna hasta tanto que nos entreguen las dhas obejas sobre que pedimos just.^{a}

El Licen^{do} de Leon

En Los Rreyes destos rreynos e prouinçias del Peru a veynte y seis dias del mes de ag^{o} de myll e qui^{o}s y ses.^{a} y siete a^{o}s ante los senores pres^{te} e oydores est^{do} en aud.^{a} pu^{ca} la presento el p.^{e} e los dhos senores mandaron lo proveydo en la otra pet.^{on}

Avendaño

/402r/ El Licen^{do} Castro — El Dr. Bravo de Saravia — El Lic^{do} don Alv^{o} Ponçe de Leon — el Lic^{do} Ser^{o} de Vigil

S^{o} Avendaño

Para que Bartolome Prol rreçetor vea las dozientas ovejas que los y^{o}s de Chacalla buelben a los y^{o}s de Canta en cunplimi^{o} de la Executoria Rreal que les esta dada sobre las tierras de Quibi e si son de dar e rresçevir e que no siendo tales los conpela a que se les entreguen de manera que sean de rresçevir y tomar.

Don Phelipe por la graçia de Dios rrey de Castilla de Leon de Aragon de las dos Seçilias de Jerusalen de Navarra de Granada de Toledo de Balençia de Galiçia de Mallorcas de Sevilla de Serdeña de Cordova de Corçega de Murçia de Jaen de los Algarves de Algezira de Gibraltar de las yslas yslas [*sic*] de Canaria de las Yndias yslas e tierra firme del mar oçeano; conde de Flandes e de Tirol etc.

A vos Br.^{me} de Prol nro rreçeptor a quien esta cometido el dar de la posesion de las tierras de Quibi a los y^{o}s de Chacalla y poner los mojones en ella / salud e graçia / Saved que por una petiçion que ante nos en la nra Rreal Audiençia e Chançilleria que por nro mandado rreside en la çiudad de Los Rreyes de los nros rreynos del Peru ante el presidente e oydores della presento don Fran^{co} Guaman Guaranga caçique de Canta por si y en nonbre de los y^{o}s a el sujetos no se apartando de la suplicaçion que tiene hecha para ante nra Rreal Persona nos hizo rrelaçion diziendo que por nos a sido mandado que una pers.^{a} vaya a cunplir la executoria rreal que por nos esta mandada dar e a partir las tierras de Quibi e las entregar a los y^{o}s de Chacalla e proveer los mojones con que les buelvan e rrestituyan las dozientas ovejas que por el conçierto presentado en la dha causa pareçe que rresçibieron dellos al tenor de la dha rreal executoria e que por quanto las dhas ovejas que las ptes contrarias pretenden pagar son ovejas pacos e biejas e llenas de carache e gusanos e las que ellos les dieron fueron obejas grandes henbras escogidas por dos españoles que estan al presente en la dha çiudad de Los Rreyes / e que pues se manda deshazer el dho conçierto que sus encomenderos hizieron entre todos ellos / e no es justo que les lleven los partos y pospartos de las dhas ovejas e sobre todo que se las den con carache e pacos / e nos pidio e suplico le mandasemos dar e diesemos nra carta e provison rreal para que las ovejas que les dieren sean de la mysma manera que las rresçivieron con los partos e pospartos e que si ellos fueren obligados a bolver los frutos de las dhas tierras que ellos los bolveran entretanto que la causa se determina en el nro Rreal Consejo de las Y^{a}s en grado de segunda suplicaçion o como la nra merçed fuese / lo qual visto por el dho pres.^{te} e oydores fue acoradado que deviamos mandar dar esta nra carta para vos en la dha rrazon e nos tubimoslo por bien por al qual vos mandamos que veys las dhas dozientas ovejas que los dhos y^{o}s de Chacalla buelven y entregan a los dhos y^{o}s de Canta en cunplimi^{o} de la dha nra carta rreal executoria que el esta mandada dar y si son buenas e de dar e rresçevir y no lo siendo conpelays a los dhos y^{o}s de Chacalla que se las den y entreguen tales e tan buenas ...(roto)... e no fagades endeal por alg^{a} manera so pena de la nra merçed e de qui^{o}s p^{o}s para la nra camara.

Dada en Los Rreyes a dos de agosto de myll e qui^{o}s e sesenta y siete años.

Yo el scr^{o} P^{o} de Avendaño escrivano de camara de la Mag.^{t} Rreal la fiz escrevyr por su mandado con acuerdo del su presidente e oydores.

Rr^{da} Juan de Murga — Chançiller Gaspar de Solis

En el pueblo de Quybe a veynte e dos dias del mes de ag^{to} de myll e quy^{o}s e sesenta e siete años; los yndios de Canta me rrequyrieron con esta provisyon para que la obedesçiese e cunpliese lo qual ansy hize e luego la yntime e notifique a los yndios de Chacalla para que traxesen las dhas obejas los quales dixeron que ya las avian enbiado a buscar y que no avian benydo e luego los dhos yndios de Canta me pidieron e rrequerieron que no les diese la posesyon de las tierras hasta tanto que les diesen y entregasen las dhas obejas e que las que tenyan para entregar por estar llenas de carache y enfermas e ser pacos chiquitos e no obejas nos las querian traer al pueblo e aun con todo eso le faltaban muchas obejas para cunplir las dozientas obejas de lo qual estaba presto a dar ynf.^{on} y luego traxo dos e tres yndios de su ayllo que lo dixeron ansy por lengua del dho Ju^{o} de Alvarado y de como lo pedia e rrequeria me pidio se lo diese ansy por fee e testimonyo seyendo t^{o}s el padre Carvajal y el padre Obiedo e Fray P^{o}.

B.^{me} de Prol scr^{o}

E despues de lo susodho veynte e tres dias del dho mes e año los dhos yndios de Canta dixeron que por q^{to} el padre Carvajal e yo andavamos dibiendo [*sic*] e amojonando las dhas tierras sobre que avian traydo pleyto y si se las entregasemos a los yndios de Chacalla no le entregarian las dhas dozientas obejas que prim^{o} estan obligados a les dar que piden e rrequyeren nos las amojonemos ny se las entreguemos hasta que cunplan con ellas y protestaron los daños e costas est^{do} presentes los yndios de Chacalla que dixeron que ya se las avian entregado e que alg.^{as} que se an muerto e otras que estaban para morir avian de ser a su costa e las demas ya las trayan para se las entregar y los dhos yndios de Canta dixeron que ellos no las avian rresçibido ny heran para rresçibir ny rresçibirian las dhas obejas sy faltaban alg.^{as} en fee de lo qual lo firme. T^{o}s: los dhos.

B.^{me} de Prol scr^{o}

/403r/ Don Fran^{co} caçique de Canta.

Muy ps^{o} s^{o}r

Pide todas las obejas.

Don Fran^{co} caçique de Canta por my y en nonbre de mis yndios en el pleito con los yndios de Chacalla sobre las tierras de Quibi / digo que los susodhos yndios de Chacalla no quieren entregar las dhas obejas y las que an traydo son çiento y tres llenas de carache y las mas dellas pacos y no es justo que pues yo les di buenas obejas que los susodhos no me las de todas juntas como yo se las di y tales como yo se las di.

A Vra Alt.^{a} pido y supp^{co} mande se me den las dhas obejas todas juntas y obejas grandes tales como yo se las di y que esten sanas pues esto es conforme a la just.^{a} la qual pido y para ello etc.

Otro si pido y sup^{co} a Vra Alt^{a} mande que las dhas obejas las saquen de nras chacaras y quando las trujeron las traygan todas juntas y sanas como tengo fecho sobre que pido just.^{a} e para ello etc.

El Licen^{do} de Leon — J. de Arrandolaça

En Los Rreyes veinte e nueve de ag^{to} de myll e quyny^{o}s e sesenta e siete a^{o}s ante los señores pres^{te} e oydores desta Rreal Audiençia en audi^{a} pu^{ca} se presento esta pet.^{on} y los dhos señores proveyeron a lo por ellos proveydo.

Avendaño

/404r/ Al muy poderoso señor presidente y oydores de la Rreal Audiençia de la çiudad de Los Rreyes.

Muy poderosos señores

No trata cosa sobre lo de don Felipe.

Sit nomen domini benedictum / Yo vine a este assiento del pueblo de Quivi por mandado de V. Alt.ª juntamente con Bartolome de Prol rreçeptor desa Audiençia Rreal de Los Rreyes y luego en cunplimyº de la Rreal Executoria señalamos los mojones de las tierras sobre que an pleiteado los indios de Canta con los yndios de Chaclla por los limites y linderos que las partio y señalo el Ynga a cada una de las partes y se hizieron los mojones a hecho de doze braças de mojon a mojon tan alto cada mojon como un estado hechos com piedra y barro y no se les dio la posesion a los indios yauyos de Chaclla hasta tanto que los indios de Canta hayan las ovejas que les fueron entregadas por Gibraleon alguazil de Su Mag.ᵗ en cunplimyº de la Rreal Executoria los quales indios de Canta maliçiosamente no an querido traer las dhas duzientas ovejas y tienen escondidas parte dellas y parte dellas se an hallado en poder de algunos prinçipales de los indios de Canta y estan juntas quatro leguas de aqui numero de çiento y treinta y siete ovejas las quales yo y el dho rreçeptor yvamos de buena voluntad a entregar a los indios de Canta las que fuesen para rreçebir y hazer buscar las demas que faltan y entregarselas ansi mismo y rrespondieron todos los prinçipales que aqui estan de Canta por lengua de don Fran^co^ hermano del señor de Canta que es ladino que no querian rreçebir las ovejas porque su señor prinçipal esta en Los Rreyes y ni ellos los quieren traer ni las dexan traer y tienen indios puestos en guarniçion de veynte en veynte para estoxuar que no vengan las ovejas por no ser entregados en ellas vea V. Alt.ª lo que //
/404v/ manda que se haga en este caso yo me vuiera ydo si no fuera porque temo que estos yndios an de rreñir unos con otros y por euitar algunas muertes me estoy esperando el mandato de lo que V. Alt.ª fuere sexuido demandar a nro señor presidente por el muy alto estado de V. Alt.ª para que nos tenga en juztiçia amen. 25 de agosto de 1567 años.

Humilde sieruo de V. Alt.ª

Frai Gaspar de Carvajal

/406r/ Los yndios de Chacalla.

Muy psº señor

Diego de Ocanpo en nonbre de los caçiques e yndios de Chacalla en la cabsa con los de Canta sobre las trras de Quivi / digo que V. Al. mando dar e dio esta sent.ª e provisyon para que los dhos yndios de Canta truxesen las ovejas que rresçibieron para que la persona que fue a poner los mojones viese sy heran de rresçibir y no lo syendo les mandase dar otras buenas porque por los abtos del proçeso consto que mys ptes se las avian entregado y syendo rrequeridos con la dha provisyon los dhos yndios de Canta no la quyeren cunplir diziendo que no an rresçibido las dhas ovejas y entre ellos ay pleytos e diferençias sobre sy an rresçibido las dhas ovejas o no y quyen las ha de traer / y se dilata la execuçion de lo probeydo por esta Rreal Abdiençia y pues mis partes tomaron y tienen posesyon de las dhas tierras por executoria desta Rreal Abdiençia y estan en posesyon dellas no es justo que con maliçias se detenga la persona que fue a poner los mojones ny se hagan tantas costas en espeçias aviendose como se an ofresçido mys partes a trocar las ovejas que les bolbyeren sy no paresçieren de dar y tomar.

A V. Al. pido y sup.^co^ probea y mande que ante todas cosas se pongan los dhos mojones y mis partes sean anparados en la dha su posesyon dando fianças de trocar las ovejas que paresçieren no ser de dar ny tomar sobre que pido just.ª y para ello etc.

El Lic^do^ Falcon — Diego de Ocanpo

/406v/ En Los Rreyes veinte y seis dias del mes de ag^to^ de myll y quiºs y sesenta y siete años ante los señores pres.^te^ y oidores desta Rreal Audiençia estando en audiençia publica de petiçiones se presento esta pet.^on^ y los dhos senores mandaron que entreguen las obejas que estan juntas siendo de dar y rresçebir y las que no ffueren para ello les hagan dar otras en su lugar y con esto entreguen la posesion de las tierras. Pres.^te^: Joan de Arrandolaça a quien se notiffico.

Avendaño

/407r/

El Liçen^do^ Castro	El D. Bravo de Saravia	El Lic^do^ don Alvº Ponçe de Leon	El Lic^do^ Serº de Vigil

Sº Avendaño

Para que las personas de Canta que rresçibieron de las de Chacalla las dozientas ovejas del conçierto las traigan ante Bartolome de Prol el qual las vea si son de dar e tomar e no lo siendo que las que dellas no lo fueren conpela a los y^{o}s de Chacalla que las den tales que se an de rreçevir e tomar.

Don Phelipe por la graçia de Dios rrey de Castilla de Leon de Aragon de las dos Seçilias de Jerusalen de Navarra de Granada de Toledo de Valençia de Galizia de Mallorcas de Sevilla de Çerdeña de Cordova de Corçega de Murçia de Jaen de los Algarves de Algezira de Gibraltar de las yslas de Canaria de las Y^{a}s yslas e tierra firme del mar oçeano; conde de Flandes e de Tirol etc. A vos Br.^{me} de Prol nro rreçeptor en la nra Rreal Audi^{a} e Chançill.^{a} que por nro mandado rresyde en la çiudad de Los Rreyes a quien esta cometido el dar de la posesion de las tierras de Quibi e poner los mojones dellas / salud e graçia / Saved que Diego de Ocanpo en nonbre de los y^{o}s de Chacalla por una petiçion que ante nos presento en la dha nra Rreal Aud.^{a} ante el pres.^{te} e oidores della nos hizo rrelaçion diziendo que a vos esta mandado que vays a anparar a sus ptes en la posesion que tomaron de las dhas tierras de Quibi por una nra Rreal Executoria e poner los mojones e que agora las ptes contrarias de Canta avian pedido que los y^{o}s de Chacalla sus ptes les entregasen las dozientas ovejas que por la dha nra Rreal Executoria le estan mandada dar buenas diziendo que no lo son las que les an entregado e porque las dhas sus ptes an cunplido con la nra Rreal Executoria e por los autos del proçeso paresçe aver sus ptes entregado las dhas dozientas ovejas e que lo que en contrario se pide es de maliçia porque se les an muerto alg.^{as} de las dhas ovejas e quieren que los dhos sus partes se las buelvan a dar bibas / nos fue pedido e suplicado mandasemos que los dhos y^{o}s de Canta traygan ante vos las ovejas que asi entregaron y que si paresçiere no ser de dar e tomar alg.^{as} dellas que los dhos sus ptes les entregaran otras que sean buenas en lugar de las que les bolvieren / e que con esto ayan cunplido / o como la nra merçed fuese / lo qual visto por el dho presidente e oydores fue acordado que deviamos mandar dar esta nra carta en la dha rrazon e nos tubimoslo por bien por la qual mandamos que los naturales del rrepartimy^{o} de Canta que rreçibieron de los y^{o}s de Chacalla las dhas dozientas ovejas en cunplimy^{o} de la dha nra Rreal Executoria las traygan ante vos e traydas veays sin son de dar e tomar e no lo siendo en las que dellas no lo fueren conpelereys a que los dhos y^{o}s de Chacalla se las dan tales que se an de rresçevir e tomar y de manera que no rresçiban en ello agravio / lo qual asi hazed e cunplid so pena de qui^{o}s p^{o}s de oro para la nra camara.

Dada en Los Rreyes a nueve de agosto de myll e qui^{o}s e sesenta y siete años.

Yo el scr^{o} P^{o} de Avendaño escrivano de camara de la Mag.^{t} Rreal la fiz escrevyr por su mandado con acuerdo del su presidente e oydores.

Rr^{da} Juan de Murga — Chançiller Gaspar de Solis

E despues de lo susodho en el dho asyento e pueblo de Quybe veynte e çinco dias del dho mes e año los yndios de Chacalla me rrequerieron con una esta provisyon de Su Mag.^{t} para que la cunpla e dixe que la obedesçia con el acatami^{o} que debia y en q^{to} al cunplimi^{o} della dixe que ya les tenya mandado que traxesen las dhas obejas con pena e que no las avian querido traer porque dizian que no las avian rresçibido y por no me constar por el proçeso averlas ellos rresçibido los unos ni los otros las quyeren traer. E luego los yndios de Chacalla me rrequerieron que les diese la posesyon de las dhas tierras que a mayor abundançia ellos estaban prestos de yr a entregar las obejas a los de Canta adonde ellas estaban pues que ellos no las querian traer como se les mandaba por esta provisyon e que si ansy lo hiziese cunpliria lo que me hera mandado / no que protestaban de se quexar de mi ante los señores presidente e oydores e de cobrar de mi las costas e daños e yo el dho escrivano rrespondi que estaba presto de yr con ellos a entregar las dhas obejas a los de Canta e hecho esto de le dar la posesyon de las dhas tierras y luego los de Canta dixeron por lengua del dho Ju^{o} de Albarado ynterpetre que ellos estan prestos de las yr a rresçibir con que sea a costa de los yndios de Chacalla que paguen el selario del rreçetor e ynterpetre no enbargante que es mal camyno y no se puede camynar a caballo seyendo obligados los de Chacalla a se las entregar en este pueblo de Quybe y otra bez les torna a mandar lo que trayan las dhas obejas como se les manda por esta probisyon so la dha pena los quales dixeron que no las avian rresçibido ny las avian de traer y lo pedieron por testimonyo en fee de lo qual lo firme de my n^{e}. T^{o}s: el dho Ju^{o} de Albarado e el padre Fray Grabiel de Obiedo. E luego los dhos yndios de Chacalla dixeron que ellos no avian de pagar las costas sino de por medio como Su Mag.^{t} lo manda. T^{o}s: los dhos.

B.^{me} de Prol scr^{o}

En el pueblo de Quybe veynte e bn dias del mes de ag.^{to} de myll e qui^{o}s e sesenta y siete años yo el presente escrivano de pedimi^{o} de los yndios de Chacalla yntime y notifique esta probisyon de suso cont.^{da} a don Fran^{co} Poma hermano de don Fran^{co} caçique prinçipal de Canta que estaba en Los Rreyes e a don Xpobal Carnabilca e a don Ant^{o} Alcapoma e a otros muchos yndios e prinçipales de Canta y les declare el hefeto de lo en ella cont.^{do} por lengua de Ju^{o} de Albarado ynterpetre los quales dixeron que ellos no avian rresçibido obejas alg.^{as} ny las tenyan que los yndios de Chacalla las tenyan e que ellos las traxesen para el hefeto en la probisyon contenydo en fee de lo qual lo firme de my n^{e}.

B.^{me} de Prol scr^{o}

E despues de lo susodho en el dho pueblo de Quybe beynte e dos dias del dho mes e año el muy rr[do] padre Fray Gaspar de Carvajal e yo el dho escrivano mandamos a los dhos yndios de Canta traxesen las obejas que abian rresçibido de los de Chacalla para ver sy heran de dar e rresçibir so p.[a] de çient p[o]s de oro y que los enbiarian presos a la çibdad de Los Rreyes los quales dixeron que no las tenian syno los de Canta [*sic*] que ellos las traxesen estando presentes los yndios de Chacalla que dizian que ellos las tenian e las avian de traer conforme a esta provisyon lo qual paso por la dha lengua. T[o]s: Gibraleon e Villarreal.

B.[me] de Prol scr[o]

E luego yncontinente visto que los bnos yndios por los otros dexaban de traer las dhas obejas les mande con asystençia del padre Carvajal que cada una de las partes juntase diez yndios e ynbiasen para las dhas obejas so pena de çient p[o]s para la camara de Su Mag.[t] los quales los juntaron e fueron por las dhas obejas. T[o]s: los dhos.

B.[me] de Prol scr[o]

E despues de lo susodho en veynte e tres dias del dho mes e año fue notificado e apresçibido [*sic*] a entranbas las dhas partes que porque no trayan las dhas obejas e los de Canta dizian que los de Chacalla las tenyan e no las querian traer porque estaban flacas e llenas de carache / e los de Chacalla dixeron que los de Canta no les querian ayudar a ello y bisto lo vno e lo otro el dho padre e yo les diximos que fuesemos adonde ellas estaban e que ally las rresçibirian los de Canta los quales dixeron que no querian que en su pueblo se las avian de entregar adonde les quytaban sus tierras e que las dhas obejas no estaban ally todas faltando algunas no las avian de rresçibir syno todas juntas. T[o]s: Ju[o] de Albarado y el padre Fray Grabiel de Obiedo e Fray P[o].

E que si alg.[as] obejas faltaban y se abian muerto abia de ser a culpa e cargo de los de Canta y no dellos que se le diese todo por ... (roto) ...

B.[me] de Prol scr[o]

/408r/ Don Felipe.

Muy p[o] señor

Mas pareçe aqui por lo que dize el don
Felipe ser sujetos a Canta que no
a Chacalla.

Don Ffelipe caçique del pueblo de Quybe/ digo que yo pedi y suplique a V. Al.[a] por my y en nonbre de mis y[o]s que no permytiese que pues todas las tierras del dho balle de Quybe fueron nuestras y de nuestros antepasados de tpo ynmemorial a esta pte y nos quytan todas las tierras del comun y de yndios muertos y del Inga para partir entre los y[o]s de Canta y los y[o]s de Chacalla que nos quyten a bueltas la tierra de las chacarrillas que cada yndio de los myos particularmente tiene para senbrar su mantenimy[o] y mantener sus hijos y muger porque el padre Caravajal por çierto enojo que tubo con vn yndio de Canta nos a quytado todas las dhas nuestras chacaras de cada yndio y a hechado los mojones por nro propio pueblo syn dexar tierra alguna en que podamos senbrar y haçemos rrelaçion del agrabio que nos hiçieron y nos haçian el dho padre Caravajal.

A Vra Alt.[a] pido y supl[co] mande que pues no fuymos oydos en la causa prençipal que se bea nra petiçion y no permyta que muramos de hanbre tan cruelmente nosotros y nuestras mugeres y hijos y seamos conpelidos a salir de nuestra propia tierra donde naçimos nosotros y nuestros antepasados / sobre que pido just.[a] para lo qual etc.

El Licen[do] de Leon

/408v/ En Los Rreyes quatro de setienbre de myll e quiny[o]s e sesenta e siete a[o]s ante los señores presydente e oydores desta Rreal Audiençia en acuerdo de justiçia / se presento esta pet.[on] y los dhos señores mandaron que se junte todo y se traya.

Avendaño

/409r/ Los y[o]s de Chacalla.

Muy p[o]s señor

Diego de Ocanpo en n[e] del caçique e yndios de Chacalla en la cabsa con los caçiques e yndios de Canta e yungas del pueblo de Quivi / sobre las tierras / digo que V. Al. mando que el padre Fray Gaspar de Carvajal con Br.[e] de Prol fuese a amojonar las dhas trras y llegados alla aunque el dho padre Carvajal como persona que tenya not.[a] de las dhas trras y avia fho otra vez ynformaçion sobre ellas declaro por donde avian de yr los mojones en los quales entravan muchas casas del dho pueblo de Quivi que las fizieron mis ptes y les pertenesçia el dho Br.[e] de Prol rreçeptor no lo quiso escryvir ny que constase de lo ny se fiziese diziendo que la executoria y provisyones desta Rreal Abdiençia no hablavan del pueblo syno de las trras y asy no podia dar trras donde oviese casa alg.[a] en lo qual fizo mal su of[o] y mis ptes quedaron agraviados.

Pido y sup[co] a V. Al. mande que el dho padre Carvajal declare lo susodho y se echen los mojones por donde an de yr y se entreguen a mys partes lo que es suyo sobre que pido just.[a] y para ello etc.

- Otro sy digo que el dho rreçeptor exçediendo de lo que le fue mandado mando que mys partes dexasen coger a las ptes contrarias el fruto de la coca y otras cosas que avian sembrado con lo qual ay grande confusyon entre los yndios y las ptes contrarias pretenden coger no solo el mayz y otras legunbres syno aun la coca que es cosa perpetua y la destruyen con la demas arboleda como cosa agena.

Pido y sup^{co} a V· Al. mande que las ptes contrarias //

/409v/ no entren a coger fruto alguno / y quando alguno ovieren de dar se tase el mayz y otras legumbres que podran coger en esta cosecha y cunplan mys partes con darlo para lo qual etc.

- Otro sy digo que mis partes an cunplido en entregar las dozientas ovejas como consta por los abtos del proçeso y las dieron todas buenas y para mas justificaçion trocaron las que dixeron que no heran buenas y aora las ptes contrarias querrian y pretenden que mys partes les paguen las que ellos an escondido o se les an muerto o perdidos por su culpa.

Pido y sup.^{co} a V. Al. declare aver mys partes cunplido e no aver lugar lo pedido por las ptes contrarias para lo qual etc.

El Lic^{do} Falcon — Diego de Ocanpo

En Los Rreyes quatro de setienbre de myll y quiny^{o}s y sesenta e siete años ante los señores pres.^{te} e oydores desta Rreal Audiençia estando en acuerdo de just.^{a} se presento esta pet.^{on} y los dhos señores mandaron que cojan el fruto de coca y arboles que oviere plantado por esta vez e que no lo arranquen ni destruyan la dha coca ni arboles que en ella oviere.

Avendaño

/410r/ Don Ant^{o} Alcapoma. Muy p^{o}s. s^{o}r

Don Antonio Alcapoma prinçipal del rrepartimi^{o} de Canta encomendado en Nicolas de Rribera por mi y en nonbre de mis yndios digo que despues de benido el padre Fray Gaspar de Carabajal de Quibi y aver dado a los yndios de Chacalla todas las trras que quisieron syn avernos pagado nras obejas porque ynsistimos en que nos las pague y por dezir que nos esta puesta pena sy no damos los mojones los dhos yndios de Chacalla fueron a nras trras y en mi propio pueblo hasta alla en la syerra / todo consultado para dezir que si nos entremetemos con ellos no nos pagaran las obejas y que V. Al. nos castigara / a lo qual V. Al. no deve dar lugar.

Por tanto a V. Al. pido y sup.^{co} mande poner rremedio en lo susodho y que vaya una persona a ver las dhas ovejas y como no son de dar ni rresçebir y se mande que se nos entreguen como V. Al. tiene mandado donde no se den los mojones por nyngunos y sy mesmo se me mande bolver las tierras que de nuebo los dhos yn^{o}s de Chacalla se entremeten a tomarme en mis propios pueblos y se quyten los mojones que por su propia abtoridad an puesto sobre que pido just.^{a} e para ello etc.

El Licen^{do} de Leon

/410v/ En Los Rreyes diez y seis de setienbre de myll e qui^{o}s y sesenta y siete años ante los señores presydente y oidores desta Rreal Audi^{a} en audi^{a} pu^{ca} se presento esta pet.^{on} y los dhos senores mandaron lo proveido para que se trayga el paresçer de Caravajal.

Avendaño

/411r/ Los y^{o}s de Canta.

Muy p^{o} s^{o}r

Los yndios de Canta en el pleito con los yauyos de Chacalla sobre las tierras de Quibi / desimos que aviendo V.Al. mandado que los y^{o}s de Quybe gozen este año del fruto que tenyan senbrado en las dhas tierras y estandolo benefiçiando fue un yndio llamado Fran^{co} Coro con çiertos y^{o}s y dixo que no entrasen nros y^{o}s en las dhas trras e los hecharon dellas y a tres yndios de Quybe asyeron de los cauellos y los lleuaron presos a su trra donde los tienen y dizen que dexen perder la coca e que no entren en ellas e asymysmo hazen otras muchas molestias.

A V. Al. pedimos e suplicamos se rremedie lo susodho mandando que nos dexen coxer librem^{te} el dho fruto que tenyamos senbrado y que nos entreguen las obejas porque hasta agora no las an dado.

En Los Rreyes treze de setienbre de myll e quiny^{o}s e sesenta e siete a^{o}s ante los señores pres^{te} e oydores desta Rreal Audi^{a} en audi^{a} pu^{ca} se presento esta pet.^{on} y los dhos señores proveyeron lo por ellos proveydo e que si por esta causa los tienen los suelten.

Avendaño

/412r/ Don Felipe prinçipal de los y^{o}s de Quibi.

Muy p^{o} s^{o}r

Aqui no dize el don Felipe ser de Chacalla
ni de Canta sino que no son sujetos.
Piden se enbie una persona que aberigue y
de las trras a cuyas fueren dexandoles
a ellos lo que puedan sembrar.

Don Felipe prinçipal de los y^{o}s yungas del pueblo de Quivi / digo que por el pleito que an traydo los y^{o}s de Chacalla con los y^{o}s de Canta sobre nras tierras V. Al. m^{de} que el padre Frai Gaspar de Caravajal las fuese a dividir el qual por un enojo que hubo con un prinçipal de Canta no solamente dio a los y^os de Chacalla las tierras sobre que traian pleito ellos entre [*sic*] pero aun todas las chacarillas de mys y^os e mias que tenyamos particulares para hazer nras comydas syn dexarnos ni aun un palmo de tierra que se pueda senbrar de ninguna cosa syno solo el pueblo y la yglesia que tenemos porque todo lo demas es pedregales y si con aquello se pasa adelante de nesçesario o se an de morir nros hijos y mugeres o nosotros buscaremos con quien estemos y pues toda aquella tierra ansi lo que se da a los y^os de Canta como a los y^os de Chacalla son myas y de mys y^os desde tienpo ynmemorial y no es justo que pues me las toman para cuyas no son que no se dizen tan desheredados que a ninguno de nosotros que de donde pueda estar.

A V. Al. pido y sup.^co lo mande rremediar mandando que vaya una persona syn paçion a dividir las dhas tierras y dexarnos siquiera las chacaras de y^os particulares sobre las quales nunca ha avido pleito pues nos toman todas mis tierras de comun y de y^os muertos puesto es conforme a just.^a la qual pido y costas sin perjuizio de la suplicaçion que tenemos fha para ante V. Persona Rreal.

El Licen^do de Leon

/412v/ En Los Rreyes doze de setienbre de myll e quinyentos y sesenta y siete a^os ante los señores pres.^te e oydores desta Rreal Audiençia en audiençia publica se presento esta pet^on y los dhos senores mandaron que se lleve esta pet.^on al acuerdo.

Avendaño

/413r/ Don Fran^co Guaman Guaranga.

Muy poderoso señor

Don Fran^co Guaman Guaranga cazique de Canta en el pleito con los yndios de Chacalla sobre las tierras de Quibi. Digo que abnque por Vra Alteza a sido muchas vezes mandado que primero que se les diese la posesion de las dhas tierras a los dhos yndios nos diesen las dozientas obejas que nosotros les dimos tales y tan buenas como se las dimos o que sean de dar y rrezebir los dhos yndios no las han querido dar sino la mitad dellas y aquellas pacos y llenas de carache y tales y tan flacas de enfermedad que es ynposible bibir / porque parezen de salir de otro ganado y no es justo que seamos despojados de nras trras y abn de las mesmas obejas que nosotros les dimos y el padre Carabajal con la mucha afiçion que tiene a los yndios de Chacala [*sic*] no solamente les ha dado las tierras sobre que no traemos pleito pero avn abiendose cometido lo de las obejas a Bartolome de Prol dize que por fuerza hemos de rrezebir las dhas obejas pacos y llenas de carache a lo qual Vra Alteza no debe dar lugar.

Por tanto a Bra Alteza pedimos y suplicamos mande yr una persona a qye bean las dhas obejas antes que se mueran para que se bea que no son de dar y rreszebir y se les mande que nos den otras tales y tan buenas como se las dimos y que en el yntretanto no se les de la posesion de las dhas tierras pues esto es conforme a justiçia la qual pedimos y para ello etc.

El Licen^do de Leon

/413v/ En Los Rreyes doze de setienbre de myll e quiny^os y sesenta y siete a^os ante los senores pres.^te e oydores desta Rreal Audiençia estando en audiençia publica se presento esta pet.^on y los dhos senores mandaron que se traya lo proveydo por el padre Caruajal ante Prol çerca desto.

Avendaño

Rrelaçion de las tierras sobre que traen pleito
los indios de Canta y de Chacalla / para los muy
poderosos señores presidente y oydores del
Audiençia Rreal de Los Rreyes.

/414r/ Quybi.

Aqui haze rrelaçion el padre Caruajal
de como el pueblo de Quibi que es de
la dyf.^a es de Chacalla.

Muy poderosos ss

Presupuesto que los Ingas señalaron y amojonaron las tierras por sus confines de cada qual de los pueblos y provincias a ellos subjectos señalaron y amojonaron las tierras de todo el confinio de Canta y de Chaqualla las quales tierras comiençan y tienen su principio con linderos de vna parte tierras de Guancayo que esta encomendado por Su Mag.^t en Gonzalo de Caçeres hijo del contador Caçeres las quales tierras dio el Inga y las partio con los de Canta y con los de Chaquella y puso sus mojones por el rio arriba de la vna vanda del rrio, tierras de yndios encomendados en Ju^o Piçarro hijo de Martin Pizarro y desde vna quebrada que esta en la sierra bien señalada que haçe el agua quando llueve començo a amojonar las tierras que perteneçen a Canta y a Chaquella, de la parte de arriba con el pueblo de Puculucha dio a Canta, y de la parte de abaxo con el pueblo de Quibe dio a Chaquella. Los mojones y linderos destas tierras que dio el Ynga a estas dos parçialidades van al sesgo derechos a la yglesia que agora esta hecha en la plaça del pueblo de Quibe //

/414v/ y van prosiguiendo hasta la sierra de Guayaringa por todo el camino de Canta dexando para los yndios de Canta toda la quebrada de Cacalarca y todas las tierras que van el rio arriba confinando con tierras de Socos yndios de Ju° Piçarro y las demas tierras de la parte de abajo con el pueblo de Quibe y las de arriba del pueblo de Quibe que lindan con la quebrada de Cacalarca consigno y dio el Ynga a la parcialidad de Chaquella. Estos dos pueblos de Puculucha que era de Canta y el de Chiquibe [*sic*] que era de Chaquella estan despoblados y los indios del un pueblo y del otro se an juntado cabe la yglesia y estan sitiados en tierras de los de Chaquella y pareçe que estan alli comodam[te] para la doctrina y como por auer poseido los yndios de Canta estas tierras por el conçierto que hicieron Fran[co] de Anpuero y Niculas de Rribera con sus indios anse pasado algunos de los yungas que eran subjectos y tributauan con los yndios de Chaquella a la parcialidad de los de Canta y como los tienen atraidos a si querrian ser subjectos a los de Canta y tributar con ellos y estarse en las tierras de los de Chaquella y esto no es justicia sino que pues sirvieron a los antepasados de Chaquella los siruan agora y se esten en sus chacaras y les ayuden a pagar el tributo y puesto que nro s[or] muchas uezes permite que los que somos religiosos y nos entremetemos en cosas fuera de nra profession no açertemos y Vra Alteza me ha mandado que yo fuesse a entender en este negocio yo he ydo por dos uezes y hecho en ello lo a mi posible predicando a los indios lo que les conuiene si en ello no he acertado remitolo a otro mejor juizio.

Sieruo de V. Alteza Fray Gaspar de Caruajal

Al muy mag° señor
Pedro de Avendaño
secretario de la Rreal Audiençia.

/416r/ Muy Mag° señor
En esta carta que escribe el p[e] Carauajal
a P° de Abendaño dize que el pueblo de
Quibe esta en las tierras de Chacalla.

Sit nomen domini benedictus amen.

Rr[bi] la carta de V. M. por la qual me manda que yo de pareçer firmado de mi nonbre que los indios de Chacalla acaben de entregar el ganado a los indios de Canta y para que los indios que tienen coca en las tierras de Chacalla gozen este año del fruto y que los indios que estan visitados por la parcialidad de Chacalla tributen con aquella parcialidad y los que estan visitados por la parcialidad de Canta tributen con Canta.

En estas tres cosas tratamos V. M. y yo en mi çelda y yo dixe a V. M. el grande mal que se siguia y los gestos y trabajos destos pobres indios de Chacalla y de Canta y que convenia que la Rreal Audiençia le pusiese silençio perpetuo porque no acabaran de pleitear y creo que en no castigarse los desacatos y osadias en desobedescer las provisiones rreales a de venir a mayor mal porque por no castigar a pocos no se enmyendan muchos y van de venir a desvergonçarse. Yo no deprendi los derechos ciuiles; la sçiencia que yo deprendi es que se de lo suyo a su dueño y que los ynocentes no padezcan. Yo no se si es justiçia que los indios de Chacalla despues de aver entregado las ovejas todas las bueluan otra vez a entregar ni se si es just.[a] que gozen este año de la coca.

Lo tercero bien entiendo que es conforme a rrazon.

Yo no cargare mi conciencia por Chacalla ni por Canta . Querria que estos indios tuviesen paz y por este bien qualquiera cosa permitiria. Esto yo lo digo y no el señor (como dize San Pablo) . Sepa V. M. que todo el interes de la coca no vale treinta pesos y que los indios de Canta les cupo en suerte y poseen mas tierras que los de Chacalla en aquel pago de trra que señalo el Inga a estas dos parcialidades. En las trras de Chacalla estava antes el pueblo de Quivi y en las tierras de Canta estava el pueblo de Puculucha donde agora tiene vna ortezuela con naranjos Nicolas de Rribera. Estos dos pueblos estan tan cercanos que con vna piedra llegara vn honbre de vn pueblo a otro. La gente destos dos pueblos esta agora cabe la yglesia y todos son tan pocos que no llegan a veinte vezinos. Estan en casas de cañizos y estan en tierras de los yndios de Chacalla y los yungas que ay entre estos moradores son tan pocos que seran siete o ocho casas en anbas parcialidades. Los demas son serranos mitimaes y a estos yungas los de Challa [*sic*] y los de Canta les rruegan con sus tierras y los querrian cada qual para si con esta condicion que les ayuden al tributo y los de Canta los hazen que pidan y pleiteen //

/416v/ que los pobres yungas no tienen causa ni ocasion de traer pleito porque en su mano y en su alvedrio esta si quieren estarse en las chacarras que tienen senbradas y en lo mejor que ellos escogieron de las tierras y para ello son rrogados. Conpelanlos a que esten donde estavan antes del conçierto que hizieron de las tierras Fran[co] de Anpuero y Nicolas de Rribera por las dozientas ovejas las quales no deven los indios de Chacalla sino Anpuero que las rreçibio y por fuerça firmo el caçique el conçierto; y dize en el conçierto que se da por entregado en dozientas ovejas las quales no rreçibio antes ante todos los españoles que se hallaron presentes al conçierto dixo el caçique de Chacalla que no queria ovejas que si Anpuero las queria que las rreçibiese; y entre Anpuero y el clerigo que era Godoy se consumieron las ovejas. Rresumome en que por el bien de paz entre estos pobre indios aunque sean agraviados los de Chacalla se les de mas de lo que les conviene a los de Canta y si no obedeçieren aya castigo exenplar por que otros teman a la Rreal Audiençia y a su justiçia.Nro señor nos de el cielo amen.

17 de setienbre 1567.

Capellan de V. M. Frai Gaspar de Caruajal

/418r/ Don Felipe caçique de Quibe.

Muy p^{o}s. s^{o}r
Aqui dize el don Felipe
no ser sujeto a los de Chacalla.

Don Felipe caçique del pueblo de Quibi por mi y en n^{e} de los dhos yn^{o}s del dho pu^{o} en el pleito con los yn^{o}s de Chacalla y los de Canta digo que no contentos los dhos yn^{o}s chacalla con avernos tomado la tierra sobre que trayan pleito con los dhos yn^{o}s de Canta y a mi y a los dhos mis yn^{o}s tomarnos las chacarillas en que senbravamos de que hemos venido a pedir socorro ante V. Alt. porque hechamos vna poca de agua en las dhas nras chacaras para coger el ffruto de que V. Alt. nos dio licen.^{a} y por que no se case aviendo preso a tres de nosotros los dhos yn^{o}s de Chacalla y llevadonos a sus pueblos dandonos de palos y haziendonos otros malos tratamy^{o}s nos tornaron aora a prender y nos traxeron dos yn^{o}s presos a esta çiudad y cada dia nos maltratan y hazen muchas molestias a lo qual V. Alt.^{a} deue poner rremedio y mandar que los dhos yn^{o}s de Chacalla no se entremetan en otra tierra ffuera de la que ffuere el dho pleito mandando que los encomenderos de los unos yn^{o}s y los otro [*sic*] que primero los conçertaron declaren y amojonen la tierra sobre que thenyan diferençia y traxeron el dho pleito sin perjuizio del dr^{o} de las ptes. Por tanto a V. Alt.^{a} pedimos y suplicamos que sin perjuizio de nro dr^{o} se tome juramento a los dhos encomenderos para que vayan a las dhas tierras y declaren sobre que tierras hera la differençia y pleito que an traydo los yn^{o}s de Canta con los de Chacalla y que aquello se amojone sin perjuizio del dr^{o} de las ptes y que ningun yn^{o}s moleste al otro sobre lo que no le pertenesçe poniendo penas bastantes para ello al que lo contr^{o} hizie [*sic*] sobre que pedimos just.^{a} y para ello etc.

El Licen^{do} de Leon

/418v/ En Los Rreyes veinte e dos de setienbre de myll e quiny^{o}s e sesenta e siete a^{o}s ante los senores pres.^{te} e oydores desta Rreal Audiençia est^{do} en acuerdo de just.^{a} se presento esta petiçion y los dhos señores proueyeron lo decretado arriba.

Avendaño

/419r/ Los y^{o}s de Chacalla.

Muy p^{o}s. señor

Querella sobre quitar los de Canta
a Chacalla las aguas.

Diego de Ocanpo en n^{e} del caçique e yndios de Chacalla en la çabsa con los caçiques e yndios de Canta sobre las trras de Quivi. Ante V. Al. me querelle e acuso criminalmente a don Fran^{co} caçique prinçipal de Canta y de los demas que por la ynformaçion paresçieren culpados y contando el caso premysas las solenydades del dr^{o} digo que es asy que no contentos los susodhos con los muchos e muy graves delictos que otras vezes an cometido contra mys ptes por les tomar las dhas trras y quedarse en ellas visto que no an sydo castigados por ellos aunque derribaron los mojones puestos por exsecutoria desta Rreal Abdiençia aora de nuevo aviendolos tornado a poner por mandado desta Rreal Abdiençia y dado a mys partes posesyon en pte de las dhas tierras los susodhos con poco temor de Dios y en grande menospreçio de vras just.^{as} an quitado las aguas con que mys ptes rregavan las dhas trras y tornado a derribar algunos de los dhos mojones y destruydo la coca en lo qual an cometido graves delictos.

Por que pido y sup^{co} a V. Al. mande aver ynformaçion de lo susodho y prender los culpados que presos yo estoy presto de ponerles acusaçion en forma sobre que pido just.^{a} e para ello etc. e juro a Dios en forma que no lo pido de maliçia.

El Lic^{do} Falcon

/419v/ En Los Rreyes veinte e dos de setienbre de myll e quiny^{o}s e sesenta e siete a^{o}s ante los señores pres^{te} e oydores desta Rreal Audiençia est^{do} en acuerdo de just.^{a} se presento esta pet.^{on} y los dhos señores mandaron que el corregidor deste rrepartimi^{o} lo vea e llamadas las ptes lo prouea e lo que p^{o}ueyere lo haga guardar.

Avendaño

/420r/ Muy p^{o} señor

Diçe que lo que se comete al corregidor
consta del proçeso.

Diego de Ocanpo en n^{e} de don Juan Baptista Caxaguaraca caçique prinçipal de Chacalla e de los prinçipales e y^{o}s a el subjetos en la cabsa con los yndios de Canta sobre las trras de Quivi / digo que estando esta cabsa mandada traer para que se declarase que mys partes an cunplido en el entrego de las ovejas y que se les deve de entregar las trras sobre que es este pleyto enteramente que no les an sydo entregadas y sobre que salgan dellas y no cojan cosa alguna las partes contrarias / a venido a mi notiçia que en el acuerdo se an dado petiçiones sobre el entrego de las ovejas / y que se cometio al corregidor y porque esto y todo lo demas consta de los abtos del proçeso y no es menester mas pleyto syno verlos y determinarlo en esta Rreal Abdiençia.

A V. Al. pido y sup.^{co} lo mande ver y que se determine y sobre todo just.^{a} y para ello etc.

El Lic^{do} Falcon

Diego de Ocanpo

/420v/ En Los Rreyes a veynte e çinco dias del mes de setienbre de mill e qui^{o}s e sesenta y siete a^{o}s ante los señores presidente e oydores en Aud.^{a} Rreal de rrelaçiones; la presento Di^{o} de Ocampo procurador en nr^{e} de su parte y los dhos señores mandaron que se traygan los autos.

Albaro Ruiz de Nabamuel

/421r/ Don Phelipe caçique de Quibi.

Muy p^{o}s. s^{o}r

Aqui no pareçe que el don Felipe
sea sujeto al rrepartimi^{o} de Chacalla.

Don Felipe caçique de Quivi por mi y en n^{e} de los demas yndios a mi sujetos y los yndios de Canta dezimos que nosotros emos dado muchas petiçiones y suplicado a V. Al. nos hiziese md de mandar que se nos entregasen las obejas que dimos syn perjuyzio de nro derecho porque los y^{o}s de Chacalla a quyen les dan nras trras no solamente no nos an querido dar las dozientas obejas que les dimos pero obra de çiento y tantas que nos truxeron no son de dar ni rresçibir por estar llenas de carache y demas desto a my el dho don Felipe y a mis yndios nos an quitado nras chacarras particulares sobre que nunca a avido pleyto entre los dhos y^{o}s de Chacalla y Canta lo qual es cosa de grande ynumanidad e ynjustiçia que nos quyten nras propias tierras sobre que nunca se traxo pleyto con persona alguna mas de por sola la boluntad del padre Fray Gaspar de Carabajal que lo quiso fazer y aviendo mandado yr persona que lo biese e viese las dhas obejas no se a hecho y nosotros padeçemos grande agravio y nesçesidad que se pierde nra sementera.

A V. Al. pido e sup.^{co} mande que vaya una persona syn sospecha a hazer la dha averiguaçion asi de las trras como del dho ganado para que no se entreguen las trras sobre que nunca a avido pleyto en las quales estan nras chacarras particulares y que se mande con brebedad //

/421v/ porque de la dilaçion rresçibimos mucho agravio y daño sobre que pedimos just.^{a} e para ello etc.

- Otro si por q^{to} V. Al. a mandado que nos dexen coger el fruto que tenemos senbrado y los dhos yndios de Chacalla nos lo destruyen y no nos lo dexan cojer y demas deso no los dexan benefiçiar. A V. Al. pedimos e suplicamos se les mande con pena que nos dexen cojer e benefiçiar el dho fruto deste año sobre que pedimos just.^{a} e para ello etc.

El Licen^{do} de Leon

En Los Rreyes a veinte y seys de setienbre de myll e qui^{o}s y sesenta y siete años ante los senores presidente y oidores desta Rreal Audiençia en audiençia se presento esta petiçion y los dhos señores mandaron lo proveido lo que se proveyo presentes Ju^{o} Sanchez de los Rrios pr. con^{o} a quien se notifico. Avendaño

/422r/ Los y^{o}s de Chacalla.

2 de otubre 1567

Muy p^{o}s. señor

Juan Yapa por mi y en nombre de don Juan Baptista Caxaguaraca caçique prinçipal de Chacalla e de los prinçipales e yndios a el subjetos en la cabsa con el caçique e yndios de Canta sobre las chacaras de Quivi / ante V. Al. me querello e acuso criminalmente a don Juan Pariguana e don Gonçalo Lucchа y a los demas que paresçieren culpados / y contando el caso premisas las solenydades del dr^{o} digo que es ansy que los susodhos no contentos con las fuerças y delictos graves que nos an fho quitandonos la posesyon que nos fue dada por exsecutoria desta Rreal Abdiençia de las dhas trras aora de nuevo aviendosenos mandado tornar a dar y anparar en ella por provisyon desta Rreal Abdiençia y estando en ella paçificamente los susodhos pospuesto el temor de Dios y en gran menospreçio de vra justiçia fueron a las dhas trras diziendo que llevavan provisyon desta Rreal Abdiençia para nos hechar dellas y enpeçaron a derribar y derribaron dos mojones de los que por vro mandado estavan puestos y a derribar y arrancar arboles y la coca / y de fho nos hecharian dellas sy no nos defendieramos y nra posesyon en lo qual cometieron graves delictos dignos de grave castigo exenplar.

Por que pido y sup^{co} a V. Al. que avida ynformaçion de lo susodho que estoy presto de dar / mande prender los cuerpos de los susodhos y secrestarles [*sic*] los bienes que presos estoy presto de ponerles acusaçion en forma sobre que pido just.^{a} y costas y para ello etc. e juro a Dios en forma que no lo pido de maliçia.

El Lic^{do} Falcon

/422v/ En Los Rreyes dose de otubre de myll e qui^{o}s e sesenta e siete años ante los señores presidente e oydores desta Rreal Audiençia estando en acuerdo de just.^{a} se presento esta pet.^{on} e proveyeron que vaya una persona a poner los dhos mojones y hazer ynformaçion sobre las personas que los quytaren e trayga presos a los caçiques e prinçipales que hallare culpados.

Avendaño

/423r/ Y^{o}s de Quibi.

Muy p^{o} s^{o}r

Los y^{o}s de Quybi en el pleyto con los yauyos sobre las trras / digo que aviendo V. Al. mandado que el corr^{o}r del rrepartimyento hiziese ynform^{on} sobre el entrego de las obejas el secretario no da la provisyon porque dize que se mandaron traer los autos y lo que se probeyo fue que se guardase lo probeydo e asy estamos detenydos. A V. Al. suplicamos mande que se nos de la dha provisyon como esta probeydo.

Joan ....

En Los Rreyes tres de otubre de myll e qui^{o}s e sesenta e siete a^{o}s ante los senores pres^{te} e oydores desta Rreal Audiençia est^{do} en audiençia pu^{ca} se presento esta pet.^{on} y los dhos señores proueyeron lo proueydo lo qual se proueyo presente Juan de Arrandolaça pr^{o}r contrario a quien se notifico.

Avendaño

/424r/ Muy pr^{o} s^{o}r

Los yndios de Canta en el pleito con los de Chacalla sobre las trras de Quibi / dezimos que V. Al. mando que fuese a averiguar todas las diferençias que ay entre nosotros ansi sobre las ovejas como sobre el poner de los mojones a Pedro de Herrazti el qual no lleba lengua.

A V. Al. pido e sup.^{co} m^{de} se le de lengua de confiança con que no sea P^{o} de Niça por ser yntimo de los de Chacalla / sobre que pido just.^{a} e para ello etc.

El Licen^{do} de Leon

En Los Rreyes catorze de otubre de myll e qui^{o}s y sesenta y siete años ante los señores presidente y oidores desta Rreal Audiençia en audi^{a} pu^{ca} se presento esta petiçion y los dhos señores mandaron que el rreçetor lleue la lengua que le paresçiere que sea sin sospecha.

Avendaño

/425r/ Los yn^{o}s de Canta.

Muy p^{o} señor

Los caçiques e yndios de Canta dezimos que sobre el entrego de las obejas y el poner de los moxones y dibision de las dhas tierras Vra Alt.^{a} m^{de} que fuese a hazello Arrasti el qual no quyere yr y los dhos y^{o}s de Chacalla entretanto nos destruyen nuestras tierras y nos hazen molestias todo a heffeto de que tengamos pasion con ellos.

A Vra Alt^{a} pedimos y suplicamos sea seruido de mandar que una persona baya a entregarnos las obexas que Vra Alt.^{a} manda que nos entreguen pues no es justo que quedemos syn las tierras que se nos mandaron quytar e sin las obejas que dimos sobre que pedimos justiçia e para ello etc.

El Licen^{do} de Leon

En Los Rreyes siete de otubre de myll e qui^{o}s y sesenta y siete años ante los señores presidente y oidores desta Rreal Audiençia en audiençia pu^{ca} se presento esta pet^{on} y los dhos senores mandaron que lo haga sin enbargo de su rrespuesta y que dentro de tres dias y que venido se le mandara pagar.

Avendaño

/426r/ Don Ju^{o} Baptista Caxaguaraca.

Muy p^{o} señor

Diego de Ocanpo en n^{e} de don Juan Baptista Caxaguaraca caçique prinçipal del rrepartimy^{o} de Chacalla y de los yndios a el subjetos en la cabsa con el caçique e yndios de Canta sobre las trras de Quivi / digo que estando muchas vezes probeydo a my pedimyento que se traiga el proçeso desta cabsa para probeer y determinar / sobre que se acaben de entregar a mys partes las trras sobre que es este pleyto y se declare aver cunplido con el entrego de las ovejas que fiziera y se mande que las ptes contrarias no entren a coger fruto algu^{o} de las dhas trras y se castiguen los que an quitado los mojones y cortado los arboles y arrancado y quemado la coca no sea visto y se libran provisyones para que una persona vaya a determinar algunos de los dhos articulos pudiendose determinar en esta Abdiençia por el proçeso y sobre otros no se probee / y sy asy se fiziese seria acabar de confundir esta cabsa y fazerla ynmortal.

Pido y sup.^{co} a V. Al. mande trayga al rrelator el proçeso y se vea y determine lo que por el proçeso se puede determinar / y se enbie persona a que exsecute lo que se probeyere / y desta manera se acabara el pleyto y se hara a las ptes justiçias la qual pido y para ello etc.

El Lic^{do} Falcon D. Ocanpo

/426v/ En Los Rreyes syete de otubre de myll e qui^{o}s e sesenta y siete años ante los senores pres.^{te} e oydores desta Rreal Aud.^{a} est.^{do} en Aud.^{a} Rreal pu^{ca} se presento esta petiçion y los dhos senores proueyeron lo p^{o}ueydo.

Avendaño

/427r/ Los y^{o}s de Chacalla.

Muy p^{o}s. señor

Diego de Ocanpo en n^{e} del caçique e yndios de Chacalla en la cabsa con los yndios de Canta sobre las trras e chacaras de Quivi / digo que las partes contrarias no contentos con las fuerças e delictos que an cometido de que estan acusados visto que no a avido castigo agora de nuevo an derribado mojones de los que estauan puestos demas de los que estauan derribados por dos vezes y an rronpido una acequia para que no pudiesemos rregar los mayzes y comidas que tenemos senbrado y para que constase por evidençia dello el caçique prinçipal fizo prender a vno de los yndios que hallo haziendo los dhos delictos y le tiene aguardando a la persona que V. Al. tiene mandado que vaya a averiguar los dhos delictos.

Pido y sup^{co} a V. Al. mande que la persona que asy fuere proçeda asymysmo sobre estos delictos y que se entregue el dho preso para ello sobre que pido just.^{a} y costas y para ello etc. e juro a Dios en forma que no lo pido de maliçia.

El Lic^{do} Falcon Diego de Ocanpo

/427v/ En Los Rreyes quinze de otubre de myll e qui^{o}s y sesenta y siete años ante los señores presidente y oidores desta Rreal Audiençia en audiençia pu^{ca} se presento esta petiçion y los dhos senores mandaron que sobre todo proçeda el que fuere.

Avendaño

/428r/ P^{o} de Arrasti.

Muy p^{o} señor

P^{o} de Herrazti digo que yo fuy por mandado de V. Al.^{a} al pueblo de Quybi a hazer çiertas ynformaçiones sobre las diferençias que los yndios de Canta y Chacalla tratan sobre el entrego de las dozientas ovejas y daños de las chacaras e açequyas de las quales hago presentaçion.

- Otro sy digo que de las dhas ynformaçiones rresulta estar culpados tres yndios de la parçialidad de Canta a los quales traxe presos y los puse en la carçel desta Corte conforme a lo que por V. Al. me fue mandado por su rreal provision.

P^{o} de Herrazti

En Los Rreyes en veinte e nuebe de otubre de myll e qui^{o}s e sesenta e siete a^{o}s ante los señores pres^{te} e oydores desta Rreal Audi^{a} estando en audi^{a} pu^{ca} se presento esta petiçion y los dhos senores mandaron que se traiga por el rrelator.

Avendaño

/429r/ El liçen^{do} Castro — El D. Bravo de Saravia — El Lic^{do} don Alv^{o} Ponçe de Leon — El Lic^{do} Ser^{o} de Vigil

Ynformaçion hecha en el pueblo de Quibi
sobre los daños que los yndios de Canta hizieron
a los de Chacalla en las chacaras e açequias.

Para que Pedro de Arrasti rreçeptor haga ynformaçion sobre lo cont.^{do} en esta prov.^{on} y prenda a los culpados y enbie a la carçel rreal desta Corte y enbie la dha ynformaçion para que se provea de justiçia a las ptes.

/429v/ Sobre lo contenydo en esta probision hize ynformaçion por la probision que esta delante / porque entranbas son sobre una mysma cosa.

Don Phelipe por la graçia de Dios rrey de Castilla de Leon de Aragon de las dos Seçilias de Jerusalen de Navarra de Granada de Toledo de Valençia de Galiçia de Mallorcas de Seuilla de Çerdeña de Cordova de Corçega de Murçia de Jaen de los Algarues de Algezira de Gibraltar de las yslas de Canaria de las Y^{as} e tierra firme del mar oçeano; conde de Flandes e de Tirol etc.

A vos Pedro de Arrasti nro rreçeptor Saued que Diego de Ocanpo en nonbre del caçique e yndios de Chacalla paresçio ante nos en la nra Rreal Audiençia e Chançilleria que por nro mandado rreside en la çiudad de Los Rreyes ante el pres.^{te} y oydores della en la causa que trata con los yndios de Canta sobre las tierras de Quibi e presento una petiçion del tenor siguiente:

"Muy poderoso señor

Diego de Ocanpo en nonbre del caçique e y^{o}s de Chacalla en la causa con los y^{o}s de Canta sobre las tierras e chacaras de Quibi digo que las partes contrarias no contentos con las fuerças e delitos que an cometido de que estan acusados visto que no a avido castigo agora de nueuo an derribado mojones de los que estauan puestos / demas de los que estauan derribados por dos vezes / y an rronpido una açequia para que no pudiesemos rregar los maizes y comidas que tenemos senbrado / y para que constase por evidençia dello el caçique prinçipal fizo prender a vno de los y^{o}s que hallo haziendo los dhos delitos y le tiene aguardando a la persona que V. Al.^{a} tiene mandado que vaya aueriguar los dhos delitos. Pido y sup.^{co} a V. Al.^{a} mande que la persona que asi fuere p^{o}çeda asimismo sobre estos delitos y que se le entregue el dho preso para ello sobre que pido just.^{a} e costas e para ello etc. e juro a Dios en forma que no lo pido de maliçia. El Liçençiado Falcon / Diego de Ocanpo."

E nos pidio //
/430r/ e sup^{co} lo mandasemos asi p^{o}ueer o como la nra merçed fuese lo qual visto por el dho pres^{te} e oidores fue acordado que deuiamos mandar dar esta nra carta para vos en la dha rrazon e nos tubimoslo por bien por la qual vos mandamos que luego que con ella fueredes rrequerido hagais ynformaçion e auerigueis como a pasado e pasa lo susodho lo qual hareis en el dho asiento de Quibi o donde bieredes que mejor podra sauer verdad y hecha la dha ynformaçion prendereis lo que della rresultaren culpados e hareis traer a la carçel rreal desta Corte y enbiareis la dha ynformaçion al dho nro presidente e oidores para que por ellos vista prouean de justiçia a las ptes que para entender en lo susodho vos damos poder cunplido tal qual para en tal caso se rrequiere.

Dada en Los Rreyes a diez y siete dias de otubre de mill e qui^{o}s e ses.^{ta} y siete años.

Yo el scr^{o} P^{o} de Avendaño escrivano de camara de la Mag.^{t} Rreal la fiz escrivyr por su mandado con acuerdo del su presidente e oydores.

Rr^{da} Juan de Murga — Chançiller Gaspar de Solis

/430v/ Don Phelipe por la graçia de Dios rrey de Castilla de Leon de Aragon de las dos Seçilias de Jerusalen de Nauarra de Granada de Toledo de Valençia de Galiçia de Mallorcas de Seuilla de Çerdeña de Cordoua de Corçega de Murçia de Jaen de los Algarues de Algezira de Gibraltar de las yslas de Canaria de las Yndias yslas e tierra firme del mar oçeano; conde de Flandes e de Tirol etc.

A vos Pedro de Arrasti nro rreçeptor salud e graçia / Saued que por parte de don Juan Bautista Caxaguaraca caçique prençipal de Chacalla y de los prençipales e yn^{o}s del rrepartimyento en el pleito que han tratado e tratan con los yn^{o}s del rrepartimyento de Canta sobre las tierras de Quibi por dos petiçiones que ante nos se presentaron en la nra Rreal Audiençia y Chançilleria que por nro mandado rreside en la çibdad de Los Rreyes se querellaron de don Fran^{co} caçique prençipal de Canta y de los demas que paresçieren culpados y asimysmo de don Juan Pariguana e don Gonçalo Lucha y demas que paresçieren culpados de las cosas y delitos contenidos en las dhas querellas cuyo tenor es el que se sigue:

"Muy poderoso señor

Diego de Ocanpo en nonbre del caçique e y^{o}s de Chacalla en la causa con los caçiques e yn^{o}s de Canta sobre las tierras de Quibi ante Vra Alteza me querello e acuso criminalmente a don Fran^{co} caçique prençipal de Canta y de los demas que por la ynformaçion paresçieren culpados y contando el caso premysas las solenidades del derecho / digo que es asi que no contentos los susodhos con los muchos e muy graues delitos que otras vezes an cometido contra mis ptes por les tomar las dhas tierras y quedarse con ellas visto que no an sido castigados por ellos aunque derribaron los mojones puestos por executoria desta Rreal Audiençia aora de nuebo auiendolos//
/431r/ tornado a poner por mandado desta Rreal Audiençia y dado a mis ptes posesion en parte de las dhas tierras / los susodhos con poco temor de Dios y en gran menospreçio de Vras Justiçias an quitado las aguas con que mis ptes rregauan las dhas tierras y tornado a derribar algunos de los dhos mojones y destruido la coca en lo qual an cometido graues delitos. Por que pido y supp^{co} a V. Alt.^{a} mande auer ynformaçion de lo susodho y prender los culpados que presos yo estoy presto de ponerles acusaçion en forma sobre que pido justiçia y para ello etc. e juro a Dios en forma que no lo pido de maliçia. El Liçençiado Falcon."

"Muy poderoso señor

Juan Yapa por mi y en nonbre de don Juan Bautista Caxaguaraca caçique prençipal de Chacalla e de los prençipales e y^{o}s a el subjetos en la causa con el caçique e y^{o}s de Canta sobre las chacaras de Quibi ante V. Al.^{a} me querello e acuso criminalmente a don Juan Pariguana e Gonçalo Luccha e a los demas que paresçieren culpados y contando el caso premisas las solenidades del derecho digo que es asi que los susodhos no contentos con las fuerças y delitos graues que nos an hecho quitandonos la posesion que nos fue dada por executoria desta Rreal Audiençia de las dhas tierras aora de nuebo auiendosenos mandado tornar a dar y anparar en ella por prouision desta Rreal Audiençia y estando en ella paçificamente los susodhos pospuesto el temor de Dios y en gran menospreçio de Vra Justiçia fueron a las dhas tierras diziendo que lleuaban proui^{on} desta Rreal Audiençia para nos hechar dellas y enpeçaron a derribar e derribaron dos mojones de los que por vro mandado estauan puestos y a derribar y arrancar arboles y la coca y de fecho nos hecharian dellas si no //

/431v/ nos defendieramos y nra posesion en lo qual cometieron graues delitos dignos de graue castigo exenplar. Por que pido e suplico a Vra Alteza que avida ynformaçion de lo susodho que estoy presto de dar mande prender los culpados de los susodhos y secrestarles los bienes que presos estoy presto de ponerles acusaçion en forma sobre que pido justiçia y costas e para ello etc. e juro a Dios en forma que no lo pido de maliçia.

El Liçençiado Falcon"

E nos fue pedido e suplicado mandasemos rresçevir ynformaçion de lo susodho que se ofresçian a dar y rresçevir y prender los cuerpos a los culpados con secresto de bienes que estando presos estauan prestos de les poner acusaçion en forma o como la nra merçed fuese lo qual visto por el presidente e oidores de la dha nra Rreal Audiençia fue acordado que os deuiamos cometer como por la presente vos cometemos la ynformaçion e aueriguacion de lo susodho y vos mandamos que luego que vos sea mostrada vais al dho asiento de Quibi partes e lugares donde os paresçiere que conuenga para aueriguar verdad donde hareys ynformaçion e aueriguareis con çitaçion de partes todo lo contenido en las dhas dos querellas que de suso van yncorporadas a los que por ella hallaredes culpados los traereys a esta Corte presos e porneis en la carçel rreal de Corte para que se haga cunplimyento de justiçia a las partes y vereis por vista de ojos los mojones que el padre Fray Gaspar de Caruajal de la Orden de Señor Santo Domingo puso en las dhas tierras de Quibi y si algu^{o}s dellos obieren quitado los hareis bolver a poner en el mesmo lugar que se pusieron por orden del dho Fray Gaspar de Caruajal y a los que hallaredes culpados en los auer quitado e mudado los traereys //
/432r/ asimismo presos segun dho es e por la presente mandamos a todas e qualesquier personas de qualquier estado condiçion que sean de quien entendieredes hazer la dha ynformaçion parezcan ante bos a dezir sus dhos e declaraçiones so las penas que de nra pte les pusieredes en las quales auemos por condenado al que lo contrario hiziere y hechas las dhas ynformaçiones las traereis ante nos para que vistas por el dho presidente e oydores se prouea en el caso lo que sea justiçia e que para entender en lo susodho cada una cosa aparte dello bos damos poder cunplido tal qual para en tal caso se rrequiere y buelto que seais de entender en lo susodho vos mandaremos pagar lo que por vro trauajo ouieredes de auer.

Dada en Los Rreyes a diez dias de otubre de mill e qui^{o}s y sesenta e siete a^{o}s.

Yo el secr^{o} P^{o} de Avendaño escrivano de camara de la Mag.^{tt} Rreal la fiz escrivyr por su mandado con acuerdo del su presidente e oydores.

Rr^{da} Juan de Murga

Chançiller
Gaspar de Solis

/432v/

Ynformaçion que hizo Arrasti
por comi.^{on} desta Rreal Audiençia de çiertos
agravios de que se quexaron los yndios de
Chacalla de los de Canta.

Aqui parece que este don Felipe
y los yndios del pueblo de Quibi son
de la parçialidad de Canta.
y tanbien lo dize a f° 436 col. 2.

En la çiudad de Los Rreyes a diez e seis dias del mes de otu^{e} año del señor de myll y quy^{o}s e senseta y siete en presençia de my P^{o} de Herrazti escrivano de Su Mag.^{t} pareçio presente Di^{o} de Ocanpo pr^{o}r del Audiençia Rreal desta dha çiudad en nonbre del caçique e yndios del rrepartimyento de Chacalla y presento la provision rreal de Su Mag.^{t} desta otra parte contenida a my dirigida y en cunplimyento della me pidio y rrequerio me fuese al pueblo de yndios de Quybi a hazer lo que por la dha rreal provision se me manda e yo el dho escrivano le rrespondi que estaua presto de lo hazer.

P^{o} de Herrazti
escry^{o} de Su Mag.^{t}

Despues de lo susodho yo P^{o} de Herrazti escr^{o} susodho en cunplimyento de lo que se me manda por la provision rreal de Su Mag.^{t} de suso contenida para hazer lo que por ella se manda sali de la dha çiudad de Los Rreyes a diez e nuebe dias del dho mes de otubre del dho año / para yr al pueblo de Quybi a veynte dias del dho mes a mediodia y luego pregunte por los caçiques prinçipales de Canta y Chacalla y don Fran^{co} Pilco y don Juan Pariona prinçipales de Canta y don Felipe Taulichunbi prinçipal de los y^{o}s del dho pueblo de Quyvi de la parçialidad de Canta me dixeron que don Fran^{co} Guaman Guaranga caçique prinçipal del rrepartimyento de Canta estaua en el dho pueblo de Canta / a los quales por lengua de Juan Diaz ynterpetre que para hazer este negoçio lleue conmygo les dixe que luego a la ora despachasen un mensagero para hazer sauer al dho don Fran^{co} caçique prinçipal como yo avia benydo a este dho pueblo de Quybi por m^{do} de los señores presidente e oydores a fazer y cunplir lo que por las dhas provisiones se me mandaua y que luego binose adonde yo estaua a ser presente porque con su çitaçion se avia de fazer el negoçio del rrepartimyento de Chacalla //
/433r/ halle en el dho pueblo de Quybi a un yndio prinçipal de Chacalla que dixo llamarse Juan de Albarado y Fran^{co} Mychuy prinçipal de

los yndios yungas del dho pueblo de Quibi de la parçialidad de Chacalla e otros yndios los quales me dixieron que don Juan Baptista su caçique prinçipal estaua ausente deste dho pueblo a los quales asymysmo les mande que enbiase a llamar al dho su caçique los quales todos los vnos y los otros me dixieron que sy lo harian y que les enbiarian mensageros para el dho hefeto lo qual todo les dixe e di a entender por lengua del dho ynterpetre e dello doy fee.

P° de Herrazti
escr° de Su Mag.ᵗ

E luego yncontinente el dho dia mes y año susodhos yo el dho Pedro de Herrazti escrivano susodho fuy a ver por vista de ojos los mojones que el padre Fray Gaspar Caruajal puso en las trras deste dho pueblo dende vn mojon grande que esta puesto junto a la yglesia del dho pueblo ençima de vn pedregal grande cabe una cruz + para arriba el valle arriba halle quarenta y nuebe mojones de piedra y trra puestos y enhiestos del altura de vn honbre y en dos mojones dellos pareçia averze quitado en cada uno dellos vna piedra e ods y los yndios de la parçialidad de Chacalla se quexaron que los yndios de la parçialidad de Canta avian quitado de los dhos dos mojones las dhas piedras y luego les mande que las tornasen a poner y asy las pusieron en my presençia los quales dhos mojones estan puestos de trecho en trecho por linde de vna barranca grande que haze el valle del dho pueblo de Quibi y por entre la dha barranca y los mojones esta el camyno rreal por donde suben a Canta / y este dia no se hizo otra cosa en el dho negoçio porque no ovo tienpo para ello.

Este mysmo dia en la tarde vino al dho pueblo de Quybi don Juan Bautista caçique prinçipal de Chacalla y par° ante my.

P° de Herrazti
escr° de Su Mag.ᵗ

/433v/ Despues de lo qual a veynte y vn dias del dho mes de otuᵉ del dho año luego en amaneçiendo sali yo el dho P° de Herrazti del dho pueblo de Quybi para yr a ver los demas mojones que el dho padre Fray Gaspar de Caruajal puso en las dhas trras de Quibi / y començando a contar desde el dho mojon grande que esta junto a la yglesia del pueblo cabe la cruz azia el rrio atrabesando el valle vn poco al sesgo hazia baxo halle puestos de trecho en trecho veynte y seis mojones de piedra y trra como los demas arriba dhos y mas vn mojon derribado y en otro mojon en lo alto del algunas piedras molidas que pareçe que las quitaron y tornaron a poner y los dhos yndios de la parçialidad de Chacalla se quexaron que los dhos y°s de la parçialidad de Canta los avian quitado / y luego les mande que tornasen a adereçar y poner los dhos mojones y asy los pusieron y hecho esto fuy a ver las chacaras de coca que los dhos yndios de Chacalla dezian que les avian quemado y arrancado los otros y halle dos pedaçitos de coca muy pequeños quemados y tanbien halle entre las dhas chacaras tres arboles de pacaes secos de averles dado fuego por abaxo a lo que pareçia y otro arbol cortado por el pie y otro desmochado por lo alto quitadas todas las rramas y començado a brotar los pinpollos de las rramas y otro quatro arboles de pacaes cortada la corteça a la rredonda del arbol pero no estanse asy otros dos arboles de pacaes tanbien chamuscados por abaxo a lo que pareçe porque no estan secos lo qual todo dixeron los dhos yndios de la parçialidad de Chacalla que lo avian hecho los otros de la parçialidad de Canta y asymysmo se quexaron que les avian quebrado la açequia por quitarles el agua para sus sementeras y que ya estaua adreçado pero no pareçe en las dhas sementeras averles hecho falta el agua asta agora y por no se aver notificado las dhas provisiones al caçique de Canta por estar ausente y no aver benydo asta agora no començe luego a hazer la ynformaçion / y asymysmo alle //
/434r/ entre las dhas chacaras de coca dos pedaçillos de coca que los arboles de la dha coca estaban mas rrasos que en las demas chacaras y los dhos y°s de la parçialidad de Chacalla se quexaron que los otros de Canta se las avian arrancado.

P° de Herrazti
escr° de Su Mag.ᵗ

Despues de lo susodho a veynte e dos dias del dho mes de otubre del dho año yo el dho P° de Herrazti fuy a ver las açequias de las trras deste dho pueblo de Quyvi donde los y°s de Chacalla me dixieron que los y°s de Canta les avian quitado el agua con que rregaban sus chacaras y me mostraron vna açequia de donde rrepartian el agua a unas chacaras e a otras y alli donde rrepartian la dha agua estaua dentro en la açequia vna piedra grande la qual dixieron los y°s de Chacalla que los yndios de Canta sus contrarios la avian hechado alli por ynpedirles que no fuese tanta agua a sus chacaras / la qual dha piedra todavia hazia algun estorbo y la hize sacar de alli y les pregunte sy les avian hecho mas daño los dhos y°s de Canta en las dhas açequias los quales dixieron que media legua de alli donde sacaban vna açequia grande del rrio que baxa del rrepartimyento de Xecos de Juan Piçarro que alli en el naçimyento del açequia vn yndio deste dho pueblo que se llama Juan Sansa de la parçialidad de Canta quito unas piedras para desbaratar la dha açequia y que luego la adreçaron y que no les //
/434v/ han quebrado la dha açequia en otra parte nynguna / y en quebrar los y°s de la parçialidad de Canta la açequia en el nacimyento della donde dizen que la quebro el dho Juan Sansa me pareçe a my que tanbien hazia el daño para si y para los demas yndios de Canta como para los de Chacalla porque para todos ellos biene el agua por la dha açequia asta junto a este dho pueblo de Quyvi donde la rreparten para todos lo qual todo declararon por lengua de Juan Diaz que dende Lima bino a este pueblo conmygo para ser ynterpetre deste negoçio del qual yo el dho escrivano tome y rre. juramento en forma de dr° de que bien y fielmente ynterpretaria so cargo del qual prometio de lo hazer asy y no firmo porque dixo que no sabia.

P° de Herrazti
escr° de Su Mag.ᵗ

Despues de lo qual a veynte y dos dias del dho mes de otubre del dho [*sic*] a las diez oras del dia poco mas o menos en el dho pueblo de Quibi paresçieron ante my el dho escr° don Fran^co Pilco e don Fran^co Chanpa y°s prinçipales del dho rrepartimyento de

Canta / e dixieron que el dho don Fran^{co} Chanpa benia de Canta por mensagero de don Fran^{co} caçique prinçipal de Canta el qual estaua enfermo y que a esta causa no podia benir a este dho pueblo si no hera en amaca y con mucho trauajo de su persona.

P^{o} de Herrazti
escr^{o} de Su Mag.^{t}

/435r/ E luego yncontinente en el dho pueblo de Quibi el dho dia veynte y dos de otu^{e} del dho año yo el dho P^{o} de Herrazti por defeto de no aver benido a este pueblo el caçique prinçipal de Canta notifique e di a entender lo contenydo en la dha Rreal provision de suso contenida a don Fran^{co} Pilco y a don Juan Pariona y a Hernando Tantavilca y^{o}s prinçipales del rrepartimi^{o} de Canta e a otros y^{o}s que con ellos estaban y tanbien a los y^{o}s yungas deste pueblo de Quibi de la parçialidad de Canta / por lengua del dho Ju^{o} Diaz ynterpetre susodho.

P^{o} de Herrazti
escr^{o} de Su Mag.^{t}

Despues de lo qual en el dho pueblo de Quibi el dho dia veynte y dos de otubre del dho año para ynformaçion de lo contenydo en la dha rreal provision el dho don Ju^{o} Baptista caçique prinçipal de Chacalla presento por testigo a M^{yn} Yauri yndio del pueblo de Llunpo del rrepartimy^{o} de Chacalla subjeto al dho don Juan Baptista al qual no le tome juramento por pareçerme que por la mucha yndignaçion y henojo questos yndios tienen vnos contra otros sobre estas trras se podria alargar en su dho y asy le encargue por lengua del dho ynterpetre que no dixese mas de la verdad donde no que la justiçia le castigaria el qual prom^{o} //
/435v/ de dezir verdad e siendo preguntado por el tenor de la dha rreal provision dixo que lo que deste caso save es que despues que el padre Fray Gaspar de Caruajal y Br^{me} de Prol mandaron poner los mojones vido este t^{o} que un yndio de Canta que se llama Alonso que es serrano y otro yndio yunga deste pueblo de la parçialidad de Canta que tanbien se llama Al^{o} que no les save los sobre nonbres mas de que los conosçe de vista / derribaron vno de los mojones que el dho padre Caruajal mando poner y este t^{o} y otros dos y^{o}s de su parçialidad como les vieron derribar el dho mojon fueron a ellos y al yndio serrano de Canta le quitaron la manta y vn prinçipal del dho rrepartimyento de Canta que se llama don Juan Pariguana les torno a quitar la dha manta y les dixo que el avia enbiado alla aquellos dos y^{o}s para que contasen los mojones pero no para que los derribasen y que tanbien vido este testigo que a otro mojon quitaron en lo alto del algunas piedras pero que no save quyen / y que asymysmo ha visto tres arboles de frutales que estan secos en las chacaras de averlos quemado / y otros dos o tres pacaes chamuscados por el pie / y otro pacae cortado por el pie / y otro quitadas todas las rramas desmochado y dos pedaçillos de chacaras de coca quemados / y otros seys arboles de pacaes cortada la corteza a la rredonda pero que no saue quyen lo hizo / y que oyo dezir este t^{o} a vn yndio deste pueblo de la parçialidad de Canta que el avia cortado aquel pacae para hazer vn horcon para su casa y que esto es lo que saue deste caso de lo que le ha sido preguntado lo qual es la verdad y en ello se afirmo por lengua del dho ynterpetre.

ante my P^{o} de Herrazti
escr^{o} de Su Mag.^{t}

/436r/ Este dia veynte e tres de otubre del dho año el dho don Juan Baptista para ynformaçion de lo susodho presento por t^{o} a vn yndio de su rrepartimy^{o} de Chacalla que se llama Alonso Quyspi del pueblo de Cullata / el qual siendo preguntado por el tenor de lo susodho dixo aviendole encargado que dixese la verdad que lo que saue es que los dias pasados despues que el padre Fray Gaspar de Caruajal les amojono las trras vn dia estando los y^{o}s deste pueblo de Quybi e rregando la chacara de mayz de don Ju^{o} Baptista su caçique prinçipal vido este t^{o} que se les yva apocando el agua del azequia y fue asta el nasçimyento del azequia a ver por donde se les yva el agua y hallo en el rrio en el nasçimyento del azequia vn yndio deste pueblo de Quibi de la parçialidad de Canta que se llama Ju^{o} Sansa junto adonde estaua quebrada la azequia / y que este t^{o} le dixo que para que avia quebrado la açequia y quitado el agua al qual le rrespondio que el la avia quebrado y que queria el para ello y este t^{o} le dixo que fuesen adonde estaua su caçique el qual no quiso benir y este t^{o} le quito la manta / y que esto es lo que saue deste caso lo qual es la verdad y en ello se afirmo y rratifico por lengua del dho ynterpetre y pareçe ser de treynta y çinco años poco mas o menos.

ante my P^{o} de Herrazti
escr^{o} de Su Mag.^{t}

Despues de lo susodho en el dho pueblo de Quibi a veynte y quatro dias del dho mes de otu^{e} del dho año para ynform^{on} de lo susodho el dho don Ju^{o} Baptista caçique de Chacalla presento por t^{o} a M^{yn} Vro yndio prinçipal que dixo ser del pueblo de Xoncomarca del rrepartimy^{o} de Chacalla al qual por lengua del dho ynterpetre le encargue que dixese la verdad el qual prometio de la dezir / e siendo preguntado por el tenor de lo susodho dixo que lo que saue es que despues que el padre Fray Gaspar de Caruajal y Br^{me} de Prol binyeron a este pueblo a amojonarles las trras vido este t^{o} que vn yndio del rrepartimi^{o} de Canta que se llama Vro que bibe en este pueblo puso dos cruzes en las trras de los y^{o}s de Chacalla mas adentro //
/436v/ de donde el dho p^{e} Caruajal mando poner los mojones y este t^{o} le dixo al dho Vro que para que ponya aquellas cruzes en las trras que no heran suyas / el qual le rrespondio que por alli avian mandado poner los mojones Gibraleon y el alguazil que bino con el / y que por eso los ponya / y que asymysmo saue que quando el otro dia fuy yo el presente escrivano a ver los daños de las chacaras dixo vn yndio deste pueblo de Quibi de la parçialidad de Canta que vn pacae que estaua cortado por el pie en la chacara que el lo avia cortado para fazer vn horcon para su casa / y que entonçes vido este t^{o} tres arboles de pacaes secos de averlos quemado por el pie y otros dos o tres chamuscados y otros arboles de pacaes cortada la corteza a la rredonda y dos pedaçitos de chacaras de coca quemados y que todos los dhos daños se an hecho despues que el dho padre Caruajal y Br^{me} de Prol se fueron de aca pero que no vido quyen los hizo mas de que le pareçe que los y^{o}s de la parçialidad de Canta por hazerles mal lo habran hecho y que esto es lo

que saue y es la verdad y en ello se afirmo por lengua del dho ynterpetre y pareçe ser de hedad de treynta y çinco años poco mas o menos.

ante my P^{o} de Herrazti
escr^{o} de Su Mag.^{t}

Aqui dize que Quibi e sus y^{o}s
es parçialidad de Canta.

Este dia para ynformaçion de lo susodho el dho don Ju^{o} Baptista presento por t^{o} a M^{yn} Caruancho yndio del pueblo de Cullata del rrepartimyento de Chacalla / al qual por lengua del dho ynterpetre le encargue que diga la verdad y prom^{o} de la dezir y siendo preguntado dixo que lo que saue es que vn dio [*sic*] deste pueblo de Quibi de la parçialidad de Canta que se llama Vina estando los y^{o}s de la parçialidad de Chacalla rregando la chacara de don Ju^{o} su caçique por quitarles el agua desbarato el azequia y hecho una piedra grande en ella para ynpedir que no fuese tanta agua a las chacaras de los y^{o}s de Chacalla / y que lo saue porque este t^{o} le hallo al dho yndio Vina estando desbaratando la açequia y este t^{o} le dixo que para que desbarataua la açequia / el qual le rrespondio que el lo avia desbaratado y que no tenya myedo de nadie y que sy benya por ventura aca algun español o saçerdote para que el le tubiese myedo y que esto es lo que saue y es la verdad y en ello se afirmo por lengua del dho ynterpetre y pareçe ser de hedad de veynte y dos años poco mas o menos.

ante my P^{o} de Herrazti
escr^{o} de Su Mag.^{t}

/437r/ Para ynform.^{on} de lo susodho el dho Ju^{o} Baptista presento por t^{o} a M^{yn} Poma yndio del pueblo de Llunpo del rrepartimy^{o} de Chacalla / el qual aviendole encargado que diga la verdad dixo que lo que saue es que vido que vn yndio deste pueblo de Quibi que se llama M^{yn} que no le saue el sobre nonbre quemo vn pedaçillo de coca en las chacaras de los y^{o}s de Chacalla / y que lo saue porque lo vido y que este t^{o} le dixo que para que lo quemaba y el dho yndio le rrespondio que aquellas trras heran suyas y que lo quemaua para senbrar alli mayz y que esto es lo que saue y es la verdad y pareçe ser de hedad de treynta a^{o}s el qual declaro por lengua del dho ynterpetre.

ante my P^{o} de Herrazti
escr^{o} de Su Mag.^{t}

Este dho dia en el dho pueblo de Quibi a veynte y quatro dias del dho mes de otu^{e} del dho año el dho don Juan Baptista para ynform^{on} de lo susodho presento por testigo a Felipe Paspon yndio deste pueblo de Quibi de la parçialidad de Chacalla al qual se le encargo que diga la verdad y prometio de la dezir / y siendo preguntado por el tenor de lo susodho dixo que lo que saue es que despues que por mandado de la Rreal Audiençia el p^{e} Caruajal y Br^{me} de Prol amojonaron las trras deste pueblo / un yndio deste dho pueblo que se llama Pedro Condor de la parçialidad de Canta corto quatro arboles que pareçian pacaes los dos dellos cortados por el pie y los otros dos descabeçados por lo alto y a lo que pareçe son cortados de poco tienpo aca y este t^{o} dixo que se los vido cortar al dho P^{o} de Condor y que esto es lo que saue para el juram^{o} que tiene hecho y pareçe ser de veynte años el qual declaro por lengua del dho ynterpetre.

ante my P^{o} de Herrazti
escr^{o} de Su Mag.^{t}

E luego yncontinente yo el dho escr^{o} dixe al dho don Ju^{o} Baptista por lengua del dho ynterpetre sy tenia mas t^{o}s que presentar sobre lo cont.^{do} en la dha rreal provision el qual dixo que no tenya mas t^{o}s que presentar.

ante my P^{o} de Herrazti

/437v/ Confesion de Xpoual Chaunyca
sobre el pacay que corto.

Despues de lo susodho en el dho pueblo de Quibi el dho dia veynte y quatro de otubre del dho año para mas aueriguaçion de lo susodho tome y rreçebi la declaraçion de Xpoual Chaunyca yndio deste dho pueblo de la parçialidad de Canta el qual siendo preguntado que por que corto vn pacae en las chacaras de los y^{o}s de Chacalla que esta cortado por el pie dixo que lo que pasa es que este t^{o} y otros y^{o}s deste pueblo yungas y serranos todos ellos de la parçialidad de Canta fueron a cortar el dho pacae para hazer vna rramada para don Fran^{co} caçique prinçipal de Canta y asy lo cortaron a este que declara dos otros y^{o}s que se llaman M^{yn} Monca y Ju^{o} Chuquy Yauri / y que otros tres pacaes que estan ya secos que este t^{o} los chamusco con el fuego porque no llebaban fruto y para que se secasen para leña porque enbaraçaban la chacara y no heran de probecho / y que otros dos pacaes que estan junto dellos que tanbien los chamusco este t^{o} que declara porque tan poco son de prouecho aquellos pero que no estan secos aquellos dos y que estos arboles de pacaes que estan en las dhas chacaras cortadas la corteza a la rredonda que un yndio del rrepartimi^{o} de Xecos que solia bibir en este pueblo les corto la corteza los quales dhos arboles avia senbrado el dho yndio de Xecos que se llama don Ju^{o} Guaman que es natural de los Lucanas y agora de poco aca a pasado a las trras de Xecos tres años ha / y que muchas vezes biene aca porque esta çerca / y que las dos cruzes que se pusieron en las trras de los de Chacalla despues que el padre Caruajal se fue de aca las pusieron dos y^{o}s de Canta que el uno se llama Vro y el otro Al^{o} Alexandre / que este que declara se las vido poner y que esto es lo que saue y es la verdad y en ello se afirmo y rratifico por lengua del dho ynterpetre.

ante my P^{o} de Herrazti
escr^{o} de Su Mag.^{t}

Declaraçion de Uro yndio de Quibi.

Este dia para la dha ynform^{on} hize pareçer ante my a vn yndio que dixo llamarse Vro y que bibe en este pueblo de Quibi en la parçialidad de Canta / el qual siendo preguntado por el tenor de lo susodho dixo que no sabia nada.

Confesion de Al^{o} Alexandre.

Para la dha ynform^{on} hize pareçer ante my a Al^{o} Alexandre Poma yndio del pueblo de Bisca del rrepartimy^{o} de Canta el qual siendo preguntado por el tenor de lo susodho por lengua del dho ynterpetre dixo que este que declara de su voluntad sin que nadie se lo manda se puso los dos mojones arribado declarados //
/438r/ en las trras de los y^{o}s de Chacalla dentro en las chacaras y que esto es la verdad y lo que saue lo qual declaro por lengua del dho ynterpetre.

ante my P^{o} de Herrazti
escr^{o} de Su Mag.^{t}

En el dho pueblo de Quibi el dho dia veynte y quatro de otu^{e} del dho año para ynformaçion de lo susodho hize pareçer ante my a don Ju^{o} Guaman yndio natural que dixo ser de los Lucanas y que ha quinze años que esta por aca y que agora çerca deste pueblo en las trras de Xecos al qual por lengua del dho ynterpetre le encargue que dixiese la verdad el qual prometio de la dezir y siendo preg.^{do} en rrazon de lo susodho dixo que estando este t^{o} en Lima su muger de henojo de que no gozaba de lo que senbraba y cultibaba en las trras deste pueblo porque estubo mucho tienpo en el corto la corteza a çiertos arboles de pacaes de que agora se quexan los y^{o}s de Chacalla y quando este t^{o} bolbio de Lima y supo lo que su muger avia hecho la rriño y castigo y le pego de lo que hizo y que esto es lo que saue y es la verdad y en ello se afirmo por lengua del dho ynterpetre.

ante my P^{o} de Herrazti
escr^{o} de Su Mag.^{t}

Ocupeme en este negoçio en la yda estada y buelta ocho dias cobre por rrazon de my trauajo treynta p^{o}s en corriente y no lleue dr ^{o}s de la escriptura y los diez dellos cobre de los y^{o}s de Chacalla y veynte de los de Canta por rrazon de estar algunos dellos culpados en los daños de las chacaras como consta por esta ynform^{on}. 30 p^{o}s.

P^{o} de Herrazti

/439r/ El Liçen^{do} Castro — El D. Bravo de Saravia — El Doctor Gz de Cuenca — El Licen. Ser^{o} de Vigil

---

Lo que dixieron el caçique e yndios de Chacalla sobre lo contenydo en esta provision.

/439v/ Don Phelipe por la graçia de Dios rrei de Castilla de Leon de Aragon de las dos Seçilias de Jerusalen de Nauarra de Granada de Toledo de Balençia de Galiçia de Mallorcas de Seuilla de Çerdeña de Cordoua de Corçega de Murçia de Jaen de los Algarues de Algeçira de Gibraltar de las yslas de Canaria de las Yndias yslas e tierra firme del mar oçeano; conde de Flandes y de Tirol etc.

A vos Pedro de Arrasti nuestro rreçevtor / Saued que Diego de Ocanpo presento una petiçion en nombre del caçique e yn^{o}s de Chacalla ante nos en la nra Audiençia y Chançilleria Rreal que por nro mandado rreside en la çibdad de Los Rreies ante el presidente e oidores della en la causa con los caçiques de Canta e yungas del pueblo de Quibi del thenor siguiente:

"Muy p^{o}s. señor

Di^{o} de Ocanpo en nombre del caçique e yn^{o}s de Chacalla en la causa con los caçiques e yn^{o}s de Canta e yungas del pueblo de Quibi sobre las tierras / digo que V. Alteza mando que el padre Ffrai Gaspar de Caruajal con Bartolome de Prol ffuese amojonar las dhas tierras y llegados alla aunque el dho padre Caruajal como persona que thenia notiçia de las dhas tierras e auia fho otra vez ynfformaçion sobre ellas declaro por donde auian de yr los mojones en los quales estauan muchas casas del dho pueblo de Quibi que las hizieron mis partes y les pertenesçia el dho Bartolome de Prol rreçetor no lo quiso escrivir ni que ni que [*sic*] constase dello ni se hiziese diziendo que la executoria y prouisiones desta Rreal Audiençia no hablauan del pueblo sino de las tierras y ansi no podria dar tierras donde obiese casas a alguna en lo qual hizo mal su off^{o} y mis ptes quedaron agrauiados. Pido e suplico a V. Alteza mande que el dho padre Caruajal declare lo susodho y se hechen los mojones por donde an de yr y se entreguen a mis partes lo que es suyo sobre que pido justiçia y para ello etc.

- Otro si digo que el dho rreçetor exçediendo de lo que le ffue mandado mando que mis partes dexasen coxer a las partes contrarias el ffruto de la coca y otras cosas que auian senbrado con lo qual ay gran conffusion entre los yn^{o}s y las ptes contrarias pretenden coxer no solo el maiz y otras legunbres sino aun la coca que es cosa perpetua y la destruyen con la demas arboleda como cosa agena.

Pido y sup^{co} a V. Alteza mande que las partes contrarias no entren a coger ffruto alguno y quando alguno ouieren de dar se tase el maiz e otras legunbres que podrian coger en esta cosecha y cun-//
/440r/ plan mis ptes con darlo para lo qual etc.

- Otro si digo que mis ptes an cunplido en entregar las dozientas ovejas como consta por los autos del proçeso y las dieron todas buenas y para mas justifficaçion trocaron las que dixeron que no heran buenas e aora las partes contrarias querrian y pretenden que mis partes les pagen [*sic*] las que ellos an escondido o se les an muerto o perdido sin culpa.

Pido e suplico a V. Alteza declare aver mis partes cunplido e no auer lugar lo pedido lo pedido [*sic*] por las partes contrarias para lo qual etc. El Liçençiado Ffalcon/Diego de Ocanpo."

E nos fue pedido e suplicado mandasemos prouer sobre ello o como la nuestra mrd ffuese lo qual visto por el dho presidente e oydores ffue acordado que deuiamos mandar dar esta nuestra carta para vos en la dha rrazon e nos touimoslo por bien por la qual vos mandamos que luego que con ella fueredes rrequerido veais la petiçion de suso yncorporada y sobre lo en ella contenido con çitaçion de las partes hagais ynformaçion de como a pasado y pasa lo uno e lo otro fecha con vro pareçer la traed o enbiad ante nos para que vista por el dho presidente y oidores se prouea sobre todo ello lo que convenga y sea justiçia e no fagades endeal por alguna manera so pena de la nra mrd y de mil p^{o}s de oro para la nuestra camara.

Dada en Los Rreyes a quinze dias de otubre de mill e qui^{o}s y sesenta y siete años.

Yo el secr^{o} P^{o} de Avendaño escrivano de camara de la Mag.^{tt} Rreal la fiz escrevyr por su mandado con acuerdo del su presidente e oydores.

Rr^{da} Juan de Murga

Chançiller
Gaspar de Solis

/440v/ En la çibdad de Los Rreyes a diez e seis dias del mes de otubre de myll y qui^{o}s e sesenta y syete años en presençia de my Pedro de Herrazti escr^{o} de Su Mag.^{t} por t^{o} presente Di^{o} de Ocanpo en nonbre del caçique e yndios del rrepartimyento de Chacalla y presento la provision rreal de Su Mag.^{t} desta otra parte contenyda a my dirigida y me pidio y rrequerio que en cunplimyento della me fuese al pueblo de yndios de Quybi a hazer y cunplir lo que por la dha rreal provision se me manda e yo el dho escr^{o} le rrespondi que estaua presto de lo hazer.

P^{o} de Herrazti
escr^{o} de Su Mag.^{t}

Despues de lo susodho en el dho pueblo de Quibi a veynte y tres dias del dho mes de otubre yo el dho P^{o} de Herrazti escr^{o} susodho por no aver benydo don Fran^{co} caçique prinçipal de Canta a este dho pueblo y por su ausiençia notifique e di a entender lo contenido en esta provision a don Fran^{co} Pilco hermano que dixo ser del dho don Fran^{co} caçique prinçipal de Canta y a don Juan Pariona y a Hernando Tantavilca prinçipales del dho rrepartimiento de Canta e a otros yndios que con ellos estaban lo qual les notifique e di a entender por lengua de Juan Diaz ynterpetre.

P^{o} de Herrazti
escr^{o} de Su Mag.^{t}

Despues de lo susodho en el dho pueblo de Quivi el dho dia mes y año susodho para ynformaçion de lo susodho hize pareçer ante my a don Juan Baptista caçique prinçipal del rrepartimi^{o} de Chacalla / y a don Juan Yapa su sobrino / y a Ju^{o} de Albarado Payco / y Al^{o} Tanta y Ju^{o} Vilca / y Juan Gabilan Chancho y M^{yn}. Yauri prinçipales del dho rrepartimyento de Chacalla / y a Juan Guaman y Fran^{co} Mychuay / y Fran^{co} Yauri y Felipe Pazpon / y Alonso Charic / y Al^{o} Xaqui //
/441r/ yndios del dho pueblo de Quivi de la parçialidad del dho don Ju^{o} Baptista y estando asy juntos les declare lo cont.^{do} en la dha provision y que me mostrasen la parte y lugar por donde mando poner los mojones el padre Fray Gaspar de Caruajal quando los dias pasados vino a este pueblo por comysion de la Rreal Avdiençia a amojonarles las trras los quales todos dixieron que el dho padre

Carvajal mando que se pusiesen los mojones en este dho pueblo dende la esquina de la yglesia a la pared de vna casa donde ençima della esta colgada vna canpana pequeña y dende alli arriba por medio del pueblo donde al presente estan poblados los yndios de la parçialidad de Canta hazia arriba y que Br^{me} de Prol dixo que no se podia dar la posesion de las trras por medio del pueblo y que lo pidiesen en la Avdiençia Rreal para que se probeyese sobre ello y que asy se pusieron los mojones por de fuera del dho pueblo por junto a las casas en lo qual se les quita vn pedaço de trra donde los dhos y^{o}s de Canta estan poblados y mas se quedan los dhos y^{o}s de Canta con siete o ocho casas que son de los y^{o}s de la parçialidad de Chacalla que las hizieron sus y^{o}s y los y^{o}s de Canta los mataron quando la rrebelion de Gonçalo Piçarro y se metieron en ellas / y que quando Gibraleon y Ant^{o} Lopez alguazil binyeron a este pueblo a amojonarles las dhas trras mandaron asymysmo poner los mojones en este dho pueblo por donde despues mando el dho padre Caruajal que se pusiesen que casy no diferieron y que en todo lo demas en el amojonar de las trras que el dho Fray Gaspar de Caruajal y Br^{me} de Prol y Gibraleon que todos estubieron conformes y estan bien puestos.

Y lo que va a dezir en esta diferençia es darseles menos detras a los y^{o}s de Chacalla treynta y ocho pasos en ancho por la parte de abaxo mediendo por el prinçipio del pueblo y por la parte de arriba quatro pasos y en largo dozientos pasos pocos mas o menos que no se puede medir derecho porque yo el dho escrivano lo medi y este pedaço de trra tienen ocupado los y^{o}s de Canta que tienen su pueblo en el y en este pedaço de trra entran asymysmo las syete o ocho casas que los yndios //
/441v/ de Chacalla dizen que les han tomado los de Canta que son al prinçipio del pueblo por la parte de abaxo / y que en lo que toca a lo contenido al segundo capitulo de la petiçion que esta ynserta en la provision rreal de suso contenida / que al tienpo que el dho padre Caruajal y Br^{me} de Prol les amojonaron las trras mandaron a los y^{o}s de Canta que por esta vez cogiesen el mayz y las demas comidas que tenyan senbradas y la coca que tenyan benefiçiada en las trras que mandaron dar a estos y^{o}s de Chacalla y no mas y que no entrasen mas a senbrar en aquellas trras ny a benefiçiar la coca que tenian plantada y que el mayz que tenyan senbrado que ya lo han cogido todo y que avnque es en su perjuizio que ellos todos como estan juntos con el dho don Juan Baptista su caçique prinçipal consienten y han por bien que los y^{o}s de la parçialidad de Canta cojan por esta vez vna vna myta de la coca que tienen benefiçiada con que la chacara de coca que vna vez cojieren no la tornen a coger ny benefiçiar mas otra vez porque son muchas chacarillas de coca y se cogen en diferentes tienpos y que todas ellas pueden acabar de coger dentro de un mes primero siguiente dende oy dho dia y que tanbien tienen senbrados vnos pocos de frisoles e yuca que tanbien consienten que por esta vez lo cojan y que esto consienten con que nyngun caçique ny prinçipal del rrepartimy^{o} de Canta se halle presente al coger desta dha cosecha sino que la cojan los yndios comunes del dho rrepartimy^{o} de Canta y los y^{o}s yungas que tienen en este pueblo porque sy los caçiques y prinçipales de Canta estubiesen aca nunca estarian en paz y cada uno dellos se iria por sy y despues dirian que ellos no lo avian visto coger y nunca faltarian enbaraços entre ellos / y que tanbien ha de ser esto con condiçion que despues que los y^{o}s de Canta ayan cogido la mytad de la coca que no arranquen la planta de la coca syno que la dexen estar asy para que estos se aprouechen della / porque quando los y^{o}s de Canta les tomaron a ellos sus trras tenyan los y^{o}s de Chacalla plantada mucha cantidad de coca mas que no la que agora dexan los yndios de Canta / y que sobre lo demas que se contiene en la dha rreal provision que ya ellos han cunplido //
/442r/ con los dhos yndios de Canta en darles las dozientas ovejas de la trra y mas les dio a rruego del dho padre Caruajal y de Br^{me} de Prol diez e siete ovejas de la trra por algunas que dezian que no heran buenas y esto fue lo que los dhos caçique prinçipal e y^{o}s declararon sobre lo cont.^{do} en la rreal provision por lengua del dho Juan Diaz ynterprete susodho.

ante my P^{o} de Herrazti
escr^{o} de Su Mag.^{t}

/443r/ El Liçen^{do} Castro — El D. Bravo de Saravia — El doctor Gz Cuenca — El Lic^{do} don Alv^{o} Ponçe de Leon

Ynformaçion sobre el entrego de las
ovejas y perdida dellas.

S^{o} Avendaño

Para que Pedro de Arrasti rreçeptor vaya al asiento de Quivi y vea por vista de ojos las ouejas que los yn^{o}s de Chacalla tienen entregadas a los y^{o}s de Canta y que tantas son y quien las tiene y si tienen enfermedades de carache o no y si son pacos o no y si son tales que se an de rresçebir e tomar e que tantas le faltan por entregar a cumplimiento de dozientas ovejas y si algunas dellas se an perdido a cuya culpa se perdieron y hecha con su pareçer lo enbie para prouer lo qual haga con çitaçion de las partes.

/443v/ Don Phelipe por la graçia de Dios rrey de Castilla de Leon de Arargon de las dos Seçilias de Jerusalen de Nauarra de Granada de Toledo de Valençia de Galiçia de Mallorcas de Seuilla de Çerdeña de Cordoua de Corçega de Murçia de Jaen de los algarues de Algezira de Gibraltar de las yslas de Canaria de las Yndias yslas e tierra firme del mar oçeano; condes [*sic*] de Flandes y de Tirol etc.

A vos Pedro de Arrasti nro rreçeptor Saued que don Fran^{co} Guaman Guaranga caçique del rrepartimyento de Canta por una petiçion que ante nos presento en la nra Rreal Audiençia y Chançilleria que por nro mandado rreside en la dha çiudad ante el presidente e oydores della en el pleito que trata con los yndios de Chacalla sobre las tierras de Quibe nos hizo rrelaçion diziendo que aunque por nos a sido muchas vezes mandado que primero que se les diese la dha prouision de las dhas tierras a los dhos yndios les diesen las dozientas ovejas que ellos les dieron tales y tan buenas como se las dieron o que sean de dar e rresebir / los dhos yndios de Chacalla no las an querido dar sino la mitad dellas y aquellas pocas y llenas de carache y tales y tan flacas de enfermedad que es ynposible bibir porque pareçen desecho de otro ganado y no es justo que sean despojados de sus tierras y aun de las mismas ovejas que ellos les dieron y que el padre Carauajal con la mucha afiçion que tiene a los yndios de Chacalla no solamente les a dado las tierras sobre que no traen pleito pero auiendose cometido lo de las ovejas a Bartolome de Prol dize que por fuerça an de rresebir las dhas ovejas pacos y llenas de carache a lo qual no deue dar lugar e nos pidio y suplico mandasemos yr una persona que vea las dhas ovejas antes que se mueran para que se vea que no son de dar ni rresebir y se les mande que les den otras tales y tan buenas como se las dieron y que en el entretanto no se les de la posesion de las dhas tierras pues es conforme a justiçia o como la nra merçed fuese lo qual visto por el dho presidente e oydores //
/444r/ fue acordado que deuiamos mandar dar esta nra carta para vos en la dha rrazon e nos touimoslo por bien por la qual vos mandamos que luego como la rreçibays vays al dho asiento de Quivi e veays por vista de ojos las ovejas que los dhos yndios de Chacalla tienen entregadas a los dhos yn^{o}s de Canta y que tantas son y si tienen enfermedad de carache o no y si son pacos o creçidas y si son de dar e rresebir y las que dellas ay bibas y quien los tiene y que tantos faltan por entregar a cumplimy^{o} de las dhas dozientas ovejas que se les mando que les entregasen y si algunas dellas se an perdido a cuya culpa se perdieron y hecha la dha aueriguaçion con çitaçion de las partes la enviad ante nos con vro pareçer para que visto por el dho presidente e oydores prouean lo que convengan e sea justiçia lo qual ansi hazed e cumplid so pena de quinientos p^{o}s para la nra camara.

Dada en Los Rreyes a diez dias de otubre de mill e quinientos y sesenta y siete años.

Yo el secr^{o} P^{o} de Avendaño escrivano de camara de la Mag.^{tt} Rreal la fiz escrivyr por su mandado con acuerdo del su presidente e oydores.

Rr^{da} Juan de Murga

Chançiller
Gaspar de Solis

/444v/ Ynformaçion sobre las ovejas.

En el pueblo de Quybi a veynte y vn dias del mes de otubre de myll y qui^{o}s e sesenta y syete años en presençia de my P^{o} de Herrazti escr^{o} de Su Mag.^{t} pareçio presente Hernando Tantavilca yndio prinçipal que dixo ser del rrepartimyento de Canta y presento la provision rreal de Su Mag.^{t} de suso contenyda y juntamente con don Fran^{co} Pilco herm^{o} que dixo ser de don Fran^{co} caçique prinçipal de Canta me pidieron que hiziese lo que por ella se mandaua lo qual dixieron por lengua de Juan Diaz ynterpetre.

P^{o} de Herrazti
escr^{o} de Su Mag.^{t}

Not^{on} al caçique e otros yndios de Chacalla.

Despues de lo susodho en el dho pueblo de Quivi a veynte y dos dias del dho mes de otubre del dho año yo el dho P^{o} de Herrazti escr^{o} susodho por lengua del dho Ju^{o} Diaz ynterpetre di a entender todo lo contenydo en la dha provision a don Juan Baptista caçique prinçipal de Chacalla y a don Juan Yapa su sobrino y a Juan de Albarado Payco y a Anton Xaguy yndios prinçipales que dixeron ser del dho rrepartimyento de Chacalla e a otros y^{o}s que con ellos estauan / y a Fran^{co} Yauri y Felipe Paspon y a Fran^{co} Mychuay y a Alonso Charic que dixieron ser yungas que biben en este dho pueblo que son de la parçialidad de Chacalla y a otros y^{o}s de su parçialidad / a la qual dha notificaçion asymysmo se hallaron presentes el dho Hernando Tantavilca y don Fran^{co} Pilco y don Juan Pariona e otros yndios de Canta / a la qual dha notificaçion el dho don Juan Baptista caçique prinçipal de Chacalla por sy e en nonbre de los demas yndios sus subjetos dixo y rrespondio por lengua del dho ynterpetre / que quando Gibraleon escrivano y Antonyo Lopez alguazil y P^{o} de Niça ynterpetre binyeron a este pueblo por mandado de los señores presidente e oydores del Avdiençia Rreal de la çiudad de Los Rreyes el les dio y entrego a los dhos yndios de Canta dozientas ovejas desta trra como por los dhos señores presidente e oydores le fue mandado y que heran buenas ovejas grandes y que los dhos Gibraleon y Ant^{o} Lopez y P^{o} de Niça las vieron y en su presençia se las entrego a los dhos yndios de Canta y que por esta rrazon no es el obligado a darles mas ovejas lo qual todo dixo y rrespondio delante de los dhos yndios de Canta de suso declarados por lengua del dho ynterpetre.

P^{o} de Herrazti
escr^{o} de Su Mag.^{t}

/445r/ E luego yncontinente en el dho pueblo de Quybi el dho dia mes e año susodho yo el dho escrivano por lengua del dho Juan Diaz les dixe e declare a los dhos don Fran^{co} Pilco y don Juan Pariona / y Hernando Tantavilca prinçipales que ya avian oydo la rrespuesta que don Juan Bautista caçique prinçipal de Chacalla avia dado a la notificaçion que le hize con la provision rreal de Su Mag.^{t} que ante my presentaron y como dezia que les avia dado y entregado todas las dhas duzientas ovejas syn faltar nynguna y que heran muy buenas y que no les debia mas / y el dho don Fran^{co} Pilco por sy y en nonbre de los demas dixo que las ovejas que los y^{o}s de

Chacalla traxieron a este pueblo de Quybi para entregar a los y^{o}s de Canta los mysmos y^{o}s de Chacalla las tornaron a llebar a su trra porque en aquella sazon llebaron presos a Lima a tres prinçipales de Canta y no avia quyen diese rrecaudo a las dhas ovejas ny estaua aca entonçes el caçique prinçipal de Canta / para rreçibirlas y que al tienpo que el padre Fray Gaspar de Caruajal y Br^{me} de Prol rreçeptor binyeron a este dho pueblo por m^{do} de la Rreal Avdiençia les dixieron que traxesen aca las dhas ovejas porque las querian ver sy heran buenas y de dar y rreçebir y para contar que tantas heran / y que entonçes fueron çiertos y^{o}s de Canta y Chacalla a buscar las dhas ovejas para traerlas aca y hallaron çiento y tres cabeças de ovejas en las trras de Chacalla y las truxieron aca y las vieron los dhos padre Caruajal y Br^{me} de Prol y que estas çiento y tres ovejas heran entre chicas y grandes porque avia algunos corderos en ellas y que las que heran grandes heran todas pacos y llenas de carache y que por esta rrazon no las quisieron rreçibir y que los dhos padre Fray Gaspar de Caruajal y Br^{me} de Prol les rrogaron e ynportunaron que las rreçibiesen y que ellos no las quysieron rreçibir por ser de la manera que dha es y que no saue que se hizieron ny han visto mas las dhas ovejas y que sy en la provision rreal de suso cont.^{da} que ellos presentaron ante my dize que rreçibieron la mytad de las dozientas ovejas que los y^{o}s de Chacalla heran obligados a les dar eso dira la provision rreal //

/445v/ por aver dho ellos que avian visto las çiento y tres ovejas que dho tiene pero no por que ellos las oviesen rreçibido y que no han rreçibido nyngunas ovejas ny sauen donde estan para traerlas aca para que no las vea / lo qual todo rrespondieron por lengua del dho ynterpetre.

P^{o} de Herrazti
escr^{o} de Su Mag.^{t}

E luego yncontinente para averiguaçion de lo susodho yo el dho escrivano por pareçerme que los t^{o}s que las partes presentasen en rrazon de lo susodho no dirian la verdad syno lo que sus caçiques les mandasen que dixiesen de ofiçio tome la declaraçion de Fran^{co} Tunqui yndio natural que dixo ser del rrepartimyento de Canta y que bibe en el pueblo de Quyçu vna legua mas arriba deste pueblo de Quibi y que es subjeto a don Fran^{co} caçique prinçipal de Canta el qual siendo preguntado por el tenor de lo susodho dixo que lo que deste caso saue es que al tienpo que el padre Fray Gaspar de Caruajal y Br^{me} de Prol binyeron a este pueblo que este t^{o} vino asymysmo aca y vio que se traxieron a este pueblo çiertas ovejas de la trra para dar a los y^{o}s de Canta los quales no las quysieron rreçibir por no estar juntas todas duzientas ovejas y que aquellas ovejas que este t^{o} vido aca que todas heran chicas porque heran pacos y algunas dellas estaban buenas y otras tenyan carache y que tanbien avia entre ellas algunos corderos / y que despues estando este t^{o} en su pueblo que como dho tiene esta vna legua de aqui poco mas o menos vido como dos y^{o}s llebaban aquellas ovejas y las llebaron a un pueblo de Chacalla que se llama Hacomarca y que no conosçio este t^{o} los y^{o}s sy heran de Canta o de Chacalla porque los vido dende lexos y que de ay a çiertos dias fue este t^{o} al dho pueblo de Hacomarca donde ay y^{o}s de Chacalla y vido alla las dhas ovejas como andaban syn pastor y que esto es lo que saue deste caso de lo que le ha sydo preguntado lo qual todo declaro por lengua del dho ynterpetre y no le tome juramento //

/446r/ por no estar satisfecho del sy diria la verdad mas de encargarle que dixiese la verdad donde no que la justiçia le castigaria.

P^{o} de Herrazti
escr^{o} de Su Mag.^{t}

E luego yncontinente en el dho pueblo de Quibi el dho dia mes y año susodho yo el dho P^{o} de Herrazti para averiguaçion de lo susodho mande a don Fran^{co} Pilco hermano del caçique prinçipal de Canta que luego enbiase quatro o çinco y^{o}s suyos al pueblo de Hacomarca del rrepartimi^{o} de Chacalla adonde el t^{o} suso escripto dixo que avia visto las ovejas y que truxiesen los y^{o}s de aquel pueblo ante my para tomarles sus dhos en rrazon de lo susodho el qual dixo que asy lo haria / y tanbien le dixe al dho don Fran^{co} que mandase a los dhos y^{o}s que myresen sy las dhas ovejas andaban en el dho pueblo o su comarca y las trugiesen aca para ver sy heran de dar y rreçibir.

P^{o} de Herrazti
escr^{o} de Su Mag.^{t}

E luego yncontinente el dho dia mes y año susodho en el dho pueblo de Quybi para ynform^{on} de lo susodho tome y rre [*sic*] la declaraçion de Fran^{co} Mychuy yndio del dho pueblo de Quibi de la parçialidad de Chacalla subjeto a don Ju^{o} Baptista caçique prinçipal de Chacalla / del qual no rreçibi juramento por entender que estos y^{o}s de Chacalla y los de Canta estan muy yndinados vnos contra otros y por no estar satisfecho del sy diria la verdad al qual le encargue que la dixiese sobre lo que le fuese preguntado / donde no que la justiçia le castigaria / el qual prometio de dezir verdad / y siendo preguntado por el tenor de lo susodho y de la dha rreal provision dixo que lo que deste caso saue es que los dias pasados oyo dezir este t^{o} como benyan de Lima a este pueblo Gibraleon y vn alguazil y P^{o} de Nyça a entregar dozientas ovejas que los y^{o}s de Chacalla avian de dar a los de Canta / y a amojonar las trras deste pueblo de Quibi / y que en aquella sazon les mando a este t^{o} y a otros quatro yndios deste dho pueblo de la parçialidad de Chacalla vn yndio prinçipal de Canta que se llama Hernando Tantavilca que se escondiesen por que no diesen alguna cosa contra los y^{o}s de Canta / y asy se escondieron de myedo //

/446v/ del de la otra parte del rrio y estando alla escondidos vido este t^{o} cantidad de ovejas de la trra junto a este pueblo de Quibi y a la rredonda del que heran muchas ovejas pero que no saue particularmente que tantas heran y que asy estubo escondido mucho tienpo avnque algunas vezes benia de noche a su casa por de comer y se bolbia a esconder de dia / y que saue quyen llevo las dhas ovejas mas de que oyo dezir a algunos y^{o}s que los y^{o}s de Canta las avian llevado / y que despues quando binyeron a este dho pueblo el padre Fray Gaspar de Caruajal y Br^{me} de Prol vido este t^{o} que vnos yndios de Canta y otros de Chacalla truxieron a este pueblo çient ovejas para dar a los y^{o}s de Canta y que las dhas çient ovejas heran algunas dellas grandes y otras heran pacos y tanbian avia algunos corderos entre ellas / pero que no tenyan carache / y que el dho padre Caruajal mando a los y^{o}s de Chacalla que truxiesen diez e siete ovejas mas y asy truxieron diez e siete cabeças machos y henbras y las unas y las otras dieron a los y^{o}s de Canta por m^{do} del dho padre Caruajal y de Br^{me} de Prol y los y^{o}s de Canta las llebaron y no saue adonde / y que este es lo que saue de lo que le a sido preguntado y que lo que dho tiene es la verdad y en ello se afirmo y rratifico por lengua del dho ynterpetre.

P^{o} de Herrazti
escr^{o} de Su Mag.^{t}

Este dia de pedimyento del dho don Juan Baptista caçique prinçipal de Chacalla mande a don Fran^{co} Pilco prinçipal de Canta que luego enbiase a llamar a don Fran^{co} Caranya y a don Fran^{co} Cundurcori prinçipales del rrepartimyento de Canta y a Fran^{co} Poma y a Arcos porque el dho don Ju^{o} dixo que estos prinçipales e y^{o}s les avian entregado las dhas dozientas cabeças de ovejas y el dho don Fran^{co} Pilco dixo que el dho Fran^{co} Poma estaua enfermo en cama y que a los otros tres que el los enbiaria a llamar luego porque estaban en Canta.

P^{o} de Herrazti
escr^{o} de Su Mag.^{t}

Este dho dia mes y año yo el dho P^{o} de Herrazti para averiguaçion de lo susodho hize pareçer ante my a Fran^{co} Yauri yndio deste dho pueblo de Quybi de la parçialidad de Chacalla subjeto al dho don Juan Baptista al qual por causas dhas no le //
/447r/ tome juramento mas de encargarle que dixiese la verdad el qual prometio de lo hazer asy e siendo preguntado por el tenor de lo susodho dixo que lo que deste caso saue es que habra dos meses poco mas o menos que este t^{o} oyo dezir en este pueblo de Quibi a vnos yndios de Canta como Gibraleon y vn alguazil benyan de Lima a este dho pueblo por m^{do} de la justiçia a entregar a los y^{o}s de Canta las ovejas que los yndios de Chacalla les avian de dar y que en aquella sazon les mandaron a este t^{o} y a otros quatro y^{o}s deste pueblo de la parçialidad de Chacalla / don Fran^{co} Yaxo y Alonso Alexandre Vilca y vn don Gonçalo prinçipales del rrepartimyento de Canta que se escondiesen por que no dixiesen los y^{o}s de Chacalla que heran naturales deste pueblo y subjetos al rrepartimy^{o} de Chacalla / y que asy se escondieron de myedo porque los dhos prinçipales de Canta les dixieron que Gibraleon les avia de dar tormento sy no se escondian y que asy andubieron sienpre escondidos asta que bino el padre Caruajal a este pueblo / y que en aquella sazon que vino el dho Gibraleon vido este t^{o} dende vn çerro donde estaua escondido muchas ovejas de la trra junto a este pueblo en la huerta de Nyculas de Rribera y entre las chacaras que andaban paçiendo / y que despues vido este t^{o} dende vn çerro que esta enfrente deste pueblo como los y^{o}s de Canta hecharon las dhas ovejas por este valle arriba la mayor parte dellas y algunas dellas se fueron por este valle abaxo sin que nadie las llebase y que dos o tres dellas se murieron junto a este pueblo porque haze aqui gran calor y no es buena trra para las ovejas de la trra y que no saue que se hizieron las otras y que quando el padre Fray Gaspar de Caruajal y Br^{me} de Prol binyerin a este pueblo vido este t^{o} como por mandado del dho padre Caruajal çiertos y^{o}s de Canta y de Chacalla truxieron a este dho pueblo çient ovejas las quales oyo dezir que las trayan de las trras de Chacalla porque los yndios de Canta no les avian dexado pasar a su trra y que las dhas çient ovejas algunas dellas heran grandes y buenas y otras heran pequeñas pacos y tanbien avia algunos corderos entre ellas pero que los corderos no entraron en la quenta de las çient ovejas sino que yvan dadas con las madres / y que por alg^{as} ovejas que tenyan carache mando el dho padre Caruajal que los dhos y^{o}s de Chacalla diesen a los de Canta otras diez e siete ovejas mas / y que asy se las dieron grandes y buenas y las vnas //
/447v/ y las otras mando llebar un don Alonso prinçipal de Canta por este valle arriba / y que asy se las vido llebar este t^{o} a çiertos y^{o}s de Canta el valle arriba por el camyno que va hazia Canta / y que no saue que las hizieron / pero que todas las ovejas que entonçes dieron los y^{o}s de Chacalla a los de Canta fueron çiento no mas porque en lugar de las diez e siete ovejas buenas que les dieron sacaron otras tantas que tenyan carache de las çiento que primero avian traydo y que esto es lo que saue de lo que se le fue preguntado y que lo que dho tiene es la verdad y en ello se afirmo y rratifico por lengua del dho ynterpetre y pareçe ser de hedad de veynte y dos años poco mas o menos / y el otro que declaro antes sera de treynta años poco mas o menos.

P^{o} de Herrazti
escr^{o} de Su Mag.^{t}

Despues de lo qual en el dho pueblo de Quybi el dho dia veynte y dos de otu^{e} del dho año para ynform^{on} de lo susodho yo el dho P^{o} de Herrazti hize pareçer ante my a Felipe Pazpon yndio del dho pueblo de Quibi de la parçialidad de Chacalla subjeto al dho don Juan Baptista al qual por lengua del dho ynterpetre encargue que dixiese la verdad donde no seria castigado por la justiçia / el qual prometio de dezir verdad y siendo preguntado por el tenor de lo susodho dixo que al tienpo que vino a este pueblo Gibraleon dende Lima este t^{o} estubo escondido como los demas t^{o}s que han declarado por mandado de los prinçipales de Canta que heran don Fran^{co} Pomacaxa y don Gonçalo de Lucha y Alonso Alexandre / y dende la otra parte del rrio que baxa por este valle vido muchas ovejas de la trra que truxieron los y^{o}s de Chacalla para entregar a los de Canta / que oyo dezir este t^{o} a un yndio deste pueblo que no se acuerda como se llama como avian buelto los y^{o}s de Chacalla a los de Canta las dozientas ovejas por m^{do} del Avdiençia Rreal / y que en aquella sazon llebaron presos a Lima çiertos prinçipales de Canta porque avian derribado los mojones y que el dho don Juan Baptista caçique prinçipal de Chacalla fue con ellos y las ovejas quedaron aca y que oyo dezir este t^{o} en este pueblo a los y^{o}s que vn prinçipal de Canta que se llama Domyngo las avia hecho llebar hasta el pueblo de Cullata que esta dos leguas de aquy poco mas o menos que es en la syerra / y que algunas fueron el valle abaxo y que vnos yndios de un prinçipal de Chacalla que se llama Estupinan las trayan hazia arriba / y los de Canta las tornaban a hechar abaxo y que asy se murieron alg.^{as} dellas porque este valle es caluroso para ellas lo qual todo lo oyo dezir a algunos yndios deste pueblo de la parçialidad de Canta //
/448r/ y que quando el padre Fray Gaspar de Caruajal y Br^{me} de Prol binyeron a este pueblo mando a los y^{o}s de Canta y Chacalla que fuesen por las ovejas y que asy fueron y traxieron çient cabeças de ovejas de la trra algunas dellas grandes y otras pequeñas que heran pacos y que tanbien avia algunos corderos pero que no entraban en la cuenta de las çient ovejas y que por algunas ovejas que

avia rruynes y sarnosas les mando dar el dho padre Caruajal diezesiete ovejas mas a los dhos y^{o}s de Canta / y se las entregaron a los dhos y^{o}s de Canta y despues las mando llebar hazia la sierra don Alonso prinçipal de Canta / y este t^{o} se las vido llebar a los y^{o}s de Canta y que esto es lo que saue deste caso lo qual es la verdad y que no saue la hedad que tiene y por su aspeto pareçe ser de dieze ocho o veynte años.

P^{o} de Herrazti
escr^{o} de Su Mag.^{t}

Despues de lo susodho en el dho pueblo de Quybi el dho dia mes e año susodho para ynformaçion de lo susodho hize pareçer ante my a Al^{o} Chaxic yndio del dho pueblo de Quibi de la parçialidad de Chacalla subjeto al dho don Juan Baptista al qual por lengua del dho ynterpetre le encargue que dixiese la verdad donde no que la justiçia le castigaria el qual por lengua del dho ynterpetre prometio de dezir verdad y siendo preguntado por el tenor de lo susodho dixo que al tienpo que Gibraleon escrivano vino a este pueblo le mandaron a este t^{o} los prinçipales de Canta que se escondiese por que su caçique no dixiese como tenya y^{o}s naturales deste pueblo y que asy se escondieron este t^{o} y otros quatro y^{o}s de la parçialidad de Chacalla y que estando este t^{o} escondido de la otra parte del rrio que pasa por este valle vido en el çimenterio de la yglesia deste pueblo mucha cantidad de ovejas las quales trugieron los y^{o}s de Chacalla para dar a los de Canta y que despues que el dho Gibraleon se bolbio a Lima estubieron en este pueblo algunos dias las ovejas y que dos prinçipales de Canta que el uno se llama don Fran^{co} Caranya y Domyngo Caruayanac las hizieron llebar hazia la syerra segun este t^{o} lo oyo dezir en este pu^{o} quando vino el padre Caruajal y que a algunos yndios de Chacalla que benian aca les oyo dezir este t^{o} que se avian muerto algunas de las dhas ovejas en el camyno llebandolas los yndios de Canta / y que quando el dho padre Caruajal Br^{me} //
/448v/ de Prol binyeron a este pueblo el dho padre Caruajal y Prol les mandaron que fuesen por las ovejas y las truxiesen aca para entregarlas a los yndios de Canta / y que asy truxieron çient ovejas y que por algunas que no heran buenas mando el dho padre Caruajal a los dhos y^{o}s de Chacalla que les diesen otras diezesiete ovejas grandes mas y que asy se las dieron y los dhos padre Caruajal y Br^{me} de Prol mandaron a los dhos y^{o}s de Canta que las rreçibiesen y ellos no las querian rreçibir y que asy estubieron algunos dias las dhas ovejas en este pueblo y despues vn prinçipal de Canta que se llama don Alonso las hizo llebar hazia la sierra y este t^{o} se las vido llebar a los y^{o}s de Canta y no saue que se an hecho las dhas ovejas y que esto es lo que saue deste caso de lo que le a sido preguntado lo qual es la verdad / y pareçe ser de dieze ocho años poco mas o menos.

ante my

P^{o} de Herrazti
escr^{o} de Su Mag.^{t}

Despues de lo susodho en el dho pueblo de Quybi a veynte y tres dias del dho mes de otubre del dho año ante my el dho P^{o} de Herrazti P^{o} de Herrazti [*sic*] pareçio el dho don Fran^{co} Pilco yndio prinçipal de Canta e dixo que el enbio ayer los quatro y^{o}s que yo le mande al pueblo de Hacomarca para que traxesen ante my los y^{o}s de aquel pueblo los quales fueron alla y han buelto y dizen que no hallaron en el dho pueblo mas de un yndio muy viejo que no puede benir y que otros dos y^{o}s moços que ay en el dho pueblo no estaban alla.

P^{o} de Herrazti
escr^{o} de Su Mag.^{t}

Despues de lo susodho en el dho pueblo de Quybi a veynte y tres dias del dho mes de otu^{e} del dho año para ynform^{on} de lo susodho tome y rreçibi juramento de Domyngo Çabac yndio natural deste pueblo de Quibi de la parçialidad de Canta el qual lo hizo segun forma de dr^{o} y siendo preguntado por el tenor de lo susodho dixo que lo que deste caso saue es que al tienpo que binyeron a este pueblo Gibraleon y Ant^{o} Lopez alguazil este t^{o} no estubo aca / pero que quando despues binyeron el padre Fray Gaspar de Caruajal y Br^{me} de Prol vino con ellos este t^{o} y que saue que despues que llegaron aca mandaron a los yndios que fuesen por las ovejas y las juntasen y //
/449r/ truxiesen a este pueblo y que asy fueron çiertos y^{o}s de Chacalla y de Canta y trugieron çiento y tres ovejas chicas y grandes pero que todas heran pacos y que algunas dellas estaban buenas y otras enfermas de carache pero que mas avia enfermas que sanas / y que despues por aquellas que estaban enfermas trugieron los y^{o}s de Chacalla çiertas ovejas mas que heran buenas y el dho padre Caruajal y Br^{me} de Prol mandaron a los y^{o}s de Canta que las rreçibiesen / y ellos dixieron que no las querian rreçibir porque no les daban todas dozientas juntas y que asy se fueron las dhas ovejas hazia la syerra hazia Chacalla pero que no saue que se an hecho mas de que oyo dezir por ay a los y^{o}s que los y^{o}s de Chacalla las avian hecho llebar deste pueblo y que esto es lo que saue de lo que se le ha preg.^{do} y que lo que dho ti^{e} es la verdad para el juram^{o} que tiene hecho y que no saue la hedad que tiene y pareçe ser de quarenta años poco mas o menos y rratificose por lengua del dho ynterpetre.

ante my

P^{o} de Herrazti
escr^{o} de Su Mag.^{t}

Este dia para ynform^{on} de lo susodho rreçibi la declaraçion de Xpoual Chaonyca yndio natural deste p^{o} de Quybi de la parçialidad de Canta del qual no rreçibi juram^{o} mas de que le encargue que dixiese la verdad el qual prom^{o} de la dezir y siendo preguntado por el tenor de lo susodho dixo que lo que deste caso saue es que al tienpo que el padre Fray Gaspar de Caruajal y Br^{me} de Prol binyeron a este pueblo trugieron los y^{o}s yauyos de Chacalla çient ovejas de la trra para dar a los de Canta las quales heran algunas dellas grandes y otras chicas y algunas dellas tenyan carache y que los dhos y^{o}s de Canta no las quysieron rreçibir y que como no las quysieron rreçibir los y^{o}s de Chacalla las llebaron hazia arriba y las dexaron alla y que no saue que se an hecho las dhas ovejas y que este es lo que saue lo qual declaro por lengua del dho ynterpetre y pareçe ser de quarenta años poco mas o menos.

ante my P^{o} de Herrazti
escr^{o} de Su Mag.^{t}

En el dho pueblo de Quybi el dho dia veynte y tres de otu^{e} del dho año para ynformaçion de lo susodho hize pareçer ante my M^{yn} Monca yndio deste pueblo de Quybi y que es mytimae aqui y que sus pasados heran naturales de Chancay al qual le encargue que dixiese la verdad y prometio de lo hazer asy y siendo preg.^{do} por el tenor de lo //
/449v/ susodho dixo que al tienpo que Gibraleon y vn alguazil y P^{o} de Nyça vinyeron a este pueblo de Quibi trugieron los y^{o}s yauyos de Chacalla vnas ovejas para dar a los de Canta y que heran ovejas pacos y que los y^{o}s de Canta nos quysieron rreçibir diziendo que heran chicas y que tres yndios yauyos de Chacalla las tornaron a llebar por ay arriba y que dos dellas se murieron aqui la una en la huerta que tiene Rribera en este pueblo y otra mas arriba / y que quando el padre Caruajal vino aca este t^{o} no estaua aqui y no saue lo que entonçes paso açerca de las ovejas ny saue donde estan las dhas ovejas y que esto es la verdad y este yndio pareçe ser de quarenta a^{o}s poco mas o menos el qual declaro por lengua del dho ynterpetre.

ante my P^{o} de Herrazti
escr^{o} de Su Mag.^{t}

Este dia para ynformaçion de lo susodho hize pareçer ante my a Ju^{o} Uru yndio natural que dixo ser del pueblo de Cullata del rrepartimy^{o} de Chacalla que esta una legua deste pueblo de Quybi poco mas o m^{o}s y que es subjeto al dho don Juan Bautista al qual le encargue que dixiese la verdad el qual prometio de la dezir y siendo preg.^{do} por el tenor de lo susodho dixo que lo que deste caso saue es que al tienpo que binyeron a este pueblo vn escrivano viejo que no saue como se llama y vn alguazil dende Lima para entregar las ovejas a los y^{o}s de Canta / saue que el dho don Ju^{o} Baptista caçique prinçipal de Chacalla hizo traer muchas ovejas de la trra para dar a los y^{o}s de Canta porque este t^{o} las vido pero que no saue que tantas heran porque no las conto pero que el dho escrivano y alguazil y caçique las contaron para entregarlas a los y^{o}s de Canta / y que a vn Fran^{co} Poma prinçipal de Canta y a otros muchos y^{o}s que con el estaban en este dho pueblo se las dieron y los dhos escrivano y alguazil les dixieron que las llebasen a su trra y ellos dixieron que no las querian rresçibir que no saue por que y que las dhas ovejas vido este t^{o} que heran buenas avnque avian muchos pacos entre ellas que mas heran los pacos que las ovejas grandes pero que heran ovejas sanas y no tenian carache / y que en este tiempo salio este t^{o} deste pueblo y se fue a su trra y las ovejas quedaron aca / y que despues desto bolbiendo este t^{o} a este pueblo de Quyvi encontro media legua de aqui a vn yndio prinçipal de Canta que se llama Caranya con otros muchos de Canta y vido como llebaba las dhas ovejas hazia la sierra y a la buelta quando este t^{o} bolbio a su trra vido las dhas ovejas çerca de donde encontro al dho Caranya //
/450r/ en vn secadal donde no avia yerba ny agua que las avian dexado alli los dhos y^{o}s de Canta / y que ellas mismas se fueron por ay arriba hazia la syerra y que muchas dellas se murieron alli en aquel secadal porque este t^{o} las vido despues muertas y que esto es lo que saue de lo que se le ha preguntado lo qual es la verdad y en ello se rratifico por lengua del dho ynterpetre el qual pareçe ser de treynta y çinco años.

ante my P^{o} de Herrazti
escr^{o} de Su Mag.^{t}

En el dho pueblo de Quibi el dho dia veynte y tres de otu^{e} del dho año para ynformaçion de lo susodho hize pareçer ante my a Juan Guaman yndio deste dho pueblo de la parçialidad de Chacalla al qual por el dho ynterpetre le encargue que dixiese la verdad donde no que la justiçia le castigaria y prom^{o} de dezir verdad y siendo preguntado por el tenor de lo susodho dixo que al tienpo que Gibraleon y Ant^{o} Lopez alguazil binyeron a este dho pueblo para entregar a los y^{o}s de Canta las ovejas que los y^{o}s de Chacalla les avian de dar este t^{o} y otros yndios deste dho pueblo de su parçialidad se escondieron por mandado de unos prinçipales de Canta que se llaman don Fran^{co} prinçipal del pueblo de Yaço y don Gonçalo prinçipal de Cocha / que les mandaron que se escondiesen por que no dixiesen que heran de la parçialidad del rrepartimy^{o} de Chacalla y que tenyan trras aqui / y que sy dixiesen que tenian aca trras se las avian de quytar todas y de myedo desto se escondieron / y que estando asy escondido dende lexos vido este t^{o} entonçes como andavan muchas [*sic*] de la trra junto a este pueblo que las avian traydo los de Chacalla para entregar a los de Canta / y que no saue quyen las llevo / y que despues quando el padre Caruajal mando a los dhos yndios de Canta que las rreçibiesen y que ellos nunca las quysieron rreçibir porque no les daban todas juntas y que despues vn yndio prinçipal de Canta que se llama don Al^{o} con otros y^{o}s de su parçialidad y otro prinçipal de Canta que se llama Caranya las llebaron por ay arriba hazia la sierra por-//
/450v/ que este t^{o} se las vido llebar y que algunas dellas se muryeron mas abaxo deste pueblo y otras mas arryba de aqui por ser trra caliente y no tener pastos y que no saue donde estan agora las dhas ovejas y esto dixo de lo que le fue preguntado lo qual es la verdad y en ello se afirmo y rratifico por lengua del dho ynterpetre y pareçe ser este yndio de çinquenta y çinco a^{o}s poco mas o menos.

ante my P^{o} de Herrazti
escr^{o} de Su Mag.^{t}

Despues de lo susodho en el dho pueblo de Quivi a veynte y quatro dias del dho mes de otubre del dho año para ynformaçion de lo susodho tome juram^{o} en forma de dr^{o} de Alonso de Ovando / so cargo del qual prometio de dezir verdad y siendo preguntado por el tenor de lo susodho dixo que lo que deste caso saue es que ayer juebes que se contaron veinte y tres dias deste dho mes binyendo juntos este t^{o} y Fran^{co} de Anpuero hijo de Fran^{co} de Anpuero v^{o} de la çiudad de Los Rreyes dende vn pueblo de Quibi se aparto este t^{o} del camyno rreal para ver sy avia alguna caça de benados por alli / y llegado a vna quebrada angosta y caliente obra de dos leguas

mas arriba deste dho pueblo de Quibi vido muchas ovejas de la trra muertas a trechos por la quebrada abaxo que a lo que pareçe segun la disposiçion de la trra se murieron de anbre y sed y de calor y començo a contar asta sesenta e quatro ovejas muertas y despues baxando por la dha quebrada abaxo vido mas ovejas muertas asta seys o syete / y a vnos y^{o}s de Chacalla que benyan con este t^{o} les pregunto que ovejas heran aquellas que estaban muertas los quales fueron a verlas y le dixieron que aquellas ovejas que estaban muertas heran de las que los y^{o}s de Chacalla avian entregado a los y^{o}s de Canta porque ellos no tenyan ovejas por alli / y que adrede las debian aver hechado alli los dhos y^{o}s de Canta para que se muriesen / y binyendo este t^{o} por el dho camyno y quebrada abaxo vido asta quatro o çinco ovejas bibas que andaban por alli y que los dhos y^{o}s de Chacalla que benian con este t^{o} dezian que aquellas que estaban bibas que tanbien heran de las que avian entregado a los y^{o}s de Canta / y que tanbien se moririan //
/451r/ presto por estar en tan mala trra donde no avia que comer y que esto es lo que saue deste caso y es la verdad para el juramento que tiene hecho y que es de hedad de veynte y çinco años y que no le tocan las g^{es} de la ley que le fueron hechas y firmolo de su n^{e}.

ante my P^{o} de Herrazti Alonso de Obando

Este dia para ynformaçion de lo susodho tome la declaraçion de M^{yn} Caxa guaman yndio que dixo ser del pueblo de Cullata del rrepartimyento de Chacalla al qual por lengua del dho ynterpetre le encargue que dixiese la verdad y siendo preguntado por el tenor de lo susodho dixo que lo que deste caso saue es que al tienpo que don Ju^{o} Baptista su caçique prinçipal entrego a los y^{o}s de Canta las dozientas ovejas de la trra en este pueblo de Quivi este t^{o} se hallo presente aqui y vido que por mandado del alguazil y escr^{o} que entonçes binyeron aca a amojonar las trras y entregar las dhas ovejas a los y^{o}s de Canta / se contaron en este dho pueblo las dhas dozientas vna a vna y este t^{o} yva atajando las que yvan contando y asy contaron dozientas ovejas y que en las dhas dozientas ovejas avia tres generos de ganado de la trra que algunas dellas heran de las grandes / y otras heran pacos / y otras mas chicas que se llaman cuyro pero que destas chicas cuyros no avia mas de çinco o seys / y que tanbien avia entre ellas dos ovejas mansas que sabian llebar carga y que todas las dhas dozientas ovejas heran sanas y sin carache / y que despues de contadas dixieron el dho escr^{o} y alguazil a los prinçipales e y^{o}s de Canta que estaban presentes que tomasen las dhas dozientas ovejas / y que dende aqui se fue este t^{o} a su trra y no saue sy los dhos y^{o}s de Canta llebaron las dhas ovejas mas de que ayer binyendo este t^{o} a este pueblo con Fran^{co} de Anpuero el moço y con Alonso de Obando vido en vna quebrada que esta mas abaxo de Hacomarca a trechos obra de setenta ovejas muertas pocas mas o menos y que le pareçe a este t^{o} y tiene por çierto que las dhas ovejas muertas son de las que entregaron a los y^{o}s de Canta porque este t^{o} vido por alli obra de diez ovejas dellas mysmas que estas estaban bibas que tanbien se moririan si no las quitan de alli porque no tienen que comer y que esto es lo que saue y es la verdad y en ello se afirmo y rratifico por lengua del dho ynterpetre. Este t^{o} pareçe ser de treinta años poco mas o menos.

ante my P^{o} de Herrazti
escr^{o} de Su Mag.^{t}

/451v/ Este dia para ynformaçion de lo susodho tome y rre. la declaraçion de M^{yn} Poma Tanta del pueblo de Xicamarca del rrepartimyento de Chacalla al qual le encargue que diga la verdad y prom^{o} de la dezir y siendo preguntado por el tenor de lo susodho dixo que lo que saue es que ayer benyendo este t^{o} a este pueblo de Quibi con Fran^{co} de Anpuero el moço y Al^{o} de Ovando vido en una quebrada que se llama Pati muchas ovejas de la trra muertas que no las conto y tanbien vido en la dha quebrada çinco ovejas bibas que estan flacas que tanbien se moriran sy no las quytan de alli porque no tienen cosa nynguna que comer que es vna quebrada seca y caliente y que esto es lo que saue y pareçe ser de veynte años y rratificose en ello por lengua del dho ynterpetre.

ante my P^{o} de Herrazti
escr^{o} de Su Mag.^{t}

E luego yncontinente este dho dia mes y año susodho yo el dho escr^{o} hize pareçer ante my a don Ju^{o} Baptista caçique prinçipal del rrepartimyento de Chacalla / y a don Antonyo Alcapoma y don Fran^{co} Pilco y don Ju^{o} Pariguana y Hern^{do} Tantavilca / y don Fran^{co} Guaraca y^{o}s prinçipales de Canta y les dixe que por quanto por la ynform^{on} de suso contenida consta que en vna quebrada que se llama Pati se an muerto muchas ovejas de las que los y^{o}s de Chacalla dizen que entregaron a los de Canta por falta de comyda por ser vna quebrada seca y sin comyda / y que algunas ovejas andan por alli bibas que enbien luego y^{o}s que las rrecojan y traygan a este pueblo para que yo las vea y las cuente y que sy no estubieren para poder las traer aca porque dizen los t^{o}s que estan muy debilitadas y flacas que poco a poco las lleben de alli adonde aya pasto y se conbalezen y que cuenten quantas son y las guarden para que aya cuenta y rrazon dellas/ y el dho don Juan Baptista caçique prinçipal de Chacalla / dixo que ya el entrego a los y^{o}s de Canta las dozientas ovejas / y que no se quyere entremeter agora en enbiar a buscar //
/452r/ ny guardar las dhas ovejas porque no quyere que despues digan los y^{o}s de Canta que en su poder se moryeron las dhas ovejas y los dhos don Ant^{o} Alcapoma y los demas prinçipales de Canta susodhos dixieron que sy las ovejas que les dieron los y^{o}s de Chacalla fueron tales como las que dieron los de Canta a los de Chacalla quando los conçertaron Nyculas de Rribera y Fran^{co} de Anpuero que los y^{o}s de Canta las rreçibieron pero que no heran buenas sino pacos y con carache y que por eso no las quysieron rreçibir / y que porque los y^{o}s de Chacalla no digan que ya rreçibieron aquellas ovejas no quyeren enbiar a buscar ny guardar esas ovejas que dizen que estan bibas en la quebrada de Pati y esto dieron por su rrespuesta los vnos y los otros por lengua del dho ynterpetre.

P^{o} de Herrazti
escr^{o} de Su Mag.^{t}

Este dia para ynform^{on} de lo susodho rreçibi la declaraçion de Marcos Vilca yndio del pueblo de Xicamarca del rrepartimy^{o} de Chacalla al qual por lengua del dho ynterpetre encargue que dixiese la verdad y prometio que la dira / y siendo preguntado por el tenor de lo susodho dixo que lo que deste caso saue es que al tienpo que Gibraleon escrivano y el alguazil binyeron a entregar las

ovejas que los y^{o}s de Chacalla avian de dar a los y^{o}s de Canta que este t^{o} se hallo presente en este pueblo de Quivi y que vido que por mandado del dho escrivano y alguazil se contaron dozientas ovejas de la trra çiento machos y çiento henbras y despues de contadas mandaron el dho escrivano y alguazil a don Fran^{co} Poma y a otros prinçipales de Canta que estaban presentes que rreçibiesen las dhas dzas ovejas los quales dixieron que no las querian rreçibir porque no heran grandes como las que ellos avian dado a los y^{o}s de Chacalla y que asy estubieron çinco dias las dhas ovejas en este pueblo de Quibi que andaban paçiendo por aqui / y despues las llebaron los dhos y^{o}s de Canta por este valle hazia la sierra y las dexaron en vna quebrada seca donde no avia agua ny yerba / y que esto lo saue porque este t^{o} se las vio llebar / y que despues desto quando vino el padre Fray Gaspar de Caruajal mando a los y^{o}s //
/452v/ de Canta y Chacalla que juntasen las dhas ovejas y asy fueron çiertos y^{o}s y truxieron çient ovejas y mas otras diez e siete ovejas que estaban aca y todas se las mando dar el dho padre Caruajal a los prinçipales de Canta que estaban presentes y les mando que las rreçibiesen y vn prinçipal que se llama don Fran^{co} Poma dixo que el no las queria rreçibir y otro prinçipal de Canta que se llama don Fran^{co} Caranya dixo: "rreçibamoslas" y asy las rreçibio y despues andubieron las dhas ovejas obra de diez dias en este pueblo y despues las llebaron los dhos y^{o}s de Canta por ay arriba a vna quebrada donde esta vn xaguey donde no ay yerba / donde se murieron muchas dellas porque este t^{o} pasando por alli las ha visto muertas.

Preguntado sy las dhas ovejas que los dhos y^{o}s de Chacalla entregaron a los y^{o}s de Canta heran buenas y grandes y sy tenyan enfermedad de carache / dixo que heran ovejas grandes y chicas y pacos y que mas cantidad avia de pacos que de ovejas grandes / y que las primeras dozientas ovejas que dieron no tenyan carache / y que las otras çiento ovejas que mandaron a traer aca por mandado del padre Caruajal que algunas dellas tenyan ya carache avnque heran de las mysmas dozientas ovejas porque estando en la dha quebrada les dio el carache porque haze mucho calor / y que esto es lo que saue y es la verdad y en ello se afirmo y rratifico por lengua del dho ynterpetre y pareçe ser de treynta años.

ante my P^{o} de Herrazti
escr^{o} de Su Mag.^{t}

Despues de lo susodho en la dha çiudad de Los Rreyes a veynte y siete dias del dho mes de otu^{e} del dho año para mas ynformaçion de lo susodho tome y rre. juramento en forma de dr^{o} de Ant^{o} Lopez alguazil so cargo del qual prometio de dezir verdad / y siendo preg^{do} por el tenor de lo susodho dixo que lo que deste caso saue es que este t^{o} como alguazil fue juntamente con Gibraleon al pueblo de Quibi a amojonar y partir las trras de aquel pueblo entre los yndios de Canta y Chacalla y a amojonarlas por virtud de vna carta executoria rreal hemanada del Audiençia Rreal desta çiudad y a entregar a los dhos yndios de Canta dozientas ovejas de la //
/453r/ trra que los yndios de Chacalla les avian de dar conforme a la dha carta executoria / y este t^{o} como tal alguazil mando contar las dhas ovejas de la trra y las conto y mando a vnos prinçipales e y^{o}s de Canta que alli estaban que las rreçibiesen / los quales rrespondieron que su caçique prinçipal no estaba alli / y que no las querian rreçibir lo qual todo lo que alli paso lo asento el dho Gibraleon escrivano y que alli se quedaron las dhas dozientas ovejas en el dho pueblo de Quibi y no saue sy las rreçibieron o no mas de que el dho ganado quedo por de los dhos y^{o}s de Canta / y que de la calidad que heran las dhas dozientas ovejas hera buen ganado y sano a lo que pareçia que no les vieron tener ningun carache ni otra enfermedad / y que le pareçe a este t^{o} que las çiento y veynte dellas pocas mas o menos heran pacos y ochenta de ovejas grandes y que esto es lo que saue deste caso de lo que se le ha preguntado y es la verdad para el juramento que tiene hecho y que es de hedad de quarenta y seis años poco mas o menos y que no le tocan las generales de la ley que le fueron hechas y no firmo porque dixo que no sabia.

ante my P^{o} de Herrazti
escr^{o} de Su Mag.^{t}

Despues de lo susodho en la çiudad de Los Rreyes en veynte y nuebe dias del dho mes y año susodho para mas ynformaçion de lo susodho tome y rre. juram^{o} en forma de dr^{o} de Alonso Diaz de Gibraleon secr^{o} de Su Mag.^{t} rresidente en esta Corte so cargo del qual prom^{o} de dezir verdad y siendo preg^{do} por el tenor de lo susodho dixo que este t^{o} como escr^{o} y Ant^{o} Lopez como alguazil en cunplimy^{o} de una executoria rreal de Su Mag.^{t} //
/453v/ hemanada desta Rreal Audiençia fueron al pueblo de Quibi a dar la posesion a don Ju^{o} Baptista caçique de Chacalla de las trras que por la dha carta executoria se le mandaban dar y en cunplimy^{o} della el dho don Ju^{o} Baptista hizo traer y traxo al dho pueblo de Quibi las ovejas que por la dha executoria se mandavan dar a los yndios de Canta las quales este t^{o} vio y conto y heran parte dellas ovejas grandes rrasas y otras ovejas pacos muy grandes y con grandes velloçinos de lana sin que nynguna dellas estubiese trasquylada / y que todas ellas heran ovejas de dar y rreçibir y sanas y sin carache nynguno / y que a lo que se acuerda podra aver entre ellas de las dhas ovejas grandes rrasas asta ochenta o noventa y las demas como dho tiene heran ovejas pacos tan grandes y mayores que las rrasas toda la mayor parte dellos y que sobre todo lo demas que paso en el dho pueblo de Quibi sobre el dar de la dha posesion de las dhas trras y entrego del dho ganado este t^{o} lo tiene dado signado que esta en el proçeso desta causa y que por ello pareçera a lo qual dixo se rreferia y rremytia y que lo que dho tiene es la verdad para el juram^{o} que ti^{e} hecho y firmolo de su n^{e}.

A^{o} Diaz de Gibr.^{on} ante my P^{o} de Herrazti
escr^{o} de Su Mag.^{t}

/454r/ Los y^{o}s de Canta.

Muy p^{o} s^{o}r

Los yndios de Canta en el pleyto con los yndios de Chacalla sobre las tierras de Quibi / digo que los dhos yndios de Chacalla nunca an querido entregar las obejas que por Vra. Alt.^{a} les a sydo tantas vezes mandado e agora diziendoles Pedro de Arrasti a quien Vra Alt.^{a} le dio comysion para que se las entregar para que se las entregasen dizen que porque nosotros no quisimos rresçibyr çiento y tres obejas que nos daban por ser la mayor pte dellas pacos y todas llenas de carache que no son obligados a entregarnoslas porque ya aquellas se les an muerto y para hazernos daño y ponernos culpa los mesmos yndios de Chacalla han quitado çiertas piedras de los mojones para dezir que nosotros las quitamos y para eso presentan t^{o}s dellos mesmos y por vyrtud de esta probança el dho P^{o} de Arrasti prendio tres yndios y los enbio asy a esta çiudad los dos yndios yungas y el uno serrano / no debiendolo hazer de dr^{o} pues aun los dhos yndios de Chacalla no an entregado las obejas que por Vra Alt.^{a} les es mandado por sent.^{a} entreguen ante todas cosas.

Por tanto a Vra Alt.^{a} pido y sup^{co} mande que ante todas cosas se nos entreguen las dhas obejas syn perjuizio de nro dr^{o} tales y tan buenas como las dimos y ansymysmo que se nos suelten los tres yndios que enbio presos Arrasti por seer yndios muy probes y meserables personas y se les mande so grabes penas a los dhos yndios de Chacalla que no ayan ynbençiones para que a nosotros no nos pongan culpa çerca de los dhos mojones para que çesen los dhos ynconbenyentes sobre que pedimos just.^{a} la qual pedimos y para en lo neçesario etc.

El Licen^{do} Falcon

/454v/ En Los Rreyes en veinte e nueue de otubre de myll e qui^{o}s e ses^{ta} e siete años ante los senores pres^{te} e oydores desta Rreal Aud^{a} estando en audi^{a} se presento esta petiçion y los dhos señores mandaron que el rrelator lo traiga todo / lo qual se proueyo presente D^{o} de Ocanpo procurador contrario a quien se notifico.

Avendaño

/455r/ En la causa entre partes de la una
el caçique e yndios de Chacalla y de la otra
el caçique e yndios de Canta e Quibi.

En la çibdad de Los Rreyes en diez y ocho dias del mes de nobiembre de mill e quinientos y sesenta e siete años los señores presidente e oidores desta Rreal Audiençia aviendo visto la dha causa mandaron guardar las sent.^{as} de vista y rrevista en ella dadas y lo hecho y proueido en execuçion dellas por el padre Carbajal y Pedro de Arrazti y que dando los yndios de Chacalla a los de Canta veinte y çinco obejas mas sobre las que les tiene dadas declararon auer cunplido con las duzientas obejas que se les mandan dar por las dhas sentençias y de lo demas pedido y demandado por las partes dieron por libres de lo pedido y acusado por la una contra la otra e asi lo proueyeron e senalaron de sus rrubricas.

El qual dho auto fue dado y pronunçiado en el dho dia mes y ano en el cont.^{do} estando presentes Joan de Arrandolaça e Ju^{o} Sanchez de los Rrios procuradores de las partes a quien se notifico e tanbien se notifico a Di^{o} de Ocanpo.

Avendaño

/456r/ Los y^{o}s de Chacalla.
suplican del abto.

Muy p^{o} señor

Diego de Ocanpo en n^{e} del caçique e yndios de Chacalla en la cabsa con el caçique e yndios de Canta sobre las trras en el articulo de las ovejas sup^{an} del abto de vros oydores en que mandaron que dando mys ptes veynte e çinco ovejas a las ptes contrarias se guardase el amojonamyento fho por el padre Carvajal y dieron por libres a las ptes contrarias / y digo que se ha de enmendar y rrebocar y fazer en todo segund por mys ptes esta pedido por lo que del proçeso rresulta en su favor que he aqui por expreso y porque constando como consta por el proçeso y paresçeres que el padre Carvajal dio por vista de ojo que los mojones se pusyeron por medio de las trras de mys ptes / y que las ptes contrarias se quedan con muchas dellas / no pudo dexarse demandar que los dhos mojones se pusyesen por donde yvan antiguamente de que a mys partes se les sygue otro daño mas que por sus trras que continuamente estaran en diferençias por estar tan vezinos / y ver mys ptes que se les an quedado con sus trras y porque constando asymysmo como consta que mys partes todas dozientas ovejas y estando como esta mandado por abto pasado en cosa juzgada que las que boluyeren a mys partes que no se an de dar y tomar se las den mys ptes buenas y de dar y tomar y aviendose asy cunplido y executado no se pudo mandar que mys partes den veynte e çinco ovejas por que las ptes contrarias las dexasen morir / pues seria que ellos sacasen prouecho de su delito y a mys partes que no an fho delicto que les vinyese daño y porque constando como consta que las ptes contrarias an derribado por dos vezes los mojones que se pusyeron por exsecutoria desta Rreal Abdiençia y quemado coca y arboles y quitado a mys partes el agua //

/456v/ no pudieron dexar de ser condenadas en las penas que tengan pedido.

Por que pido y sup[co] a V. Al. rreboque el dho abto para fazer en todo segund que pedido tengo sobre que pido just.[a] y las costas y para ello etc.

El Licen[do] Falcon Diego de Ocanpo

En Los Rreyes en beynte e dos dias de nouiembre de myll e qui[o]s e ses.[ta] e siete a[o]s ante los senores presidente e oydores desta Rreal Audi[a] estando en audi[a] pu[ca] se presento esta petiçion y los dhos señores mandaron que se guarde lo proueydo sin enbargo de la sup[on] lo qual se proueyo presente el dho Di[o] de Ocanpo procurador a quien se notifico.

Avendaño

Abreviaciones más frecuentes

abundami^{o}	abundamiento
acreçentami^{o}	acrecentamiento
acreçentami^{to}	acrecentamiento
admi^{on}	administración
ag^{o}	agosto
ag^{to}, agot^{o}, ago^{to}	agosto
Al^{o}	Alonso
Al^{a}	Alteza
alcal	alcalde
alcald	alcalde
alçami^{o}	alcamiento
alg^{a}	alguna
alg^{a}s	algunas
alg^{as}	algunas
alg^{o}	alguno
alg^{s}	algunas
algu^{a}	algunas
Alua,Alu^{o}	Aluaro
Alvo,Alv^{o}	Alvaro
alld	alcalde
allde	alcalde
alldes	alcaldes
allds	alcaldes
amonest^{on}	amonestación
an	ante
Ant	Antonio
Ant^{o}	Antonio
a^{o}	año
aos, a^{o}s	años
apçebimi^{o}	apercebimiento
apçbimi^{o}	apercebimiento
apon, ap^{on}	apelación
ap^{o}vechado	aprovechado
ap^{o}vechar	aprovechar
asst^{o}, ass^{to}	assiento
asto, as^{to}	asiento
At^{o}	Antonio
au^{a}	ausencia
auçia, au^{çia}	audiencia
audia, aud^{ia}	audiencia
aueg^{o}, aueg^{on}	aueriguación
avda, avd^{a}	avdiencia
avdia, avd^{ia}	avdiencia
ayuntami^{o}, ayuntami^{to}	ayuntamiento
b^{a}	buena
baste, bast^{e}	bastante
bast^{e}s	bastantes
benefio, benef^{io}	beneficio
Ber^{do}	Bernardo
b^{e}s, b^{es}	bienes
b md	buestra merced
Bme, B^{me}	Bartolomé
bna	buena
b^{r}	barchiller
bs	bienes
b^{te}	beinte
bu^{a}	buena
bu^{a}s, bu^{as}	buenas
bu^{o}	bueno
bu^{o}s	buenos
cama, cam^{a}	cámara
cant^{d}, can^{td}	cantidad
cedu^{a}	cédula
çib^{a}	çibdad (ciudad)
çinq	cinquenta
çinta, cinqt^{a}	cinquenta
çinq^{ta}, çinq^{ua}	cinquenta
çiud, çivd	ciudad
c^{o}	consejo
c^{o}	con
coms^{on}	comisión
comp^{a}, compi^{a}	compañia
con^{a}	contra
conbi^{e}	conbiene
con^{do}	contenido
conf^{e}	conforme
conocimi^{o}, conocim^{to}	conocimiento
conoci^{o}	conocimiento
cont^{a}	contra
conthe, conthe^{do}	conthenido
cont^{o}	contrato
cont^{or}	contador
contr^{o}	contrario
cump^{do}	cumplido
cumplimi^{o}, cumplim^{to}	cumplimiento
cumpp^{do}, cunplimi^{o}	cumplimiento
Chancell^{ia}	Chancellería
delig^{a}	deligencia
dem^{da}	demanda
denças, denç^{as}	denuncias
derco, der^{co}	derecho
derc^{o}s	derechos
derc^{o}s	derechos
der^{o}	derecho
dgo, dg^{o}	Diego
dha	dicha
dhas	dichas
dho	dicho

dhos	dichos
di^{e}	diciembre
di^{ze}	diziembre
difficul^{d}, dificul^{d}	dificultad
dif^{o}	difunto
diliga, dilig^{a}, diligen^{a}	diligencia
Di^{o}	Diego
dist^{a}	distancia
dizdo, dizd^{o}	diziendo
diz^{e}, dizi^{e}	diziembre
dn, d^{n}	don
doc^{ta}	doctrina
doct^{o}, doc^{to}	documento
docum^{to}	documento
dom^{o}	domingo
d^{o}s	doscientos
doz^{o}s	dozeientos
dr	derecho
dr^{co}, dreco, dro, dr^{o}, dre^{co}	derecho
enpedimi^{o}, enpedimi^{to}	enpedimiento
enr^{o}	enero
ent^{do}	enterado
ent^{o}	entero
entr^{o}, ent^{ro}	entero
esçpto	escripto
escrbo	escribano
escrptura	escriptura
esct^{a}	escritura
est^{do}	estando
est^{o}	estado
estui^{a}	estuviera
ex^{a}	excelencia
ffa, ff^{a}	fecha
ffo, ff^{o}	fecho
ffr	fray
fha, fh^{a}	fecha
fho, fh^{o}	fecho
fielm^{te}	fielmente
firm^{on}	firmaron
for^{a}	forma
fr	fray
fr^{da}	firmada
gen^{a}lmente, gen^{a}lm^{te}	generalmente
Germo, Ger^{mo}	Gerónimo
G^{o}	Gonzalo
Gr^{mo}	Gerónimo
grra, grr^{a}	guerra
haci^{do}	haciendo
haz^{da}	hazienda
Hernez	Hernández
Hez	Hernández
hordenami^{os}, hordenami^{tos}	hordenamientos
hordin^{o}	hordinario
Hrrdz	Hernandez
h^{ta}	hasta
ilmo, il^{mo}	Ilustrísimo
ille, ill^{e}	Ilustre
inform^{on}, infor^{on}, infor	información
in^{o}, ynd^{o}	indio, yndio
inst^{a}	instancia
jui^{o}	juicio
juram^{o}, juramto, juram^{to}, jur^{mo}	juramento
just^{a}	justicia
l^{do}, lic^{do}, licen^{do}	licenciado
ma, m^{a}	memoria
m^{a}	materia
mag^{co}	magnífico
mag^{co}s^{or}	magnífico señor
mag^{d}, mag^{t}	magestad
mandami^{o}, mandami^{to}	mandamiento
mçd, md	merced
mda, m^{da}	manda
m^{de}	mande
md^{o}, m^{do}	mandado
mds	mercedes
M^{yn}.	Martyn
Min	Martín
mio, mi^{o}	mandamiento
mmi^{co}	magnifico
m^{o}	mandamiento
m^{o}s	mandamos
mrd, mr^{d}	merced
m^{t}	majestad
n^{a}	nueva
nas	nuestras
nas	necesarias
n^{e}	nombre
neces^{d}	necesidad
neç^{o}	necesario
neg^{o}	negocio
ness^{o}	necesario
ning^{a}, ning^{o}	ninguna, ninguno
n^{o}	número
no^{e}	nombre
not^{a}	noticia
notablem^{e}	notablemente
not^{da}	notificada
noti^{a}	noticia
noti^{on}	notificación
nra, nro	nuestra, nuestro
nu^{a}	nueva
num^{o}	número
nva	nueva
oblig^{o}	obligado
oblig^{o}, oblig^{on}, obl^{on}	obligación
offio	offio
ofo, of^{o}	oficio
ordenami^{o}, ordenam^{to}	ordenamiento
o^{te}	otubre
otorgami^{o}, otorgami^{to}	otorgamiento
pa, p^{a}	para
p^{a}	pena
pag^{do}	pagado
pal, p^{a}l	principal
par^{e}	parece
par^{o}	pareció
pbydo	probeydo
pcipal, pcp^{a}	principal
p^{e}, pd, p^{d}	padre
pedimi^{o}, pedimi^{to}	pedimiento
pe^{on}, pet^{on}	petición
plicos, pli^{cos}	públicos
pnte, pn^{te}	presente
P^{o}	Pedro
p^{o}	pero
p^{o}, po^{s}	peso, pesos
po^{dr}, pod^{res}	poderoso, poderosos
pos^{on}	posesión
pra, pr^{a}	primera
pral, pr^{a}l, pr^{al}	principal
preg^{a}, preg^{a}s	pregunta, preguntas
preg^{do}, preg^{o}	preguntado
prese, pres^{e}	presente
pres^{on}	presentación

pres^{to}	presentó
preste, prest^{e}	presente
pria, pr^{ia}	propia
pr^{ia}, pri^{o}, pr^{o}	primera, primero
prom^{o}	prometió
pr^{on}	provisión
proua	prouanza
prs^{a}	persona
p^{so}, p^{o}	poderoso
psona, psn^{a}	persona
pate, p^{te}	parte
pton	petición
pu^{co}, pu^{ca}	público, pública
pueyo	proueyó
pu^{e}	pueblo
qdo	quando
qnto, qnt^{o}, qu^{to}	quanto
Ral, R^{a}l, R^{al}	Real
ra^{o}	razón
rbi, eb^{i}	recibí
rdo, rd^{o}	recibido
rd^{o}	ratificado
redo, re^{do}	recibido
rel^{on}	relación
requerimi^{o}, requerimi^{to}	requerimiento
resdo, res^{do}	rescibido
rezudo, rezv^{do}, rez^{o}	rezivido
rl, r^{l}	real
rmo, rm^{o}, r^{mo}	reverendísimo
r^{o}	registro
r^{o}, r^{o}s	recibió, recibidos
s^{a}	señoría
secre^{o}, secretr^{o}, secret^{ro}	secretario
sencia, sent^{a}, senti^{a}	sentencia
serui^{o}, servi^{o}	seruicio, servicio
ses, s^{es}, s^{e}s	señores
sig^{e}, sigt^{e}	siguiente
s^{o}	suso
s^{or}, s^{r}	señor
sup^{a}, sup^{ca}	súplica
sup^{co}	suplico
syg^{e}	syguiente
testimi^{o}, testio, testi^{o}	testimonio
tgos, tg^{o}s	testigos
ti^{e}	tiene
t^{o}, t^{o}s	testigo, testigos
tpo, tp^{o}	tiempo
tra, trrn, trs	tierra, tierras
tr^{do}	traslado
trm^{o}, trn^{o}, tr^{no}	término
tsimi^{o}	testimonio
tuui^{a}, tuvi^{a}	tuuuiera, tuviera
v, v^{a}	vuestro, vuestros, vuestra
vad, va^{d}, v^{ad}	verdad
v^{e}	viere
v^{e}	veinte
veçi^{o}, veç^{o}, veç^{o}s	vecino, vecinos
V S^{a}	Vuestra Señorío
V Mcd	Vuestra merced
vta, v^{ta}	venta
vte	veinte
vz^{o}	vezino
X^{o}bal, X^{o}ual, X^{o}u^{a}l	Christobal, Christoual
ygalm^{te}	ygualmente
Ylmo, Ylm^{e}	yllustre
Ynd^{a}, Ynd^{a}s	Yndia, Yndias
ynform^{on}	Ynformación
yn^{o}, yn^{o}s	yndio, yndios

Indice por folios*

ESPAÑOLES

Encomenderos

*The numbers in this index refer to folio numbers; they can be found along the left-hand margin of the *Justicia 413* document in Part III. When the number is italicized in this index, it means that there is more than one folio with the same number.

*Los números que están en este índice se refieren a los números de los folios. Se encuentran en el margen izquierdo del documento *Justicia 413* en Parte III. Cuando se encuentra un número en este índice en letra itálica, indica que hay más de un folio con el mismo número.

Escrivanos y Testigos

Intérpretes

GEOGRÁFICO

INDIOS

Incas, Orejones, Capitanes

PALABRAS QUECHUAS

PLANTAS

PUEBLOS Y PROVINCIAS